Handbook of Hydrometry

Handbook of Hydrometry

Edited by Danny Fuller

SYRAWOOD
PUBLISHING HOUSE

New York

Published by Syrawood Publishing House,
750 Third Avenue, 9th Floor,
New York, NY 10017, USA
www.syrawoodpublishinghouse.com

Handbook of Hydrometry
Edited by Danny Fuller

International Standard Book Number: 978-1-64740-145-0 (Hardback)

Cataloging-in-Publication Data

Handbook of hydrometry / edited by Danny Fuller.
    p. cm.
Includes bibliographical references and index.
ISBN 978-1-64740-145-0
1. Hydraulic measurements. 2. Physical measurements. 3. Hydraulic engineering--Instruments. I. Fuller, Danny.
TC177 .H36 2022
681.28--dc23

TABLE OF CONTENTS

PREFACE

The purpose of the book is to provide a glimpse into the dynamics and to present opinions and studies of some of the scientists engaged in the development of new ideas in the field from very different standpoints. This book will prove useful to students and researchers owing to its high content quality.

Hydrometry observes the components of the hydrological cycle such as rainfall, groundwater characteristics and water quality. It also studies the flow characteristics of surface waters. It is concerned with the measurement of the components of the hydrological cycle, such as the quantification of water resources. Hydrometry encompasses concepts from various areas of engineering such as hydrology, control systems, computer sciences, data management, and communications. It uses the hydrometer to measure the characteristics such as density and specific gravity of a liquid. Hydrometry is primarily concerned with measuring the river flows. These measurements are useful in predicting and managing the river flows that can control floods. It also helps in water supply, agriculture and energy production. This book discusses the fundamentals as well as modern approaches to hydrometry. It presents researches and studies performed by experts across the globe. It will serve as a valuable source of reference for graduate and post-graduation students.

At the end, I would like to appreciate all the efforts made by the authors in completing their chapters professionally. I express my deepest gratitude to all of them for contributing to this book by sharing their valuable works. A special thanks to my family and friends for their constant support in this journey.

Editor

Divergence of actual and reference evapotranspiration observations for irrigated sugarcane with windy tropical conditions

R. G. Anderson[1,*], **D. Wang**[1], **R. Tirado-Corbalá**[1,**], **H. Zhang**[1,***], **and J. E. Ayars**[1]

[1]USDA, Agricultural Research Service, San Joaquin Valley Agricultural Sciences Center, Water Management Research Unit, Parlier, California, USA
[*]USDA, Agricultural Research Service, U.S. Salinity Laboratory, Contaminant Fate and Transport Unit, Riverside, California, USA
[**]Crops and Agro-Environmental Science Department,University of Puerto Rico, Mayagüez, Puerto Rico, USA
[***]USDA, Agricultural Research Service, Water Management Research Unit, Fort Collins, Colorado, USA

Correspondence to: R. G. Anderson (ray.anderson@ars.usda.gov)

Abstract. Standardized reference evapotranspiration (ET) and ecosystem-specific vegetation coefficients are frequently used to estimate actual ET. However, equations for calculating reference ET have not been well validated in tropical environments. We measured ET (ET_{EC}) using eddy covariance (EC) towers at two irrigated sugarcane fields on the leeward (dry) side of Maui, Hawaii, USA in contrasting climates. We calculated reference ET at the fields using the short (ET_0) and tall (ET_r) vegetation versions of the American Society for Civil Engineers (ASCE) equation. The ASCE equations were compared to the Priestley–Taylor ET (ET_{PT}) and ET_{EC}. Reference ET from the ASCE approaches exceeded ET_{EC} during the mid-period (when vegetation coefficients suggest ET_{EC} should exceed reference ET). At the windier tower site, cumulative ET_r exceeded ET_{EC} by 854 mm over the course of the mid-period (267 days). At the less windy site, mid-period ET_r still exceeded ET_{EC}, but the difference was smaller (443 mm). At both sites, ET_{PT} approximated mid-period ET_{EC} more closely than the ASCE equations ((ET_{PT}-ET_{EC}) < 170 mm). Analysis of applied water and precipitation, soil moisture, leaf stomatal resistance, and canopy cover suggest that the lower observed ET_{EC} was not the result of water stress or reduced vegetation cover. Use of a custom-calibrated bulk canopy resistance improved the reference ET estimate and reduced seasonal ET discrepancy relative to ET_{PT} and ET_{EC} in the less windy field and had mixed performance in the windier field. These divergences suggest that modifications to reference ET equations may be warranted in some tropical regions.

1 Introduction

Accurate estimates of evapotranspiration (ET) are needed for numerous purposes, including efficient irrigation scheduling (Davis and Dukes, 2010), parameterizing and running different classes of biogeochemical and hydrologic models (Fisher et al., 2005; Zhao et al., 2013), assessing changes in regional hydrology under different cultivation systems (Ferguson and Maxwell, 2011; Holwerda et al., 2013; Waterloo et al., 1999), and evaluating the impacts of agricultural production on regional and global climate (Kueppers et al., 2007; Lo and Famiglietti, 2013; Puma and Cook, 2010) and hydrology (Anderson et al., 2012; Vörösmarty et al., 1998). In irrigated agriculture, underestimation of required ET can lead to suboptimal yield due to water stress (Kang et al., 2002), whereas overestimation of ET can lead to excessive applied water, thus reducing water available for other uses or additional acreage (Perry, 2005), degrading water quality (Smith, 2000), and decreasing economic competitiveness (Hargreaves and Samani, 1984).

While accurate ET estimates are essential, ET can be challenging to measure. Numerous approaches have been developed to measure or estimate ET, including lysimeters (Meissner et al., 2010), micrometeorological methods (Anderson and Goulden, 2009; Baldocchi, 2003; Hemakumara et al., 2003), satellite remote sensing (Bastiaanssen et al., 2005; Tang et al., 2009), and water balance methods. While these approaches vary in their spatial/temporal scale and methodological assumptions and accuracy, most require significant observational costs, technical expertise, or have operational difficulties that are too high for most farmers.

Because of the difficulties in actual ET measurement, the vegetation coefficient/reference ET approach (Jensen, 1968) has gained widespread acceptance for estimating actual ET for varied applications (e.g., Arnold et al., 1998; Cristea et al., 2013). This approach involves calculating a reference ET for a standard land surface, usually grass or alfalfa, using meteorological data and relating the reference surface to the ecosystem/land cover of interest with empirical coefficient(s):

$$ET_A = K_C * ET_0, \qquad (1)$$

where ET_A is actual ET, ET_0 is reference ET, and K_c is the coefficient for the specific land cover type. Two of the most commonly used standard methods include the Food and Agricultural Organization (FAO) approach presented in Irrigation and Drainage paper 56, hereafter referred to as FAO-56 (Allen et al., 1998), and the American Society of Civil Engineers approach, hereafter referred to as ASCE (Allen et al., 2005). Both approaches are based on the combination Penman–Monteith (PM) formula (Monteith, 1965) and account for ET from both solar irradiation and advectively driven ET due to wind and vapor pressure deficit (VPD). Both the FAO-56 and ASCE approaches assume standard measurement conditions and surface parameters (canopy height, surface resistance, albedo, etc.), thus allowing canopy and atmospheric resistance terms to be condensed into constants. Both methods also provide scaling procedures to account for variation in meteorological measurements as well as missing or erroneous data.

Validation work of standardized reference ET equations against large weighing lysimeters with reference surfaces has been done primarily in the western continental USA with low atmospheric humidity (Evett et al., 2000; Jensen et al., 1990). Internationally, most other reference ET validation has been done in Mediterranean climates with similar low humidity (Lecina et al., 2003; Ventura et al., 1999). Relatively little evaluation of these equations has been done in areas with higher relative humidity, presumably because of the perceived lack of use for reference ET equations in these areas. However, reference ET equations are used in more humid regions for applications such as watershed modeling (Rao et al., 2011), forecasting water demand (Tian and Martinez, 2012), and determining irrigation needs (Suleiman and Hoogenboom, 2007). As such, it is necessary to test these reference ET equations in regions with high relative humidity to ensure accurate ET parameterization.

One major tropical and subtropical crop that has generally high ET is sugarcane. Sugarcane is a good crop to test reference ET parameterizations because of its longer full-canopy period, when actual crop ET should be at its maximum relative to reference ET equations, and its high crop coefficient (K_c) that generally exceeds 1. Previous research in irrigated sugarcane has found full-canopy ET rates that equal or exceed evaporation rates from open-water pans (Campbell et al., 1960; Thompson and Boyce, 1967). Since the development and implementation of reference ET equations, researchers have generally found irrigated sugarcane to have a K_c greater than 1 in Australia and Swaziland (Inman-Bamber and McGlinchey, 2003), Brazil (Da Silva et al., 2012), and Texas (Salinas and Namken, 1977). However, all of these studies found variable and differing K_c values, with Inman-Bamber and McGlinchey noting a correspondence between meteorological events and outlying daily K_c values. Sugarcane's high water use, the potential for expanded irrigation to reduce yield deficits and increase production in tropical regions (Inman-Bamber et al., 1999), and the potential for sugarcane irrigation to stress water resources during dry periods in tropical areas (Ramjeawon, 1994) make it a good case study for evaluating reference ET equations in tropical regions.

To evaluate the performance of standardized reference ET equations, we established two eddy covariance (EC) towers over irrigated sugarcane fields in Hawaii, USA to measure ET (ET_{EC}). We calculated reference ET using the ASCE approach for short (ET_0) and tall (ET_r) reference vegetation. The FAO-56 ET_0 was not used as it is identical to ASCE ET_0 for calculations on a daily time step (Irmak et al., 2006; Suleiman and Hoogenboom, 2009). We also compared ET_{EC} to the Priestley–Taylor (PT) ET equation (ET_{PT}). Our objectives were to (1) determine if standardized reference ET equations adequately parameterized actual ET across differing microclimates, (2) determine the meteorological conditions that contribute to discrepancies in the standardized equations, and (3) examine corrections to improve estimates of reference ET under relatively more humid conditions.

2 Methods

2.1 Study region

We evaluated reference ET approaches in two sugarcane (*Saccharum officinarum* L.) fields with identical cultivars (Heinz et al., 1981) at a commercial farm on Maui, Hawaii (Fig. 1 and Table 1). Climatic conditions vary across the farm, with changes in precipitation, wind, solar irradiation, and air temperature due to orographic effects. Normal annual precipitation ranges from 275 to 1275 mm year^{-1} from the leeward (south) side to the windward (northeast) side of the plantation (Giambelluca et al., 2013). Elevations on the plantation range from near sea level to ~ 340 m. The western side of the plantation is generally windier (Table 1). Drip irrigation is used to maximize limited surface and ground water resources (Moore and Fitschen, 1990). Drip tape spacing is 2.70 m with sugarcane rows planted 45 cm away from the tape on both sides; the tape irrigates at 1.58 L^{-1} h^{-1} m^{-1} and is regulated to 83 kPa of pressure at the head of the row. Irrigation amounts were recorded by the farm; rainfall was

Table 1. Eddy covariance field site information.

Micrometeorological site information		
Field	Lee	Windy
Latitude (°N)	20.784664	20.824633
Longitude (°W)	156.403869	156.491278
Elevation (m)	203	44
Date field planted	28 March 2011	11 May 2011
Date tower established	21 July 2011	23 July 2011
Begin of mid-period (cover > 80 %)	3 November 2011	5 December 2011
End of analysis	26 July 2012	27 August 2012
Natural Resource Conservation Service (NRCS) soil series	Waiakoa very stony, silty clay loam	Pulehu cobbly silt loam
Bulk density*	1.22	1.35
Porosity (%)	54	49
Soil texture classification**	Clay	Sandy clay loam
Soil texture – sand (%)	31	51
Soil texture – silt (%)	15	16
Soil texture – clay (%)	54	33
Soil volumetric water content (VWC) at saturation (mm/40 cm depth)	216	196
Soil water storage (water content at 30 % VWC-wilting point) (mm)	60	72
Wilting point (% VWC)	15	12
Matric potential at 30 % VWC (MPa)	NA***	−0.01
Matric potential at 24 % VWC (MPa)	NA	−0.033
Field size (ha)	99.1	62.6
Field length (m) (predominant wind)	> 500	415
Field length (m) (shortest direction)	220	150
Mean meteorological observations (1 August 2011–31 July 2012)		
Mean daily air temperature (°C)	22.3	23.4
Mean minimum daily air temperature (°C)	17.8	20.4
Mean maximum daily air temperature (°C)	27.3	26.9
Mean daily wind speed (m s^{-1})	2.0	4.6
Mean daily net radiation (MJ m^{-2} day^{-1})	10.7	11.3
Mean daily relative humidity (%)	65	62

* All reported soil properties averaged/summed over the first 40 cm of soil depth. ** Soil texture was determined in the lab using the hydrometer method. ** Matric potential not available for Lee because of extreme logistical difficulty in obtaining intact Tempe cell samples at depth for determination of water retention characteristics.

recorded at nearby weather stations (Supplement S1). As is typical for Hawaii (Heinz and Osgood, 2009), sugarcane is grown on a 24-month rotation with planting and harvesting throughout most of the year. Peak ET, as determined by the length of the mid-season period, lasts significantly longer (330 days) than for sugarcane in other regions (190–220 days) (Doorenbos and Pruitt, 1977; Inman-Bamber and McGlinchey, 2003).

2.2 EC measurements and data analysis

We installed two micrometeorological towers in contrasting micro-climates (Fig. 1 and Table 1). These towers are at the "Windy" site (lower elevation, higher wind velocity, more constant wind direction, and sandy clay loam soil) and the "Lee" site (higher elevation, lower wind velocity, and clay soil). Field fetch in the prevailing wind directions was over 200 m for both towers. The slope in both fields, as determined using the 1/3 arcsec (∼ 10 m) digital elevation model from the US Geological Survey's National Elevation Dataset (http://ned.usgs.gov/index.html), is less than 3 %. Beyond the edge of each field, Windy was surrounded by sugarcane fields on all sides for over 1500 m; Lee was bordered by non-irrigated rangeland in the non-prevailing wind directions (east and south) and contiguous sugarcane fields to the north and east.

Tower instrumentation included an integrated EC system (EC150 – Campbell Scientific, Logan, Utah, USA[1]) with an open-path infrared gas analyzer, aspirated temperature probe,

---

[1]Mention of trade names or commercial products in this publication is solely for the purpose of providing specific information

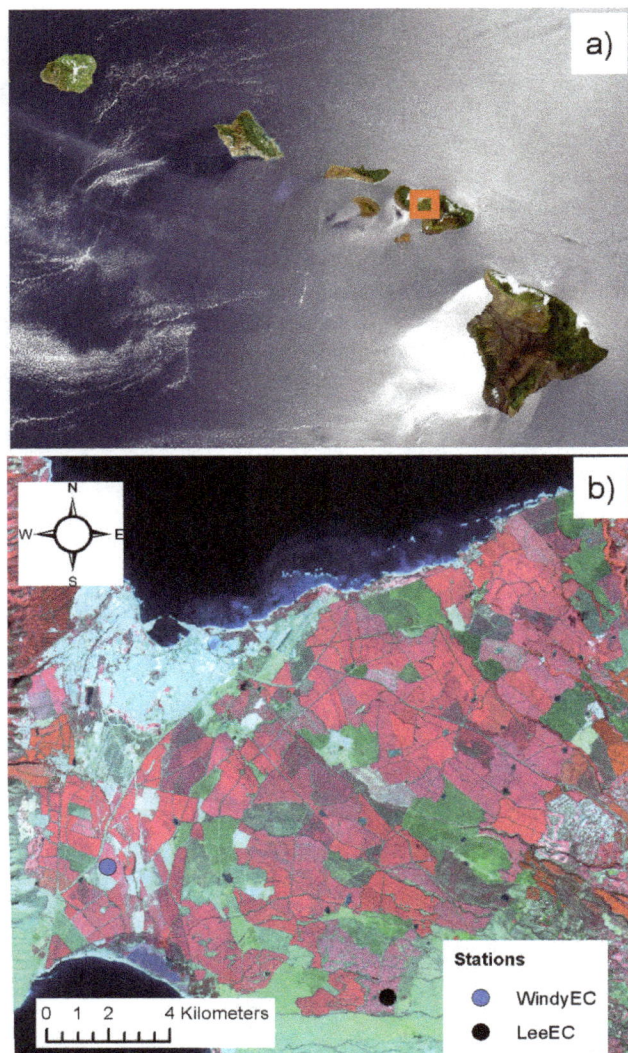

Figure 1. (a) True color image of the main Hawaiian islands from the moderate resolution imaging spectroradiometer (250 m resolution – image date: 27 May 2003). Study region is outlined in red box. **(b)** The study region on central Maui showing the location of the eddy covariance towers (Windy and Lee) used in this study. Image is false color Landsat 7 (30 m resolution – image date: 5 February 2000).

attached 3-D sonic anemometer head (CSAT3A – Campbell Scientific), and enhanced barometer (PTB110 – Vaisala, Vantaa, Finland). Relative humidity and air temperature were measured by a combined temperature and relative humidity probe (HMP45C – Vaisala). Net radiation was measured with a single component net radiometer (NR-Lite2 – Kipp and Zonen, Delft, Netherlands). We corrected the single component net radiometer for the effect of wind following Cobos and Baker (2003). Ground heat flux was measured as the average

and does not imply recommendation or endorsement by the US Department of Agriculture.

of four self-calibrating heat flux plates (HFP01SC – Huskeflux, Delft, Netherlands). The plates were installed at 5 cm depth at four lateral locations perpendicular to the irrigation drip line (Sect. 2.1): 0 cm (drip line), 45 cm (sugarcane row), 75 cm, and 135 cm (mid-point between drip lines). All instruments were factory calibrated to ISO 9001:2008 standards prior to deployment; data were recorded and processed on solid-state data loggers (CR3000, Campbell Scientific).

Two water content reflectometry probes (CS616 – Campbell Scientific) were installed at 20 cm depth at two locations perpendicular to the drip line (45 and 135 cm) to measure soil volumetric water content (VWC). These locations were chosen to correspond with the sugarcane row (center of root zone) and halfway between sugarcane rows. VWC was measured to independently assess potential water stress in both fields. VWC was calculated using a quadratic equation with empirically determined coefficients specific to each field following the manufacturer's recommendation. Soil water retention and permanent wilting point were also determined for Windy but, due to rockiness at the Lee site, could not be determined for Lee because of the logistical difficulty and equipment risk in obtaining intact Tempe cell samples below the surface. More technical details on soil calibrations are provided in Supplement S1.

The EC150 system measured CO_2, H_2O, wind velocity, and sonic temperature at 10 Hz. Other variables were averaged to 30 min fluxes. We processed raw covariances on the data logger and post-processed high-frequency time series data with commercial software (Eddy Pro Advanced V 3.0 and 4.0 – LI-COR, Lincoln, Nebraska USA). Data logger flux calculations were downloaded daily via a cellular modem. High-frequency (10 Hz) data and half-hourly fluxes were transferred monthly via a data card. Raw time series data were checked following the tests of Vickers and Mahrt (1997). Sonic anemometer tilt was corrected using double rotation (Kaimal and Finnigan, 1994); lags between the infrared gas analyzer and sonic anemometer were determined using maximum covariance. We corrected for density fluctuations (Webb et al., 1980), low pass filtering (Moncrieff et al., 1997), and high pass filtering (Moncrieff et al., 2004). Flux footprint lengths were calculated following Kljun et al. (2004), and quality flags were assigned following the CarboEurope standard (Mauder and Foken, 2011). We independently calculated stability (Obukhov, 1971). After installation, tower heights were periodically adjusted to keep meteorological instrumentation ~ 3.0–3.3 m above the zero-plane displacement height, which was assumed to be 67 % of canopy height (Arya, 2001). Canopy height was measured biweekly, concurrent with the vegetation cover observations (Sect. 2.4). Additional detailed EC cross-validation activities are described in Supplement S1.

Half-hourly fluxes with instrumentation errors flagged by the EC150 system, rainfall, or lack of turbulence (friction velocity $< 0.1 \, \mathrm{m\,s^{-1}}$) were excluded. Excluded fluxes were gap-filled as a function of fluxes measured from similar meteorological periods using the Max Planck Institute tool (http://www.bgc-jena.mpg.de/~MDIwork/eddyproc/index.php) (Reichstein et al., 2005). Gap-filled fluxes were used to calculate daily and cumulative fluxes but were excluded from half-hourly analyses. We corrected fluxes for energy budget closure by regressing daily EC-observed available energy against measured available energy (net radiation minus ground heat flux) and forcing the regression through the origin, preserving the daily mean Bowen ratio and adjusting each day's ET by the regression slope for the entire study period (Anderson and Wang, 2014; Leuning et al., 2012).

2.3 Reference ET equations, corrections, and evaluation of controls

At each tower, daily and hourly reference ET was calculated using the ASCE short (ET_0) and tall (ET_r) reference equations, where short and tall refer to parameterized surfaces similar to well-watered fescue grass (short) and alfalfa (tall), with differences in the equations due to assumed leaf area index (LAI) and bulk canopy resistance to ET:

$$ET_{sz} = \frac{0.408\Delta\,(R_n - G) + \gamma\frac{C_n}{T+273}u_2\,(e_s - e_a)}{\Delta + \gamma\,(1 + C_d u_2)}. \quad (2)$$

As shown in Eq. (2), ET_{sz} is the reference ET type (ET_r or ET_0 in mm day^{-1} or mm h^{-1} depending on time step); R_n and G are net radiation and ground heat flux (MJ m^{-2} day^{-1} or MJ m^{-2} h^{-1}), respectively; γ is the psychrometric constant (kPa °C^{-1}); T is mean daily or hourly air temperature (°C); u_2 is mean daily or hourly wind speed measured at or scaled to 2 m height; e_s and e_a are mean saturation and actual vapor pressure (kPa), respectively; and C_n and C_d are empirical numerator and denominator constants that change with the reference surface and time step (Table 1 in Allen et al., 2005). We scaled all meteorological variables from 3 m above the zero-plane displacement to 2 m, following the ASCE procedure for adjusting meteorological measurements at non-standard heights. Following ASCE, mean daily meteorological values were calculated as an average of daily minimum and maximum values as opposed to averaging all 24 h of measurements. Differences between these averaging approaches were small (mean T difference of 0.26 and 0.27 °C in Windy and Lee, respectively). Measured net radiation and ground heat fluxes were used for all calculations.

We also calculated another reference using the PT equation (Priestley and Taylor, 1972). PT was chosen as a comparison because of its different treatment of advection in comparison to the PM-type equations, its wide usage, and the relative simplicity of its meteorological inputs compared to PM. The PT equation is

$$ET_{PT} = \frac{\alpha}{\lambda}*\frac{\Delta\,(R_n - G)}{\Delta + \gamma}. \quad (3)$$

ET_{PT} is the PT ET (mm day^{-1}); Δ, γ, R_n, and G are the same as in Eq. (2); λ is the latent heat of vaporization; and α is an empirical constant. We assumed that λ is 2.45 MJ mm^{-1}, which is the same as the ASCE/FAO-56 approach. We used an α of 1.26, which is widely, but not universally, representative of a well-watered surface across a variety of climates (e.g., Eichinger et al., 1996; McAneney and Itier, 1996).

To examine the discrepancies between the ASCE equations (ET_0 and ET_r), the PT equation (ET_{PT}), and measured ET_{EC} we inverted the PM equation to calculate bulk canopy resistance (r_c) from ET_{EC} and ET_{PT} and compared the calculated r_c to the constant r_c used to calculate ET_0 and ET_r during the mid-period. The ASCE parameterization to calculate atmospheric resistance (r_a) was used in the inverted PM equation. Days with available energy (net radiation (R_n) − ground heat flux (G)) of < 5 MJ day^{-1} were excluded because low radiation values would result in extreme r_c values and to avoid including days with precipitation, which would bias the net radiation measurement of the NR-Lite2.

Once the discrepancies between reference and measured ET became apparent (see Sects. 3.2 and 3.3), we attempted two corrections to the ASCE reference ET approach to better parameterize sugarcane water use. One was a climatological correction to the ET coefficient (K_{C-adj}). Following the FAO-56 approach (Allen et al., 1998), an adjustment term (K_{adj}) was calculated:

$$K_{adj} = 0.04*(U_{2avg} - 2) - 0.004*(RH_{avg} - 45)*h_{avg}^{0.3}, \quad (4)$$

$$K_{C-adj} = K_{C-FAO} + K_{adj}. \quad (5)$$

In Eqs. (4) and (5), K_{C-FAO} is the literature mid-canopy K_C value, U_{2avg} is mean location wind speed (m s^{-1}) at 2 m height, RH_{avg} is mean location relative humidity, and h_{avg} is average vegetation height. For our study we used average wind speed, relative humidity, and vegetation height over the mid-period to calculate these parameters in the absence of longer-term climate data. The FAO-56 provides a range of mid-period K_C values for sugarcane (1.25–1.40) for short reference ET. For adjustment, we chose the lowest end of the range (1.25) for K_{C-FAO} to enable the most conservative estimate of parameterized ET.

The second correction was to parameterize the ASCE–PM equation with a custom constant r_c. To estimate a r_c value, an intermediate bulk canopy resistance of 165 s m^{-1} was used, which was chosen as the weighted average of the r_c calculated by inverting the ET_{PT} at Windy and Lee. We then ran the full-form PM equation to calculate a new reference ET (ET_{r-cane}).

Along with corrections to the reference ET equations, we examined potential controls on the discrepancies between reference and measured ET values. Daytime and nighttime r_c were investigated by inverting the full PM equation with measured ET to see if there was a systematic time-of-day difference between the fields and to see if errors in daytime- or nighttime-parameterized r_c were disproportionally contributing to discrepancies in reference ET. Daily daytime and nighttime r_c were calculated for days that had at least eight (daytime) and four (nighttime) non-gap-filled half-hourly flux measurements. For these calculations, daytime was defined as $R_n > 50\,\mathrm{Wm}^{-2}$ and nighttime as $R_n < -10\,\mathrm{Wm}^{-2}$. We used this definition to avoid including periods with near zero Rn that would blow up the inverted PM equation. Finally, we evaluated the correlation between meteorological observations and discrepancies between the ASCE tall reference ET equation ($\mathrm{ET_r}$) and $\mathrm{ET_{EC}}$ to assess the importance of the advective and radiation terms in the PM equation.

2.4 Canopy cover and determination of mid-period

We measured fractional canopy cover with an optical camera to obtain an independent, conservative determination of the mid-season period (mid-period) for intercomparison of measured and reference ET. The mid-period is one of the growth/ET stages in the FAO/ASCE methodology and corresponds to maximum plant transpiration and the highest ecosystem coefficient (K_{c-mid}). In unstressed sugarcane, the mid-period coefficient should exceed 1 (Allen et al., 1998), thus measured ET should exceed reference ET. The camera (TetraCam ADC multispectral camera, TetraCam Inc., Chatsworth, California, USA) contains a single precision 3.2 megapixel image sensor optimized for capturing green, red, and near-infrared wavebands of reflected light. A telescoping pole tripod system (GeoData Systems Management Inc., Berea, Ohio, USA) was used to suspend the camera directly above the plant at a height of 7 m and aim vertically downward at nadir view. Each field was photographed every $\sim 16 \pm 2$ days. Ten images were taken in two lines perpendicular to the irrigation line at pre-selected sampling locations in each field at solar noon $\pm$ 2 hours; sampling locations were identical throughout the study. Each image was preprocessed with image processing software (LView Pro 2006 – CoolMoom Corp., Hallandale, Florida, USA) to paint out the pixels of soil, grass, shadow, and other background. The preprocessed image was then analyzed using proprietary software (PixelWrench, TetraCam Inc.) to classify fractional vegetation cover based on threshold analysis, and the cover readings from the 10 locations were averaged to determine mean and standard error of field vegetation cover. We considered the beginning of the mid-period to be the latter of the beginning date of mid-period from the FAO-56 K_C curve (Allen et al., 1998) or the date when canopy cover clearly exceeded 80 %, which has been shown to coincide with the start of mid-period (Carr and Knox, 2011). The end of the K_{C-mid} period was set to 27 August 2012, which was the last date of irrigation data prior to the end of the FAO-56 mid-period. Finally, we further restricted the end of the mid-period in the earlier planted field (Lee) to ensure that the length of the

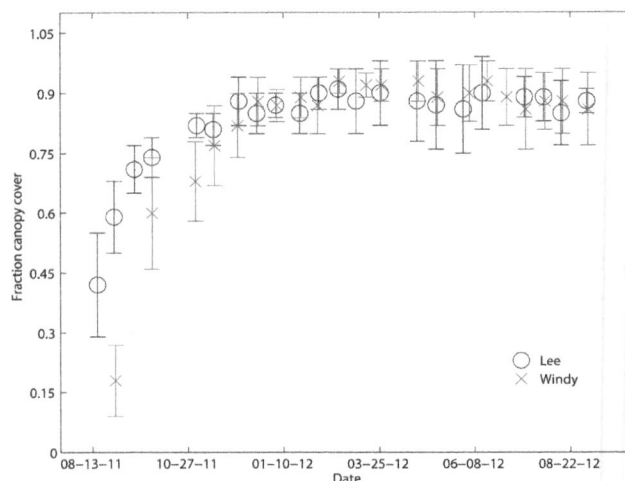

Figure 2. Measured mean and standard deviation of fractional vegetation cover from TetraCam for Windy and Lee fields.

mid-period was identical in both fields for intercomparison purposes.

2.5 LAI and stomatal resistance measurements

We measured LAI and leaf stomatal resistance in a field campaign during the mid-period for both EC fields (July 2012). LAI was measured using a nondestructive, optical plant canopy analyzer (LAI 2200, LI-COR Inc.) on 13 July in the Lee field and 16 July in the Windy field. At each of the 10 TetraCam sampling locations in each field (Sect. 2.4), we made 10 below-canopy and 5 above-canopy measurements with the optical canopy analyzer; we then used the manufacturer's software (FV2200, LI-COR Inc.) to determine mean and standard error of LAI for both fields. To observe leaf level stomatal resistance, we used a steady-state diffusion porometer (SC-1, Decagon Devices Inc.), which has been used to observe response to different irrigation regimes in multiple agronomic crops (e.g., Ballester et al., 2013; Hirich et al., 2014; Mabhaudhi et al., 2013; Mendez-Costabel et al., 2014). At each TetraCam point nine leaves were measured: three fully sunlit upper-canopy leaves near (< 20 vertical cm away from) the top visible dewlap (TVD) point (Glaz et al., 2008), three mid-level leaves that were attached to the cane stalk below the TVD height but were still mostly sunlit, and three lower canopy leaves that were partially to mostly shaded. Porometry measurements were made in a 30 s measurement window using the porometer's automatic mode. We also repeated the stomatal resistance measurements at five of the TetraCam points in the Windy field to evaluate the larger discrepancies in reference ET observed in that field.

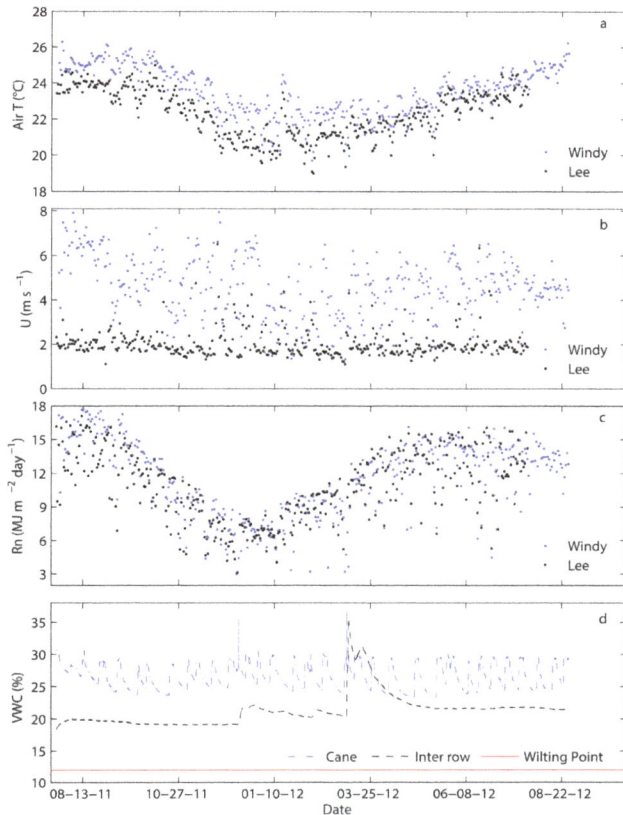

Figure 3. Meteorological and soil observations during the study period: (**a**) mean daily air temperature; (**b**) mean 24 h wind velocity; (**c**) cumulative daily net radiation; and (**d**) soil volumetric water content (VWC) data from Windy field at 20 cm depth underneath cane row (45 cm away from drip line) and inter-row or midway between drip lines (137 cm away from drip line). Wilting point noted as solid red line (12 % VWC).

3　Results

3.1　Fractional vegetation cover, LAI, and leaf stomatal resistance

Fractional vegetation cover increased rapidly in both fields after the beginning of the EC measurements (Fig. 2). Initial cover was $< 20\%$ in Windy and $< 45\%$ in Lee (112 and 142 days after planting (DAP), respectively). Some early TetraCam sampling dates were missed due to initial equipment failures. Vegetation cover exceeded 80 % in Lee on 3 November 2011 and 5 December 2011 in Windy (220 and 208 DAP, respectively), which we considered the onset of the mid-period. Both of these dates are later than the onset of mid-period according to the FAO-56 curve (180 DAP). Variation in cover was largest at the beginning of the study period (standard deviation of $\sim 10\%$) (Fig. 2). Vegetation cover was least variable near the onset of the mid-period (standard de-

Table 2. A summary of cumulative irrigation, rain, actual measured evapotranspiration-ET_{EC}, and reference evapotranspiration values (ASCE short-ET_0 and tall-ET_r, Priestley–Taylor ET_{PT}, and a custom cane reference ET-ET_{r-cane}) for the entire study period and the mid-period. All values are in mm.

	Lee		Windy	
	Whole study	Mid-period	Whole study	Mid-period
Irrigation	1599	1348	1928	1221
Rain	58	58	140	122
ET_{EC}	1191	843	1389	1001
ET_0	1487	1042	2099	1367
ET_r	1828	1292	2861	1861
ET_{PT}	1470	1008	1707	1096
ET_{r-cane}	1317	947	1662	1128

viation $< 5\%$). Mean canopy height reached 3.97 m in Lee and 4.09 m in Windy by the end of the study.

Mean ± standard error of measured LAI was 4.9 ± 0.2 in Windy on 13 July 2012 and 4.7 ± 0.3 in Lee on 16 July 2012. Midday leaf stomatal resistance (r_s) observations of fully sunlit leaves in Windy ($n = 32$) and Lee ($n = 21$) showed substantial variation, ranging from 45 to $259\,s\,m^{-1}$ in Windy and 40 to $640\,s\,m^{-1}$ in Lee. Median r_s in Windy and Lee were 112 and $114\,s\,m^{-1}$, respectively. Mean ± standard deviations of r_s in Windy and Lee were $125 \pm 57\,s\,m^{-1}$ and $161 \pm 157\,s\,m^{-1}$, respectively. There were two observations in Lee of sunlit stomatal resistance of $> 500\,s\,m^{-1}$. Excluding these two observations resulted in a revised mean and median r_s in Lee of 114 and $104\,s\,m^{-1}$, respectively. Mean sunlit stomatal resistance was not significantly different ($p < 0.01$) from $100\,s\,m^{-1}$ in either Windy ($p = 0.02$) or Lee ($p = 0.09$).

3.2　Meteorological observations

Air temperature and net radiation were similar in both Windy and Lee (Fig. 3a and c; Table 1). In Windy, mean daily air temperature ranged from 19.0 to 25.0 °C over the study period, whereas in Lee, mean daily air temperature ranged from 19.7 to 26.3 °C. Mean air temperature was higher in Windy than Lee (23.5 and 22.3 °C, respectively) with a similar, low day-to-day variability (standard deviation of 1.3 °C for both fields). Daily net radiation (Rn) was also similar between fields; Rn was slightly higher in Windy versus Lee (11.5 and $10.9\,MJ\,m^{-2}\,day^{-1}$; Fig. 3c and Table 1). Both fields showed larger relative variations in R_n ($\sim 10\,MJ\,m^{-2}\,day^{-1}$) than in other meteorological observations. Wind velocities were sharply divergent between the two fields. Mean wind velocity was more than twice as high ($4.6\,m\,s^{-1}$ versus $2.0\,m\,s^{-1}$) in Windy compared to Lee (Fig. 3b; Table 1). Wind velocities were also more variable in Windy than Lee (standard deviation of 1.4 and $0.7\,m\,s^{-1}$, respectively).

VWC observations in the Windy field underneath the sugar cane row/line varied from 23 to 30 % during the mid-

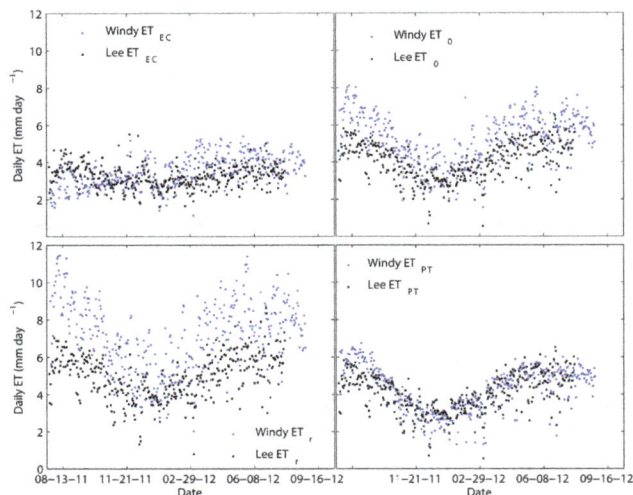

Figure 4. Daily measured and reference ETs for EC tower fields from the tower's establishment until the end of the study period for each field.

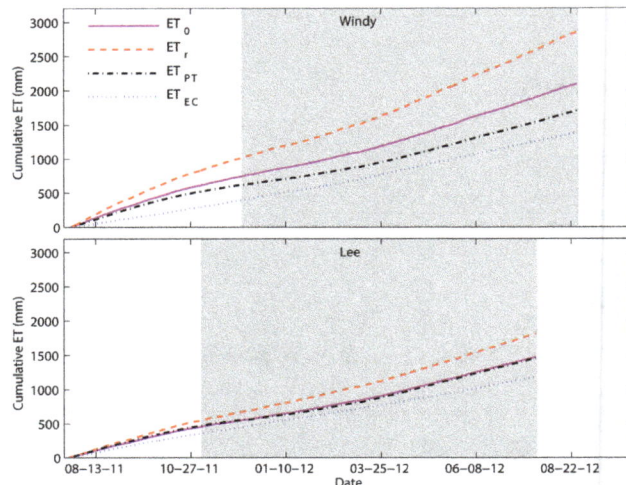

Figure 5. Cumulative measured and reference ET for Windy and Lee. Shaded background indicates mid-period when ground canopy cover exceeded 80 %.

period except after major rain events in December 2011 and March 2012 when they spiked to 36–37 % (Fig. 3d). At all times, VWC remained well above wilting point (12 %) for both sensors (Table 1). Available plant water in the top 40 cm of the soil at minimum VWC was ~ 40 mm. Soil matric potentials in Windy near typical maximum (30 %) and minimum (24 %) soil VWC were −0.01 and −0.033 MPa, respectively (Table 1). Shallow VWC observations underneath the cane row are likely indicative of plant water stress due to the majority of drip-irrigated Hawaiian sugarcane roots being at less than 50 cm depth (Evensen et al., 1997). VWC observations between drip lines showed relatively little periodicity compared to underneath the cane row, indicating that neither irrigation events nor root depletion was impacting VWC at this location. Due to difficulties with instrument installation and instrument failure we were not able to obtain a reliable time series of soil VWC observations in the Lee field. Precipitation at both fields was less than 150 mm over the course of the study, with irrigation providing more than 90 % of the water input (Table 2).

From the tower's establishment until the end of the study period, daily EC measured ET (ET_{EC}) ranged from 1.6 to 5.5 mm day^{-1} with a mean of 3.2 mm day^{-1} in Lee and 1.6 to 5.5 mm day^{-1} with a mean 3.8 mm day^{-1} in Windy (Fig. 4). ET_{EC} showed relatively little seasonal variation (< 3 mm day^{-1} from summer maxima to winter minima) and greater day-to-day variations of 1–2 mm day^{-1}. Cumulatively, mid-period ET_{EC} was 158 mm higher in Windy than in Lee (Fig. 5; Table 2). Factors contributing to higher ET_{EC} in Windy include higher wind speed, slightly higher Rn, a higher mean air temperature, and lower mean daily relative humidity. However, maximum daily air temperature is higher near Lee than Windy. Ground heat flux was minimal (< 3 %

of R_n during daytime periods) at both sites during the mid-period.

Quality control checks on the EC data indicated no significant issues with ET measurements. Energy closure varied significantly between the sites, with daily energy closure of the turbulent fluxes of 75 % at Lee and 97 % at Windy. As data processing and instrumentation were identical between sites, the difference in energy closure is very likely due to the differences in topography and turbulence between the two fields, particularly nighttime turbulence (Anderson and Wang, 2014). Friction velocity at Windy rarely dropped below the critical threshold (0.1 m s^{-1}) at night (2.5 % of the half-hourly fluxes). Mean 90 % footprint lengths during the study period, determined following Kljun et al. (2004), were 158 m in Windy and 124 m in Lee, which indicate that our EC towers were observing the field of interest even during the rare periods (~ 7 % of record) when we were observing in the short fetch direction (Table 1), such as during Kona winds (winds from the south and west). During the predominant trade wind flows (prevailing winds from the northeast), our fetch in both fields was > 200 m.

3.3   Reference ET at EC tower sites

Daily short (ET_0) and tall (ET_r) ASCE reference ET were significantly different between the two sites (Fig. 4). In Windy, ET_0 ranged from 1.6 to 8.1 mm day^{-1} over the study period with a mean of 5.2 mm day^{-1} (5.1 mm day^{-1} over the mid-period). ET_r ranged from 2.0 to 12.3 mm day^{-1} with a mean of 7.14 mm day^{-1} (7.0 mm day^{-1} for mid-period). For Lee, ET_0 varied from 0.6 to 6.5 mm day^{-1} with a mean of 4.0 mm day^{-1} (3.9 mm day^{-1} for mid-period). ET_r ranged from 0.8 to 8.6 mm day^{-1} with a mean of 5.0 mm day^{-1} (4.8 mm day^{-1} mid-period). The PT ET (ET_{PT}) showed less

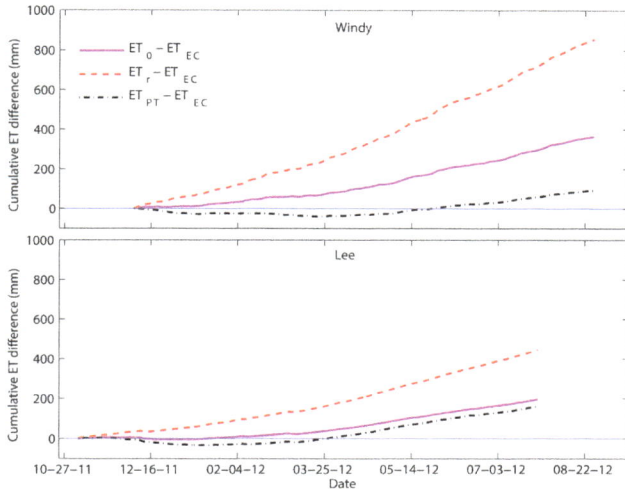

Figure 6. Cumulative difference between reference and measured ET since the beginning of the mid-period in each EC tower field.

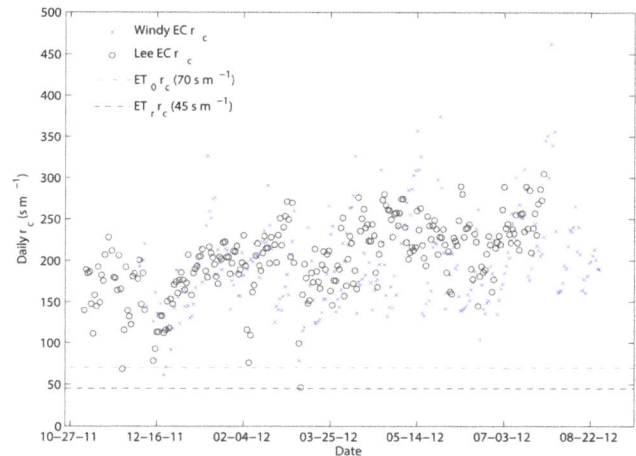

Figure 7. Calculated daily bulk canopy resistance at Windy and Lee from the EC towers for the mid-period. Dotted lines show daily time step resistances from short-canopy ($ET_0 - 70\,s\,m^{-1}$) and tall-canopy ($ET_r - 50\,s\,m^{-1}$) reference surfaces.

difference between the two fields. Mean ET_{PT} was slightly higher at Windy (4.3 and 4.1 mm day^{-1} mid-period) than at Lee (4.0 and 3.8 mm day^{-1} mid-period).

Over the course of the study, the cumulative ET_0 in Windy was 612 mm higher than in Lee, and cumulative ETr was 1032 mm higher (Fig. 5; Table 2). Similar to the daily values, cumulative ET_{PT} values were considerably closer, with Windy exceeding Lee by 237 mm. As expected, the cumulative difference between reference equations and ET_{EC} grew in the early portion of the study period prior to the mid-period (Fig. 5). During the mid-period, the difference between ET_r and ET_{EC} grew significantly larger in both EC fields. Windy also saw increasing differences between ET_0, ET_{PT}, and ET_{EC}, whereas in Lee, cumulative ET_0 and ET_{PT} tracked quite closely with each other.

To further evaluate these discrepancies between reference and ET_{EC}, we calculated the cumulative difference between the three reference ET equations and ET_{EC} during the mid-period (Fig. 6). ET_{PT} was the only equation with near zero cumulative difference for a substantial amount of the mid-period for both fields; ET_0 was near 0 for the Lee field from October 2011 to February 2012 but not for the Windy field. Over the mid-period in Windy, the difference between cumulative ET_{EC} and ET_{PT} ranged from −40 mm in March 2012 to 92 mm at the end of the study period (August 2012) with cumulative differences of < 40 mm until July 2012. In Lee, the differences were greater, varying between −33 and 161 mm. The difference with ET_0 ranged from 0 (at the beginning of mid-period) to 362 and 195 mm in Windy and Lee, respectively. ET_r showed the greatest cumulative differences of 854 and 443 mm in Windy and Lee.

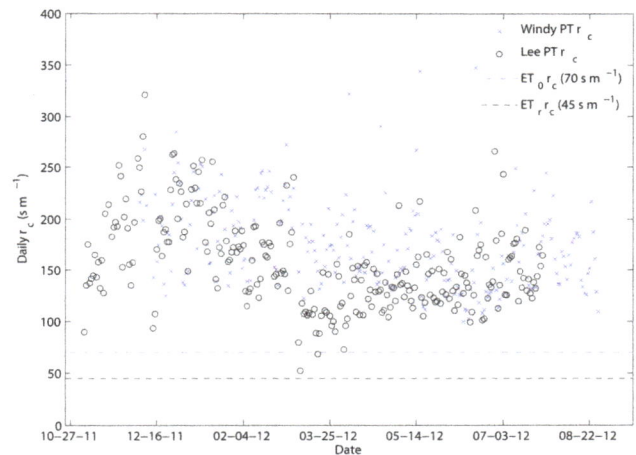

Figure 8. Calculated daily bulk canopy resistances at Windy and Lee from inverting the Priestley–Taylor (PT) ET for the mid-period. Dotted lines again show daily time step resistances from short and tall canopy for comparison.

3.4 Bulk canopy resistances at EC towers, soil observations, and patterns in ET discrepancies

r_c varied considerably between Windy and Lee for ET_{EC}. For the mid-period, mean $\pm$ standard deviations of daily r_c at Lee and Windy were 201 ± 47 and $145 \pm 36\,s\,m^{-1}$, respectively (Fig. 7). With respect to ET_{PT}, mean $\pm$ standard deviations of daily r_c at Lee and Windy during the mid-period were $146 \pm 28\,s\,m^{-1}$ and $175 \pm 42\,s\,m^{-1}$, respectively (Fig. 8). In all cases, mean r_c values were significantly higher ($> 75\,s\,m^{-1}$) than the daily r_c values used to parameterize the ET_0 and ET_r equations.

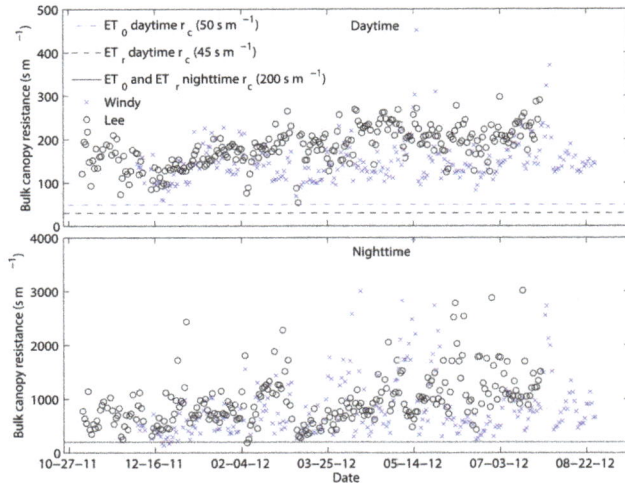

Figure 9. Calculated mean nighttime and daytime bulk canopy resistances (following Fig. 6) compared to assumed resistances.

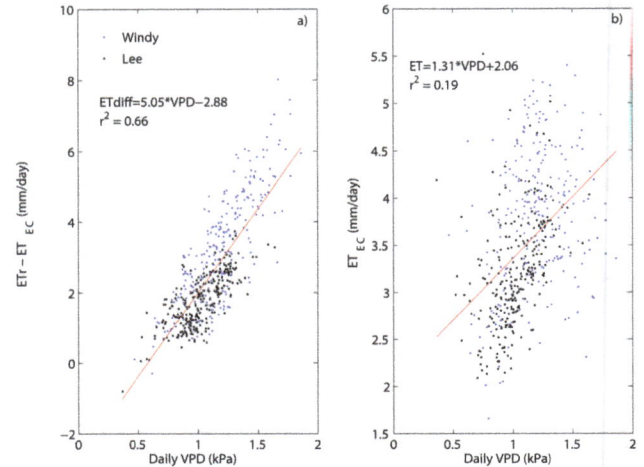

Figure 10. (a) Relationship between daily ET discrepancy (ET_r-ET_{EC}) and daily vapor pressure deficit (VPD) from the beginning of the mid-period to the end of the study period. Regression equation is fitted to the entire pool of data from Lee and Windy. **(b)** Relationship between measured ET and daily VPD; time period and regression approach are the same as in **(a)**.

Daily daytime and nighttime r_c are shown in Fig. 9. Nighttime r_c shows greater difference between towers, with mean $\pm$ standard deviation in Windy and Lee of 675 ± 289 and $808 \pm 445 \, s\,m^{-1}$ and substantially larger absolute and relative standard deviation in r_c. For both fields, daytime and nighttime r_c were larger than the ASCE r_c parameterizations for almost all days. One other notable feature of the resistance terms was the low atmospheric resistance (r_a); in Windy and Lee, mean daily r_a were 17.7 and $38.6 \, s\,m^{-1}$, respectively, over the study period.

With respect to meteorological controls on the discrepancies between ET_r and ET_{EC}, the only parameter that was highly correlated to ET discrepancy (ET_r-ET_{EC}) was VPD with a coefficient of determination (r^2) of 0.66 (Fig. 10a). VPD showed a much stronger correlation with ET discrepancy than ET_{EC} ($r^2 = 0.19$) (Fig. 10b). Available energy was moderately correlated with ET discrepancy ($r^2 = 0.37$) while all other tested parameters (daily minimum, mean and maximum wind speed and temperature) had weak or no correlation with ET discrepancy ($r^2 < 0.1$).

3.5 Corrections to better parameterize sugarcane water use

The climatological K_C adjustment (K_{adj}) had relatively little impact on calculated water use. In the Windy field K_{adj} was -0.0126, and in Lee K_{adj} was -0.0359. For both fields, the wind adjustment offset the relative humidity/vegetation height adjustment as all three parameters were greater than zero. The magnitude of the K_{adj} term was insufficient to account for the observed discrepancies between reference ET and ET_{EC}.

Cumulative differences between ET_{r-cane} and ET_{EC} are shown in Fig. 11 along with the differences between ET_{PT} and ET_{EC}. ET_{r-cane} showed some improvements over ET_{PT}

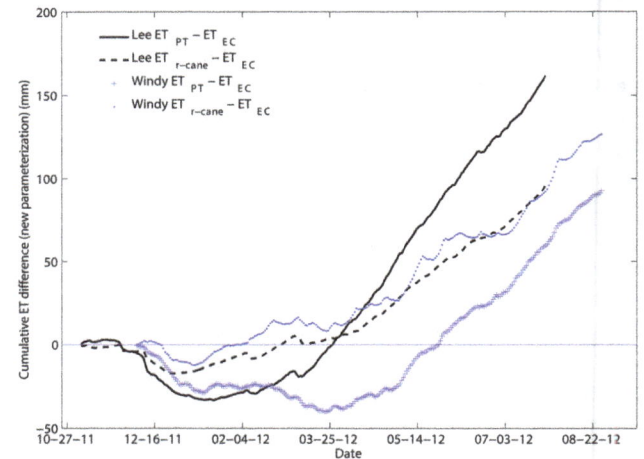

Figure 11. Cumulative difference between new reference ET (custom bulk canopy resistance of $165 \, s\,m^{-1}$) and measured ET for both EC tower fields during the mid-period.

in predicting measured ET between October 2011 and March 2012; in particular ET_{r-cane} had less underestimation of ET (15 to 27 mm improvement) in winter and spring for both fields and had consistently better performance in the Lee field. ET_{r-cane} had worse performance than ET_{PT} during the summer in the Windy field (40 mm). The minimum cumulative difference between ET_{r-cane} and ET_{EC} was -12 and -18 mm in Windy and Lee, respectively. The maximum cumulative difference between ET_{r-cane} and ET_{EC} was 132 and 164 mm at the end of the study period in Windy and Lee, respectively.

4 Discussion

4.1 Is Hawaiian sugarcane representative of a fully transpiring reference ET surface?

Well-irrigated, full-canopy sugarcane has generally been reported to have an ET rate 1.1 to 1.4 times the ASCE/FAO-56 reference ET_0 equation (Da Silva et al., 2012; Inman-Bamber and McGlinchey, 2003), and rain-fed sugarcane has been reported to have an ET rate approaching ET_0 (Cabral et al., 2012). Furthermore, a reference PM–ET equation designed specifically for sugarcane, created by McGlinchey and Inman-Bamber (1996), has a bulk canopy resistance that is slightly lower than the daily ASCE ET_r equation (40 versus $45\,s\,m^{-1}$ for ASCE ET_r). Therefore, the significant overestimation of measured ET (ET_{EC}) by the ET_0 and ET_r equations found in this study was quite surprising. Although Windy and Lee fields had slight differences in planting dates, available soil water capacity, and fetch (Table 1), we do not believe these account for the observed ET/reference ET differences between the fields. Seasonal variation in temperature in Hawaii is quite small; wind speeds appeared to be uncorrelated to seasonality. Wind fields in central Maui are generally very strong, and our separate calculations of reference ET using independent farm weather station observations (Supplement S1) and publicly available airport weather data from Kahului airport (http://mesonet.agron.iastate.edu/request/download.phtml?networkHI_ASOS – station ID PHOG) show higher-than-typical values of reference ET for a tropical region.

The quality of EC observations was good, especially at the Windy tower where high turbulence, flux footprints that were well within field boundaries, low proportion of time periods requiring gap filling, and excellent energy budget closure ($H + LE$ was $> 95\,\%$ of daily R_n-G) indicated that the methodological requirements of the EC method were well satisfied (Anderson and Wang, 2014). At the Lee tower, EC measurements showed a more typical pattern, with a larger number of gaps during still nighttime periods when ET is low. Furthermore, seasonal and annual totals of ET have been shown to be relatively insensitive to gap-filling methodologies (Alavi et al., 2006). Finally, while the gap-filling method of Reichstein et al. (2005) may systematically underestimate wet canopy evaporation due to exclusion of all EC periods during and immediately after rain, this bias is likely to be insignificant at our sites due to the low precipitation (Table 2) and drip irrigation that would minimize wetting of the leaves.

One hypothesis is that portions of the fields measured by our EC towers were under significant water stress or had less-than-optimal cover and thus were not representative of a reference ET type surface. Uniformity of irrigation is a major concern with drip irrigation, particularly with sub- and near-surface drip lines where root development can plug or pinch drip lines, leading to insufficient irrigation (e.g., Soopramanien et al., 1990). At our field with higher ET (Windy),

visible dry lines arising from pinched drip tubes appeared in parts of the field at and after the end of the study period. However, there are multiple independent lines of evidence against this hypothesis.

With respect to canopy cover, the TetraCam observations of cover (Fig. 2) show that fractional cover remained above $80\,\%$, a threshold for the mid-period K_C (Carr and Knox, 2011; Inman-Bamber and McGlinchey, 2003). More evidence for full canopy comes from the LAI measurements made in July 2012 toward the end of the mid-period. In both Lee and Windy, mean LAI (4.7 and 4.9) were slightly higher than the LAI (4.5) parameterized in the ET_r equation (Allen et al., 2005). These two types of data indicate that incomplete cover is not an issue with our study sites.

Another possibility is that the sugarcane leaves are under significant water stress and thus are transpiring at a lower rate. Four factors show that the sugarcane is unlikely to be water stressed. First, porometer measurements from the July 2012 campaign of midday, sunlit leaf stomatal resistance were not significantly $> 100\,s\,m^{-1}$. The $100\,s\,m^{-1}$ comes from the mean leaf level stomatal resistance of a sunlit leaf on a well-watered plant, as measured by Szeicz and Long (1969), and which is used as a basis for scaling bulk canopy resistance in the ASCE and FAO-56 approaches (Allen et al., 1998, 2005). Second, we compared the daily observed ET coefficient (K_C) from the day immediately preceding a substantial irrigation or rain event (defined as $> 8\,mm\,day^{-1}$) during the mid-period with daily K_C 2 and 3 days after the irrigation event using a paired t test ($n = 106$ in Windy and $n = 98$ in Lee). We reasoned that stressed full-canopy sugarcane would respond to irrigation within 3 days, but that 3 days were short enough to avoid confounding changes due to variations in field water budgets. Neither field showed significantly greater daily ET_{EC} following an irrigation during the mid-period ($p > 0.40$ for all tests). Third, the soil VWC data from the Windy field indicate relatively high soil moisture content; available soil water underneath the cane row in the middle of the root zone always remained at $> 50\,\%$ of available capacity. Windy's soils were also near field capacity (and far above permanent wilting point) based on matric potential at typical maximum and minimum soil VWC (Table 1). The VWC content also argues against severe water stress that might persist after irrigation relieves the soil moisture deficit; thus if the ASCE reference ET equations and coefficients were applicable to this situation, we should see at least some days with ET_{EC} in the range of ET_0 and ET_r (6–$10\,mm\,day^{-1}$ in Windy) when soil moisture was near or above field capacity. Fourth, measured irrigation plus precipitation as recorded by the plantation was compared to measured cumulative ET_{EC}, with cumulative mid-period irrigation and precipitation exceeding ET_{EC} by $342\,mm$ in Windy (Table 2). At all times in the Windy field, cumulative ET_{EC} was significantly less than irrigation plus precipitation. In Lee, by early January 2012 cumulative precipitation and irrigation exceeded ET_{EC}; by the end of the mid-period (July 2012), cumulative irrigation and precipitation exceeded cumulative ET_{EC} by $> 500\,mm$ (Table 2). In summary, the evidence of full canopy and the lack of evidence of water stress indicated that the mid-period sugarcane at our study fields should be fully transpiring.

4.2 Why do the standardized ASCE reference ET equations differ between similar sites?

Without clear evidence of water stress or lack of canopy cover over the study sites, we examine some explanations for the overestimation of the ASCE ET_0 and ET_r compared to ET_{EC} and ET_{PT}. Four hypotheses include (1) scaling of leaf level stomatal resistance to whole canopy bulk resistance, (2) incorrect parameterization of daytime leaf level resistance, (3) underestimation of nighttime bulk canopy resistance, and (4) underestimation of atmospheric resistance. Scaling up leaf level resistance measurements has long been recognized as a major challenge (Bailey and Davies, 1981; Furon et al., 2007; Sprintsin et al., 2012) due to heterogeneity of environmental variables. The ASCE/FAO reference ET methods take a single layer "big leaf" approach to scaling to convert non-stressed leaf resistances (r_s) into whole canopy bulk resistances (r_c) by using an "effective LAI" where r_c is calculated by dividing r_s by effective LAI. ASCE assumes that effective LAI is equivalent to 0.5 times measured LAI, which is assumed to be 2.9 for ET_0 and 4.5 for ET_r, thus resulting in effective LAIs of 1.4 and 2.3, respectively. Studies of well-watered crops have found effective LAIs which vary quite significantly from those assumed for the reference surface. Tolk et al. (1996) found an effective LAI of 1.3 for irrigated maize in Texas that was only 30 % of the maximum measured LAI. Other studies (Alfieri et al., 2008; Mehrez et al., 1992) have assumed effective LAI as a linear function of LAI, with effective LAI equaling 50 % of LAI when LAI is 6. Ultimately, the effective LAI concept is only a presumed distribution of leaves with differing r_s (Bailey and Davies, 1981); there is a possibility that the relatively unique production system in our study fields results in a different, distinctive leaf distribution with a lower effective LAI. Along with effective LAI, another leaf parameter that could be different is leaf level resistance (r_s). Although we did not find a highly significant difference between measured r_s and the r_s assumed in the ASCE parameterizations ($100\,s\,m^{-1}$), we were able to measure r_s in only one field campaign during the mid-period, where r_s observations were limited by clouds and other logistical limitations. A large number of r_s observations are needed to accurately characterize r_c (Denmead, 1984); more than we could feasibly measure during our field campaign. We also note that other researchers (e.g., Zhang et al., 2008) have found non-stressed r_s values greater than $100\,s\,m^{-1}$.

Two other nonbiological factors could help explain the discrepancy between ASCE reference and mid-period ET_{EC}. One is nighttime r_c. Both ASCE approaches assume a nighttime r_c of $200\,s\,m^{-1}$, which is based on measurements of damp soil beneath a grass lysimeter (Allen et al., 2006). Measured nighttime r_c at our fields was significantly higher. We suspect that the taller sugarcane canopy and substantial layer of trash and lodged cane minimizes bare soil water evaporation, thus increasing nighttime r_c. Oliver and Singels (2012) found a significant decrease in soil evaporation in sugarcane with surfaces covered by crop residue. Furthermore, the minimal daytime ground heat flux (< 5 %) further reduces nighttime ET. Another factor is the canopy energy storage that is considerable in high-biomass systems (Anderson and Wang, 2014). Finally, we note that nighttime r_c is likely to be a locally specific value; $200\,s\,m^{-1}$ is too low for our study region but too high for other regions with significant advection (Evett et al., 2012).

Along with nighttime r_c, we examined the role of atmospheric resistance (r_a) in parameterizing ET, given the low observed mean r_a at Windy ($< 20\,s\,m^{-1}$) and the demonstrated importance of atmospheric resistance/conductance parameterizations in coastal tropical regions for accurate ET parameterization (e.g., Holwerda et al., 2012). Given the canopy architecture of mid-period sugarcane in our study fields, we were not certain about the equations that are commonly used to parameterize zero plane displacement height and roughness lengths, which are also used in the ASCE reference ET equations. To test the effect of r_a uncertainty, a sensitivity analysis was conducted. We used r_a that was 200 and 50 % of the original r_a and recalculated r_c for both EC towers. In all cases, the new r_a changed the r_c values by $< 10\,s\,m^{-1}$, with most r_c values changed by $< 5\,s\,m^{-1}$. These values are too small to explain the discrepancy between observed and parameterized r_c. The presence of r_a in both the numerator and denominator of the PM equation limits the impact of variation in r_a on r_c.

Finally, we note that the ASCE and FAO reference ET and PT ET equations show varying sensitivity to meteorological variables depending upon climate. Multiple studies have shown spatial, seasonal, and interannual variation in the sensitivity of reference ET to meteorological inputs, with the most sensitive input (air temperature, wind velocity, relative humidity, etc.) changing depending upon season and location (e.g., Bandyopadhyay et al., 2009; Estévez et al., 2009; Gong et al., 2006; Huo et al., 2013; Irmak et al., 2006; Liang et al., 2008; Liu et al., 2014). Irmak et al. (2006) and Estévez et al. (2009) found increased sensitivity to reference ET parameterization at locations with higher wind velocities in the United States and Spain, respectively. Bandyopadhyay et al. (2009) and Huo et al. (2013) reported that decreased wind velocities accounted for the largest proportion of decreased reference ET in climatically differing regions in India and China. Across a large river basin in China (Chiang Jiang), Gong et al. (2006) showed that sensitivities of reference ET to other meteorological variables (air temperature and relative humidity) depended significantly on the spatial pattern of wind sensitivity. With respect to the PT equation, variability in the PT coefficient (α) has been found at lower to middle LAI (LAI less than 3) depending upon the soil wetness and covering (Ding et al., 2013). This may be particularly relevant for our system in early growth stages with fractional soil wetness and partial cover from sugarcane detritus (trash). Conversely, at mid- to full canopy (LAI greater than 3) or when soil moisture was greater than 50 % of the available field capacity, α showed little sensitivity.

5 Summary and conclusion

We investigated discrepancies between two standardized reference ET equations and EC-measured ET at two field sites over irrigated sugarcane on Maui. At both fields, measured daily ET during the mid-period should have approached the tall reference ET equation and exceeded the short reference ET equation. At both fields, both ASCE reference ET equations significantly overestimated mid-period ET compared to EC observations of ET. The PT equation performed substantially better at the Windy field than the short reference ET, while the short reference ET equation and PT were more closely matched at the Lee field. We used a custom bulk canopy resistance derived from inverting PT ET; the custom cane reference ET equation had less seasonal variation in ET discrepancy. Multiple independent field observations did not indicate insufficient canopy cover or plant water stress reducing ET_{EC} significantly.

This study indicated nighttime bulk canopy resistance, leaf stomatal resistance, and effective LAI as possible causes for the discrepancy in bulk canopy resistance (and reference ET estimates) between the ASCE reference equations and mid-period ET_{EC}. The higher bulk canopy resistances and relationship between ET discrepancies and vapor pressure deficit indicated that the ASCE equations overestimated the advective component of ET. Ultimately, validation with field methods, including micrometeorology and water balance methods, is needed to establish the accuracy of the ASCE equations in a region where they have not been tested previously. Adjusting the bulk canopy resistance to local climate to reduce the advective component of ET may make the full ASCE PM equation a more appropriate equation in this region.

The PT equation performs better than ET_r or ET_0 in our study region. The PT equation likely provides a more robust estimation of reference ET in regions with high humidity. The simplicity of the PT equation also makes it attractive for use in larger scale project planning as it has been parameterized in satellite-based ET models (e.g., Choi et al., 2011; Jin et al., 2011) and can be used in regions with a relative paucity of surface meteorological data, unlike the ASCE/FAO equations that require near-surface wind speed and humidity data, which are currently supplied by surface meteorological stations and interpolated in satellite-based approaches (Allen et al., 2007; Hart et al., 2009).

The results illustrate the importance of the careful use of reference ET equations and coefficients for assessing actual ET in hydrologic applications. Our finding of high bulk canopy resistance and low atmospheric resistance supports Widmoser's (2009) recommendation for research of the canopy resistance/atmospheric resistance ratio. Many areas with changing hydrology (Elison Timm et al., 2011) and areas that currently and may soon use irrigation in previously non-irrigated fields (Baker et al., 2012; Salazar et al., 2012) are outside of the semi-arid areas where reference ET methods have been primarily developed and tested. As such, it will be important to ensure that the appropriate reference equation is used to parameterize evaporative demand.

Acknowledgements. We thank Ilja van Meerveld, Lixin Wang, Maarten Waterloo, and two anonymous reviewers for their constructive feedback on this manuscript. Don Schukraft discussed previous meteorological investigations and observations at the farm. Jim Gartung, ARS-Parlier, assisted with the establishment of the EC tower and TetraCam measurements. David Grantz provided insight into the historical evaluation of ET data for Hawaiian sugarcane. Adel Youkhana, Neil Abranyi, Jason Drogowski, and the farm crew assisted with data collection and field logistical support. This research was supported by USDA Agricultural Research Service, national program 211: Water Availability and Watershed Management and by the US Navy, Office of Naval Research. Note: the US Department of Agriculture (USDA) prohibits discrimination in all its programs and activities on the basis of race, color, national origin, age, disability, and where applicable, sex, marital status, familial status, parental status, religion, sexual orientation, genetic information, political beliefs, reprisal, or because all or part of an individual's income is derived from any public assistance program (not all prohibited bases apply to all programs.) Persons with disabilities who require alternative means of communication of program information (Braille, large print, audiotape, etc.) should contact USDA's TARGET Center at (202) 720-2600 (voice and TDD). To file a complaint of discrimination, write to USDA, Director, Office of Civil Rights, 1400 Independence Avenue, S.W., Washington, D.C. 20250-9410, or call (800) 795-3272 (voice) or (202) 720-6382 (TDD). USDA is an equal opportunity provider and employer.

Edited by: L. Wang

References

Alavi, N., Warland, J. S., and Berg, A. A.: Filling gaps in evapotranspiration measurements for water budget studies: Evaluation of a Kalman filtering approach, Agr. Forest Meteorol., 141, 57–66, doi:10.1016/j.agrformet.2006.09.011, 2006.

Alfieri, J. G., Niyogi, D., Blanken, P. D., Chen, F., LeMone, M. A., Mitchell, K. E., Ek, M. B., and Kumar, A.: Estimation of the Minimum Canopy Resistance for Croplands and Grasslands Using Data from the 2002 International H2O Project, Mon. Weather Rev., 136, 4452–4469, doi:10.1175/2008MWR2524.1, 2008.

Allen, R. G., Pereira, L. S., Raes, D., and Smith, M.: Crop evapotranspiration?: guidelines for computing crop water requirements, Food and Agriculture Organization of the United Nations, Rome, 1998.

Allen, R. G., Walter, I. A., Elliott, R. L., Howell, T. A., Itenfisu, D., Jensen, M. E., and Snyder, R. L: The ASCE standardized reference evapotranspiration equation, American Society of Civil Engineers, Reston, Va., 2005.

Allen, R. G., Pruitt, W. O., Wright, J. L., Howell, T. A., Ventura, F., Snyder, R., Itenfisu, D., Steduto, P., Berengena, J., Yrisarry, J. B., Smith, M., Pereira, L. S., Raes, D., Perrier, A., Alves, I., Walter, I., and Elliott, R.: A recommendation on standardized surface resistance for hourly calculation of reference ETo by the FAO56 Penman-Monteith method, Agr. Water Manage., 81, 1–22, doi:10.1016/j.agwat.2005.03.007, 2006.

Allen, R. G., Tasumi, M., and Trezza, R.: Satellite-Based Energy Balance for Mapping Evapotranspiration with Internalized Calibration (METRIC)-Model, J. Irrig. Drain. Eng., 133, 380–394, doi:10.1061/(ASCE)0733-9437(2007)133:4(380), 2007.

Anderson, R. G. and Goulden, M. L.: A mobile platform to constrain regional estimates of evapotranspiration, Agr. Forest Meteorol., 149, 771–782, doi:10.1016/j.agrformet.2008.10.022, 2009.

Anderson, R. G. and Wang, D.: Energy budget closure observed in paired Eddy Covariance towers with increased and continuous daily turbulence, Agr. Forest Meteorol., 184, 204–209, doi:10.1016/j.agrformet.2013.09.012, 2014.

Anderson, R. G., Lo, M.-H., and Famiglietti, J. S.: Assessing surface water consumption using remotely-sensed groundwater, evapotranspiration, and precipitation, Geophys. Res. Lett., 39, L16401, doi:10.1029/2012GL052400, 2012.

Arnold, J. G., Srinivasan, R., Muttiah, R. S., and Williams, J. R.: Large Area Hydrologic Modeling and Assessment Part I: Model Development, J. Am. Water Resour. Assoc., 34, 73–89, doi:10.1111/j.1752-1688.1998.tb05961.x, 1998.

Arya, S. P.: Introduction to micrometeorology, Academic Press, San Diego, 2001.

Bailey, W. G. and Davies, J. A.: Bulk stomatal resistance control on evaporation, Boundary-Lay. Meteorol., 20, 401–415, doi:10.1007/BF00122291, 1981.

Baker, J. M., Griffis, T. J., and Ochsner, T. E.: Coupling landscape water storage and supplemental irrigation to increase productivity and improve environmental stewardship in the US Midwest, Water Resources Res., 48, W05301, doi:10.1029/2011WR011780, 2012.

Baldocchi, D. D.: Assessing the eddy covariance technique for evaluating carbon dioxide exchange rates of ecosystems: past, present and future, Glob. Change Biol., 9, 479–492, doi:10.1046/j.1365-2486.2003.00629.x, 2003.

Ballester, C., Jiménez-Bello, M. A., Castel, J. R., and Intrigliolo, D. S.: Usefulness of thermography for plant water stress detection in citrus and persimmon trees, Agr. Forest Meteorol., 168, 120–129, doi:10.1016/j.agrformet.2012.08.005, 2013.

Bandyopadhyay, A., Bhadra, A., Raghuwanshi, N. S., and Singh, R.: Temporal Trends in Estimates of Reference Evapotranspiration over India, J. Hydrol. Eng., 14, 508–515, doi:10.1061/(ASCE)HE.1943-5584.0000006, 2009.

Bastiaanssen, W. G. M., Noordman, E. J. M., Pelgrum, H., Davids, G., Thoreson, B. P., and Allen, R. G.: SEBAL Model with Remotely Sensed Data to Improve Water-Resources Management under Actual Field Conditions, J. Irrig. and Drain. Eng., 131, 85–93, doi:10.1061/(ASCE)0733-9437(2005)131:1(85), 2005.

Cabral, O. M. R., Rocha, H. R., Gash, J. H., Ligo, M. A. V., Tatsch, J. D., Freitas, H. C., and Brasilio, E.: Water use in a sugarcane plantation, GCB Bioenergy, 4, 555–565, doi:10.1111/j.1757-1707.2011.01155.x, 2012.

Campbell, R., Chang, J.-H., and Cox, D.: Evapotranspiration of Sugar Cane in Hawaii as Measured by In-Field Lysimeters in Relation to Climate, in: Proceedings of the International Society of Sugarcane Technologists; 10th Congress, Hawaii, 3–22 May, 1959, 637–645, 1960.

Carr, M. K. V. and Knox, J. W.: The Water Relations and Irrigation Requirements of Sugar Cane (Saccharum Officinarum): A Review, Exp. Agric., 47, 1–25, doi:10.1017/S0014479710000645, 2011.

Choi, M., Woong Kim, T., and Kustas, W. P.: Reliable estimation of evapotranspiration on agricultural fields predicted by the Priestley–Taylor model using soil moisture data from ground and remote sensing observations compared with the Common Land Model, Int. J. Remote Sens., 32, 4571–4587, doi:10.1080/01431161.2010.489065, 2011.

Cobos, D. R. and Baker, J. M.: Evaluation and Modification of a Domeless Net Radiometer, Agron. J., 95, 177–183, 2003.

Cristea, N. C., Kampf, S. K., and Burges, S. J.: Revised Coefficients for Priestley-Taylor and Makkink-Hansen Equations for Estimating Daily Reference Evapotranspiration, J. Hydrol. Eng., 18, 1289–1300, doi:10.1061/(ASCE)HE.1943-5584.0000679, 2013.

Da Silva, T. G. F., de Moura, M. S. B., Zolnier, S., Soares, J. M., Vieira, V. J. S., and Júnior, W. G. F.: Water requirement and crop coefficient of irrigated sugarcane in a semi-arid region, Revista Brasileira de Engenharia Agricola e Ambiental, 16, 64–71, 2012.

Davis, S. L. and Dukes, M. D.: Irrigation scheduling performance by evapotranspiration-based controllers, Agr. Water Manage., 98, 19–28, doi:10.1016/j.agwat.2010.07.006, 2010.

Denmead, O. T.: Plant physiological methods for studying evapotranspiration: Problems of telling the forest from the trees, Agr. Water Manage., 8, 167–189, doi:10.1016/0378-3774(84)90052-0, 1984.

Ding, R., Kang, S., Li, F., Zhang, Y., and Tong, L.: Evapotranspiration measurement and estimation using modified Priestley–Taylor model in an irrigated maize field with mulching, Agr. Forest Meteorol., 168, 140–148, doi:10.1016/j.agrformet.2012.08.003, 2013.

Doorenbos, J. and Pruitt, W.: Crop water requirements. FAO irrigation and drainage paper 24, Land and Water Development Division, FAO, Rome, 1977.

Eichinger, W. E., Parlange, M. B., and Stricker, H.: On the Concept of Equilibrium Evaporation and the Value of the Priestley-Taylor Coefficient, Water Resour. Res., 32, 161–164, doi:10.1029/95WR02920, 1996.

Elison Timm, O., Diaz, H. F., Giambelluca, T. W., and Takahashi, M.: Projection of changes in the frequency of heavy rain events over Hawaii based on leading Pacific climate modes, J. Geophys. Res., 116, D04109, doi:10.1029/2010JD014923, 2011.

Estévez, J., Gavilán, P., and Berengena, J.: Sensitivity analysis of a Penman-Monteith type equation to estimate reference evapotranspiration in southern Spain, Hydrol. Process., 23, 3342–3353, doi:10.1002/hyp.7439, 2009.

Evensen, C. I., Muchow, R. C., El-Swaify, S. A., and Osgood, R. V.: Yield Accumulation in Irrigated Sugarcane: I. Effect of Crop Age and Cultivar, Agron. J., 89, 638–646, doi:10.2134/agronj1997.00021962008900040016x, 1997.

Evett, S. R., Howell, T. A., Todd, R. W., Schneider, A. D., and Tolk, J. A.: Alfalfa reference ET measurement and prediction, in National irrigation symposium. Proceedings of the 4th Decennial Symposium, Phoenix, Arizona, USA, 14–16 November 2000, 266–272, American Society of Agricultural Engineers, St. Joseph Mich., 2000.

Evett, S. R., Lascano, R. J., Howell, T. A., Tolk, J. A., O'Shaughnessy, S. A., and Colaizzi, P. D.: Single- and dual-surface iterative energy balance solutions for reference ET, Trans. ASABE, 55, 533–541, doi:10.13031/2013.41388, 2012.

Ferguson, I. M. and Maxwell, R. M.: Hydrologic and land–energy feedbacks of agricultural water management practices, Environ. Res. Lett., 6, 014006, doi:10.1088/1748-9326/6/1/014006, 2011.

Fisher, J. B., DeBiase, T. A., Qi, Y., Xu, M., and Goldstein, A. H.: Evapotranspiration models compared on a Sierra Nevada forest ecosystem, Environ. Model. Soft., 20, 783–796, doi:10.1016/j.envsoft.2004.04.009, 2005.

Furon, A. C., Warland, J. S., and Wagner-Riddle, C.: Analysis of Scaling-Up Resistances from Leaf to Canopy Using Numerical Simulations, Agron. J., 99, 1483, doi:10.2134/agronj2006.0335, 2007.

Giambelluca, T. W., Chen, Q., Frazier, A. G., Price, J. P., Chen, Y.-L., Chu, P.-S., Eischeid, J. K., and Delparte, D. M.: Online Rainfall Atlas of Hawai'i, B. Am. Meteorol. Soc., 94, 313–316, doi:10.1175/BAMS-D-11-00228.1, 2013.

Glaz, B., Reed, S. T., and Albano, J. P.: Sugarcane Response to Nitrogen Fertilization on a Histosol with Shallow Water Table and Periodic Flooding, J. Agron. Crop Sci., 194, 369–379, doi:10.1111/j.1439-037X.2008.00329.x, 2008.

Gong, L., Xu, C., Chen, D., Halldin, S., and Chen, Y. D.: Sensitivity of the Penman–Monteith reference evapotranspiration to key climatic variables in the Changjiang (Yangtze River) basin, J. Hydrol., 329, 620–629, doi:10.1016/j.jhydrol.2006.03.027, 2006.

Hargreaves, G. H. and Samani, Z. A.: Economic Considerations of Deficit Irrigation, J. Irrig. Drain. Eng., 110, 343–358, doi:10.1061/(ASCE)0733-9437(1984)110:4(343), 1984.

Hart, Q. J., Brugnach, M., Temesgen, B., Rueda, C., Ustin, S. L., and Frame, K.: Daily reference evapotranspiration for California using satellite imagery and weather station measurement interpolation, Civil Eng. Environ. Systems, 26, 19–33, 2009.

Heinz, D. J. and Osgood, R. V.: A History of the Experiment Station: Hawaiian Sugar Planters' Association, Hawaiian Planters' Record, 61, 1–108, 2009.

Heinz, D. J., Tew, T. L., Meyer, H. K., and Wu, K. K.: Registration of H65-7052 Sugarcane (Reg. No. 51), Crop Sci., 21, 634–635, doi:10.2135/cropsci1981.0011183X002100040050x, 1981.

Hemakumara, H., Chandrapala, L., and Moene, A. F.: Evapotranspiration fluxes over mixed vegetation areas measured from large aperture scintillometer, Agr. Water Manage., 58, 109–122, doi:10.1016/S0378-3774(02)00131-2, 2003.

Hirich, A., Choukr-Allah, R., and Jacobsen, S.-E.: Deficit Irrigation and Organic Compost Improve Growth and Yield of Quinoa and Pea, J. Agronom. Crop Sci., 200, 390–398, doi:10.1111/jac.12073, 2014.

Holwerda, F., Bruijnzeel, L. A., Scatena, F. N., Vugts, H. F., and Meesters, A. G. C. A.: Wet canopy evaporation from a Puerto Rican lower montane rain forest: The importance of realistically estimated aerodynamic conductance, J. Hydrol., 414–415, 1–15, doi:10.1016/j.jhydrol.2011.07.033, 2012.

Holwerda, F., Bruijnzeel, L. A., Barradas, V. L., and Cervantes, J.: The water and energy exchange of a shaded coffee plantation in the lower montane cloud forest zone of central Veracruz, Mexico, Agr. Forest Meteorol., 173, 1–13, doi:10.1016/j.agrformet.2012.12.015, 2013.

Huo, Z., Dai, X., Feng, S., Kang, S., and Huang, G.: Effect of climate change on reference evapotranspiration and aridity index in arid region of China, J. Hydrol., 492, 24–34, doi:10.1016/j.jhydrol.2013.04.011, 2013.

Inman-Bamber, N. G. and McGlinchey, M. G.: Crop coefficients and water-use estimates for sugarcane based on long-term Bowen ratio energy balance measurements, Field Crop Res., 83, 125–138, doi:10.1016/S0378-4290(03)00069-8, 2003.

Inman-Bamber, N. G., Robertson, M. J., Muchow, R. C., Wood, A. W., Pace, R., and Spillman, M. F.: Boosting yields with limited irrigation water, Proc. Aust. Soc. Sugar Cane Techol., 21, 203–211, 1999.

Irmak, S., Payero, J. O., Martin, D. L., Irmak, A., and Howell, T. A.: Sensitivity Analyses and Sensitivity Coefficients of Standardized Daily ASCE-Penman-Monteith Equation, J. Irrig. Drain. Eng., 132, 564–578, doi:10.1061/(ASCE)0733-9437(2006)132:6(564), 2006.

Jensen, M. E.: Water consumption by agricultural plants, in: Water deficit and plant growth, Vol. 1. Development, control and measurement, Vol. II, 1–22, New York, London: Academic Press, 1968.

Jensen, M. E., Burman, R. D., and Allen, R. G.: Evapotranspiration and irrigation water requirements: a manual, American Society of Civil Engineers, New York, N.Y., 1990.

Jin, Y., Randerson, J. T., and Goulden, M. L.: Continental-scale net radiation and evapotranspiration estimated using MODIS satellite observations, Remote Sens. Environ., 115, 2302–2319, doi:10.1016/j.rse.2011.04.031, 2011.

Kaimal, J. C. and Finnigan, J. J.: Atmospheric boundary layer flows: their structure and measurement, Oxford University Press, USA, 1994.

Kang, S., Zhang, L., Liang, Y., Hu, X., Cai, H., and Gu, B.: Effects of limited irrigation on yield and water use efficiency of winter wheat in the Loess Plateau of China, Agr. Water Manage., 55, 203–216, doi:10.1016/S0378-3774(01)00180-9, 2002.

Kljun, N., Calanca, P., Rotach, M. W., and Schmid, H. P.: A Simple Parameterisation for Flux Footprint Predictions, Boundary-Lay. Meteorol., 112, 503–523, doi:10.1023/B:BOUN.0000030653.71031.96, 2004.

Kueppers, L. M., Snyder, M. A., and Sloan, L. C.: Irrigation cooling effect: Regional climate forcing by land-use change, Geophys. Res. Lett., 34, L03703, doi:10.1029/2006GL028679, 2007.

Lecina, S., Martínez-Cob, A., Pérez, P. J., Villalobos, F. J., and Baselga, J. J.: Fixed versus variable bulk canopy resistance for reference evapotranspiration estimation using the Penman–Monteith equation under semiarid conditions, Agr. Water Manage., 60, 181–198, doi:10.1016/S0378-3774(02)00174-9, 2003.

Leuning, R., van Gorsel, E., Massman, W. J., and Isaac, P. R.: Reflections on the surface energy imbalance problem, Agr. Forest Meteorol., 156, 65–74, doi:10.1016/j.agrformet.2011.12.002, 2012.

Liang, L., Li, L., Zhang, L., Li, J., and Li, B.: Sensitivity of penman-monteith reference crop evapotranspiration in Tao'er River Basin of northeastern China, Chinese Geogr. Sci., 18, 340–347, doi:10.1007/s11769-008-0340-x, 2008.

Liu, H., Zhang, R., and Li, Y.: Sensitivity analysis of reference evapotranspiration (ETo) to climate change in Beijing, China, Desal. Water Treat., 52, 2799–2804, doi:10.1080/19443994.2013.862030, 2014.

Lo, M.-H. and Famiglietti, J. S.: Irrigation in California's Central Valley Strengthens the Southwestern U.S. Water Cycle, Geophys. Res. Lett., 6, 301–306, doi:10.1002/grl.50108, 2013.

Mabhaudhi, T., Modi, A. T., and Beletse, Y. G.: Response of taro (Colocasia esculenta L. Schott) landraces to varying water regimes under a rainshelter, Agr. Water Manage., 121, 102–112, doi:10.1016/j.agwat.2013.01.009, 2013.

Mauder, M. and Foken, T.: Documentation and Instruction Manual of the Eddy Covariance Software Package TK2, Universitätsbibliothek Bayreuth, Bayreuth, available at: https://epub.uni-bayreuth.de/342/1/ARBERG046.pdf (last access: 25 October 2014), 2011.

McAneney, K. J. and Itier, B.: Operational limits to the Priestley-Taylor formula, Irrig. Sci., 17, 37–43, doi:10.1007/s002710050020, 1996.

McGlinchey, M. G. and Inman-Bamber, N.G.: Predicting sugarcane water use with the Penman–Monteith equation, in Proceedings of the International Conference on Evapotranspiration and Irrigation Scheduling, San Antonio, 592–598, ASAE, St. Joseph Mich., 1996.

Mehrez, M. B., Taconet, O., Vidal-Madjar, D., and Valencogne, C.: Estimation of stomatal resistance and canopy evaporation during the HAPEX-MOBILHY experiment, Agr. Forest Meteorol., 58, 285–313, doi:10.1016/0168-1923(92)90066-D, 1992.

Meissner, R., Rupp, H., Seeger, J., Ollesch, G., and Gee, G. W.: A comparison of water flux measurements: passive wick-samplers versus drainage lysimeters, Eur. J. Soil Sci., 61, 609–621, doi:10.1111/j.1365-2389.2010.01255.x, 2010.

Mendez-Costabel, M. P., Wilkinson, K. L., Bastian, S. E. P., Jordans, C., McCarthy, M., Ford, C. M., and Dokoozlian, N. K.: Effect of increased irrigation and additional nitrogen fertilisation on the concentration of green aroma compounds in V itis vinifera L. Merlot fruit and wine: Green aroma compounds in Merlot, Aust. J. Grape Wine Res., 20, 80–90, doi:10.1111/ajgw.12062, 2014.

Moncrieff, J. B., Massheder, J. M., de Bruin, H., Elbers, J., Friborg, T., Heusinkveld, B., Kabat, P., Scott, S., Soegaard, H., and Verhoef, A.: A system to measure surface fluxes of momentum, sensible heat, water vapour and carbon dioxide, J. Hydrol., 188–189, 589–611, doi:10.1016/S0022-1694(96)03194-0, 1997.

Moncrieff, J., Clement, R., Finnigan, J., and Meyers, T.: Averaging, Detrending, and Filtering of Eddy Covariance Time Series, in: Handbook of Micrometeorology, Vol. 29, edited by: Lee, X., Massman, W., and Law, B., 7–31, Kluwer Academic Publishers, Dordrecht, 2004.

Monteith, J. L.: Evaporation and environment, Symp. Soc. Exp. Biol., 19, 205–234, 1965.

Moore, R. C. and Fitschen, J. C.: The drip irrigation revolution in the Hawaiian sugarcane industry, in Visions of the future?: proceedings of the 3. National Irrigation Symposium, held in conjunction with the 11. Annual International Irrigation Exposition, 1990, Phoenix Civic Plaza, Phoenix, Arizona, ASAE, St. Joseph, Mich., 1990.

Obukhov, A. M.: Turbulence in an atmosphere with a non-uniform temperature, Boundary-Lay. Meteorol., 2, 7–29, doi:10.1007/BF00718085, 1971.

Olivier, F. and Singels, A.: The effect of crop residue layers on evapotranspiration, growth and yield of irrigated sugarcane, Water SA, 38, 78–86, doi:10.4314/wsa.v38i1.10, 2012.

Perry, C.: Irrigation reliability and the productivity of water: A proposed methodology using evapotranspiration mapping, Irrig. Drain. Systems, 19, 211–221, doi:10.1007/s10795-005-8135-z, 2005.

Priestley, C. H. B. and Taylor, R. J.: On the Assessment of Surface Heat Flux and Evaporation Using Large-Scale Parameters, Mon. Weather Rev., 100, 81–92, doi:10.1175/1520-0493(1972)100<0081:OTAOSH>2.3.CO;2, 1972.

Puma, M. J. and Cook, B. I.: Effects of irrigation on global climate during the 20th century, J. Geophys. Res., 115, D16120, doi:10.1029/2010JD014122, 2010.

Ramjeawon, T.: Water resources management on the small Island of Mauritius, Int. J. Water Resour. D., 10, 143–155, doi:10.1080/07900629408722619, 1994.

Rao, L., Sun, G., Ford, C., and Vose, J.: Modeling potential evapotranspiration of two forested watersheds in the southern Appalachians, Trans. ASABE, 54, 2067–2078, doi:10.13031/2013.40666, 2011.

Reichstein, M., Falge, E., Baldocchi, D., Papale, D., Aubinet, M., Berbigier, P., Bernhofer, C., Buchmann, N., Gilmanov, T., Granier, A., Grunwald, T., Havrankova, K., Ilvesniemi, H., Janous, D., Knohl, A., Laurila, T., Lohila, A., Loustau, D., Matteucci, G., Meyers, T., Miglietta, F., Ourcival, J.-M., Pumpanen, J., Rambal, S., Rotenberg, E., Sanz, M., Tenhunen, J., Seufert, G., Vaccari, F., Vesala, T., Yakir, D., and Valentini, R.: On the separation of net ecosystem exchange into assimilation and ecosystem respiration: review and improved algorithm, Glob. Change Biol., 11, 1424–1439, doi:10.1111/j.1365-2486.2005.001002.x, 2005.

Salazar, M. R., Hook, J. E., Garcia y Garcia, A., Paz, J. O., Chaves, B., and Hoogenboom, G.: Estimating irrigation water use for maize in the Southeastern USA: A modeling approach, Agr. Water Manage., 107, 104–111, doi:10.1016/j.agwat.2012.01.015, 2012.

Salinas, F. and Namken, L. N.: Irrigation scheduling for sugarcane in the Lower Rio Grande Valley of Texas, Proc. Am. Soc. Sugar Cane Technol., 6, 186–191, 1977.

Smith, M.: The application of climatic data for planning and management of sustainable rainfed and irrigated crop production, Agr. Forest Meteorol., 103, 99–108, doi:10.1016/S0168-1923(00)00121-0, 2000.

Soopramanien, G. C., Berthelot, B., and Batchelor, C. H.: Irrigation research, development and practice in Mauritius, Agr. Water Manage., 17, 129–139, doi:10.1016/0378-3774(90)90060-C, 1990.

Sprintsin, M., Chen, J. M., Desai, A., and Gough, C. M.: Evaluation of leaf-to-canopy upscaling methodologies against carbon flux data in North America, J. Geophys. Res., 117, G011023, doi:10.1029/2010JG001407, 2012.

Suleiman, A. A. and Hoogenboom, G.: Comparison of Priestley-Taylor and FAO-56 Penman-Monteith for Daily Reference Evapotranspiration Estimation in Georgia, J. Irrig. Drain. Eng., 133, 175–182, doi:10.1061/(ASCE)0733-9437(2007)133:2(175), 2007.

Suleiman, A. A. and Hoogenboom, G.: A comparison of ASCE and FAO-56 reference evapotranspiration for a 15-min time step in humid climate conditions, J. Hydrol., 375, 326–333, doi:10.1016/j.jhydrol.2009.06.020, 2009.

Szeicz, G. and Long, I. F.: Surface Resistance of Crop Canopies, Water Resources Res., 5, 622–633, doi:10.1029/WR005i003p00622, 1969.

Tang, Q., Peterson, S., Cuenca, R. H., Hagimoto, Y., and Lettenmaier, D. P.: Satellite-based near-real-time estimation of irrigated crop water consumption, J. Geophys. Res., 114, D05114, doi:10.1029/2008JD010854, 2009.

Thompson, G. D. and Boyce, J. P.: Daily measurements of potential evapotranspiration from fully canopied sugarcane, Agric. Meteorol., 4, 267–279, doi:10.1016/0002-1571(67)90027-1, 1967.

Tian, D. and Martinez, C. J.: Forecasting Reference Evapotranspiration Using Retrospective Forecast Analogs in the Southeastern United States, J. Hydrolmeteorol., 13, 1874–1892, doi:10.1175/JHM-D-12-037.1, 2012.

Tolk, J. A., Howell, T. A., Steiner, J. L., and Krieg, D. R.: Corn canopy resistance determined from whole plant transpiration, in: Proceedings of the International Conference on Evapotranspiration and Irrigation Scheduling, San Antonio, 347–351, ASAE, St. Joseph Mich., 1996.

Ventura, F., Spano, D., Duce, P., and Snyder, R. L.: An evaluation of common evapotranspiration equations, Irrig. Sci., 18, 163–170, doi:10.1007/s002710050058, 1999.

Vickers, D. and Mahrt, L.: Quality Control and Flux Sampling Problems for Tower and Aircraft Data, J. Atmos. Ocean. Technol., 14, 512–526, doi:10.1175/1520-0426(1997)014<0512:QCAFSP>2.0.CO;2, 1997.

Vörösmarty, C. J., Federer, C. A., and Schloss, A. L.: Potential evaporation functions compared on US watersheds: Possible implications for global-scale water balance and terrestrial ecosystem modeling, J. Hydrol., 207, 147–169, doi:10.1016/S0022-1694(98)00109-7, 1998.

Waterloo, M. J., Bruijnzeel, L. A., Vugts, H. F., and Rawaqa, T. T.: Evaporation from Pinus caribaea plantations on former grassland soils under maritime tropical conditions, Water Resour. Res., 35, 2133–2144, doi:10.1029/1999WR900006, 1999.

Webb, E. K., Pearman, G. I., and Leuning, R.: Correction of flux measurements for density effects due to heat and water vapour transfer, Q. J. Roy. Meteorol. Soc., 106, 85–100, doi:10.1002/qj.49710644707, 1980.

Widmoser, P.: A discussion on and alternative to the Penman–Monteith equation, Agr. Water Manage., 96, 711–721, doi:10.1016/j.agwat.2008.10.003, 2009.

Zhang, B., Kang, S., Li, F., and Zhang, L.: Comparison of three evapotranspiration models to Bowen ratio-energy balance method for a vineyard in an arid desert region of northwest China, Agr. Forest Meteorol., 148, 1629–1640, doi:10.1016/j.agrformet.2008.05.016, 2008.

Zhao, L., Xia, J., Xu, C., Wang, Z., Sobkowiak, L., and Long, C.: Evapotranspiration estimation methods in hydrological models, J. Geograph. Sci., 23, 359–369, doi:10.1007/s11442-013-1015-9, 2013.

Dead Sea evaporation by eddy covariance measurements vs. aerodynamic, energy budget, Priestley–Taylor, and Penman estimates

Jutta Metzger[1], **Manuela Nied**[1], **Ulrich Corsmeier**[1], **Jörg Kleffmann**[2], **and Christoph Kottmeier**[1]

[1]Institute of Meteorology and Climate Research, Karlsruhe Institute of Technology (KIT),
P.O. Box 3640, 76021 Karlsruhe, Germany
[2]Physikalische und Theoretische Chemie, Fakultät für Mathematik und Naturwissenschaften,
Bergische Universität Wuppertal, 42097 Wuppertal, Germany

Correspondence: Jutta Metzger (jutta.metzger@kit.edu)

Abstract. The Dead Sea is a terminal lake, located in an arid environment. Evaporation is the key component of the Dead Sea water budget and accounts for the main loss of water. So far, lake evaporation has been determined by indirect methods only and not measured directly. Consequently, the governing factors of evaporation are unknown. For the first time, long-term eddy covariance measurements were performed at the western Dead Sea shore for a period of 1 year by implementing a new concept for onshore lake evaporation measurements. To account for lake evaporation during offshore wind conditions, a robust and reliable multiple regression model was developed using the identified governing factors wind velocity and water vapour pressure deficit. An overall regression coefficient of 0.8 is achieved. The measurements show that the diurnal evaporation cycle is governed by three local wind systems: a lake breeze during daytime, strong downslope winds in the evening, and strong northerly along-valley flows during the night. After sunset, the strong winds cause half-hourly evaporation rates which are up to 100 % higher than during daytime. The median daily evaporation is $4.3 \, \mathrm{mm \, d^{-1}}$ in July and $1.1 \, \mathrm{mm \, d^{-1}}$ in December. The annual evaporation of the water surface at the measurement location was $994 \pm 88 \, \mathrm{mm \, a^{-1}}$ from March 2014 until March 2015. Furthermore, the performance of indirect evaporation approaches was tested and compared to the measurements. The aerodynamic approach is applicable for sub-daily and multi-day calculations and attains correlation coefficients between 0.85 and 0.99. For the application of the Bowen ratio energy budget method and the Priestley–Taylor method, measurements of the heat storage term are inevitable on timescales up to 1 month. Otherwise strong seasonal biases occur. The Penman equation was adapted to calculate realistic evaporation, by using an empirically gained linear function for the heat storage term, achieving correlation coefficients between 0.92 and 0.97. In summary, this study introduces a new approach to measure lake evaporation with a station located at the shoreline, which is also transferable to other lakes. It provides the first directly measured Dead Sea evaporation rates as well as applicable methods for evaporation calculation. The first one enables us to further close the Dead Sea water budget, and the latter one enables us to facilitate water management in the region.

1 Introduction

Since several years, the lake level of the Dead Sea declines by over $1 \, \mathrm{m \, a^{-1}}$ (approx. 600–$700 \times 10^6 \, \mathrm{m^3 \, a^{-1}}$), meaning that the balance of the Dead Sea water budget is no longer sustained. The main water inflow to the Dead Sea is the Jordan River, but through anthropogenic interferences the discharge of the Jordan River into the Dead Sea decreased by 90 % down to 60–$400 \times 10^6 \, \mathrm{m^3 \, a^{-1}}$ (Asmar and Ergenzinger, 2002; Holtzman et al., 2005) compared to its natural discharge before 1955. Further natural inflow by groundwater discharge and surface runoff is in the range of 235–$243 \times 10^6 \, \mathrm{m^3 \, a^{-1}}$ (Siebert et al., 2014). As the Dead Sea is a terminal lake, no natural outflow exists, but water is withdrawn from the lake for mineral and potash production. The loss of water is about 250×10^6 m^3 a^{-1} (Lensky et al., 2005). Thus, evaporation has to be the main loss of water from the Dead Sea. Even though evaporation is of particular importance for the Dead Sea water budget, the variation in evaporation estimates is high. The spread of the evaporation estimates ranges from 1.05 to 2.00 m a^{-1}, which is comparable to a volume loss of 700–1334×10^6 m^3 a^{-1} (Stanhill, 1994; Salameh and El-Naser, 1999). It is important to reduce these uncertainties and assess the water budget components of the Dead Sea for a climatological purpose, but it is also of importance for the people in the area and the socio-economic development of the region to anticipate the evolution of these components and the resulting consequences for the environment. For instance, the lake level decline causes severe environmental problems. It influences the adjacent aquifers, their groundwater tables, and flow paths (Siebert et al., 2016), and it results in a shifting of the fresh–saline groundwater interface (Yechieli et al., 2006), which is connected to the development of sinkholes (Yechieli et al., 2006; Abelson et al., 2006). Since the 1980s, over 4000 sinkholes have formed at the western shore of the Dead Sea, which affect industrial, agricultural, and environmentally protected areas, leading to a substantial economic loss (Arkin and Gilat, 2000). Furthermore, evaporation influences the climatic conditions through a considerable change in the fraction of land and water surface. The changing fraction of water and land surfaces leads to a changing partitioning of the net radiation into sensible and latent heat flux. This results in a weaker horizontal gradient of the air temperature between the air masses over the water and land surface, resulting in a weaker pressure gradient, and thus weakens the lake breeze. As the lake breeze has an attenuating effect on the diurnal temperature amplitude and advects humidity towards the land, a weaker lake breeze results in higher maximum temperatures and decreasing humidity in the southern part of the valley (Alpert et al., 1997). Furthermore, it increases the diurnal penetration of the westerly winds into the valley in the afternoon. These westerly winds have often high wind velocities enhancing the evaporation and thus accelerating the lake level decline. Alpert et al. (1997) showed that in the 1940s, before the lake level and thus the water surface started to decrease, the much stronger easterly lake breeze delayed the penetration of the westerly winds considerably. The changing atmospheric conditions, together with the changing groundwater tables, result in a severe dieback of vegetation and the drying up of springs, endangering the unique flora and fauna in the Dead Sea region, such as the unique fish population of the Ein Feshkha reserve (Goren and Ortal, 1999; Lipchin et al., 2009).

In view of these environmental changes, resulting from the lake level decline, more accurate estimates of the Dead Sea evaporation are required (Kottmeier et al., 2016). Previous studies on the Dead Sea evaporation used indirect methods, such as water budget calculations (Salameh, 1996; Salameh and El-Naser, 2000); the energy budget approach (Stanhill, 1994; Lensky et al., 2005); aerodynamic methods (Salhotra et al., 1985; Oroud, 1994); or the combination of the latter two methods, called the combination approach (Calder and Neal, 1984; Asmar and Ergenzinger, 1999; Oroud, 2011). Variations in evaporation estimates between the studies result from assumptions on single water budget components such as groundwater inflow, different lengths of the time series of input variables, different measurement locations, and measurement uncertainties. To minimise the spread of 1.05 to 2.00 m a^{-1} in the evaporation estimates (Stanhill, 1994; Salameh and El-Naser, 1999) and reduce uncertainties, direct measurements of the Dead Sea evaporation are required. The eddy covariance technique is the only method to obtain direct evaporation measurements. Thus, it is considered the most accurate and reliable method to estimate evaporation (Rimmer et al., 2009; Tanny et al., 2008). All other methods assess evaporation indirectly, which means that all measurement errors accumulate into the estimated evaporation (Assouline and Mahrer, 1993). With the high temporal resolution of the measurements, the data can also be linked to meteorological variables afterwards. However, it is quite expensive and difficult to perform such measurements as it requires highly accurate instruments and their continuous maintenance. Various studies using eddy covariance measurements have been conducted around the world and also in Israel. Assouline and Mahrer (1993) measured evaporation from Lake Kinneret, a freshwater lake north of the Dead Sea, crossed by the Jordan River, and Tanny et al. (2008) measured evaporation from a small reservoir also north of the Dead Sea using eddy covariance systems. However, to the authors' knowledge, no eddy covariance measurements were performed at the Dead Sea, where the environmental problems are severe. That is why, in the framework of the international DESERVE project (DEad SEa Research VEnue; Kottmeier et al., 2016), a new concept for assessing lake evaporation from onshore measurements was applied. Long-term eddy covariance measurements were conducted at the Dead Sea shore, which provided evaporation data of the water surface for onshore wind conditions. These measurements were combined with a statistical model to calculate evaporation for offshore wind conditions. The comprehensive data set is analysed in this paper with the following aims: to (i) provide an applicable method for measuring lake evaporation, using a station located at the shoreline; (ii) evaluate the actual evaporation rates of the Dead Sea at the measurement location and their diurnal and intra-annual variability; and (iii) evaluate the applicability of the commonly used indirect methods to calculate evaporation from the Dead Sea, and assess the capacity of the methods to retrieve the evaporation term, in the future, when eddy covariance measurements are not available any more.

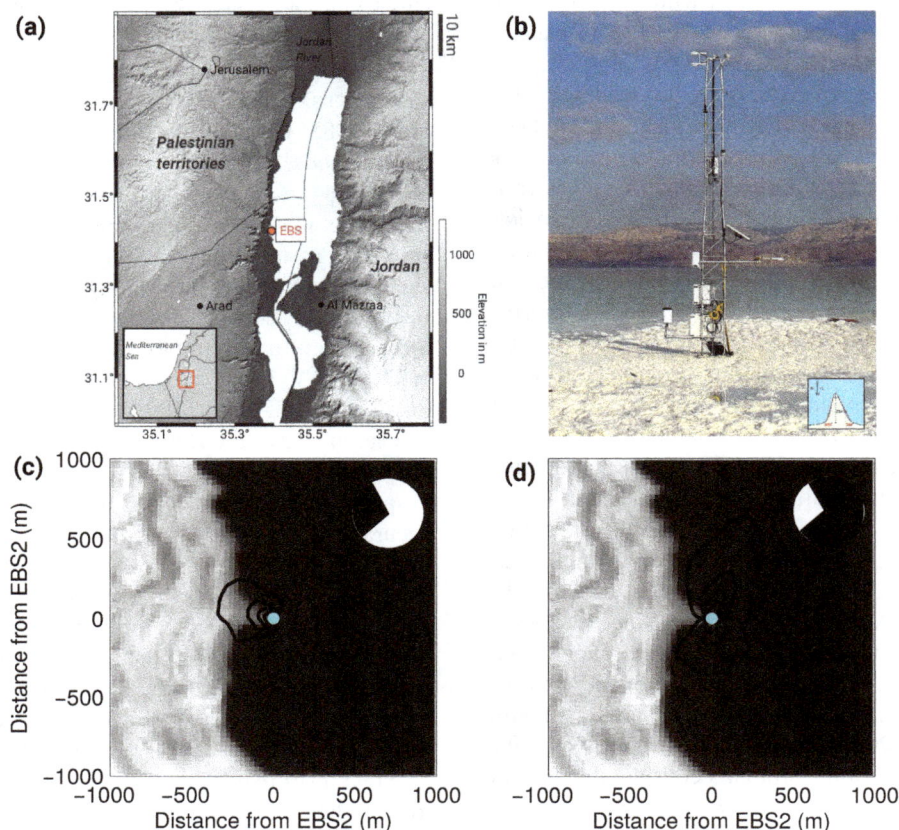

Figure 1. Map of the research area and location of the measurement site **(a)**; image of the measurement site and sketch of the headland (inlet) **(b)**; Landsat 8 images of the headland with location of the EBS (blue dot); and with the results of the footprint analysis **(c, d)**. Contour lines (from inside to outside) represent 20, 40, 60, and 80 % of the flux footprint area calculated with the footprint model of Kljun et al. (2015) for offshore wind conditions with wind direction between 230 and 330° **(c)** and for the other wind directions **(d)**. Satellite data were provided by the US Geological Survey.

2   Measurement site and instrumentation

The Dead Sea is a hypersaline terminal lake, located at the lowest point of the Jordan rift valley. It is surrounded by the Judean Mountains to the west and the Moab mountains to the east (Fig. 1a). Nowadays, the Dead Sea consists of two basins, the northern basin with approximately $600 \, km^2$ and the shallow artificial evaporation ponds in the south with approximately $280 \, km^2$, which are used for potash and mineral production. Since the 1950s, the lake level of the northern basin dropped by over 30 m, from $-395 \, m$ a.m.s.l. to the current $-429.9 \, m$ a.m.s.l. (Givati and Tal, 2016). The southern basin is held on a constant level by pumping water from the northern basin to the south. The area between the lake and the eastern and western mountain chains is rocky desert. When freshwater springs emerge along the shoreline, sufficient water is available for plants to grow. Although the total area of these vegetated areas is very small compared to the area covered by water or desert, these vegetated areas are very important for the diversity of the local ecosystems.

To measure the energy balance components of the water surface, an energy balance station (EBS) was installed directly at the shoreline (Fig. 1b). The station, was located 3 km south of Ein Gedi on the tip of a headland at the western shore of the Dead Sea (Fig. 1a). At the time of the measurements, the station was located at $-428 \, m$ a.m.s.l.; the headland was 214 m long and was surrounded by water from 300 to 260° (insert in Fig. 1b).

At the station the following meteorological variables were measured and averaged over 10 min: temperature and humidity at 2 m height (HC2S3, Rotronic), temperature at 6 m (100KGA1A, BetaTherm), longwave and shortwave radiation components of the upper and lower half space (CNR4, Kipp&Zonen) at 2 m height, precipitation (tipping bucket rain gauge 552202, Young), and atmospheric pressure (PTR330, Vaisala) at 1 m height. With a temporal resolution of 20 Hz, water vapour, CO_2 concentration, sonic temperature, and the three wind components were measured with an open path integrated gas analyser and sonic anemometer (IRGASON) from Campbell Scientific at 6 m height.

As the station was located at the shoreline, the radiation measurements of the lower half space represented the land surface conditions. For the water surface they have to be calculated. The applied method is explained in Sect. 3.1. Furthermore, the heat storage of the lake was not measured and was therefore calculated as the residuum of the energy balance equation ($R_n = LE + H + \Delta Q$) using half-hourly measurements. Notable hereby is that ΔQ also contains the possible non-closure of the energy balance. Considering the values of common energy balance closure studies (Foken, 2008; Wilson et al., 2002) the heat storage is thus most likely about 20 % smaller than calculated.

3 Data and methods

Measurement data from March 2014 until March 2015 were analysed. To achieve the research aims, the following calculations and methods were applied. The shortwave and longwave radiation components of the lower half space were calculated. This is presented in Sect. 3.1. The latent and sensible heat flux were calculated from the 20 Hz data using the eddy covariance method. The principle of the method, the post-processing, and data quality control steps are presented in Sect. 3.2. Furthermore, a multiple regression model was used to calculate evaporation for offshore wind conditions and it was validated using the Monte Carlo cross validation (MCCV) technique, which is explained in Sect. 3.3. The indirect methods which are evaluated for calculating evaporation from the Dead Sea water surface and the performed sensitivity studies are presented in Sect. 3.4.

3.1 Calculation of radiation components

The measurements of the radiation components of the lower half space were not conducted directly over the water surface, but over the land surface. Therefore, these two components had to be calculated for the water surface. The reflected shortwave radiation was calculated using literature values of the Dead Sea albedo. Stanhill (1987) calculated the albedo of the Dead Sea surface from ship measurements and reported values of 0.06 in the summer months, 0.09 in the winter months, and an annual average of 0.07. He also reported albedo values from Kondrat'Ev (1969) for the latitude of the Dead Sea and the cloud cover observed in the northern part of the Dead Sea, which was 0.08 for November and 0.07 as an average annual albedo value. To confirm the validity of the literature values for our site, a short-term experiment was conducted in November 2014. The measured albedo values of 0.08 to 0.09 concur well with the literature values for winter. As the literature values for summer could not be compared to measurements, the annual average of 0.07 was used for all calculations. The longwave outgoing radiation was calculated using the Stefan–Boltzmann equation

$$RI \uparrow = \epsilon \cdot k_B \cdot T_S^4, \tag{1}$$

with the water surface emissivity $\epsilon = 0.98$ (e.g. Konda et al., 1994) and the Stefan–Boltzmann constant, k_B. For the surface water temperature, T_S, no in situ measurements were available. Also, remotely sensed surface water temperature products from satellites could not be used, as operational algorithms are calibrated to mean sea level and do not take the additional 421 m atmospheric layer in the Dead Sea valley into account. Nehorai et al. (2009) showed that a calibration of satellite data with in situ measurements is necessary. Furthermore, Nehorai et al. (2009) raised concerns that enhanced water vapour input into the atmosphere through evaporation causes stronger absorption of thermal IR radiation, leading to a screening of the Dead Sea surface and, thus, incorrect estimates of the surface water temperature. Based on the results of Nehorai et al. (2013), which showed that "surface water temperature is highly correlated to air temperature ($R^2 = 0.93$–0.98) in all seasons", the Monin–Obukhov similarity approach was used to calculate surface water temperature from the measured air temperature (see Appendix A) and is further on referred to as T_{MO}.

3.2 Calculation of sensible and latent heat flux

To calculate the sensible and latent heat flux from the wind, temperature, and humidity data measured by the IRGASON, the eddy covariance technique was used. This method uses the fluctuations of the vertical wind velocity and temperature around a temporal mean, here 30 min, to calculate the sensible heat flux,

$$H = c_p \cdot \rho_a \cdot \overline{w'T'_{sonic}}, \tag{2}$$

and of the vertical wind velocity and the absolute humidity to calculate the latent heat flux:

$$LE = L_v \cdot \overline{w'a'} \cdot 1000. \tag{3}$$

The overbar represents the time average over 30 min; c_p is the specific heat at constant pressure in $J\,K^{-1}\,kg^{-1}$, ρ_a is the density of the air in $kg\,m^{-3}$, w' is the deviation of the vertical wind speed from the mean vertical wind speed in $m\,s^{-1}$, T'_{sonic} is the deviation of the sonic temperature from the mean sonic temperature in K, and a' is the deviation of absolute humidity from the mean absolute humidity in $kg\,m^{-3}$. L_v is the latent heat of vaporisation in $kJ\,kg^{-1}$,

$$L_v = 3148.4 - 2.37 \cdot T_w, \tag{4}$$

which depends on water temperature T_w in K. For salt water, L_v increases with increasing salinity (Steiner, 1948). Therefore, for the calculation of the latent heat flux of the Dead Sea water, the salinity of the water has to be considered. To get the dependency of L_v on water temperature for the Dead Sea, respective measurements were undertaken. The vapour pressure of the Dead Sea water was measured as a function of water temperature with a calibrated capacitance manometer (see Appendix B). The following equation for the Dead

Sea water was derived with the same units as in Eq. (4):

$$L_v = 5150.6561 - 13.9530 \cdot T_w + 0.0162 \cdot T_w^2. \qquad (5)$$

As evaporation takes place directly at the water surface of the lake, in a layer of approximately $10\,\mu m$ (Emery et al., 2001), surface water temperature should be used for the calculation of L_v. Thus, the introduced T_{MO} was used for this purpose.

3.2.1 Post-processing of eddy covariance data

Post-processing of eddy covariance data is essential as field measurements generally do not fulfil all the theoretical concepts and assumptions of the eddy covariance theory. In particular, measurement limitations of the sensors, non-stationary conditions over the averaging period, and horizontal heterogeneity have to be considered (Foken et al., 2012). Therefore, the following post-processing steps were applied to the data set using the software package TK3 (Mauder and Foken, 2011). First, data were checked on plausibility using individual thresholds for each meteorological variable. Then, a spike detection, using the algorithm after Mauder et al. (2013), was applied. No fluxes were calculated if more than 10 % of the data in the corresponding 30 min interval were missing. To account for a not perfectly levelled sonic anemometer, meaning that the vertical axis is not perpendicular to the surface and thus the vertical wind measurements are affected by the horizontal wind components, the coordinate system of the sonic anemometer was rotated using the planar fit method after Wilczak et al. (2001). It rotates the coordinate system to the main wind direction and then rotates the system around the y axis, such that the z axis is positioned perpendicular to the horizontal plan and that the mean vertical wind over the period that is used to define the plane is $0\,m\,s^{-1}$.

Spectral corrections were performed to account for the loss of energy for high frequencies, due to path-length averaging and limited sensor frequency response, following the approach after Mauder and Foken (2011). The influence of humidity on sonic temperature plays an important role for the calculation of the sensible heat flux. To account for this influence, the Schotanus correction (Schotanus et al., 1983) was applied. This correction is particularly important for flux calculations at sites with high humidity fluctuations, such as over the water surface. The water vapour measurements are influenced by temperature and humidity changes, as only the molar density of water vapour is measured and not the mass mixing ratio. To consider the density fluctuations, corrections after Webb et al. (1980) were applied.

3.2.2 Quality control and data coverage

The overall performance of the system was very good, and only 2.1 % of the sensible heat flux data and 2.4 % of the latent heat flux data were missing. To assure data quality of the flux measurements, several quality criteria were applied. Latent heat flux data were rejected when the signal strength of the radiation source to measure the water vapour was below 50 %, when the variability of the signal from one 10 min average to the next one was higher than 0.6 % within the 30 min time interval, and during precipitation events, as a disturbance of the water vapour measurements was expected for these conditions. Due to these quality criteria 10 % of the latent heat flux data were rejected. Further quality control was performed using the steady state test after Foken and Wichura (1996), which analyses each 30 min time interval on stationarity and the integral turbulence characteristics (ITC) test after Foken et al. (2012), which checks data on fully developed turbulent conditions. A combined quality flag considering the steady state test and the ITC test (Foken, 1999) was used to classify the data into nine classes. Class 1 to 6 describe data, which can be used for the analysis, and classes 7 to 9 were rejected. After the quality control, data availability was 86.3 % for sensible and 78.5 % for latent heat flux data. Furthermore, the flux footprint had to be considered. A footprint analysis was performed, using the model after Kljun et al. (2015). Results show that flux data for wind directions between 230 and 330° had to be rejected as the fetch was over land, while the aim of this station was to measure evaporation from the water surface (Fig. 1c). For northerly to southerly wind directions, the fetch was over water and the average fetch contributing to 80 % of the flux ranged from 0 to 300 m and 0 to 600 m, respectively. (Fig. 1d). The amount of flux data rejected due to the footprint was about 19 %. The total available flux data from the water surface was thus 67.1 % for sensible and 59.2 % for latent heat flux. This was reasonably good compared to other eddy covariance studies at other lakes, where a data availability between 36 and 56 % was reported (e.g. Jonsson et al., 2008; Mammarella et al., 2015; Bouin et al., 2012).

3.3 Multiple regression model for the latent heat flux

Through the installation of the EBS at the shoreline, flux data from the water surface are only available for onshore wind conditions, and all data for offshore wind conditions, i.e. wind directions between 230 and 330°, are rejected for the analysis (Fig. 1c). However, for the analysis of the diurnal and intra-annual variability of the evaporation rates, estimates of the fluxes for these wind directions are important, as otherwise evaporation rates in the afternoon, when westerly downslope winds prevail, would be missing. Therefore, a multiple regression model is applied to find a suitable relationship between the turbulent fluxes and governing meteorological variables, such as wind speed, vapour pressure deficit, net radiation, and surface water temperature. The vapour pressure deficit is calculated using the surface water temperature T_{MO} (see Appendix C). A Monte Carlo cross validation, first introduced by Picard and Cook (1984), is performed to test the model's robustness and get an estimate of the model error. The work flow is as follows: (i) data be-

Table 1. Selection of commonly used equations to calculate evaporation (Ev) in mm d^{-1}. The original version and the default version (V0) used in Sect. 3.4 and 4.4 are presented.

Method	Name	Original equation	Default version (V0)
Aerodynamic/mass transfer	Aerodynamic	[a]$Ev = \frac{0.622}{\rho_w p} C_e \rho_a \, v_a \, (E_w - e_a)$	$C_e = \dfrac{\kappa^2}{\left(\ln\left(\frac{z_m - z_d}{z_0}\right)\right)^2}$
Energy budget	BREB	[b]$Ev = \frac{R_n - G - F_n - \Delta Q}{\rho_w \cdot L_v \cdot (1 + Bo)}$	$\cancel{F_n}, \cancel{G}, \cancel{\Delta Q}$
Combination	Priestley–Taylor	[c]$Ev = c_{PT} \frac{\Delta \cdot (R_n - G)}{\rho_w L_v (\Delta + \gamma)}$	$\cancel{G}$
Combination	Penman	[d]$Ev = \frac{\Delta \cdot R_n + \gamma C_e v_a \rho_w L_v (E_a - e_a)}{\rho_w L_v (\Delta + \gamma)}$	orig. Eq. used
	Bowen ratio	$Bo = \frac{H}{LE} \approx \frac{c_p \cdot p}{0.622 \cdot L_v} \cdot \frac{T_S - T_a}{E_w - e_a} = \gamma \cdot \frac{T_S - T_a}{E_w - e_a}$	

C_e = transfer coefficient for evaporation
c_p = specific heat capacity at constant pressure
$c_{PT} = 1.26$ = Priestley–Taylor coefficient
e_a = vapour pressure at air temperature
E_a = saturation vapour pressure at air temperature
E_w = saturation vapour pressure at surface water temperature
Ev = evaporation
F_n = net advected heat flux
G = ground heat flux
L_v = latent heat of vaporisation
p = air pressure

R_n = net radiation
T_S = surface water temperature
T_a = air temperature
v_a = wind velocity
Bo = Bowen ratio
γ = psychometric constant
Δ = slope of the saturation vapour pressure vs. temperature curve
ΔQ = heat storage of the lake
ρ_a = air density
ρ_w = water density

[a] Brutsaert (1982), [b] Dingman (2002), [c] Priestley (1972), [d] Van Bavel (1966)

tween 230 and 330° are removed from the data set. (ii) Two approaches are used to divide the data in a training and validation data set. The first approach uses randomly chosen data points of about 15 % of the total data set as validation data and the second approach uses a randomly chosen wind sector of 45° as validation data. The usage of these two approaches allows the general test of the model's robustness but also its sensitivity on a certain wind sector. (iii) After each division a regression model is built with the training data set and then applied on the data of the validation group. The deviation of the calculated from the measured flux values yields the model error of one realisation. (iv) After multiple applications, in this case 500 times, the model error is averaged and results in the prediction error of the regression model. A large prediction error indicates a dependency of the model on the choice of the training data set and therefore has to be rejected.

3.4 Indirect methods to estimate evaporation

For the calculation of evaporation, several equations, based on different physical approaches, exist. Each approach connects evaporation to different meteorological parameters and is designed for different time intervals, ranging from sub-daily calculations to a time interval of at least 7 days. Four commonly used indirect methods to estimate evaporation (Table 1) will be tested in this paper by comparing their results to the eddy covariance measurements. An aerodynamic

approach also known as mass transfer approach, the energy budget method, and two combination approaches, namely the Priestley–Taylor and Penman equation, will be evaluated on time intervals of 1, 7, 14, and 28 days. The aerodynamic approach is the only approach which is also designed for sub-daily time intervals and will thus also be tested for 30 min time intervals. Additionally, sensitivity studies are performed to quantify the influence of simplification within the approaches, which are often made in literature.

The first method is the aerodynamic approach after Brutsaert (1982), where only wind speed and vapour pressure deficit are required. With the assumption of equal transfer coefficients for evaporation and momentum ($C_e = C_d$) under neutral conditions, the logarithmic wind profile can be used (Van Bavel, 1966) (Table 1, V0). This is the default version of the aerodynamic method for the sensitivity studies. The second method is the energy budget method expressed as the Bowen ratio energy budget (BREB) (Table 1). For this approach several of the input variables are difficult to obtain. The amount of net advected heat into the water body, F_n, meaning the heat advected into the lake by water inflow and precipitation, and the loss of heat by water outflow have to be known. If the in- and outflows are small compared to the size of the water body, or water temperatures are similar, the term can be neglected (Dingman, 2002; Rosenberry et al., 2007). Moreover, the ground heat flux G, meaning the heat exchange at the bottom of the lake, is required. It can usually

Table 2. Overview of the sensitivity studies performed for the evaporation equations. Sensitivity studies applied to a method are marked with an X.

Version	Explanation	Aerodynamic	BREB	Priestley–Taylor	Penman
0	Default (see Table 1)	X	X	X	X
1	Atmospheric Stability	X	–	–	X
2	Heat storage term derived with hysteresis approach	–	X	X	X
3	Heat storage term derived as a linear function of R_n from V0	–	X	X	X
4	Removal of T_{MO} from R_n calculation	–	–	–	X
5	Removal of T_{MO} from R_n calculation and heat storage term from hysteresis approach	–	–	–	X
6	Removal of T_{MO} from R_n calculation and heat storage term derived as a linear function of R_n from V4	–	–	–	X

be neglected, as the amount for deep lakes is small compared to the other components (Henderson-Sellers, 1986). Another component difficult to obtain is the heat storage of the lake, ΔQ. It requires measurements of lake temperature at different depths from a raft station or a ship. On longer timescales it can often be neglected. Because of the aforementioned reasons and the difficulty in obtaining these three terms, the net advected heat, the ground heat flux, and the heat storage term are neglected in many studies. Thus, for the default version (V0) of the BREB method these three terms are neglected (Table 1). Even though neglecting the heat storage on the timescales investigated is a coarse assumption, it serves as a basis for the sensitivity studies V1 and V2. Using V0, only net radiation, surface water temperature, air temperature, and the vapour pressure deficit have to be known, which are relatively easy to obtain and thus an easy approach to calculate evaporation. The third method to calculate evaporation is the combination approach, considering the energy budget and the aerodynamic influence. Priestley (1972) proposed an equation which considers the aerodynamic influence by using an empirically gained coefficient of $c_{PT} = 1.26$ (Table 1). Because of the same reason as mentioned above, the ground heat flux is neglected in the default version (V0) of the Priestley–Taylor equation. Method four is a combination of the energy budget equation with the aerodynamic approach first developed by Penman (1948). In his approach he already neglected net advected heat, the ground heat flux, and the heat storage. Van Bavel (1966) further generalised Penman's equation by replacing the empirical wind function through the logarithmic wind profile, assuming neutral conditions (Table 1). This equation will be used as the default version (V0) for testing the Penman approach.

In total, six sensitivity studies were performed. An overview of the sensitivity studies and to which of the methods they are applied to is given in Table 2. Sensitivity study V1 considers non-neutral atmospheric conditions, by incorporating stability correction factors into C_e. As only the aerodynamic and the Penman approach are based on mass-transfer, V1 is applied to these two equations only. Studies V2 and V3 consider the heat storage of the lake ΔQ and are

applied to the BREB, Priestley–Taylor, and Penman method. For this purpose R_n is replaced with $(R_n - \Delta Q)$. Duan and Bastiaanssen (2015) proposed a hysteresis approach to calculate the heat storage term, depending only on the net radiation ($\Delta Q = a + b \cdot R_n + c \cdot dR_n/dt$). This approach is applied to the measurement data and the resulting coefficients (a, b, c) are used in sensitivity version V2 to calculate ΔQ. To avoid the use of the calculated heat storage from the measurements, in V3 it is assumed that the heat storage term is directly proportional to the net radiation and that the deviation of the default version (V0) from the measurements equals the heat storage term. The last three sensitivity tests were applied to the Penman approach only. In V4 the uncertainty caused by the calculated longwave outgoing radiation with T_{MO} was eliminated by using an approximation from Kohler and Parmele (1967) where they calculated the longwave net radiation and the psychrometric constant using air temperature only. This further reduces the amount of necessary input parameters, which makes the equation more easily applicable, when net radiation of the water surface is not directly measured. In version V5 the approximation after Kohler and Parmele (1967) is applied together with the hysteresis model for the heat storage term (V2). The last sensitivity test (V6) combines the approximation after Kohler and Parmele (1967) with a linear function for the heat storage term, derived from the deviation of V4 from the measurements.

4 Results

4.1 Meteorological conditions

In the Dead Sea valley the measured average annual air temperature was 26.5 °C for the measurement period, which was slightly higher than the long-term annual mean of 25.9 °C found by Hecht and Gertman (2003) for the period 1992 to 2002. Maximum daily air temperatures regularly exceeded 40 °C in summer (Fig. 2) and the annual precipitation was 273 mm. (Fig. 2). The precipitation amount for the observation period is high compared to the mean annual precipitation of the standard normal period 1961 to 1990 of 80 mm

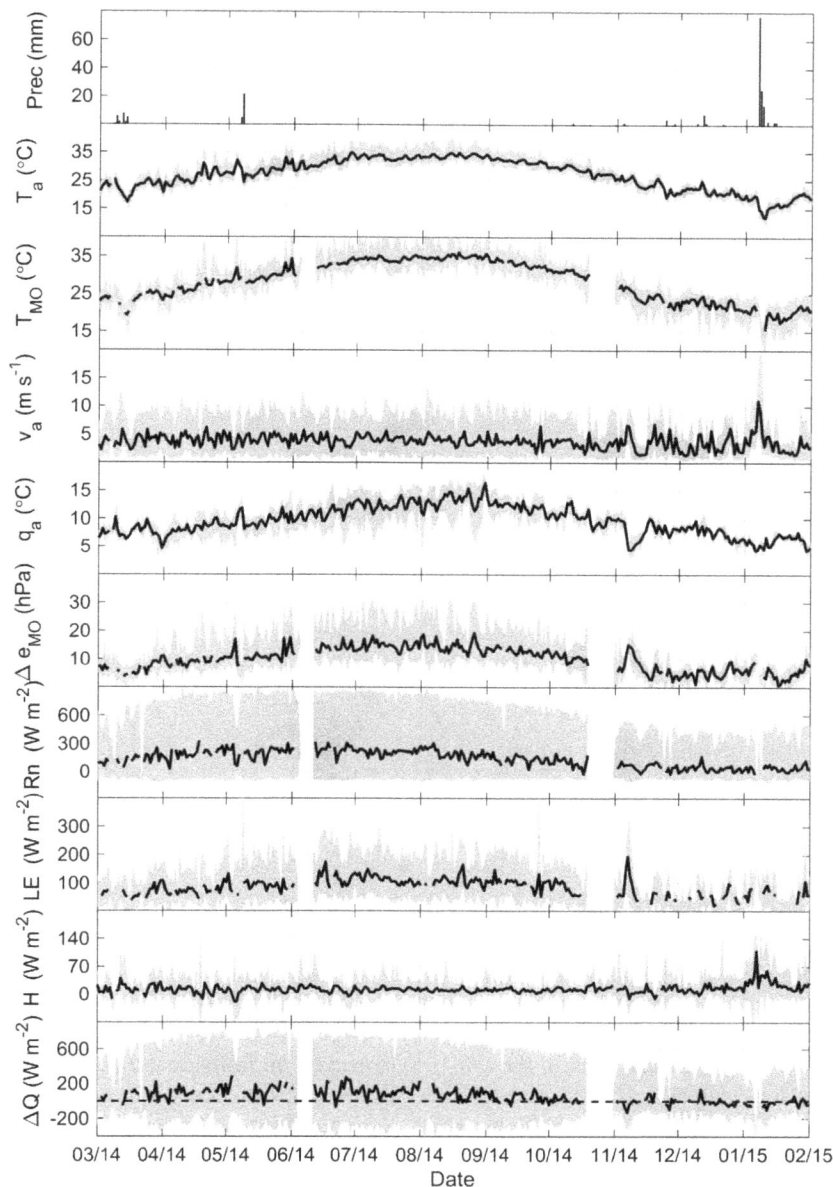

Figure 2. Daily precipitation (prec), 24 h running mean of air temperature (T_a), surface water temperature (T_{MO}), wind velocity (v_a), specific humidity (q_a), vapour pressure deficit (Δe_{MO}), net radiation (R_n), latent heat flux (LE), sensible heat flux (H), and heat storage (ΔQ). The grey shaded area represents the range between daily minimum and maximum values of the respective variable.

(Goldreich, 2003). It resulted from a few heavy precipitation events in January 2015, which made the observation period 2014–2015 a relatively wet year for the area. The wind velocity did not show a clear annual cycle. From March until October, mean, maximum, and minimum were relatively similar. However, during winter, a different behaviour was found when the wind increased in connection with the stronger large-scale activity (Fig. 2). The relative uniform wind velocities from spring until autumn resulted from periodic local wind systems, governing the conditions in the valley. Between sunrise and sunset a lake breeze prevailed, leading to

north-easterly winds at the station with a median wind velocity of $3\,\mathrm{m\,s^{-1}}$ (Figs. 2 and 3a). The lake breeze occurred throughout the year, with an occurrence rate exceeding 70 % of the days in summer 2014, and 58 and 48 % of the days in spring and autumn 2014, respectively. In winter, the synoptic conditions gained more influence and often superimposed the local wind field such that a north-easterly lake breeze was only observed on about 32 % of the days and a south-easterly flow on 26 % of the days in winter 2014–2015. In the evening, north-westerly downslope winds, often enhanced by the Mediterranean Sea breeze (Alpert et al., 1997; Naor

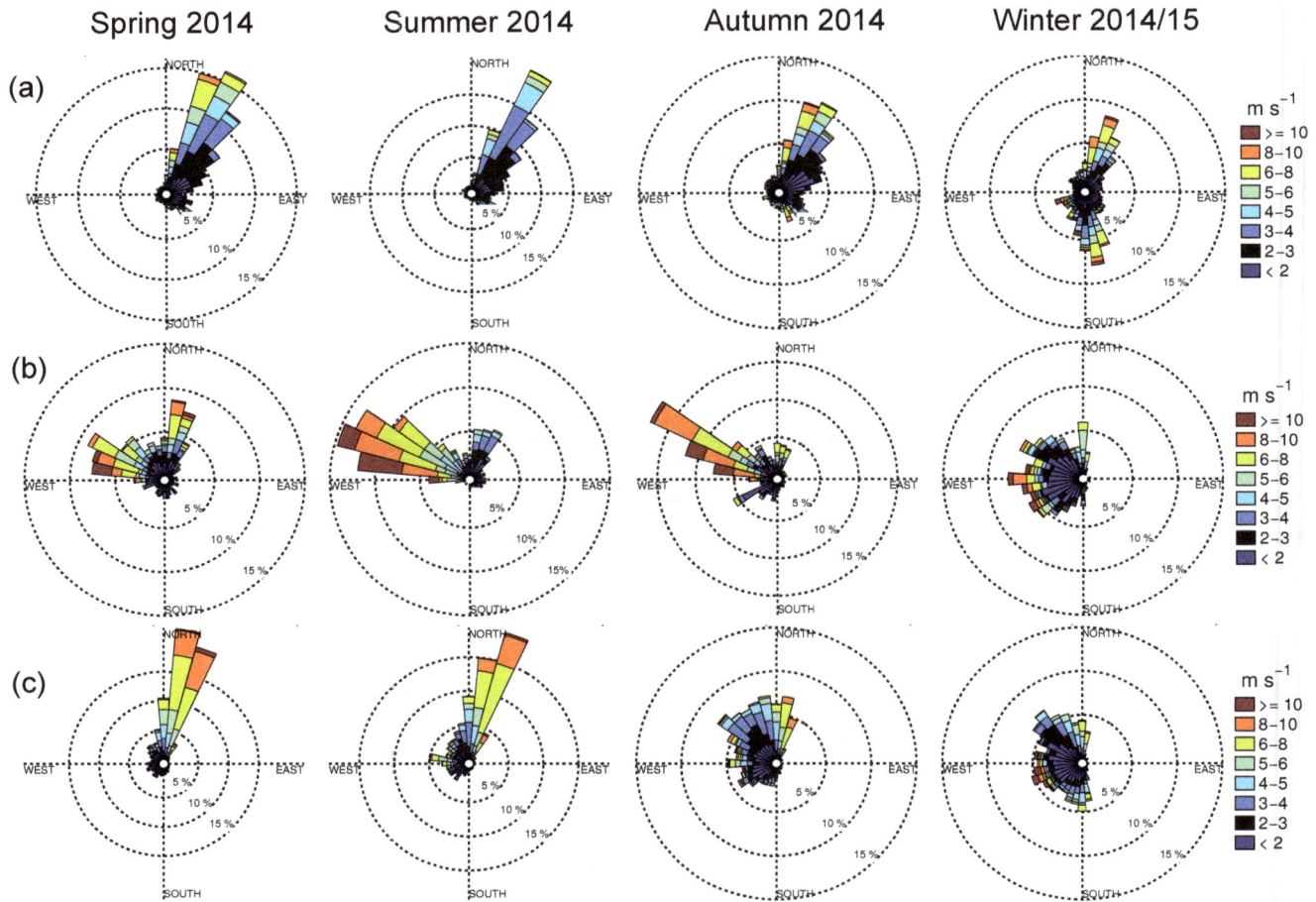

Figure 3. Wind conditions between **(a)** 06:30 and 17:30 LT, **(b)** 17:30 and 20:30 LT, and **(c)** 20:30 and 06:30 LT. Data are shown for spring, summer, autumn, and winter 2014–2015.

et al., 2017), lead to accelerated wind velocities in the valley (Fig. 3b). These downslope winds occurred on about 57 % of the days in summer and still 28 % of the days in spring and 45 % of the days in autumn. The downslope winds regularly reached mean wind velocities exceeding $10\,\mathrm{m\,s^{-1}}$ (Fig. 3b). During the night, a northerly along-valley flow prevailed mainly in spring and summer. The along-valley flow also reached wind velocities exceeding $10\,\mathrm{m\,s^{-1}}$ (Fig. 3c). The difference between the saturation vapour pressure at the water surface and the actual vapour pressure of the air (Δe) had a mean value of 9.75 hPa. It had a clear annual cycle with maximum values above 30 hPa in summer. Individual peaks in winter were related to special synoptic conditions, e.g. in the beginning of November, when a Red Sea Trough with a central axis advected dry and warm air into the valley over the course of several days. The annual cycles of the energy balance components are also shown in Fig. 2. The net radiation reaches maximum values exceeding $900\,\mathrm{W\,m^{-2}}$ in summer and about $500\,\mathrm{W\,m^{-2}}$ in winter. The sensible heat flux is small throughout the year. The mean latent heat flux values are higher in summer compared to the winter months.

However, on individual days in winter some latent heat flux values even exceeded the summer values. The heat storage shown in Fig. 2 shows that a considerable amount of energy is stored but also released over the course of the day. However, this term also contains the possible non-closure of the energy balance. Assuming common literature values of the non-closure (Wilson et al., 2002; Foken, 2008), the actual heat storage is most likely 20 % smaller than shown here. On a seasonal basis the sensible heat flux accounts for about 5 to 10 % of the net radiation in spring, summer, and autumn, whereas it accounts for nearly 40 % in winter. The latent heat flux accounts for 43 and 53 % of the net radiation in spring and summer, leading to a high heat storage amount of 51 and 42 %, respectively. In autumn over 74 % of the net radiation is transformed into latent heat flux, such that the heat storage amount is small. In winter, the latent heat flux is in the range of 92 % of the net radiation, meaning that the heat storage term is negative, releasing the heat to the atmosphere, represented through the higher sensible heat flux. Similar behaviour of the flux components was found for other lakes (e.g. Giadrossich et al., 2015).

Table 3. Correlation coefficients for latent heat flux (LE) with wind speed (v_a), net radiation (R_n), surface water temperature (T_{MO}), and vapour pressure deficit calculated with surface water temperature (Δe_{MO}). Correlation coefficients over 0.5 are bold. Data are shown for the meteorological seasons 2014–2015 and the entire data set.

	v_a	R_n	T_{MO}	$\Delta e_{T_{MO}}$
Spring	**0.68**	−0.19	0.07	0.06
Summer	**0.73**	−0.16	0.00	−0.12
Autumn	**0.53**	0.16	0.36	0.46
Winter	**0.81**	0.27	0.19	**0.56**
Total	**0.59**	0.03	0.42	0.38

4.2 Multiple regression model for the latent heat flux

The footprint model showed that the fetch of the fluxes is over land for wind directions between 230 and 330°. The affected amount of latent heat flux data is 19 %. Through the predominant local wind systems, these wind directions occur almost exclusively in the evening between 17:30 and 20:30 LT (LT = UTC + 2) from spring until autumn (Fig. 3) and, thus, most of the data within this time frame are excluded. For the analysis of the diurnal variability of the latent heat flux from the water surface, and also for the intra-annual and annual amounts, it is important to close these gaps. A multiple regression model was applied to calculate the latent heat flux for offshore wind conditions. The choice of the input variables for the multiple regression model was based on the analysis of the linear correlation between the latent heat flux and different meteorological variables. The correlation coefficients for the variables are shown in Table 3. For the latent heat flux, the highest correlation is achieved with wind speed, with correlation coefficients between 0.53 and 0.81 for the different seasons, followed by the vapour pressure deficit, and finally surface water temperature and net radiation. This is different from cooler climates where the highest correlation was found with vapour pressure deficit (Blanken et al., 2000; Nordbo et al., 2011) and also from lakes in Mediterranean climate, where vapour pressure deficit had the same impact as wind speed (e.g. Bouin et al., 2012). The influence of the vapour pressure deficit varies strongly between the different seasons. In spring and summer no correlation exists between latent heat flux and the vapour pressure deficit, but in autumn, winter, and for the total data set correlation coefficients are between 0.38 and 0.56. Although correlation with individual meteorological variables is already good, none of the variables can fully explain the latent heat flux. A stepwise multiple regression model was applied with the following variables to find the best fitting solution for the latent heat flux:

$$X_{LE} = (v_a, \Delta e_{T_{MO}}, R_n, T_{MO}). \tag{6}$$

The model X_{LE} gave the same dependency for all seasons. The latent heat flux depended on a linear combination of

Table 4. Results of the stepwise linear regression model X_{LE} for the latent heat flux. The corresponding correlation coefficient (R) of the model after a variable is added to the model is shown. For the model with $v_a \cdot \Delta e_{MO}$, the correlation coefficient (R) is given. The prediction errors yielded by the MCCV with randomly chosen validation data points (er_r) and randomly chosen validation sectors (er_s) are shown for both models. Results are shown for the meteorological seasons and for the entire data set.

	X_{LE}				$v_a \cdot \Delta e_{T_{MO}}$		
	v_a	$\Delta e_{T_{MO}}$	er_r (%)	er_s (%)	R	er_r (%)	er_s (%)
Spring	0.68	0.77	0.32	8.61	0.79	−0.01	10.57
Summer	0.73	0.77	0.17	2.31	0.76	−0.17	1.60
Autumn	0.53	0.82	−0.16	1.25	0.84	0.42	6.17
Winter	0.81	0.85	2.94	0.02	0.85	4.72	−0.31
Total	0.59	0.80	0.96	4.79	0.83	0.80	6.78

Table 5. Coefficients of the model equations to calculate latent heat flux (LE). The equations have the general form: LE $= a + b \cdot v_a + c \cdot \Delta e$. Coefficients are shown for the meteorological seasons 2014–2015 and the entire data set.

	Spring	Summer	Autumn	Winter	Total
a	−32.52	−25.41	−58.91	−15.29	−36.92
b	13.33	18.41	16.21	11.07	14.31
c	5.51	4.61	7.56	4.46	6.13

wind speed and vapour pressure deficit. The correlation coefficient ranged from 0.77 in spring and summer to 0.85 in winter (Table 4). The aerodynamic approach to estimate evaporation is based on the product of wind speed and vapour pressure deficit (Table 1), instead of a linear combination. For comparison, the correlation of the product of wind speed and vapour pressure deficit with the latent heat flux was calculated additionally and resulted in nearly the same correlation coefficients (Table 4). The results of the Monte Carlo cross validation analysis reveal that the model X_{LE} results in small prediction errors. The prediction error varies between −0.16 and 2.94 % for randomly chosen data points and for randomly chosen control sectors between 0.02 and 8.61 %. The model with $v_a \cdot \Delta e_{T_{MO}}$ results in higher model errors varying between −0.17 and 4.72 % for randomly chosen data points and between −0.31 and 10.57 % for randomly chosen control sectors. Even though the correlation coefficients are similar for both models, model X_{LE} was chosen for the calculation of the latent heat flux, instead of the commonly used $\Delta e \cdot v_a$, because of the robustness and the smaller prediction error. The model coefficients are shown in Table 5.

In summary, the regression model X_{LE} provides a suitable and robust method to calculate the latent heat flux for offshore wind conditions. To ensure that the model is not applied outside the conditions for which it has been constructed, the extreme values of offshore wind velocity and vapour pressure deficit are not considered to calculate evaporation and it is checked that data are always within the model

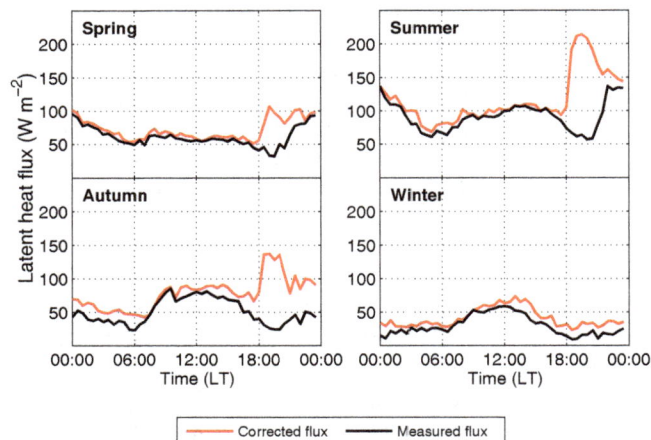

Figure 4. Median diurnal cycles of the measured latent heat flux (black lines) and the latent heat flux corrected with the multiple regression model for wind directions between 230 and 330° (red lines).

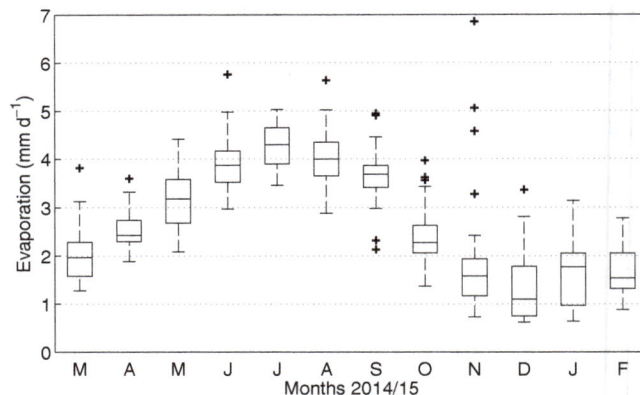

Figure 5. Box plot of daily evaporation rates. Red lines indicate medians, the edges of the boxes are the 25th and 75th percentiles, the whiskers extend to the most extreme data points not considered outliers, and outliers are plotted individually by red crosses.

boundaries. Evaporation values, which can not be calculated because wind velocity or vapour pressure deficit are outside the boundaries are treated as missing values. With this method 90 % of the originally rejected latent heat flux data due the fetch criteria can be calculated with the model. The total data availability is increased from 59.2 to 76.8 %.

The calculation of the latent heat flux for offshore wind conditions is especially important for the analysis of the diurnal cycle of the latent heat flux, and also for its intra-annual variation. The comparison of the mean diurnal cycles of the measured fluxes with the cycles including the calculated values for offshore wind conditions (corrected fluxes) shows that during the day the differences are small (Fig. 4). As the prevailing wind direction is north-east, caused by the lake breeze, nearly no calculations are necessary, as the flux footprint is located over water (Fig. 1d). However, in the evening, when downslope winds prevail in spring, summer, and autumn, the differences are quite large (Fig. 4). During this time period, the measured values represent the latent heat flux from the land surface, with values around or below $50\,\mathrm{W\,m^{-2}}$. In contrast, the calculated values represent the latent heat flux from the water surface, with values up to $200\,\mathrm{W\,m^{-2}}$ in summer. Hence, the regression model allows a detailed analysis of the diurnal cycle of the fluxes, even though the station is located at the shoreline.

4.3 Diurnal and intra-annual variability

The latent heat flux is the dominating turbulent flux at the water surface (Fig. 2). It has a strong diurnal cycle. During daytime, the latent heat flux reaches values of $100\,\mathrm{W\,m^{-2}}$ in summer and autumn, and $70\,\mathrm{W\,m^{-2}}$ in spring and winter (Fig. 4). The maximum values are reached after sunset around 19:00 LT in spring, summer, and autumn. In spring

about $105\,\mathrm{W\,m^{-2}}$ are reached, in summer $213\,\mathrm{W\,m^{-2}}$, and in autumn $136\,\mathrm{W\,m^{-2}}$. During the night, the latent heat flux continues to be higher than during daytime and reaches minimum values shortly before sunrise. In winter, this late maximum is not observable and values during nighttime are lower than during daytime. The unusual diurnal cycle with highest latent heat flux values after sunset and during the night are clearly connected to the diurnal cycle of wind speed and vapour pressure deficit and thus to the wind systems. This is most pronounced in summer. During the day, the lake breeze with relatively low wind velocities, (Fig. 3a) causes moderate latent heat flux rates. The downslope winds in the evening have generally high wind velocities (Fig. 3b) and advect drier air into the valley, which results in high vapour pressure deficits and thus high latent heat flux values. The high values during night result from accelerated wind velocities (Fig. 3c), rather than high vapour pressure deficits.

For the calculation of daily and yearly evaporation, still-existing data gaps were closed, using the median evaporation rate of the corresponding time step of the respective month. The uncertainty due to this gap-filling method was estimated using the median absolute deviation, which is the median of the absolute deviations from the data's median.

In spring, evaporation rates steadily increase until a maximum median evaporation of $4.3\,\mathrm{mm\,d^{-1}}$ is reached in July (Fig. 5). Afterwards, evaporation rates decrease until a minimum median evaporation of $1.1\,\mathrm{mm\,d^{-1}}$ is reached in December (Fig. 5). The annual cycle of evaporation follows the solar cycle with a time lag of about 1 month. Summing the evaporation values over the whole measurement period results in an annual evaporation of $994.5 \pm 88.2\,\mathrm{mm}$, where $81.2\,\mathrm{mm}$ of the uncertainty result from the gap-filling method and $7.0\,\mathrm{mm}$ from to the regression model. Also visible in Fig. 5 is the higher variation of the daily evaporation rates between November and February. This is the so-called wet season when synoptic patterns gain more influence on the atmospheric conditions in the valley (Bitan, 1974, 1976). The governing factors of evaporation, i.e. wind speed and vapour pressure deficit, are very variable during this time. On the one hand, winter storms with rain and high air humidity can reach the region, which decreases the evaporation rate. On the other hand, winter storms without rain but high wind velocities, which advect very dry air to the Dead Sea, can significantly increase the evaporation rate (Shafir and Alpert, 2011). The highest variability (not considering outliers) can be seen in January, with daily evaporation rates between 0.6 and 3.1 mm d^{-1}. In November, daily evaporation rates vary between 0.7 and 2.4 mm d^{-1}, but on 3 consecutive days evaporation rates exceed these values. Evaporation rates of 5.1, 6.9, and 4.6 mm d^{-1} are measured, which is the absolute maximum of the whole measurement period. These extreme evaporation rates are caused by a Red Sea Trough with a central axis and a dominant high to the east, which causes south-easterly winds above the valley. It can be observed that through the complex orography a pressure-driven channelling occurs along the valley axis, resulting in a near-surface northerly wind with constantly high averaged wind speed exceeding 10 m s^{-1} (not shown). This leads to the advection of warm and very dry air over the lake, which, together with the high wind velocities, increases the evaporation dramatically. This case was also used to test the performance of the regression model as on these 3 consecutive days only 3 out of 72 evaporation values had to be calculated due to the fetch criteria. Applying the regression model to calculate evaporation on these 3 days completely yields good results for day 1 and three were the difference was only 4–5 %, but it also shows the potential underestimation of extreme evaporation rates as the model underestimated the daily evaporation on the second day by 18 %.

4.4 Indirect methods to estimate evaporation

With the comprehensive data set of the measurements, it is possible, for the first time, to evaluate four of the commonly used evaporation equations for their applicability for Dead Sea evaporation on different timescales (30 min, 1, 7, 14 and 28 d) and perform a sensitivity analysis on simplifications and assumption used for the equations. The main goal is the identification of the best fitting equation, by using measurements purely made on land, as data from raft stations or buoys are often difficult to obtain (Giadrossich et al., 2015). The calculated evaporation rates are compared to the eddy covariance measurements and evaluated in terms of their correlation coefficient, slope and offset of the regression line, mean difference, and monthly differences between the estimates and the measurements. Additionally, the relative over- or underestimation of the annual evaporation is compared to the measured amount of 994 ± 88.2 mm.

The first equation is the aerodynamic approach after Brutsaert (1982) (Table 1). This equation uses wind speed and vapour pressure deficit as governing factors. In the default version (V0) the stability of the atmosphere is not considered. The aerodynamic approach is the only approach designed for sub-daily time intervals. The correlation coefficient for 30 min averages is 0.85 and it tends to overestimate evaporation rates. The slope of the regression line is 1.26 (Table 6) and the mean difference is 0.92 ± 0.54 mm d^{-1}. For time intervals of 1 d and longer, the aerodynamic approach yields better results. The correlation coefficients vary between 0.94 for 1 d intervals and 0.99 for 28 d intervals, mean differences are smaller, 0.02 ± 0.54 mm d^{-1} for the 1 d interval, and the slopes of the regression lines vary around 1.10 (Table 6, Fig. 6a,V0). The mean differences are evenly distributed throughout the year, showing no seasonal bias, and the annual evaporation is well represented (Fig. 7a,V0). A sensitivity study was performed, considering the near-surface stability (V1), using the stability factors after Cline (1997). However, the comparison with V0 shows that the inclusion of the stability has a negligible effect (Table 6).

The BREB method is first used in the simplified version shown in Table 1, neglecting net advected heat fluxes, the ground heat flux, and the heat storage term. With this version (V0), only net radiation, surface water temperature, and air temperature have to be known. These variables are relatively easy to obtain and it would therefore be an easy approach to calculate evaporation. However, neglecting the heat storage term results in a strong bias of the evaporation rates. The correlation coefficients range from only 0.67 for 1 d time intervals to 0.87 for 28 d intervals, the slope varies from 1.27 to 1.72, respectively, and the largest offset is −1.35 mm d^{-1} (Table 6). This indicates a strong overestimation of high evaporation rates in spring and summer and an underestimation of small evaporation rates mainly in winter (Fig. 6b,V0), resulting in a clear seasonal bias. From April until September daily evaporation rates are overestimated by up to 3 mm d^{-1} and underestimated during the rest of the year (Fig. 7b,V0). This seasonal bias was also observed in other studies, e.g. Winter et al. (1995); Rosenberry et al. (2007). Compared to the measured values, this results in an overestimation of the annual evaporation by 22 %, calculated from the 28 d averages. For the other time intervals the overestimation of the annual evaporation was comparable and is therefore not shown. The sensitivity study V2, considers the heat storage of the lake using a hysteresis model. Correlation coefficients are better and the mean differences are reduced (Table 6). However, the slope and offset shows that the heat storage term is still not represented correctly. The slopes and the offsets indicate an overestimation of small evaporation rates and an underestimation of high evaporation rates (Fig. 6b,V2). The intra-annual performance improved slightly and evaporation estimates between November and April are quite good; however, evaporation rates are underestimated in summer and autumn by about 1 mm d^{-1} (Fig. 7b,V2). This results in a underestimation of the annual evaporation by about 11 %. Sensitivity study V3 also

Table 6. Slope and offset of the regression lines between the evaporation estimates calculated with the different equations and the evaporation measurements and the corresponding correlation coefficient (R), for averaging periods of 30 min, 1, 7, 14, and 28 days. Mean difference (MD) and standard deviation (SD) in mm d^{-1} are shown for 1 and 28 d as no relevant differences for the other time intervals exist. V0 to V6 indicate the different sensitivity studies (see Table 2). The best fitting solutions are indicated with bold numbers.

		Slope					Offset					R					MD $\pm$ SD	
		30 min	1 d	7 d	14 d	28 d	30 min	1 d	7 d	14 d	28 d	30 min	1 d	7 d	14 d	28 d	1 d	28 d
Aero-	**V0**	**1.26**	**1.13**	**1.08**	**1.10**	**1.12**	**−0.01**	**−0.33**	**−0.24**	**−0.29**	**−0.34**	**0.85**	**0.94**	**0.94**	**0.98**	**0.99**	**0.02 ± 0.54**	**−0.02 ± 0.24**
dynamic	V1	1.27	1.16	1.10	1.13	1.14	−0.01	−0.30	−0.17	−0.23	−0.26	0.85	0.94	0.94	0.98	0.99	0.13 ± 0.54	0.11 ± 0.24
BREB	V0	–	1.27	1.51	1.63	1.72	–	−0.13	−0.78	−1.11	−1.35	–	0.67	0.78	0.84	0.87	0.60 ± 1.78	0.61 ± 1.26
	V2	–	0.45	0.57	0.63	0.67	–	1.21	0.89	0.70	0.59	–	0.69	0.83	0.90	0.96	−0.30 ± 0.89	−0.30 ± 0.40
	V3	–	1.17	1.39	1.50	1.58	–	−0.12	−0.72	−1.02	−1.24	–	0.67	0.78	0.84	0.87	0.33 ± 1.62	0.33 ± 1.11
Priestley–	V0	–	1.35	1.61	1.74	1.84	–	−0.28	−0.98	−1.33	−1.58	–	0.69	0.80	0.86	0.89	0.69 ± 1.81	0.70 ± 1.30
Taylor	V2	–	0.49	0.62	0.70	0.74	–	1.17	0.81	0.59	0.47	–	0.73	0.87	0.93	0.98	−0.24 ± 0.84	−0.23 ± 0.32
	V3	–	1.17	1.39	1.51	1.59	–	−0.24	−0.85	−1.15	−1.37	–	0.69	0.80	0.86	0.89	0.23 ± 1.53	−0.24 ± 1.04
Penman	V0	–	1.44	1.58	1.69	1.76	–	0.19	−0.20	−0.49	−0.69	–	0.78	0.83	0.88	0.91	1.38 ± 1.52	1.38 ± 1.17
	V1	–	1.44	1.57	1.68	1.76	–	0.24	−0.12	−0.41	−0.61	–	0.78	0.83	0.88	0.91	1.44 ± 1.52	1.45 ± 1.16
	V2	–	0.75	0.80	0.85	0.89	–	1.34	1.21	1.04	0.94	–	0.87	0.89	0.94	0.97	0.65 ± 0.61	0.64 ± 0.25
	V3	**–**	**0.94**	**0.99**	**1.05**	**1.09**	**–**	**0.29**	**0.16**	**0.00**	**−0.11**	**–**	**0.82**	**0.84**	**0.90**	**0.92**	**0.13 ± 0.82**	**0.13 ± 0.51**
	V4	–	1.54	1.73	1.80	1.87	–	0.45	−0.09	−0.28	−0.48	–	0.89	0.91	0.94	0.96	1.92 ± 1.16	1.91 ± 1.07
	V5	–	0.96	1.01	1.04	1.06	–	1.22	1.10	1.00	0.93	–	0.92	0.93	0.96	0.97	1.10 ± 0.51	1.10 ± 0.27
	V6	**–**	**0.78**	**0.80**	**0.81**	**0.84**	**–**	**0.59**	**0.54**	**0.49**	**0.42**	**–**	**0.92**	**0.92**	**0.95**	**0.97**	**−0.02 ± 0.50**	**−0.02 ± 0.29**

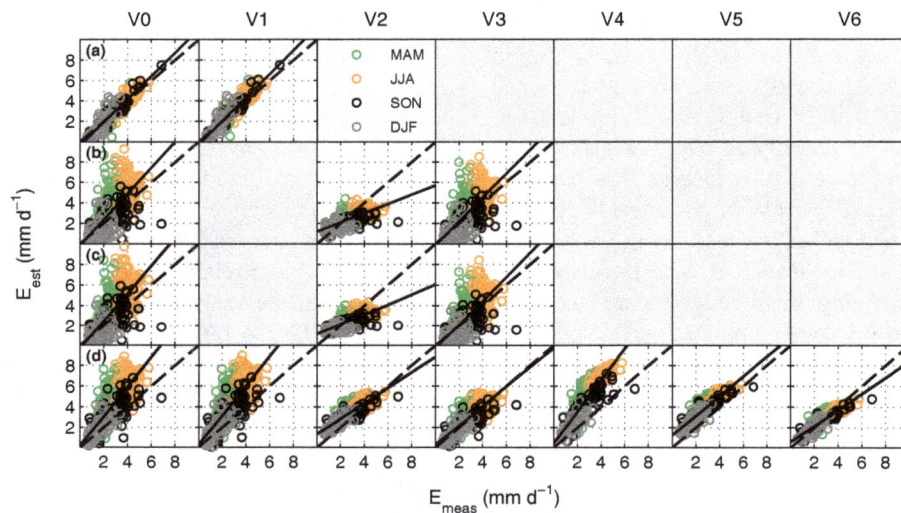

Figure 6. Correlation between estimated and measured daily evaporation rates for **(a)** the aerodynamic approach, **(b)** the BREB method, **(c)** the Priestley–Taylor equation, and **(d)** the Penman equation and their sensitivity studies (Table 2) calculated from 1 d averages. The colours indicate the meteorological seasons spring (MAM), summer (JJA), autumn (SON), and winter (DJF). The regression line is shown in black and the 1 : 1 line as a dashed red line.

accounts for the heat storage term by using $\Delta Q = 0.08 \cdot R_n$, derived from the deviation of V0 from the measurements. This approach can only slightly improve the correlation coefficient, slope, offset, and mean difference in comparison to V0 (Table 6). Only the annual evaporation improves compared to the default version and overestimates evaporation by only 13 % instead of 22 % (Fig. 7b,V3).

The Priestley–Taylor equation, as described in (Table 6), results in correlation coefficients between 0.69 for 1 d and 0.89 for 28 d time intervals. Like the BREB equation, slopes are too high with values between 1.35 and 1.84 and offsets vary between −0.28 and −1.58 mm d^{-1} (Table 6). By neglecting the heat storage term small evaporation rates are

underestimated and large ones overestimated (Fig. 6c,V0), resulting in a strong seasonal bias and an overestimation of the annual evaporation by 26 % (Fig. 7c,V0). Sensitivity test V2 yields similar results as for the BREB equation (Table 6). With the hysteresis model the seasonal bias shifts to an underestimation of evaporation in summer and autumn and relatively good results for winter and spring, resulting in a total underestimation of the annual evaporation by 8 % (Fig. 7c,V2). In V3 the heat storage is considered as a linear function of R_n and, thus, results in a new Priestley–Taylor coefficient of 1.09. With V3 the seasonal bias is reduced but high evaporation rates are still overestimated and low ones

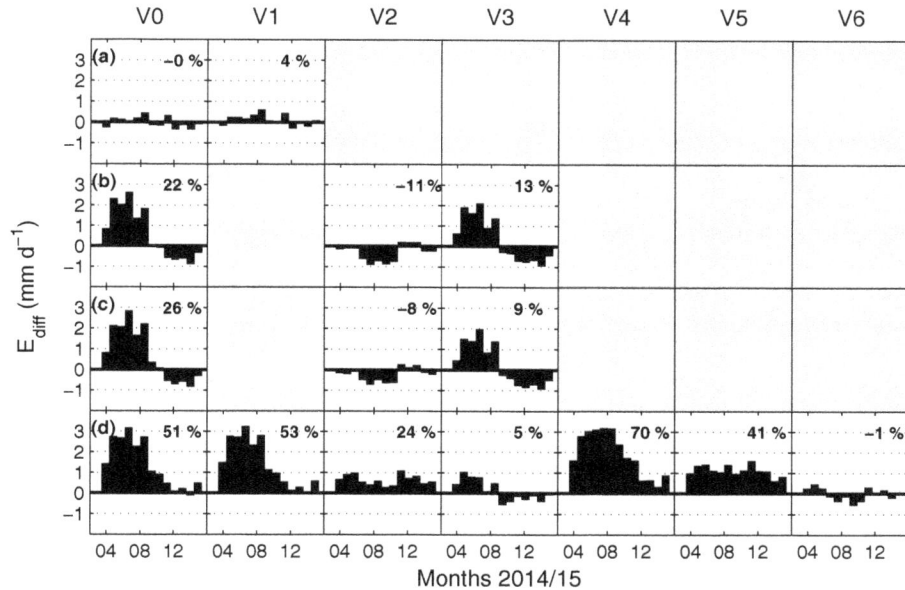

Figure 7. Differences between the estimated daily evaporation rates calculated from the 28 d time averages and the measured daily evaporation rates for **(a)** the aerodynamic approach, **(b)** the BREB method, **(c)** the Priestley–Taylor equation, **(d)** the Penman equation, and their sensitivity studies (Table 2). The red numbers show the total deviation of the accumulated calculated annual evaporation (28 d averages) from the accumulated measured evaporation.

underestimated (Table 6). The annual evaporation is overestimated by 9 % (Fig. 7c,V3).

The last equation tested is the Penman equation. In its original form (Table 6, V0) it results in correlation coefficients of 0.78 for time averages of 1 d to 0.91 for 28 d (Table 6). However, the slopes of the regression lines vary between 1.44 and 1.76, respectively, and indicate an overestimation. The mean differences also show a strong variability. Evaporation rates are strongly overestimated from spring until autumn (Fig. 6d,V0), exceeding the measured daily evaporation rates by up to 100 % (compare Fig. 7d,V0 to Fig. 5). The annual evaporation is thus also overestimated by 51 % showing that the original Penman equation is not applicable for the investigation of intra-annual variations. The consideration of the heat storage using the hysteresis model (V2) yields considerable improvements regarding the correlation coefficient. Its value varies between 0.87 and 0.97, and the mean difference and its standard deviation is reduced, meaning that the spread of the calculated values is smaller (Table 6). The slopes for V2 are all below unity and the offsets above $0.94 \, \mathrm{mm \, d^{-1}}$, meaning that small evaporation rates are overestimated (Fig. 6d,V2). This is also apparent in the intra-annual deviation of the estimated evaporation rates from the measured ones. Deviations are below or around $1 \, \mathrm{mm \, d^{-1}}$ for all months, resulting in a total overestimation of the annual evaporation by 24 % (Fig. 7d,V2).

The calculation of the heat storage term as a linear function of the net radiation results in $\Delta Q = 0.46 \cdot R_\mathrm{n}$. Using this function for the heat storage term in V3, the results are strongly improved. The slopes of the regression lines are close to one, offsets are small, and also the mean differences are smaller (Table 6). Correlation coefficients vary between 0.82 and 0.92 and the annual evaporation is only overestimated by 5 %, which is in the range of the measurement uncertainties. However, the results show a seasonal bias with an overestimation in spring and summer and a underestimation in autumn and winter (Fig. 7d,V3).

Another commonly used variation of the Penman equation is the removal of the surface water temperature from the calculation of the net radiation. This is tested in V4. However, in V4 the heat storage term is still missing an thus does not result in reliable evaporation rates (Figs. 6 and 7d,V4). The combination of the hysteresis model with the removal of the surface water temperature (V5) yields an improvement of the correlation coefficients, the slope of the regression lines, and also the standard deviations, but the calculated rates show an offset of over $0.93 \, \mathrm{mm \, d^{-1}}$ (Table 6). This results in a constant overestimation of evaporation rates throughout the year and results in an annual evaporation which is 41 % higher than the measured one (Fig. 7d,V5). The last test for the Penman equation combines the removal of the surface water temperature with a derived linear function for the heat storage term from V4. With a heat storage term $\Delta Q = 0.77 \cdot R_\mathrm{n}$ the discrepancy of the calculated from the measured rates can be minimised. The regression line is still slightly tilted (Fig. 6d,V6), small evaporation rates are overestimated, and large ones underestimated, but the mean difference is nearly zero and the standard deviation is in the range of 0.29 to

$0.5\,\mathrm{mm\,d^{-1}}$ (Table 6). The annual evaporation is well represented with this adjustments of the equation (Fig. 7d,V6).

5 Discussion and conclusion

The eddy covariance method is used for the first high-resolution, direct evaporation measurements of the Dead Sea. The first aim of this study was to present an applicable method to measure evaporation with a shoreline station. The measurement strategy is based on the installation of the station on a headland, surrounded by water from $320°$. The advantage of this setup at the shoreline is the avoidance of raft motion and sea spray influencing the measurements, where the latter one leads to a serious soiling of the instrument and influences data quality strongly. The major drawback of land-based eddy covariance measurements is the limited data availability of measured lake evaporation as part of the flux footprint is located over land. In this study 19 % and in other works 15–25 % (e.g. Mammarella et al., 2015; Nordbo et al., 2011) of the data were rejected due to the fetch criteria. This was overcome by a novel approach. A multiple regression model was trained with the onshore wind and vapour pressure deficit data. With this model lake evaporation for offshore wind conditions was calculated and, thus, data availability was increased from 59.2 to 76.8 %. The uncertainty introduced by this method is small with a prediction error of the calculated values of 4.8 %, making it a very reliable method. However, there is still some uncertainty due to this method which cannot be accounted for directly. On the one hand, extreme values of wind velocity and water vapour pressure deficit were not used to calculate evaporation when they were outside the model boundaries. This leads most likely to an underestimation of the actual evaporation rate. On the other hand, wind velocity and vapour pressure deficit could decrease with increasing distance from the shoreline, which would lead to an overestimation of evaporation. However, the comparison with results from measurements in the middle of the lake (Weiss et al., 1988; Hecht and Gertman, 2003) shows that even in the middle of the lake westerly winds with hourly averaged velocities between 8 and $12\,\mathrm{m\,s^{-1}}$ were observed. Wind lidar measurements confirmed that the westerly winds regularly reach several km over the lake without loosing their strength (Metzger, 2017). In conclusion, offshore wind measurements seem representative for lake conditions and reasonable for the calculation of evaporation. A decrease in vapour pressure deficit has to be considered but is most likely small for the following reasons. Firstly, the fetch of the station is limited to $600\,\mathrm{m}$, meaning that the distance the air mass passes over the water is short. Secondly, the westerly winds are connected with high turbulence and, thus, strong vertical mixing (Metzger, 2017). From these results we conclude that the approach is also applicable to other lakes, in the case where the measured onshore wind velocity and vapour pressure deficit values are representative for offshore

conditions to appropriately train the model, and the fetch of the flux measurements is small enough that the meteorological measurements at the shoreline are representative for the fetch.

The second aim was to evaluate the diurnal and intra-annual variability of Dead Sea evaporation. The annual Dead Sea evaporation was found to be $994 \pm 88.2\,\mathrm{mm}$ for the measurement period. The uncertainty of 8.8 % results mostly from the gap-filling procedure ($81.2\,\mathrm{mm}$) and not from the regression model. As gaps result from system malfunction or bad data quality, the uncertainty can be reduced by improving the system performance or by finding a better method to fill the gaps. The annual evaporation coincides well with previous findings such as Stanhill (1994) with $1005\,\mathrm{mm\,a^{-1}}$ and is close to the results from Lensky et al. (2005) (1100–$1200\,\mathrm{mm\,a^{-1}}$), which both estimated the evaporation based on theoretical energy budget approaches. A certain degree of differences between the results is inevitable as the studies considered different data sets and different time periods, meaning different water salinities and different weather conditions. However, the measurements are far away from the $2000\,\mathrm{mm}$ from Salameh and El-Naser (1999), who estimated evaporation based on water budget calculations, which could indicate uncertainties in the assessment of the water budget components. Therefore, the results could be implemented into hydrological models to study the uncertain water budget components and the development of the water budget in the future. Furthermore, the results show that the diurnal cycle of evaporation is in phase with the wind velocity, which corresponds to findings of other studies in the Jordan Valley (e.g. Assouline, 1993; Assouline et al., 2008). As a result the strong westerly winds in the evening double evaporation compared to midday values. These findings are also important for other lakes, where strong and dry wind systems are observed, e.g. bora, tramontane, mistral. Bouin et al. (2012) showed that the tramontane in France trebles evaporation from a lagoon compared to non-tramontane conditions. In respect of ongoing climate change our results could motivate a regional study on the impact of climate change on the future evolution of thermally and orographically induced wind systems in the Mediterranean region. So far, there is little information, although it is important for the future development of the water bodies. As expected, Dead Sea evaporation is lower compared to other less and non-saline lakes. The ratio to Lake Kinneret, which is located only $100\,\mathrm{km}$ north, is 0.68 in summer but only 0.83 in winter. This difference is most likely caused by the different climatic conditions in winter. Lake Kinneret receives a considerable amount of rainfall due to more humid air masses as it is located within a Mediterranean climate zone (Goldreich, 2003), whereas the Dead Sea has arid climate, where, even in winter, very little rainfall occurs.

For the prospective affordable long-term assessment of evaporation, different equations to calculate evaporation were tested for their applicability for the Dead Sea. The best suitable, and also the only method applicable on sub-daily timescales, is the aerodynamic approach. It is shown that the consideration of the atmospheric stability in the calculations has a negligible effect on the results. These results coincide with results for Lake Kinneret (Shilo et al., 2015; Rimmer et al., 2009) and make this method easily applicable for evaporation calculations applying data from a shoreline station. The other approaches are developed for longer time intervals and are not applicable for sub-daily calculations. The results also confirm the findings from various other studies (Rimmer et al., 2009; Giadrossich et al., 2015; Tanny et al., 2008; Rosenberry et al., 2007) that for the BREB, Priestley–Taylor, and Penman method, the knowledge of the heat storage term is essential to achieve reliable results, as neglecting the heat storage results in a strong seasonal bias. Using estimates of the heat storage term does not provide acceptable results for the BREB or for the Priestley–Taylor method. For the Penman equation, an applicable solution is achieved when using the empirically gained function for the heat storage. Thus, we conclude that the BREB and Priestley–Taylor method are not applicable with data from a shoreline station, but the aerodynamic and the adapted Penman method can be used, making expensive raft measurements expendable. For future application it is advisable to use the Penman method only for longer time intervals as its prediction skill improves with increasing time interval and to use the aerodynamic method for short time intervals. The use of low-maintenance, cost-efficient measurements to estimate evaporation on short timescales is beneficial for economic purposes, such as the production of minerals from the saline water, as well as for further investigations of the water budget of the lake. For instance, pumping rates for mineral production can be adjusted according to the evaporation rates.

Appendix A: Calculation of surface water temperature

The surface water temperature T_s was not measured and could also not be retrieved from satellite data. Therefore, it was calculated following Monin–Obukhov's similarity approach:

$$T_{MO} = T_s = T(z_m) - \frac{\theta^*}{\kappa} \cdot \left(\ln \frac{z_m}{z_0} - \Psi_H(\zeta_m, \zeta_0) \right). \quad (A1)$$

T_{MO} is the calculated surface water temperature at the height of the roughness length z_0, which is assumed as 0.001 m; z_m is the measurement height in m; $\zeta_m = z_m L_*^{-1}$ and $\zeta_0 = z_0 L_*^{-1}$ are independent dimensionless parameters using the Monin–Obukhov length L_*; and $\frac{\theta^*}{\kappa}$ is a scaling parameter defined as

$$\frac{\theta^*}{\kappa} = -\frac{1}{\kappa u^*} \frac{H}{\rho_0 c_p}, \quad (A2)$$

with $\kappa = 0.4$, which is the Kármán constant; sensible heat flux H in $W\,m^{-2}$; specific heat capacity $c_p = 1004\,J\,K^{-1}\,kg^{-1}$; and density of the air ρ_0 in $kg\,m^{-3}$. Ψ_H is the integral over the empirical gained functions φ_H:

$$\Psi_H(\zeta_m, \zeta_0) = \int_{z_0}^{z_m} = \frac{1 - \varphi_H}{z} dz. \quad (A3)$$

In this work the φ functions from Dyer (1974) are used:

$$\varphi_H = 1 + 5\zeta \qquad \qquad \zeta > 0, \quad (A4)$$
$$\varphi_H = (1 - 16\zeta)^{-1/2} \qquad -1 < \zeta < 0. \quad (A5)$$

Appendix B: Measurement of the latent heat of vaporisation

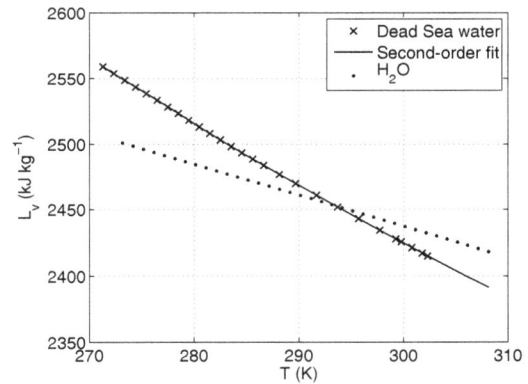

Figure B1. Dependency of the specific latent heat of vaporisation (L_v) on temperature. Measurements of L_v for the saline water of the Dead Sea, a second-order polynomial fit, and literature values for pure water (H_2O) are shown.

The latent heat of vaporisation and the activity of water β for the highly saline water of the Dead Sea were measured using a water probe taken at the measurement site of the EBS at the end of 2014. First, the saturation vapour pressure of pure water E_w was measured with a capacitance manometer, which was calibrated by a linear regression to literature values (Weast et al., 1987). Afterwards, the saturation vapour pressure of the saline water, E_s, was measured as a function of water temperature with the calibrated manometer. Through this approach possible measurement uncertainties of the manometer could be minimised. The activity of water can then be calculated as

$$\beta = \frac{E_s}{E_w}. \tag{B1}$$

The averaged activity for the Dead Sea water is $\beta = 0.65$.

The molar latent heat of vaporisation, ΔH_{vap} ($J\,mol^{-1}$), can be derived by using the general form of the Clausius–Clapeyron equation, assuming that the molar volume of the liquid can be neglected against the molar volume of the gas, and by using the ideal gas law:

$$\Delta H_{vap} = -R\frac{d(\ln E_s)}{d\left(\frac{1}{T_w}\right)}. \tag{B2}$$

$R = 8.314\,J\,mol^{-1}\,K^{-1}$ is the universal gas constant, the corrected saturation vapour pressure of the saline water is E_s in hPa, and water temperature is T_w in K. With the molar mass of water $m_{H_2O} = 0.018\,kg\,mol^{-1}$, the specific latent heat of vaporisation L_v can be calculated,

$$L_v = \frac{\Delta H_{vap}}{m_{H_2O} \cdot 1000}, \tag{B3}$$

in $kJ\,kg^{-1}$, and can then be fitted to the water temperature T_w (Fig. B1). The regression formula is

$$L_v = 5150.6561 - 13.9530 \cdot T_w + 0.0162 \cdot T_w^2. \tag{B4}$$

Appendix C: Calculation of vapour pressure deficit

The vapour pressure deficit for the regression approach is calculated as follows. The vapour pressure deficit is defined as the difference between the saturation vapour pressure above the saline water, E_s, and the atmospheric vapour pressure at 2 m height, $e_{a,2m}$:

$$\Delta e = E_s - e_{a,2m}. \tag{C1}$$

The saturation vapour pressure of saline water is lower than that of freshwater, E_w, by a factor β, caused by the vapour pressure depression by dissolved salts (Raoult's law) (Atkins, 2014).

$$E_s = \beta \cdot E_w. \tag{C2}$$

The activity β depends on the composition of the dissolved salts and is determined to 0.65 for the Dead Sea water in this

study (Appendix B). Saturation vapour pressure over water can be calculated using the Magnus equation after Bolton (1980):

$$E_w(T_S) = 6.112 \cdot \exp\left(\frac{17.67 \cdot (T_S - 273.15)}{T_S - 29.65}\right), \tag{C3}$$

with surface water temperature T_S in K. As surface water temperature is not directly measured at the station, vapour pressure deficit is calculated using surface water temperature obtained by the Monin–Obukhov theory, T_{MO}, in K:

$$\Delta e_{MO} = \beta \cdot E_w(T_{MO}) - e_{a,2m}. \tag{C4}$$

Competing interests. The authors declare that they have no conflict of interest.

Special issue statement. This article is part of the special issue "Environmental changes and hazards in the Dead Sea region (NHESS/ACP/HESS/SE inter-journal SI)". It is not associated with a conference.

Acknowledgements. The current study was carried out in the framework of the Dead Sea Research Venue (DESERVE) (http://www.deserve-vi.net), an international project funded by the Helmholtz Association of German Research Centres as a Virtual Institute (VH-VI-527). We would like to thank Bernhard Deny and Philipp Gasch for maintaining the stations. We thank Pinhas Alpert and Eduard Karat from Tel Aviv University for their support. We also want to thank David Seveloff and Yael Maor from the Dead Sea and Arava Science Center for their support with the measurements and Ein Gedi Spa for the provision of the measurement location. The authors also thank the editor and four anonymous referees for their insightful remarks which helped to improve the paper.

Edited by: Matthew Hipsey

References

Abelson, M., Yechieli, Y., Crouvi, O., Baer, G., Wachs, D., Bein, A., and Shtivelman, V.: Evolution of the Dead Sea sinkholes, Geol. Soc. Spec. Pap., 401, 241–253, 2006.

Alpert, P., Shafir, H., and Issahary, D.: Recent changes in the climate at the Dead Sea – a preliminary study, Climatic Change, 37, 513–537, https://doi.org/10.1023/A:1005330908974, 1997.

Arkin, Y. and Gilat, A.: Dead Sea sinkholes – an ever-developing hazard, Environ. Geol., 39, 711–722, 2000.

Asmar, B. and Ergenzinger, P.: Dynamic simulation of the Dead Sea, Adv. Water Resour., 25, 263–277, 2002.

Asmar, B. N. and Ergenzinger, P.: Estimation of evaporation from the Dead Sea, Hydrol. Process., 13, 2743–2750, https://doi.org/10.1002/(SICI)1099-1085(19991215)13:17<2743::AID-HYP845>3.0.CO;2-U, 1999.

Assouline, S.: Estimation of lake hydrologic budget terms using the simultaneous solution of water, heat, and salt balances and a Kalman Filtering Approach: Application to Lake Kinneret, Water Resour. Res., 29, 3041–3048, https://doi.org/10.1029/93WR01181, 1993.

Assouline, S. and Mahrer, Y.: Evaporation from Lake Kinneret: 1. Eddy correlation system measurements and energy budget estimates, Water Resour. Res., 29, 901–910, 1993.

Assouline, S., Tyler, S., Tanny, J., Cohen, S., Bou-Zeid, E., Parlange, M., and Katul, G.: Evaporation from three water bodies of different sizes and climates: Measurements and scaling analysis, Adv. Water Resour., 31, 160–172, 2008.

Atkins, P.: Physical Chemistry, 10 edn., Oxford Univ. Press, Oxford, 2014.

Bitan, A.: The wind regime in the north-west section of the Dead-Sea, Arch. Meteor. Geophy. B, 22, 313–335, https://doi.org/10.1007/BF02246585, 1974.

Bitan, A.: The influence of the special shape of the Dead-Sea and its environment on the local wind system, Arch. Meteor. Geophy. B, 24, 283–301, 1976.

Blanken, P. D., Rouse, W. R., Culf, A. D., Spence, C., Boudreau, L. D., Jasper, J. N., Kochtubajda, B., Schertzer, W. M., Marsh, P., and Verseghy, D.: Eddy covariance measurements of evaporation from Great Slave Lake, Northwest Territories, Canada, Water Resour. Res., 36, 1069–1077, https://doi.org/10.1029/1999WR900338, 2000.

Bolton, D.: The computation of equivalent potential temperature, Mon. Weather Rev., 108, 1046–1053, 1980.

Bouin, M.-N., Caniaux, G., Traulle, O., Legain, D., and Le Moigne, P.: Long-term heat exchanges over a Mediterranean lagoon, J. Geophys. Res.-Atmos., 117, https://doi.org/10.1029/2012JD017857, 2012.

Brutsaert, W.: Evaporation into the Atmosphere. Theory, History, and Applications, Reidel, Dordrecht, the Netherlands, 1982.

Calder, I. and Neal, C.: Evaporation from saline lakes: a combination equation approach, Hydrolog. Sci. J., 29, 89–97, 1984.

Cline, D. W.: Snow surface energy exchanges and snowmelt at a continental, midlatitude Alpine site, Water Resour. Res., 33, 689–701, 1997.

Dingman, S.: Physical Hydrology, Physical Hydrology, 2nd edn., Prentice Hall, Upper Saddle River, New Jersey, 2002.

Duan, Z. and Bastiaanssen, W.: A new empirical procedure for estimating intra-annual heat storage changes in lakes and reservoirs: review and analysis of 22 lakes, Remote Sens. Environ., 156, 143–156, 2015.

Dyer, A.: A review of flux-profile relationships, Bound.-Lay. Meteorol., 7, 363–372, https://doi.org/10.1007/BF00240838, 1974.

Emery, W., Castro, S., Wick, G., Schluessel, P., and Donlon, C.: Estimating sea surface temperature from infrared satellite and in situ temperature data, B. Am. Meteorol. Soc., 82, 2773, https://doi.org/10.1175/1520-0477(2001)082<2773:ESSTFI>2.3.CO;2, 2001.

Foken, T.: Der Bayreuther Turbulenzknecht, Tech. Rep. 1, Univ. Bayreuth, Abt. Mikrometeorol., Bayreuth, 16 pp., 1999.

Foken, T.: The energy balance closure problem: an overview, Ecol. Appl., 18, 1351–1367, 2008.

Foken, T. and Wichura, B.: Tools for quality assessment of surface-based flux measurements, Agr. Forest Meteorol., 78, 83–105, https://doi.org/10.1016/0168-1923(95)02248-1, 1996.

Foken, T., Leuning, R., Oncley, S., Mauder, M., and Aubinet, M.: Corrections and data quality control, in: Eddy Covariance. A Practical Guide to Measurements and Data Analysis, edited by: Aubinet, M., Vesala, T., and Papale, D., Springer, Dordrecht, 85–132, 2012.

Giadrossich, F., Niedda, M., Cohen, D., and Pirastru, M.: Evaporation in a Mediterranean environment by energy budget and Penman methods, Lake Baratz, Sardinia, Italy, Hydrol. Earth Syst. Sci., 19, 2451–2468, https://doi.org/10.5194/hess-19-2451-2015, 2015.

Givati, A. and Tal, A.: Monthly hydrological status: upper water in major drainage areas and water tables in the country system, The Hydrological Service and Water Authority, Israel, 2016.

Goldreich, Y.: The Climate of Israel: Observation, Research, and Application, Kluwer Academic/Plenum Publishers, New York (u.a.), 2003.

Goren, M. and Ortal, R.: Biogeography, diversity and conservation of the inland water fish communities in Israel, Biol. Conserv., 89, 1–9, https://doi.org/10.1016/S0006-3207(98)00127-X, 1999.

Hecht, A. and Gertman, I.: Dead Sea meteorological climate, in: Fungal Life in the Dead Sea, edited by: Nevo, E., Oren, A., and Wasser, S., International Center for Cryptogamic Plants and Fungi, Haifa, 68–114, 2003.

Henderson-Sellers, B.: Calculating the surface energy balance for lake and reservoir modeling: a review, Rev. Geophys., 24, 625–649, 1986.

Holtzman, R., Shavit, U., Segal-Rozenhaimer, M., Gavrieli, I., Marei, A., Farber, E., and Vengosh, A.: Quantifying ground water inputs along the lower Jordan River, J. Environ. Qual., 34, 897–906, 2005.

Jonsson, A., Åberg, J., Lindroth, A., and Jansson, M.: Gas transfer rate and CO_2 flux between an unproductive lake and the atmosphere in northern Sweden, J. Geophys. Res.-Biogeo., 113, https://doi.org/10.1029/2008JG000688, 2008.

Kljun, N., Calanca, P., Rotach, M. W., and Schmid, H. P.: A simple two-dimensional parameterisation for Flux Footprint Prediction (FFP), Geosci. Model Dev., 8, 3695–3713, https://doi.org/10.5194/gmd-8-3695-2015, 2015.

Kohler, M. and Parmele, L.: Generalized estimates of free-water evaporation, Water Resour. Res., 3, 997–1005, 1967.

Konda, M., Imasato, N., Nishi, K., and Toda, T.: Measurement of the sea surface emissivity, J. Oceanogr., 50, 17–30, 1994.

Kondrat'Ev, K. Y.: Radiation in the Atmosphere, International Geophysics Series, Vol. 12, Academic Press, New York, 1969.

Kottmeier, C., Agnon, A., Al-Halbouni, D., Alpert, P., Corsmeier, U., Dahm, T., Eshel, A., Geyer, S., Haas, M., Holohan, E., Kalthoff, N., Kishcha, P., Krawczyk, C., Lati, J., Laronne, J. B., Lott, F., Mallast, U., Merz, R., Metzger, J., Mohsen, A., Morin, E., Nied, M., Rödiger, T., Salameh, E., Sawarieh, A., Shannak, B., Siebert, C., and Weber, M.: New perspectives on interdisciplinary earth science at the Dead Sea: The DESERVE project, Sci. Total Environ., 544, 1045–1058, 2016.

Lensky, N. G., Dvorkin, Y., Lyakhovsky, V., Gertman, I., and Gavrieli, I.: Water, salt, and energy balances of the Dead Sea, Water Resour. Res., 41, 1–13, https://doi.org/10.1029/2005WR004084, 2005.

Lipchin, C., Sandler, D., and Cushman, E.: The Jordan River and Dead Sea Basin: Cooperation Amid Conflict, Springer Science and Business Media, Dordrecht, the Netherlands, 2009.

Mammarella, I., Nordbo, A., Rannik, Ü., Haapanala, S., Levula, J., Laakso, H., Ojala, A., Peltola, O., Heiskanen, J., Pumpanen, J., and Vesala, T.: Carbon dioxide and energy fluxes over a small boreal lake in Southern Finland, J. Geophys. Res.-Biogeo., 120, 1296–1314, 2015.

Mauder, M. and Foken, T.: Documentation and instruction manual of the eddy-covariance software package TK3, vol. 46, Univ. Bayreuth, Abt. Mikrometeorologie, Bayreuth, 2011.

Mauder, M., Cuntz, M., Drüe, C., Graf, A., Rebmann, C., Schmid, H. P., Schmidt, M., and Steinbrecher, R.: A strategy for quality and uncertainty assessment of long-term eddy-covariance measurements, Agr. Forest Meteorol., 169, 122–135, https://doi.org/10.1016/j.agrformet.2012.09.006, 2013.

Metzger, J. V.: Wind systems and energy balance in the Dead Sea Valley, Wissenschaftliche Berichte des Instituts für Meteorologie und Klimaforschung des Karlsruher Instituts für Technologie; Band 74, KIT Scientific Publishing, Karlsruhe, 2017.

Naor, R., Potchter, O., Shafir, H., and Alpert, P.: An observational study of the summer Mediterranean Sea breeze front penetration into the complex topography of the Jordan Rift Valley, Theor. Appl. Climatol., 127, 275–284, 2017.

Nehorai, R., Lensky, I. M., Lensky, N. G., and Shiff, S.: Remote sensing of the Dead Sea surface temperature, J. Geophys. Res.-Oceans, 114, https://doi.org/10.1029/2008JC005196, c05021, 2009.

Nehorai, R., Lensky, N., Brenner, S., and Lensky, I.: The dynamics of the skin temperature of the Dead Sea, Adv. Meteorol., 2013, 1–9, 2013.

Nied, M.: DESERVE_measurement_data_20160421, available at: https://www.deserve-vi.net/index.php/publications (last access: 7 February 2018), 2016.

Nordbo, A., Launiainen, S., Mammarella, I., Leppäranta, M., Huotari, J., Ojala, A., and Vesala, T.: Long-term energy flux measurements and energy balance over a small boreal lake using eddy covariance technique, J. Geophys. Res.-Atmos, 116, https://doi.org/10.1029/2010JD014542, 2011.

Oroud, I.: Evaluation of saturation vapor pressure over hypersaline water bodies at the southern edge of the Dead Sea, Jordan, Sol. Energy, 53, 497–503, 1994.

Oroud, I.: Evaporation estimates from the Dead Sea and their implications on its water balance, Theor. Appl. Climatol., 106, 523–530, https://doi.org/10.1007/s00704-011-0452-6, 2011.

Penman, H. L.: Natural evaporation from open water, bare soil and grass, P. Roy. Soc. Lond. A Mat., 193, 120–145, 1948.

Picard, R. R. and Cook, R. D.: Cross-validation of regression models, J. Am. Stat. Assoc., 79, 575–583, 1984.

Priestley, C. H. B.: On the assessment of surface heat flux and evaporation using large scale parameters, Mon. Weather Rev., 100, 81–92, 1972.

Rimmer, A., Samuels, R., and Lechinsky, Y.: A comprehensive study across methods and time scales to estimate surface fluxes from Lake Kinneret, Israel, J. Hydrol., 379, 181–192, 2009.

Rosenberry, D. O., Winter, T. C., Buso, D. C., and Likens, G. E.: Comparison of 15 evaporation methods applied to a small mountain lake in the northeastern USA, J. Hydrol., 340, 149–166, 2007.

Salameh, E.: Water quality degradation in Jordan (Impacts on Environment, Economy and Future Generations Resources Base), Friedrich Ebert Stiftung, Royal Society for the Conservation of Nature, Amman, 1996.

Salameh, E. and El-Naser, H.: Does the actual drop in Dead Sea level reflect the development of water sources within its drainage basin?, Acta Hydroch. Hydrob., 27, 5–11, 1999.

Salameh, E. and El-Naser, H.: Changes in the Dead Sea Level and their Impacts on the Surrounding Groundwater Bodies, Acta Hydroch. Hydrob., 28, 24–33, https://doi.org/10.1002/(SICI)1521-401X(200001)28:1<24::AID-AHEH24>3.0.CO;2-6, 2000.

Salhotra, A. M., Adams, E. E., and Harleman, D. R. F.: Effect of salinity and ionic composition on evaporation: analysis of Dead Sea evaporation pans, Water Resour. Res., 21, 1336–1344, https://doi.org/10.1029/WR021i009p01336, 1985.

Schotanus, P., Nieuwstadt, F., and De Bruin, H.: Temperature measurement with a sonic anemometer and its application to heat and moisture fluxes, Bound.-Lay. Meteorol., 26, 81–93, https://doi.org/10.1007/BF00164332, 1983.

Shafir, H. and Alpert, P.: Regional and local climatic effects on the Dead-Sea evaporation, Climatic Change, 105, 455–468, https://doi.org/10.1007/s10584-010-9892-8, 2011.

Shilo, E., Ziv, B., Shamir, E., and Rimmer, A.: Evaporation from Lake Kinneret, Israel, during hot summer days, J. Hydrol., 528, 264–275, 2015.

Siebert, C., Rödiger, T., Mallast, U., Gräbe, A., Guttman, J., Laronne, J. B., Storz-Peretz, Y., Greenman, A., Salameh, E., Al-Raggad, M., Vachtman, D., Zvi, A. B., Ionescu, D., Brenner, A., Merz, R., and Geyer, S.: Challenges to estimate surface- and groundwater flow in arid regions: The Dead Sea catchment, Sci. Total Environ., 485–486, 828–841, https://doi.org/10.1016/j.scitotenv.2014.04.010, 2014.

Siebert, C., Rödiger, T., Geyer, S., Laronne, J. B., Hillel, N., Sauter, M., and Mallast, U.: Multidisciplinary investigations of the transboundary Dead Sea basin and its water resources, in: Integrated Water Resources Management: Concept, Research and Implementation, Springer, 107–127, Cham, 2016.

Stanhill, G.: The radiation climate of the dead sea, J. Climatol., 7, 247–265, https://doi.org/10.1002/joc.3370070305, 1987.

Stanhill, G.: Changes in the rate of evaporation from the Dead Sea, Int. J. Climatol., 14, 465–471, https://doi.org/10.1002/joc.3370140409, 1994.

Steiner, L. E.: Introduction to Chemical Thermodynamics, McGraw-Hill Book Co., New York, 510 pp., 1948.

Tanny, J., Cohen, S., Assouline, S., Lange, F., Grava, A., Berger, D., Teltch, B., and Parlange, M.: Evaporation from a small water reservoir: direct measurements and estimates, J. Hydrol., 351, 218–229, https://doi.org/10.1016/j.jhydrol.2007.12.012, 2008.

Van Bavel, C.: Potential evaporation: the combination concept and its experimental verification, Water Resour. Res., 2, 455–467, 1966.

Weast, R. C., Astle, M. J., and Beyer, W. H.: CRC Handbook of Chemistry and Physics, 68th Edition, CRC Press, Boca Raton, USA, 1987.

Webb, E. K., Pearman, G. I., and Leuning, R.: Correction of flux measurements for density effects due to heat and water vapour transfer, Q. J. Roy. Meteor. Soc., 106, 85–100, https://doi.org/10.1002/qj.49710644707, 1980.

Weiss, M., Cohen, A., and Mahrer, Y.: Upper atmosphere measurements and meteorological measurements on the Dead Sea, Tech. rep., Ministry of Energy and Infrastructure, 19 pp., Israel, 1988 (in Hebrew).

Wilczak, J., Oncley, S., and Stage, S.: Sonic anemometer tilt correction algorithms, Bound.-Lay. Meteorol., 99, 127–150, https://doi.org/10.1023/A:1018966204465, 2001.

Wilson, K., Goldstein, A., Falge, E., Aubinet, M., Baldocchi, D., Berbigier, P., Bernhofer, C., Ceulemans, R., Dolman, H., Field, C., Grelle, A., Ibrom, A., Law, B., Kowalski, A., Meyers, T., Moncrieff, J., Monson, R., Oechel, W., Tenhunen, J., Valentini, R., and Verma, S.: Energy balance closure at FLUXNET sites, Agr. Forest Meteorol., 113, 223–243, 2002.

Winter, T., Rosenberry, D., and Sturrock, A.: Evaluation of 11 equations for determining evaporation for a small lake in the north central United States, Water Resour. Res., 31, 983–993, 1995.

Yechieli, Y., Abelson, M., Bein, A., Crouvi, O., and Shtivelman, V.: Sinkhole swarms along the Dead Sea coast: reflection of disturbance of lake and adjacent groundwater systems, Geol. Soc. Am. Bull., 118, 1075–1087, 2006.

3

Comparing statistical and process-based flow duration curve models in ungauged basins and changing rain regimes

M. F. Müller and S. E. Thompson

Department of Civil and Environmental Engineering, Davis Hall, University of California, Berkeley CA, USA

Correspondence to: M. F. Müller (marc.muller@berkeley.edu)

Abstract. The prediction of flow duration curves (FDCs) in ungauged basins remains an important task for hydrologists given the practical relevance of FDCs for water management and infrastructure design. Predicting FDCs in ungauged basins typically requires spatial interpolation of statistical or model parameters. This task is complicated if climate becomes non-stationary, as the prediction challenge now also requires extrapolation through time. In this context, process-based models for FDCs that mechanistically link the streamflow distribution to climate and landscape factors may have an advantage over purely statistical methods to predict FDCs.

This study compares a stochastic (process-based) and statistical method for FDC prediction in both stationary and non-stationary contexts, using Nepal as a case study. Under contemporary conditions, both models perform well in predicting FDCs, with Nash–Sutcliffe coefficients above 0.80 in 75 % of the tested catchments. The main drivers of uncertainty differ between the models: parameter interpolation was the main source of error for the statistical model, while violations of the assumptions of the process-based model represented the main source of its error. The process-based approach performed better than the statistical approach in numerical simulations with non-stationary climate drivers. The predictions of the statistical method under non-stationary rainfall conditions were poor if (i) local runoff coefficients were not accurately determined from the gauge network, or (ii) streamflow variability was strongly affected by changes in rainfall. A Monte Carlo analysis shows that the streamflow regimes in catchments characterized by frequent wet-season runoff and a rapid, strongly non-linear hydrologic response are particularly sensitive to changes in rainfall statistics. In these cases, process-based prediction approaches are favored over statistical models.

1 Introduction

The flow duration curve (FDC) provides a compact summary of the variability of daily streamflow by indicating what proportion of the flow regime exceeds a given flow rate. FDCs have considerable practical relevance, particularly in supporting decisions that are affected by the availability and reliability of surface water. Common applications of FDCs include the design and management of hydropower infrastructure (e.g., Basso and Botter, 2012; Müller, 2015), the determination of environmental flow standards for ecosystem protection (e.g., Lazzaro et al., 2013), the allocation of water resources for consumptive uses (e.g., Alaouze, 1989), or the prediction of streamflow time series in ungauged or poorly gauged catchments (e.g., Hughes and Smakhtin, 1996; Westerberg et al., 2014).

Despite their utility, empirical FDCs are unavailable for many basins, primarily because they require extensive on-site observations of daily streamflow (Vogel and Fennessey, 1994). Globally, the majority of catchments remain ungauged (or the gauge data that exist are subject to significant quality assurance and data availability constraints). Furthermore, the global number of stream gauges continues to decline because of ongoing budgetary constraints faced by water monitoring agencies (Stokstad, 1999; United States Geological Survey, 2015). Therefore FDCs must typically be estimated in data-scarce areas. The most widely used techniques for FDC estimation are simple, graphical methods. Such empirical methods are easy to implement but often rely on overly simplistic assumptions that lead to substantial prediction errors. For instance, in Nepal, the regionalization method prescribed in official design manuals (e.g., Chitrakar, 2004; Alternative Energy Promotion Center, 2014) relies on *one* in situ observation of streamflow during the dry season to scale standardized regional indices for monthly flows. The procedure neglects the inter-annual variability of low flows, which leads to important biases in the predicted flow distributions (see Sect. S1 of the Supplement). Even in gauged catchments, FDCs constructed from historical observations may not represent current flow conditions well, because flow regimes are impacted by climate change and anthropogenic alterations of the catchments (e.g., Botter et al., 2013; Mu et al., 2007). Predicting streamflow in ungauged basins, particularly in the context of environmental change, remains both a fundamental necessity for water managers and a major research challenge (Blöschl et al., 2013; Montanari et al., 2013).

Recent efforts to predict FDCs in ungauged catchments focus on statistical approaches that predict the flow distribution based on the catchment's similarity to nearby, gauged watersheds (Castellarin et al., 2013). Index flow approaches, which regionalize specific index flows (typically the mean flow), and use those indices to rescale empirical FDCs from similar catchments, are particularly popular (e.g., Chalise et al., 2003; Castellarin et al., 2004b; Sauquet and Catalogne, 2011; Arora et al., 2005). While differing in methodological details, all index flow approaches assume that FDCs do not vary within homogeneous regions, except by a scaling factor. Because they do not assume any specific runoff-generating process, statistical methods are versatile. They have been successfully applied globally to predict FDCs in a variety of climates and catchment types (Blöschl et al., 2013). However, methods are also insensitive to the diversity of controls on the shape of the FDC exerted by climate processes and catchment characteristics. This may affect their reliability under non-stationary conditions (Milly et al., 2008). Finally, the calibration of statistical methods relies on extensive streamflow observations from a large number of representative and well-characterized catchments (e.g., Cheng et al., 2012; Coopersmith et al., 2012). Their performance is therefore sensitive to the spatial density of available gauges (Blöschl et al., 2013), and their reliability in regions where streamflow data are truly scarce is uncertain.

Stochastic, process-based models that mechanistically link the drivers, state, and response of the system are a promising avenue to address these issues. In these models, basic assumptions about the stochastic structure of rainfall and the (deterministic) response of catchments allow the analytic derivation of streamflow probability density functions (PDFs). (Note that because the FDC can be obtained directly by transforming the PDF, a predictive technique that yields the streamflow PDF will also allow the FDC to be estimated.) Botter et al. (2007b) show that runoff follows a gamma distribution if catchments behave as a linear reservoir, forced by stochastic rainfall that follows a marked Poisson process. The resulting gamma distribution depends on two parameters that are determined by the recession characteristics of the catchment, and by the frequency and intensity of effective rain. This process-based approach to the streamflow PDF has been extended to include the fast flow component of streamflow (Muneepeerakul et al., 2010), non-linearities in subsurface storage–runoff relationships (Botter et al., 2009), the effects of short-term snowmelt (Schaefli et al., 2013), and the carryover of subsurface storage between seasons in seasonally dry climates (Müller et al., 2014). Although the stochastic framework allows the effects of changes in climate or landscape to be independently modeled, it relies on strong simplifying assumptions about the spatial homogeneity of catchments. These assumptions make the existing process models less versatile than statistical methods. Nonetheless, the approach has low calibration requirements because it relies on a small number of parameters, which can be determined using rainfall, climate, and geomorphological characteristics of the catchments (Doulatyari et al., 2015). This information is increasingly available in ungauged basins, thanks to remote-sensing technologies, even when ground-based measurements are sparse.

Process-based models successfully reproduce streamflow PDFs in numerous gauged catchments worldwide (Botter et al., 2007a; Ceola et al., 2010), including Nepal (Müller et al., 2014). Yet their predictive performance in ungauged basins remains largely unassessed, particularly in regions where the local gauge density is globally representative (as opposed to densely monitored catchments in developed countries such as, e.g., France and Austria in Castellarin et al., 2013). For lower gauge densities, it is unclear whether the advantages of the process-based approaches, which are derived from an explicit representation of flow-generating processes, are outweighted by the limitations imposed by the restrictive assumptions underlying these methods – and whether this trade-off is altered by non-stationarity in climate drivers.

Using Nepal as a test case, this study compares the process-based and statistical approaches on the basis of (i) their ability to predict FDCs in ungauged basins, (ii) their sensitivity to data scarcity, represented both by the spatial density of the stream gauge network and by the temporal extent (length) of the available streamflow records, and (iii) their ability to accommodate changes in the rainfall regime.

Nepal provides an ideal setting to compare the two approaches, for four reasons. First, the country is representative of global availability of streamflow data, as measured by the density of its stream gauge network (Fig. 1a). Second, methods drawn from both statistical and process-based approaches have been developed and validated in Nepal. Here we compare the stochastic–dynamic framework developed in Müller et al. (2014) with the index flow model described in Chalise et al. (2003). Third, flow generation processes in Nepalese Himalayan catchments are complex, particularly with respect to the spatial and temporal properties of precipitation. Rainfall derives from the Indian summer monsoon and is strongly affected by topography. As a result, local rainfall is temporally autocorrelated, spatially

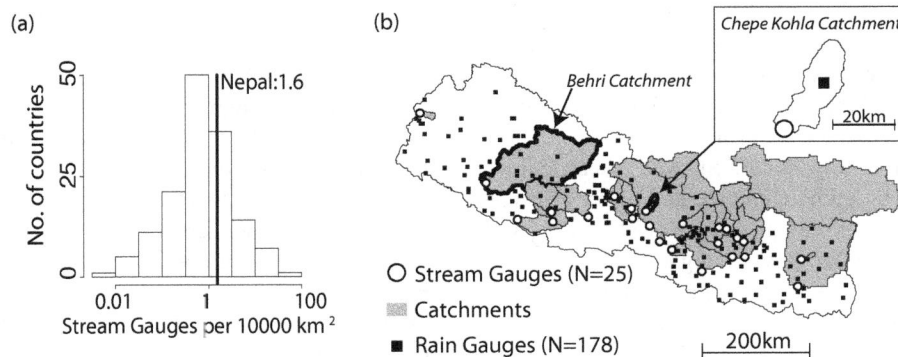

Figure 1. (a) Global histogram of the approximate spatial density of streamflow gauges by nation, represented by the sample of 8540 gauges indexed by the Global Runoff Data Center for 146 countries (Global Runoff Data Center, 2014). With a density of 1.6 gauges per $10\,000\,\text{km}^2$, Nepal falls close to the mode of the global distribution. (b) Location of the rain gauges, streamflow gauges, and corresponding Nepalese catchments used in the analysis.

heterogeneous, and highly seasonal. There is also significant carryover of groundwater storage between the wet and dry seasons, so that dry-season discharge reflects the features of the antecedent wet season. These characteristics violate many of the assumptions that underlie the process-based method. The analysis in Nepal is therefore likely to provide a conservative estimate of the potential performance of the process-based method in ungauged basins. Finally, developing reliable methods for FDC prediction in Nepal represents an opportunity for "use-inspired science" (Thompson et al., 2013b). Nepal has an enormous untapped hydropower potential and is in dire need of electrical power, particularly in rural areas. A reliable method to estimate FDCs in ungauged catchments would be a valuable tool to support the development of micro-hydropower, a sustainable technology for rural electrification (Müller, 2015).

Section 2 describes the two models and the procedures used to estimate their parameters from streamflow and rainfall observations. Section 3 presents the results of the comparative analysis in Nepal. Section 4 examines the key sources of errors for both models and discusses implications for both "Prediction in ungauged catchments" (PUB) and "Predictions under change" (PUC) beyond Nepal.

2 Methods

2.1 Compared approaches

2.1.1 Process-based model

The process-based approach models daily streamflow as a random variable. Subject to strong simplifying assumptions about rainfall stochasticity and runoff generation, the streamflow PDF can be analytically derived. During the wet season, daily rainfall is represented as a stationary marked Poisson process with exponentially distributed depths. Assuming lin-

ear evapotranspiration losses, Botter et al. (2007b) showed that effective rain, that is, the portion of the total rainfall that contributes to streamflow generation, also follows a stationary marked Poisson process. For a spatially homogenous catchment with an exponentially distributed response time (i.e., a catchment that behaves as a linear reservoir), this effective rainfall will produce gamma-distributed streamflow. The parameters of the gamma distribution are derived from the frequency (λ_P) and mean depth (α_P) of rainfall, and from the recession constant (k) of the catchment. If rainfall in the dry season is sufficiently minimal that effective rainfall does not contribute to runoff generation, then dry-season streamflow represents only the discharge of groundwater stored during the previous wet season. This discharge is modeled as a single seasonal recession with stochastic initial conditions that depend on the wet-season properties. Because groundwater is not replenished during the dry season, the water table is subject to a large transient drawdown, resulting in a non-linear discharge behavior and a power-law relation between recession rate and discharge (Brutsaert and Nieber, 1977). We showed in Müller et al. (2014) that the distributions of streamflow, and therefore the FDC, in seasonally dry climates that meet the assumptions above, can be expressed analytically as a function of seven independent parameters: the frequency (λ_P) and mean intensity (α_P) of wet-season rainfall, maximum daily evapotranspiration during the wet season (ET), the water storage capacity of the soil in the root zone (SSC), the (linear) wet-season recession constant (k), the duration of the dry season (T_d), and the exponent of the power-law recession during the dry season (b). The model admits an additional input parameter, the scale a of the power-law seasonal recession, which we showed in Müller et al. (2014) can be expressed as a function of k, b, λ_P, and α_P. The formal derivation of the model is summarized in Appendix A.

The model was successfully validated in a variety of regions with seasonally dry climates worldwide, including Nepal, where observed FDCs were predicted in 24 gauged catchments with a median Nash–Sutcliffe coefficient of 0.90 on log-transformed flow quantiles (Müller et al., 2014). The approach successfully reproduced both the rain-driven distribution of flows during the wet season and the release of stored monsoon water during the dry-season recession. In this study, we assess the operational performance of the process-based approach as a tool to predict streamflow in *ungauged* catchments. Therefore, we do not further attempt to attribute model errors to parameters versus the model structure in the results presented in Sect. 3, since in practice these errors are confounded in any real application. The relative significance of these two error sources is nonetheless discussed in Sect. 4.1.1.

In ungauged catchments, the process-based model is implemented as follows. Three of the seven parameters of the model (T_d, λ_P, α_P) are rainfall characteristics that can be estimated in ungauged basins using meteorological observations. Recession parameters (k and b) describe aquifer properties that are challenging to observe at the catchment scale. They can be estimated using observed streamflow time series in nearby gauged basins and subsequently interpolated from nearby gauges using the geostatistical approach described in Müller and Thompson (2015), which accounts for the topology of the stream network. The last two parameters (ET) and (SSC) describe catchment-scale soil moisture dynamics that are arduous to determine empirically. Previous applications of the model relied on reasonable values of ET and SSC, based on land use, soil, and climate characteristics of the catchment (e.g., Botter et al., 2007a; Ceola et al., 2010). Alternatively, runoff coefficients can be used to directly relate rainfall statistics to streamflow increments (Doulatyari et al., 2015). Runoff coefficients describe the ratio of mean discharge to mean precipitation, and can be predicted in ungauged basins using water balance models and meteorological observations. This approach circumvents the need to estimate ET and SSC, but the accuracy of predicted runoff coefficients in ungauged catchments is critically dependent on the type of water balance model used and on the availability of appropriate calibration data (Doulatyari et al., 2015). Instead, this study follows the former procedure and uses reasonable estimates of ET and SSC for Nepal.

2.1.2 Statistical model

The statistical approach is entirely driven by observation data and does not assume any specific runoff generation process. Instead, it identifies and exploits statistical correlations that may occur between streamflow observed at existing gauges and the geology, topography, and climate of the corresponding catchments. The index flow model used in this study was developed by Chalise et al. (2003) to regionalize FDCs in Nepal to assess the potential for small hydropower develop-

ment. The model is based on local flow indices for mean ($Q_m = E[Q]$) and low ($q_{95} = Q_{95}/Q_m$, where Q_{95} is the 95th streamflow percentile) flows, and uses a non-parametric approach to represent the shape of the FDC. Empirical FDCs from available gauges are normalized by Q_m and pooled into equally sized groups based on the q_{95} index of the gauge. A standardized curve is determined for each group by taking the average of the normalized flows corresponding to each duration, in order to represent the average catchment response in the group. The chosen statistical approach is considerably less complex than many alternative state-of-the-art methods using multiple (often non-linear) equations to relate multiple flow quantiles to a variety of observed covariates (see Castellarin et al., 2013, for a review). However, Chalise et al. (2003) is, to our knowledge, the most recent statistical method specifically developed and validated in the study region. The approach is parsimonious and adapted to situations, where in situ observations of catchment characteristics are scarce. The method is therefore representative of the level of complexity of statistical approaches likely to be implemented in developing countries for practical hydrological engineering purposes.

Predictions in ungauged catchments are obtained by first using linear regressions to predict Q_m and q_{95}. Although the original method calls for a stepwise multiple regression approach to determine regression covariates inductively, we used the regression models obtained in Chalise et al. (2003): Q_m is regressed against annual rainfall (R_y) and gauge elevation (z_{min}) as a proxy for evapotranspiration, and q_{95} is regressed against the ratios of catchment area occupied by each of the considered geological units. The two regressions loosely represent the long-term water balance and short-term response of the catchment. The predicted low-flow index is then used to determine the standardized FDC shape, which is finally multiplied by the predicted mean flow to obtain the FDC. An important assumption, inherent to the linear regression models, is that the dependent variable (here Q_m and q_{95}) is not spatially correlated when controlling for the considered covariates. This assumption is reasonable in Nepal, where the typical distance between stream gauges is much larger than the correlation scale of runoff (Müller and Thompson, 2015). In more densely gauged areas (or if runoff is correlated over larger distances), streamflow observations at neighboring or flow-connected gauges are likely to be correlated. In these regions, accounting for the effect of distance and stream network topology when interpolating flow indices (e.g., using TopREML Müller and Thompson, 2015) will improve predictions.

2.2 Study region and data

The two methods were evaluated using observed streamflow data from 25 Nepalese catchments mapped in Fig. 1b. The gauges in this data set (HKH-FRIEND, 2004; Department of Hydrology and Meteorology, 2011) have at least 10 years

Table 1. Catchment characteristics. Median values and interquartile distances (IQD) are given for the whole sample of 25 gauges. The table also presents characteristics of the Chepe Kohla watershed considered in the analysis as a case study.

	Streamflow			Topography			Climate							Recession	
	Q_m	q_{95}	N_y	A	z_m	z_M	P_y	T_{mons}	λ_P	α_P	AR	CV	ET	k	b
All gauges															
Median	76.1	0.14	22	1355	481	5209	1952	99	0.71	18.8	0.29	0.92	2.5	0.17	2.38
Min	7.3	0.06	10	130	116	1913	1260	88	0.54	12.1	0.09	0.61	0.40	0.07	1.99
Max	1462.4	0.25	41	32817	1641	8369	4030	152	0.91	33.0	0.51	1.53	3.27	0.32	2.99
Chepe Kohla	23.0	0.14	31	277	475	4711	3050	100	0.84	26.5	0.09	1.03	2.1	0.20	2.41

Q_m is mean annual flow in $m^3 s^{-1}$; q_{95} is the 95th flow percentile normalized by Q_m; N_y indicates the number of observation years; A is the catchment area in km^2; z_m and z_M are, respectively, the minimum and maximum elevation of the basin's meters; P_y is mean precipitation in $mm\,yr^{-1}$; T_{mons} is the estimated duration of the monsoon in days; λ_P is rainfall frequency during the monsoon (in d^{-1}); α_P is mean rainfall intensity in $mm\,d^{-1}$; AR is the first-order autocorrelation coefficient of rainfall occurrence (AR = 0 if rainfall follows a Poissonian process), CV is the coefficient of variation of rainfall intensity on rainy days (CV = 1 if depths are exponentially distributed); ET ($mm\,d^{-1}$) is the reference evapotranspiration during the rainy season (Lambert and Chitrakar, 1989); k is the linear recession constant estimated during the monsoon (in d^{-1}) and b is the non-linear exponent of the seasonal recession. A soil moisture capacity of 16 mm is assumed throughout the country (Müller et al., 2014).

of daily streamflow records. They were checked for consistency, using double mass plots (Searcy and Hardison, 1960) and bias: we discarded non-glaciated catchments that had a precipitation deficit in their long-term water balance. Watersheds were delineated using the ASTER GDEM v2 digital elevation model (NASA Land Processes Distributed Active Archive Center, LP DAAC). The study watersheds are located in central Nepal but cover a wide variety of catchment sizes, elevation ranges, precipitation characteristics, and geological units (Table 1).

We focused on the Chepe Kohla catchment in central Nepal (Fig. 1b, insert) as a case study for analyses requiring resampling (Sect. 2.3.1) or simulation (Sect. 2.3.2) of streamflow time series. The Chepe Kohla watershed has a long (by Nepalese standards) record of daily streamflow observations (31 years) and is representative of the full sample of gauges in terms of topography and recession behavior (Table 1). The catchment is also small (i.e., close to spatially homogenous), and local rainfall is well approximated by a marked Poisson process (first-order autocorrelation coefficient of rainfall occurrence (AR): 0.09; coefficient of variation of rainfall depths (CV): 1.09), echoing the underlying assumptions of the process-based model.

Rainfall characteristics over the sampled catchments were obtained from 178 precipitation gauges (HKH-FRIEND, 2004; Department of Hydrology and Meteorology, 2011), also mapped in Fig. 1b. The average duration of the dry season (T_d) was estimated at each precipitation gauge by fitting a step function to the corresponding rainfall time series (Müller and Thompson, 2013), and wet-season precipitation records were used to compute the frequency and mean intensity of rainfall (λ_P and α_P). Rainfall characteristics were then aggregated at the catchment level by assuming that the rain process aggregates linearly within the basins. For rainfall occurrence, we assumed that the duration between rain events caused by two consecutive storms can be estimated as the average of

the inter-arrival times measured at the rain gauges within the catchment. This allows us to compute catchment-level rainfall frequency as

$$\lambda_P = \left(\frac{1}{N_g} \sum_i^{N_g} \frac{1}{\lambda_P^{(i)}} \right)^{-1},$$

where $\lambda_P^{(i)}$ designates rainfall frequency observed at gauge i and N_g the number of rain gauges within the catchment. Similarly, the catchment-level duration between rainy seasons is assumed to be the average of the durations observed within the catchment:

$$T_d = \frac{1}{N_g} \sum_i^{N_g} T_d^{(i)}.$$

Finally, the precipitation depth received on any given day by a catchment is assumed to be the average of the precipitation depths observed by individual rain gauges. It follows that the aggregated mean rainfall intensity can be expressed as

$$\alpha_P = \lambda_P^{-1} \frac{1}{N_g} \sum_i^{N_g} \lambda_P^{(i)} \alpha_P^{(i)}.$$

If no precipitation station is located within the catchment, rainfall characteristics observed at the rain station closest to the catchment centroid were considered. Although aggregating rainfall *time series* before computing their statistics would better account for spatial correlation in rainfall, aggregating rainfall *statistics* instead allows for non-overlapping observation periods (assuming rainfall is stationary). This is important in the context of Nepal, where rain gauges are scarce with sporadic observations. Unfortunately, the low density of rain gauges within the considered basins prevents a formal treatment of spatial correlation when aggregating frequencies. However, in a previous study (Müller and Thompson, 2013) we observed large spatial correlation ranges on rainfall occurrence in Nepal (125 km during the monsoon). Under these conditions the selected method stands out as the most parsimonious approach to utilize multiple, yet sparse, rainfall observations.

Recession characteristics were estimated using streamflow observations as described in Müller et al. (2014). We computed wet-season recession constants (k) by regressing the logarithm of streamflow against time for each period of consecutively decreasing streamflow during the wet season. The recession constant was then obtained by taking the median value of the regression coefficients of recessions lasting more than 4 days. The power-law exponent of dry-season recessions (b) was obtained by fitting a non-linear recession curve

$$Q(t) = (Q_0^{1-b} - a(1-b)t)^{\frac{1}{1-b}} \qquad (1)$$

to base flow, which was computed from observed streamflow time series using the Lyne–Hollick algorithm (Nathan and McMahon, 1990). The last streamflow peak of the wet season was taken as initial flow condition Q_0, and we used a stochastic optimization algorithm (simulated annealing, Bélisle, 1992) to minimize least square fitting errors. In ungauged catchments, the scale exponent of the seasonal recession was approximated as (Müller et al., 2014)

$$a \approx \frac{\lambda}{-r} \left(e^{\frac{-r}{m}} - 1 \right) \left(\alpha_Q \cdot (m+1) \right), \qquad (2)$$

where $r = 1 - b$; m is the ratio between the frequency λ of effective rain events and the linear recession constant k, and α_Q is the average depth of effective rain events (see Appendix A).

Potential evapotranspiration was approximated by applying the empirical relation estimated by Lambert and Chitrakar (1989) for Nepal during the rainy season (July–September):

$$\text{ET} \approx 4.0 - 0.0008 \cdot z_{\text{mean}},$$

where ET is given in mm d^{-1} and z_{mean} is the average elevation of the catchment in meters. The formula provides daily average evapotranspiration estimates for each month. It accounts for elevation but assumes a spatially homogenous elevation gradient. A uniform soil moisture capacity of 50 mm was assumed throughout the country, based on empirical observations reported in Shrestha (1997). By neglecting local variation in soil characteristics, this produces conservative estimates of the performance of the process-based model in ungauged basins.

2.3 Comparative analyses

2.3.1 Predictions in ungauged basins

We used three cross-validation techniques to evaluate the predictive ability of both methods in ungauged basins.

Firstly, a leave-one-out analysis was carried out to assess predictive performances in a realistic situation, where FDCs are predicted in Nepal using all streamflow gauges available in the region. Secondly, we examined the sensitivity of the methods to decreasing data availability by reducing the number of gauges available to calibrate the models. Finally, we performed a similar data-degradation procedure, but in this case we reduced the number of daily streamflow observations, while holding the number of gauges constant. This final analysis accounts for the challenges posed by recent or temporary installation of stream gauges, which introduce uncertainties into the estimation of model parameters due to the short streamflow records used. These errors can propagate through the model and affect the prediction of FDCs.

In a leave-one-out analysis, one gauge is "left out" of the data set, and streamflow is predicted at the "missing" location using observations from the remaining gauges. The predicted FDC is then compared to observations from the omitted gauge. The resulting error between observation and prediction yields the prediction performance of the method at that catchment if it was not gauged. Repeating the procedure for all gauges offers an approximation to the overall prediction error of the method. To measure this error, we constructed error duration curves (Müller et al., 2014), where the relative prediction error at each flow quantile is plotted against the corresponding duration. Error duration curves allow the partitioning of prediction errors across flow quantiles to be visualized. General prediction performances (across all durations) at individual gauges were also determined using the Nash–Sutcliffe coefficient (NSC) on log streamflow quantiles (Müller et al., 2014):

$$\text{NSC} = 1 - \frac{\sum_{t=1}^{364} \left(\ln Q_t^{(\text{emp})} - \ln Q_t^{(\text{mod})} \right)^2}{\sum_{t=1}^{364} \left(\ln Q_t^{(\text{emp})} - E \ln Q_t^{(\text{emp})} \right)^2}, \qquad (3)$$

where $Q_t^{(\text{emp})}$ and $Q_t^{(\text{mod})}$ are the empirical and modeled streamflow quantiles of duration t.

The effect of the number of calibration gauges was assessed using a jackknife cross-validation analysis (Shao and Tu, 2012; Müller and Thompson, 2013). At each of 10 000 iterations, a selected fraction of the available gauges was randomly sampled (without replacement) and used to predict the FDC at one (randomly selected) remaining gauge. Prediction accuracies for flow duration curves (given by the NSC) and uncertainties on the spatial interpolation of model parameters were reported for each iteration. The procedure was repeated for decreasing numbers of selected "training" gauges.

The available streamflow data did not allow a direct evaluation of the effects of time-series length through cross-validation, because such an analysis requires substantial overlaps in the monitoring periods of all gauges. Therefore we focused the final analysis on the Chepe Kohla catchment, which has the longest observation record in our data set. We evaluated the effect of the length of the available observation

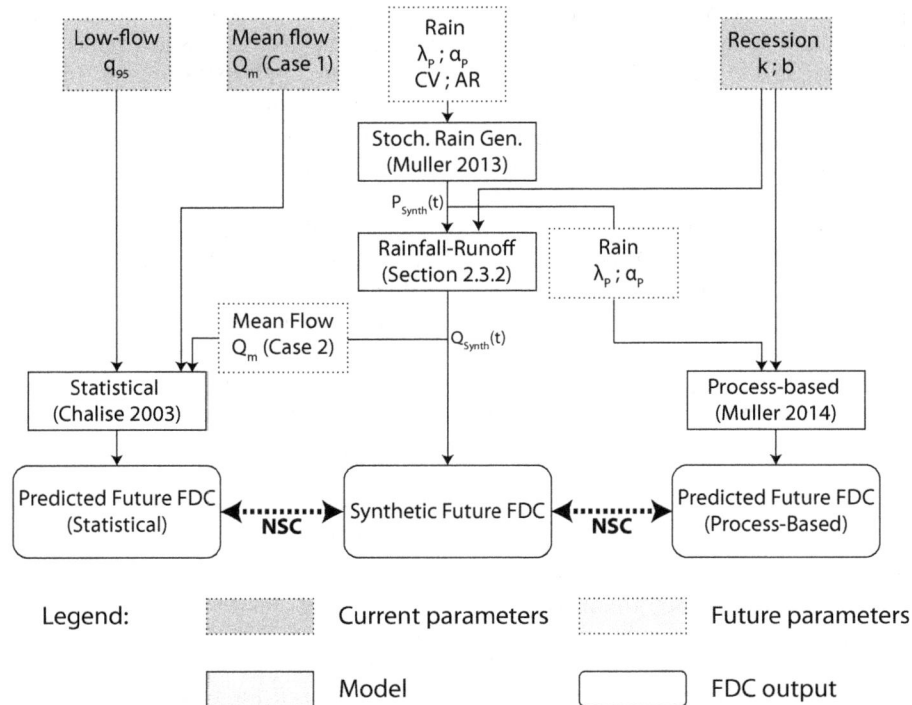

Figure 2. Numerical simulation analysis to assess predictions under change. Future rainfall characteristics (frequency λ_P, mean intensity α_P, auto-correlation coefficient AR and coefficient of variation CV) are determined according to expected changes in rain regimes in Nepal (see Sect. 2.3.2) and fed into a stochastic rainfall generator. The resulting 1000 years of synthetic daily rainfall values ($P_{\text{Synth}}(t)$) are fed into a rainfall–runoff model that simulates the processes described in Sect. 2.1.1. The rainfall–runoff model uses current recession, soil, and evapotranspiration conditions observed at the Chepe Kohla catchment. The resulting 1000 years of synthetic daily flow values ($Q_{\text{Synth}}(t)$) are then reordered to construct an empirical synthetic (future) FDC, which was compared (in terms of the Nash–Sutcliffe coefficient) to modeled FDCs predicted by the statistical and process-based models. The process-based model admits *current* recession conditions but *future* estimates for rainfall frequency (λ_P) and mean intensity (α_P). Note that unlike the numerically generated empirical FDC, the process-based model assumes Poissonian rainfall with exponentially distributed depths, that is, CV = 1 and AR = 0. Current low-flow characteristics (q_{95}) are fed into the statistical model, as well as the current or future (i.e., computed from synthetic streamflow time series) mean flow, depending on the extent to which mean rainfall is an unbiased predictor of mean flow (Cases 1 and 2 described in Sect. 2.3.2).

records on parameter estimation, and propagated the ensuing uncertainty in the parameters to the FDCs predicted by each model. To do this, we selected a fixed number of full years of streamflow observations, estimated the parameters, predicted the FDC using these parameters, and compared the results to the empirical FDC obtained from the full observation record. The procedure was repeated 10 000 times. The estimation errors in the model parameters and the resulting FDC prediction performances (NSC) were recorded as a function of the number of sampled years. This analysis is not intended to describe the models' ability to predict FDCs at catchments with short observation records: in this case, constructing an empirical FDC using the available (however short) observation record is likely to be the best course of action (Castellarin et al., 2004a). Instead, the analysis is intended to simulate the effect of short observation records on FDC prediction at nearby, *ungauged* catchments. The underlying assumptions behind this analysis are that (i) the error associated with inter-

polation is independent of the flow record length, and (ii) the Chepe Kohla catchment is representative of Nepalese basins.

2.3.2 Predictions under change

We used numerical simulations to assess the ability of both models to predict streamflow when subject to changing rainfall regimes, as described in Fig. 2.

Synthetic streamflow time series were generated by coupling the stochastic rainfall generator described in Müller and Thompson (2013) to a rainfall–runoff model. The generated wet-season rainfall is a first-order Markov process (i.e., rainfall occurrence on a given day is correlated with rainfall occurrence on the previous day) with gamma-distributed rainfall intensities, and as such produces a rainfall record that explicitly violates the assumptions under-pinning the process-based model. The duration of the rainy season was assumed constant, and no rainfall was generated during the dry season. Wet-season streamflow was simulated by feeding syn-

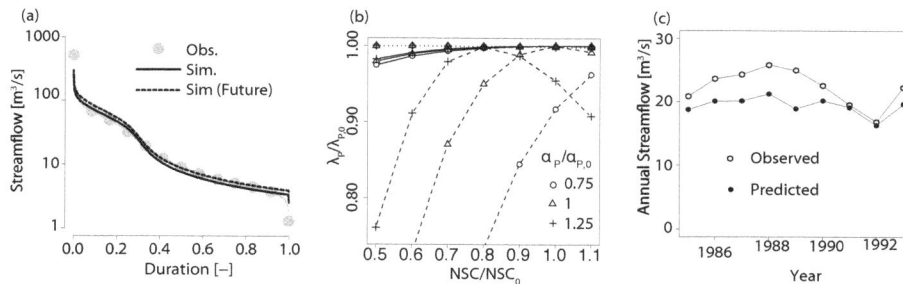

Figure 3. Sensitivity of models to changes in the precipitation regime. (**a**) Empirical and simulated flow duration curves at Chepe Kohla. The simulated FDC obtained from the stochastic rainfall generator and the bucket watershed model (solid) reproduce the empirical FDC constructed from the observed streamflow well (grey dots). Rainfall changes expected in Nepal ($\alpha_P/\alpha_{P,0} = 1.2$, $\lambda_P/\lambda_{P,0} = 0.98$) do not have a substantial influence on the simulated flow distribution (dashed). α_P and λ_P designate the mean depth and frequency of wet-season rainfall, respectively. (**b**) Sensitivities to relative changes in rainfall frequency and intensity over the Chepe Kohla catchment. The performance of the process-based model is not affected by rainfall changes (dotted). The sensitivity of the statistical model depends on its ability to predict changes in mean flow from annual rainfall. The model is highly sensitive to rain changes if average streamflow cannot be predicted (dashed), and is robust to moderate changes if average flow is perfectly predicted (solid). (**c**) The linear regression of the statistical model underestimates annual flows at the Chepe Kohla when using a cross-sectional sample (25 gauges) to estimate the local relation between average rainfall and average runoff.

thetic rainfall into a linear reservoir (with a recession constant k) with linear evapotranspiration losses, as in Müller et al. (2014). Dry-season discharge was obtained by simulating non-linear seasonal recessions of duration T_d starting at randomly selected runoff peaks in the (previously generated) wet-season streamflow. These assumptions are close to the observed reality in Nepal, as seen in Fig. 3a, where the FDC constructed from the simulated streamflow is a close approximation to the empirical FDC in the Chepe Kohla watershed. We translated the effect of shifts in precipitation regimes into changed streamflow for the Chepe Kohla catchment by considering a range of future combinations for rainfall frequencies and intensities. In line with what is expected in Nepal (Turner and Slingo, 2009; Turner and Annamalai, 2012), we considered negative changes in the frequency and positive changes in the mean daily rainfall depth. We neglected changes in soil moisture capacity, evapotranspiration, rainfall autocorrelation, and the duration of the rainy season. These parameters are explicit in the process-based model, so we expect differences in the sensitivity of the process-based and statistical models to climate change to be underestimated by this procedure. For each rainfall scenario, we evaluated the performance of the models in a changing climate by generating 1000 years of daily streamflow using future rainfall frequencies and intensities.

We compared the synthetic FDCs to model predictions that were made with *future* rainfall statistics but *contemporary* recession and low-flow parameters (Fig. 2). The statistical method in Chalise et al. (2003) uses a linear regression over a cross-sectional sample of observations to predict mean flow based on mean rainfall and altitude. The regression may fail to capture a variety of unobserved characteristics affecting both rainfall and streamflow (e.g., local topographic fea-

tures), and hence may not capture the causal relation between the two variables. The extent of this bias cannot be quantified a priori, so we considered two extreme cases: infinite and zero bias. The infinite bias case (Case 1 in Fig. 2) represents the case where no effective relationship can be determined between rainfall and mean flow. The best estimator of future mean flow is then the *current* flow condition. Conversely, if regression coefficients perfectly describe the effect of annual rainfall on average flow (Case 2 in Fig. 2), then the future flow conditions can be perfectly estimated using the (known) future annual rainfall. We modeled this situation by estimating Q_m directly from the (simulated) *future* flow conditions. While the two cases differed in the determination of mean flow (Q_m), the low-flow parameter (q_{95}) was determined from current flow conditions in both cases. In Chalise et al. (2003), q_{95} is normalized by Q_m and represents recession behavior, which is assumed independent of rainfall. The process-based predictions were obtained by inserting future rainfall statistics and contemporary recession constants into the analytical FDC equation described in Appendix A. The two models were compared by plotting prediction performances (NSC) against the relative change in the frequency and intensity of synthetic rainfall.

Although the recession assumptions of the process-based model are taken to generate the synthetic streamflow used as a control, we believe that the analysis is not biased against the statistical approach for three reasons. Firstly, the only parameter of the statistical approach that is influenced by rainfall (Q_m) is also computed from synthetic streamflow (Case 2 in Fig. 2). Secondly, although based on identical recession assumptions, the process-based model and the synthetic streamflow generator are driven by different stochastic rainfall processes (i.e., Poisson and Markov, respectively).

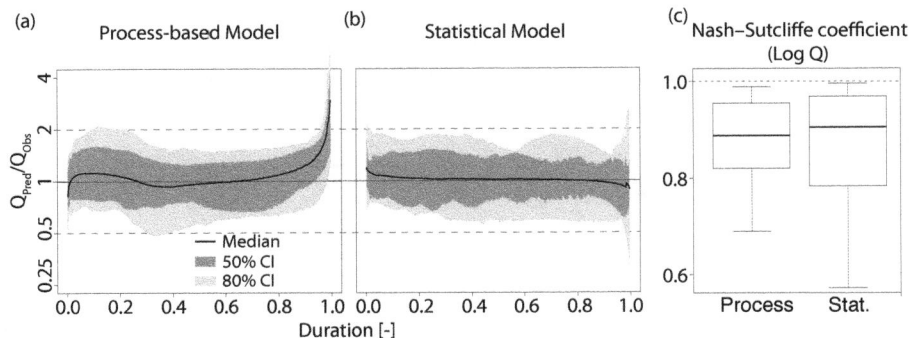

Figure 4. Flow duration curve prediction performance in ungauged basins. The error duration curves of the leave-one-out cross-validation analysis using the process-based and statistical models are presented in panels **(a)** and **(b)**, respectively. Relative errors are plotted on a log scale in order to allow the graphs to be balanced on the y axis: a relative prediction error of 2 (the model predicts double the observed value) is at the same distance from $y = 1$ (perfect prediction) as a relative error of $1/2$ (the model predicts half the observed value). Durations are plotted on the x axis, with $x = 0$ and $x = 1$ for the highest and lowest flow quantiles, respectively. Panel **(c)** shows box plots of Nash–Sutcliffe coefficients computed from log-transformed flow quantiles.

Lastly and most importantly, empirical observations reveal that synthetic streamflow distributions generated under contemporaneous rainfall conditions reproduce closely FDCs constructed from gauge records (Fig. 3a), showing that the underlying recession assumptions are, in fact, representative of runoff processes actually occurring in Nepal.

3 Results

3.1 Prediction in ungauged basins

Results from the leave-one-out cross-validation analysis are presented in Fig. 4 and show that both methods perform similarly in the prediction of FDCs in ungauged basins. Error duration curves (Fig. 4a and b) show comparable streamflow prediction uncertainties: 75 % of the predicted flow quantiles are between half and double the observed streamflow for both models, although the low flows in the process-based model display an increasing upwards bias (Fig. 4b). Considering the Nash–Sutcliffe coefficients computed at the individual basin level, the mean and median performances are again comparable for both models, but the accuracy of the statistical model predictions is more variable across sites than the process model predictions, as indicated by the larger spread of the Nash–Sutcliffe coefficients (Fig. 4c).

Figure 5a (top) shows prediction performances of both models as the number of streamflow gauges available for predictions decreases, and indicates that the performance of both models is relatively insensitive to the gauge density, until it declines to less than approximately 0.6 gauges per $10\,000\,\mathrm{km}^2$. For such situations, which represent discarding of more than half the available gauges in Nepal, the statistical model performance declines rapidly compared to the process-based model. Prediction performances are strongly affected by uncertainties on the interpolation of model pa-

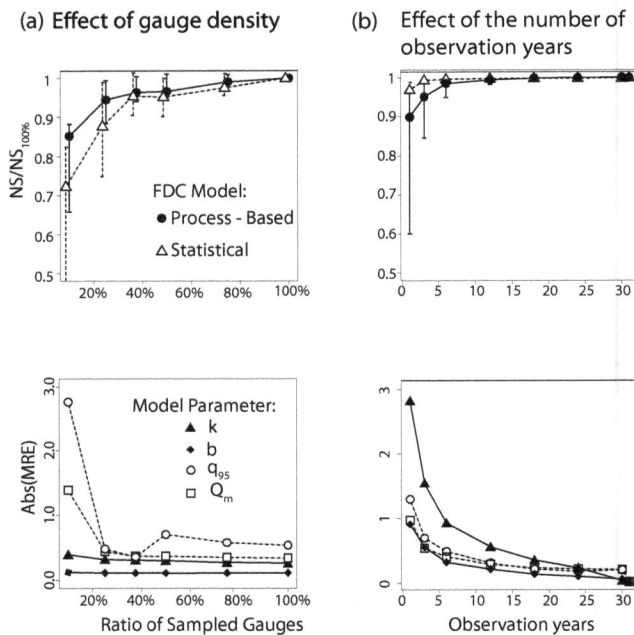

Figure 5. Sensitivity of models to data scarcity. **(a)** Cross-validation analysis showing the sensitivity of both models to a decreasing number of calibration gauges. **(b)** Resampling analysis of streamflow observations in the Chepe Kohla ($N = 10\,000$) catchment showing the effect of the number of observation years. In panels **(a)** and **(b)**, the effects on FDC prediction performances (top) are shown by plotting the ratio of calibration gauges sampled (or the number of observation years) against the relative Nash–Sutcliffe coefficient (with the NSC for the full set of available data as reference). The plot shows the median value for all iterations, and the error bars indicate the interquartile (25–75 %) range. The prediction uncertainties of model parameters (bottom) are given in absolute values of relative prediction errors.

rameters, as seen in Fig. 5a (bottom). Interpolation uncertainties are generally larger for the flow indices of the statistical model (Q_m and q_{95}) than for the recession parameters of the process-based model (k and b). This explains the larger spread in prediction performances of the former (Fig. 4c and error bars in Fig. 5a (top)). The parameter uncertainties are also relatively insensitive to the total gauge density until about 60 % of the originally available gauges are discarded. At this point, the uncertainties associated with estimation of the flow indices increase significantly, while the process-based model parameters remain more reasonably estimated.

When considering short observation windows, parameter uncertainties also drive the performance of the models. Figure 5b (top) shows the prediction performance of both models at the Chepe Khola watershed, as the number of observation years used to estimate the model parameters is reduced. In this case, the statistical model outperforms the process-based model when less than 10 years of streamflow observations are available. The parameter uncertainties associated with the short time-series estimates (Fig. 5b, bottom) suggest that a longer time series of streamflow observations is needed to accurately estimate the wet-season recession parameter (k), resulting in the lower performance of the process-based model for short streamflow records.

3.2 Prediction under change

Simulation results presented in Fig. 3b show both models' ability to predict a simulated future flow duration curve of the Chepe River under a range of different possible changes in rainfall regimes. In all simulations, parameters describing the hydrological response of the basin (k, b, and q_{95}) are determined using current flow conditions, and evapotranspiration is assumed constant. The results show that explicitly modeling rainfall–runoff processes allows the process-based model to accommodate the effects of the changing precipitation regime. In contrast, the performance of the statistical model is affected to various degrees by shifts in rainfall regimes, depending on how the model translates changes in annual precipitation to changes in average flows. If these shifts are perfectly represented by the model, then prediction errors arise solely from changes in the shape of the FDC, and the process and statistical models perform similarly in the Chepe Kohla watershed across the full range of considered rainfall scenarios (Fig. 3b, dashed curve). If, however, average (future) streamflows cannot be reliably predicted from the predicted changes in annual rainfall, the statistical model does not accommodate flow regime changes at all. In this case, future FDCs are modeled using current streamflow observations, and the ensuing prediction errors can be substantial (Fig. 3b, dotted curve). The simulated cases provide upper and lower bounds for the actual performance of the statistical model in future rainfall regimes. We evaluated the model's ability to predict Q_m by using cross-sectional data (i.e., aver-

age streamflow and annual rainfall from the 25 catchments) to estimate the linear relation between Q_m and annual rainfall R_y. Applied to the Chepe Kohla watershed, the estimated regression coefficients allowed the annual streamflow to be estimated from annual precipitation with a bias of -13 % and a coefficient of determination of $R^2 = 0.57$ (Fig. 3c). Regardless, prediction errors remained negligible for both bounds (NSC > 0.95) for the range of changes actually anticipated in Nepal (e.g., $\Delta\lambda_P/\lambda_P \approx 0.98$ and $\Delta\alpha_P/\alpha_P \approx 1.20$ for the $2 \cdot CO_2$ scenario – Turner and Slingo, 2009).

4 Discussion

4.1 Predictions in ungauged basins

The analysis suggests that both statistical and process-based methods to estimate FDCs in ungauged basins perform comparably in Nepal, over a wide range of gauge densities and observation durations. Yet prediction performances varied significantly between the models as data became increasingly sparse. The statistical method is more sensitive to spatially sparse data, which degrades the interpolation accuracy of Q_m. In contrast, the estimation method for recession parameters makes the process-based approach more sensitive to temporally restricted observations, which reduce the accuracy with which recession parameters can be estimated. This suggests that the performance of the two models in ungauged basins is affected by different sources of uncertainty. In this section, we investigate the source of prediction error in each method and discuss the implications for their application in ungauged basins beyond Nepal.

4.1.1 Sources of uncertainty

The statistical model relies on two assumptions about the correlations of observed data. The first assumption is that catchments with similar low-flow indices (q_{95}) have identical hydrological responses, and therefore identical FDC shapes. Second, the model assumes that the flow indices (Q_m and q_{95}) at ungauged catchments can be best predicted using linear regressions against observable covariates (annual rainfall, elevation, and geology). The latter assumption does not hold if the flow indices are spatially auto-correlated, or if the posited linear relations are spatially heterogeneous or, in fact, non-linear. Furthermore, "omitted variable" biases (Greene, 2003) will arise if an unobserved variable is correlated with both a covariate and a flow index. For instance, local topographic features may affect both the annual rainfall and the average streamflow in mountainous regions. Violation of the second assumption leads to substantial uncertainty in the interpolation of the flow indices in Nepal and drives the prediction errors of the statistical approach, as shown in Sect. S2 of the Supplement.

While the performance of the process-based model is also driven by parameter estimation uncertainties, these errors arise from simplifying assumptions about local hydrological processes (rather than uncertainties from their statistical interpolation from neighboring gauges). Additional cross-validation analyses (shown in Sect. S2 of the Supplement) suggest that uncertainties caused by the aggregation of observed point-rainfall statistics at the catchment level drive prediction errors of high-flow quantiles. While increasingly accurate remote sensing rainfall data will progressively allow such spatial heterogeneities to be resolved, current precipitation products (e.g., TRMM 3B42) remain substantially biased in mountainous regions like Nepal, where they do not outperform available rain gauges in predicting the frequency and intensity of areal rainfall (Müller and Thompson, 2013). A second source of error arises from the simplifying assumptions made about streamflow recession that do not hold perfectly in the observed catchments. Because they describe the same watershed, the wet and dry recession parameters are assumed to be physically related. In Müller et al. (2014), the scale parameter of the non-linear seasonal recession (a) is expressed as an explicit function of the two recession parameters (k and b) for sufficiently short recession times, where power-law recessions can be approximated by exponential functions. We show in the Supplement (Sect. S2) that, although this approach provides more accurate estimates of a than would be obtained through spatial interpolation, estimation uncertainties remain, propagate through the model, and result in prediction errors during the dry season.

4.1.2 Applicability beyond Nepal

This study compares two specific methods on their ability to predict FDCs in the particular context of ungauged Nepalese basins. Results are thus not necessarily representative of the relative performance of process-based and statistical methods in general, particularly in regions where abundant field data allow more advanced statistical approaches to be implemented. Yet fundamentally, the statistical model relies on observed correlations rather than assumptions about hydrologic mechanisms. Because FDC shapes are modeled non-parametrically, the approach is applicable to regions with highly variable catchment responses. However, prediction performance in ungauged basins is constrained by interpolation errors in the mean flow. This makes the method unsuitable for regions where the local determinants of mean flow (i.e., rainfall, evapotranspiration, glacial melt) cannot be accurately monitored at the catchment level. In contrast, a key advantage of the process-based model is its ability to exploit characteristics of the stochastic structure of rainfall that can be estimated from daily rainfall observations. The model is appropriate for regions where the spatial heterogeneity of runoff is driven by rainfall, and where the frequency and intensity of rainfall depths at the catchment level can be readily estimated (i.e., small catchments with numerous rain gauges, or places where satellite observations provide a good representation of rainfall statistics). Unlike rainfall, recession

behavior arises from lumped and complex interactions between climate, vegetation, and groundwater processes that typically cannot be monitored in a spatially explicit manner. The process-based model is therefore inappropriate for regions where the hydrologic response of the catchment is the main source of runoff heterogeneity, or where the assumed recession behavior (in particular the relation between a, k, and b) does not occur.

Conveniently, the appropriate implementation contexts for both methods appear to be complementary, and the optimal method in a given region is determined by the driving source of runoff heterogeneity in the catchments. Ultimately, the performance of both methods is constrained by their ability to estimate their parameters in ungauged basins. This relation is apparent in Fig. 5, where drops in prediction performances correspond to increases in the estimation uncertainty of model parameters. Under these conditions, the performance of each method is driven by the ability of the available observations to capture the variability of the model parameters. When interpolated from neighboring gauges, uncertainties are governed by the interplay between the layout of the gauges and the spatial correlation range of the considered model parameter. When estimated from short observation records, accuracy is determined by the extent to which the available record is representative of the temporal variability of the parameter. These interactions between data availability and runoff variability are inherently local and will affect the determination of the most appropriate method for any given region.

4.2 Prediction under change

Expected shifts in the frequency and intensity of monsoon rainfall over Nepal only have a marginal impact on the streamflow distributions in the Chepe Kohla catchment, as shown by the numerical simulation presented in Fig. 3a (dashed curve). Consequently, changes in rainfall regime do not appear to affect the performance of either model (Fig. 3b), unless they are significantly larger than expected. Climate change may nonetheless affect flow predictions elsewhere. It is therefore helpful to consider the conditions under which FDCs can be reliably predicted in a changing climate.

Although rainfall stationarity is an inherent assumption of the process-based approach, climate change can be incorporated by updating the relevant parameters to their future value to predict the (pseudo-)stationary future state of the system. The method accounts for otherwise confounding changes in the frequency and intensity of rainfall, which are expected in Nepal. By explicitly accounting for soil moisture dynamics and recession behavior, the model emulates the (causal) effect of rainfall on streamflow. As a result, the method reliably predicts the distribution of future streamflow, provided that governing flow generation processes are in line with the basic assumptions listed in Sect. 2.1.1.

In contrast, the statistical model is solely based on observed correlations, leading to two important sources of errors for predictions under change. First, the model only accommodates rainfall changes to the extent that the estimated statistical relation between rainfall and runoff is representative of local runoff coefficients. The model will not reliably predict future streamflows if runoff coefficients are strongly spatially heterogeneous, or if the cross-sectional sample of gauges fails to capture important processes governing mean flow. This source of uncertainty appears to be significant in Nepal, as illustrated by the substantial bias in annual flow predictions in Fig. 3c. Secondly, the statistical model only considers the effect of *average* rainfall on *average* flow: the effect of rainfall *distribution* on streamflow *distribution* is ignored. As a result, the model cannot predict changes in the shape of FDCs that are brought about by changing rainfall. The prediction performance of the statistical approach is therefore determined by the resilience of the flow regime, that is, the extent to which streamflow distribution is affected by shifting rain signals (Botter et al., 2013): the method will perform poorly in catchments with non-resilient flow regimes. The Monte Carlo analysis presented in the Supplement (Sects. S3 and S4) shows that streamflow resilience in seasonally dry catchments depends on two distinct seasonal effects: a "direct" effect driven by the ratio between λ_P and k during the wet season, and an "indirect" effect during the dry season, when resilience is determined by the interplay between Q_0 (i.e., wet-season rainfall) and b. In seasonally dry climates, we expect the statistical method to be most reliable in regions where wet seasons are short with limited total rainfall but persistent flow regimes, and where the recession behavior during the dry season is close to linear.

Lastly, a key assumption in this study is that catchment response (in terms of low-flow or recession characteristics) is independent of climate. It is possible that shifts in climate have an effect on catchment response by affecting the partitioning of effective rainfall between storage and runoff. Although not quantitatively assessed in this study, we expect that this effect would negatively affect the performance of both approaches.

5 Conclusions

Stochastic, process-based models predicted the FDCs for ungauged catchments in Nepal well, with a performance that was comparable to that of statistical models. It suggests that in regions with globally representative gauge densities, and under seasonally dry climates, the advantages of the statistical approaches relative to stochastic models noted in previous analyses (Blöschl et al., 2013) may not apply. Fundamentally, the performances of both approaches are strongly affected by the method chosen to estimate model parameters in ungauged basins, so this conclusion comes with the caveat that this study cannot be interpreted as a general benchmark to compare these approaches at a global level. Although we believe that the selected models are appropriate to compare process-based and statistical approaches for practical PUB application in Nepal, their relative performance may be different in other regions, where more abundant information on catchment characteristics allow more complex (and presumably more accurate) regionalization approaches to be applied. Thus, substantial research remains to be done to compare these approaches in other parts of the world, where locally appropriate methods should be carefully considered.

Nonetheless, this study finds a complementarity between the different sources of uncertainty in the stochastic and statistical methods. This suggests that model selection should be driven by a consideration of the main drivers of heterogeneity in any study catchment: process-based models are advisable if climate is likely to be the main source of runoff heterogeneity. Conversely, statistical methods are more appropriate for regions with substantially different recession behaviors across catchments. These distinctions provide a potentially robust basis for model selection in any given application.

The results also suggest that the sensitivity of statistical approaches to changes in rainfall statistics is dependent on the "resilience" of the flow regime as defined by Botter et al. (2013). Overall, the process-based models are more reliable in projecting FDCs into new rainfall regimes. This is particularly true for catchments characterized by a strong wet-season runoff and a rapid, strongly non-linear hydrologic response, because their flow regime is particularly vulnerable to rainfall changes, making the assumptions of the statistical model inappropriate.

The excellent performance of both process-based and statistical models for the FDC and PDF in ungauged basins suggests that extending probabilistic analyses in such basins to also include flow-derived variables such as hydropower capacity (Basso and Botter, 2012) or ecological responses (Thompson et al., 2013a) may be feasible. While these prospects are enticing, we note that a model's ability to predict an FDC with high fidelity is not necessarily indicative of prediction performances on all derived stochastic properties. For instance, Dralle et al. (2015) demonstrate that the crossing properties of streamflow can be very poorly estimated by stochastic process-based models, even in applications where the same models predict the PDF of flow well. Further exploration of the potential opportunities and limitations afforded by use of probabilistic models in ungauged basins offers a promising avenue for future study.

Appendix A: Process-based streamflow distribution model for seasonally dry climates

This appendix presents the analytical expression of FDC in seasonal climates derived in Müller et al. (2014). The approach assumes that rainfall can be represented as a marked Poisson process with exponentially distributed depths. Catchments are modeled as spatially homogenous linear reservoirs with linear evapotranspiration losses. Under these conditions, wet-season streamflow can be represented as a gamma-distributed random variable (Botter et al., 2007b):

$$Q_w \sim \Gamma(m, \alpha_Q^{-1}),$$

with $m = \lambda/k$ and $\alpha_Q = \alpha_P k A$, and where k is the linear recession constant, A the area of the contributing catchment and α_P the mean intensity of wet-season rainfall. The frequency λ of runoff events can be expressed as a function of the frequency (λ_P) and intensity of rainfall (Botter et al., 2007b):

$$\lambda = \eta \frac{\exp(-\gamma)\gamma^{\frac{\lambda_P}{\eta}}}{\Gamma_L(\lambda_P/\eta, \gamma)}, \tag{A1}$$

where $\Gamma_L(\cdot, \cdot)$ is the lower incomplete gamma function, and where $\eta = \text{ET}/\text{SSC}$ and $\gamma = \text{SSC}/\alpha_P$ are, respectively, the ratio between maximum evapotranspiration and soil storage capacity, and the ratio between soil storage capacity and mean rainfall intensity.

Dry-season streamflow is modeled as a seasonal recession starting at the last discharge peak of the wet season. Because wet-season streamflow is a gamma-distributed variable, streamflow at discharge peaks, and therefore the initial condition of the seasonal recession, is itself a gamma-distributed variable (Müller et al., 2014):

$$Q_{\text{peak}} \sim \Gamma(m + 1, \alpha_Q^{-1}).$$

Assuming a power-law relation between discharge and recession rate, the cumulative distribution function of dry-season streamflow can be expressed as (Müller et al., 2014)

$$P_{Q_d}(Q) = \begin{cases} 1 + \frac{q_d^r \Gamma_1 - \alpha_Q^r \Gamma_2}{a r T_d \Gamma(m+1)}, & \text{if } Q > -(a r T_d)^{\frac{1}{r}} \\ & \text{and } r < 0 \\ 1 + \frac{q_d^r \Gamma_1 - \alpha_Q^r \Gamma_2}{a r T_d \Gamma(m+1)} & \text{otherwise} \\ \quad + \frac{\alpha_Q^r \Gamma_4 + (Q^r - a r T_d)\Gamma_3}{a r T_d \Gamma(m+1)}, \end{cases}$$

with

$$\Gamma_1 = \Gamma_U(m + 1, \alpha_Q^{-1} Q),$$

$$\Gamma_2 = \Gamma_U\left(r + m + 1, \alpha_Q^{-1} Q\right),$$

$$\Gamma_3 = \Gamma_U\left(m + 1, \alpha_Q^{-1}(Q^r + a r T_d)^{\frac{1}{r}}\right),$$

$$\Gamma_4 = \Gamma_U\left(r + m + 1, \alpha_Q^{-1}(Q^r + a r T_d)^{\frac{1}{r}}\right).$$

$\Gamma(\cdot)$ and $\Gamma_U(\cdot, \cdot)$ denote the complete and upper incomplete gamma functions; T_d is the duration of the dry season;

$r = 1 - b$ and a are the parameters of the non-linear recession, which are assumed stationary. Because they describe the same watershed, recession parameters for the wet and dry seasons are related. If power-law recessions can be approximated by an exponential function for sufficiently short recession times, we can express a as a function of k and b (Müller et al., 2014):

$$a \approx \frac{\lambda}{-r}\left(e^{\frac{-r}{m}} - 1\right)\left(\alpha_Q \cdot (m + 1)\right). \tag{A2}$$

The law of total probability can finally be used to combine seasonal streamflow distributions and derive the cumulative distribution function of streamflow for the whole year:

$$P_Q(Q) = \left(1 - \frac{T_d}{365}\right) \cdot P_{Q_w}(Q) + \frac{T_d}{365} \cdot P_{Q_d}(Q). \tag{A3}$$

The FDC for seasonally dry climates is finally obtained by plotting the streamflow quantiles Q against $1 - P_Q(Q)$, the complement of the cumulative distribution function of streamflow.

Acknowledgements. The Swiss National Science Foundation is gratefully acknowledged for funding (M. F. Müller).

Edited by: S. Archfield

References

Alaouze, C. M.: Reservoir releases to uses with different reliability requirements, AWRA Water Resour. Bull., 25, 1163–1168, 1989.

Alternative Energy Promotion Center: Construction and Installation Manual for Micro Hydropower Project Installers, Government of Nepal, 2014 (in Nepalese).

Arora, M., Goel, N., Singh, P., and Singh, R.: Regional flow duration curve for a Himalayan river Chenab, Nord. Hydrol., 36, 193–206, 2005.

Basso, S. and Botter, G.: Streamflow variability and optimal capacity of run-of-river hydropower plants, Water Resour. Res., 48, W10527, doi:10.1029/2012WR012017, 2012.

Bélisle, C. J.: Convergence theorems for a class of simulated annealing algorithms on Rd, J. Appl. Probab., 29, 885–895, 1992.

Blöschl, G., Sivapalan, M., Wagener, T., Viglione, A., and Savenije, H.: Runoff Prediction in Ungauged Basins: Synthesis across Processes, Places and Scales, Cambridge University Press, 2013.

Botter, G., Peratoner, F., Porporato, A., Rodriguez-Iturbe, I., and Rinaldo, A.: Signatures of large-scale soil moisture dynamics on streamflow statistics across U.S. climate regimes, Water Resour. Res., 43, W11413, doi:10.1029/2007WR006162, 2007a.

Botter, G., Porporato, A., Rodriguez-Iturbe, I., and Rinaldo, A.: Basin-scale soil moisture dynamics and the probabilistic characterization of carrier hydrologic flows: slow, leachingprone components of the hydrologic response, Water Resour. Res., 43, W02417, doi:10.1029/2006WR005043, 2007b.

Botter, G., Porporato, A., Rodriguez-Iturbe, I., and Rinaldo, A.: Nonlinear storage–discharge relations and catchment streamflow regimes, Water Resour. Res., 45, W10427, doi:10.1029/2008WR007658, 2009.

Botter, G., Basso, S., Rodriguez-Iturbe, I., and Rinaldo, A.: Resilience of river flow regimes, P. Natl. Acad. Sci. USA, 110, 12925–12930, doi:10.1073/pnas.1311920110, 2013.

Brutsaert, W. and Nieber, J. L.: Regionalized drought flow hydrographs from a mature glaciated plateau, Water Resour. Res., 13, 637–643, doi:10.1029/WR013i003p00637, 1977.

Castellarin, A., Galeati, G., Brandimarte, L., Montanari, A., and Brath, A.: Regional flow-duration curves: reliability for ungauged basins, Adv. Water Resour., 27, 953–965, doi:10.1016/j.advwatres.2004.08.005, 2004a.

Castellarin, A., Vogel, R., and Brath, A.: A stochastic index flow model of flow duration curves, Water Resour. Res., 40, W03104, doi:10.1029/2003WR002524, 2004b.

Castellarin, A., Botter, G., Hughes, D., Liu, S., Ouarda, T., Parajka, J., Post, D., Sivapalan, M., Spence, C., Viglione, A., and Vogel, R.: Prediction of flow duration curves in ungauged basins, chapt. 7, in: Runoff Prediction in Ungauged Basins: Synthesis Across Processes, Places and Scales, edited by: Blöschl, G., Sivapalan, M., Wagener, T., Viglione, A., and Savenije, H., Cambridge University Press, 135–162, 2013.

Ceola, S., Botter, G., Bertuzzo, E., Porporato, A., Rodriguez-Iturbe, I., and Rinaldo, A.: Comparative study of ecohydrological streamflow probability distributions, Water Resour. Res., 46, W09502, doi:10.1029/2010WR009102, 2010.

Chalise, S., Kansakar, S., Rees, G., Croker, K., and Zaidman, M.: Management of water resources and low flow estimation for the Himalayan basins of Nepal, J. Hydrol., 282, 25–35, doi:10.1016/S0022-1694(03)00250-6, 2003.

Cheng, L., Yaeger, M., Viglione, A., Coopersmith, E., Ye, S., and Sivapalan, M.: Exploring the physical controls of regional patterns of flow duration curves –Part 1: Insights from statistical analyses, Hydrol. Earth Syst. Sci., 16, 4435–4446, doi:10.5194/hess-16-4435-2012, 2012.

Chitrakar, P.: Micro-Hydropower Design Aids Manual, Small Hydropower Promotion Project (GTZ) and Mini-Grid Support Program, Alternate Energy Promotion Center, Government of Nepal, 2004.

Coopersmith, E., Yaeger, M. A., Ye, S., Cheng, L., and Sivapalan, M.: Exploring the physical controls of regional patterns of flow duration curves – Part 3: A catchment classification system based on regime curve indicators, Hydrol. Earth Syst. Sci., 16, 4467–4482, doi:10.5194/hess-16-4467-2012, 2012.

Department of Hydrology and Meteorology: Daily Streamflow and Precipitation Data, Kathmandu, 2011.

Doulatyari, B., Betterle, A., Basso, S., Biswal, B., Schirmer, M., and Botter, G.: Predicting streamflow distributions and flow duration curves from landscape and climate, Adv. Water Resour., 83, 285–298, doi:10.1016/j.advwatres.2015.06.013, 2015.

Dralle, D. N., Karst, N., and Thompson, S.: Dry season streamflow persistence in seasonal climates, Water Resour. Res., doi:10.1002/2015WR017752, online first, 2015.

Global Runoff Data Center: Global Runoff Data Base, Global Runoff Data Centre, Koblenz, Federal Institute of Hydrology (BfG), 2014.

Greene, W. H.: Econometric Analysis, Prentice Hall, Upper Saddle River, NJ, 2003.

HKH-FRIEND: Hindu Kush Himalayan – Flow Regimes from International Experimental and Network Data, UNESCO International Hydrological Programme, UNESCO Paris, France, 2004.

Hughes, D. and Smakhtin, V.: Daily flow time series patching or extension: a spatial interpolation approach based on flow duration curves, Hydrol. Sci. J., 41, 851–871, 1996.

Lambert, L. and Chitrakar, B.: Variation of potential evapotranspiration with elevation in Nepal, Mt. Res. Dev., 9, 145–152, 1989.

Lazzaro, G., Basso, S., Schirmer, M., and Botter, G.: Water management strategies for run-of-river power plants: profitability and hydrologic impact between the intake and the outflow, Water Resour. Res., 49, 8285–8298, doi:10.1002/2013WR014210, 2013.

Milly, P., Julio, B., Malin, F., Robert, M., Zbigniew, W., Dennis, P., and Ronald, J.: Stationarity is dead, Science, 319, 573–574, doi:10.1126/science.1151915, 2008.

Montanari, A., Young, G., Savenije, H., Hughes, D., Wagener, T., Ren, L., Koutsoyiannis, D., Cudennec, C., Toth, E., and Grimaldi, S.: Panta Rhei – Everything Flows: Change in hydrology and society – the IAHS Scientific Decade 2013–2022, Hydrolog. Sci. J., 58, 1256–1275, doi:10.1080/02626667.2013.809088, 2013.

Mu, X., Zhang, L., McVicar, T. R., Chille, B., and Gau, P.: Analysis of the impact of conservation measures on stream flow regime in catchments of the Loess Plateau, China, Hydrol. Process., 21, 2124–2134, doi:10.1002/hyp.6391, 2007.

Müller, M. F.: Bridging the Information Gap: Remote Sensing and Micro Hydropower Feasibility in Data-Scarce Regions, Doctoral Dissertation, University of California, Berkeley, CA, 2015.

Müller, M. F. and Thompson, S. E.: Bias adjustment of satellite rainfall data through stochastic modeling: methods development and application to Nepal, Adv. Water Resour., 60, 121–134, doi:10.1016/j.advwatres.2013.08.004, 2013.

Müller, M. F. and Thompson, S. E.: TopREML: a topological restricted maximum likelihood approach to regionalize trended runoff signatures in stream networks, Hydrol. Earth Syst. Sci., 19, 2925–2942, doi:10.5194/hess-19-2925-2015, 2015.

Müller, M. F., Dralle, D. N., and Thompson, S. E.: Analytical model for flow duration curves in seasonally dry climates, Water Resour. Res., 50, 5510–5531, doi:10.1002/2014WR015301, 2014.

Muneepeerakul, R., Azaele, S., Botter, G., Rinaldo, A., and Rodriguez-Iturbe, I.: Daily streamflow analysis based on a two-scaled gamma pulse model, Water Resour. Res., 46, W11546, doi:10.1029/2010WR009286, 2010.

NASA Land Processes Distributed Active Archive Center (LP DAAC): ASTER GDEM v2, NASA Land Processes Distributed Active Archive Center (LP DAAC), ASTER L1 B, USGS/Earth Resources Observation and Science (EROS) Center, Sioux Falls, 2011.

Nathan, R. and McMahon, T.: Evaluation of automated techniques for base flow and recession analyses, Water Resour. Res., 26, 1465–1473, 1990.

Sauquet, E. and Catalogne, C.: Comparison of catchment grouping methods for flow duration curve estimation at ungauged sites in France, Hydrol. Earth Syst. Sci., 15, 2421–2435, doi:10.5194/hess-15-2421-2011, 2011.

Schaefli, B., Rinaldo, A., and Botter, G.: Analytic probability distributions for snow-dominated streamflow, Water Resour. Res., 49, 2701–2713, doi:10.1002/wrcr.20234, 2013.

Searcy, J. and Hardison, C.: Double-Mass Curves. Manual of Hydrology: Part I, General Surface Water Techniques, US Geological Survey Water-Supply Paper 1541-B, United States Government Printing Office, Washington, 1960.

Shao, J. and Tu, D.: The Jackknife and Bootstrap, Springer-Verlag, New York, 2012.

Shrestha, D. P.: Assessment of soil erosion in the Nepalese Himalaya: a case study in Likhu Khola Valley, Middle Mountain Region, Land Husbandry, 2, 59–80, 1997.

Stokstad, E.: Scarcity of rain, stream gages threatens forecasts, Science, 285, 1199–1200, doi:10.1126/science.285.5431.1199, 1999.

Thompson, S., Levin, S., and Rodriguez-Iturbe, I.: Linking plant disease risk and precipitation drivers: a dynamical systems framework, Am. Nat., 181, E1–E16, doi:10.1086/668572, 2013a.

Thompson, S. E., Sivapalan, M., Harman, C. J., Srinivasan, V., Hipsey, M. R., Reed, P., Montanari, A., and Blöschl, G.: Developing predictive insight into changing water systems: use-inspired hydrologic science for the Anthropocene, Hydrol. Earth Syst. Sci., 17, 5013–5039, doi:10.5194/hess-17-5013-2013, 2013b.

Turner, A. G. and Annamalai, H.: Climate change and the south Asian summer monsoon, Nature Climate Change, 2, 587–595, doi:10.1038/nclimate1495, 2012.

Turner, A. G. and Slingo, J. M.: Subseasonal extremes of precipitation and active-break cycles of the Indian summer monsoon in a climate-change scenario, Q. J. Roy. Meteor. Soc., 135, 549–567, 2009.

United States Geological Survey: USGS Threatened and Endangered Streamgages, available at: http://streamstats09.cr.usgs.gov/ThreatenedGages/ThreatenedGages_str.html (last access: 5 August 2015), 2015.

Vogel, R. and Fennessey, N.: Flow-duration curves. I: New interpretation and confidence intervals, J. Water Res. Pl.-ASCE, 120, 485–504, 1994.

Westerberg, I. K., Gong, L., Beven, K. J., Seibert, J., Semedo, A., Xu, C.-Y., and Halldin, S.: Regional water balance modelling using flow-duration curves with observational uncertainties, Hydrol. Earth Syst. Sci., 18, 2993–3013, doi:10.5194/hess-18-2993-2014, 2014.

4

Regionalizing nonparametric models of precipitation amounts on different temporal scales

Tobias Mosthaf and András Bárdossy

Institute for Modeling Hydraulic and Environmental Systems, Universität Stuttgart, Stuttgart, Germany

Correspondence to: Tobias Mosthaf (tobias.mosthaf@iws.uni-stuttgart.de)

Abstract. Parametric distribution functions are commonly used to model precipitation amounts corresponding to different durations. The precipitation amounts themselves are crucial for stochastic rainfall generators and weather generators. Nonparametric kernel density estimates (KDEs) offer a more flexible way to model precipitation amounts. As already stated in their name, these models do not exhibit parameters that can be easily regionalized to run rainfall generators at ungauged locations as well as at gauged locations. To overcome this deficiency, we present a new interpolation scheme for nonparametric models and evaluate it for different temporal resolutions ranging from hourly to monthly. During the evaluation, the nonparametric methods are compared to commonly used parametric models like the two-parameter gamma and the mixed-exponential distribution. As water volume is considered to be an essential parameter for applications like flood modeling, a Lorenz-curve-based criterion is also introduced. To add value to the estimation of data at sub-daily resolutions, we incorporated the plentiful daily measurements in the interpolation scheme, and this idea was evaluated. The study region is the federal state of Baden-Württemberg in the southwest of Germany with more than 500 rain gauges. The validation results show that the newly proposed nonparametric interpolation scheme provides reasonable results and that the incorporation of daily values in the regionalization of sub-daily models is very beneficial.

1 Introduction

Rainfall time series of differing temporal resolutions are needed for various applications like water engineering design, flood modeling, risk assessments and ecosystem and hydrological impact studies (Wilks and Wilby, 1999; Burton et al., 2008). As many precipitation records are too short and contain erroneous measurements, stochastic precipitation models can be used to generate synthetic time series instead. Starting from single-site models (summarized in Wilks and Wilby, 1999) and multisite models for simultaneous time series at various sites (e.g., Wilks, 1998; Buishand and Brandsma, 2001; Bárdossy and Plate, 1992), models that allow for gridded simulations are finally developed (e.g., Wilks, 2009; Burton et al., 2008).

For modeling precipitation, one crucial variable is the precipitation amount, which follows a certain distribution. Distributions of the daily precipitation amounts are strongly right skewed, with many small values and few large values (Wilks and Wilby, 1999; Li et al., 2012; Chen and Brissette, 2014). This also holds true for different temporal resolutions with increasing skewness for higher temporal resolutions and vice versa. This means that rainfall intensity distributions depend on the temporal scale of the observed values. Applying single-site or multisite precipitation models at ungauged locations requires the regionalization of the precipitation amount distributions. This can be done in two different ways.

1. Interpolate the precipitation amounts from observation points for every time step to the target location(s) and set up a distribution with the interpolated values.

2. Fit a distribution function to the precipitation amounts separately for each gauge and interpolate the distribution functions to the target location(s).

The first approach seems more straightforward, but exhibits several deficiencies, such as the overestimation of the rainfall probability, the underestimation of the variance and the underestimation of the maximum rainfall value. In the Supplement Sect. S1, an example demonstrates these problems. Due to the relative inefficiency of the first interpolation approach, the second is preferred.

In most stochastic rainfall models, the theoretical parametric distribution functions are fitted to the empirical values using, e.g., the exponential distribution or the two-parametric gamma distribution (Wilks and Wilby, 1999; Papalexiou and Koutsoyiannis, 2012). It is possible to either interpolate the parameters of the theoretical distribution or to interpolate the moments (e.g., the mean and standard deviation) of the rainfall intensities (Wilks, 2008; Haberlandt, 1998). Lall et al. (1996) introduced a more flexible nonparametric single-site rainfall model, for which they used nonparametric KDEs with a prior logarithmic transformation to model daily rainfall intensities. They addressed the problem of regionalization by using nonparametric estimates of distribution functions. However, a different interpolation scheme is required for nonparametric estimations, as they do not use any parameter that can be simply interpolated.

In the present work, we introduce a regionalization strategy for nonparametric distributions and compare it to the traditional regionalization of parametric distributions for varying temporal resolutions from hourly to monthly scales. The common procedure to interpolate parametric distribution functions is outlined as follows.

1. Fit a parametric distribution (e.g., a gamma or exponential distribution) at each sampling site to the empirical distribution function (EDF).

2. Interpolate the moment(s) or parameter(s) of the fitted parametric distribution.

3. Set up the theoretical cumulative distribution function (CDF) at every interpolation target with the interpolated moment(s) or parameter(s).

The newly proposed procedure for the nonparametric distribution functions is the following.

1. Fit a nonparametric distribution to log-transformed rainfall values using a Gaussian kernel.

2. Estimate the interpolation (kriging) weights with the precipitation values of a certain quantile.

3. Apply these weights to the values of certain discrete quantiles.

4. Linearly interpolate the remaining quantile values to receive a continuous CDF for all target locations.

In Arns et al. (2013), a similar approach is used to interpolate the quantile value differences of water levels for a bias correction between the empirical distributions of the observed and modeled values at the German North Sea coast. In contrast to their work, entire theoretical distribution functions are estimated in our work through interpolation. Goulard and Voltz (1993) introduced a curve kriging procedure to regionalize fitted functions, which was further developed by Giraldo et al. (2011). Based on their work, Menafoglio et al. (2013) developed a universal kriging approach for the nonstationary interpolation of functional data, which was applied in Menafoglio et al. (2016) for the simulation of soil particle distribution functions. As CDF curves are special functions that are monotonically non-decreasing between 0 and 1, the curve kriging procedure additionally needs to be constrained to these conditions. Our approach can deal with these conditions directly.

After describing the study region Baden-Württemberg in Sect. 2, the concept of precipitation amount models is introduced in Sect. 3. The data selection in Sect. 4 is followed by an investigation of the spatial dependence of the precipitation amount models in Sect. 5. The theory of precipitation amount models is addressed in Sect. 6, and the basis of the proposed interpolation procedure for nonparametric models is established in Sect. 7. The application of different regionalization procedures for precipitation amount models is explained in Sect. 8. The implementation of daily rainfall observations within the interpolation of sub-daily distribution functions is outlined in Sect. 9. The resulting performance of the different precipitation amount models at the point locations and their regionalization is depicted in Sect. 10.

2 Study region and data

The study region is the federal state of Baden-Württemberg, which is located in the southwest of Germany. The Black Forest mountain range, in the west, and the Swabian Alps, extending from southwest to northeast, exhibit the highest elevations in Baden-Württemberg. The rising of large-scale moist air masses across the mountainous regions causes higher rainfall amounts on the windward side and lower amounts on the leeward side. In the summer months, slopes with differing inclinations lead to a warming of the air that triggers convection currents, leading to a greater number of showers and thunderstorms over the mountainous regions. This shows a dependence of rainfall on elevation with seasonal differences. The rain-bearing westerly winds lead to high rainfall amounts in the Black Forest. The relatively lower altitude of the Swabian Alps results in lower rainfall amounts as they lie in the shadow of the Black Forest (Landesanstalt für Umwelt, Messungen und Naturschutz Baden-Württemberg , LUBW).

The years from 1997 to 2011 are chosen as the investigation period, as the German Meteorological Service (DWD) set up many new rain gauges in 1997. A relatively homogeneous data set is obtained by only choosing gauges with observation periods greater than or equal to 5 years, which also provide rainfall measurements for at least 80 % of the

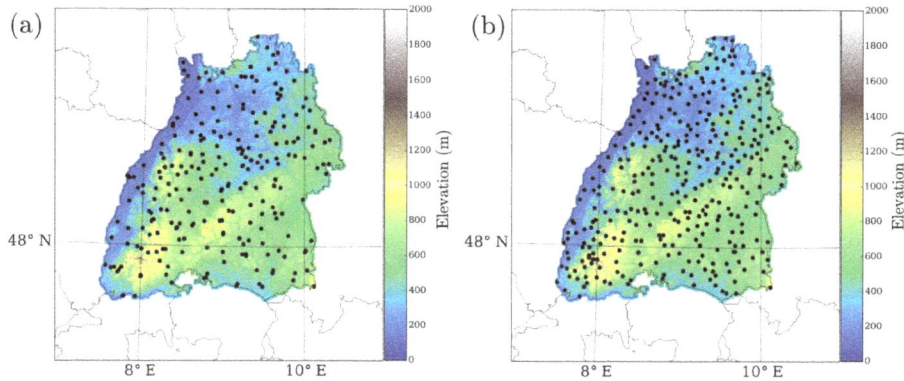

Figure 1. The locations of the high-resolution (hourly and 5 min; **a**) and daily rain gauges (**b**) in Baden-Württemberg.

time steps within their observation period. We had access to (i) 242 hourly and 5 min resolution and (ii) 347 daily gauges available in the study region, with 80 sites having both daily and high-resolution instruments. The observations are provided by the DWD and the Environmental Agency of Baden-Württemberg (LUBW). The high-resolution rain gauges are mostly equipped with tipping buckets and gravimetric measurement devices (Beck, 2013). Figure 1 shows the study region with the locations of the two sets of rain gauges.

3 Modeling precipitation amounts at point locations

Modeling precipitation amounts in our context means estimating the distribution functions. The usage of these distribution functions includes the implicit assumption of temporally independent and identically distributed (i.i.d.) variables. This assumption is generally accepted for daily rainfall as the autocorrelation of consecutive nonzero daily precipitation is relatively small and usually of less importance. For higher temporal resolutions, such as hourly, autocorrelation needs to be incorporated in the model (Wilks and Wilby, 1999). In practice, different methods exist to take such a correlation into account. One approach is to include autocorrelation prior to the sampling procedure by using conditional distributions. These conditions may be event statistics, like the duration of a rainfall event (e.g., Acreman, 1990), or varying statistical moments depending on the hour of the day (e.g., Katz and Parlange, 1995). Another approach is introducing autocorrelation after the sampling procedure. Bárdossy (1998) uses the empirical distributions of hourly rainfall intensities to sample values for which the random order is subsequently changed within a simulated annealing scheme to consider autocorrelation. In Bárdossy et al. (2000), the theoretical representations (CDFs) of the empirical distributions are used to allow for the regionalization of the distributions and enable simulations at ungauged locations. The non-exceedance probabilities of a CDF are referred to as *quantiles* in this work and their corresponding rainfall values are called *quantile values*.

4 Data selection

For the applications of rainfall estimates, like hydrological or hydraulic modeling, the correct representation of small rainfall values is not necessary as their contribution to decisively high discharge rates is rather small. Furthermore, tipping bucket gauges lead to wrong estimates, especially for low rainfall values (Habib et al., 2001). Relative estimation errors increase for decreasing rainfall rates (Nystuen et al., 1996; Ciach, 2003), and they only represent a small part of the total water volume, but the number of smaller rainfall values is rather high. To avoid the negative effect of this high number of inaccurate values and due to their minor importance for further applications, this study focuses on medium and high rainfall values.

Therefore, the quantile threshold (Q_{th}) for hourly (1 h) values is set to 0.95. This means that values smaller than the quantile value at $Q_{th} = 0.95$ are excluded. To investigate the total water volume represented by rainfall values above this quantile at point locations, the Lorenz curve (Lorenz, 1905) is used. We considered a water volume analysis for varying quantiles as important, to show that high quantiles not only represent the decisively higher rainfall intensities, but also a large proportion of the total water volume. Focusing on these quantiles during the model setup is likely to lead to a better model, as lower quantiles would disturb the model estimation due to measurement errors and the higher quantiles already represent a high percentage of the total water volume. The volume of the lower quantiles can then be modeled by simple and robust methods, as they do not require a very precise estimation due to their high inaccuracy and minor importance.

After arranging the n observations x_i in non-decreasing order, the Lorenz curve L_i can be calculated from a population (in our case, the rainfall values at a single gauge) with the following formula:

$$L(i) = \frac{\sum_{j=1}^{i} x_j}{\sum_{j=1}^{n} x_j}. \tag{1}$$

Table 1. The basic rainfall information of the study region for different aggregations (agg): P_0 is the probability of 0 mm rainfall, Q_{th} stands for the defined quantile thresholds or the threshold ranges and QV_{th} represents the corresponding quantile values (rainfall) for the defined Q_{th}.

Agg	P_0 (–)	Q_{th} (–)	QV_{th} (mm)
1 h	0.82–0.93	0.95	0.2–1.6
2 h	0.76–0.9	0.93	0.3–2.3
3 h	0.71–0.87	0.92	0.4–3.1
6 h	0.61–0.81	0.9	0.7–5.1
12 h	0.46–0.72	0.86	1.2–7.7
1 d	0.38–0.6	0.72	1.0–6.4
5 d	0.1–0.22	0.29	1.0–7.2
m	0.0–0.02	0.0–0.02	0

The hourly threshold quantile values (QV_{th}) range between 0.2 and 1.6 mm for $Q_{th} = 0.95$ depending on the location of the gauge (see Table 1). The Lorenz curve in Fig. 2a shows that the hourly values above $Q_{th} = 0.95$ represent between 70 and 95 % of the total water volume (100 % minus the cumulative share of the water volume).

Based on the hourly values (1 h) of the high-resolution data set, the aggregated rainfall values of different temporal resolutions are obtained: 2-hourly (2 h), 3-hourly (3 h), 6-hourly (6 h) and 12-hourly (12 h). Through the aggregation of the daily values (1 d) in the daily data set, 5-daily (5 d) and monthly (m) values are obtained. In order to exclude small values and still consider the values producing a high percentage of the water volume, the Q_{th} for the sub-daily resolutions are defined with the mean Lorenz curves in Fig. 2b. The mean hourly Lorenz curve yields 0.15 as the cumulative share of the water volume for $Q_{th} = 0.95$ (85 % of the total water volume is represented by larger values), which is also defined as the target share for the remaining sub-daily resolutions. This target share of 0.15 results in the following values of Q_{th} for the sub-daily resolutions: 0.93, 0.92, 0.9 and 0.86 (see Table 1). For aggregations greater than or equal to 1 day, the number of values is rather small and their estimation errors are lower due to an increasing accumulation time (Ciach, 2003; McMillan et al., 2012). Nevertheless, only the values above the highest quantile of 1 mm in the study region are used for the daily (1 d) and 5-daily (5 d) resolution (see Table 1), as the smaller values may still exhibit measurement errors.

For the estimation of the basic statistics in Table 1 and for the following calculations, the rain values of the investigated aggregations smaller than 0.1 mm are set to 0 mm. The reason is to achieve the homogenization of the data sets for different years and gauges, as the discretization ranges from 0.01 to 0.1 mm depending on the gauge.

5 Probability distributions of precipitation amounts in a spatial context

This section focuses on the spatial dependence of the precipitation amount distributions, as the applied interpolation technique of ordinary kriging (OK) is based on the assumption that the variable of interest (the CDF) is more likely to be dissimilar with increasing distances. For the purpose of describing the development of the distribution functions in space, the test statistic T of the two-sample Cramér–von Mises criterion is used (Anderson, 1962). It evaluates the similarity of two CDFs, in our case the similarity of the CDFs from the observations at two different point locations. The test statistic T is defined according to Anderson (1962) as

$$T = \frac{U}{NM(N+M)} - \frac{4MN-1}{6(M+N)}, \quad (2)$$

where

$$U = N \cdot \sum_{i=1}^{N} (r_i - i)^2 + M \cdot \sum_{j=1}^{M} (s_j - j)^2, \quad (3)$$

with N as the number of observations in the first sample and M as number of observations in the second sample. Both observations are joined together in one pooled data set, and the ranks are determined in ascending order of all observations in the pooled data set. The r_i values are the ranks of the N observations for the first sample in the pooled data set and s_j are the sorted ranks of the M observations for the second sample in the pooled data set. T can be interpreted as the mean difference in the CDF values (quantiles) of the observed rainfall intensities between the two data sets. So, if T increases for increasing distances, the CDFs are less similar for increasing distances.

For the calculations of T, only the rainfall values above the different Q_{th} (see Table 1) are used. The graphs in Fig. 3 show a decreasing similarity in the distribution functions with increasing distances over all temporal resolutions, as the values of T increase with increasing distances. Note that the average T values of the hourly (1 h) data in Fig. 3a are shown as the highest dashed line in Fig. 3b. The continuity of the whole distribution changes in space, and not only the continuity of the values in a single quantile. This shows the applicability of interpolation techniques like OK.

6 Precipitation amount models

In the following subsections, nonparametric and parametric models for precipitation amounts at single sites are introduced. Before estimating the nonparametric or parametric distributions at each observation gauge, the observations smaller than QV_{th} are censored from the sample of each gauge and QV_{th} is subtracted from the values above them to fit to the support of the theoretical distribution functions

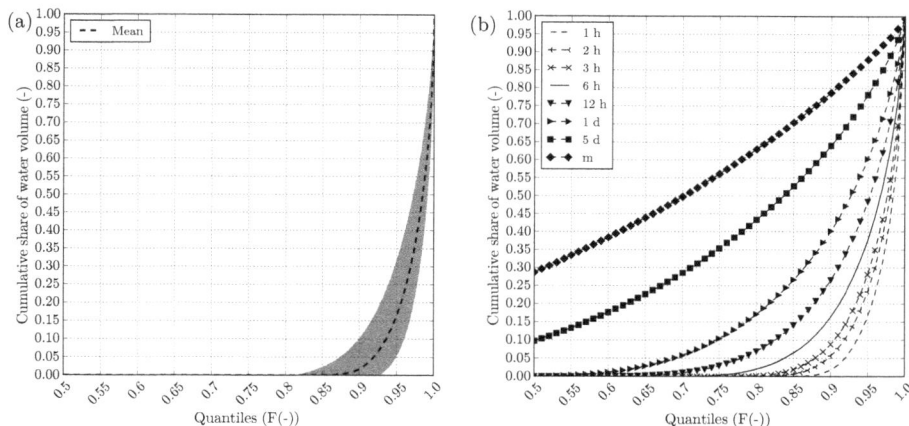

Figure 2. The **(a)** range of the Lorenz curves and the mean Lorenz curve for the hourly rainfall values of all the rainfall gauges inside the study region. The **(b)** mean Lorenz curves for different temporal resolutions.

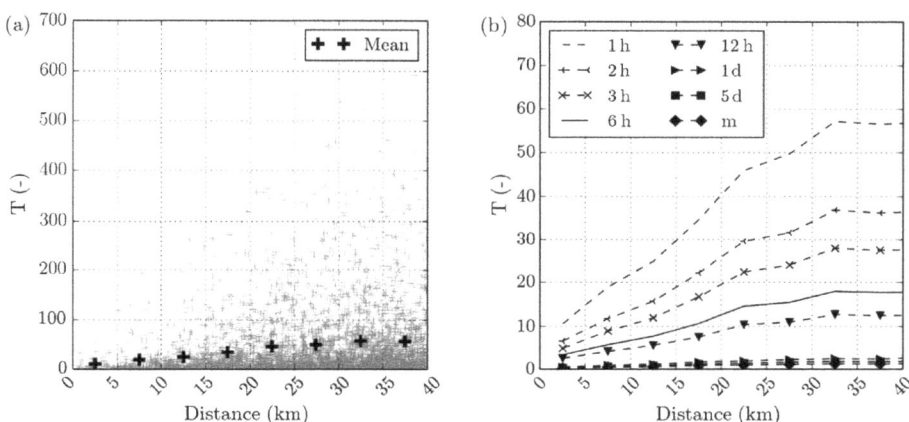

Figure 3. The T statistic over distance: **(a)** the results for the hourly distribution functions of all gauge pairs (the gray crosses) and their mean calculated for 5 km classes. **(b)** The mean values of the T statistic for different temporal resolutions (for more detail on the temporal resolutions of 1 d, 5 d and m, see Fig. S2).

$[0, \infty)$. QV_{th} varies from gauge to gauge for different temporal resolutions (see Table 1). After estimating the theoretical CDFs, the quantiles F are scaled with Q_{th}

$$F_{sc} = F \cdot (1 - Q_{th}) + Q_{th}, \qquad (4)$$

and QV_{th} is added to the quantile values. Only the monthly resolution is excluded from the whole scaling procedure, as all monthly rainfall values are used.

6.1 Nonparametric models

The nonparametric KDEs for the precipitation amount distributions were previously used and are described for the daily precipitation amounts in Rajagopalan et al. (1997) and Peel and Wilson (2008). Using this nonparametric method means that no theoretical distribution needs to be preassigned; only a kernel and its bandwidth need to be chosen. That is why they are assumed to be more flexible. A kernel in this context

is a function which is centered over each observation value and is itself a probability density function with a variance controlled by its bandwidth (Bowman and Azzalini, 1997). The probability density function (PDF) or KDE $f(x)$ of every data set is then constructed through a linear superposition of these kernels (Peel and Wilson, 2008), where n is the number of observed values, K is the kernel function, h is the bandwidth of the kernel, x are discrete kernel supporting points and x_i are observed rainfall values:

$$f(x) = \frac{1}{n} \sum_{i=1}^{n} K(x - x_i; h). \qquad (5)$$

The estimation of $f(x)$ is performed with an R (R Core Team, 2015) implementation of Wand (2015). However, since our nonparametric interpolation scheme is based on CDFs and not on PDFs, the CDF is needed. In order to obtain a CDF from the KDEs, an integration is required, which is done numerically with the composite trapezoidal rule (e.g.,

Atkinson, 1989). For numerical reasons, quantiles slightly greater than 1 are sometimes obtained, which are simply set to 1 so that they remain in the correct range.

To model the right-skewed precipitation amounts with their bounded support on $[0, \infty)$, either an asymmetric kernel like the Gamma kernel (Chen, 2000) or a symmetric kernel with a prior logarithmic transformation of the values (Rajagopalan et al., 1997) can be used to avoid boundary bias. A boundary bias occurs when kernels with infinite support are used for data with bounded support, as this would lead to a leakage of probability mass (Rajagopalan et al., 1997).

In this work, the symmetric Gaussian kernel with a prior transformation of data to logarithms is chosen, as this is an implicit adaptive kernel method with increasing bandwidths for increasing values and therefore alleviates the need to choose variable bandwidths with skewed data (Lall et al., 1996; Charpentier and Flachaire, 2014). The Gaussian kernel is chosen as it is straightforward and its application is facilitated through several software implementations (Sheather, 2004). The Gaussian kernel $K(t)$ is described in Eq. (6):

$$K(t) = \frac{1}{h\sqrt{2\pi}} \cdot \exp\left(\frac{-t^2}{2h^2}\right). \tag{6}$$

If the density of the logarithmically transformed observed values $y = \log(x)$ is f_Y and a Gaussian kernel is used for this density estimation, the density estimation f_X of the original values x according to Charpentier and Flachaire (2014) is

$$f_X(x) = f_Y(\log(x)) \frac{1}{x}. \tag{7}$$

Finally, the bandwidth h needs to be chosen, which is commonly indicated as the key step for KDEs (e.g., Bowman, 1984; Harrold et al., 2003; Sheather, 2004; Charpentier and Flachaire, 2014) as a poor bandwidth selection may result in a peakedness or an over-smoothing of the density estimation. Due to this great importance of the bandwidth selection, the performances of different selection methods are investigated.

1. The simplest and most widely used selection method is Silverman's rule of thumb (Silverman, 1986), which is defined as

$$h_{\text{opt, SRT}} = 0.9 \cdot \min\left(s; \frac{q_3 - q_1}{1.349}\right) n^{-1/5} \tag{8}$$

to obtain the optimal kernel bandwidth $h_{\text{opt, SRT}}$ with n sample values, where s is the standard deviation and $q_3 - q_1$ is the interquartile range. Silverman's rule of thumb (SRT) is deduced by minimizing an approximation of the mean integrated squared error between the estimated and the true densities, where the Gaussian distribution is referred to as the true distribution (Charpentier and Flachaire, 2014).

2. The second method is a plug-in approach developed by Sheather and Jones (1991), which is widely recommended due to its good performance (Jones et al., 1996;

Rajagopalan et al., 1997; Sheather, 2004). Instead of using a Gaussian reference distribution, it uses a prior non-parametric estimate in the approximation of the mean integrated square error and therefore requires a numerical calculation (Charpentier and Flachaire, 2014) to find the optimal bandwidth $h_{\text{opt, SJ}}$. This is performed with the R implementation of Wand (2015) within this work.

Instead of minimizing the mean integrated squared error, Bowman (1984) recommended minimizing the integrated squared error through a least squares cross-validation (LSCV), which is applied using the R package of Duong (2015). Another common cross-validation method is the maximum likelihood cross-validation (MLCV). Cross-validation methods tend to produce small bandwidths and therefore tend to produce a peakedness in the density (Rajagopalan et al., 1997; Sheather, 2004; Peel and Wilson, 2008), which we also observed in our applications. Due to this deficiency, neither cross-validation method is considered in what follows.

6.2 Parametric models

Within the parametric procedure, five different parametric distributions are used to model the precipitation amounts of all aggregations in this study. The most commonly used models are the exponential distribution and the two-parameter gamma distribution (Wilks and Wilby, 1999). The mixed-exponential distribution was recommended in Wilks and Wilby (1999) and was first used for daily precipitation amounts by Woolhiser and Pegram (1979). Another common and efficient distribution to model precipitation amounts, especially with daily temporal resolution, is the generalized Pareto distribution (Chen and Brissette, 2014; Li et al., 2012). In addition to these models, the Weibull distribution is used, which showed good performance for modeling monthly precipitation amounts in Baden-Württemberg (Beck, 2013). The CDF $F(x)$ and the PDF $f(x)$ of each parametric distribution used here are listed in the following.

1. For the exponential distribution with the parameter λ, these functions are

$$f(x; \lambda) = \lambda e^{-\lambda x}, \tag{9}$$

$$F(x; \lambda) = 1 - e^{-\lambda x}. \tag{10}$$

2. For the two-parameter gamma distribution, they are

$$f(x; \theta, k) = \frac{x^{k-1} e^{-\frac{x}{\theta}}}{\Gamma(k)\theta^k}, \tag{11}$$

$$F(x; \theta, k) = \frac{\gamma\left(k, \frac{x}{\theta}\right)}{\Gamma(k)}, \tag{12}$$

where Γ is the gamma function and γ is the incomplete gamma function.

3. For the two-parameter Weibull distribution, $F(x)$ and $f(x)$ are

$$f(x; \lambda, k) = \frac{k}{\lambda}\left(\frac{x}{\lambda}\right)^{(k-1)} e^{-(x/\lambda)^k}, \tag{13}$$

$$F(x; \lambda, k) = 1 - e^{-(x/\lambda)^k}. \tag{14}$$

4. The mixed-exponential distribution exhibits the following functions:

$$f(x; \lambda_1, \lambda_2, \alpha) = \alpha \lambda_1 e^{-\lambda_1 x} + (1 - \alpha)\lambda_2 e^{-\lambda_2 x}, \tag{15}$$

$$F(x; \lambda_1, \lambda_2, \alpha) = 1 - \alpha e^{-\lambda_1 x} - (1 - \alpha)e^{-\lambda_2 x}. \tag{16}$$

5. The generalized Pareto distribution exhibits the following PDF

$$\begin{aligned} f(x; k, \alpha) &= \alpha^{-1}(1 + kx/\alpha)^{-1-\frac{1}{k}}, &k \neq 0 \\ &= \alpha^{-1} e^{-x/\alpha}, &k = 0 \end{aligned} \tag{17}$$

and CDF

$$\begin{aligned} F(x; k, \alpha) &= 1 - (1 + kx/\alpha)^{-1/k}, &k \neq 0 \\ &= 1 - e^{-x/\alpha}, &k = 0. \end{aligned} \tag{18}$$

The parametric distributions with more than two parameters are not considered, as this would complicate the regionalization of the distributions due to the dependencies among the parameters. For the three-parameter mixed-exponential distribution, the parameter α is fixed for the whole study region (Wilks, 2008), transforming it into a two-parameter distribution.

In order to estimate the optimal parameter sets of the presented parametric distributions for each rainfall gauge and temporal resolution, the method of moments (MOM) and the maximum likelihood method (MLM) using a numerical maximization via a simplex algorithm are applied. The MLM is applied to all mentioned parametric distributions. In the special case of the mixed-exponential distribution, the parameter α is varied between 0.01 and 0.5 within the parameter estimation. For each value of α, the sum of the log-transformed likelihoods is calculated over all gauges with varying values of the remaining parameters, while the maximum sum defines the parameter set. To apply MOM, the mean $\overline{x}$ and standard deviation s_x of the sample values need to be calculated for the gamma and generalized Pareto (Hosking and Wallis, 1987) distribution. In order to use MOM for the Weibull distribution, the method described in Cohen (1965) is applied. For the estimation of the mixed-exponential distribution parameters, MOM is not applied due to its shortcomings as described in Rider (1961). MOM is also not applied to the one-parameter exponential distribution, as it would yield the same results as those from the MLM.

7 Nonparametric distributions in a spatial context

In order to establish the basis of the proposed regionalization procedure for nonparametric models and to get a more detailed idea of the spatial relationship of the distribution functions, the EDFs of the hourly and monthly rainfall intensities from the gauge at Stuttgart/Schnarrenberg and its five closest gauges are plotted in Fig. 4. It is therefore not of importance which EDF belongs to which gauge, but rather the relationship that the EDFs have with each other. The two graphs in Fig. 4 show that the order of the EDFs stays quite persistent over different quantiles for both aggregations, as the EDFs do not cross each other very often. In other words, if one gauge exhibits the highest rainfall values for a certain quantile, it also exhibits the highest rainfall values for the other quantiles and vice versa. The red and purple EDFs on the left graph illustrate this quite nicely.

A more global look at the spatial relation between different EDFs can be obtained with the Spearman's rank correlation ρ_{xy} of the quantile values of all gauges for different quantile pairs. As we want to investigate the persistence of EDFs for the whole study region, we are only interested in the ranks, or rather the order, of the different quantile values for differing quantiles, which can be identified by calculating ρ_{xy}. In our application the two input data sets for calculating ρ_{xy} represent the quantile values of two different pairs of quantiles over all gauges in the study region. These pairwise rank correlations of the quantile values of all gauge pairs are calculated starting from Q_{th} until 1 in 0.001 steps for sub-daily aggregations and in 0.005 steps for aggregations greater than or equal to 1 day. This procedure is repeated until the rank correlation of every quantile with every other quantile is obtained. Finally the mean values of the rank correlation belonging to each quantile are calculated (see the dotted gray lines in Fig. 5). The highest mean rank correlation is indicated with a red cross in this figure, which also defines the control quantile (Q_c) with the highest mean rank correlation. The rank correlations of Q_c with the remaining quantiles lead to the dashed lines in Fig. 5.

Figure 5 demonstrates that most of the rank correlations are greater than 0.85, indicating a persistence of quantile values over a large interval of quantiles as well as over the whole study region for hourly through monthly data. Lower correlations can be observed for the highest and lowest quantiles, which indicates nonpersistent behavior for these quantiles. This behavior is similar for all temporal resolutions. Therefore, the quantile values of Q_c can be used to set up the interpolation weights. Applying these weights to the remaining quantiles from Q_{th} until 1 should lead to good regionalization results for nonparametric CDFs.

In Table 2, the control quantiles Q_c with the highest mean correlations are summarized for all temporal resolutions. As the precipitation mechanisms are different in summer and winter in Baden-Württemberg, the rainfall data sets are also analyzed separately for summer (from May to August) and

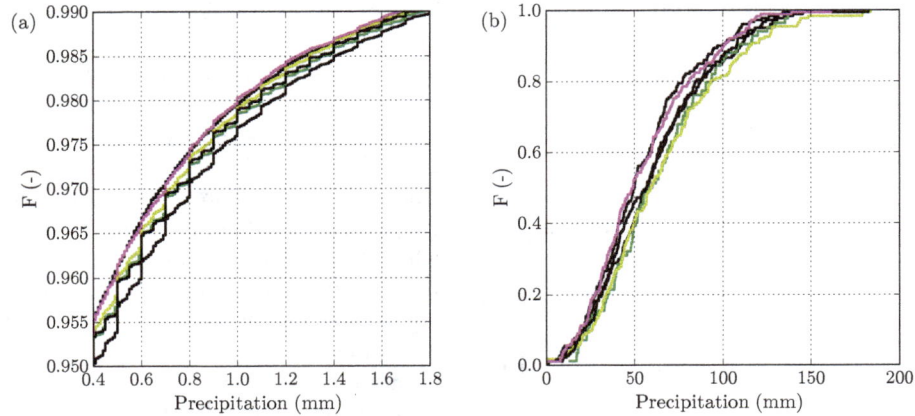

Figure 4. The EDFs of the hourly **(a)** and monthly **(b)** precipitation amounts for the gauge at Stuttgart/Schnarrenberg and its five closest gauges for a quantile interval. This shows that the order of the EDFs is quite persistent over a wide quantile range for low and high resolutions. Note: As the daily and hourly data sets are not the same, the colors in the two graphs do not correspond to the same gauges.

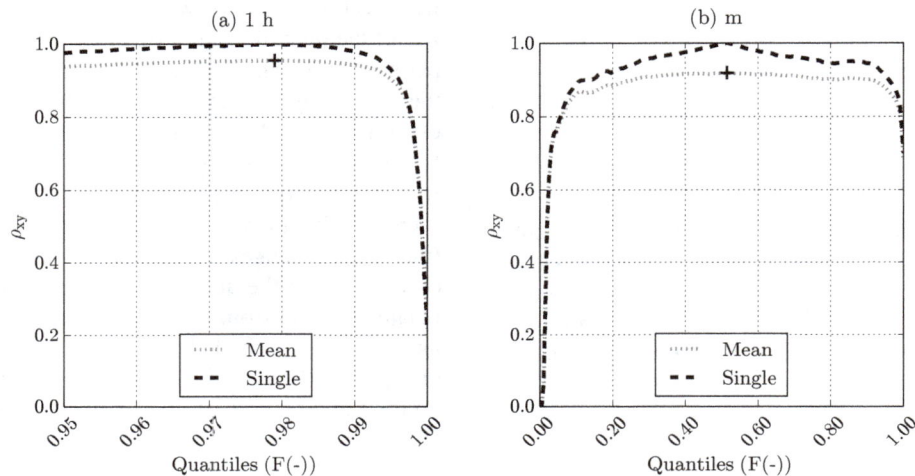

Figure 5. The mean rank correlations ρ_{xy} of the **(a)** hourly (1 h) and **(b)** monthly (m) quantile values for all gauge pairs of discrete quantiles in 0.001 (1 h) and 0.005 steps (m) ranging from Q_{th} to 1 (the gray dotted line). They are calculated to define the control quantile (Q_{c}), which exhibits the highest mean rank correlation ρ_{xy} (the red cross). The black dashed line shows the (single) rank correlations ρ_{xy} of the quantile values at Q_{c} (the red cross) with the quantile values of the remaining quantiles.

winter (from September to April). Q_{c} is mostly close to the center of the considered quantile ranges, which are also shown in Table 2. Nevertheless, the strong similarity in the winter and summer control quantiles Q_{c} is worth noting. The proposed procedure to interpolate nonparametric distribution functions using the same interpolation weights for different quantiles seems feasible, as the persistence of the order for the quantile values of the spatially distributed rain gauges is evident. Only the values of the very high and low quantiles show nonpersistent behavior. Therefore, quality measures that focus on the difference in these values will be introduced.

8   Regionalizing precipitation amount models

In the following, the regionalization of the point models in order to obtain the precipitation amount models at ungauged locations is described. The regionalization method OK is introduced first. Then, the approaches to regionalize the parametric and nonparametric distributions are explained.

As only a short overview of OK will be given, the interested reader is referred to the common geostatistical literature, like Kitanidis (1997), for further information. The empirical variogram $\gamma_e(h)$ is calculated using Eq. (19)

$$\gamma_e(h) = \frac{1}{2n(h)} \sum_{i=1}^{n(h)} (z(\boldsymbol{x}_i) - z(\boldsymbol{x}_i + h))^2, \tag{19}$$

Table 2. The control quantiles (Q_c) that exhibit the highest mean pairwise rank correlations with the other quantiles. They are shown for different temporal aggregations (agg) and separately for summer and winter. Additionally, the (center) quantile in the middle of the investigated quantile range is shown.

Agg	Season		Center quantile
---	Winter	Summer	
1 h	0.977	0.979	0.975
2 h	0.963	0.967	0.965
3 h	0.959	0.966	0.96
6 h	0.949	0.953	0.95
12 h	0.924	0.922	0.93
1 d	0.835	0.865	0.86
5 d	0.615	0.575	0.645
m	0.545	0.46	0.5

where $n(h)$ is the number of gauge pairs for distance h, x_i represents the position of gauge i and $z(x_i)$ is the variable value at gauge i. As the distances between the rainfall gauges are never a continuous set of distances, the h in Eq. (19) represents the different distance intervals. For the following applications, the width of the interval of h is 10 km and the maximum distance is 100 km. For the theoretical variogram $\gamma_t(h)$, one single model from the following four is chosen based on the least squares criterion. The s parameters represent the sills, and the r parameters represent the ranges of the variograms.

1. Gaussian model:

$$\gamma_t(h) = s_1\left(1 - e^{-\frac{h^2}{r_1^2}}\right).$$ (20)

2. Spherical model:

$$\gamma_t(h) = s_2\left(1.5\frac{h}{r_2} - 0.5\left(\frac{h}{r_2}\right)^3\right).$$ (21)

3. Exponential model:

$$\gamma_t(h) = s_3\left(1 - e^{-\frac{h}{r_3}}\right).$$ (22)

4. Matern model (Pardo-Iguzquiza and Chica-Olmo (2008) with K_v as the modified bessel function of the second kind):

$$\gamma_t(h) = s_4\left(1 - \frac{1}{2^{v-1}\Gamma(v)}\left(\frac{h}{r_4}\right)^v K_v\left(\frac{h}{r_4}\right)\right).$$ (23)

The next step within OK is solving the corresponding equation system to estimate an interpolated value at an unobserved location x_0:

$$\sum_{j=1}^{n}\phi_j\gamma_t(x_i - x_j) + \mu = \gamma_t(x_i - x_0)\ i = 1, \ldots, n,$$
$$\sum_{j=1}^{n}\phi_j = 1,$$ (24)

where n is the number of gauges included in the interpolation (10 within this work) and μ is the Lagrange multiplier.

As already outlined in the Introduction, either the parameters (Kleiber et al., 2012) or the moments (Haberlandt, 1998; Wilks, 2008) of the parametric distributions can be interpolated to regionalize the parametric models. Within this work, the moments are interpolated, when MOM is used for fitting the parametric distributions. If MLM is used, the parameters are interpolated. Since only the rainfall values above QV_{th} (see Table 1) are used, QV_{th} also needs to be interpolated within the parametric approach.

Kernel-smoothed distribution functions do not provide a parameter that can be interpolated; thus, a procedure other than that for the parametric distributions needs to be applied. When the spatial relation of the rainfall EDFs in Sect. 7 are analyzed, a persistent order of quantile values over a wide range of quantiles is observed. Therefore, the interpolation weights of the quantile values for the control quantile Q_c (see Table 2) can be applied to the remaining quantiles.

For all gauges, the quantile values QV_c of the control quantile Q_c are estimated with the inverse of the gauge-wise numerically integrated nonparametric CDF F_{np}:

$$QV_c = F_{np}^{-1}(Q_c).$$ (25)

With these QV_c at the observation points, the interpolation weights ϕ_j for the target locations are estimated with OK (see Eq. 24). Then, these weights are applied to the quantile values of the quantiles between Q_{th} and 1 in 0.0001 steps. Finally, the remaining quantile values are linearly interpolated to receive a continuous CDF for all target locations. In order to ensure a monotonically increasing CDF, only positive interpolation weights are allowed. This makes the use of OK problematic. It can only be used if the equation system (see Eq. 24) is solved with positive weights, which leads to additional constraints:

$$\phi_j \geq 0\ j = 1, \ldots, n.$$ (26)

Considering these additional constraints, the OK equation system is solved with a SCIPY implementation (Jones et al., 2001) of a FORTRAN algorithm by Lawson and Hanson (1987), which solves the Karush–Kuhn–Tucker conditions for the nonnegative least squares problem. In the following, this kriging procedure will be called positive kriging (PK). Another way to solve this extended optimization problem with an application of the Lagrange method is presented in Szidarovszky et al. (1987). The persistence of the quantile values described in Sect. 7 also implies the persistence of the quantiles. The interpolation of the quantiles for the discrete rainfall values would therefore also be an option. However, this would complicate the regionalization as not only the monotonicity needs to be preserved, but also the value range of the quantiles from 0 to 1.

9 Dependence of sub-daily on daily values

As the high-resolution rain gauge monitoring network in the study area is quite sparse and the corresponding time series are often incomplete, it would be useful to include more dense and complete secondary information in the interpolation of the sub-daily distributions. Therefore, the applicability of the daily values to improve their interpolation is investigated, as the daily monitoring network has a higher density. The simple disaggregation strategy (rescaled nearest neighbor) of Bárdossy and Pegram (2016) is applied to all days to obtain the distributions of the sub-daily resolutions at the locations of the daily gauges by allocating the sub-daily values from the closest high-resolution gauge to the daily target gauge. The procedure to incorporate the daily values in the interpolation of the sub-daily values should be the following.

1. Choose a daily target gauge and allocate the sub-daily rainfall values of the closest (concerning the horizontal distance) high-resolution gauge to it.

2. Aggregate the sub-daily values of the high-resolution gauge to the daily values $p_{\text{sub-daily}}(t)$ and calculate a scaling factor for every day t by additionally using the values of the daily target gauge $p_{\text{daily}}(t)$:

$$\text{sc}(t) = \frac{p_{\text{daily}}(t)}{p_{\text{sub-daily}}(t)}. \quad (27)$$

3. Multiply all of the sub-daily values of the nearest gauge by this scaling factor. The scaling factor changes from day to day and simply ensures that the daily sums of the disaggregated sub-daily values at the target gauge equal the daily values measured at the target.

4. Repeat steps 1 to 3 for all daily gauges.

5. Calculate the sub-daily statistic of interest from these scaled values at every daily gauge and incorporate them in the interpolation procedure.

The applicability of this procedure is tested with a cross-validation, which is described in Sect. S3. For the incorporation of the daily values within the regionalization of the parametric and nonparametric sub-daily distributions, a special regionalization technique is not needed. The rescaling method (NNS) is applied to all available daily gauges. If for a certain day no hourly values are available for the closest gauge, the next closest gauge is used for the rescaling of that day in order to increase the sub-daily sample size at the daily gauge. After obtaining the sub-daily values at the daily gauges, they are simply treated as additional control points for the regionalization.

10 Performance

This section is divided into three parts. In Sect. 10.1, the quality measures are introduced. In Sect. 10.2, the performance of the precipitation amount models for the pointwise estimations are compared for all temporal resolutions. The regionalization of the precipitation amount models is addressed in Sect. 10.3. The precipitation amount models are fitted and regionalized separately for winter (from September to April) and summer (from May to August), as the rain-producing weather processes are different in these two seasons.

10.1 Quality measures

The validation of the precipitation amount models at point locations and their regionalization is evaluated with two different quality measures. These quality measures need to be measures considering the CDF and not the PDF, as the interpolation of the nonparametric distributions only provides CDFs for ungauged locations.

The most common goodness-of-fit test to estimate the quality of fitted distributions is the Kolmogorov–Smirnov test. As distributions of precipitation amounts are positively skewed, most of the values are small or medium values, which leads to the highest gradient of the CDF for these values. Therefore, a greater difference in the corresponding CDF quantiles would be more likely and would govern the Kolmogorov–Smirnov test. However, these medium values are less important than the higher precipitation amounts for most of the precipitation model applications.

For this reason, the Cramér–von Mises criterion as a more integral measure and a Lorenz-curve-based measure, which allows for conclusions about the representation of the water volume, are used. The Cramér–von Mises criterion W^2 for single samples is (Stephens, 1974)

$$W^2 = \frac{1}{12n} \sum_{i=1}^{n} \left(\frac{2i-1}{2n} - F(x_i) \right)^2, \quad (28)$$

where $F(x_i)$ represents the theoretical distribution (nonparametric or parametric) of the observed values x_i in ascending order. For sub-monthly resolutions, the Cramér–von Mises criterion is slightly modified, as only quantiles above Q_{th} (see Table 1) are used:

$$W^2 = \frac{1}{12n} \sum_{i=1}^{n} \left(\left(\frac{2i-1}{2n} \cdot (1 - Q_{\text{th}}) + Q_{\text{th}} \right) - F(x_i) \right)^2. \quad (29)$$

As already mentioned in Sect. 7, a quality measure that describes the representation of high quantiles is needed. For Lorenz curves, high vertical differences are supposed to appear more frequently for high quantiles as the slope increases with increasing quantiles. Therefore, a measure respecting the vertical differences of the Lorenz curves is suitable. In Sect. 4, the estimation of the Lorenz curve with the observed rainfall values was described. However, the Lorenz curve $L(F(x))$ can also be estimated from the theoretical CDF $F(x)$, which is a preferable approach, as random rainfall values do not need to be generated from the CDF previous to

Table 3. The mean and median of the two quality measures W^2 and L_d for the 10 precipitation amount models over the study region for hourly values (1 h) in the winter season. The bold numbers indicate the lowest (best) value of the corresponding measure.

	W^2		L_d	
	Mean	Median	Mean	Median
P-Exp-MLM	0.009718	0.008104	0.2399	0.2004
P-Gamma-MLM	0.00263	0.002146	0.0752	0.04835
P-Mixed-Exp-MLM	0.0007967	0.0004331	0.02026	0.007648
P-Pareto-MLM	0.0006701	0.0003277	0.008036	**0.001959**
P-Weibull-MLM	0.001578	0.0012	0.03891	0.02249
P-Gamma-MOM	0.03089	0.01897	0.1656	0.04182
P-Pareto-MOM	0.001074	0.0005668	**0.004482**	0.002213
P-Weibull-MOM	0.01418	0.00827	0.08677	0.04182
NP-SRT	0.0003752	0.0001995	0.01815	0.01448
NP-SJ	**0.0003485**	**0.0001954**	0.01492	0.01156

the Lorenz curve estimation:

$$L(F(x)) = \frac{\int_0^F x(F)\,\mathrm{d}F}{\int_0^1 x(F)\,\mathrm{d}F}, \tag{30}$$

where $x(F)$ is the gauge-wise quantile function (the inverse of the CDF). The integrals of the quantile functions are estimated numerically, because the nonparametrically estimated distribution functions are not analytically invertible. The Lorenz curve criterion L_d used here is the squared difference of the observed $L(F_n(x))$ and the modeled Lorenz curve $L(F(x))$:

$$L_d = \sum_{i=1}^{n} (L(F_n(x)) - L(F(x)))^2. \tag{31}$$

The differences in the Lorenz curves are only estimated for values greater than $\mathrm{QV_{th}}$ (see Table 1). Within the validation of the regionalization, only the values above the highest $\mathrm{QV_{th}}$ among the observed and regionalized values for each gauge are evaluated, as they may differ for the different techniques.

10.2 Point models

To determine an overall performance ranking for the remaining models, the arithmetic mean and the median over the number of gauges for both measures of quality (the Cramér–von Mises criterion W^2 and the Lorenz curve criterion L_d) are first calculated for each precipitation amount model. This leads to four different measures, which are shown for hourly values in the winter season in Table 3. Note that the mean values reflect the robustness and the median values represent a good average performance of one precipitation model for the whole study region.

To combine the four statistics (the mean and median of W^2 and L_d, respectively) into one single performance measure, every value in Table 3 is then divided by the smallest (best)

value (the bold numbers) of its corresponding quality measures, indicating the relative performance with respect to the best model. This leads to one number for each statistic and precipitation model starting from 1 for the best-performing model for each statistic. The bigger this number, the worse its relative performance. These four numbers are then added together, which results in a single number for each precipitation amount model to define the performance ranking for each temporal resolution. A ranking number of 4 is the lowest possible number and implies that the related model shows the best performance for all four quality measures. In Table 4, the ranking numbers for all temporal resolutions and both seasons are shown.

With the ranking numbers, the best-performing precipitation amount model is estimated for each season and temporal resolution. Among the nonparametric methods (NP), Silverman's rule of thumb (SRT) and the plug-in approach of Sheather and Jones (1991) (SJ) show very similar results. The generalized Pareto distribution with an MLM parameter estimation (Pareto-MLM) exhibits the best performance among the (P) parametric models for the hourly resolution. The mixed-exponential distribution with an MLM parameter estimation (Mixed-Exp-MLM) leads to the best results for the remaining sub-daily and daily resolutions. For temporal resolutions greater than 1 d the Weibull distribution with a MOM parameter estimation (Weibull-MOM) leads to the best results, except for the daily resolution in the winter season, where the Pareto-MOM combination is better. The best performance of the Weibull distribution for the monthly values coincides with the results of Beck (2013) for the same study region.

The performance ranking of the different methods is quite similar in winter and summer. The nonparametric methods always lead to better performances concerning the Cramér–von Mises criterion W^2. The parametric estimations lead to better results regarding the Lorenz curve criterion L_d (for details, see Tables S2 and S3). Figure 6 may provide an explanation for the differences in performance regarding these two quality measures. The graphs show the CDFs and Lorenz curves for the hourly (1 h) and 12-hourly (12 h) resolution for a chosen gauge. For the hourly resolution, the nonparametric SRT method leads to better results for both measures. An equally good performance regarding the W^2 for the parametric and nonparametric method can be observed for the 12-hourly resolution. However, the nonparametric method performs worse regarding the L_d measure, as it overestimates the water volume represented by the higher quantiles. The reason for this can already be observed in the CDF, where the nonparametric method systematically overestimates the values of the high quantiles. The parametric method can lead to overestimations and underestimations. This influences the W^2 criterion in the same way as a constant overestimation (see the squared differences in Eq. 28), but it seems to lead to better results regarding the L_d criterion.

Table 4. The performance ranking numbers of the precipitation amount models for the pointwise estimations. The underlined numbers indicate the best parametric (P) and nonparametric (NP) models. The bold numbers indicate the best overall model.

				Winter season				
	1 h	2 h	3 h	6 h	12 h	1 d	5 d	m
P-Exp-MLM	225.18	145.85	129.44	69.16	37.67	41.63	14.60	742.32
P-Gamma-MLM	59.99	30.67	22.79	10.43	6.98	11.51	9.03	13.23
P-Mixed-Exp-MLM	12.93	**5.72**	**5.38**	**5.09**	**4.79**	**5.15**	15.65	742.67
P-Pareto-MLM	**6.39**	6.12	7.29	7.09	6.38	6.40	6.19	978.92
P-Weibull-MLM	30.83	14.01	9.74	5.75	4.93	7.51	6.45	24.33
P-Gamma-MOM	265.89	138.15	94.77	41.47	20.24	23.48	6.64	5.50
P-Pareto-MOM	8.11	8.8	10.23	9.25	7.96	8.08	**5.71**	51.16
P-Weibull-MOM	123.72	61.72	44.64	21.28	12.37	13.71	5.82	**5.11**
NP-SRT	13.54	22.06	35.43	42.16	36.22	31.21	17.33	12.40
NP-SJ	11.23	22.12	33.82	43.21	38.15	29.15	16.77	17.35
				Summer season				
	1 h	2 h	3 h	6 h	12 h	1 d	5 d	m
P-Exp-MLM	245.24	233.50	188.82	70.63	26.14	25.83	21.67	850.52
P-Gamma-MLM	51.51	40.53	30.14	12.82	7.43	7.98	10.00	7.22
P-Mixed-Exp-MLM	10.37	**5.93**	**5.31**	**4.71**	**4.77**	**4.79**	23.10	850.52
P-Pareto-MLM	7.58	13.44	12.79	7.13	5.55	5.58	6.67	709.69
P-Weibull-MLM	21.08	13.92	10.42	6.63	5.48	6.02	6.61	25.62
P-Gamma-MOM	289.15	145.27	87.73	33.09	15.51	13.44	6.45	7.27
P-Pareto-MOM	16.48	14.46	11.55	7.28	5.99	6.08	5.82	47.81
P-Weibull-MOM	98.40	51.03	35.23	16.38	9.91	8.48	**5.46**	**5.05**
NP-SRT	**6.15**	19.05	31.54	37.41	36.15	30.76	19.00	9.27
NP-SJ	6.91	22.09	36.11	46.41	41.17	33.14	17.64	12.32

The parameter estimation through MOM in combination with the Weibull distribution performs better for the higher aggregations, which exhibit more symmetric distributions. For the daily and sub-daily aggregations, the MLM parameter estimation in combination with the mixed-exponential distribution mostly leads to the best results.

The overall performance is best with the mixed-exponential distribution for temporal resolutions between 2 hours (2 h) and 1 day (1 d) in both seasons. For the hourly distribution (1 h), the nonparametric models show the best overall performance in the summer season and the third-best performance after the generalized Pareto (Pareto-MLM and Pareto-MOM) distribution in the winter season. For the monthly resolution (m), the Weibull distribution exhibits the best overall performance in both seasons. For the 5-daily resolution, the MOM estimation provides the best result in winter (Pareto-MOM) and summer (Weibull-MOM).

10.3 Regionalization

In order to estimate the quality of the regionalized precipitation amount models, a 2-fold cross-validation (split sampling) is used. Two equally sized samples of observation points are randomly generated (Fig. 7). The simplest regionalization method is using the estimates of the nearest neighbor (NN) in the calibration set, which are therefore used as benchmarks for the quality of the regionalization procedure. Additionally, the daily rescaled nearest neighbors (NNS) are used as a benchmark. In this case, all daily gauges are used for the rescaling except for the daily observations at the locations of the respective validation sample.

Following the results of the pointwise estimation in the previous section, only the Weibull-MOM and the Mixed-Exp-MLM models among the parametric models are investigated for the regionalization, as they show good performance for differing aggregations. They are both investigated for all aggregations to test the difference for interpolated moments or parameters, except for the monthly aggregation, for which only the Weibull distribution is investigated. In order to regionalize the Weibull-MOM model, the mean and standard deviation are spatially interpolated. For the regionalization of the Mixed-Exp-MLM model, the parameters λ_1 and λ_2 are interpolated, while its parameter α is kept constant for the whole study region.

As the two nonparametric approaches SRT and SJ show very similar results during the pointwise estimation, only

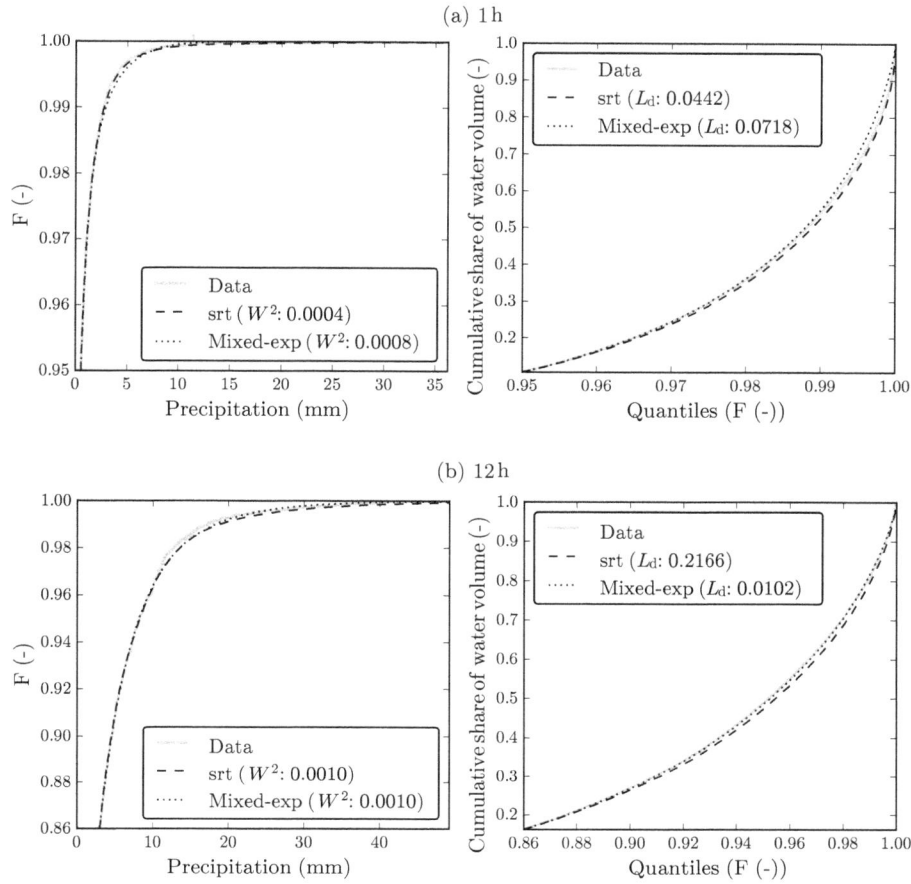

Figure 6. Empirical (data), nonparametric (SRT) and parametric (Mixed-Exp) CDF and Lorenz curve examples for the hourly (1 h, **a**) and 12-hourly (12 h, **b**) resolution of a chosen gauge. The values of the two quality measures L_d and W^2 are also indicated.

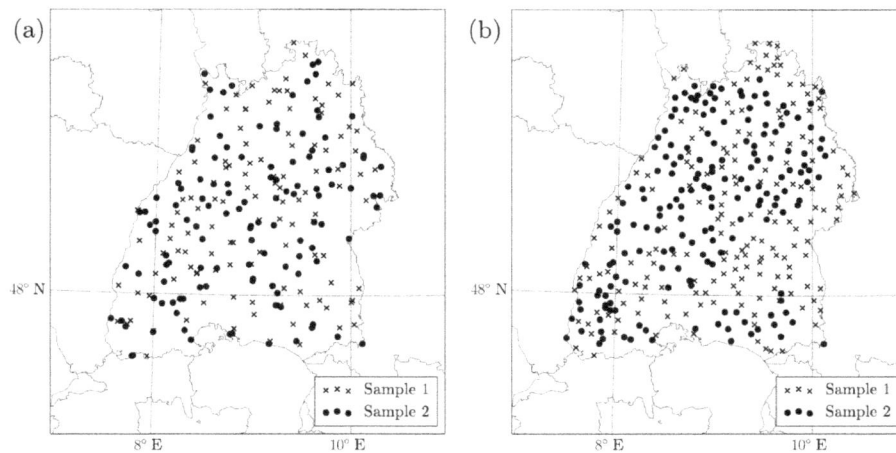

Figure 7. The locations of the two 2-fold cross-validation samples for the sub-daily (**a**) and daily gauges (**b**).

the SRT approach is interpolated. For the regionalization of the nonparametric model, the QV_c values (see Table 2 and Eq. 25) are used to estimate the interpolation weights, which are further applied to the remaining quantiles. Following the conclusions in Sect. 9, the daily gauges can be used to set up distribution functions for the sub-daily values with a scaled nearest neighbor approach (NNS).

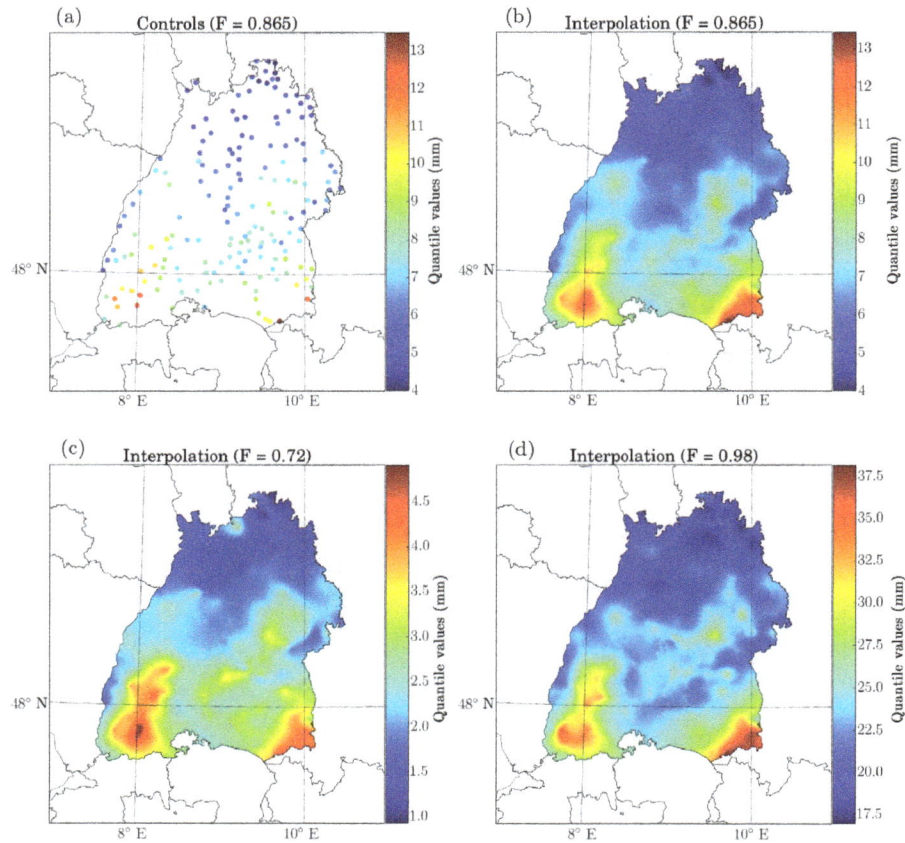

Figure 8. Illustrations for the kriging procedure of the nonparametric distributions with the daily values (1 d) in the summer season using calibration sample 1 (see Fig. 7). The **(a)** nonparametric QV_c of $Q_c = 0.865$ at the gauges, which then lead to the interpolated values **(b)** using the interpolation weights ϕ_j resulting from PK. The same interpolation weights ϕ_j are used for the remaining quantiles, for which example results are shown for the quantiles **(c)** F = 0.72 and **(d)** F = 0.98. An exponential variogram with a range of 41 km and a sill of 2.2 mm^2 is used.

10.3.1   Variogram estimation

The first step during the regionalization procedure is the estimation of the theoretical variograms. The interpolation variables of the three precipitation amount models, for which theoretical variograms need to be estimated for the two seasons and eight temporal resolutions, are as follows.

1. P-Mixed-Exp-MLM: λ_1, λ_1.

2. P-Weibull-MOM: mean, standard deviation.

3. NP-SRT: QV_c values (see Table 2 and Eq. 25).

During the estimation of the parameters of the Weibull distribution with MOM, QV_{th} is subtracted from the rainfall values prior to the estimation of the mean and the standard deviation. As the mean of these values shows lower spatial dependencies than the mean of the censored values without subtraction, QV_{th} is added to the mean values of the parameter estimation before the regionalization. After the regionalization, they are subtracted again to determine the parameters

of the Weibull distribution. The variogram models are also fitted to QV_{th}, as the corresponding values serve as starting points for the parametric models at the ungauged locations. Figures S4 to S7 in the Supplement show example theoretical variograms for the different parameters for temporal resolutions of 1 and 12 h for the winter and summer seasons of calibration sample 2.

It is difficult to compare the spatial persistence of T (see Fig. 3) with the spatial persistence of the different distribution parameters (see Figs. S4 to S7), as T considers the whole distribution function and the distribution parameters only describe the properties of the distribution. However, the range of T was about 35 km, which can also be observed for some of the parameters, especially the mean of P-Weibull-MOM, the QV_c of NP-SRT and QV_{th}.

10.3.2   Precipitation amount models

The regionalization of the precipitation amount models is evaluated with the same quality measures as the pointwise es-

Table 5. The performance ranking numbers for the 2-fold cross-validation of the regionalized precipitation amount models in the winter season. The underlined numbers indicate the best parametric (P) and nonparametric (NP) models. The bold numbers indicate the best overall model for each validation sample and temporal resolution.

	Calibration sample 1							
	1 h	2 h	3 h	6 h	12 h	1 d	5 d	m
OK-MOM	10.08	9.85	9.62	6.95	6.36	**4.56**	**4.37**	**4.14**
OK-MLM	7.04	7.22	7.11	10.52	6.70	7.60	6.35	–
OK-MOM DAILY	7.45	5.97	5.72	4.21	5.32	–	–	–
OK-MLM DAILY	4.46	**4.05**	**4.02**	**4.07**	**4.39**	–	–	–
PK-NP	7.18	8.21	8.94	8.98	9.21	5.56	7.24	6.94
PK-NP DAILY	**4.09**	5.74	6.00	5.68	10.09	–	–	–
NNS-MOM	7.59	6.76	6.73	5.79	7.01	–	–	–
NN-MOM	13.78	13.27	13.68	10.45	9.85	6.05	6.09	6.48
NNS-MLM	6.48	5.88	6.44	5.61	5.69	–	–	–
NN-MLM	10.04	9.81	10.39	10.19	10.51	5.41	7.40	288.07
NNS-NP	5.82	7.09	7.75	7.27	11.53	–	–	–
NN-NP	10.65	12.22	13.19	13.10	13.81	7.65	9.03	9.76
	Calibration sample 2							
	1 h	2 h	3 h	6 h	12 h	1 d	5 d	m
OK-MOM	9.82	9.81	9.41	7.46	6.59	**4.00**	**4.69**	**4.14**
OK-MLM	5.19	6.19	6.87	10.29	6.73	5.45	6.85	–
OK-MOM DAILY	5.90	5.83	6.39	**4.58**	6.26	–	–	–
OK-MLM DAILY	4.33	**4.16**	**4.39**	5.62	**4.38**	–	–	–
PK-NP	5.67	8.37	10.86	11.70	9.54	6.37	9.73	7.59
PK-NP DAILY	**4.14**	6.00	7.49	8.49	11.15	–	–	–
NNS-MOM	6.24	7.06	7.09	5.51	7.42	–	–	–
NN-MOM	11.37	13.40	12.31	11.44	9.17	5.10	5.91	5.23
NNS-MLM	4.94	5.08	4.90	4.67	5.62	–	–	–
NN-MLM	7.25	9.25	9.52	9.78	8.47	4.82	6.90	283.64
NNS-NP	4.80	6.96	8.52	8.52	12.37	–	–	–
NN-NP	8.36	11.34	12.90	14.37	11.81	7.35	11.12	8.96

timation, the Cramér–von Mises criterion W^2 and the Lorenz curve criterion L_d. The investigated interpolation approaches for the parametric distributions are listed in the following.

1. OK-MOM: OK of the Weibull distribution fitted with MOM.

2. OK-MLM: OK of the mixed-exponential distribution fitted with MLM.

3. OK-MOM Daily: OK of the Weibull distribution including the scaled NNS values of the daily gauges (only for the sub-daily aggregations).

4. OK-MLM Daily: OK of the mixed-exponential distribution including the scaled NNS values of the daily gauges (only for the sub-daily aggregations).

The interpolation approaches for the nonparametric models are as follows.

1. PK-NP: PK of the nonparametric models, which are estimated using SRT.

2. PK-NP Daily: PK of the nonparametric models including the scaled NNS values of the daily gauges (only for the sub-daily aggregations).

In Fig. 8, parts of the interpolation procedure for PK-NP are shown for the daily aggregation; the nonparametric QV_c at the calibration gauges and three interpolation fields are shown.

In Tables 5 and 6, the performance ranking numbers of the regionalized precipitation amount models are summarized for the winter season and for the summer season. The differences between the two cross-validation samples are quite small, so the performances result not from the positioning of the gauges in the samples, but from the interpolation approaches. Among the parametric methods, the MOM approaches mostly perform better than the MLM approaches

Table 6. The performance ranking numbers for the 2-fold cross-validation of the regionalized precipitation amount models in the summer season. The underlined numbers indicate the best parametric (P) and nonparametric (NP) models. The bold numbers indicate the best overall model for each validation sample and temporal resolution.

	1 h	2 h	3 h	6 h	12 h	1 d	5 d	m
				Calibration sample 1				
OK-MOM	34.02	13.42	10.23	7.15	**4.73**	**4.22**	**4.00**	**4.14**
OK-MLM	10.96	7.15	7.00	17.75	9.54	4.49	11.24	–
OK-MOM DAILY	22.16	13.63	10.57	<u>5.74</u>	9.48	–	–	–
OK-MLM DAILY	<u>10.46</u>	**4.19**	**4.22**	7.10	13.18	–	–	–
PK-NP	5.37	10.40	12.16	11.70	<u>9.71</u>	<u>9.83</u>	<u>11.34</u>	5.89
PK-NP DAILY	<u>4.30</u>	<u>8.27</u>	<u>9.97</u>	<u>10.42</u>	18.88	–	–	–
NNS-MOM	20.70	12.94	10.29	6.39	10.29	–	–	–
NN-MOM	30.16	19.43	16.53	11.26	7.72	6.18	<u>6.41</u>	5.55
NNS-MLM	10.37	<u>4.36</u>	<u>4.48</u>	**4.41**	7.51	–	–	–
NN-MLM	11.29	11.69	11.80	9.98	<u>7.35</u>	<u>5.19</u>	11.57	269.85
NNS-NP	**4.10**	8.79	10.26	10.72	20.24	–	–	–
NN-NP	6.26	14.98	16.41	15.95	12.41	11.59	12.70	7.84
				Calibration sample 2				
	1 h	2 h	3 h	6 h	12 h	1 d	5 d	m
OK-MOM	29.60	9.95	8.78	6.61	**4.11**	**4.05**	**4.05**	**4.10**
OK-MLM	6.42	5.66	5.99	83.02	7.01	5.56	24.15	–
OK-MOM DAILY	24.89	11.54	9.23	<u>6.02</u>	7.11	–	–	–
OK-MLM DAILY	<u>4.58</u>	**4.00**	**4.00**	61.46	5.66	–	–	–
PK-NP	5.67	<u>6.82</u>	<u>8.11</u>	<u>8.53</u>	<u>6.62</u>	<u>9.63</u>	<u>9.79</u>	6.99
PK-NP DAILY	**4.27**	7.01	8.30	9.75	13.60	–	–	–
NNS-MOM	24.66	12.74	10.38	6.98	8.54	–	–	–
NN-MOM	27.71	14.10	11.90	9.10	5.81	5.81	<u>4.82</u>	4.90
NNS-MLM	5.53	<u>5.09</u>	<u>4.91</u>	**4.43**	6.30	–	–	–
NN-MLM	8.90	8.15	7.94	7.63	<u>5.63</u>	<u>5.25</u>	8.69	261.23
NNS-NP	<u>5.41</u>	7.80	9.35	10.34	14.38	–	–	–
NN-NP	9.18	10.34	11.78	9.90	7.90	10.83	10.41	8.03

for aggregations greater than or equal to 1 day during the winter season. In the summer season, the MOM approaches perform mostly worse than the MLM approaches for aggregations smaller than 6 h and vice versa for higher aggregations. Interpolating moments, therefore, seems to be more robust than interpolating parameters of the distributions as the performance ranking changed in favor of the MOM approaches compared to the pointwise results (see Table 4). Only for the more strongly skewed distributions of the smaller aggregations does the MLM approach still outperform the MOM approach.

Comparing the nonparametric interpolation approaches with the parametric interpolation approaches shows that the nonparametric approach performs best for hourly (1 h) values for both calibration samples in both seasons. This is in line with the pointwise estimations, for which the nonparametric approaches also produced very good results for the hourly resolution in both seasons.

It is obvious that using the scaled values of the daily gauges is very beneficial, as the approaches incorporating these values almost always include the best-performing method, except for the 12 h aggregation in the summer season.

As a benchmark, the interpolation results are also shown for the parametric and nonparametric estimates of the nearest neighbors (NN) and additionally using scaled daily gauges for the sub-daily aggregations (NNS). Among the benchmark methods, the NNS approaches perform better than the simpler NN approaches for the sub-daily aggregations, except for the 12-hourly (12 h) resolution in summer. Since the best interpolation approach almost always, with only three exceptions, performs better than the best nearest neighbor approach, the regionalization of the distributions seems to be worthwhile.

11 Conclusions

Comparing different modeling schemes for the precipitation amounts at point locations (see Table 4) over different temporal resolutions has revealed several findings. The nonparametric estimates only perform better for the hourly resolution in the summer season. They have problems, especially in reproducing the volume correctly, as they seem to have difficulties with high quantiles. The causes for this deficiency could be the numeric interpolation or the small number of rainfall values at high quantiles. For temporal resolutions between 2 h and 1 month, the parametric distributions outperform the nonparametric distributions for both seasons. Among the parametric methods, the MLM parameter estimation (Mixed-Exp-MLM and Pareto-MLM) performs better for the sub-daily and daily aggregations, whereas the MOM parameter estimation (Weibull-MOM and Pareto-MOM) has the advantage for higher aggregations.

The regionalization of the precipitation amount models showed (see Tables 5 and 6) that the proposed interpolation scheme for the nonparametric distributions is useful as it does not worsen its performance ranking compared to the estimation at point locations. Among the parametric methods, the interpolation of moments turned out to be more robust than the interpolation of parameters. The proposed regionalization scheme for nonparametric models could also be tested in different research fields whenever nonparametric distributions may provide good representations of pointwise models and the order of the quantiles is persistent over spatially distributed locations. Especially for applications in which multimodal distributions are common, this interpolation scheme may be of great interest because kernel density estimates, in contrast to parametric models, can easily model multimodal distributions.

As auxiliary variables, the use of daily gauges for sub-daily resolutions is very beneficial, as was suggested by our data analysis in Sect. S3 and is also proven by the evaluation of the regionalization.

In general, the regionalization of the distributions seems to be worthwhile as it nearly always performs better than the nearest neighbor (horizontal distance) approaches, which would be the simplest estimate. As lower rainfall values were excluded from this study due to their minor importance and measurement errors, the results are not directly comparable to those of most of the other publications within this research field.

The difficulty for nonparametric distributions in representing water volumes may be reduced by using the Epanechnikov kernel with finite support as proposed by Rajagopalan

et al. (1997). However, the use of an Epanechnikov kernel instead of a Gaussian kernel reduces the ability to model precipitation beyond the range of historical data. Additionally, ways of incorporating elevation within the regionalization of the nonparametric distributions need to be tested. Mamalakis et al. (2017) used kriged two-component parametric distributions (a generalized Pareto distribution for higher, and an exponential distribution for lower daily precipitation amounts) for the bias correction and downscaling of rainfall resulting from a climate model. They applied a parameter estimation through probability-weighted moments, which could also be compared to the presented estimation approaches for the regionalization of distributions on varying temporal resolutions. Finally, the nonparametric interpolation approach could also be applied to parametric or empirical distributions and should be tested for various study regions.

Competing interests. The authors declare that they have no conflict of interest.

Acknowledgements. The work of many people developing different libraries in the PYTHON programming language (Jones et al., 2001; van der Walt et al., 2011; McKinney, 2010) made the work on the presented research much more convenient and helped a lot to illustrate (Hunter, 2007) its findings. The authors appreciate the valuable comments on an earlier draft of the manuscript by Geoffrey Pegram (University of KwaZulu-Natal, Durban, South Africa). Three anonymous reviewers helped a lot to improve the quality of the manuscript. The authors also want to thank the German Federal Ministry of Education and Research (BMBF) for funding this work through the funding measure *Smart and Multifunctional Infrastructural Systems for Sustainable Water Supply, Sanitation and Stormwater Management (INIS)*. Furthermore, the authors would like to thank the Environmental Agency of Baden-Württemberg (LUBW) and the German Meteorological Service (DWD) for the provision of rainfall data.

Edited by: D. Koutsoyiannis

References

Acreman, M. C.: A simple stochastic model of hourly rainfall for Farnborough, England, Hydrolog. Sci. J., 35, 119–148, doi:10.1080/02626669009492414, 1990.

Anderson, T. W.: On the Distribution of the Two-Sample Cramervon Mises Criterion, Ann. Math. Statist., 33, 1148–1159, doi:10.1214/aoms/1177704477, 1962.

Arns, A., Wahl, T., Dangendorf, S., Mudersbach, C., and Jensen, J.: Ermittlung regionalisierter Extremwasserstände für die Schleswig-Holsteinische Nordseeküste, Hydrologie und Wasserbewirtschaftung, HW57, doi:10.5675/HyWa_2013,6_1, http://www.hywa-online.de/ermittlung-regionalisierter-extremwasserstaende-fuer-die-schleswig-holsteinische-nordseekueste/ (last access: 28 April 2017), 2013.

Atkinson, K. E.: An Introduction to Numerical Analysis (2nd ed.), John Wiley & Sons, New York, 1989.

Bárdossy, A.: Generating precipitation time series using simulated annealing, Water Resour. Res., 34, 1737–1744, doi:10.1029/98WR00981, 1998.

Bárdossy, A. and Pegram, G. G. S.: Space-time conditional disaggregation of precipitation at high resolution via simulation, Water Resour. Res., 52, 920–937, doi:10.1002/2015WR018037, 2016.

Bárdossy, A. and Plate, E. J.: Space-time model for daily rainfall using atmospheric circulation patterns, Water Resour. Res., 28, 1247–1259, doi:10.1029/91WR02589, 1992.

Bárdossy, A., Giese, H., Haller, B., and Ruf, J.: Erzeugung synthetischer Niederschlagsreihen in hoher zeitlicher Auflösung für Baden-Württemberg, Wasserwirtschaft, 90, 548–553, 2000.

Beck, F.: Generation of spatially correlated synthetic rainfall time series in high temporal resolution: a data driven approach, PhD thesis, Universität Stuttgart, available at : http://elib.uni-stuttgart. de/opus/volltexte/2013/8216 (last access: 28 April 2017), 2013.

Bowman, A. and Azzalini, A.: Applied Smoothing Techniques for Data Analysis: The Kernel Approach with S-Plus Illustrations: The Kernel Approach with S-Plus Illustrations, OUP Oxford, 1997.

Bowman, A. W.: An Alternative Method of Cross-Validation for the Smoothing of Density Estimates, Biometrika, 71, 353–360, doi:10.1093/biomet/71.2.353, 1984.

Buishand, T. A. and Brandsma, T.: Multisite simulation of daily precipitation and temperature in the Rhine Basin by nearest-neighbor resampling, Water Resour. Res., 37, 2761–2776, doi:10.1029/2001WR000291, 2001.

Burton, A., Kilsby, C., Fowler, H., Cowpertwait, P., and O'Connell, P.: RainSim: A spatial-temporal stochastic rainfall modelling system, Environ. Modell. Softw,, 23, 1356–1369, doi:10.1016/j.envsoft.2008.04.003, 2008.

Charpentier, A. and Flachaire, E.: Log-Transform Kernel Density Estimation of Income Distribution, doi:10.2139/ssrn.2514882, 2014.

Chen, J. and Brissette, F.: Stochastic generation of daily precipitation amounts: review and evaluation of different models, Clim. Res., 59, 189–206, doi:10.3354/cr01214, 2014.

Chen, S.: Probability Density Function Estimation Using Gamma Kernels, Ann. I. Stat. Math., 52, 471–480, doi:10.1023/A:1004165218295, 2000.

Ciach, G.: Local random errors in tipping-bucket rain gauge measurements, J. Atmos. Ocean. Tech., 20, 752–759, doi:10.1175/1520-0426(2003)20<752:LREITB>2.0.CO;2, 2003.

Cohen, A. C.: Maximum Likelihood Estimation in the Weibull Distribution Based on Complete and on Censored Samples, Technometrics, 7, 579–588, doi:10.1080/00401706.1965.10490300, 1965.

Duong, T.: ks: Kernel Smoothing, available at: http://CRAN. R-project.org/package=ks (last access: 28 April 2017), r package version 1.10.0, 2015.

DWD (Deutscher Wetterdienst): WebWerdis (Web-based Weather Request and Distribution System), available at: http://www. dwd.de/DE/leistungen/webwerdis/webwerdis.html, last access: 28 April 2017a.

DWD (Deutscher Wetterdienst): CDC (Climate Data Center), available at: http://www.dwd.de/DE/klimaumwelt/cdc/cdc_ node.html, last access: 28 April 2017b.

Giraldo, R., Delicado, P., and Mateu, J.: Ordinary kriging for function-valued spatial data, Environ. Ecol. Stat., 18, 411–426, doi:10.1007/s10651-010-0143-y, 2011.

Goulard, M. and Voltz, M.: Geostatistical Interpolation of Curves: A Case Study in Soil Science, 805–816, Springer, Netherlands, Dordrecht, doi:10.1007/978-94-011-1739-5_64, 1993.

Haberlandt, U.: Stochastic Rainfall Synthesis Using Regionalized Model Parameters, J. Hydrol. Eng., 3, 160–168, doi:10.1061/(ASCE)1084-0699(1998)3:3(160), 1998.

Habib, E., Krajewski, W., and Kruger, A.: Sampling errors of tipping-bucket rain gauge measurements, J. Hydrol. Eng., 6, 159–166, doi:10.1061/(ASCE)1084-0699(2001)6:2(159), 2001.

Harrold, T. I., Sharma, A., and Sheather, S. J.: A nonparametric model for stochastic generation of daily rainfall occurrence, Water Resour. Res., 39, doi:10.1029/2003WR002182, 2003.

Hosking, J. R. M. and Wallis, J. R.: Parameter and Quantile Estimation for the Generalized Pareto Distribution, Technometrics, 29, 339–349, doi:10.2307/1269343, 1987.

Hunter, J. D.: Matplotlib: A 2D Graphics Environment, Comput. Sci. Eng., 9, 90–95, doi:10.1109/MCSE.2007.55, 2007.

Jones, E., Oliphant, T., Peterson, P., et al., SciPy: Open source scientific tools for Python, available at: http://www.scipy.org/ (last access: 28 April 2017), 2001.

Jones, M. C., Marron, J. S., and Sheather, S. J.: A brief survey of bandwidth selection for density estimation, J. Am. Stat. Assoc., 91, 401–407, 1996.

Katz, R. W. and Parlange, M. B.: Generalizations of Chain-Dependent Processes: Application to Hourly Precipitation, Water Resour. Res., 31, 1331–1341, doi:10.1029/94WR03152, 1995.

Kitanidis, P. K.: Introduction to geostatistics: Application to hydrogeology, Cambridge University Press, 1997.

Kleiber, W., Katz, R. W., and Rajagopalan, B.: Daily spatiotemporal precipitation simulation using latent and transformed Gaussian processes, Water Resour. Res., 48, 1–17, doi:10.1029/2011WR011105, 2012.

Lall, U., Rajagopalan, B., and Tarboton, D. G.: A nonparametric wet/dry spell model for resampling daily precipitation, Water Resour. Res., 32, 2803–2823, doi:10.1029/96WR00565, 1996.

Landesanstalt für Umwelt, Messungen und Naturschutz Baden-Württemberg (LUBW): Klimaatlas Baden-Württemberg, available at: http://www4.lubw.baden-wuerttemberg.de/servlet/ is/244295/ (last access: 28 April 2017), 2006.

Lawson, C. and Hanson, R.: Solving Least Squares Problems, SIAM, 1987.

Li, C., Singh, V. P., and Mishra, A. K.: Simulation of the entire range of daily precipitation using a hybrid probability distribution, Water Resour. Res., 48, doi:10.1029/2011WR011446, 2012.

Lorenz, M. O.: Methods of Measuring the Concentration of Wealth, Publ. Am. Stat. Assoc., 9, 209–219, doi:10.2307/2276207, 1905.

Mamalakis, A., Langousis, A., Deidda, R., and Marrocu, M.: A parametric approach for simultaneous bias correction and high-resolution downscaling of climate model rainfall, Water Resour. Res., 53, 2149–2170, doi: 10.1002/2016WR019578, 2017.

McKinney, W.: Data Structures for Statistical Computing in Python, in: Proceedings of the 9th Python in Science Conference, edited by: van der Walt, S. and Millman, J., 51–56, 2010.

McMillan, H., Krueger, T., and Freer, J.: Benchmarking observational uncertainties for hydrology: rainfall, river discharge and water quality, Hydrol. Process., 26, 4078–4111, doi:10.1002/hyp.9384, 2012.

Menafoglio, A., Secchi, P., and Dalla Rosa, M.: A Universal Kriging predictor for spatially dependent functional data of a Hilbert Space, Electron. J. Statist., 7, 2209–2240, doi:10.1214/13-EJS843, 2013.

Menafoglio, A., Guadagnini, A., and Secchi, P.: Stochastic simulation of soil particle-size curves in heterogeneous aquifer systems through a Bayes space approach, Water Resour. Res., 52, 5708–5726, doi:10.1002/2015WR018369, 2016.

Nystuen, J., Proni, J., Black, P., and Wilkerson, J.: A comparison of automatic rain gauges, J. Atmos. Ocean. Tech., 13, 62–73, doi:10.1175/1520-0426(1996)013<0062:ACOARG>2.0.CO;2, 1996.

Papalexiou, S. M. and Koutsoyiannis, D.: Entropy based derivation of probability distributions: A case study to daily rainfall, Adv. Water Resour., 45, 51–57, doi:10.1016/j.advwatres.2011.11.007, 2012.

Pardo-Iguzquiza, E. and Chica-Olmo, M.: Geostatistics with the Matern semivariogram model: A library of computer programs for inference, kriging and simulation, Comput. Geosci., 34, 1073–1079, doi:10.1016/j.cageo.2007.09.020, 2008.

Peel, S. and Wilson, L. J.: Modeling the Distribution of Precipitation Forecasts from the Canadian Ensemble Prediction System Using Kernel Density Estimation, Weather Forecast., 23, 575–595, doi:10.1175/2007WAF2007023.1, 2008.

R Core Team: R: A Language and Environment for Statistical Computing, R Foundation for Statistical Computing, Vienna, Austria, availabale at: https://www.R-project.org/ (last access: 28 April 2017), 2015.

Rajagopalan, B., Lall, U., and Tarboton, D. G.: Evaluation of kernel density estimation methods for daily precipitation resampling, Journal of Stochastic Hydrology and Hydraulics, 11, 523–547, 1997.

Rider, P. R.: The Method of Moments Applied to a Mixture of Two Exponential Distributions, Ann. Math. Stat., 32, 143–147, doi:10.1214/aoms/1177705147, 1961.

Sheather, S. J.: Density Estimation, Statist. Sci., 19, 588–597, doi:10.1214/088342304000000297, 2004.

Sheather, S. J. and Jones, M. C.: A Reliable Data-Based Bandwidth Selection Method for Kernel Density Estimation, J. Roy. Stat. Soc. B, 53, 683–690, 1991.

Silverman, B. W.: Density estimation for statistics and data analysis, Chapman and Hall, London, 1. publ. edn., 1986.

Stephens, M. A.: EDF Statistics for Goodness of Fit and Some Comparisons, J. Am. Stat. Assoc., 69, 730–737, doi:10.1080/01621459.1974.10480196, 1974.

Szidarovszky, F., Baafi, E. Y., and Kim, Y. C.: Kriging without negative weights, Math. Geol., 19, 549–559, doi:10.1007/BF00896920, 1987.

van der Walt, S., Colbert, S. C., and Varoquaux, G.: The NumPy Array: A Structure for Efficient Numerical Computation, Comput. Sci. Eng., 13, 22–30, doi:10.1109/MCSE.2011.37, 2011.

Wand, M.: KernSmooth: Functions for Kernel Smoothing Supporting Wand & Jones (1995), available at: http://CRAN.R-project.org/package=KernSmooth (last access: 28 April 2017), r package version 2.23-15, 2015.

Wilks, D. S.: Multisite generalization of a daily stochastic precipitation generation model, J. Hydrol., 210, 178–191, doi:10.1016/S0022-1694(98)00186-3, 1998.

Wilks, D. S.: High-resolution spatial interpolation of weather generator parameters using local weighted regressions, Agr. Forest Meteorol., 148, 111–120, doi:10.1016/j.agrformet.2007.09.005, 2008.

Wilks, D. S.: A gridded multisite weather generator and synchronization to observed weather data, Water Resour. Res., 45, doi:10.1029/2009WR007902, 2009.

Wilks, D. S. and Wilby, R. L.: The weather generation game: a review of stochastic weather models, Prog. Phys. Geog., 23, 329–357, doi:10.1177/030913339902300302, 1999.

Woolhiser, D. A. and Pegram, G. G. S.: Maximum Likelihood Estimation of Fourier Coefficients to Describe Seasonal Variations of Parameters in Stochastic Daily Precipitation Models, J. Appl. Meteor., 18, 34–42, doi:10.1175/1520-0450(1979)018<0034:MLEOFC>2.0.CO;2, 1979.

5

Measurement and interpolation uncertainties in rainfall maps from cellular communication networks

M. F. Rios Gaona[1], **A. Overeem**[1,2], **H. Leijnse**[2], **and R. Uijlenhoet**[1]

[1]Hydrology and Quantitative Water Management Group, Department of Environmental Sciences, Wageningen University, 6708 PB Wageningen, the Netherlands
[2]R&D Observations and Data Technology, Royal Netherlands Meteorological Institute, 3731 GA De Bilt, the Netherlands

Correspondence to: M. F. Rios Gaona (manuel.riosgaona@wur.nl)

Abstract. Accurate measurements of rainfall are important in many hydrological and meteorological applications, for instance, flash-flood early-warning systems, hydraulic structures design, irrigation, weather forecasting, and climate modelling. Whenever possible, link networks measure and store the received power of the electromagnetic signal at regular intervals. The decrease in power can be converted to rainfall intensity, and is largely due to the attenuation by raindrops along the link paths. Such an alternative technique fulfils the continuous effort to obtain measurements of rainfall in time and space at higher resolutions, especially in places where traditional rain gauge networks are scarce or poorly maintained.

Rainfall maps from microwave link networks have recently been introduced at country-wide scales. Despite their potential in rainfall estimation at high spatiotemporal resolutions, the uncertainties present in rainfall maps from link networks are not yet fully comprehended. The aim of this work is to identify and quantify the sources of uncertainty present in interpolated rainfall maps from link rainfall depths. In order to disentangle these sources of uncertainty, we classified them into two categories: (1) those associated with the individual microwave link measurements, i.e. the errors involved in link rainfall retrievals, such as wet antenna attenuation, sampling interval of measurements, wet/dry period classification, dry weather baseline attenuation, quantization of the received power, drop size distribution (DSD), and multi-path propagation; and (2) those associated with mapping, i.e. the combined effect of the interpolation methodology and the spatial density of link measurements.

We computed ∼ 3500 rainfall maps from real and simulated link rainfall depths for 12 days for the land surface of the Netherlands. Simulated link rainfall depths refer to path-averaged rainfall depths obtained from radar data. The ∼ 3500 real and simulated rainfall maps were compared against quality-controlled gauge-adjusted radar rainfall fields (assumed to be the ground truth). Thus, we were able to not only identify and quantify the sources of uncertainty in such rainfall maps, but also test the actual and optimal performance of one commercial microwave network from one of the cellular providers in the Netherlands. Errors in microwave link measurements were found to be the source that contributes most to the overall uncertainty.

1 Introduction

Accurate rainfall estimates are crucial inputs for hydrological models, especially those employed for forecasting flash floods, due to the short timescales in which they develop. Rainfall rates can be retrieved from microwave links because rain droplets attenuate the electromagnetic signal between transmitter and receiver along the microwave link path. The principles behind rainfall estimates from microwave attenuation were investigated by Atlas and Ulbrich (1977). They established the nearly linear relationship between the rainfall intensity and the specific attenuation of the signal for frequencies between 10 and 35 GHz.

Messer et al. (2006) and Leijnse et al. (2007) used commercial microwave links to estimate rainfall rates. Note that networks of such links have not been designed for that purpose. In the last decade several studies have developed methods to improve rainfall estimates from microwave link measurements (Leijnse et al., 2008, 2010; Overeem et al., 2011; Schleiss et al., 2013; Chwala et al., 2014). In addition, Goldshtein et al. (2009) and Zinevich et al. (2008, 2009, 2010) proposed methods to estimate rainfall fields via commercial microwave networks. Giuli et al. (1991) had previously reconstructed rainfall fields from simulated microwave attenuation measurements. Overeem et al. (2011) developed an algorithm to estimate rainfall from minimum and maximum received signal levels over 15 min intervals, in which the wet antenna effect is corrected for, and where wet and dry spells are identified from the removal of signal losses not related to rainfall by using nearby links.

Rainfall fields can generally be retrieved from commercial microwave link networks at a higher resolution than rain gauge networks. This holds not only for the spatial resolution (usually microwave links outnumber rain gauges) but also for the temporal resolution (microwave link measurements can be obtained for 1 s, 1 min, 15 min, or daily intervals at either instantaneous or minimum and maximum samples of received signal level (RSL) measurements; Messer et al., 2012). The massive deployment of microwave links provides a complementary network to measure rainfall, especially in countries where rain gauges are scarce or poorly maintained, and where ground-based weather radars are not (yet) deployed (Doumounia et al., 2014).

Recently, Overeem et al. (2013) obtained 15 min and daily rainfall depths from one commercial microwave link network for 12 days for the land surface of the Netherlands ($\sim 35\,000\,\mathrm{km}^2$; ~ 1750 links). They interpolated these rainfall depths to obtain rainfall fields to be compared against gauge-adjusted radar rainfall maps. Although the associated biases were small, the corresponding uncertainties were not. The coefficient of determination, i.e. the square of the correlation coefficient, between link-based and gauge-adjusted radar rainfall maps was 0.49 for the 15 min timescale, and 0.73 for the daily timescale. They did not explore the sources of error that impeded these correlations to reach higher values, though. Here, we address this issue with the aim to unravel and understand the sources of error (and their uncertainties) present in the methodology proposed by Overeem et al. (2013) to estimate rainfall fields. We split the overall uncertainty in rainfall maps from commercial microwave networks into two main sources of error: (1) those associated with the individual microwave link measurements, such as wet antenna attenuation, sampling interval of measurements, wet/dry period classification, dry weather baseline attenuation, drop size distribution (DSD), and multi-path propagation; and (2) those associated with mapping, that is, the combined effect of the interpolation methodology and the spatial density of microwave link measurements. Note that not all the links in the network continuously report data. Only the overall effects of measurement and interpolation errors are addressed here, not all measurement errors separately.

This paper is organized as follows: Sect. 2 describes the data sets and methodology developed by Overeem et al. (2013) to estimate rainfall maps, jointly with the methodologies for this work to derive rainfall maps to identify and quantify error sources. Section 3 compares the results obtained here with those presented in Overeem et al. (2013). Section 4 highlights our major findings. Finally, Sects. 5 and 6 provide a summary, conclusions, and recommendations.

2 Materials and methods

2.1 Data

Two categories of data were used: link data and radar data. These two data sets are fully independent given that each one originates from a different source: microwave link measurements, and a combination of radar and rain gauge measurements, respectively. Link and radar data contain rainfall depths from the 12-day validation period studied by Overeem et al. (2013), which is spread across the months of June, August, and September 2011. This validation period was selected because of its large number of rainfall events. Figure 1 conceptually illustrates the steps we followed to quantify uncertainties in rainfall maps from link networks.

2.1.1 Link data

Link data refer to rainfall depths retrieved from measurements of the attenuation of electromagnetic signals from one commercial microwave link network in the Netherlands. Overeem et al. (2011, 2013) thoroughly explained the methodology to convert measurements of the decrease in the received power to rainfall depths, with reference to a level representative of dry weather. Briefly explained, their methodology is based on four steps: (1) a link is considered to be affected by rainfall if the received power jointly decreases with that of nearby links; (2) a reference signal level representative of dry weather, i.e. the median signal level of all dry periods in the previous 24 h, is determined, and the signal is subtracted from this reference level; the result is the attenuation estimate; (3) microwave links for which accumulated (over 1 day) specific attenuation deviates too much (from that of nearby links) are excluded from the analysis; and (4) 15 min average rainfall intensities are computed from a weighted average of minimum and maximum rainfall intensities obtained by a power-law correlation of specific attenuation (Atlas and Ulbrich, 1977). These rainfall intensities are expressed as path-averaged rainfall depths, and are assumed to be representative of the rainfall across the link path. Full details of the algorithm can be found in Overeem et al. (2011, 2013).

Data from up to 1751 link paths are available, with path lengths from 0.13 to 20.26 km, and frequencies from 12.8 to 39.4 GHz (Fig. 2). It is also clear that the network is designed such that the link frequency decreases as path length in-

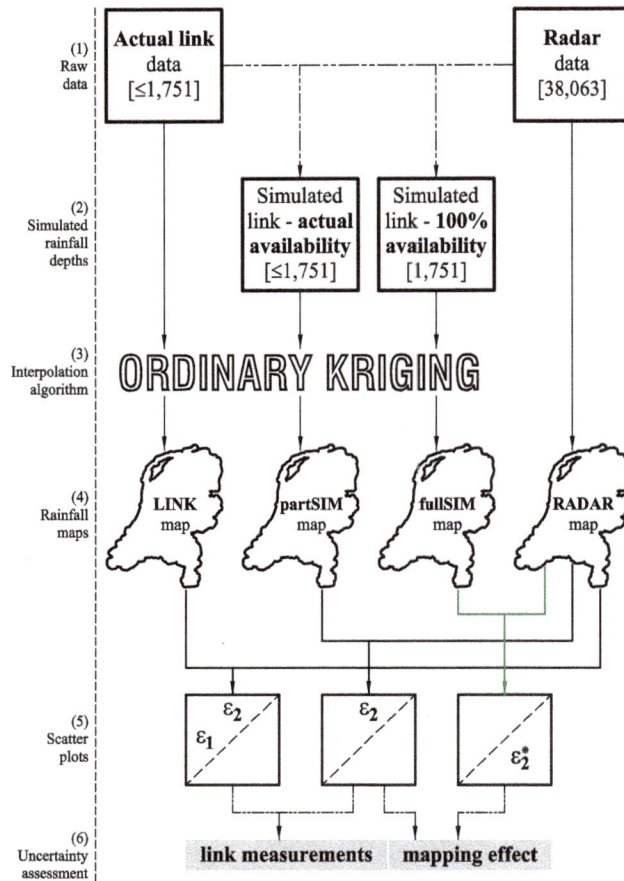

Figure 1. Flowchart to visualize the hierarchical process to identify and quantify uncertainties in rainfall maps from link networks. From top to bottom: (1–2) raw data are selected and rainfall depths simulated; (3–4) through the interpolation methodology rainfall maps are obtained; (5) from the comparison between rainfall maps scatter plots are created; and (6) from the comparison between these scatter plots (and their metrics), the error sources are quantified. ε_1 and ε_2 represent the categories in which the sources of error are classified. Specifically, ε_1 indicates the error from microwave link rainfall retrievals, and ε_2 indicates the error related to mapping. ε_2^* indicates the best case for the mapping-related error (i.e. all links are available all of the time). The number between brackets (1–2) indicates the number of data for every single map or data set.

creases, mainly because low-frequency links suffer less from rain attenuation.

Figure 3 presents the spatial distribution of one commercial link network from one of the providers in the Netherlands, as well as the temporal availability for each link path. Due to data storage problems, wet/dry classification, and outlier removal, it is not feasible to have link data for all the possible link paths in the network (1751) for every time step. The temporal availability per link varies from 0.9 to 99.9 %, with a global average over the entire 12-day data set of 83.5 %.

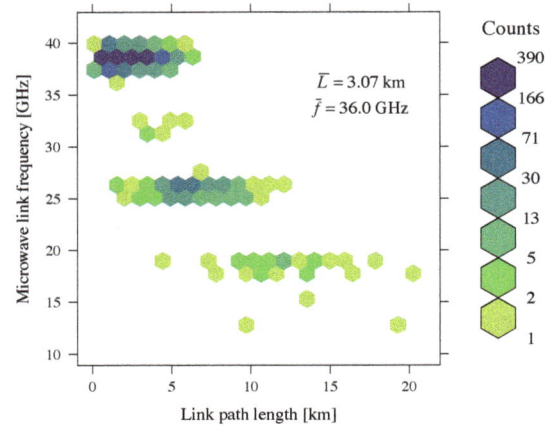

Figure 2. Scatter density plot of microwave link frequencies vs. link-path lengths for the 12-day validation period. The colour scale is logarithmic.

The spatial distribution of the network has two characteristics: (1) there is a strong contrast between urban and rural areas with regard to the spatial distribution of the network; and (2) there are gaps in the network, because of a complete absence of link data or low data availability. Analyses of the link-path orientations show no preferred orientations, i.e. a uniform distribution (such analyses are not presented in this paper).

2.1.2   Radar data

Radar data are taken from the climatological rainfall data set[1] of two C-band Doppler weather radars operated by the Royal Netherlands Meteorological Institute (KNMI) (Overeem et al., 2009a, b, 2011). The composite image of rainfall depths has a temporal resolution of 5 min, and a spatial resolution (pixel size) of $0.92\,\mathrm{km}^2$ (rounded to $1\,\mathrm{km}^2$ in figures, tables, and subsequent analyses), for the entire land surface of the Netherlands (38 063 pixels). This composite image is adjusted with rainfall depths from one automatic and one manual rain gauge network (32 and 325 gauges, respectively) also operated by KNMI. The spatial and temporal resolution, and its accuracy, make this data set a reliable source of rainfall data. We used the same radar data set as that used in Overeem et al. (2013).

2.2   Simulated link rainfall depths

Simulated link rainfall depths are averages of radar data based on the topology and time-availability features of the link network. The purpose of simulated link rainfall depths is twofold: (1) to evaluate the performance of the link net-

---

[1]KNMI climatological rainfall data sets are freely available at the IS-ENES climate4impact portal: http://climate4impact.eu/impactportal/data/catalogbrowser.jsp?catalog=http://opendap.knmi.nl/knmi/thredds/./radarprecipclim.xml.

work assuming that all links provide perfect measurements of path-averaged rainfall at the 15 min interval, and (2) to evaluate the performance of the link network if all links would be available all the time.

Because link data were obtained in intervals of 15 min, sets of three consecutive 5 min radar composite images were summed up on a pixel-by-pixel basis. The simulation allows us to separate mapping errors from other errors. For detailed studies on the effects of link length and frequency, temporal sampling, power resolution, and wet antenna attenuation in link measurements see Leijnse et al. (2008, 2010). After the addition of 5 min radar composite images, the link network topology was overlaid on the 15 min radar composite image, and all pixels under every link path were selected. Then, for every link path and its associated pixels, rainfall depths were averaged. This was a weighted average in which the weight was taken as the fraction of the total link path that overlaps one radar pixel. For instance, if a 1 km link path was located 0.6 km over one pixel and 0.4 km over a contiguous pixel, the average rainfall depth was the sum of 60 % of the first pixel's rainfall depth plus 40 % of the second pixel's rainfall depth.

Not all link data are available for all the possible link paths in the network (1751) at every time step. In addition to the performance of the actual topology of the network, the complete availability of radar data allowed us to simulate the optimal performance of the link network, i.e. the performance that could theoretically be achieved if all links (1751) would be available all the time.

2.3 Rainfall maps

The rainfall depths from actual link measurements and both types of simulations (actual and 100 % network availability) were spatially interpolated to obtain 15 min rainfall maps with a spatial resolution of 1 km^2. In all rainfall maps the land surface of the Netherlands was represented by 38 063 pixels. For any given time step, interpolated rainfall maps were compared on a pixel-by-pixel basis against the radar rainfall fields. Hence, 15 min rainfall maps were obtained for the 12-day validation period, i.e. 1152 rainfall maps in total for each of the four sets of rainfall maps considered (namely, radar, actual links, simulated links with partial availability, and simulated links with 100 % availability). In subsequent figures and tables, these four data sets will be identified as "RADAR", "LINK", "partSIM", and "fullSIM", respectively (see Fig. 1); 15 min rainfall maps were accumulated to daily rainfall maps, i.e. 12 daily rainfall maps per data set.

Ordinary kriging (OK) was employed to generate rainfall maps, because it is the simplest and most straightforward method that accounts for the local variability of the stochastic process, rainfall in this case (Cressie, 1990; Haining et al., 2010). Kriging is ideally suited for interpolation of highly irregularly spaced data points. Nevertheless, this method comes with its own limitations, and a number of assumptions should be made for the method to be valid, e.g.

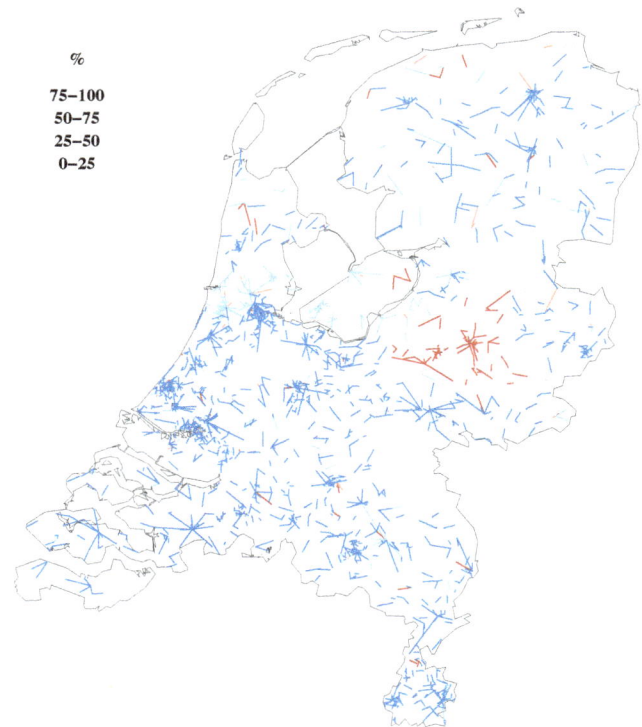

Figure 3. Topology of the T-Mobile NL microwave link network used for this study. The colour scale of the microwave network represents the temporal availability of the link data for the 12-day validation period. The average availability is 83.5 %.

isotropy and statistical stationarity. These assumptions are further explained in Sect. 6. The path-averaged link rainfall estimates are assigned to the point at the centre of the link, so that these point data can be used in the OK interpolation. This conversion from line-scale to point-scale data is part of our mapping method, and hence errors resulting from this conversion are part of the mapping uncertainty.

Any kriging method heavily relies on the function that describes the spatial covariance, i.e. the semi-variogram. The semi-variogram is a continuous function that describes how the spatial dependence of a random variable changes with distance and direction (Isaaks and Srivastava, 1989, chap. 7). Like Overeem et al. (2013), we chose the semi-variogram approach of van de Beek et al. (2011) because it is a simple isotropic spherical model developed for the Netherlands on the basis of a 30-year climatological rainfall data set. van de Beek et al. (2011) concluded that the seasonality in range and sill of the semi-variogram can be described by cosine-function models with the day-of-year as the independent variable. Note that they assumed the nugget to be zero. van de Beek et al. (2012) also developed two methodologies that allowed for the spherical semi-variogram to be downscaled from daily to hourly time steps. We chose their second methodology, namely, power-law scaling of cosine-function parameters, because it was shown to perform better. This downscaling methodology was based on hourly rainfall data aggregated to 2, 3, 4, 6, 8, 12, and 24 h. Here we extended this power-law downscaling to smaller timescales, namely, 0.25 h, i.e. 15 min.

For the LINK, partSIM, and fullSIM data sets, 15 min rainfall maps were obtained as follows: first, the spherical semivariogram parameters were computed and downscaled for the given day of the year. Hence, a single semi-variogram is applied to all 15 min time steps within that given day. The nugget was defined as 10 % of the sill. Second, rainfall depths were assigned to the coordinates of the link paths' middle points. Third, rainfall depths were interpolated over the spatial grid of the radar data set. The interpolation algorithm always selects the closest 100 rainfall depths to the pixel for which the interpolation is carried out. This selection was established to speed up the interpolation process; 24 h rainfall maps were obtained from the aggregation of 15 min rainfall maps.

2.4 Error and uncertainty metrics

To quantify the uncertainty in rainfall maps from microwave link networks, we used three metrics: (1) the relative bias, (2) the coefficient of variation, and (3) the coefficient of determination.

The relative bias is a relative measure of the average error between the interpolated and radar rainfall fields (considered to be the ground truth):

$$\text{relative bias} = \frac{\overline{R}_{\text{res}}}{\overline{R}_{\text{radar}}} = \frac{\sum_{i=1}^{n} R_{\text{res},i}}{\sum_{i=1}^{n} R_{\text{radar},i}}, \quad (1)$$

where $R_{\text{res},i} = R_{\text{int},i} - R_{\text{radar},i}$. In Eq. (1), n represents all possible pixels and time steps for the 12-day validation period.

The coefficient of variation is a dimensionless measure of dispersion, which is defined as the standard deviation divided by the mean (Haan, 1977). In this case we took the standard deviation of the residuals divided by the mean of the reference field (i.e. the mean of the radar rainfall field):

$$\text{CV} = \frac{\sqrt{\frac{1}{n-1} \sum_{i=1}^{n} \left(R_{\text{res},i} - \overline{R}_{\text{res}}\right)^2}}{\overline{R}_{\text{radar}}}. \quad (2)$$

The coefficient of variation is a measure of uncertainty (similar to the root mean squared error). For instance, a CV = 0 would indicate a hypothetical case with no bias and no uncertainty, i.e. a case in which all data points would fall exactly on the 1 : 1 line.

The coefficient of determination is a measure of the strength of the linear dependence between two random variables, interpolated and radar rainfall depths, in this case. It

is simply defined as the square of the correlation coefficient between the interpolated and radar rainfall depths:

$$r^2 = \frac{\left[\sum_{i=1}^{n} \left(R_{\text{radar},i} - \overline{R}_{\text{radar}}\right) \cdot \left(R_{\text{int},i} - \overline{R}_{\text{int}}\right)\right]^2}{\left[\sum_{i=1}^{n} \left(R_{\text{radar},i} - \overline{R}_{\text{radar}}\right)^2\right] \cdot \left[\sum_{i=1}^{n} \left(R_{\text{int},i} - \overline{R}_{\text{int}}\right)^2\right]}. \quad (3)$$

The coefficient of determination represents the fraction of the variance of the reference variable that can be explained by a linear regression. In a case of perfect linear correlation, i.e. $r^2 = 1$, all data points would fall on a straight line without any scatter. Hence, the linear regression would be able to explain 100 % of the variance of the reference variable in that case. However, perfect linearity does not imply unbiased estimation because the regression line could not necessarily coincide with the 1 : 1 line, even if it captures all variability.

3 Results

From the actual and simulated link rainfall depths, rainfall maps were obtained for three cases: (1) 15 min rainfall maps from interpolation of 15 min rainfall depths; (2) 24 h rainfall maps from the sum of these 15 min rainfall maps; and (3) 15 min rainfall maps from interpolation of 15 min rainfall depths, in which each pixel (interpolated rainfall depth) was averaged with the surrounding pixels within a 9×9 pixel square. The reason for this posterior average of the rainfall depths was to limit representativeness errors in time (Overeem et al., 2013). Incidentally, this area (~ 81 km^2) roughly corresponds to the spatial extent of typical water management units in the Netherlands.

Appendix A presents five examples of 24 h and 15 min rainfall maps. Overeem et al. (2013, Supplement) showed daily comparisons between actual link rainfall maps and radar rainfall fields for the 12-day validation period. Here, we present 5 of those 12 cases for reference. These comparisons are extended to both types of simulated link rainfall maps (actual and 100 % network availability) (Fig. A1). Five comparisons of 15 min rainfall maps are also presented (Fig. A2). These examples provide information on the improvement in rainfall fields when the sources of error studied here are removed.

For any given time step, interpolated rainfall maps were compared on a pixel-by-pixel basis against radar rainfall fields. This pixel-by-pixel comparison was done via scatter density plots of interpolated against radar rainfall depths (ground truth). Figure 4 presents an array of scatter plots, for the three cases of spatiotemporal aggregation, for the actual and both types of simulated link rainfall depths (actual and 100 % network availability). Each of the scatter plots in Fig. 4 corresponds to all 15 min (or 24 h) rainfall maps within the 12-day validation period. These plots show paired rainfall

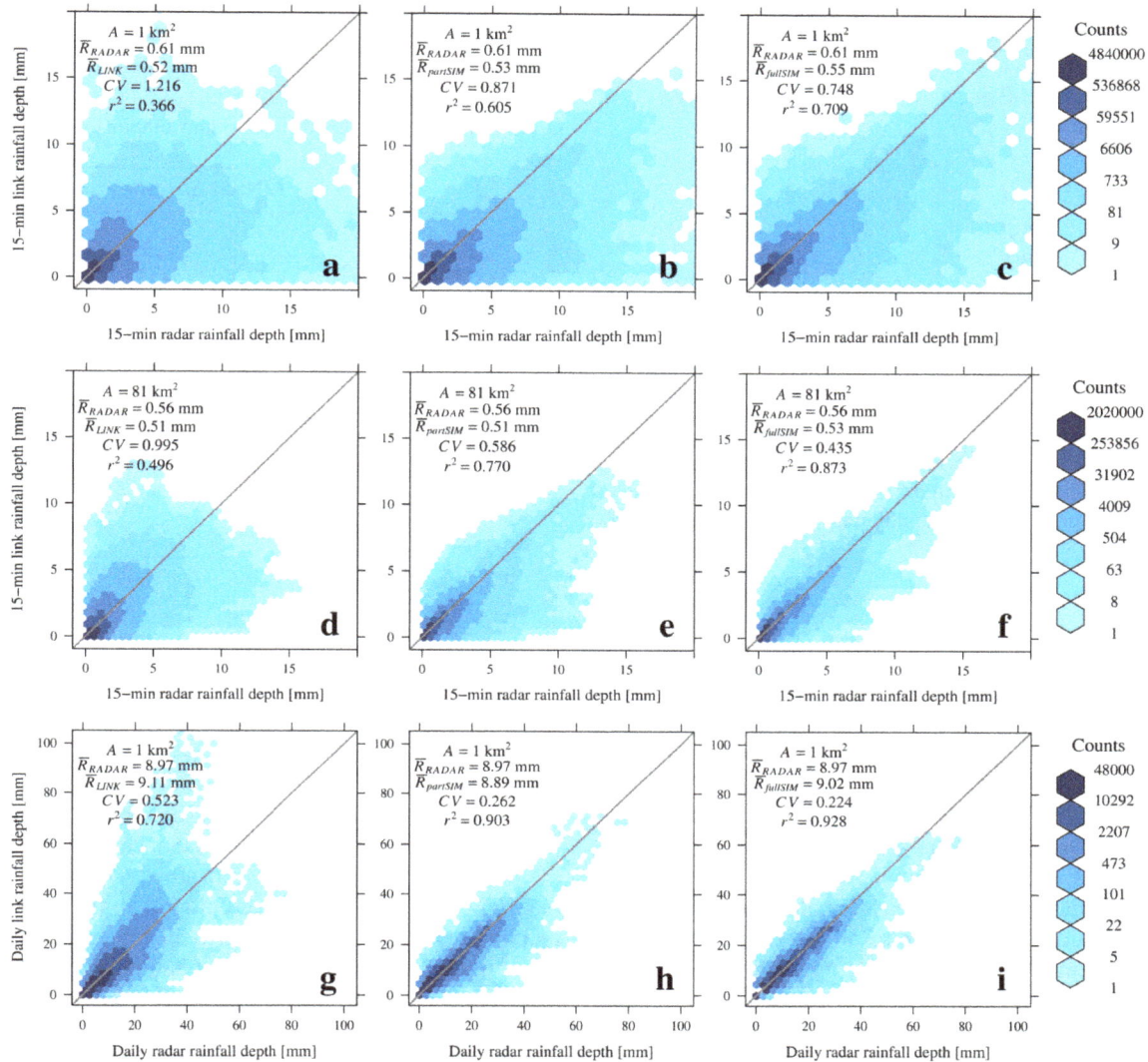

Figure 4. Scatter density plots of interpolated link rainfall depths vs. radar rainfall depths for 15 min and 24 h. Top row (**a, b, c**): 15 min rainfall depths; middle row (**d, e, f**): 15 min rainfall depths averaged with the surrounding pixels within a 9×9 pixel square; bottom row (**g, h, i**): daily sum of 15 min rainfall depths. Left column (**a, d, g**): actual link rainfall maps vs. radar rainfall fields; centre column (**b, e, h**): simulated link rainfall maps (actual availability) vs. radar rainfall fields; right column (**c, f, i**): simulated link rainfall maps (100 % availability) vs. radar rainfall maps. (**d**) and (**g**) are comparable to Overeem et al. (2013). The colour scale is logarithmic.

depths of interpolated and radar rainfall maps, for any pair in which the radar rainfall depth is larger than 0.1 mm.

The scatter density plot of Fig. 5 corresponds to the actual and simulated link rainfall depths (actual availability) at the locations of the links, i.e. before any interpolation was applied. Only those pairs for which at least one rainfall depth exceeded 0.1 mm were plotted.

Table 1 summarizes the values of the relative bias, the coefficient of variation (of the residuals), and the coefficient of determination (i.e. the squared correlation coefficient) for the three cases of spatiotemporal aggregation, for the actual and both types of simulated link rainfall depths.

4 Discussion

From left to right and from top to bottom, the general picture that arises from Fig. 4 and Table 1 is (1) a reduced systematic error (relative bias); (2) a smaller random error (CV); and (3) a stronger linear dependence (r^2). This suggests a general improvement of the interpolated link rainfall depths with respect to the radar rainfall depths, as more sources of error are removed from the analysis.

Figure 4a, d, and g show the relation between the actual link and radar rainfall depths, for the three cases of spatiotemporal aggregation. The scatter in these plots can be attributed to all possible sources of error in rainfall maps from microwave link measurements, i.e. those associated with the

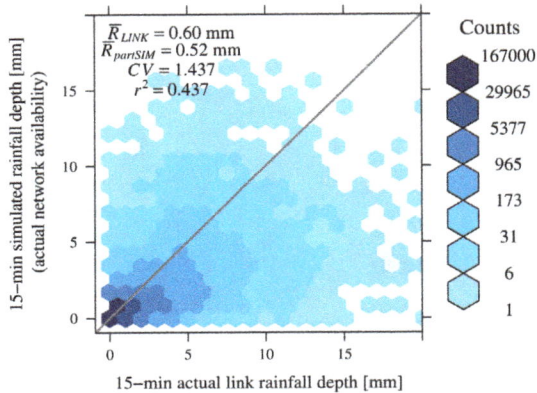

Figure 5. Scatter density plot of simulated link rainfall depths (actual availability) vs. actual link rainfall depths for all 15 min time steps in the 12-day validation period. Both simulated and actual link rainfall depths are path-averaged rainfall depths. The colour scale is logarithmic.

Table 1. Relative bias, and coefficients of variation and determination for the three cases of spatiotemporal aggregation (15 min [1 km^2], 15 min [81 km^2], 24 h [1 km^2]), for the three sets of link measurements, i.e. the actual and both types of simulated link rainfall depths (actual and 100 % network availability).

	LINK	partSIM	fullSIM
Relative bias [%]			
15 min [1 km^2]	−14.3	−13.0	−9.3
15 min [81 km^2]	−9.1	−9.1	−5.6
24 h [1 km^2]	+1.6	−0.8	+0.7
Coefficient of variation − CV			
15 min [1 km^2]	1.216	0.871	0.748
15 min [81 km^2]	0.995	0.586	0.435
24 h [1 km^2]	0.523	0.262	0.224
Coefficient of determination − r^2			
15 min [1 km^2]	0.366	0.605	0.709
15 min [81 km^2]	0.496	0.770	0.873
24 h [1 km^2]	0.720	0.903	0.928

link measurements themselves and those associated with the interpolation of individual measurements (mapping).

The dark blue shading close to the 1 : 1 line for small rainfall depths in all panels of Fig. 4 indicates a good agreement between rainfall estimates from microwave links and radar (note that the colour scale is logarithmic). Conversely, for larger rainfall depths the scatter seems to relatively increase for the actual link measurements (panels a, d, g), while it decreases for the simulated link measurements (all other panels). Such deviations must be the result of errors in individual link measurements as well as the combination of limited spatial coverage of the link network (Fig. 3) with the strong variability of rainfall in space. The relative contribution of the measurement errors to the total error hence increases with rainfall amounts.

From Fig. 4 and Table 1, it is clear as well that the relative bias is most sensitive to the spatial and temporal aggregation level. If all paired rainfall accumulations would have been used (and not only those in which at least the radar rainfall depth exceeds 0.1 mm) one would expect the relative bias to be exactly the same for all aggregation levels, because both aggregation and computation of the bias are linear operators (Eq. 1).

There is a limited improvement in terms of the coefficients of variation and determination, when the scatter plots in the second column of Fig. 4 are compared to those in the third column, as well as their respective statistics in Table 1. This means that the main reduction of uncertainty is achieved when the actual link measurements are replaced with the simulated microwave link measurements, rather than to increase the actual link network availability to 100 % for all links. This implies that a significant fraction of the overall uncertainty must be due to errors and uncertainties in the link measurements themselves, rather than due to errors and un-

certainties associated with mapping, at which rainfall maps are reconstructed.

Figure 4c, f, and i and the last column of Table 1 indicate the best possible performance that can be achieved with the employed link network (if all links would yield perfect measurements of path-averaged rainfall all the time). The remaining scatter can be attributed to the interpolation methodology (including the assignment of path-averaged rainfall intensities to the link's centre point), the spatial variability of rainfall, and the effect of other factors such as the variable and limited density of the link network (more links in urban than in rural areas).

When 15 min rainfall depths at the 1 km^2 spatial scale (Fig. 4a–c) are summed to daily rainfall depths (Fig. 4g–i), the discrepancies in rainfall estimates at 15 min tend to cancel each other. This explains the sharp decrease in the coefficient of variation, and the sharp increase in the coefficient of determination between 15 min and 24 h rainfall accumulations, which implies a certain degree of independence among the errors in the 15 min accumulations.

Figure 5 compares simulated against actual link rainfall depths, before any interpolation was applied. This indicates the performance of the 1751 individual links in terms of rainfall retrieval, regardless of the errors and uncertainties introduced by interpolation (mapping). Note that the coefficient of variation is larger than that of the 1 km^2, 15 min rainfall accumulations presented in panel a of Fig. 4; and that the coefficient of determination is between those coefficients presented in panels a and d of Fig. 4. If we would assume that rainfall retrieval and mapping errors are independent, we would expect the CV in Fig. 4 to be greater than that in Fig. 5.

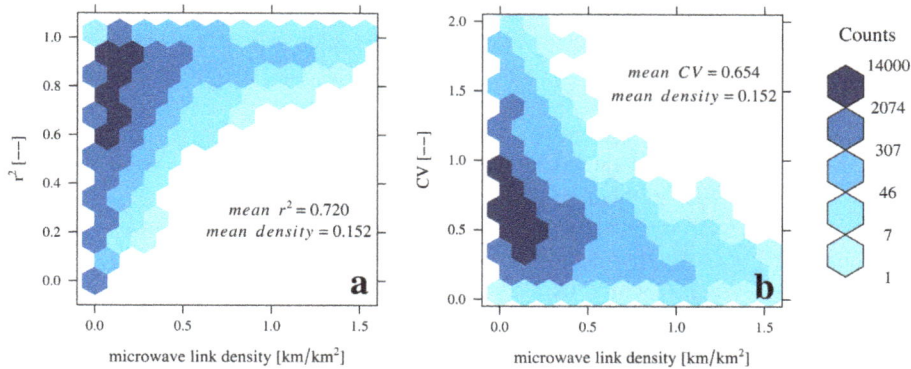

Figure 6. Scatter density plots of coefficient of determination (r^2) and coefficient of variation (CV) vs. microwave link density (averaged over $155\,km^2$), for the fullSIM case at 15 min and $1\,km^2$ spatial scale. The colour scale is logarithmic.

This means that there is a clear interplay between these two types of errors, and that the assumption of independence does not hold. This may be explained by the fact that we use kriging with a variogram that includes a nugget. In areas with a dense link network, the weight of each individual link is relatively small in the computation of the interpolated rainfall field. This reduces the effect of large errors in a given link. In areas with lower link densities the nugget of the employed variogram has a similar reducing effect on large errors.

From Fig. 6 it can be seen that a higher density in the link network guarantees good correlation between the estimated values of rainfall and the ground truth, and a low coefficient of variation of the residuals. From the left panel (Fig. 6a), it can be concluded that lower link densities also contribute (and in large proportion) to higher correlation coefficients. This means that without considering errors in link measurements, these latter being the largest source of uncertainty in country-wide rainfall fields, the network density and the mapping methodology considered here are, respectively, high and good enough to retrieve accurate rainfall fields at such country-wide scales (at least in the Netherlands).

5 Summary and conclusions

Our goal was to quantify the errors and uncertainties in rainfall maps from commercial microwave link networks. In general, these errors can be attributed to different sources like wet antenna attenuation, sampling interval of measurements, wet/dry period classification, dry weather baseline attenuation, drop size distribution (DSD), multi-path propagation, interpolation methodology and algorithm, the availability of microwave link measurements, and the variability of rainfall itself across time and space. For the purpose of this paper we classified all possible sources of error into two categories: (1) those associated with the link measurements themselves (retrieval algorithm included), and (2) those associated with mapping. Only the overall effects of physical and interpola-

tion errors were addressed here, not all physical errors separately.

To quantify the errors and uncertainties that can be attributed to these two categories, rainfall maps created from three sets of link rainfall depths were compared: actual link measurements, simulated link measurements with the actual network availability, and simulated link measurements with 100 % network availability assumed. Simulated link rainfall depths are not affected by errors and uncertainties attributed to actual link measurements; therefore, we could estimate uncertainties attributed to mapping. Based on a pixel-by-pixel comparison, interpolated rainfall maps of the Netherlands were compared against radar rainfall fields (considered to be the ground truth). These comparisons were carried out on the basis of scatter density plots and three metrics: relative bias, coefficient of variation (CV), and coefficient of determination (r^2).

We found that measurement errors themselves are the source of error that contributes most to the overall uncertainty in rainfall maps from commercial microwave link networks.

In a standard operational framework, data from commercial microwave link networks may not be continuously available for the entire network. Such data gaps affect the accuracy of the retrieved rainfall intensities. Because we were able to simulate rainfall depths on the basis of radar composites, we could investigate the hypothetical case in which data from a commercial link network would be available for all time steps, and for all possible link paths in the network. This best-case scenario could explain an additional 10 % of the variance explained by error-free link measurements with actual network availability for the 15 min accumulation (3 % for the 24 h accumulation). Note that these percentages are particular for the region and period considered in this study. Nevertheless, even the best-case scenario showed a remaining and significant amount of uncertainty that could not be removed in rainfall maps. This means that the space–time variability of rainfall is such that it would require an even more dense and robust network of microwave links to generate more accurate rainfall maps at country-wide scales. The uncertainties in link rainfall retrievals found in this paper are partly explained by the combined effects of rainfall space variability along the link, non-linearity of the retrieval relation, imperfect temporal sampling strategy, quantization of the received power (data stored in integer number of dBs), and wet antenna attenuation (and correction) investigated by Leijnse et al. (2008, in particular Fig. 13, upper right panel on p. 1487). They reported a CV of ~ 1.0, which explains a significant part of the CV (1.44) given in Fig. 5. Daily rainfall maps from microwave links showed less uncertainty compared to 15 min rainfall maps, because errors present in 15 min rainfall maps tend to cancel each other when 15 min rainfall maps are aggregated.

6 Constraints and recommendations

The kriging algorithm we used was that of Pebesma (1997) and Pebesma and Wesseling (1998). The interpolated maps from simulated link rainfall depths represent the outcome of a process in which a linear feature (link path) obtained from the average of volume samples (radar data) is assigned to a point (link-path middle point). Each of these features (area, line, volume, point) represents what in geostatistics is referred to as support, i.e. the spatial resolution at which the random variable is analyzed (Cressie and Wikle, 2011, chap 4.1). The arbitrary change from line to point support introduces a source of error that is implicitly included in the errors related to mapping.

Apart from its simplicity and the 30-year rainfall data set on which it is based, we also chose the isotropic spherical semi-variogram of van de Beek et al. (2011), because a consistent semi-variogram model estimated from link data was not feasible for 15 min rainfall intensities. Isotropic semi-variograms assume equal spatial dependence in all possible directions. Rainfall is generally a phenomenon that exhibits anisotropy in time and space (Lepioufle et al., 2012; Velasco-Forero et al., 2009; Guillot and Lebel, 1999; Amani and Lebel, 1997). Nevertheless, it is reasonable to assume isotropy for the Netherlands given its relative small area and flat topography. OK assumes the mean to be constant and unknown within the region of interpolation. When this unknown mean presents substantial changes over short distances, the assumption of statistical stationarity is no longer valid. Universal kriging, kriging with external drift, and regression kriging are more sophisticated interpolation techniques that incorporate trends to account for non-stationarity (e.g. Schuurmans et al., 2007). The performance of these geostatistical techniques to retrieve link rainfall maps was beyond the scope of this research.

If a similar study were to be carried out in a country with different conditions than those present in the Netherlands, three issues should be considered: (1) the spatial and operational configuration of the link network, (2) the climatology of the region where the link network operates, and (3) the spatial scale at which the analysis is carried out.

The first issue, the spatial and operational configuration of the link network, refers to the distribution of link frequencies, lengths, and densities of link networks around the world. For instance, the commercial microwave link network used in this study has an average link-path length of 3.1 km, a mean frequency of 36.0 GHz, and a global average availability of 83.5 % across the Netherlands (Figs. 2 and 3). Other regions may have more extensive urban and/or rural areas. In particular, for rural areas one expects to find longer link paths, and therefore lower microwave frequencies. Another issue related to the lower frequencies, e.g. 7 GHz, is the low sensitivity to rainfall and the non-linearity of the R–k relationship, mostly in tropical regions (Doumounia et al., 2014). This non-linearity will lead to biases in rainfall intensities in cases of large rainfall variability along the link path (positive biases at lower frequencies where the exponent of the R–k power law is smaller than 1; see Leijnse et al., 2010). Thus, the performance of the rainfall retrieval algorithm for such link networks will differ from the performance found in this study. For instance, in places where link paths are longer (tens of kilometres) the error due to spatial variability of rainfall along the link path becomes more important (Berne and Uijlenhoet, 2007; Leijnse et al., 2008, 2010). Moreover, less dense networks with long link paths will provide less detailed information about rainfall.

The second issue, the climatology of the region refers to the local pattern of rainfall that characterizes different regions around the world. The rainfall characteristics of the Netherlands are different from the ones encountered in, e.g., (sub-)tropical regions. For instance, the spherical semi-variogram model applied here was derived from climatological rain gauge data for the Netherlands. Furthermore, rainfall characteristics such as raindrop size distributions or the distribution of rainfall intensities will affect the optimal values of the parameters of the retrieval algorithm. Therefore, for regions with different rainfall climatologies than the Netherlands, variations should be considered not only in the interpolation methodology but also in the algorithms and their parameters to retrieve rainfall intensities.

The third issue refers to the spatial scale at which rainfall maps are reconstructed. The analyses presented here focussed on 15 min (and 24 h) maps at 1 and 81 km^2, and the differences in error characteristics are significant. For larger regions, for instance, the uncertainty attributed to mapping could play a major role in the overall error distribution. Still, the scale at which rainfall can effectively be retrieved depends greatly on the density of the underlying link network. This means that in regions with a much lower link density than in the Netherlands, the effective spatial resolution for which rainfall maps can be derived will be lower.

Appendix A: Comparison of 24 h and 15 min rainfall maps

In Fig. A1, the LINK column (top and bottom rows – 20110907_08:00 and 20110819_08:00) shows how daily rainfall depths are greatly overestimated by link data, especially in places where there is intense rainfall, and the density of the network is higher. Simulated rainfall depths (actual availability) show improvement of rainfall fields with regard to link-based rainfall fields. Conversely, to actual link rainfall maps, simulated rainfall fields based on the actual availabil-

ity of the network present a slight underestimation of rainfall depths. Simulated link rainfall fields (actual and 100 % network availability) are similar because the effect of actual or 100 % availability among 15 min intervals is smoothed out by the sum of 15 min rainfall fields.

Figure A2 shows how accurate rainfall events are captured across the Netherlands at 15 min intervals. Note how the accuracy is improved for the best-case scenario of 100 % network availability (fullSIM column).

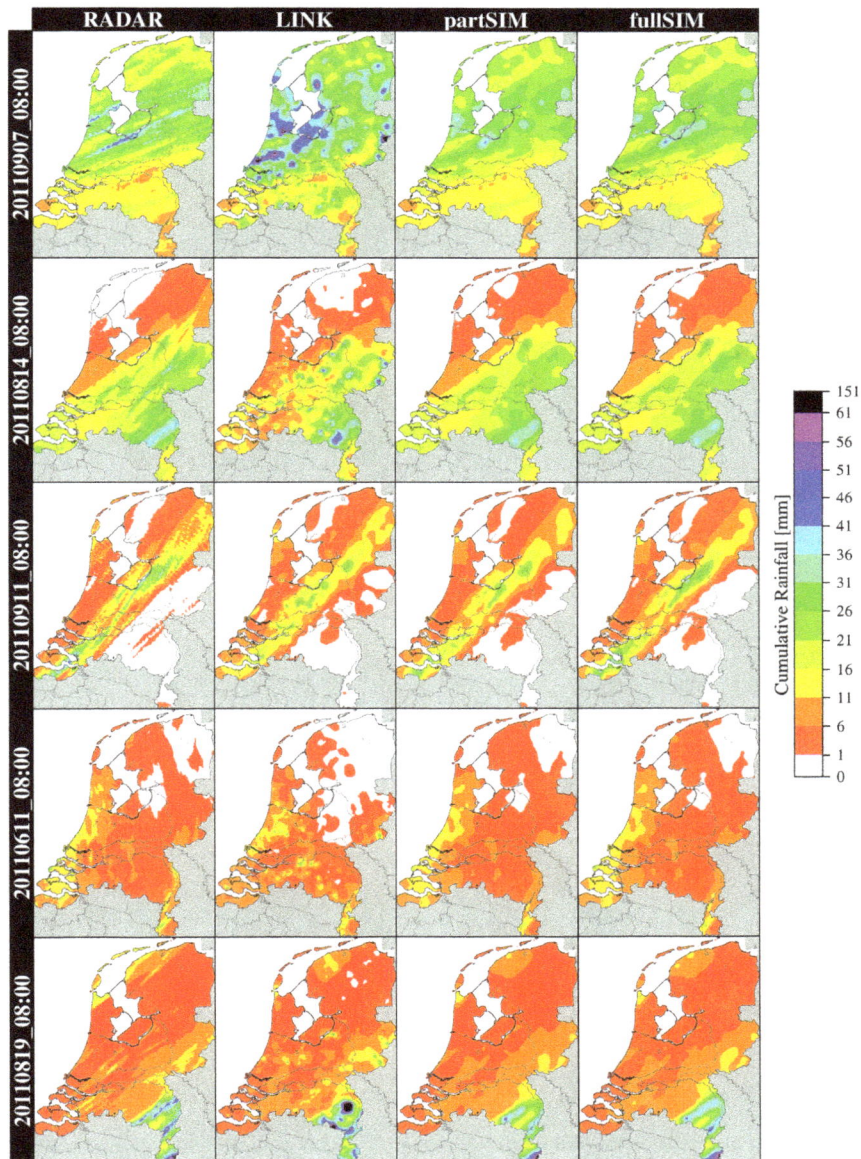

Figure A1. Comparison of daily interpolated rainfall maps with regard to radar rainfall fields (ground truth; left column). The rows show five of the 12 days of the validation period. Daily rainfall maps were aggregated from 15 min rainfall maps. The row labels indicate the end UTC for which the maps were obtained.

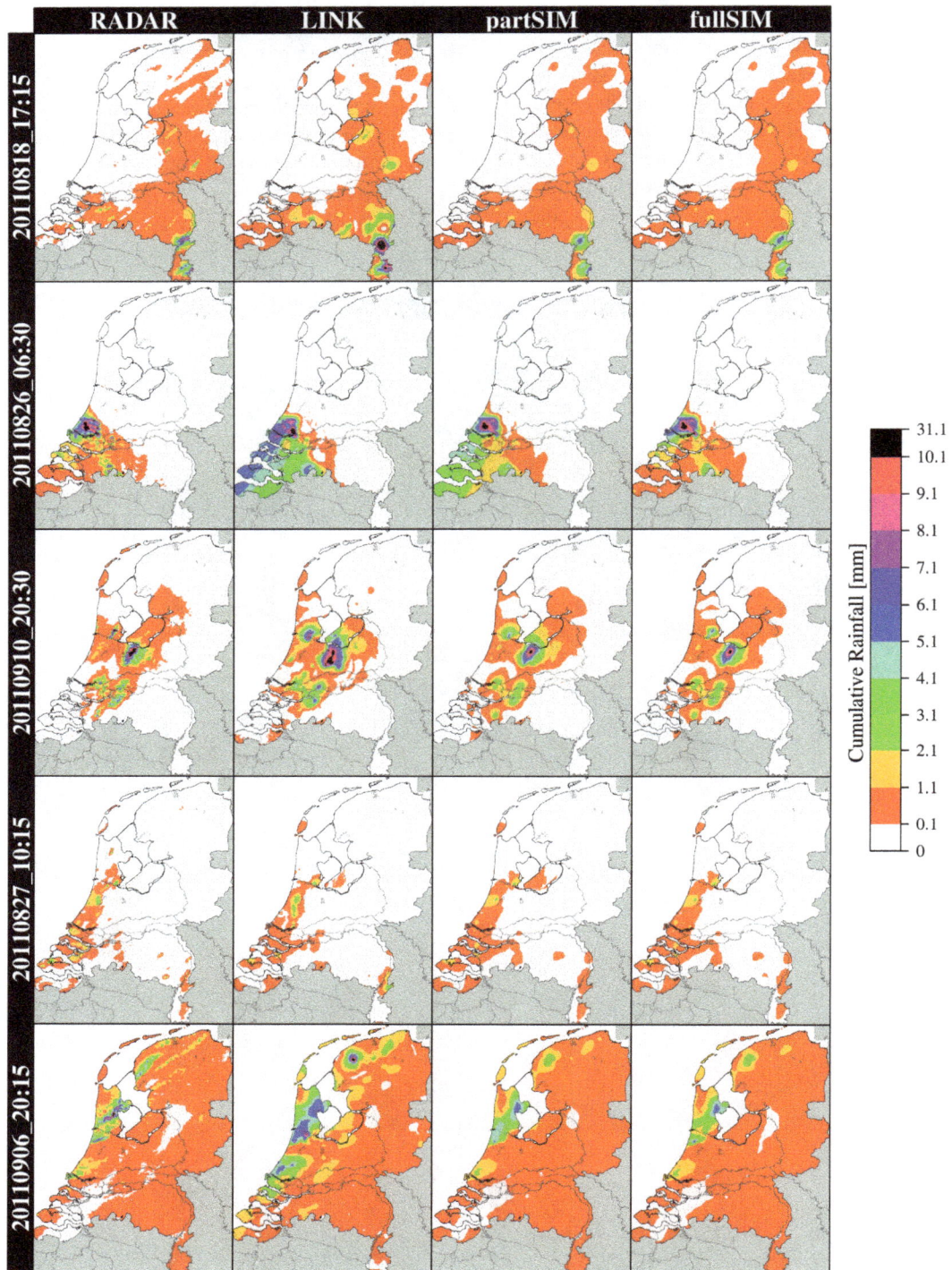

Figure A2. Comparison of 15 min interpolated rainfall maps with regard to radar rainfall fields (ground truth, left column). The rows show five of the 1152 time steps (cases) present in the 12-day validation period. The row labels indicate the start UTC for which the maps were obtained.

Acknowledgements. We gratefully acknowledge Ronald Kloeg and Ralph Koppelaar from T-Mobile NL who provided us with the cellular telecommunication link data. We would like to thank Marc Bierkens from Utrecht University for the fruitful discussions. This work was financially supported by the Netherlands Organisation for Scientific Research NWO (project ALW-GO-AO/11-15), and the Netherlands Technology Foundation STW (project 11944).

Edited by: P. Gentine

References

Amani, A. and Lebel, T.: Lagrangian kriging for the estimation of Sahelian rainfall at small time steps, J. Hydrol., 192, 125–157, doi:10.1016/S0022-1694(96)03104-6, 1997.

Atlas, D. and Ulbrich, C. W.: Path- and area-integrated rainfall measurement by microwave attenuation in the 1–3 cm band, J. Appl. Meteorol., 16, 1322–1331, doi:10.1175/1520-0450(1977)016<1322:PAAIRM>2.0.CO;2, 1977.

Berne, A. and Uijlenhoet, R.: Path-averaged rainfall estimation using microwave links: uncertainty due to spatial rainfall variability, Geophys. Res. Lett., 34, L07403, doi:10.1029/2007GL029409, 2007.

Chwala, C., Kunstmann, H., Hipp, S., and Siart, U.: A monostatic microwave transmission experiment for line integrated precipitation and humidity remote sensing, Atmos. Res., 144, 57–72, doi:10.1016/j.atmosres.2013.05.014, 2014.

Cressie, N.: The origins of kriging, Math. Geol., 22, 239–252, doi:10.1007/BF00889887, 1990.

Cressie, N. and Wikle, C. K.: Statistics for Spatio-Temporal Data, Wiley series in Probability and Statistics, Wiley, Oxford, 2011.

Doumounia, A., Gosset, M., Cazenave, F., Kacou, M., and Zougmore, F.: Rainfall monitoring based on microwave links from cellular telecommunication networks: first results from a West African test bed, Geophys. Res. Lett., 41, 6016–6022, doi:10.1002/2014GL060724, 2014.

Giuli, D., Toccafondi, A., Gentili, G. B., and Freni, A.: Tomographic reconstruction of rainfall fields through microwave attenuation measurements, J. Appl. Meteorol., 30, 1323–1340, doi:10.1175/1520-0450(1991)030<1323:TRORFT>2.0.CO;2, 1991.

Goldshtein, O., Messer, H., and Zinevich, A.: Rain rate estimation using measurements from commercial telecommunications links, IEEE T. Signal Proces., 57, 1616–1625, doi:10.1109/TSP.2009.2012554, 2009.

Guillot, G. and Lebel, T.: Approximation of Sahelian rainfall fields with meta-Gaussian random functions, Stoch. Env. Res. Risk. A., 13, 113–130, doi:10.1007/s004770050035, 1999.

Haan, C. T.: Statistical Methods in Hydrology, Iowa State University Press, Ames, 1977.

Haining, R. P., Kerry, R., and Oliver, M. A.: Geography, spatial data analysis, and geostatistics: an overview, Geogr. Anal., 42, 7–31, doi:10.1111/j.1538-4632.2009.00780.x, 2010.

Isaaks, E. H. and Srivastava, R.: Applied Geostatistics, Oxford University Press, available at: http://books.google.nl/books?id=vC2dcXFLI3YC (last access: 23 March 2015), 1989.

Leijnse, H., Uijlenhoet, R., and Stricker, J. N. M.: Hydrometeorological application of a microwave link: 2. Precipitation, Water Resour. Res., 43, WR004989, doi:10.1029/2006WR004989, 2007.

Leijnse, H., Uijlenhoet, R., and Stricker, J.: Microwave link rainfall estimation: Effects of link length and frequency, temporal sampling, power resolution, and wet antenna attenuation, Adv. Water Resour., 31, 1481–1493, doi:10.1016/j.advwatres.2008.03.004, 2008.

Leijnse, H., Uijlenhoet, R., and Berne, A.: Errors and uncertainties in microwave link rainfall estimation explored using drop size measurements and high-resolution radar data, J. Hydrometeorol., 11, 1330–1344, doi:10.1175/2010JHM1243.1, 2010.

Lepioufle, J.-M., Leblois, E., and Creutin, J.-D.: Variography of rainfall accumulation in presence of advection, J. Hydrol., 464-465, 494–504, doi:10.1016/j.jhydrol.2012.07.041, 2012.

Messer, H., Zinevich, A., and Alpert, P.: Environmental monitoring by wireless communication networks, Science, 312, p. 713, doi:10.1126/science.1120034, 2006.

Messer, H., Zinevich, A., and Alpert, P.: Environmental sensor networks using existing wireless communication systems for rainfall and wind velocity measurements, IEEE Instru. Meas. Mag., 15, 32–38, doi:10.1109/MIM.2012.6174577, 2012.

Overeem, A., Buishand, T. A., and Holleman, I.: Extreme rainfall analysis and estimation of depth-duration-frequency curves using weather radar, Water Resour. Res., 45, WR007869, doi:10.1029/2009WR007869, 2009a.

Overeem, A., Holleman, I., and Buishand, A.: Derivation of a 10-year radar-based climatology of rainfall, J. Appl. Meteorol. Clim., 48, 1448–1463, doi:10.1175/2009JAMC1954.1, 2009b.

Overeem, A., Leijnse, H., and Uijlenhoet, R.: Measuring urban rainfall using microwave links from commercial cellular communication networks, Water Resour. Res., 47, WR010350, doi:10.1029/2010WR010350, 2011.

Overeem, A., Leijnse, H., and Uijlenhoet, R.: Country-wide rainfall maps from cellular communication networks, P. Natl. Acad. Sci. USA, 110, 2741–2745, doi:10.1073/pnas.1217961110, 2013.

Pebesma, E. J.: Gstat user's manual, available at: http://www.gstat.org/gstat.pdf (last access: 23 March 2015), 1997.

Pebesma, E. J. and Wesseling, C. G.: Gstat: a program for geostatistical modelling, prediction and simulation, Comput. Geosci., 24, 17–31, doi:10.1016/S0098-3004(97)00082-4, 1998.

Schleiss, M., Rieckermann, J., and Berne, A.: Quantification and modeling of wet-antenna attenuation for commercial microwave links, IEEE Geosci. Remote S., 10, 1195–1199, doi:10.1109/LGRS.2012.2236074, 2013.

Schuurmans, J. M., Bierkens, M. F. P., Pebesma, E. J., and Uijlenhoet, R.: Automatic Prediction of High-Resolution Daily Rainfall Fields for Multiple Extents: The Potential of Operational Radar, J. Hydrometeorol., 8, 1204–1224, doi:10.1175/2007JHM792.1, 2007.

van de Beek, C. Z., Leijnse, H., Torfs, P. J. J. F., and Uijlenhoet, R.: Climatology of daily rainfall semi-variance in The Netherlands, Hydrol. Earth Syst. Sci., 15, 171–183, doi:10.5194/hess-15-171-2011, 2011.

van de Beek, C. Z., Leijnse, H., Torfs, P., and Uijlenhoet, R.: Seasonal semi-variance of Dutch rainfall at hourly to daily scales, Adv. Water Resour., 45, 76–85, doi:10.1016/j.advwatres.2012.03.023, 2012.

Velasco-Forero, C. A., Sempere-Torres, D., Cassiraga, E. F., and Gómez-Hernández, J. J.: A non-parametric automatic blending methodology to estimate rainfall fields from rain gauge and radar data, Adv. Water Resour., 32, 986–1002, doi:10.1016/j.advwatres.2008.10.004, 2009.

Zinevich, A., Alpert, P., and Messer, H.: Estimation of rainfall fields using commercial microwave communication networks of variable density, Adv. Water Resour., 31, 1470–1480, doi:10.1016/j.advwatres.2008.03.003, 2008.

Zinevich, A., Messer, H., and Alpert, P.: Frontal rainfall observation by a commercial microwave communication network, J. Appl. Meteorol. Clim., 48, 1317–1334, doi:10.1175/2008JAMC2014.1, 2009.

Zinevich, A., Messer, H., and Alpert, P.: Prediction of rainfall intensity measurement errors using commercial microwave communication links, Atmos. Meas. Tech., 3, 1385–1402, doi:10.5194/amt-3-1385-2010, 2010.

6

Relative effects of statistical preprocessing and postprocessing on a regional hydrological ensemble prediction system

Sanjib Sharma[1], **Ridwan Siddique**[2], **Seann Reed**[3], **Peter Ahnert**[3], **Pablo Mendoza**[4], **and Alfonso Mejia**[1]

[1]Department of Civil and Environmental Engineering, The Pennsylvania State University, University Park, PA, USA
[2]Northeast Climate Science Center, University of Massachusetts, Amherst, MA, USA
[3]National Weather Service, Middle Atlantic River Forecast Center, State College, PA, USA
[4]Advanced Mining Technology Center (AMTC), Universidad de Chile, Santiago, Chile

Correspondence: Sanjib Sharma (sanjibsharma66@gmail.com) and Alfonso Mejia (amejia@engr.psu.edu)

Abstract. The relative roles of statistical weather preprocessing and streamflow postprocessing in hydrological ensemble forecasting at short- to medium-range forecast lead times (day 1–7) are investigated. For this purpose, a regional hydrologic ensemble prediction system (RHEPS) is developed and implemented. The RHEPS is comprised of the following components: (i) hydrometeorological observations (multisensor precipitation estimates, gridded surface temperature, and gauged streamflow); (ii) weather ensemble forecasts (precipitation and near-surface temperature) from the National Centers for Environmental Prediction 11-member Global Ensemble Forecast System Reforecast version 2 (GEFSRv2); (iii) NOAA's Hydrology Laboratory-Research Distributed Hydrologic Model (HL-RDHM); (iv) heteroscedastic censored logistic regression (HCLR) as the statistical preprocessor; (v) two statistical postprocessors, an autoregressive model with a single exogenous variable (ARX(1,1)) and quantile regression (QR); and (vi) a comprehensive verification strategy. To implement the RHEPS, 1 to 7 days weather forecasts from the GEFSRv2 are used to force HL-RDHM and generate raw ensemble streamflow forecasts. Forecasting experiments are conducted in four nested basins in the US Middle Atlantic region, ranging in size from 381 to 12 362 km^2.

Results show that the HCLR preprocessed ensemble precipitation forecasts have greater skill than the raw forecasts. These improvements are more noticeable in the warm season at the longer lead times (> 3 days). Both postprocessors, ARX(1,1) and QR, show gains in skill relative to the raw ensemble streamflow forecasts, particularly in the cool season, but QR outperforms ARX(1,1). The scenarios that implement preprocessing and postprocessing separately tend to perform similarly, although the postprocessing-alone scenario is often more effective. The scenario involving both preprocessing and postprocessing consistently outperforms the other scenarios. In some cases, however, the differences between this scenario and the scenario with postprocessing alone are not as significant. We conclude that implementing both preprocessing and postprocessing ensures the most skill improvements, but postprocessing alone can often be a competitive alternative.

1 Introduction

Both climate variability and climate change, increased exposure from expanding urbanization, and sea level rise are increasing the frequency of damaging flood events and making their prediction more challenging across the globe (Dankers et al., 2014; Wheater and Gober, 2015; Ward et al., 2015). Accordingly, current research and operational efforts in hydrological forecasting are seeking to develop and implement enhanced forecasting systems, with the goals of improving the skill and reliability of short- to medium-range streamflow forecasts (0–14 days) and providing more effective early warning services (Pagano et al., 2014; Thiemig et al., 2015; Emerton et al., 2016; Siddique and Mejia, 2017). Ensemble-based forecasting systems have become the preferred paradigm, showing substantial improvements over single-valued deterministic ones (Schaake et al., 2007; Cloke and Pappenberger, 2009; Demirel et al., 2013; Fan et al., 2014; Demargne et al., 2014; Schwanenberg et al., 2015; Siddique and Mejia, 2017). Ensemble streamflow forecasts can be generated in a number of ways, being the most common approach to the use of meteorological forecast ensembles to force a hydrological model (Cloke and Pappenberger, 2009; Thiemig et al., 2015). Such meteorological forecasts can be generated by multiple alterations of a numerical weather prediction model, including perturbed initial conditions and/or multiple model physics and parameterizations.

A number of ensemble prediction systems (EPSs) are being used to generate streamflow forecasts. In the United States (US), the NOAA's National Weather Service River Forecast Centers are implementing and using the Hydrological Ensemble Forecast Service to incorporate meteorological ensembles into their flood forecasting operations (Demargne et al., 2014; Brown et al., 2014). Likewise, the European Flood Awareness System from the European Commission (Alfieri et al., 2014) and the Flood Forecasting and Warming Service from the Australia Bureau of Meteorology (Pagano et al., 2016) have adopted the ensemble paradigm. Furthermore, different regional EPSs have been designed and implemented for research purposes, to meet specific regional needs, and/or for real-time forecasting applications. Two examples, among several others (Zappa et al., 2008, 2011; Hopson and Webster, 2010; Demuth and Rademacher, 2016; Addor et al., 2011; Golding et al., 2016; Bennett et al., 2014; Schellekens et al., 2011), are the Stevens Institute of Technology's Stevens Flood Advisory System for short-range flood forecasting (Saleh et al., 2016) and the National Center for Atmospheric Research (NCAR) System for Hydromet Analysis, Research, and Prediction for medium-range streamflow forecasting (NCAR, 2017). Further efforts are underway to operationalize global ensemble flood forecasting and early warning systems, e.g., through the Global Flood Awareness System (Alfieri et al., 2013; Emerton et al., 2016).

EPSs are comprised of several system components. In this study, a regional hydrological ensemble prediction system (RHEPS) is used (Siddique and Mejia, 2017). The RHEPS is an ensemble-based research forecasting system, aimed primarily at bridging the gap between hydrological forecasting research and operations by creating an adaptable and modular forecast emulator. The goal with the RHEPS is to facilitate the integration and rigorous verification of new system components, enhanced physical parameterizations, and novel assimilation strategies. For this study, the RHEPS is comprised of the following system components: (i) precipitation and near-surface temperature ensemble forecasts from the National Centers for Environmental Prediction 11-member Global Ensemble Forecast System Reforecast version 2 (GEFSRv2), (ii) NOAA's Hydrology Laboratory-Research Distributed Hydrologic Model (HL-RDHM) (Reed et al., 2004; Smith et al., 2012a, b), (iii) a statistical weather preprocessor (hereafter referred to as preprocessing), (iv) a statistical streamflow postprocessor (here-

after referred to as postprocessing), (v) hydrometeorological observations, and (vi) a verification strategy. Recently, Siddique and Mejia (2017) employed the RHEPS to produce and verify ensemble streamflow forecasts over some of the major river basins in the US Middle Atlantic region. Here, the RHEPS is specifically implemented to investigate the relative roles played by preprocessing and postprocessing in enhancing the quality of ensemble streamflow forecasts.

The goal with statistical processing is to use statistical tools to quantify the uncertainty of and remove systematic biases in the weather and streamflow forecasts in order to improve the skill and reliability of forecasts. In weather and hydrological forecasting, a number of studies have demonstrated the benefits of separately implementing preprocessing (Sloughter et al., 2007; Verkade et al., 2013; Messner et al., 2014a; Yang et al., 2017) and postprocessing (Shi et al., 2008; Brown and Seo, 2010; Madadgar et al., 2014; Ye et al., 2014; Wang et al., 2016; Siddique and Mejia, 2017). However, only a very limited number of studies have investigated the combined ability of preprocessing and postprocessing to improve the overall quality of ensemble streamflow forecasts (Kang et al., 2010; Zalachori et al., 2012; Roulin and Vannitsem, 2015; Abaza et al., 2017). At first glance, in the context of medium-range streamflow forecasting, preprocessing seems necessary and beneficial since meteorological forcing is often biased and its uncertainty is more dominant than the hydrological one (Cloke and Pappenberger, 2009; Bennett et al., 2014; Siddique and Mejia, 2017). In addition, some streamflow postprocessors assume unbiased forcing (Zhao et al., 2011) and hydrological models can be sensitive to forcing biases (Renard et al., 2010).

The few studies that have analyzed the joint effects of preprocessing and postprocessing on short- to medium-range streamflow forecasts have mostly relied on weather ensembles from the European Centre for Medium-range Weather Forecasts (ECMWF) (Zalachori et al., 2012; Roulin and Vannitsem, 2015; Benninga et al., 2017). Kang et al. (2010) used different forcing but focused on monthly, as opposed to daily, streamflow. The conclusions from these studies have been mixed (Benninga et al., 2017). Some have found statistical processing to be useful (Yuan and Wood, 2012), particularly postprocessing, while others have found that it contributes little to forecast quality. Overall, studies indicate that the relative effects of preprocessing and postprocessing depend strongly on the forecasting system (e.g., forcing, hydrological model, statistical processing technique), and conditions (e.g., lead time, study area, season), underscoring the research need to rigorously verify and benchmark new forecasting systems that incorporate statistical processing.

The main objective of this study is to verify and assess the ability of preprocessing and postprocessing to improve ensemble streamflow forecasts from the RHEPS. This study differs from previous ones in several important respects. The assessment of statistical processing is done using a spatially distributed hydrological model whereas previous studies have tended to emphasize spatially lumped models. Much of the previous studies have used ECMWF forecasts; here we rely on GEFSRv2 precipitation and temperature outputs. Also, we test and implement a preprocessor, namely heteroscedastic censored logistic regression (HCLR), which has not been used before in streamflow forecasting. We also consider a relatively wider range of basin sizes and longer study period than in previous studies. In particular, this paper addresses the following questions:

- What are the separate and joint contributions of preprocessing and postprocessing over the raw RHEPS outputs?

- What forecast conditions (e.g., lead time, season, flow threshold, and basin size) benefit potential increases in skill?

- How much skill improvement can be expected from statistical processing under different uncertainty scenarios (i.e., when skill is measured relative to observed or simulated flow conditions)?

The remainder of the paper is organized as follows. Section 2 presents the study area. Section 3 describes the different components of the RHEPS. The main results and their implications are examined in Sect. 4. Lastly, Sect. 5 summarizes key findings.

2 Study area

The North Branch Susquehanna River (NBSR) basin in the US Middle Atlantic region (MAR) is selected as the study area (Fig. 1), with an overall drainage area of $12\,362\,\text{km}^2$. The NBSR basin is selected as flooding is an important regional concern. This region has a relatively high level of urbanization and high frequency of extreme weather events, making it particularly vulnerable to damaging flood events (Gitro et al., 2014; MARFC, 2017). The climate in the upper MAR, where the NBSR basin is located, can be classified as warm, humid summers and snowy, cold winters with frozen precipitation (Polsky et al., 2000). During the cool season, a positive North Atlantic Oscillation phase generally results in increased precipitation amounts and occurrence of heavy snow (Durkee et al., 2007). Thus, flooding in the cool season is dominated by heavy precipitation events accompanied by snowmelt runoff. In the summer season, convective thunderstorms with increased intensity may lead to greater variability in streamflow. In the NBSR basin, we select four different US Geological Survey (USGS) daily gauge stations, representing a system of nested subbasins, as the forecast locations (Fig. 1). The selected locations are the Ostelic River at Cincinnatus (USGS gauge 01510000), Chenango River at Chenango Forks (USGS gauge 01512500), Susquehanna River at Conklin (USGS gauge 01503000), and Susquehanna River at Waverly (USGS gauge 01515000) (Fig. 1).

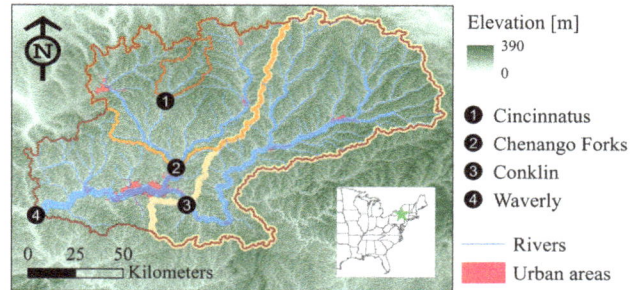

Figure 1. Map illustrating the location of the four selected river basins in the US middle Atlantic region.

The drainage area of the selected basins ranges from 381 to $12\,362\,\text{km}^2$. Table 1 outlines some key characteristics of the study basins.

3 Approach

In this section, we describe the different components of the RHEPS, including the hydrometeorological observations, weather forecasts, preprocessor, postprocessors, hydrological model, and the forecasting experiments and verification strategy.

3.1 Hydrometeorological observations

Three main observation datasets are used: multisensor precipitation estimates (MPEs), gridded near-surface air temperature, and daily streamflow. MPEs and gridded near-surface air temperature are used to run the hydrological model in simulation mode for parameter calibration purposes and to initialize the RHEPS. Both the MPEs and gridded near-surface air temperature data at $4 \times 4\,\text{km}^2$ resolution were provided by the NOAA's Middle Atlantic River Forecast Center (MARFC) (Siddique and Mejia, 2017). Similar to the NCEP stage-IV dataset (Moore et al., 2015; Prat and Nelson, 2015), the MARFC's MPEs represent a continuous time series of hourly, gridded precipitation observations at $4 \times 4\,\text{km}^2$ cells, which are produced by combining multiple radar estimates and rain gauge measurements. The gridded near-surface air temperature data at $4 \times 4\,\text{km}^2$ resolution were developed by the MARFC by combining multiple temperature observation networks as described by Siddique and Mejia (2017). Daily streamflow observations for the selected basins were obtained from the USGS. The streamflow observations are used to verify the simulated flows, and the raw and postprocessed ensemble streamflow forecasts.

3.2 Meteorological forecasts

GEFSRv2 data are used for the ensemble precipitation and near-surface air temperature forecasts. The GEFSRv2 uses

Table 1. Main characteristics of the four study basins.

Location of outlet	Cincinnatus, New York	Chenango Forks, New York	Conklin, New York	Waverly, New York
NWS id	CINN6	CNON6	CKLN6	WVYN6
USGS id	01510000	01512500	01503000	01515000
Area (km^2)	381	3841	5781	12 362
Latitude	42°32′28″	42°13′05″	42°02′07″	41°59′05″
Longitude	75°53′59″	75°50′54″	75°48′11″	76°30′04″
Minimum daily flow[a] (m^3 s^{-1})	0.31 (0.11)	4.05 (2.49)	6.80 (5.32)	13.08 (6.71)
Maximum daily flow[a] (m^3 s^{-1})	172.73 (273.54)	1248.77 (1401.68)	2041.64 (2174.734)	4417.42 (4417.42)
Mean daily flow[a] (m^3 s^{-1})	8.89 (9.17)	82.36 (81.66)	122.93 (121.99)	277.35 (215.01)
Climatological flow ($Pr = 0.95$)[b] (m^3 s^{-1})	29.45	266.18	382.28	843.84

[a] The numbers in parentheses are the historical (based on entire available record, as opposed to the period 2004–2012 used in this study) daily minimum, maximum, or mean recorded flow. [b] $Pr = 0.95$ indicates flows with exceedance probability of 0.05.

the same atmospheric model and initial conditions as the version 9.0.1 of the Global Ensemble Forecast System and runs at T254L42 ($\sim 0.50°$ Gaussian grid spacing or ~ 55 km) and T190L42 ($\sim 0.67°$ Gaussian grid spacing or ~ 73 km) resolutions for the first and second 8 days, respectively (Hamill et al., 2013). The reforecasts are initiated once daily at 00:00 Coordinated Universal Time. Each forecast cycle consists of 3-hourly accumulations for day 1 to day 3 and 6-hourly accumulations for day 4 to day 16. In this study, we use 9 years of GEFSRv2 data, from 2004 to 2012, and forecast lead times from 1 to 7 days. The period 2004 to 2012 is selected to take advantage of data that were previously available to us (i.e., GEFSRv2 and MPEs for the MAR) from a recent verification study (Siddique et al., 2015). Forecast lead times of up to 7 days are chosen since we previously found that the GEFSRv2 skill is low after 7 days (Siddique et al., 2015; Sharma et al., 2017). The GEFSRv2 data are bilinearly interpolated onto the 4×4 km^2 grid cell resolution of the HL-RDHM model.

3.3 Distributed hydrological model

NOAA's HL-RDHM is used as the spatially distributed hydrological model (Koren et al., 2004). Within HL-RDHM, the Sacramento Soil Moisture Accounting model with Heat Transfer (SAC-HT) is used to represent hillslope runoff generation, and the SNOW-17 module is used to represent snow accumulation and melting.

HL-RDHM is a spatially distributed conceptual model, where the basin system is divided into regularly spaced, square grid cells to account for spatial heterogeneity. Each grid cell acts as a hillslope capable of generating surface, interflow, and groundwater runoff that discharges directly into the streams. The cells are connected to each other through

the stream network system. Further, the SNOW-17 module allows each cell to accumulate snow and generate hillslope snowmelt based on the near-surface air temperature. The hillslope runoff, generated at each grid cell by SAC-HT and SNOW-17, is routed to the stream network using a nonlinear kinematic wave algorithm (Koren et al., 2004; Smith et al., 2012a). Likewise, flows in the stream network are routed downstream using a nonlinear kinematic wave algorithm that accounts for parameterized stream cross-section shapes (Koren et al., 2004; Smith et al., 2012a). In this study, we run HL-RDHM using a 2 km horizontal resolution. Further information about the HL-RDHM can be found elsewhere (Koren et al., 2004; Reed et al., 2007; Smith et al., 2012a; Fares et al., 2014; Rafieeinasab et al., 2015; Thorstensen et al., 2016; Siddique and Mejia, 2017).

To calibrate HL-RHDM, we first run the model using a priori parameter estimates previously derived from available datasets (Koren et al., 2000; Reed et al., 2004; Anderson et al., 2006). We then select 10 out of the 17 SAC-HT parameters for calibration based upon prior experience and preliminary sensitivity tests. During the calibration process, each a priori parameter field is multiplied by a factor. Therefore, we calibrate these factors instead of the parameter values at all grid cells, assuming that the a priori parameter distribution is true (e.g., Mendoza et al., 2012). The multiplying factors are adjusted manually first; once the manual changes do not yield noticeable improvements in model performance, the factors are tuned up using stepwise line search (SLS; Kuzmin et al., 2008; Kuzmin, 2009). This method is readily available within HL-RDHM, and has been shown to provide reliable parameter estimates (Kuzmin et al., 2008; Kuzmin, 2009). With SLS, the following objective function is optimized:

$$OF = \sqrt{\sum_{i=1}^{m} \left[q_i - s_i(\Omega) \right]^2}, \tag{1}$$

where q_i and s_i denote the daily observed and simulated flows at time i, respectively; Ω is the parameter vector being estimated; and m is the total number of days used for calibration. A total of 3 years (2003–2005) of streamflow data are used to calibrate the HL-RDHM for the selected basins. The first year (year 2003) is used to warm up HL-RDHM. To assess the model performance during calibration, we use the percent bias (PB), modified correlation coefficient (R_m), and Nash–Sutcliffe efficiency (NSE) (see Appendix for details). Note that these metrics are used during the manual phase of the calibration process, and to assess the final results from the implementation of the SLS. However, the actual implementation of the SLS is based on the objective function in Eq. (1).

3.4　Statistical weather preprocessor

Heteroscedastic censored logistic regression (Messner et al., 2014a; Yang et al., 2017) is implemented to preprocess the ensemble precipitation forecasts from the GEFSRv2. HCLR is selected since it offers the advantage, over other regression-based preprocessors (Wilks, 2009), of obtaining the full, continuous predictive probability density function (pdf) of precipitation forecasts (Messner et al., 2014b). Also, HCLR has been shown to outperform other widely used preprocessors, such as Bayesian model averaging (Yang et al., 2017). In principle, HCLR fits the conditional logistic probability distribution function to the transformed (here the square root) ensemble mean and bias corrected precipitation ensembles. Note that we tried different transformations (square root, cube root, and fourth root) and found a similar performance between the square and cube root, both outperforming the fourth root. In addition, HCLR uses the ensemble spread as a predictor, which allows the use of uncertainty information contained in the ensembles.

The development of the HCLR follows the logistic regression model initially proposed by Hamill et al. (2004) as well as the extended version of that model proposed by Wilks (2009). The extended logistic regression of Wilks (2009) is used to model the probability of binary responses such that

$$P(y \leq z|x) = \Lambda[\omega(z) - \delta(x)], \tag{2}$$

where $\Lambda(.)$ denotes the cumulative distribution function of the standard logistic distribution, y is the transformed precipitation, z is a specified threshold, x is a predictor variable that depends on the forecast members, $\delta(x)$ is a linear function of the predictor variable x, and the transformation $\omega(.)$ is a monotone nondecreasing function. Messner et al. (2014a) proposed the heteroscedastic extended logistic regression (HELR) preprocessor with an additional predictor

variable φ to control the dispersion of the logistic predictive distribution,

$$P(y \leq z|x) = \Lambda \left\{ \frac{\omega(z) - \delta(x)}{\exp[\eta(\varphi)]} \right\}, \tag{3}$$

where $\eta(.)$ is a linear function of φ. The functions $\delta(.)$ and $\eta(.)$ are defined as follows:

$$\delta(x) = a_0 + a_1 x \tag{4}$$

and

$$\eta(\varphi) = b_0 + b_1\varphi, \tag{5}$$

where a_0, a_1, b_0, and b_1 are parameters that need to be estimated; $x = \frac{1}{K} \sum_{k=1}^{K} f_k^{\frac{1}{2}}$, i.e., the predictor variable x is the mean of the transformed, via the square root, ensemble forecasts f; K is the total number of ensemble members; and φ is the standard deviation of the square-root-transformed precipitation ensemble forecasts.

Maximum likelihood estimation with the log-likelihood function is used to estimate the parameters associated with Eq. (3) (Messner et al., 2014a, b). One variation of the HELR preprocessor that can easily accommodate nonnegative variables, such as precipitation amounts, is HCLR. For this, the predicted probability or likelihood π_i of the ith observed outcome is determined as follows (Messner et al., 2014b):

$$\pi_i = \begin{cases} \Lambda \left[\dfrac{\omega(0) - \delta(x)}{\exp[\eta(\varphi)]} \right] & y_i = 0 \\ \lambda \left[\dfrac{\omega(y_i) - \delta(x)}{\exp[\eta(\varphi)]} \right] & y_i > 0 \end{cases}, \tag{6}$$

where $\lambda[.]$ denotes the likelihood function of the standard logistic function. As indicated by Eq. (6), HCLR fits a logistic error distribution with point mass at zero to the transformed predictand.

HCLR is applied here to each GEFSRv2 grid cell within the selected basins. At each cell, HCLR is implemented for the period 2004–2012 using a leave-one-out approach. For this, we select 7 years for training and the 2 remaining years for verification purposes. This is repeated until all the 9 years have been preprocessed and verified independently of the training period. This is done so that no training data are discarded and the entire 9-year period of analysis can be used to generate the precipitation forecasts. HCLR is employed for 6-hourly precipitation accumulations for lead times from 6 to 168 h. To train the preprocessor, we use a stationary training period, as opposed to a moving window, for each season and year to be forecasted, comprised of the seasonal data from all the 7 training years. Thus, to forecast a given season and specific lead time, we use ~6930 forecasts (i.e., 11 members × 90 days per season × 7 years). We previously tested using a moving window training approach and found that the results were similar to the stationary window approach

(Yang et al., 2017). To make the implementation of HCLR as straightforward as possible, the stationary window is used here. Finally, the Schaake shuffle method as applied by Clark et al. (2004) is implemented to maintain the observed space–time variability in the preprocessed GEFSRv2 precipitation forecasts. At each individual forecast time, the Schaake shuffle is applied to produce a spatial and temporal rank structure for the ensemble precipitation values that is consistent with the ranks of the observations.

3.5 Statistical streamflow postprocessors

To statistically postprocess the flow forecasts generated by the RHEPS, two different approaches are tested, namely a first-order autoregressive model with a single exogenous variable, ARX(1,1), and quantile regression (QR). We select the ARX(1,1) postprocessor since it has been suggested and implemented for operational applications in the US (Regonda et al., 2013). QR is chosen because it is of similar complexity to the ARX(1,1) postprocessor but for some forecasting conditions it has been shown to outperform it (Mendoza et al., 2016). Furthermore, the ARX (1,1) and QR postprocessors have not been compared against each other for the forecasting conditions specified by the RHEPS. The postprocessors are implemented for the years 2004–2012, using the same leave-one-out approach used for the preprocessor. For this, the 6-hourly precipitation accumulations are used to force the HL-RDHM and generate 6-hourly flows. Note that we use 6-hourly accumulations since this is the resolution of the GEFSRv2 data after day 4 and this is a temporal resolution often used in operational forecasting in the US. Since the observed flow data are mean daily, we compute the mean daily flow forecast from the 6-hourly flows. The postprocessor is then applied to the mean daily values from day 1 to 7.

3.5.1 First-order autoregressive model with a single exogenous variable

To implement the ARX(1,1) postprocessor, the observation and forecast data are first transformed into standard normal deviates using the normal quantile transformation (NQT) (Krzysztofowicz, 1997; Bogner et al., 2012). The transformed observations and forecasts are then used as predictors in the ARX(1,1) model (Siddique and Mejia, 2017). Specifically, for each forecast lead time, the ARX(1,1) postprocessor is formulated as follows:

$$q_{i+1}^T = (1 - c_{i+1}) q_i^T + c_{i+1} f_{i+1}^T + \xi_{i+1}, \tag{7}$$

where q_i^T and q_{i+1}^T are the NQT-transformed observed flows at time steps i and $i + 1$, respectively; c is the regression coefficient; f_{i+1}^T is the NQT transformed forecast flow at time step $i + 1$; and ξ is the residual error term. In Eq. (7), assuming that there is significant correlation between ξ_{i+1} and q_i^T, ξ_{i+1} can be calculated as follows:

$$\xi_{i+1} = \frac{\sigma_{\xi_{i+1}}}{\sigma_{\xi_i}} \rho (\xi_{i+1}, \xi_i) \xi_i + \vartheta_{i+1}, \tag{8}$$

where σ_{ξ_i} and $\sigma_{\xi_{i+1}}$ are the standard deviation of ξ_i and ξ_{i+1}, respectively; $\rho(\xi_{i+1}, \xi_i)$ is the serial correlation between ξ_{i+1} and ξ_i; and ϑ_{i+1} is a random Gaussian error generated from $N(0, \sigma_{\vartheta_{i+1}}^2)$. To estimate $N(0, \sigma_{\vartheta_{i+1}}^2)$, the following equation is used:

$$\sigma_{\vartheta_{i+1}}^2 = \left[1 - \rho^2 (\xi_{i+1}, \xi_i)\right] \sigma_{\xi_{i+1}}^2. \tag{9}$$

To implement Eq. (7), 10 equally spaced values of c_{i+1} are selected from 0.1 to 0.9. For each value of c_{i+1}, $\sigma_{\vartheta_{i+1}}^2$ is determined from Eq. (9), using the training data to determine the other variables in Eq. (9). Then, ϑ_{i+1} is generated from $N(0, \sigma_{\vartheta_{i+1}}^2)$ and ξ_{i+1} is calculated from Eq. (8). The result from Eq. (8) is used with Eq. (7) to generate a trace of q_{i+1}^T which is transformed back to real space using the inverse NQT. These steps are repeated to generate multiple traces for each value of c_{i+1}. For each value of c_{i+1}, the ARX(1,1) model is trained and used to generate ensemble streamflow forecasts, which are in turn used to compute the mean continuous ranked probability score (CRPS) for the 7-year training period under consideration. Thus, the mean CRPS is computed for each value of c_{i+1}, and the value of c_{i+1} that produces the smallest mean CRPS is then selected for use in the 2-year verification period under consideration. This is repeated until all the years (2004–2012) have been postprocessed and verified independently of the training period. The ARX(1,1) postprocessor is applied at each individual lead time. Thus, at each forecast lead time, an optimal value of c_{i+1} is estimated by minimizing the mean CRPS following the steps previously outlined. For lead times beyond the initial one (day 1), 1-day-ahead predictions are used as the observed streamflow. For the cases in which q_{i+1}^T falls beyond the historical maxima, extrapolation is used by modeling the upper tail of the forecast distribution as hyperbolic (Journel and Huijbregts, 1978).

3.5.2 Quantile regression

Quantile regression (Koenker and Bassett Jr., 1978; Koenker, 2005) is employed to determine the error distribution, conditional on the ensemble mean, resulting from the difference between observations and forecasts (Dogulu et al., 2015; López López et al., 2014; Weerts et al., 2011; Mendoza et al., 2016). QR is applied here in streamflow space, since it has been shown that, in hydrological forecasting applications, QR has similar skill performance in streamflow space as well as normal space (López López et al., 2014). Another advantage of QR is that it does not make any prior assumptions regarding the shape of the distribution. Further, since QR results in conditional quantiles rather than conditional means, QR is less sensitive to the tail behavior of the streamflow dataset and, consequently, less sensitive to outliers. Note that although QR is here implemented separately for each

Table 2. Summary and description of the verification scenarios.

Scenario	Description
S1	Verification of the raw ensemble precipitation forecasts from the GEFSRv2
S2	Verification of the preprocessed ensemble precipitation forecasts from the GEFSRv2: GEFSRv2 + HCLR
S3	Verification of the raw ensemble flood forecasts: GEFSRv2 + HL-RDHM
S4	Verification of the preprocessed ensemble flood forecasts: GEFSRv2 + HCLR + HL-RDHM
S5	Verification of the postprocessed ensemble flood forecasts: GEFSRv2 + HL-RDHM + QR
S6	Verification of the preprocessed and postprocessed ensemble flood forecasts: GEFSRv2 + HCLR + HL-RDHM + QR

lead time, the mathematical notation does not reflect this for simplicity.

The QR model is given by

$$\varepsilon_\tau' = d_\tau + e_\tau \overline{f}, \tag{10}$$

where ε_τ' is the error estimate at quantile interval τ, $\overline{f}$ is the ensemble mean, and d_τ and e_τ are the linear regression coefficients a τ. The coefficients are determined by minimizing the sum of the residuals based on the training data as follows:

$$\min \sum_{i=1}^{N} w_\tau \left[\varepsilon_{\tau,i} - \varepsilon_\tau' \left(i, \overline{f}_i \right) \right]. \tag{11}$$

Here, $\varepsilon_{\tau,i}$ and $\overline{f}_i$ are the paired samples from a total of N samples; $\varepsilon_{\tau,i}$ is computed as the observed flow minus the forecasted one, $q_\tau - f_\tau$; and w_τ is the weighting function for the τth quantile defined as follows:

$$w_\tau (\zeta_i) = \begin{cases} (\tau - 1)\zeta_i & \text{if } \zeta_i \leq 0 \\ \tau \zeta_i & \text{if } \zeta_i > 0 \end{cases}, \tag{12}$$

where ζ_i is the residual term defined as the difference between $\varepsilon_{\tau,i}$ and $\varepsilon_\tau'(i, \overline{f}_i)$ for the quantile τ. The minimization in Eq. (11) is solved using linear programming (Koenker, 2005).

Lastly, to obtain the calibrated forecast, f_τ, the following equation is used:

$$f_\tau = \overline{f} + \varepsilon_\tau'. \tag{13}$$

In Eq. (13), the estimated error quantiles and the ensemble mean are added to form a calibrated discrete quantile relationship for a particular forecast lead time and thus generate an ensemble streamflow forecast.

3.6 Forecast experiments and verification

The verification analysis is carried out using the Ensemble Verification System (Brown et al., 2010). For the verification, the following metrics are considered: Brier skill score (BSS), mean continuous ranked probability skill score (CRPSS), and the decomposed components of the CRPS (Hersbach, 2000), i.e., the CRPS reliability ($CRPS_{rel}$) and CRPS potential ($CRPS_{pot}$). The definition of each of these metrics is provided in the appendix. Additional details about the verification metrics can be found elsewhere (Wilks, 2011; Jolliffe and Stephenson, 2012). Confidence intervals for the verification metrics are determined using the stationary block bootstrap technique (Politis and Romano, 1994), as done by Siddique et al. (2015). All the forecast verifications are done for lead times from 1 to 7 days.

To verify the forecasts for the period 2004–2012, six different forecasting scenarios are considered (Table 2). The first (S1) and second (S2) scenario verify the raw and preprocessed ensemble precipitation forecasts, respectively. Scenarios 3 (S3), 4 (S4), and 5 (S5) verify the raw, preprocessed, and postprocessed ensemble streamflow forecasts, respectively. The last scenario, S6, verifies the combined preprocessed and postprocessed ensemble streamflow forecasts. In S1 and S2, the raw and preprocessed ensemble precipitation forecasts are verified against the MPEs. For the verification of S1 and S2, each grid cell is treated as a separate verification unit. Thus, for a particular basin, the average performance is obtained by averaging the verification results from different verification units. The streamflow forecast scenarios, S3–S6, are verified against mean daily streamflow observations from the USGS. The quality of the streamflow forecasts is evaluated conditionally upon forecast lead time, season (cool and warm), and flow threshold.

4 Results and discussion

This section is divided into four subsections. The first subsection demonstrates the performance of the spatially distributed model, HL-RDHM. The second subsection describes the performance of the raw and preprocessed GEFSRv2 ensemble precipitation forecasts (forecasting scenarios S1 and S2). In the third subsection, the two statistical postprocessing techniques are compared. Lastly, the verification of different ensemble streamflow forecasting scenarios is shown in the fourth subsection (forecasting scenarios S3–S6).

4.1 Performance of the distributed hydrological model

To assess the performance of HL-RDHM, the model is used to generate streamflow simulations which are verified against daily observed flows, covering the entire period of analysis

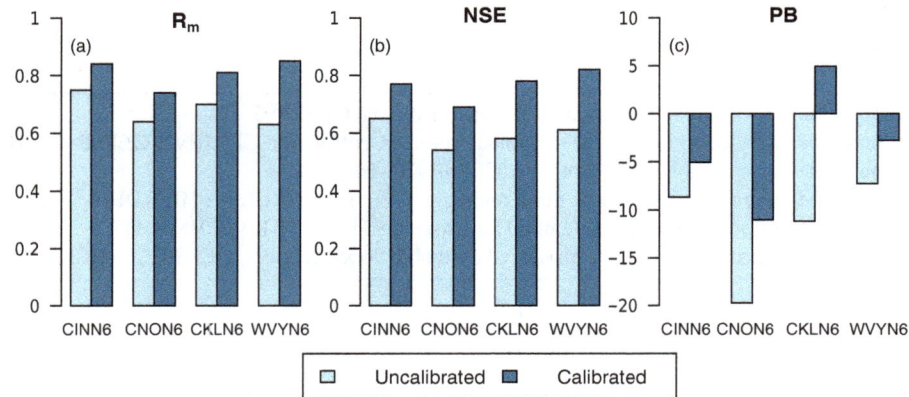

Figure 2. Performance statistics for the uncalibrated and calibrated simulation runs for the entire period of analysis (years 2004–2012): **(a)** R_{m}, **(b)** NSE, and **(c)** PB.

(years 2004–2012). Note that the simulated flows are obtained by forcing HL-RDHM with gridded observed precipitation and near-surface temperature data. The verification is done for the four basin outlets shown in Fig. 1. To perform the verification and assess the quality of the streamflow simulations, the following statistical measures of performance are employed: modified correlation coefficient, Nash–Sutcliffe efficiency, and percent bias. The mathematical definition of these metrics is provided in the appendix. The verification is done for both uncalibrated and calibrated simulation runs for the entire period of analysis. The main results from the verification of the streamflow simulations are summarized in Fig. 2.

The performance of the calibrated simulation runs is satisfactory, with R_{m} values ranging from ~ 0.75 to 0.85 (Fig. 2a). Likewise, the NSE, which is sensitive to both the correlation and bias, ranges from ~ 0.69 to 0.82 for the calibrated runs (Fig. 2b), while the PB ranges from ~ 5 to $-11\,\%$ (Fig. 2c). Relative to the uncalibrated runs, the R_{m}, NSE, and PB values improve by ~ 18, 29, and 47 %, respectively. Further, the performance of the calibrated simulation runs is similar across the four selected basins, although the largest basin, WVYN6 (Fig. 2), shows slightly higher performance with R_{m}, NSE, and PB values of 0.85, 0.82, and $-3\,\%$ (Fig. 2), respectively. The lowest performance is seen in CNON6 with R_{m}, NSE, and PB values of 0.75, 0.7, and $-11\,\%$ (Fig. 2), respectively. Nonetheless, the performance metrics for both the uncalibrated and calibrated simulation runs do not deviate widely from each other in the selected basins, with perhaps the only exception being PB (Fig. 2c).

4.2 Verification of the raw and preprocessed ensemble precipitation forecasts

To examine the skill of both the raw and preprocessed GEFSRv2 ensemble precipitation forecasts, we plot in Fig. 3 the CRPSS (relative to sampled climatology) as a function of the

forecast lead time (day 1 to 7) and season for the selected basins. Two seasons are considered: cool (October–March) and warm (April–September). Note that a CRPSS value of zero means no skill (i.e., same skill as the reference system) and a value of 1 indicates maximum skill. The CRPSS is computed using 6-hourly precipitation accumulations.

The skill of both the raw and preprocessed ensemble precipitation forecasts tends to decline with increasing forecast lead time (Fig. 3). In the warm season (Fig. 3a–d), the CRPSS values vary overall, across all the basins, in the range from ~ 0.17 to 0.5 and from ~ 0.0 to 0.4 for the preprocessed and raw forecasts, respectively; while in the cool season (Fig. 3e–h) the CRPSS values vary overall in the range from ~ 0.2 to 0.6 and from ~ 0.1 to 0.6 for the preprocessed and raw forecasts, respectively. The skill of the preprocessed ensemble precipitation forecasts tends to be greater than the raw ones across basins, seasons, and forecast lead times. Comparing the raw and preprocessed forecasts against each other, the relative skill gains from preprocessing are somewhat more apparent in the medium-range lead times (> 3 days) and warm season. That is, the differences in skill seem not as significant in the short-range lead times (≤ 3 days). This seems particularly the case in the cool season, where the confidence intervals for the raw and preprocessed forecasts tend to overlap (Fig. 3e–h).

Indeed, seasonal skill variations are noticeable in all the basins. Even though the relative gain in skill from preprocessing is slightly greater in the warm season, the overall skill of both the raw and preprocessed forecasts is better in the cool season than the warm one. This may be due, among other potential factors, to the greater uncertainty associated with modeling convective precipitation, which is more prevalent in the warm season, by the NWP model used to generate the GEFSRv2 outputs (Hamill et al., 2013; Baxter et al., 2014). Nonetheless, the warm season preprocessed forecasts show gains in skill across all the lead times and basins. For a particular season, the forecast ensembles across the different

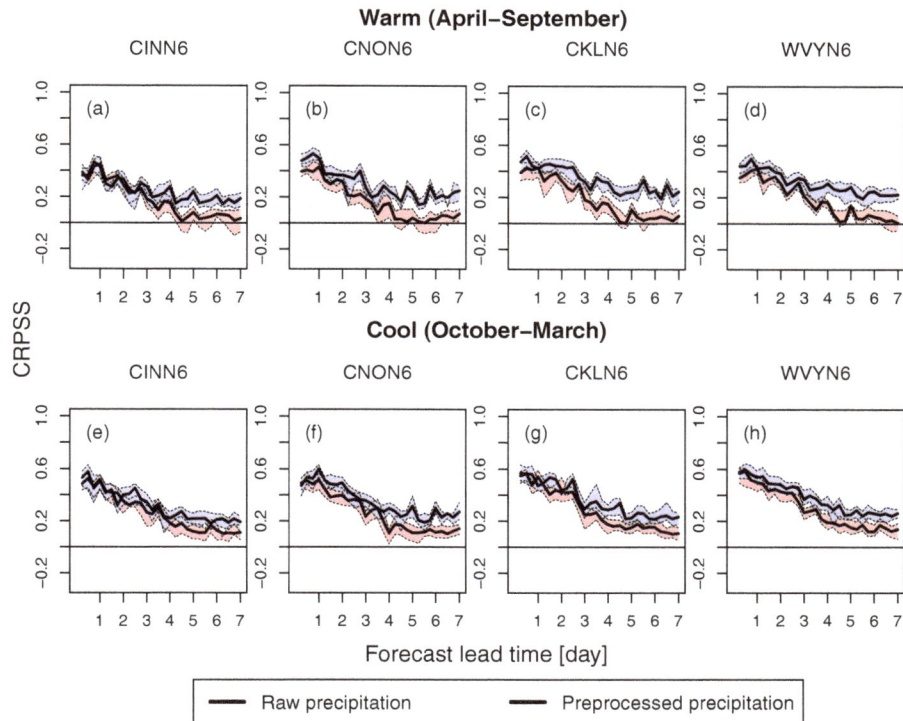

Figure 3. CRPSS (relative to sampled climatology) of the raw (red curves) and preprocessed (blue curves) ensemble precipitation forecasts from the GEFSRv2 vs. the forecast lead time during the **(a)–(d)** warm (April–September) and **(e)–(h)** cool season (October–March) for the selected basins.

basins tend to display similar performance; i.e. the analysis does not reflect skill sensitivity to the basin size as in other studies (Siddique et al., 2015; Sharma et al., 2017). This is expected here since the verification is performed for each GEFSRv2 grid cell, rather than verifying the average for the entire basin. That is, the results in Fig. 3 are for the average skill performance obtained from verifying each individual grid cell within the selected basins.

Based on the results presented in Fig. 3, we may expect some skill contribution to the streamflow ensembles from forcing the HL-RDHM with the preprocessed precipitation, as opposed to using the raw forecast forcing. It may also be expected that the contributions are greater for the medium-range lead times and warm season. This will be examined in Sect. 4.4; prior to that we compare next the two postprocessors, namely ARX(1,1) and QR.

4.3 Selection of the streamflow postprocessor

The ability of the ARX(1,1) and QR postprocessors to improve ensemble streamflow forecasts is investigated here. The postprocessors are applied to the raw streamflow ensembles at each forecast lead time from day 1 to 7. To examine the skill of the postprocessed streamflow forecasts, Fig. 4 displays the CRPSS (relative to the raw ensemble streamflow forecasts) versus the forecast lead time for all the selected

basins, for both warm (Fig. 4a–d) and cool (Fig. 4e–h) seasons. In the cool season (Fig. 4e–h), the tendency is for both postprocessing techniques to demonstrate improved forecast skill across all the basins and lead times. The skill can improve as much as 40 % at the later lead times (Fig. 4f). The skill improvements, however, from the ARX(1,1) postprocessor are not as consistent for the warm season (Fig. 4a–d), displaying negative skill values for some of the lead times in all the basins. The latter underscores an inability of the ARX(1,1) postprocessor to enhance the raw streamflow ensembles for the warm season. In some cases (Fig. 4b and e–f), the skill of the postprocessors shows an increasing trend with the lead time. This is the case since the skill is here measured relative to the raw streamflow forecasts, which is done to better isolate the effect of the postprocessors on the streamflow forecasts.

The gains in skill from QR vary from $\sim 0\,\%$ (Fig. 4b at the day 1 lead time) to $\sim 40\,\%$ (Fig. 4f at lead times > 4 days) depending upon the season and lead time. The gains from ARX(1,1), however, vary from $\sim 0\,\%$ (Fig. 4g at the day 1 lead time) to a much lower level of $\sim 28\,\%$ (Fig. 4f at the day 4 lead time) during the cool season, while there are little to no gains in the warm season. In the cool season (Fig. 4e–h), both postprocessors exhibit somewhat similar performance at different lead times, with the exception of

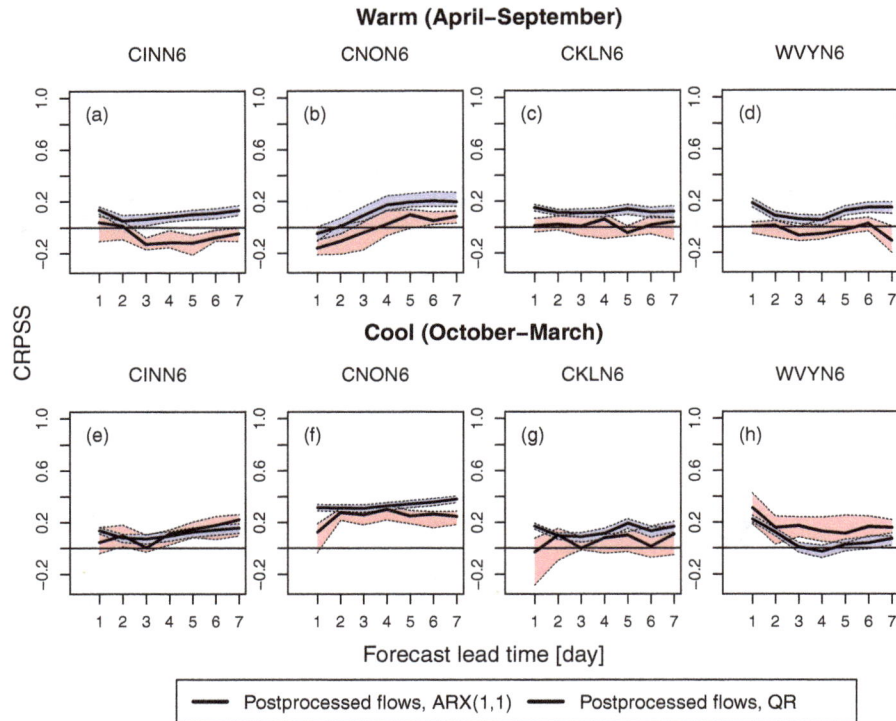

Figure 4. CRPSS (relative to the raw forecasts) of the ARX(1,1) (red curves) and QR (blue curves) postprocessed ensemble flood forecasts vs. the forecast lead time during the **(a)–(d)** warm (April–September) and **(e)–(h)** cool season (October–March) for the selected basins.

Fig. 4h, but in the warm season QR tends to consistently perform better than ARX(1,1). The overall trend in Fig. 4 is for QR to mostly outperform ARX(1,1), with the difference in performance being as high as 30 % (Fig. 4d at the day 7 lead time). This is noticeable across all the basins (except WVYN6 in Fig. 4h) for most of the lead times and for both seasons.

As discussed and demonstrated in Fig. 4, QR performs better than ARX(1,1). We also computed reliability diagrams, as determined by Sharma et al. (2017), for the two postprocessors (plots not shown) and found that QR tends to display better reliability than ARX(1,1) across lead times, basins, and seasons. Therefore, we select QR as the statistical streamflow postprocessor to examine the interplay between preprocessing and postprocessing in the RHEPS.

4.4   Verification of the ensemble streamflow forecasts for different statistical processing scenarios

In this subsection, we examine the effects of different statistical processing scenarios on the ensemble streamflow forecasts from the RHEPS. The forecasting scenarios considered here are S3–S6 (Table 2 defines the scenarios). To facilitate presenting the verification results, this subsection is divided into the following three parts: CRPSS, CRPS decomposition, and BSS.

4.4.1   CRPSS

The skill of the ensemble streamflow forecasts for S3–S6 is assessed using the CRPSS relative to the sampled climatology (Fig. 5). The CRPSS in Fig. 5 is shown as a function of the forecast lead time for all the basins, and the warm (Fig. 5a–d) and cool (Fig. 5e–h) seasons. The most salient feature of Fig. 5 is that the performance of the streamflow forecasts tends for the most part to progressively improve from S3 to S6. This means that the forecast skill tends to improve across lead times, basin sizes, and seasons as additional statistical processing steps are included in the RHEPS forecasting chain. Although there is some tendency for the large basins to show better forecast skill than the small ones, the scaling (i.e., the dependence of skill on the basin size) is rather mild and not consistent across the four basins.

In Fig. 5, the skill first increases from the raw scenario (i.e., S3, where no statistical processing is done) to the scenario where only preprocessing is performed, S4. The gain in skill between S3 and S4 is generally small at the short lead times (< 3 days) but increases for the later lead times; this is somewhat more evident for the cool season than the warm one. This skill trend between S3 and S4 is not entirely surprising, as we previously saw (Fig. 3) that differences between the raw and preprocessed precipitation ensembles are more significant at the later lead times. The skill in Fig. 5

Warm (April–September)

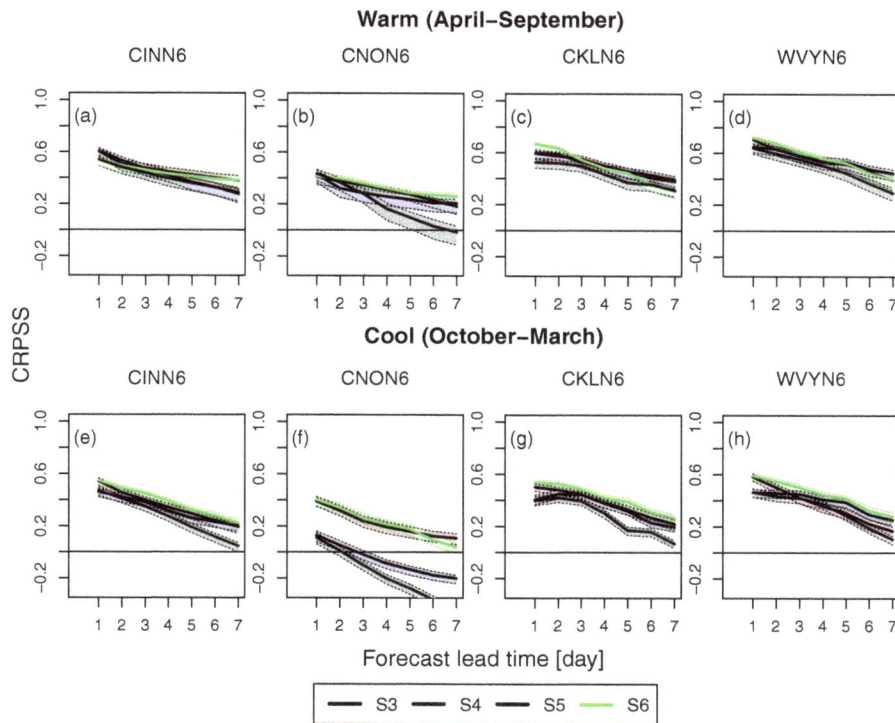

Cool (October–March)

Figure 5. Continuous ranked probability skill score (CRPSS) of the mean ensemble flood forecasts vs. the forecast lead time during the **(a)**–**(d)** warm (April–September) and **(e)**–**(h)** cool season (October–March) for the selected basins. The curves represent the different forecasting scenarios S3–S6. Note that S3 consists of GEFSRv2 + HL-RDHM, S4 of GEFSRv2 + HCLR + HL-RDHM, S5 of GEFSRv2 + HL-RDHM + QR, and S6 of GEFSRv2 + HCLR + HL-RDHM + QR.

then shows further improvements for both S5 and S6, relative to S4. Although S6 tends to outperform both S4 and S5 in Fig. 5, the differences in skill among these three scenarios are not as significant; their confidence intervals tend to overlap in most cases, with the exception of Fig. 5f, in which S4 underperforms relative to both S5 and S6. Figure 5 shows that S6 is the preferred scenario in that it tends to more consistently improve the ensemble streamflow forecasts across basins, lead times, and seasons than the other scenarios. It also shows that postprocessing alone, S5, may be slightly more effective than preprocessing alone, S4, in correcting the streamflow forecast biases.

There are also seasonal differences in the forecast skill among the scenarios. The skill of the streamflow forecasts tends to be slightly greater in the warm season (Fig. 5a–d) than in the cool one (Fig. 5e–h) across all the basins and lead times. In the warm season (Fig. 5a–d), all the scenarios tend to show similar skill, except CNON6 (Fig. 5b), with S5 and S6 only slightly outperforming S3 and S4. In the cool season (Fig. 5e–h), with the exception of CNON6 (Fig. 5f), the performance is similar among the scenarios for the short lead times, but S3 tends to consistently underperform for the later lead times relative to S4–S6. There is also a skill reversal between the seasons when comparing the ensemble precipitation (Fig. 3) and streamflow (Fig. 5) forecasts. That is, the

skill tends to be higher in the cool season than the warm one in Fig. 3, but this trend reverses in Fig. 5. The reason for this reversal is that in the cool season hydrological conditions are strongly influenced by snow dynamics, which can be challenging to represent with HL-RDHM, particularly when specific snow information or data are not available. In any case, this could be a valuable area for future research since it appears here to have a significant influence on the skill of the ensemble streamflow forecasts.

The underperformance of S4 in the CNON6 basin (Fig. 5f), relative to the other scenarios, is in part due to the unusually low skill of the raw ensemble streamflow forecasts of S3, so that even after preprocessing the skill improvement attained with S4 is not comparable to that associated with S5 and S6. This is also the case for CNON6 in the warm season (Fig. 5b). However, in addition, during the cool season it is likely that streamflows in CNON6 are affected by a reservoir just upstream from the main outlet of CNON6. The reservoir is operated for flood control purposes. During the cool season the reservoir affects low flows by maintaining them at a somewhat higher level than in natural conditions. Since we do not account for reservoir operations in our hydrological modeling, it is likely that one of the benefits of postprocessing is, in this case, that it corrects for this modeling bias. In fact, this is also reflected in the calibration results (e.g.,

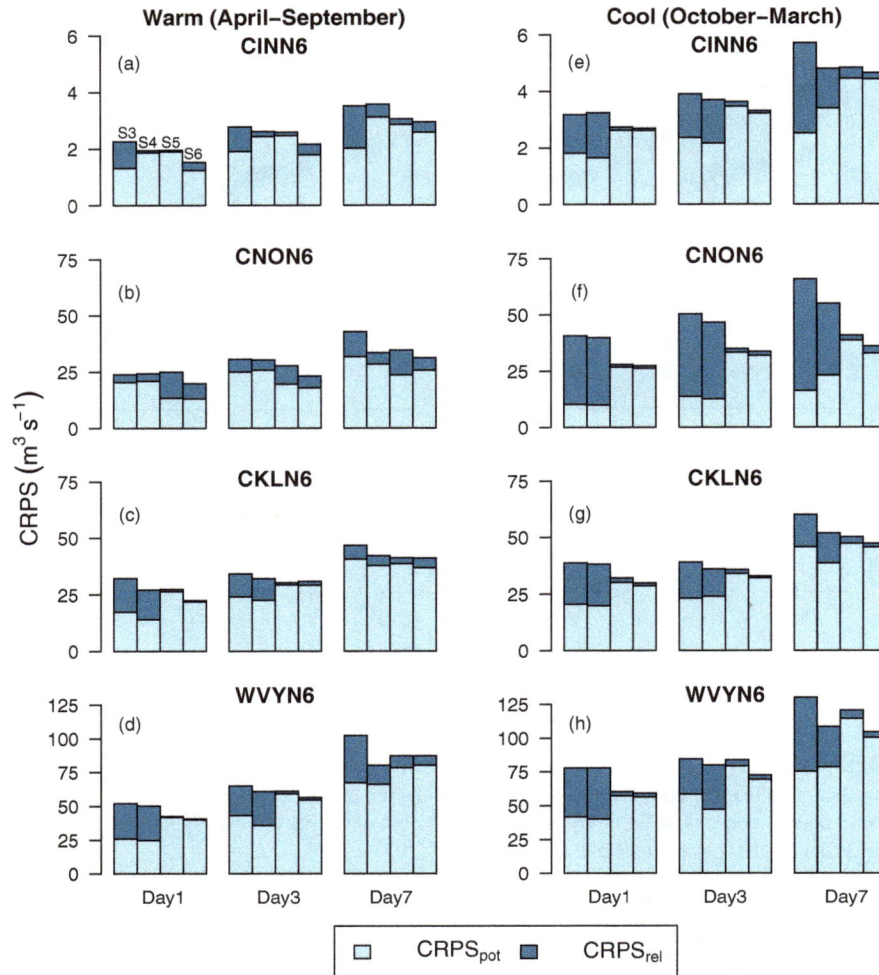

Figure 6. Decomposition of the CRPS into CRPS potential ($CRPS_{pot}$) and CRPS reliability ($CRPS_{rel}$) for forecasts lead times of 1, 3, and 7 days during the warm **(a)–(d)** (April–September) and cool season **(e)–(h)** (October–March) for the selected basins. The four columns associated with each forecast lead time represent the forecasting scenarios S3–S6 (from left to right). Note that S3 consists of GEFSRv2 + HL-RDHM, S4 of GEFSRv2 + HCLR + HL-RDHM, S5 of GEFSRv2 + HL-RDHM + QR, and S6 of GEFSRv2 + HCLR + HL-RDHM + QR.

in Fig. 2c), where the performance of CNON6 is somewhat lower than in the other basins. Interestingly, after postprocessing (S5 in Fig. 5f), the skill of CNON6 is as good as that of CINN6, even though at the day 1 lead time the skill for S3 is ~ 0.1 for CNON6 (Fig. 5f) and ~ 0.4 for CINN6 (Fig. 5e). Hence, the postprocessor seems capable of compensating somewhat for the lesser performance of CNON6 in both calibration or after preprocessing in the cool season.

4.4.2 CRPS decomposition

Figure 6 displays different components of the mean CRPS against lead times of 1, 3, and 7 days for all the basins according to both the warm (Fig. 6a–d) and cool (Fig. 6e–h) seasons. The components presented here are reliability ($CRPS_{rel}$) and potential CRPS ($CRPS_{pot}$) (Hersbach,

2000). $CRPS_{rel}$ measures the average reliability of the ensemble forecasts across all the possible events, i.e., it examines whether the fraction of observations that fall below the jth of n ranked ensemble members is equal to j/n on average. $CRPS_{pot}$ represents the lowest possible CRPS that could be obtained if the forecasts were made perfectly reliable (i.e., $CRPS_{rel} = 0$). Note that the CRPS, $CRPS_{rel}$, and $CRPS_{pot}$ are all negatively oriented, with perfect score of zero. Overall, as was the case with the CRPSS (Fig. 5), the CRPS decomposition reveals that forecast reliability tends mostly to progressively improve from S3 to S6.

Interestingly, improvements in forecast quality for S5 and S6, relative to the raw streamflow forecasts of S3, are mainly due to reductions in $CRPS_{rel}$ (i.e., by making the forecasts more reliable), whereas for S4 better forecast quality is achieved by reductions in both $CRPS_{rel}$ and $CRPS_{pot}$.

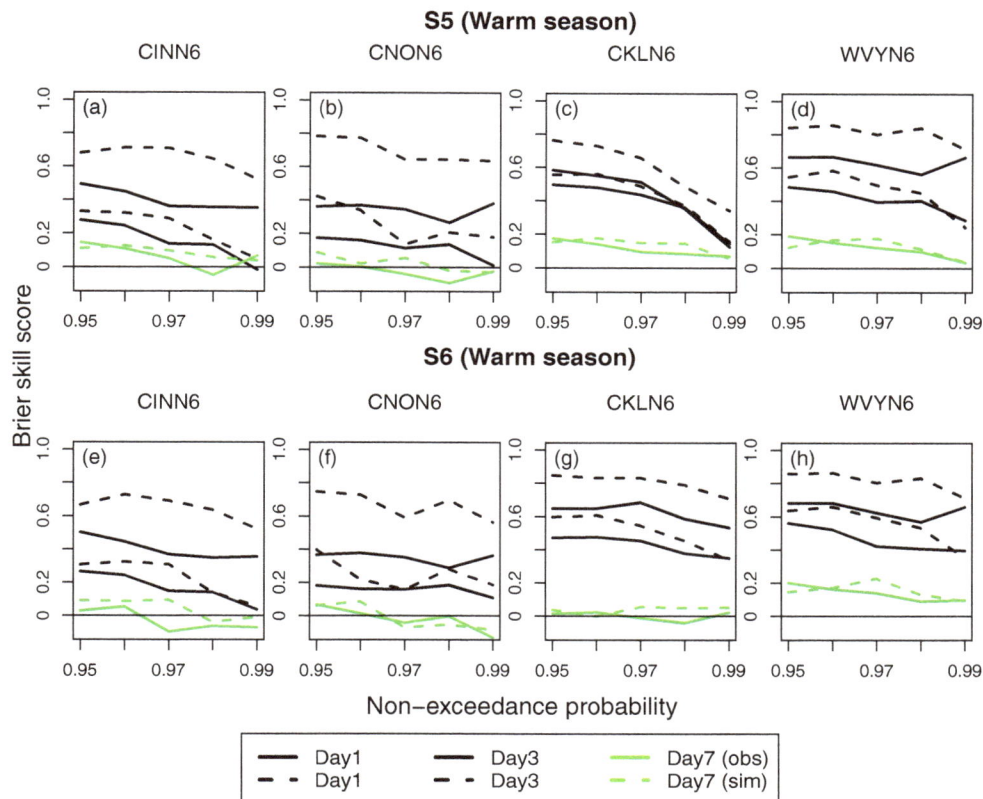

Figure 7. Brier skill score (BSS) of the mean ensemble flood forecasts for S5 **(a–d)** and S6 **(e–h)** vs. the flood threshold for forecast lead times of 1, 3, and 7 days during the warm (April–September) season for the selected basins. The BSS is shown relative to both observed (solid lines) and simulated floods (dashed lines).

$CRPS_{pot}$ appears to play a bigger role in S4 than in the other scenarios, since in many cases in Fig. 6 the $CRPS_{pot}$ value for S4 is the lowest among all the scenarios. The explanation for this lies in the implementation of the HCLR preprocessor, which uses the ensemble spread as a predictor of the dispersion of the predictive pdf and the $CRPS_{pot}$ is sensitive to the spread (Messner et al., 2014a). In terms of the warm and cool seasons, the warm season tends to show a slightly lower CRPS than the cool one for all the scenarios. There are other, more nuanced differences between the two seasons. For example, S5 is more reliable than S4 in several cases in Fig. 6, such as for the day 1 lead time in the cool season. The CRPS decomposition demonstrates that the ensemble streamflow forecasts for S5 and S6 tend to be more reliable than for S3 and S4. It also shows that the forecasts from S5 and S6 tend to exhibit comparable reliability. However, the ensemble streamflow forecasts generated using both preprocessing and postprocessing, S6, ultimately result in lower CRPS than the other scenarios. The latter is seen across all the basins, lead times, and seasons, except in one case (Fig. 6d at the day 7 lead time).

4.4.3 BSS

In our final verification comparison, the BSS values of

the ensemble streamflow forecasts for S5 (Fig. 7a–d) and S6 (Fig. 7e–h) are plotted against the non-exceedance probability associated with different streamflow thresholds ranging from 0.95 to 0.99. The BSS is computed for all the basins, for the warm season, and at lead times of 1, 3 and 7 days. In addition, the BSS is computed relative to both observed (solid lines in Fig. 7) and simulated (dashed lines in Fig. 7) flows. When the BSS is computed relative to observed flows, it considers the effect on forecast skill of both meteorological and hydrological uncertainties. While the BSS relative to simulated flows is mainly affected by meteorological uncertainties. The difference between the two, i.e., the BSS relative to observed flows minus the BSS relative to simulated ones, provides an estimate of the effect of hydrological uncertainties on the skill of the streamflow forecasts. Similar to the CRPSS, the BSS value of zero means no skill (i.e., same skill as the reference system) and a value of 1 indicates perfect skill.

In general, the skill of streamflow forecasts tends to decrease with lead time across the flow thresholds and basins. In contrast to the CRPSS (Fig. 5), where S6 tends for the majority of cases to slightly outperform S5, the BSS values for the different flow thresholds appear similar for S5 (Fig. 7a–d) and S6 (Fig. 7e–h). The only exception is CKLN6 (Fig. 7c and g), where S6 has better skill than S5 at the day 1 and 3 lead times, particularly at the highest flow thresholds considered. With respect to the basin size, the skill tends to improve somewhat from the small to the large basin. For instance, for non-exceedance probabilities of 0.95 and 0.99 at the day 1 lead time, the BSS values for the smallest basin (Fig. 7a), measured relative to the observed flows, are ~ 0.49 and 0.35, respectively. For the same conditions, both values increase to ~ 0.65 for the largest basin (Fig. 7d).

The most notable feature in Fig. 7 is that the effect of hydrological uncertainties on forecast skill is evident at the day 1 lead time, while meteorological uncertainties clearly dominate at the day 7 lead time. With respect to the latter, notice that the solid and dashed green lines for the day 7 lead time tend to be very close to each other in Fig. 7, indicating that hydrological uncertainties are relatively small compared to meteorological ones. Hydrological uncertainties are largest at the day 1 lead time, particularly for the small basins (Fig. 7a–b and e–f). For example, for a non-exceedance probability of 0.95 and at a day 1 lead time (Fig. 7b), the BSS value relative to the simulated and observed flows are ~ 0.79 and 0.38, respectively, suggesting a reduction of $\sim 50\%$ skill due to hydrological uncertainties.

5   Summary and conclusion

In this study, we used the RHEPS to investigate the effect of statistical processing on short- to medium-range ensemble streamflow forecasts. First, we assessed the raw precipitation forecasts from the GEFSRv2 (S1), and compared them with the preprocessed precipitation ensembles (S2). Then, streamflow ensembles were generated with the RHEPS for four different forecasting scenarios involving no statistical processing (S3), preprocessing alone (S4), postprocessing alone (S5), and both preprocessing and postprocessing (S6). The verification of ensemble precipitation and streamflow forecasts was done for the years 2004–2012, using four nested gauge locations in the NBSR basin of the US MAR. We found that the scenario involving both preprocessing and postprocessing consistently outperforms the other scenarios. In some cases, however, the differences between the scenario involving preprocessing and postprocessing, and the scenario with postprocessing alone are not as significant, suggesting for those cases (e.g., warm season) that postprocessing alone can be effective in removing systematic biases. Other specific findings are as follows:

- The HCLR preprocessed ensemble precipitation forecasts show improved skill relative to the raw forecasts. The improvements are more noticeable in the warm season at the longer lead times (> 3 days).

- Both postprocessors, ARX(1,1) and QR, show gains in skill relative to the raw ensemble streamflow forecasts in the cool season. In contrast, in the warm season, ARX(1,1) shows little or no gains in skill. Overall, for the majority of cases analyzed, the gains with QR tend to be greater than with ARX(1,1), especially during the warm season.

- In terms of the forecast skill (i.e., CRPSS), in the warm season the scenarios including only preprocessing and only postprocessing have a comparable performance to the more complex scenario consisting of both preprocessing and postprocessing, while in the cool season, the scenario involving both preprocessing and postprocessing consistently outperforms the other scenarios but the differences may not be as significant.

- The skill of the postprocessing-alone scenario and the scenario that combines preprocessing and postprocessing was further assessed using the Brier skill score for different streamflow thresholds and the warm season. This assessment suggests that for high flow thresholds the similarities in skill between both scenarios, S5 and S6, become greater.

- Decomposing the CRPS into reliability and potential components, we observed that the scenario that combines preprocessing and postprocessing results in slightly lower CRPS than the other scenarios. We found that the scenario involving only postprocessing tends to demonstrate similar reliability to the scenario consisting of both preprocessing and postprocessing across most of the lead times, basins, and seasons. We also found that in several cases the postprocessing-alone scenario displays improved reliability relative to the preprocessing-alone scenario.

These conclusions are specific to the RHEPS forecasting system, which is mostly relevant to the US research and operational communities as it relies on a weather and a hydrological model that are used in this domain. However, the use of a global weather forecasting system illustrates the potential of applying the statistical techniques tested here in other regions worldwide.

The emphasis of this study has been on benchmarking the contributions of statistical processing to the RHEPS. To accomplish this, our approach required that the quality of ensemble streamflow forecasts be verified over multiple years (i.e., across many flood cases) to obtain robust verification statistics. Future research, however, could be focused on studying how distinct hydrological processes contribute or constrain forecast quality. This effort could be centered around specific flood events rather than on the statistical, many-cases approach taken here. To further assess the relative importance of the various components of the RHEPS, additional tests involving the uncertainty to initial hydrologic conditions and hydrological parameters could be performed. For instance, the combined use of data assimilation and post-processing has been shown to produce more reliable and sharper streamflow forecasts (Bourgin et al., 2014). The potential for the interaction of preprocessing and postprocessing with data assimilation to significantly enhance streamflow predictions, however, has not been investigated. This could be investigated in the future with the RHEPS, as the pairing of data assimilation with preprocessing and postprocessing could facilitate translating the improvements in the preprocessed meteorological forcing down the hydrological forecasting chain.

Data availability. Daily streamflow observation data for the selected forecast stations can be obtained from the USGS website (https://waterdata.usgs.gov/nwis/). Multisensor precipitation estimates are obtained from the NOAA's Middle Atlantic River Forecast Center. Precipitation and temperature forecast datasets can be obtained from the NOAA Earth System Research Laboratory website (https://www.esrl.noaa.gov/psd/forecasts/reforecast2/download.html).

Appendix A: Verification metrics

Modified correlation coefficient. The modified version of the correlation coefficient, called a modified correlation coefficient R_m, compares event-specific observed and simulated hydrographs (McCuen and Snyder, 1975). In the modified version, an adjustment factor based on the ratio of the observed and simulated flow is introduced to refine the conventional correlation coefficient R. The modified correlation coefficient R_m is defined as follows:

$$R_\mathrm{m} = R \frac{\min\{\sigma_\mathrm{s}, \sigma_q\}}{\max\{\sigma_\mathrm{s}, \sigma_q\}}, \tag{A1}$$

where σ_s and σ_q denote the standard deviation of the simulated and observed flows, respectively.

Percent bias. PB measures the average tendency of the simulated flows to be larger or smaller than their observed counterparts. Its optimal value is 0.0 where positive values indicate model overestimation bias, and negative values indicate model underestimation bias. The PB is estimated as follows:

$$PB = \frac{\sum_{i=1}^{N} (s_i - q_i)}{\sum_{i=1}^{N} q_i} \times 100, \tag{A2}$$

where s_i and q_i denote the simulated and observed flow, respectively, at time i.

Nash–Sutcliffe efficiency. the NSE (Nash and Sutcliffe, 1970) is defined as the ratio of the residual variance to the initial variance. It is widely used to indicate how well the simulated flows fit the observations. The range of NSE can vary between negative infinity to 1.0, with 1.0 representing the optimal value and values should be larger than 0.0 to indicate minimally acceptable performance. The NSE is computed as follows:

$$NSE = 1 - \frac{\sum_{i=1}^{N} (s_i - q_i)^2}{\sum_{i=1}^{N} (q_i - \overline{q}_i)^2}, \tag{A3}$$

where s_i, q_i, and $\overline{q}_i$ are the simulated, observed, and mean observed flow, respectively, at time i.

Brier skill score. the Brier score (Brier, 1950) is analogous to the mean squared error, but where the forecast is a probability and the observation is either a 0.0 or 1.0. The BS is given by

$$BS = \frac{1}{n} \sum_{i=1}^{n} \left[F_{f_i}(z) - F_{q_i}(z) \right]^2, \tag{A4}$$

where the probability of f_i to exceed a fixed threshold z is

$$F_{f_i}(z) = P_\mathrm{r}\left[f_i > z \right]. \tag{A5}$$

Here, n is again the total number of forecast-observation pairs, and

$$F_{q_i}(z) = \begin{cases} 1, & q_i > z \\ 0, & \text{otherwise} \end{cases}. \tag{A6}$$

In order to compare the skill score of the main forecast system with respect to the reference forecast, it is convenient to define the Brier Skill Score (BSS):

$$BSS = 1 - \frac{BS_\mathrm{main}}{BS_\mathrm{reference}} \tag{A7}$$

where BS_main and $BS_\mathrm{reference}$ are the BS values for the main forecast system (i.e. the system to be evaluated) and reference forecast system, respectively. Any positive values of the BSS, from 0 to 1, indicate that the main forecast system performs better than the reference forecast system. Thus, a BSS of 0 indicates no skill and a BSS of 1 indicates perfect skill.

Mean continuous ranked probability skill score. Continuous ranked probability score (CRPS) quantifies the integrated square difference between the cumulative distribution function (cdf) of a forecast, $F_\mathrm{f}(z)$, and the corresponding cdf of the observation, $F_q(z)$. The CRPS is given by

$$CRPS = -\int_{-\infty}^{\infty} \left[F_\mathrm{f}(z) - F_q(z) \right]^2 \mathrm{d}z. \tag{A8}$$

To evaluate the skill of the main forecast system relative to the reference forecast system, the associated skill score, the mean continuous ranked probability skill score (CRPSS) is defined as follows:

$$CRPSS = 1 - \frac{CRPS_{main}}{CRPS_{reference}}, \qquad (A9)$$

where the CRPS is averaged across n pairs of forecasts and observations to calculate the mean CRPS of the main forecast system ($CRPS_{main}$) and reference forecast system ($CRPS_{reference}$). The CRPSS varies from $-\infty$ to 1. Any positive values of the CRPSS, from 0 to 1, indicate that the main forecast system performs better than the reference forecast system.

To further explore the forecast skill, the $CRPS_{main}$ is decomposed into the CRPS reliability ($CRPS_{rel}$) and potential ($CRPS_{pot}$) such that $CRPS_{main}$ can be calculated as follows (Hersbach, 2000):

$$CRPS_{main} = CRPS_{rel} + CRPS_{pot}. \qquad (A10)$$

The $CRPS_{rel}$ measures the average reliability of the precipitation ensembles similarly to the rank histogram, which shows whether the frequency that the verifying analysis was found in a given bin is equal for all bins (Hersbach, 2000). The $CRPS_{pot}$ measures the CRPS that one would obtain for a perfect reliable system. It is sensitive to the average ensemble spread and outliers.

Competing interests. The authors declare that they have no conflict of interest.

Acknowledgements. We acknowledge the funding support provided by the NOAA/NWS through award NA14NWS4680012.

Edited by: Shraddhanand Shukla

References

Abaza, M., Anctil, F., Fortin, V., and Perreault, L.: On the incidence of meteorological and hydrological processors: effect of resolution, sharpness and reliability of hydrological ensemble forecasts, J. Hydrol., 555, 371–384, 2017.

Addor, N., Jaun, S., Fundel, F., and Zappa, M.: An operational hydrological ensemble prediction system for the city of Zurich (Switzerland): skill, case studies and scenarios, Hydrol. Earth Syst. Sci., 15, 2327–2347, https://doi.org/10.5194/hess-15-2327-2011, 2011.

Alfieri, L., Burek, P., Dutra, E., Krzeminski, B., Muraro, D., Thielen, J., and Pappenberger, F.: GloFAS – global ensemble streamflow forecasting and flood early warning, Hydrol. Earth Syst. Sci., 17, 1161–1175, https://doi.org/10.5194/hess-17-1161-2013, 2013.

Alfieri, L., Pappenberger, F., Wetterhall, F., Haiden, T., Richardson, D., and Salamon, P.: Evaluation of ensemble streamflow predictions in Europe, J. Hydrol., 517, 913–922, 2014.

Anderson, R. M., Koren, V. I., and Reed, S. M.: Using SSURGO data to improve Sacramento Model a priori parameter estimates, J. Hydrol., 320, 103–116, 2006.

Baxter, M. A., Lackmann, G. M., Mahoney, K. M., Workoff, T. E., and Hamill, T. M.: Verification of quantitative precipitation reforecasts over the southeastern United States, Weather Forecast., 29, 1199–1207, 2014.

Bennett, J. C., Robertson, D. E., Shrestha, D. L., Wang, Q., Enever, D., Hapuarachchi, P., and Tuteja, N. K.: A System for Continuous Hydrological Ensemble Forecasting (SCHEF) to lead times of 9days, J. Hydrol., 519, 2832–2846, 2014.

Benninga, H.-J. F., Booij, M. J., Romanowicz, R. J., and Rientjes, T. H. M.: Performance of ensemble streamflow forecasts under varied hydrometeorological conditions, Hydrol. Earth Syst. Sci., 21, 5273–5291, https://doi.org/10.5194/hess-21-5273-2017, 2017.

Bogner, K., Pappenberger, F., and Cloke, H. L.: Technical Note: The normal quantile transformation and its application in a flood forecasting system, Hydrol. Earth Syst. Sci., 16, 1085–1094, https://doi.org/10.5194/hess-16-1085-2012, 2012.

Bourgin, F., Ramos, M.-H., Thirel, G., and Andreassian, V.: Investigating the interactions between data assimilation and postprocessing in hydrological ensemble forecasting, J. Hydrol., 519, 2775–2784, 2014.

Brier, G. W.: Verification of forecasts expressed in terms of probability, Mon. Weather Rev., 78, 1–3, 1950.

Brown, J. D. and Seo, D.-J.: A nonparametric postprocessor for bias correction of hydrometeorological and hydrologic ensemble forecasts, J. Hydrometeorol., 11, 642–665, 2010.

Brown, J. D., Demargne, J., Seo, D.-J., and Liu, Y.: The Ensemble Verification System (EVS): A software tool for verifying ensemble forecasts of hydrometeorological and hydrologic variables at discrete locations, Environ. Model. Softw., 25, 854–872, 2010.

Brown, J. D., He, M., Regonda, S., Wu, L., Lee, H., and Seo, D.-J.: Verification of temperature, precipitation, and streamflow forecasts from the NOAA/NWS Hydrologic Ensemble Forecast Service (HEFS): 2. Streamflow verification, J. Hydrol., 519, 2847–2868, 2014.

Clark, M., Gangopadhyay, S., Hay, L., Rajagopalan, B., and Wilby, R.: The Schaake shuffle: A method for reconstructing space–time variability in forecasted precipitation and temperature fields, J. Hydrometeorol., 5, 243–262, 2004.

Cloke, H. and Pappenberger, F.: Ensemble flood forecasting: a review, J. Hydrol., 375, 613–626, 2009.

Dankers, R., Arnell, N. W., Clark, D. B., Falloon, P. D., Fekete, B. M., Gosling, S. N., Heinke, J., Kim, H., Masaki, Y., Satoh, Y., Stacke, T., Wada, Y., and Wisser, D.: First look at changes in flood hazard in the Inter-Sectoral Impact Model Intercomparison Project ensemble, P. Natl. Acad. Sci. USA, 111, 3257–3261, https://doi.org/10.1073/pnas.1302078110, 2014.

Demargne, J., Wu, L., Regonda, S. K., Brown, J. D., Lee, H., He, M., Seo, D.-J., Hartman, R., Herr, H. D., and Fresch, M.: The science of NOAA's operational hydrologic ensemble forecast service, B. Am. Meteorol. Soc., 95, 79–98, 2014.

Demirel, M. C., Booij, M. J., and Hoekstra, A. Y.: Effect of different uncertainty sources on the skill of 10 day ensemble low flow forecasts for two hydrological models, Water Resour. Res., 49, 4035–4053, 2013.

Demuth, N. and Rademacher, S.: Flood Forecasting in Germany – Challenges of a Federal Structure and Transboundary Cooperation, Flood Forecasting: A Global Perspective, Elsevier, 125–151, 2016.

Dogulu, N., López López, P., Solomatine, D. P., Weerts, A. H., and Shrestha, D. L.: Estimation of predictive hydrologic uncertainty using the quantile regression and UNEEC methods and their comparison on contrasting catchments, Hydrol. Earth Syst. Sci., 19, 3181–3201, https://doi.org/10.5194/hess-19-3181-2015, 2015.

Durkee, D. J., Frye, D. J., Fuhrmann, M. C., Lacke, C. M., Jeong, G. H., and Mote, L. T.: Effects of the North Atlantic Oscillation on precipitation-type frequency and distribution in the eastern United States, Theor. Appl. Climatol., 94, 51–65, 2007.

Emerton, R. E., Stephens, E. M., Pappenberger, F., Pagano, T. C., Weerts, A. H., Wood, A. W., Salamon, P., Brown, J. D., Hjerdt, N., and Donnelly, C.: Continental and global scale flood forecasting systems, Wiley Interdisciplin. Rev.: Water, 3, 391–418, 2016.

Fan, F. M., Collischonn, W., Meller, A., and Botelho, L. C. M.: Ensemble streamflow forecasting experiments in a tropical basin: The São Francisco river case study, J. Hydrol., 519, 2906–2919, 2014.

Fares, A., Awal, R., Michaud, J., Chu, P.-S., Fares, S., Kodama, K., and Rosener, M.: Rainfall-runoff modeling in a flashy tropical watershed using the distributed HL-RDHM model, J. Hydrol., 519, 3436–3447, 2014.

Gitro, C. M., Evans, M. S., and Grumm, R. H.: Two Major Heavy Rain/Flood Events in the Mid-Atlantic: June 2006 and September 2011, J. Operat. Meteorol., 2, 152–168, https://doi.org/10.15191/nwajom.2014.0213, 2014.

Golding, B., Roberts, N., Leoncini, G., Mylne, K., and Swinbank, R.: MOGREPS-UK convection-permitting ensemble products for surface water flood forecasting: Rationale and first results, J. Hydrometeorol., 17, 1383–1406, 2016.

Hamill, T. M., Whitaker, J. S., and Wei, X.: Ensemble reforecasting: Improving medium-range forecast skill using retrospective forecasts, Mon. Weather Rev., 132, 1434–1447, 2004.

Hamill, T. M., Bates, G. T., Whitaker, J. S., Murray, D. R., Fiorino, M., Galarneau Jr., T. J., Zhu, Y., and Lapenta, W.: NOAA's second-generation global medium-range ensemble reforecast dataset, B. Am. Meteorol. Soc., 94, 1553–1565, 2013.

Hersbach, H.: Decomposition of the continuous ranked probability score for ensemble prediction systems, Weather Forecast., 15, 559–570, 2000.

Hopson, T. M. and Webster, P. J.: A 1–10-day ensemble forecasting scheme for the major river basins of Bangladesh: Forecasting severe floods of 2003–07, J. Hydrometeorol., 11, 618–641, 2010.

Jolliffe, I. T. and Stephenson, D. B.: Forecast verification: a practitioner's guide in atmospheric science, Wiley, West Sussex, England, 2012.

Journel, A. G. and Huijbregts, C. J.: Mining geostatistics, Academic Press, London, 1978.

Kang, T. H., Kim, Y. O., and Hong, I. P.: Comparison of pre-and post-processors for ensemble streamflow prediction, Atmos. Sci. Lett., 11, 153–159, 2010.

Koenker, R.: Quantile regression, Cambridge University Press, Cambridge, 38, https://doi.org/10.1017/CBO9780511754098, 2005.

Koenker, R. and Bassett Jr., G.: Regression quantiles, Econometrica, 46, 33–50, 1978.

Koren, V., Smith, M., Wang, D., and Zhang, Z.: 2.16 Use of soil property data in the derivation of conceptual rainfall-runoff model parameters, in: Proceedings of the 15th Conference on Hydrology, American Meteorological Society, Long Beach, California, 103–106, 2000.

Koren, V., Reed, S., Smith, M., Zhang, Z., and Seo, D.-J.: Hydrology laboratory research modeling system (HL-RMS) of the US national weather service, J. Hydrol., 291, 297–318, 2004.

Krzysztofowicz, R.: Transformation and normalization of variates with specified distributions, J. Hydrol., 197, 286–292, 1997.

Kuzmin, V.: Algorithms of automatic calibration of multi-parameter models used in operational systems of flash flood forecasting, Russ. Meteorol. Hydrol., 34, 473–481, 2009.

Kuzmin, V., Seo, D.-J., and Koren, V.: Fast and efficient optimization of hydrologic model parameters using a priori estimates and stepwise line search, J. Hydrol., 353, 109–128, 2008.

López López, P., Verkade, J. S., Weerts, A. H., and Solomatine, D. P.: Alternative configurations of quantile regression for estimating predictive uncertainty in water level forecasts for the upper Severn River: a comparison, Hydrol. Earth Syst. Sci., 18, 3411–3428, https://doi.org/10.5194/hess-18-3411-2014, 2014.

Madadgar, S., Moradkhani, H., and Garen, D.: Towards improved post-processing of hydrologic forecast ensembles, Hydrol. Process., 28, 104–122, 2014.

MARFC: http://www.weather.gov/marfc/Top20, last access: 1 April 2017.

McCuen, R. H. and Snyder, W. M.: A proposed index for comparing hydrographs, Water Resour. Res., 11, 1021–1024, 1975.

Mendoza, P. A., McPhee, J., and Vargas, X.: Uncertainty in flood forecasting: A distributed modeling approach in a sparse data catchment, Water Resour. Res., 48, W09532, https://doi.org/10.1029/2011wr011089, 2012.

Mendoza, P. A., Wood, A., Clark, E., Nijssen, B., Clark, M., Ramos, M. H., and Voisin, N.: Improving medium-range ensemble streamflow forecasts through statistical postprocessing, Presented at 2016 Fall Meeting, AGU, 11–15 December 2016, San Francisco, California, 2016.

Messner, J. W., Mayr, G. J., Zeileis, A., and Wilks, D. S.: Heteroscedastic extended logistic regression for postprocessing of ensemble guidance, Mon. Weather Rev., 142, 448–456, 2014a.

Messner, J. W., Mayr, G. J., Wilks, D. S., and Zeileis, A.: Extending extended logistic regression: Extended versus separate versus ordered versus censored, Mon. Weather Rev., 142, 3003–3014, 2014b.

Moore, B. J., Mahoney, K. M., Sukovich, E. M., Cifelli, R., and Hamill, T. M.: Climatology and environmental characteristics of extreme precipitation events in the southeastern United States, Mon. Weather Rev., 143, 718–741, 2015.

Nash, J. E. and Sutcliffe, J. V.: River flow forecasting through conceptual models part I – A discussion of principles, J. Hydrol., 10, 282–290, 1970.

NCAR: https://ral.ucar.edu/projects/system-for-hydromet-analysis-research-and-prediction-sharp, last access: 1 April 2017.

Pagano, T. C., Elliott, J., Anderson, B., and Perkins, J.: Australian Bureau of Meteorology Flood Forecasting and Warning, in: Flood Forecasting, Elsevier, 3–40, 2016.

Pagano, T. C., Wood, A. W., Ramos, M.-H., Cloke, H. L., Pappenberger, F., Clark, M. P., Cranston, M., Kavetski, D., Mathevet, T., and Sorooshian, S.: Challenges of operational river forecasting, J. Hydrometeorol., 15, 1692–1707, 2014.

Politis, D. N. and Romano, J. P.: The stationary bootstrap, J. Am. Stat. Assoc., 89, 1303–1313, 1994.

Polsky, C., Allard, J., Currit, N., Crane, R., and Yarnal, B.: The Mid-Atlantic Region and its climate: past, present, and future, Clim. Res., 14, 161–173, 2000.

Prat, O. P. and Nelson, B. R.: Evaluation of precipitation estimates over CONUS derived from satellite, radar, and rain gauge data sets at daily to annual scales (2002–2012), Hydrol. Earth Syst. Sci., 19, 2037–2056, https://doi.org/10.5194/hess-19-2037-2015, 2015.

Rafieeinasab, A., Norouzi, A., Kim, S., Habibi, H., Nazari, B., Seo, D.-J., Lee, H., Cosgrove, B., and Cui, Z.: Toward high-resolution flash flood prediction in large urban areas – Analysis of sensitivity to spatiotemporal resolution of rainfall input and hydrologic modeling, J. Hydrol., 531, 370–388, 2015.

Reed, S., Koren, V., Smith, M., Zhang, Z., Moreda, F., and Seo, D. J.: Overall distributed model intercomparison project results, J. Hydrol., 298, 27–60, 2004.

Reed, S., Schaake, J., and Zhang, Z.: A distributed hydrologic model and threshold frequency-based method for flash flood forecasting at ungauged locations, J. Hydrol., 337, 402–420, 2007.

Regonda, S. K., Seo, D. J., Lawrence, B., Brown, J. D., and Demargne, J.: Short-term ensemble streamflow forecasting using operationally-produced single-valued streamflow forecasts – A Hydrologic Model Output Statistics (HMOS) approach, J. Hydrol., 497, 80–96, 2013.

Renard, B., Kavetski, D., Kuczera, G., Thyer, M., and Franks, S. W.: Understanding predictive uncertainty in hydrologic modeling: The challenge of identifying input and structural errors, Water Resour. Res., 46, W05521, https://doi.org/10.1029/2009wr008328, 2010.

Roulin, E. and Vannitsem, S.: Post-processing of medium-range probabilistic hydrological forecasting: impact of forcing, initial conditions and model errors, Hydrol. Process., 29, 1434–1449, 2015.

Saleh, F., Ramaswamy, V., Georgas, N., Blumberg, A. F., and Pullen, J.: A retrospective streamflow ensemble forecast for an extreme hydrologic event: a case study of Hurricane Irene and on the Hudson River basin, Hydrol. Earth Syst. Sci., 20, 2649–2667, https://doi.org/10.5194/hess-20-2649-2016, 2016.

Schaake, J. C., Hamill, T. M., Buizza, R., and Clark, M.: HEPEX: the hydrological ensemble prediction experiment, B. Am. Meteorol. Soc., 88, 1541–1547, 2007.

Schellekens, J., Weerts, A., Moore, R., Pierce, C., and Hildon, S.: The use of MOGREPS ensemble rainfall forecasts in operational flood forecasting systems across England and Wales, Adv. Geosci., 29, 77–84, https://doi.org/10.5194/adgeo-29-77-2011, 2011.

Schwanenberg, D., Fan, F. M., Naumann, S., Kuwajima, J. I., Montero, R. A., and Dos Reis, A. A.: Short-term reservoir optimization for flood mitigation under meteorological and hydrological forecast uncertainty, Water Resour. Manage., 29, 1635–1651, 2015.

Sharma, S., Siddique, R., Balderas, N., Fuentes, J. D., Reed, S., Ahnert, P., Shedd, R., Astifan, B., Cabrera, R., Laing, A., Klein, M., and Mejia, A.: Eastern U.S. Verification of Ensemble Precipitation Forecasts, Weather Forecast., 32, 117–139, 2017.

Shi, X., Andrew, W. W., and Dennis, P. L.: How essential is hydrologic model calibration to seasonal streamflow forecasting, J. Hydrometeorol., 9, 1350–1363, 2008.

Siddique, R. and Mejia, A.: Ensemble Streamflow Forecasting across the US Mid-Atlantic Region with a Distributed Hydrological Model Forced by GEFS Reforecasts, J. Hydrometeorol., 18, 1905–1928, 2017.

Siddique, R., Mejia, A., Brown, J., Reed, S., and Ahnert, P.: Verification of precipitation forecasts from two numerical weather prediction models in the Middle Atlantic Region of the USA: A precursory analysis to hydrologic forecasting, J. Hydrol., 529, 1390–1406, 2015.

Sloughter, J. M. L., Raftery, A. E., Gneiting, T., and Fraley, C.: Probabilistic quantitative precipitation forecasting using Bayesian model averaging, Mon. Weather Rev., 135, 3209–3220, 2007.

Smith, M. B., Koren, V., Reed, S., Zhang, Z., Zhang, Y., Moreda, F., Cui, Z., Mizukami, N., Anderson, E. A., and Cosgrove, B. A.: The distributed model intercomparison project – Phase 2: Motivation and design of the Oklahoma experiments, J. Hydrol., 418, 3–16, 2012a.

Smith, M. B., Koren, V., Zhang, Z., Zhang, Y., Reed, S. M., Cui, Z., Moreda, F., Cosgrove, B. A., Mizukami, N., and Anderson, E. A.: Results of the DMIP 2 Oklahoma experiments, J. Hydrol., 418, 17–48, 2012b.

Thiemig, V., Bisselink, B., Pappenberger, F., and Thielen, J.: A pan-African medium-range ensemble flood forecast system, Hydrol. Earth Syst. Sci., 19, 3365–3385, https://doi.org/10.5194/hess-19-3365-2015, 2015.

Thorstensen, A., Nguyen, P., Hsu, K., and Sorooshian, S.: Using Densely Distributed Soil Moisture Observations for Calibration of a Hydrologic Model, J. Hydrometeorol., 17, 571–590, 2016.

Verkade, J., Brown, J., Reggiani, P., and Weerts, A.: Post-processing ECMWF precipitation and temperature ensemble reforecasts for operational hydrologic forecasting at various spatial scales, J. Hydrol., 501, 73–91, 2013.

Wang, Q., Bennett, J. C., and Robertson, D. E.: Error reduction and representation in stages (ERRIS) in hydrological modelling for ensemble streamflow forecasting, Hydrol. Earth Syst. Sci., 20, 3561–3579, https://doi.org/10.5194/hess-20-3561-2016, 2016.

Ward, P. J., Jongman, B., Salamon, P., Simpson, A., Bates, P., De Groeve, T., Muis, S., De Perez, E. C., Rudari, R., and Trigg, M. A.: Usefulness and limitations of global flood risk models, Nat. Clim. Change, 5, 712–715, 2015.

Weerts, A. H., Winsemius, H. C., and Verkade, J. S.: Estimation of predictive hydrological uncertainty using quantile regression: examples from the National Flood Forecasting System (England and Wales), Hydrol. Earth Syst. Sci., 15, 255–265, https://doi.org/10.5194/hess-15-255-2011, 2011.

Wheater, H. S. and Gober, P.: Water security and the science agenda, Water Resour. Res., 51, 5406–5424, 2015.

Wilks, D. S.: Extending logistic regression to provide full-probability-distribution MOS forecasts, Meteorol. Appl., 16, 361–368, 2009.

Wilks, D. S.: Statistical methods in the atmospheric sciences, Academic Press, Diego, California, 2011.

Yang, X., Sharma, S., Siddique, R., Greybush, S. J., and Mejia, A.: Postprocessing of GEFS Precipitation Ensemble Reforecasts over the US Mid-Atlantic Region, Mon. Weather Rev., 145, 1641–1658, 2017.

Ye, A., Qingyun, D., Xing, Y., Eric, F. W., and John, S.: Hydrologic post-processing of MOPEX streamflow simulations, J. Hydrol., 508, 147–156, 2014.

Yuan, X. and Wood, E. F.: Downscaling precipitation or bias-correcting streamflow? Some implications for coupled general circulation model (CGCM)-based ensemble seasonal hydrologic forecast, Water Resour. Res., 48, W12519, https://doi.org/10.1029/2012WR012256, 2012.

Zalachori, I., Ramos, M.-H., Garçon, R., Mathevet, T., and Gailhard, J.: Statistical processing of forecasts for hydrological ensemble prediction: a comparative study of different bias correction strategies, Adv. Sci. Res., 8, 135–141, https://doi.org/10.5194/asr-8-135-2012, 2012.

Zappa, M., Rotach, M. W., Arpagaus, M., Dorninger, M., Hegg, C., Montani, A., Ranzi, R., Ament, F., Germann, U., and Grossi, G.: MAP D-PHASE: real-time demonstration of hydrological ensemble prediction systems, Atmos. Sci. Lett., 9, 80–87, 2008.

Zappa, M., Jaun, S., Germann, U., Walser, A., and Fundel, F.: Superposition of three sources of uncertainties in operational flood forecasting chains, Atmos. Res., 100, 246–262, 2011.

Zhao, L., Duan, Q., Schaake, J., Ye, A., and Xia, J.: A hydrologic post-processor for ensemble streamflow predictions, Adv. Geosci., 29, 51–59, https://doi.org/10.5194/adgeo-29-51-2011, 2011.

7

Area-averaged evapotranspiration over a heterogeneous land surface: aggregation of multi-point EC flux measurements with a high-resolution land-cover map and footprint analysis

Feinan Xu[1,2], Weizhen Wang[1], Jiemin Wang[1], Ziwei Xu[3], Yuan Qi[1], and Yueru Wu[1]

[1]Key Laboratory of Remote Sensing of Gansu Province, Heihe Remote Sensing Experimental Research Station, Northwest Institute of Eco-Environment and Resources, Chinese Academy of Sciences, Lanzhou, 730000, China
[2]University of Chinese Academy of Sciences, Beijing, 100049, China
[3]State Key Laboratory of Earth Surface Processes and Resource Ecology, Faculty of Geographical Science, Beijing Normal University, Beijing, 100875, China

Correspondence to: Weizhen Wang (weizhen@lzb.ac.cn)

Abstract. The determination of area-averaged evapotranspiration (ET) at the satellite pixel scale/model grid scale over a heterogeneous land surface plays a significant role in developing and improving the parameterization schemes of the remote sensing based ET estimation models and general hydro-meteorological models. The Heihe Watershed Allied Telemetry Experimental Research (HiWATER) flux matrix provided a unique opportunity to build an aggregation scheme for area-averaged fluxes. On the basis of the HiWATER flux matrix dataset and high-resolution land-cover map, this study focused on estimating the area-averaged ET over a heterogeneous landscape with footprint analysis and multivariate regression. The procedure is as follows. Firstly, quality control and uncertainty estimation for the data of the flux matrix, including 17 eddy-covariance (EC) sites and four groups of large-aperture scintillometers (LASs), were carefully done. Secondly, the representativeness of each EC site was quantitatively evaluated; footprint analysis was also performed for each LAS path. Thirdly, based on the high-resolution land-cover map derived from aircraft remote sensing, a flux aggregation method was established combining footprint analysis and multiple-linear regression. Then, the area-averaged sensible heat fluxes obtained from the EC flux matrix were validated by the LAS measurements. Finally, the area-averaged ET of the kernel experimental area of HiWATER was estimated. Compared with the formerly used and rather simple approaches, such as the arithmetic average and area-weighted methods, the present scheme is not only with a much better database, but also has a solid grounding in physics and mathematics in the integration of area-averaged fluxes over a heterogeneous surface. Results from this study, both instantaneous and daily ET at the satellite pixel scale, can be used for the validation of relevant remote sensing models and land surface process models. Furthermore, this work will be extended to the water balance study of the whole Heihe River basin.

1 Introduction

Land surface evapotranspiration (ET) is not only a key component in the regional water circulation, but is also essential in the surface energy balance and land surface process. Under the condition of increasing shortage of water resources, high-precision estimation of ET at regional scale is essential for such applications, as the management of river basin water resources, regional planning, and the sustainable development of agriculture (Wang et al., 2003). Currently, the commonly used methods for acquisition of regional ET are ground-based observation, remote sensing based estimation, and model simulation, respectively.

The Earth's surface is always characterized by spatial heterogeneity. Large land surface heterogeneity affects greatly the exchanges of momentum, heat, and water between the

Area-averaged evapotranspiration over a heterogeneous land surface: aggregation of multi-point EC flux...

105

land surface and atmosphere (Mengelkamp et al., 2006). Indeed, the surface heterogeneity caused either by the contrast in soil moisture or vegetation type generates a large spatial variability of fluxes, which limit the use of the eddy-covariance (EC) system, unless one deploys a network of EC devices (Ezzahar et al., 2009b). The flux tower group can quantify the turbulent exchange of energy and mass between the atmosphere and a variety of surface types (Sellers et al., 1995), and these local point measurements need to be aggregated to provide meaningful area-averaged fluxes (André et al., 1986). If special aggregation rules for local flux measurements are applied, measurements can provide averaged fluxes at model grid scale (Beyrich et al., 2006; Mahrt et al., 2001). But given the EC network's high price and the requirement for their continuous maintenance, the large-aperture scintillometer (LAS) is a useful alternative method for direct measurements of area-averaged sensible heat fluxes on the scale of 1–5 km (Ezzahar et al., 2009b; Ezzahar and Chehbouni, 2009).

Satellite has been considered a promising data source for deriving regional ET with the development of remote sensing techniques (Ezzahar et al., 2009a). In response to increasing demand for spatially distributed hydrologic information, many satellite-based approaches have been developed for routine monitoring of ET at a regional scale (Anderson et al., 2012). Nevertheless, the effectiveness of the remote sensing based methods for estimating ET must be fully assessed by ground-based area-averaged flux measurements, mainly due to the uncertainties of model inputs and parameterization schemes (Wang et al., 2003). Furthermore, there may be a bias in directly comparing a remote sensing based ET estimation with in situ measurements, because of their spatial-scale mismatch and spatial heterogeneity at the sub-pixel scale (Jia et al., 2012).

General atmospheric–hydrological models (e.g., numerical weather prediction) can adequately describe the interaction between the atmosphere and the underlying surface using complex parameterization schemes. The development and validation of these models are usually based on measurements performed over homogeneous land surfaces. While the assumption of homogeneity might be justified at the local scale ($10–10^3$ m), it is often violated at the scale of the grid resolution of current regional atmospheric models (about 10^4 m) (Beyrich et al., 2006; Beyrich and Mengelkamp, 2006). Therefore, it is significantly important to determine the area-averaged surface fluxes at the satellite pixel scale/model grid scale ($10^3–10^4$ m) for the evaluation of general hydro-meteorological models and relevant remote sensing models. The estimation of area-averaged fluxes is usually not straightforward, especially for heterogeneous land surfaces.

A number of international field experiments have been performed over heterogeneous land surfaces in different geographical and climate regions of the Earth in recent decades (Mengelkamp et al., 2006; Beyrich et al., 2006; Wang, 1999),

such as HAPEX–MOBILHY (André et al., 1986), FIFE (Sellers et al., 1988), HAPEX-SAHEL (Goutorbe et al., 1994), BOREAS (Sellers et al., 1995), NOPEX (Halldin et al., 1998), and LITFASS-2003 (Mengelkamp et al., 2006). In these experiments, surface fluxes at the model grid scale, estimated from multi-point flux observations using various flux aggregation techniques, were compared with those obtained from LAS systems and remote sensing estimation methods. In the former studies, the most commonly used and rather simple flux aggregation methods mainly include the arithmetic average method, the area-weighted method, and the footprint-weighted method (Liu et al., 2016). These studies revealed, under careful data processing and quality control (Charuchittipan et al., 2014) as well as analysis of the energy balance closure for flux data (Foken et al., 2006, 2010), that the integration of the multi-site EC flux measurements and area-averaged fluxes from scintillometers and aircraft observations can provide reasonable estimates over a heterogeneous landscape (Mahrt et al., 2001; Beyrich et al., 2006; Liu et al., 2016).

However, the integration schemes of the aforementioned methods are applicable for relatively uniform sites, of which the local flux measurements are representative of the individual surface types. For the interpretation of tower flux measurements over a heterogeneous land surface, operational footprint analysis is an essential approach (Schmid, 2002). The development of footprint models provides diagnostic tools to quantify the representativeness of tower flux measurements for selected sites (Horst and Weil, 1992; Kim et al., 2006). Besides, it had been demonstrated that the footprint climatology can be combined with information on the spatial variability of vegetation types provided by satellite images (Kim et al., 2006; Chen et al., 2008). Land cover reflects the combined effects of vegetation, climate, soil, and topography; some relationship could be found between land cover and measured surface fluxes (Ogunjemiyo et al., 2003). Ran et al. (2016) proposed four indicators with footprint analysis and a land-cover map to improve the representativity of EC towers and correct the EC flux measurements. But this method did not obtain the surface fluxes of individual land-cover types, but just corrected the EC observations with some prior coefficients. Some previous studies have successfully related the aircraft observed fluxes to surface cover types with the integration of footprint analysis and a satellite-based land-cover map (Ogunjemiyo et al., 2003; Kirby et al., 2008; Hutjes et al., 2010). Among these works, a flux disaggregation method (Hutjes et al., 2010), developed from a former study presented by Ogunjemiyo et al. (2003), would be a promising method for integrating multiple tower-based flux measurements to satellite pixel or grid scale on account of its theoretical framework. The application of this method in attributing heterogeneous EC flux measurements to separate land-cover classes will hopefully be a way to have insight into the component fluxes from various land-cover types. It also provides a chance to develop a flux aggregation scheme

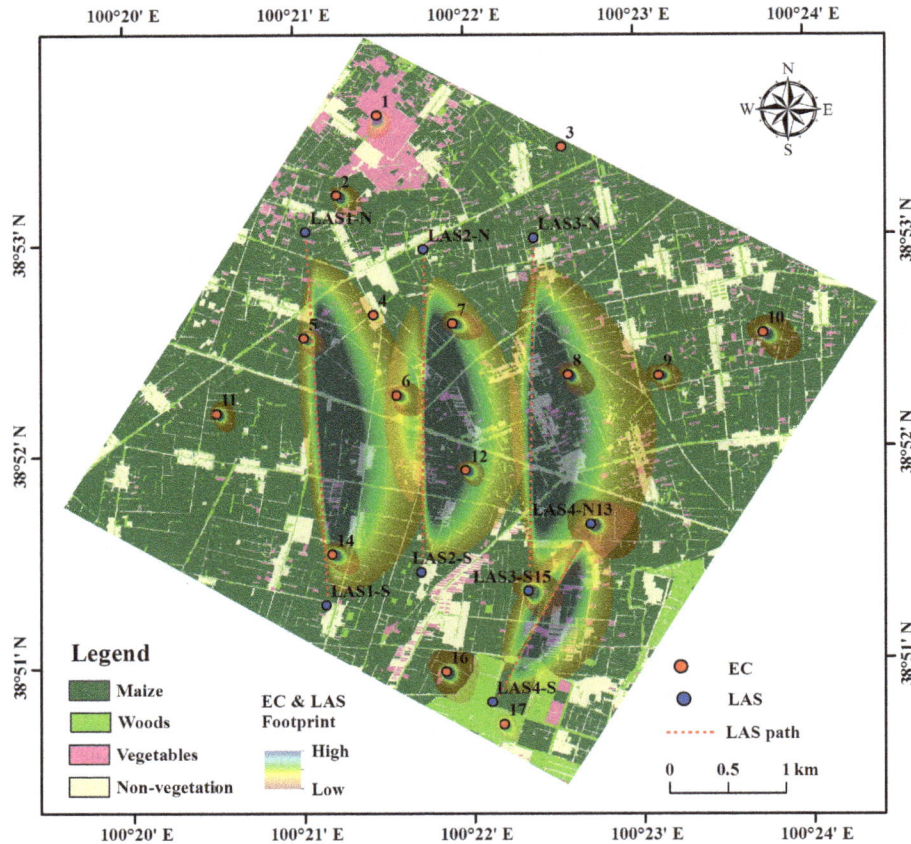

Figure 1. The land-cover map of the kernel experiment area of HiWATER 2012 and the daytime-averaged (8:30–15:30, BST) footprint distribution for EC and LAS (90 % flux contribution) for 29–30 June 2012.

for exploring the extension of multiple EC flux observations to satellite pixel/gird scale.

A multi-scale observation experiment on evapotranspiration over a heterogeneous land surface was conducted in the middle reaches of the Heihe River basin during the project of HiWATER (Heihe Watershed Allied Telemetry Experimental Research) in 2012 (Li et al., 2013; Liu et al., 2016). A comprehensive flux matrix, consisting of 17 EC sites and four groups of LAS systems within a $5 \times 5\,\mathrm{km}^2$ area, was specifically designed to capture the multi-scale characteristics of ET over a heterogeneous landscape during the experiment. The HiWATER flux matrix, with abundant multi-scale flux data, provided a unique opportunity to build an aggregation scheme for area-averaged fluxes over a heterogeneous land surface. The objective of this study is to integrate multi-point EC flux measurements to area-average with high-resolution land-cover map and footprint analysis. The main issues were as follows: (1) the representativeness of the EC flux matrix was quantitatively evaluated; (2) a flux aggregation scheme was established and used for estimating area-averaged sensible heat fluxes from the EC flux matrix, taking LAS measurements as a reference to check the integration algorithm; and (3) the developed scheme was applied to determine the

area-averaged evapotranspiration over a heterogeneous land surface.

2   Study sites and data

2.1   Site description

This study was based on ground-based observation dataset, collected from the multi-scale flux matrix of HiWATER from May to September 2012. The kernel experimental area ($5 \times 5\,\mathrm{km}^2$) of the multi-scale observation experiment was located in the Yingke and Daman irrigation district within Zhangye oasis. The land-cover types in the kernel experimental area were dominated by maize (72 %), vegetables (5 %), orchard and shelterbelt (8 %) as well as residential area and roads (15 %). The small square maize fields were staggered with windbreak trees, roads and irrigation ditches; the surface status of this oasis was actually very heterogeneous. As shown by the numbers 1–17 in Fig. 1, 17 sites were installed according to the distribution of crop planting structure and land cover. Each of them was equipped with an eddy-covariance system (with two layers in site 15) and an automatic weather station (AWS) to capture the exchange process

Table 1. Details of eddy-covariance systems in the HiWATER flux matrix.

Site no.	Longitude (°)	Latitude (°)	Elevation (m)	Turbulence sensors	Sensor height (m)	Surface type
1	100.3582	38.8932	1552.75	Gill/Li7500A	3.8	Vegetables
2	100.35406	38.88695	1559.09	CSAT3/Li7500	3.7	Maize
3	100.37634	38.89053	1543.05	Gill/Li7500A	3.8	Maize
4	100.35753	38.87752	1561.87	CSAT3/Li7500A	4.2/ 6.2 after 19 Aug	Residential area
5	100.35068	38.87574	1567.65	CSAT3/Li7500	3.0	Maize
6	100.3597	38.87116	1562.97	CSAT3/Li7500A	4.6	Maize
7	100.36521	38.87676	1556.39	CSAT3/Li7500A	3.8	Maize
8	100.37649	38.87254	1550.06	CSAT3/Li7500	3.2	Maize
9	100.38546	38.87239	1543.34	Gill/Li7500A	3.9	Maize
10	100.39572	38.87567	1534.73	CSAT3/Li7500	4.8	Maize
11	100.34197	38.86991	1575.65	CSAT3/Li7500	3.5	Maize
12	100.36631	38.86515	1559.25	CSAT3/Li7500	3.5	Maize
13	100.37841	38.86076	1550.73	CSAT3/Li7500A	5.0	Maize
14	100.3531	38.85867	1570.23	CSAT3/Li7500	4.6	Maize
15	100.37223	38.85555	1556.06	CSAT3/Li7500A	4.5/34	Maize
16	100.36411	38.84931	1564.31	Gill/Li7500	4.9	Maize
17	100.36972	38.8451	1559.63	CSAT3/EC150	7.0	Orchard

Table 2. Details of large-aperture scintillometers in the HiWATER flux matrix.

Site	Longitude (°)	Latitude (°)	LAS type, manufacturers	Path length (m)	Effective height (m)
LAS 1	North 100.35090	38.88413	BLS900, Scintec, Germany	3256	33.45
	South 100.35285	38.85470	RR9340, Rainroot, China	3256	33.45
LAS 2	North 100.36236	38.88256	BLS900, Scintec, Germany	2841	33.45
	South 100.36171	38.85717	BLS450, Scintec, Germany	2841	33.45
LAS 3	North 100.37319	38.88338	BLS900, Scintec, Germany	3111	33.45
	South 100.37223	38.85555	LAS, Kipp & Zonen, Netherlands	3111	33.45
LAS 4	North 100.37841	38.86076	BLS450, Scintec, Germany	1854	22.45
	South 100.36840	38.84682	RR9340, Rainroot, China	1854	22.45

of surface water and energy budget at the local scale. Spatial distribution of EC/AWS systems is shown in Fig. 1, with site 1 of vegetable (pepper) field, site 4 of residential area, site 17 of apple orchard, and the other 14 sites are spatially distributed on the dominated maize fields. Key micrometeorological observations at each AWS included four-component radiation, one or two levels wind/temperature/relative humidity, soil temperature/moisture and soil heat flux. Among these sites, site 15 was a superstation equipped with two levels of EC system, seven-level wind speed/direction and air temperature/humidity. Four groups of large-aperture scintillometers were installed across the experimental district (see Fig. 1). Details of the EC and LAS systems in HiWATER flux matrix are given in Tables 1 and 2, respectively.

2.2 Data collection, processing and quality control

2.2.1 Flux data processing and quality control

Data on the typical clear days of 29–30 June 2012 were se-

lected for the following analysis, including EC data from 16 towers (except site 3 and the highest level (34 m) of site 15) and four groups of LAS data as well as the multi-point micrometeorological data listed above. The last round of irrigation in each plot was done before 26 June; during the 2 days, there was almost no irrigation in the flux matrix. Firstly, AWS data sampled at 10 min were averaged to a 30 min period. Careful data processing and quality control for EC and LAS raw data were then performed so as to ensure a high-quality flux dataset.

The EddyPro software developed by LI-COR (Lincoln, Nebraska USA, www.licor.com/eddypro) was used to process the 10 Hz raw EC data into a half-hourly averaged flux data, by procedures including spike removal, angle of attack correction (for Gill), time lag correction, coordinate rotation (2-D rotation), frequency response correction, sonic virtual temperature correction, and corrections for density fluctuation (Webb–Pearman–Leuning, WPL). Data quality assessment was performed for the turbulent flux every 30 min using the flagging system with three different levels (0, 1, and 2) (Mauder and Foken, 2015). Detailed information on the processing steps can be found in Wang et al. (2015) and Xu et al. (2013). For this study, only the flux data of flag 0 (the best) were used. Flux data of flag 2, as well as the data at night when the friction velocity was below $0.1 \, \mathrm{m \, s^{-1}}$, were discarded (Blanken, 1998; Liu et al., 2011). To obtain daily ET, at first, a gap-filling method, based on the nonlinear regression (establishing the relationship between the latent heat flux and net radiation), for the 30 min latent heat fluxes (LE) was used. Then, the daily ET was calculated by summing the half-hourly gap-filled ET to 24 h totals.

For the EC systems used in the data analysis, we have tried to reduce the systematic errors to a minimum with a pre-observation inter-comparison, and careful maintenances during the observation period (Xu et al., 2013). The random errors were also analyzed by a separate research, which can be minimized in an ensemble average (Wang et al., 2015). The energy balance closure ratio (EBR) for the EC data of the flux matrix was also carefully assessed. Generally, the EBR during the 3 and half months was good. For the 17 EC stations in the intensive observation area, the average EBR was about 0.92. Except the lowest EBR (0.78) in orchard site, values in other sites were scattered without clear relation to the surface status. For site 15 with two heights of EC system, the relevant EBR were 0.89 (at 4.5 m) and 1.03 (for 34 m), respectively (Xu et al., 2017). For daytime conditions (global radiation $Rg > 20 \, \mathrm{W \, m^{-2}}$), the systematic error of the scalar fluxes (δ_F^{sys}) can be quantified indirectly through the surface energy balance closure (EBR), which is defined as $\delta_F^{sys} = F \cdot (1/\mathrm{EBR} - 1)$ (Mauder et al., 2013). So the systematic error of the turbulent flux for HiWATER flux matrix was generally about 8 %.

The LAS system provided a measurement of the structure parameter for the refractive index of air (C_n^2) with an output period of 1 min. The raw data were firstly checked, mainly rejecting the saturated cases when $C_n^2 < 0.193 R^{-8/3} \lambda^{1/3} D^{5/3}$ (where R is the path length, D the optical aperture, and λ the wavelength) (Ochs and Wilson, 1993). Then, the data were averaged to 30 min, and the path-average sensible heat fluxes were iteratively calculated based on Monin–Obukhov similarity theory (MOST) (Andreas, 1988). The parameters used in this calculation, like the roughness height and zero-plane displacement, were obtained following Martano (2000); other parameters, including wind speed, air pressure and temperature, Obukhov length, and Bowen ratio, were directly obtained from relevant EC measurements. For this study, the sensible heat fluxes from the LAS systems at daytime (8:30–15:30, Beijing Standard Time, BST; the time difference between local time and BST is approximately +1 h 18 min) were selected.

As for the eddy-covariance systems, quality control for the flux data from the four groups of LAS was also done. The systematic errors from data processing, e.g., the larger effects of Bowen-ratio correction in this oasis area, were carefully minimized. We checked the sensible heat fluxes (H) from the four groups of LAS with that from the nearer ECs. Except for LAS3 (under its path there were clearly some village buildings, so the H_LAS is higher), others agreed very well with that of ECs. These are consistent with the other two studies in the Heihe River basin: Liu et al. (2011) have reported that the LAS measured sensible heat flux over an alpine meadow was consistent with that of EC, and Xu et al. (2013) have concluded that both the EC and LAS system measurements were reliable and comparable during the intensive observation period of HiWATER.

2.2.2 High-resolution land-cover map from remote sensing

Based on the airborne hyper-spectral images acquired by the Compact Airborne Spectrographic Imager (CASI) on 29 June 2012 and the Canopy Height Model (CHM) data from the lidar data collected on 9 July 2012, a land-cover classification map with 1 m spatial resolution was derived using an object-based classification method. This was done mainly for the kernel experimental area. The overall accuracy of the land-cover classification map is up to 90 %, and the Kappa coefficient is approximately 0.9. The detailed classification process of the map can be found in Liu and Bo (2015).

Land-cover misclassification still occurred on this map because of spectral similarity, especially on the edges of different surface-cover types. To obtain a more accurate land-cover map, the misclassified patches of the land-cover map were visually and manually revised, according to the high-resolution Charge-Coupled Device (CCD) images (acquired on 26 July) and the Google Earth imagery (on 3 September 2012). Finally, for the aim of this study, the refined 12 kinds of land classification types, of which most were different vegetables of small areas, were merged into four kinds (maize, woods, vegetables, and non-vegetation types) in accordance with crop species and surface types, as shown in Fig. 1. Among the four land-cover types, the non-vegetation type mainly contains the buildings, roads, as well as plastic film covered greenhouses, while the woods type consists of orchard and shelterbelts.

3 Methodology

3.1 Aggregation method combining footprint analysis and multivariate regression

It is generally accepted that an area-averaged flux equals the area-weighted sum of the component fluxes from individual land-cover classes (Hutjes et al., 2010).

$$F = \sum_{k=1}^{n} A_k F_k, \qquad (1)$$

where F is the total flux of any scalar (e.g., the sensible and latent heat flux in our case) for a specified area, A_k is the fractional coverage of an individual land-cover class k within that area, and F_k is the flux from the individual land-cover class k; n is the number of land-cover classes that is distinguished in the specified area.

The observed flux (F_{obs}) at height z_m can be closely related to the true surface flux from upwind measurement point through the footprint function, in continuous form (Leclerc and Foken, 2014):

$$F_{obs}(x_{obs}, y_{obs}, z_m) = \qquad (2)$$
$$\int_{-\infty}^{\infty} \int_{-\infty}^{\infty} F(x, y, 0) w(x, y, z_m) \, dx \, dy$$

Here x_{obs}, y_{obs} are the site coordinates, z_m is the effective observation height, defined as $z_m = z - d$ (where z is the sensor height, d the zero-plane displacement). The footprint function $w(x, y, z_m)$ describes the flux portion seen at the sensor position (x_{obs}, y_{obs}, z_m). Equation (2) can be discretized for a uniform grid over a landscape, as in a land-cover classification map based on a remote sensing image. Leaving out the height dependence for simplification, it becomes

$$F_{obs} = \sum_{k=1}^{n} F_k \sum_{i=1}^{N} \sum_{j=1}^{M} w_{ij} \Delta x \Delta y, \qquad (3)$$

where each pixel $\Delta x \Delta y$ of the map is assumed to be homogeneous, which is uniquely classified as belonging to class k. The fraction of the kth land-cover type in the footprint (fp) is then defined as

$$X_{fp,k} = \sum_{i=1}^{N} \sum_{j=1}^{M} w_{ij} \Delta x \Delta y. \qquad (4)$$

Combining Eqs. (3) and (4), the multi-linear model for the flux becomes

$$F_{obs} = \sum_{k=1}^{n} F_k X_{fp,k}. \qquad (5)$$

A critical assumption under the flux aggregation method is that each land cover k (area A_k) has a constant source strength (F_k). Thus, as in Eq. (1), the flux (F) for a specific area is a weighted aggregation of its various land-cover classes. Based on multi-point tower flux measurements (F_{obs}), multiple-linear regression equations can be formulated by overlaying the flux footprint with a land-cover map ($X_{fp,k}$), as follows in Eq. (5). In this study, the multiple-linear

regression method (using the "regress" and "robustfit" algorithms from the Matlab® statistical toolbox) determined the regression coefficients (estimates of the specific flux for each land-cover class in the case F_k) by minimizing the squared residuals, and the standard error of each coefficient estimate was also quantified. So the specific flux (F_k) not only contains estimates, but also its uncertainties in regression. For each LAS path, the measured flux (i.e., sensible heat flux) can also be disaggregated into the component flux by the relevant footprint function as in Eq. (5). This can be taken as a validation of the former step.

The accuracy of this method is highly dependent on four most important aspects: (1) better flux data for all EC sites; (2) a better land-cover classification map; (3) more precise flux footprint analysis; and (4) good flux and footprint data for LAS. So properly processed flux data, an accurate high-resolution land-cover map, and appropriate footprint models are the foundation of formulating a better multiple-linear regression. Sometimes the multi-linear regression equations may not be solvable. When this problem happens, the classification accuracy of the land-cover map should be carefully checked, and the selected footprint model should also be verified with an alternative method.

3.2 Footprint models

The Eulerian analytical footprint model, which developed by Kormann and Meixner (2001), was used for estimating the single time flux footprint of EC measurements, due to its ease of use and wide range of stability as well as its numerical stability (Leclerc and Foken, 2014). Besides, as we have checked, its footprint estimates were in good agreement with the calculations of more sophisticated backward Lagrangian footprint models, such as the Kljun scheme (Kljun et al., 2002; Kljun et al., 2015). The footprint function $w(x, y, z)$ can be expressed in terms of a crosswind integrated flux footprint function $f^y(x, z)$ and a Gaussian crosswind distribution function $D_y(x, y)$. The analytic solution of Kormann and Meixner (2001) is depicted by Eq. (6). More details on the derivation of $f^y(x, z)$ and $D_y(x, y)$ as well as the relevant parameters can be seen in Kormann and Meixner (2001).

$$w(x, y, z) = f^y(x, z) D_y \qquad (6)$$
$$= \frac{1}{\Gamma(\mu)} \frac{\xi^\mu}{x^{1+\mu}} e^{-\xi/x} \frac{1}{\sqrt{2\pi}\sigma} e^{-\frac{y^2}{2\sigma^2}}$$

The flux contribution source area of LAS measurements was estimated by combining the footprint function $w(x, y, z)$ for point flux measurement with the path-weighting function $W(x)$ of LAS (Meijninger et al., 2002). For equal transmitter and receiver apertures, this path-weighting function is symmetrical bell-shaped having a center maximum and tapering to zero at the transmitter and receiver end. For the LAS footprint calculation, the approach of Korman and Meixner (2001) was still used for the single-point footprint

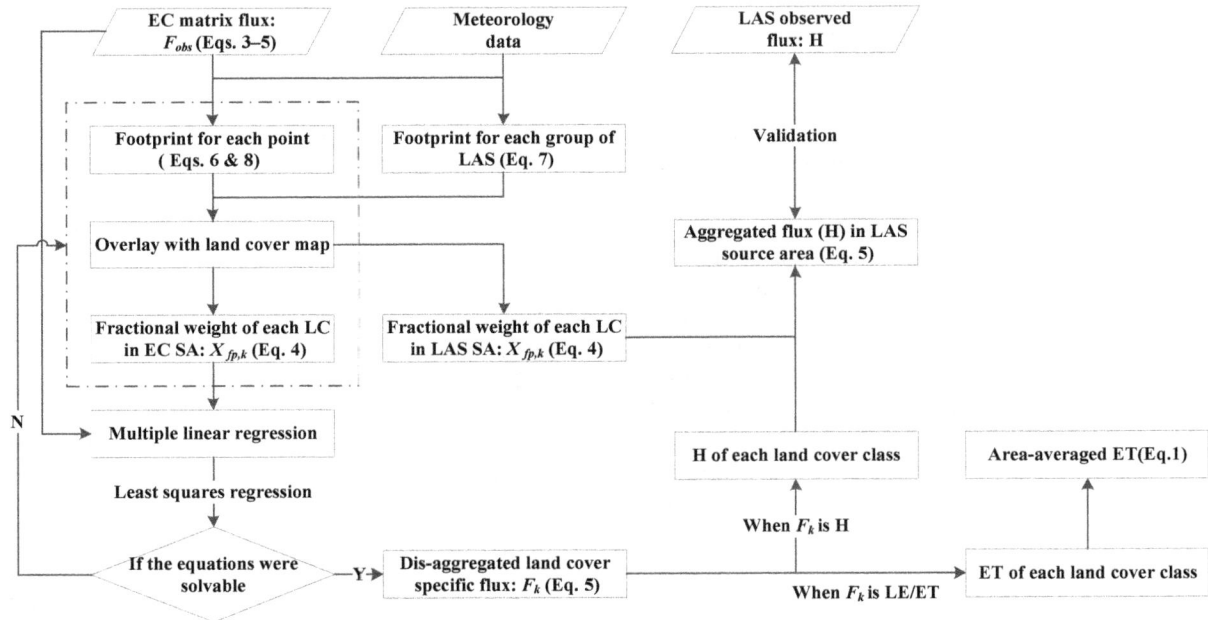

Figure 2. Schematic illustration of data processing steps; LC: land-cover class; SA: source area; H: sensible heat flux; LE: latent heat flux; ET: evapotranspiration.

estimation. The equation of the LAS footprint function is

$$f_{\mathrm{LAS}} = \int_{x_2}^{x_1} W(x)w(x - x', y - y', z_{\mathrm{LAS}})\mathrm{d}x, \qquad (7)$$

where x_1, x_2 are the positions of LAS transmitter and receiver, respectively. x, y represent the locations of points along the path of LAS. x', y' are the coordinates of each of the upwind points. z_{LAS} is the effective height of LAS measurements.

To obtain the averaged footprint of EC flux measurements (such as daily and monthly), the flux-weighted footprint climatology method was applied for each pixel (Liu et al., 2016). The equation of the weighted footprint climatology is

$$w_c(x, y, z) = \sum_i^N w_i(x, y, z)\mathrm{Flux}(i) / \sum_i^N \mathrm{Flux}(i). \qquad (8)$$

Here i denotes the time step (e.g., 30 min), N is the total number of 30 min periods within the selected time frame (such as daily scale), $\mathrm{Flux}(i)$ is the EC observed flux at the i time step (half-hourly LE in our case), and $w_i(x, y, z)$ represents a half-hourly footprint estimate calculated via Eq. (6).

The inputs of the analytical footprint model mainly include the measurement height, wind direction, wind speed, and the Obukhov length, which can be easily derived from flux tower measurements. The daily-averaged flux footprint of the EC observations was calculated by Eq. (8). The flux contribution of the total source area was set to 90 % for both

EC and LAS measurements. The normalized daily-averaged footprint of ECs and half-hourly footprint estimates of LASs were separately overlaid with the land-cover map to determine the footprint-weighted contribution of each land-cover class to the measured flux from the EC and LAS systems.

3.3 Data processing flow for the determination of area-averaged fluxes

The overall data processing flow for determining the area-averaged evapotranspiration over a heterogeneous land surface mainly includes three steps (Fig. 2).

Firstly, the spatial representativeness of 16 EC sites within the $5 \times 5 \, \mathrm{km}^2$ experimental area was quantitatively assessed by overlaying flux footprint climatology with a high-resolution land-cover map. Detailed analyses on this aspect are going to be presented in the following section.

Secondly, the sensible heat fluxes we aggregated were evaluated via the LAS measured path-weighted areal flux. Specifically speaking, based on footprint analysis and a high-resolution land-cover map, the sensible heat flux for individual land-cover types was firstly disaggregated from multiple EC flux measurements by performing a multiple-linear regression analysis (Eq. 5). To obtain area-averaged fluxes representative of the LAS source area, the EC-disaggregated fluxes for all land-cover classes were aggregated again according to the fractional weight of each land-cover class in the LAS footprint (Eq. 4). Finally, the EC-aggregated fluxes were compared with LAS observations.

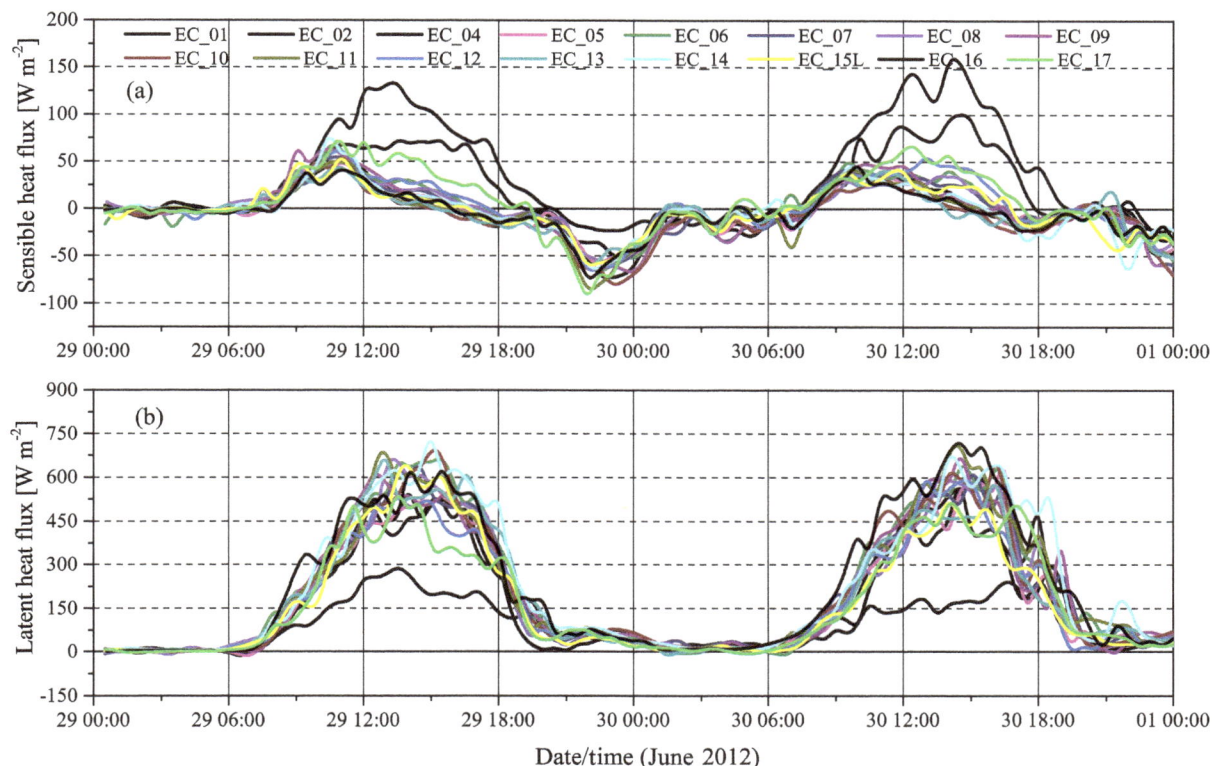

Figure 3. Diurnal cycle of the sensible heat fluxes (**a**) and latent heat fluxes (**b**) between different sites during 29 and 30 June 2012.

At last, the area-averaged evapotranspiration over a heterogeneous land surface was estimated with EC measurements and the flux integration scheme proposed and verified in the last step (Eq. 1).

4 Results and discussion

4.1 The characteristics of the surface heat and water vapor fluxes

Figure 3 depicts the diurnal cycle of the sensible and latent heat fluxes at different sites during two clear days. It not only reveals the energy exchange of different sites but also the significant differences in the magnitude of the sensible and latent heat fluxes between different surface types during the growing season.

The sensible heat flux over residential area (EC04) was higher than that over the vegetated surfaces (Fig. 3a) and reached a maximum of about 150 W m^{-2} at afternoon, while the latent heat flux was smallest among all sites (Fig. 3b).

Over the vegetated surfaces (maize, orchard, vegetable), the sensible heat flux was normally less than 100 W m^{-2} because of sufficient irrigation (Fig. 3a), and it was significantly different between different vegetated surfaces (Fig. 4a). Besides, difference in latent heat flux was also clear.

The maize fields showed higher latent heat fluxes and lower sensible heat fluxes than that of other two vegetated surfaces. One of the possible reasons is that both of the orchard area and vegetable field are comparatively sparsely vegetated than the maize field during the crop growing period. Over maize field, the sensible heat flux was relatively small and even negative in the midafternoon (known as the "oasis effect") (Figs. 3a, 4a); while the latent heat flux was quite large, with a maximum value up to 600 W m^{-2} (Figs. 3b, 4b). There was also a difference in sensible and latent heat fluxes among maize sites (Fig. 4). The values of standard deviation (SD) of LE and H for 13 maize sites were about 43 and 8 W m^{-2}, respectively.

4.2 Analysis of the representativeness of the multi-point EC flux measurements

To further understand the variability of surface energy fluxes between different sites in a heterogeneous landscape, footprint analyses for the spatial representativeness of 16 EC sites were performed by superimposing flux footprint with a land-cover map derived from aircraft images (Fig. 1). The fraction of all land-cover classes present in the daytime-averaged footprint is shown in Fig. 5.

Due to the differences of observation height and some variations of wind direction/speed, the EC source areas of differ-

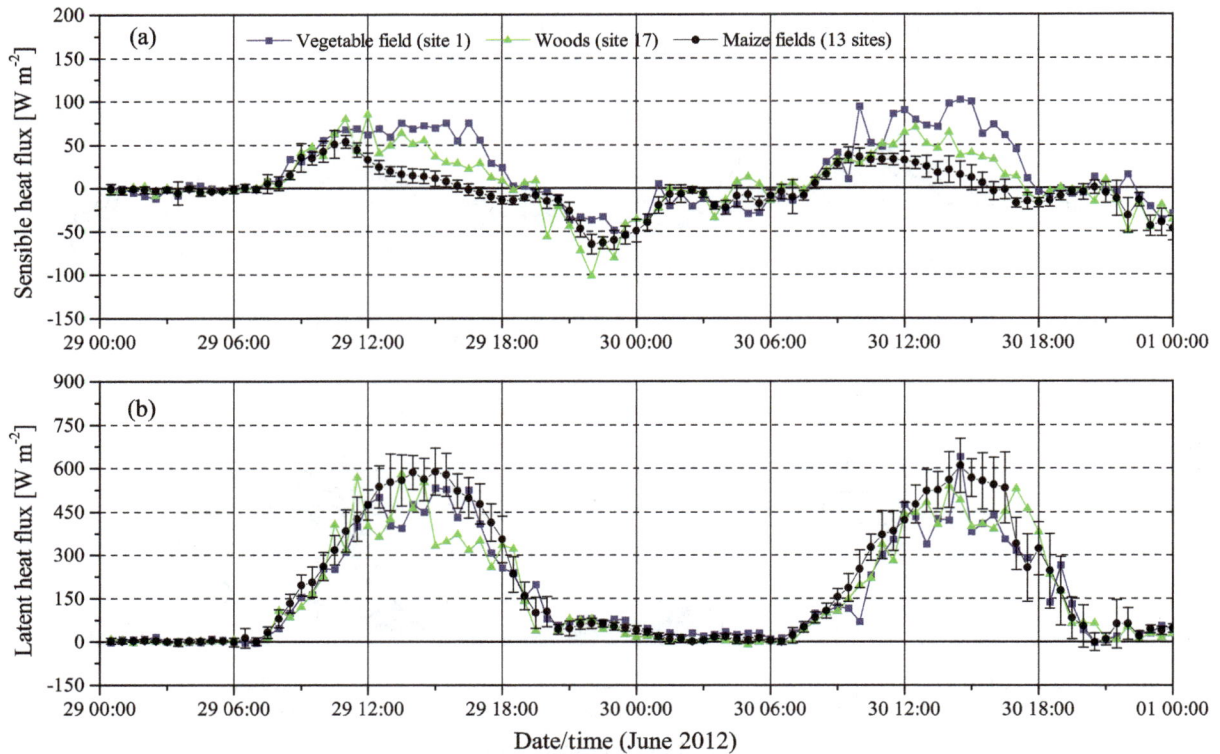

Figure 4. Diurnal cycle of the mean sensible **(a)** and latent **(b)** heat fluxes for 13 maize field sites and different types of vegetation; the error bar is the standard deviation.

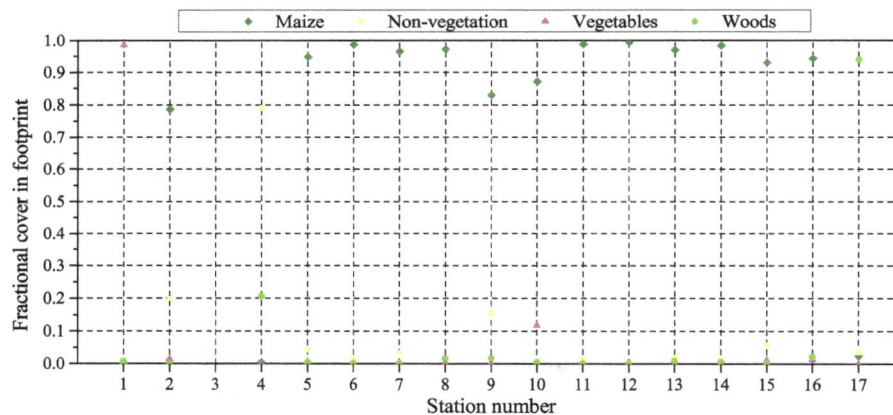

Figure 5. The fractional weight of each land-cover class in the daytime-averaged flux footprint of each EC flux measurement for 29–30 June 2012.

ent sites were distinctly different (Fig. 1). For each EC flux measurements, the relative contribution of each land cover to the total measured flux was presented in Fig. 5. For sites 1 and 17, the dominated surface types in the source area were vegetable and orchard, respectively. For site 4, the fractional weight of the non-vegetation type in the footprint was about 0.79, and that for woods was around 0.21. For other sites, the relative contribution of maize field to the EC measured flux was generally higher than 0.9. However, for site 2 and site 9, the non-vegetated surfaces covered 0.15 to 0.20 of the footprint area; while at site 10, the contribution of vegetable type to the flux measurements reached about 0.12.

The above analysis shows that the EC-matrix measurements are generally representative of the dominated surface types in the HiWATER intensive observation area. The "averaged" fluxes determined by weighting the upwind surface

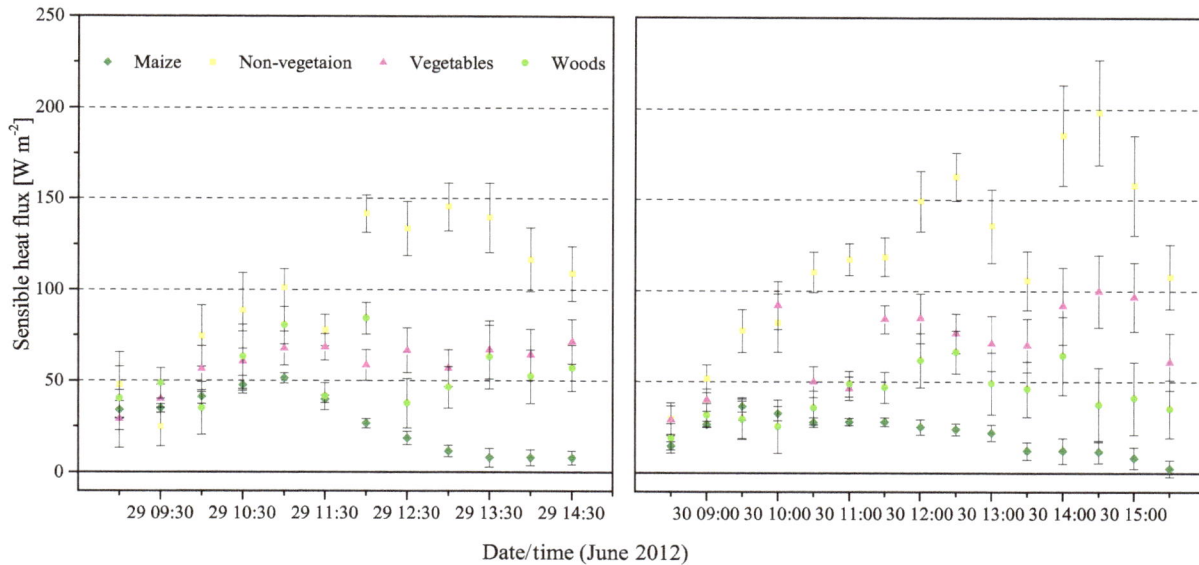

Figure 6. The diurnal cycle of the sensible heat flux for each land-cover class for 29–30 June 2012; the error bars are standard errors of sensible heat flux estimated in the regression.

flux by individual land-cover classes and the relevant flux footprints will be representative of the target area.

4.3 Evaluation of the EC-aggregated fluxes

Firstly, the sensible heat fluxes for each land-cover type were disaggregated from EC flux observations via the scheme mentioned in Sect. 3.1. Figure 6 gives the variation of the EC-disaggregated component fluxes and their uncertainties in the regression. The diurnal cycle of the sensible heat fluxes for each land-cover class was highly significant, and the difference in fluxes between different land-cover types was also distinct. The mean standard error (SE) of estimates of sensible heat flux for maize fields was the lowest, with a value of about 4 W m^{-2}, mainly because of its dominant land-cover type and the majority of EC sites, while the SE values for land-cover types of non-vegetation, vegetable, and woods were larger, with values of about 15, 12, and 13 W m^{-2}, respectively, mainly due to the heterogeneity of the underlying surfaces, especially for site 4.

Then, the EC-disaggregated fluxes for all land-cover classes were aggregated again to obtain area-averaged fluxes representative of the LAS source area. Figure 7 illustrates a scatterplot of half-hourly area-averaged sensible heat fluxes estimated from the EC flux matrix (hereafter referred to as H_ECagg) versus LAS measurements (H_LAS), as well as the linear regression parameters (equations and R^2). The different statistics between them are also listed in Table 3.

For LAS1 (see Fig. 7a and Table 3), a good agreement is found between EC aggregated fluxes and LAS measurements, with high correlation and a low root mean square error (RMSE) value ($R^2 = 0.75$, RMSE $= 1$ W m^{-2}). The scatter

Table 3. Different statistics between LAS observed flux and EC aggregated flux at LAS sites.

LAS sites	RMSE (W m^{-2})	MBE (W m^{-2})	MAPE (%)
LAS1	1	4	10
LAS2	7	2	16
LAS3	18	−18	32
LAS4	13	10	34

Remarks: RMSE $= \sqrt{\sum_{i=1}^{n}(P_i - O_i)^2/n}$; MAPE $= \frac{100}{n}\sum_{i=1}^{n}\frac{|P_i - O_i|}{\overline{O}}$;

MBE $= \sum_{i=1}^{n}(P_i - O_i)/n$; P_i is the EC aggregated value; O_i is the LAS observed value; $\overline{O}$ is the mean measured value; n is the number of samples. RMSE is root mean square error, MAPE is mean absolute percentage error, and MBE is the mean bias error.

points in the graph are nearly close to the 1 : 1 line. The MBE and MAPE values were almost 4 W m^{-2} and 10 %, respectively. For LAS2 (Fig. 7b), there was a little more scatter and a little higher RMSE ($R^2 = 0.59$, RMSE $= 7$ W m^{-2}), probably owing to a slight parts of village buildings under its path center (Fig. 1). For LAS4 (Fig. 7d), the sensible heat flux aggregated from the EC matrix was a little larger than that from LAS measurements ($R^2 = 0.65$, RMSE $= 13$ W m^{-2}).

For LAS3 (Fig. 7c, Table 3), there was a slightly weak relationship between LAS measured sensible heat fluxes and that from multi-point EC measurements. The scatter points were overall below the 1 : 1 line, and both RMSE and MBE were larger than those of LAS1, LAS2, and LAS4. This would be from a larger area of residential buildings distributed under the central part of the LAS path (Fig. 1), and their representative sensible heat flux derived from a single site 4 would

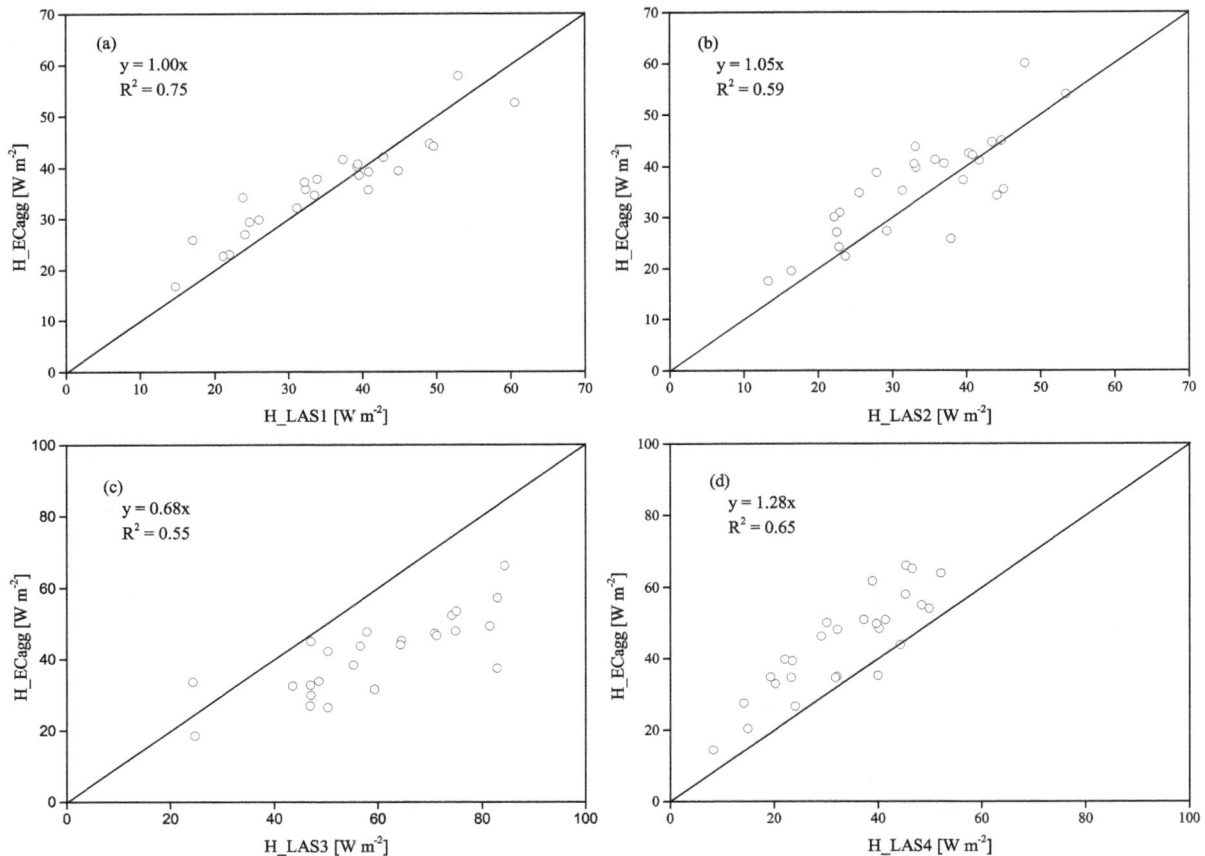

Figure 7. The comparison between LAS observed fluxes (x axis) and EC aggregated fluxes (y axis).

be insufficient. Besides, as the closure ratio of the surface energy balance (EBR) is comparatively lower over heterogeneous areas, such as for site 4 (Xu et al., 2017), the energy flux from large eddies or secondary circulations induced from surface heterogeneities may not be captured by a single-point EC, but may be measured via the LAS system. Thus, the LAS observations might be able to close the surface energy balance better than the EC method (Foken, 2008; Foken et al., 2010).

The flux uncertainties of LAS, induced in its data processing by use of EC measurements (such as Obukhov length and Bowen ratio), also have considerable influence on the comparison. When using a constant Bowen ratio as in former processing of the LAS data (Xu et al., 2013), the MAPE of the comparison between LAS and EC aggregation would be increased by about 20 %, and the value of the RMSE would be more than 10 W m^{-2}.

The present flux aggregation method relies on the using of footprint model and high-resolution land-cover map (Eq. 5). We have checked the footprint results between the Kormann and Meixner (2001) model we used with that of Kljun's

scheme (Kljun et al., 2002, 2015), the differences were apparently minor. Hutjes et al. (2010) also reported that the uncertainties about the real size of the footprint do not affect the disaggregating of flux estimates too much as long as the uncertainty does not lead to significantly different fractional covers. Besides, the error in the land-cover classification map we used is no more than 3 %. So the effect of uncertainties in $X_{\mathrm{fp},k}$ on estimates of F_k is rather small. As mentioned in Sect. 2.2.1, the systematic error of EC fluxes was also small. Therefore, disaggregation from fluxes over a heterogeneous area into individual land-cover types is feasible. The present scheme is also practical over a truly complex landscape or a simple "binary" land (Hutjes et al., 2010).

In the comparison with LAS measurements, both the arithmetic-averaged and area-averaged methods (Liu et al., 2016) displayed a relatively larger scatter and larger values of RMSE and MAPE than the present aggregation method, even for a relatively homogeneous land surface (e.g., near the path of LAS1). The present scheme not only has a much better database, but also has a solid grounding in physics and mathematics in the aggregation of area-averaged fluxes over a heterogeneous surface.

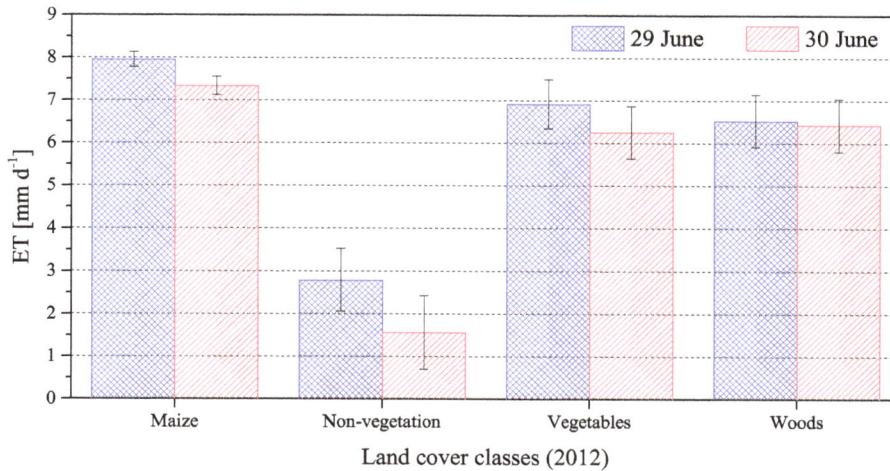

Figure 8. The disaggregated daily ET for each land-cover class in the kernel experimental area of HiWATER on 29 and 30 June 2012; the error bar is the standard error of ET estimates in the regression.

Table 4. ET for each land-cover class and their proportion of the kernel experimental area ET.

Land-cover class	Area (km^2)	29 June 2012		30 June 2012	
		ET (10^3 m^3 day^{-1})	ET proportion of total ET (%)	ET (10^3 m^3 day^{-1})	ET proportion of total ET (%)
Maize	17.42	138.56	81.6	127.86	83.2
Woods	1.96	12.78	7.5	12.58	8.2
Vegetables	1.20	8.28	4.9	7.48	4.9
Non-vegetation	3.62	10.11	6.0	5.67	3.7

4.4 Estimation of area-averaged evapotranspiration (ET)

The present flux aggregation scheme, as evaluated in Sect. 4.3, was adopted to determine the area-averaged ET over the study area with multi-point EC flux measurements and high-resolution land-cover map. The EC disaggregated daily ET for four land-cover classes on 2 typical clear days is shown in Fig. 8; the standard errors of the ET estimates from least squares regression are also plotted as error bars. As can be seen, the daily ET values for maize field were the highest (7–8 mm) during the crop growing season. For woods areas the value of daily ET was about 6.5 mm; and for vegetable field it was 6–7 mm. On the contrary, the daily ET for non-vegetation type varied largely, with values about 2.8 mm on 29 June and 1.6 mm on 30 June, respectively.

Table 4 lists the total ET for different land-cover classes and their proportions of the total area. The total ET for the study area was approximately 169.7×10^3 m^3 day^{-1} on 29 June and 153.6×10^3 m^3 day^{-1} on 30 June. The ET of maize fields occupied more than 80 % of the total area, due to its domination in the study area, while that for woods, vegetables, and non-vegetation fields occupied about 8, 5, and 5 %, respectively.

The area-averaged ET over the kernel experimental area of HiWATER was finally estimated, with values of approximately 7 mm day^{-1} on 29 June and 6 mm day^{-1} on 30 June, respectively.

5 Summary and conclusions

On the basis of an accurate high-resolution land-cover classification map and multi-point ground-based flux measurements from 16 EC systems and four groups of LAS systems during the intensive observation period of HiWATER, a flux aggregation method for determining area-averaged flux was established through the combination of footprint analysis and multiple-linear regression. The method was applied to estimate the area-averaged surface fluxes over a heterogeneous surface from multi-point EC flux measurements, and its results were verified by the LAS measurements. Ultimately, the integration method was applied to estimate area-averaged ET over the study area.

Robust quality control and uncertainty estimation for the EC and LAS data, done through careful data processing and inter-comparison as well as assessment of the energy balance closure, ensure the accuracy of the flux dataset used in data analysis. For the deep interpretation of the surface fluxes over different land surfaces, the combination of footprint analyses for the representativeness of EC flux measurements and high-resolution land-cover map can be a practical way, and it is also the foundation for the establishment of the flux aggregation algorithm.

With a high-quality flux dataset (EC & LAS), precise flux footprint estimates, and an accurate land-cover classification map, a flux aggregation method can be successfully established by multiple linear regression analysis. It achieves the goal of determining the area-averaged fluxes over heterogeneous areas from the EC flux matrix. However, the agreement between the results of the flux integration method and LAS measurements partly relates to the heterogeneity of the land surface, as it influences greatly the energy balance closure of the EC flux dataset. On the other hand, the bias may be partly attributed to the insufficient distribution of flux stations on some dominant land covers; the uncertainty of LAS measurements was also a factor that affects the result of the comparison.

In spite of the limitations mentioned above, the current flux integration scheme provides a unique opportunity to discompose the heterogeneous land surface fluxes into their single components, and the disaggregation process has the potential to scale up multiple EC measurements to an oasis landscape, even to a whole river basin. Besides, compared with the formerly used and rather simple approaches (e.g., the arithmetic average and area-weighted methods), the present scheme is not only with a much better database, but also has a solid grounding in physics and mathematics in the integration of area-averaged fluxes over a heterogeneous surface. Results from this study, such as daily ET at the satellite pixel scale, can be applied for the validation of flux estimates of meso-γ-scale (1–20 km) models. Furthermore, this work will be extended to the water balance study of the whole Heihe River basin, which is quite practical for hydrological modeling and basin water resource management.

Competing interests. The authors declare that they have no conflict of interest.

Acknowledgements. We thank Thomas Foken and another three, anonymous, reviewers for their constructive comments that significantly improved the presentation of this paper. We also appreciate all the scientists, engineers, and students who participated in the HiWATER field campaigns. This study was supported by the National Natural Science Foundation of China (grant numbers: 41671373; 41271359) and the Science and Technology Service Network Initiative Project of the Chinese Academy of Sciences.

Edited by: Bob Su

References

Anderson, M. C., Kustas, W. P., Alfieri, J. G., Gao, F., Hain, C., Prueger, J. H., Evett, S., Colaizzi, P., Howell, T., and Chávez, J. L.: Mapping daily evapotranspiration at Landsat spatial scales during the BEAREX'08 field campaign, Adv. Water Resour., 50, 162–177, 2012.

André, J.-C., Goutorbe, J.-P., and Perrier, A.: HAPEX-MOBLIHY: A Hydrologic Atmospheric Experiment for the Study of Water Budget and Evaporation Flux at the Climatic Scale, B. Am. Meteorol. Soc., 67, 138–144, 1986.

Andreas, E. L.: Estimating Cn^2 over snow and sea ice from meteorological data, J. Opt. Soc. Am. A, 5, 481–495, 1988.

Beyrich, F. and Mengelkamp, H.-T.: Evaporation over a heterogeneous land surface: EVA_GRIPS and the LITFASS-2003 experiment – an overview, Bound.-Lay. Meteorol., 121, 5–32, 2006.

Beyrich, F., Leps, J.-P., Mauder, M., Bange, J., Foken, T., Huneke, S., Lohse, H., Lüdi, A., Meijninger, W. M., and Mironov, D.: Area-averaged surface fluxes over the LITFASS region based on eddy-covariance measurements, Bound.-Lay. Meteorol., 121, 33–65, 2006.

Blanken, P.: Turbulent flux measurements above and below the overstory of a boreal aspen forest, Bound.-Lay. Meteorol., 89, 109–140, 1998.

Charuchittipan, D., Babel, W., Mauder, M., Leps, J. P., and Foken, T.: Extension of the Averaging Time in Eddy-Covariance Measurements and Its Effect on the Energy Balance Closure, Bound.-Lay. Meteorol., 152, 303–327, 2014.

Chen, B., Black, T. A., Coops, N. C., Hilker, T., Trofymow, J. A., and Morgenstern, K.: Assessing Tower Flux Footprint Climatology and Scaling Between Remotely Sensed and Eddy Covariance Measurements, Bound.-Lay. Meteorol., 130, 137–167, https://doi.org/10.1007/s10546-008-9339-1, 2008.

Ezzahar, J. and Chehbouni, A.: The use of scintillometry for validating aggregation schemes over heterogeneous grids, Agr. Forest Meteorol., 149, 2098–2109, 2009.

Ezzahar, J., Chehbouni, A., Er-Raki, S., and Hanich, L.: Combining a large aperture scintillometer and estimates of available energy to derive evapotranspiration over several agricultural fields in a semi-arid region, Plant Biosyst., 143, 209–221, https://doi.org/10.1080/11263500802710036, 2009a.

Ezzahar, J., Chehbouni, A., Hoedjes, J., Ramier, D., Boulain, N., Boubkraoui, S., Cappelaere, B., Descroix, L., Mougenot, B., and Timouk, F.: Combining scintillometer measurements and an aggregation scheme to estimate area-averaged latent heat flux during the AMMA experiment, J. Hydrol., 375, 217–226, 2009b.

Foken, T.: The energy balance closure problem: An overview, Ecol. Appl., 18, 1351–1367, 2008.

Foken, T., Wimmer, F., Mauder, M., Thomas, C., and Liebethal, C.: Some aspects of the energy balance closure problem, Atmos. Chem. Phys., 6, 4395–4402, https://doi.org/10.5194/acp-6-4395-2006, 2006.

Foken, T., Mauder, M., Liebethal, C., Wimmer, F., Beyrich, F., Leps, J.-P., Raasch, S., DeBruin, H. A., Meijninger, W. M., and Bange, J.: Energy balance closure for the LITFASS-2003 experiment, Theor. Appl. Climatol., 101, 149–160, 2010.

Goutorbe, J., Lebel, T., Tinga, A., Bessemoulin, P., Brouwer, J., Dolman, A., Engman, E., Gash, J., Hoepffner, M., and Kabat, P.: HAPEX-Sahel: a large-scale study of land-atmosphere interactions in the semi-arid tropics, Ann. Geophys., 12, 53–64, 1994.

Halldin, S., Gottschalk, L., van de Griend, A. A., Gryning, S.-E., Heikinheimo, M., Högström, U., Jochum, A., and Lundin, L.-C.: NOPEX – a northern hemisphere climate processes land surface experiment, J. Hydrol., 212, 172–187, 1998.

Horst, T. and Weil, J.: Footprint estimation for scalar flux measurements in the atmospheric surface layer, Bound.-Lay. Meteorol., 59, 279–296, 1992.

Hutjes, R., Vellinga, O., Gioli, B., and Miglietta, F.: Dis-aggregation of airborne flux measurements using footprint analysis, Agr. Forest Meteorol., 150, 966–983, 2010.

Jia, Z., Liu, S., Xu, Z., Chen, Y., and Zhu, M.: Validation of remotely sensed evapotranspiration over the Hai River Basin, China, J. Geophys. Res.-Atmos., 117, D13113, https://doi.org/10.1029/2011JD017037, 2012.

Kim, J., Guo, Q., Baldocchi, D., Leclerc, M., Xu, L., and Schmid, H.: Upscaling fluxes from tower to landscape: Overlaying flux footprints on high-resolution (IKONOS) images of vegetation cover, Agr. Forest Meteorol., 136, 132–146, 2006.

Kirby, S., Dobosy, R., Williamson, D., and Dumas, E.: An aircraft-based data analysis method for discerning individual fluxes in a heterogeneous agricultural landscape, Agr. Forest Meteorol., 148, 481–489, 2008.

Kljun, N., Rotach, M., and Schmid, H.: A three-dimensional backward Lagrangian footprint model for a wide range of boundary-layer stratifications, Bound.-Lay. Meteorol., 103, 205–226, 2002.

Kljun, N., Calanca, P., Rotach, M. W., and Schmid, H. P.: A simple two-dimensional parameterisation for Flux Footprint Prediction (FFP), Geosci. Model Dev., 8, 3695–3713, https://doi.org/10.5194/gmd-8-3695-2015, 2015.

Kormann, R. and Meixner, F. X.: An analytical footprint model for non-neutral stratification, Bound.-Lay. Meteorol., 99, 207–224, 2001.

Leclerc, M. Y. and Foken, T.: Footprints in Micrometeorology and Ecology, Springer, Heidelberg, New York, Dordrecht, London, 239 pp., 2014.

Li, X., Cheng, G., Liu, S., Xiao, Q., Ma, M., Jin, R., Che, T., Liu, Q., Wang, W., Qi, Y., Wen, J., Li, H., Zhu, G., Guo, J., Ran, Y., Wang, S., Zhu, Z., Zhou, J., Hu, X., and Xu, Z.: Heihe Watershed Allied Telemetry Experimental Research (HiWATER): Scientific Objectives and Experimental Design, B. Am. Meteorol. Soc., 94, 1145–1160, 2013.

Liu, S. M., Xu, Z. W., Wang, W. Z., Jia, Z. Z., Zhu, M. J., Bai, J., and Wang, J. M.: A comparison of eddy-covariance and large aperture scintillometer measurements with respect to the energy balance closure problem, Hydrol. Earth Syst. Sci., 15, 1291–1306, https://doi.org/10.5194/hess-15-1291-2011, 2011.

Liu, S., Xu, Z., Song, L., Zhao, Q., Ge, Y., Xu, T., Ma, Y., Zhu, Z., Jia, Z., and Zhang, F.: Upscaling evapotranspiration measurements from multi-site to the satellite pixel scale over heterogeneous land surfaces, Agr. Forest Meteorol., 230, 97–113, 2016.

Liu, X. and Bo, Y.: Object-Based Crop Species Classification Based on the Combination of Airborne Hyperspectral Images and Li-DAR Data, Remote Sens., 7, 922–950, 2015.

Mahrt, L., Vickers, D., Sun, J., and McCaughey, J. H.: Calculation of area-averaged fluxes: Application to BOREAS, J. Appl. Meteorol., 40, 915–920, 2001.

Martano, P.: Estimation of surface roughness length and displacement height from single-level sonic anemometer data, J. Appl. Meteorol., 39, 708–715, 2000.

Mauder, M. and Foken, T.: Documentation and instruction manual of the eddy covariance software package TK3 (update), Arbeitsergebnisse, Universität Bayreuth, Abt. Mikrometeorologie, 64 pp., 2015.

Mauder, M., Cuntz, M., Drüe, C., Graf, A., Rebmann, C., Schmid, H. P., Schmidt, M., and Steinbrecher, R.: A strategy for quality and uncertainty assessment of long-term eddy-covariance measurements, Agr. Forest Meteorol., 169, 122–135, 2013.

Meijninger, W., Hartogensis, O., Kohsiek, W., Hoedjes, J., Zuurbier, R., and De Bruin, H.: Determination of area-averaged sensible heat fluxes with a large aperture scintillometer over a heterogeneous surface–Flevoland field experiment, Bound.-Lay. Meteorol., 105, 37–62, 2002.

Mengelkamp, H.-T., Beyrich, F., Heinemann, G., and Ament, F., Bange, J., Berger, F. H., Bösenberg, J., Foken, T., Hennemuth, B., Heret, C., Huneke, S., Johnsen, K.-P., Kerschgens, M., Kohsiek, W., Leps, J.-P., Liebethal, C., Lohse, H., Mauder, M., Meijninger, W. M. L., Raasch, S., Simmer, C., Spieß, T., Tittebrand, A., Uhlenbrook, S., and Zittel, P.: Evaporation over a heterogeneous land surface: the EVA-GRIPS project, B. Am. Meteorol. Soc., 87, 775–786, 2006.

Ochs, G. and Wilson, J.: A Second-generation Large-aperature Scintillometer, US Department of Commerce, National Oceanic and Atmospheric Administration, Environmental Research Laboratories, Wave Propagation Laboratory, 1993.

Ogunjemiyo, S. O., Kaharabata, S. K., Schuepp, P. H., MacPherson, I. J., Desjardins, R. L., and Roberts, D. A.: Methods of estimating CO_2, latent heat and sensible heat fluxes from estimates of land cover fractions in the flux footprint, Agr. Forest Meteorol., 117, 125–144, 2003.

Ran, Y., Li, X., Sun, R., Kljun, N., Zhang, L., Wang, X., and Zhu, G.: Spatial representativeness and uncertainty of eddy covariance carbon flux measurements for upscaling net ecosystem productivity to the grid scale, Agr. Forest Meteorol., 230, 114–127, 2016.

Schmid, H. P.: Footprint modeling for vegetation atmosphere exchange studies: a review and perspective, Agr. Forest Meteorol., 113, 159–183, 2002.

Sellers, P., Hall, F., Asrar, G., Strebel, D., and Murphy, R.: The first ISLSCP field experiment (FIFE), B. Am. Meteorol. Soc., 69, 22–27, 1988.

Sellers, P., Hall, F., Ranson, K. J., Margolis, H., Kelly, B., Baldocchi, D., den Hartog, G., Cihlar, J., Ryan, M. G., and Goodison, B.: The boreal ecosystem-atmosphere study (BOREAS): an overview and early results from the 1994 field year, B. Am. Meteorol. Soc., 76, 1549–1577, 1995.

Wang, J.: Land surface process experiments and interaction study in China: From HEIFE to IMGRASS and GAME-Tibet/TIPEX, Plateau Meteorol., 18, 280–294, 1999.

Wang, J., Gao, F., and Liu, S.: Remote sensing retrieval of evapotranspiration over the scale of drainage basin, Remote Sens. Technol. Appl., 18, 332–338, 2003.

Wang, J., Zhuang, J., Wang, W., Liu, S., and Xu, Z.: Assessment of Uncertainties in Eddy Covariance Flux Measurement Based on Intensive Flux Matrix of HiWATER-MUSOEXE, IEEE Geosci. Remote S., 12, 259–263, 2015.

Xu, Z., Liu, S., Li, X., Shi, S., Wang, J., Zhu, Z., Xu, T., Wang, W., and Ma, M.: Intercomparison of surface energy flux measurement systems used during the HiWATER-MUSOEXE, J. Geophys. Res.-Atmos., 118, 13140–13157, 2013.

Xu, Z., Ma, Y., Liu, S., Shi, W., and Wang, J.: Assessment of the Energy balance closure under advective conditions and its impact using remote sensing data, J. Appl. Meteorol. Climatol., 56, 127–140, 2017.

8

Inundation mapping based on reach-scale effective geometry

Cédric Rebolho[1], Vazken Andréassian[1], and Nicolas Le Moine[2]

[1]Irstea, UR HYCAR, 1 Rue Pierre-Gilles de Gennes, 92160 Antony, France
[2]Sorbonne Universités, UPMC Univ Paris 06, CNRS, EPHE, UMR 7619 Metis, 4 Place Jussieu, 75005 Paris, France

Correspondence: Cédric Rebolho (cedric.rebolho@irstea.fr)

Abstract. The production of spatially accurate representations of potential inundation is often limited by the lack of available data as well as model complexity. We present in this paper a new approach for rapid inundation mapping, MHYST, which is well adapted for data-scarce areas; it combines hydraulic geometry concepts for channels and DEM data for floodplains. Its originality lies in the fact that it does not work at the cross section scale but computes effective geometrical properties to describe the reach scale. Combining reach-scale geometrical properties with 1-D steady-state flow equations, MHYST computes a topographically coherent relation between the "height above nearest drainage" and streamflow. This relation can then be used on a past or future event to produce inundation maps. The MHYST approach is tested here on an extreme flood event that occurred in France in May–June 2016. The results indicate that it has a tendency to slightly underestimate inundation extents, although efficiency criteria values are clearly encouraging. The spatial distribution of model performance is discussed and it shows that the model can perform very well on most reaches, but has difficulties modelling the more complex, urbanised reaches. MHYST should not be seen as a rival to detailed inundation studies, but as a first approximation able to rapidly provide inundation maps in data-scarce areas.

1 Introduction

Floods are a recurring phenomenon in France: in September 2014, intense rainfall affected the south of the country, leading to several deaths and about EUR 0.6 billion worth of damage. The following year, in October, about 20 people died in the south-east due to massive flooding, which caused a loss of EUR 0.5 billion. Then, in June 2016, large-scale flooding occurred over the Seine and Loire catchments, mainly affecting their tributaries and resulting in four deaths at a cost of EUR 1.4 billion. These are only examples which underline the value of flood inundation mapping to anticipate the impact of such events. Public authorities and insurance companies are showing a growing interest in the field of rapid inundation modelling, and for the development of simple methods, that would work for any river with easily available data.

Flood hazard assessment usually combines rainfall observations or simulations, a hydrological model, streamflow simulations or observations, and an inundation model in order to generate inundation extents, height maps and sometimes other information (e.g. velocities). Traditionally, flood inundation models are derived from the shallow water equations (SWEs) in one or two dimensions (the so-called hydraulic models), with various simplifications that have proved to give satisfying results. For instance, the Regional Flood Model (RFM), probably one of the most comprehensive approaches published so far, is made of four parts (Falter et al., 2014): a daily distributed rainfall–runoff model, a 1-D hydraulic model for channel routing, a 2-D hydraulic model for floodplain mapping and a flood loss estimation model. Its application on the Mulde catchment in Germany (Falter et al., 2015) showed mixed results concerning inundation extents, correctly predicting only 50 % of the flooded area for the August 2002 event. This underestimation was explained by dike breaches that were not accounted for within the model. A lack of observed data did not allow validation on other events.

Not all hydraulic models need to have this degree of complexity. It is indeed possible to neglect specific parts of the SWEs depending on the situation. Usually, 2-D models use the complete Saint-Venant equations while 1-D models often disregard one or several terms, leading, for instance, to the diffusive wave or kinematic wave approximations (e.g. Moussa and Cheviron, 2015). Some methods choose to couple 1-D and 2-D models, the former for streamflow routing and the latter for overbank flow (Morales-Hernández et al., 2016). Despite the accuracy of such models, studies often try to further simplify them because of the large computing time to simulate small areas and the lack of precise data required to run these models.

LISFLOOD-FP (Bates and De Roo, 2000), a hydraulic model developed to simulate floodplain inundation, was used in several studies (Horritt and Bates, 2001; Hunter et al., 2005; Biancamaria et al., 2009). The model offers different possibilities: using 2-D equations or 1-D equations decoupled on a 2-D grid with kinematic, diffusive or inertial approximations (Bates et al., 2010). Horritt and Bates (2002) published a comparison between different models with gradually increasing complexities (1-D, 1-D on 2-D grid and 2-D) and, surprisingly, showed that the 1-D model had a better ability to reproduce the two events that were used in validation. The subsequent analysis concluded that the reach studied was relatively narrow and could easily be modelled using simple methods, and the authors argued that the other models would be more appropriate for more complex reaches.

However, these examples concern relatively small and well-instrumented reaches and assessing flood hazard at a larger scale may require different approaches. Alfieri et al. (2014) applied LISFLOOD-ACC, an inertial version of LISFLOOD-FP with decoupled 1-D equations on a $100\,\mathrm{m}$ resolution grid over Europe in order to map flood hazards for a 100-year return period, assuming a constant return period along the reaches. Broadly speaking, the model splits rivers into small reaches, to apply the hydraulic models independently and to merge simulated maps together, but only for rivers with a catchment larger than $500\,\mathrm{km}^2$. The model was then validated against regional and national hazard maps for six catchments in Germany and the United Kingdom and showed a general over-prediction. Another variation of LISFLOOD-FP for large-scale flood inundation modelling was introduced by Neal et al. (2012), including a new sub-grid representation of channel networks for improved model accuracy (Neal et al., 2012; Schumann et al., 2013).

Le Bihan et al. (2017) developed an approach aimed at the forecasting context, in order to cope with excessive computing times. The solution chosen was to run a simple 1-D hydraulic model during a "pre-analysis phase" and create a catalogue of inundation extents corresponding to various return periods. These maps are then used, in a forecasting context, to give an estimate of the level of flooding, depending on the forecast discharge.

The lack of precise data (especially for channel cross sections) and the computing time required by numerical methods for solving the SWE motivated the development of potentially alternative methods, mostly based on DEM analysis. For instance, the rapid flood spreading method (RFSM,

Gouldby et al., 2008) was chosen to divide floodplains into impact zones of different elevations in order to explore the effects of dike breaches using a spilling algorithm based on water depth. Other methods derive inundation maps from topographic information only: one can cite EXZECO (Pons et al., 2010), which introduces elevation noise in the DEM in order to create a single map of "maximum flow accumulation" that can be seen as a potential inundation area, and HAND ("height above nearest drainage"), a descriptor originally used for terrain classification (Rennó et al., 2008; Nobre et al., 2011), which has recently been adapted to static flood inundation mapping (Nobre et al., 2016) and is increasingly used to produce flood maps (e.g. Afshari et al., 2018; Speckhann et al., 2018; McGrath et al., 2018). HAND calculates the difference between river cells' elevation and that of the connected floodplain cells, thus giving relative height information which can be compared to observed flood depths and the corresponding inundation extent.

MHYST, the method presented in this paper, is a simplified approach developed with the aim of rapidly producing inundation maps in data-scarce areas. It combines (i) concepts of hydraulic geometry to characterise channel geometry and (ii) DEM-derived relative elevations to characterise the floodplain; it does not work at the cross section scale but computes effective geometrical properties representative of the reach scale. Combining reach-scale geometrical properties with simplified steady-state hydraulic laws allows one to rapidly generate flood inundation maps while ensuring reach-scale coherence. After describing the method and the calibration dataset, MHYST is compared against the inundation extent observed for the major event that occurred in May–June 2016 in France. The last section discusses the spatial distribution of performance and the impact of uncertainties on the results obtained.

2 MHYST: a simplified steady-state hydraulic approach

The MHYST model stands for *Modélisation HYdraulique simplifiée en écoulement STationnaire*, i.e. Simplified Steady-state Hydraulic Modelling. It is a flood inundation model which aims to map inundation extents at the reach scale. Where classic hydraulic models use cross sections, this method is based on an effective geometry representative of each river reach. Since no detailed geometric data were available to describe the shape and roughness of the channel river bed for this study, a sub-grid representation of the channel was derived from hydraulic geometry relationships linking the drainage area with bankfull width and height (Leopold and Maddock, 1953). When discharge exceeds bankfull capacity, the model computes a reach-scale relation between streamflow and the HAND defined by Nobre et al. (2016). This relation can finally be used to assess which height corresponds to the given streamflow, and thus to derive the corresponding inundation map.

2.1 Processing of DEM: from elevations to height above nearest drainage

The initial step consists of processing the digital elevation model (DEM) in order (i) to obtain a flowing drainage direction map, (ii) to identify the subcatchments (corresponding to the river reaches), and (iii) to compute the height above nearest drainage (HAND) in each subcatchment. This initial processing is the basis of the floodplain analysis in MHYST. To compute the drainage direction map, we used the D8 method from the Flow Direction function provided by ArcGIS 10.3. It computes the drainage direction by calculating the steepest slope from the eight possible directions for a given cell.

Figure 1 shows the procedure used to compute HAND values: for a given floodplain cell, it is the difference between its elevation and that of the closest river cell in terms of drainage direction. For instance, the cell of elevation 25 at the top of the figure is linked to (i.e. flows towards) the most upstream red river cell which has an elevation of 18: thus, its HAND value is $25 - 18 = 7$. This relative height has been used as a proxy for inundation height by various studies (Nobre et al., 2016; Afshari et al., 2018). To derive an inundation map from HAND values, we must define a threshold height H_T: the flooded area corresponds to all the cells whose HAND value is strictly lower than H_T (Jafarzadegan and Merwade, 2017).

2.2 Model description

MHYST is mostly based on a DEM and its derivatives (drainage map and drainage areas) and on the hydraulic equations describing a steady uniform flow at the reach scale. This means that for a given time step (day in this case), at a given reach, we make the approximation that the flow is constant over time and space (this is obviously a strong simplification that we will discuss later). Table 1 sums up the variables used in the following equations as well as their respective units and interpretations. Table 2 describes the two free parameters of the model. The following equations show the path to build a reach-scale relation between H_T and the streamflow Q by calculating, with hydraulic formulas, the discharge value corresponding to a given H_T. Once this relation is known, the model can easily simulate a hydrological event by inverting the relation, and by searching for the H_T which corresponds to the given Q (Fig. 2).

Other variables can be directly calculated from the DEM (Fig. 4): for a given threshold height H_T at a reach of length L (L is a fixed parameter of the model), $V(H_T)$ is the sum of volumes above all flooded pixels and $S(H_T)$ is the area occupied by the flooded cells. $A(H_T)$, in Eq. (1), is the average cross section area of the flooded reach and it depends on $V(H_T)$ and on the bankfull cross section area of the channel ($A_b = h_b \cdot W_b$, Fig. 3). This variable can also be defined as the

sum of the channel cross section area ($A_{ch} = A_b + H_T \cdot W_b$) and the floodplain cross section area $\left(A_{fp} = A(H_T) - A_{ch} \right)$. $B(H_T)$, in Eq. (2), is the average surface width of the flooded reach, defined similarly from $S(H_T)$ and L.

$$A(H_T) = \frac{1}{L} \cdot V(H_T) + A_b = A_{ch} + A_{fp} \tag{1}$$

$$B(H_T) = \frac{1}{L} \cdot S(H_T) \tag{2}$$

The only unknown variables in these equations are subgrid parameters h_b (bankfull water level) and W_b (bankfull width), i.e. the bankfull geometry, which cannot be obtained from usual DEMs and are only available from detailed surveys for a small number of rivers. Indeed, in situ bathymetric data are quite scarce and red lasers cannot penetrate the water surface more than a few centimetres, which means that the real elevation of the river bed is mostly not correctly represented in the DEMs. This is why we chose to use downstream hydraulic geometry equations to estimate these geometric parameters, assuming a rectangular channel, the size of which depends on the upstream drainage area (Eqs. 3 and 4). To assess the coefficients α and β, we used satellite images from the French platform Géoportail in order to link observed bankfull widths and drainage areas. The values found for the Loing catchment are $\alpha = 0.053$ and $\beta = 0.822$. The other coefficients, δ and ω, were taken from a study by Blackburn-Lynch et al. (2017), which attempted to regionalise these parameters in the US. We used the general values found for the whole set of catchments: $\delta = 0.27$ and $\omega = 0.21$. Although these values probably add uncertainties in the model, they are an accessible way to assess bankfull channel geometry and could still be improved by local bankfull studies when available.

$$W_b = \alpha \cdot A_D^{\beta} \tag{3}$$

$$h_b = \delta \cdot A_D^{\omega} \tag{4}$$

The fundamental equations of the MHYST model come from an experimental study by Nicollet and Uan (1979) which defines the DEBORD formulation as in Eqs. (5) to (7). Building on the Manning–Strickler formula, these authors proposed an empirical parameterisation of turbulent momentum exchange between the channel and the floodplain. This formulation expresses the conveyance capacity depending on channel-related and floodplain-related variables. The coefficient C takes into account the interaction of flows between the fast-flowing channel and the slow-flowing floodplain, as well as the corresponding head losses.

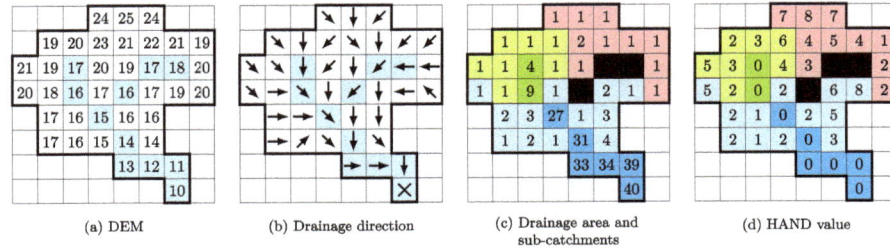

(a) DEM (b) Drainage direction (c) Drainage area and sub-catchments (d) HAND value

Figure 1. Processing of DEM and calculation of the HAND value for a hypothetical catchment.

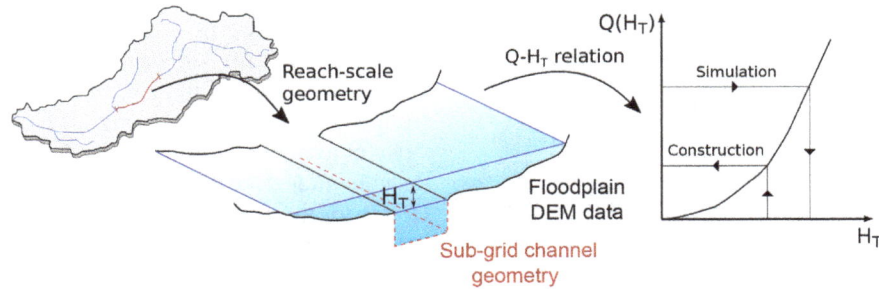

Figure 2. Representation of the model structure: the reach-scale geometry is derived from hydraulic geometry relationships and DEM data and is then used to compute a relation between the threshold height H_T and the discharge Q. L is a fixed characteristic of the reach.

$$D_e = K_{ch} \cdot C \cdot A_{ch} \cdot R_{ch}^{2/3} + K_{fp}$$
$$\cdot \sqrt{A_{fp}^2 + A_{ch} \cdot A_{fp} \cdot \left(1 - C^2\right)} \cdot R_{fp}^{2/3} \qquad (5)$$

$$C = \begin{cases} C_0 = 0.9 \cdot \left(\dfrac{K_{fp}}{K_{ch}}\right)^{1/6} & \text{if } r = \dfrac{R_{fp}}{R_{ch}} > 0.3 \\[2mm] \dfrac{1 - C_0}{2} \cdot \cos\left(\dfrac{\pi \cdot r}{0.3}\right) + \dfrac{1 + C_0}{2} & \text{if } 0 \le r \le 0.3 \end{cases} \qquad (6)$$

$$Q = D_e \cdot \sqrt{I_f} \qquad (7)$$

The streamflow Q is finally defined from the conveyance capacity and the channel slope, since we hypothesise a uniform flow. R_{ch} and R_{fp} can easily be calculated from the assumed reach geometry (Eqs. 8 and 9), which only leaves the Strickler coefficients as unknown variables.

$$R_{ch} = \frac{A_{ch}}{W_b + 2 \cdot h_b} \qquad (8)$$

$$R_{fp} = \frac{A_{fp}}{B(H_T) - W_b} \qquad (9)$$

The two Strickler coefficients add 2 degrees of freedom, and K_{ch} is additionally used to calculate the bankfull flow from the Manning–Strickler formula (Eq. 10).

$$Q_b = K_{ch} \cdot \left(\frac{W_b \cdot h_b}{L_b + 2 \cdot h_b}\right)^{2/3} \cdot \sqrt{I_f} \cdot h_b \cdot W_b \qquad (10)$$

Here, we sum up the procedure, which operates at the reach scale:

1. For a given threshold height H_T, we use the DEBORD formulation to calculate the corresponding discharge Q.

2. By repeating the operation for all possible H_T, we obtain a reach-specific table matching values of H_T and Q.

3. When working on an event where only Q is known, when it is greater than Q_b (which means that the river overflowed), the model looks for the corresponding H_T value in the table, by calculating a linear interpolation between two values if necessary, and then assigns to each cell in the subcatchment a flooded height $h_{flood} = \max(0; H_T - \text{HAND}_{cell})$.

Although this method and that of Zheng et al. (2018) were developed independently, they share a lot of similarities, both using HAND to derive a reach-scale geometry which is used as input for a simplified hydraulic model. However, in addition to HAND, MHYST uses downstream hydraulic geometry relationships to evaluate a sub-grid representation of the channel geometry. The hydraulic model is also different: Zheng et al. (2018) use the Manning–Strickler formula, while MHYST computes streamflow values from the DEBORD formulation.

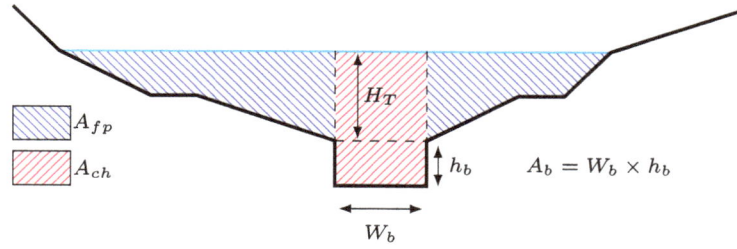

Figure 3. Typical cross section segmentation, with the cross section area of the channel (A_{ch}), that of the floodplains (A_{fp}) and the bankfull cross section area (A_b) which is calculated from the average bankfull height (h_b) and width (W_b) computed from downstream hydraulic geometry relationships.

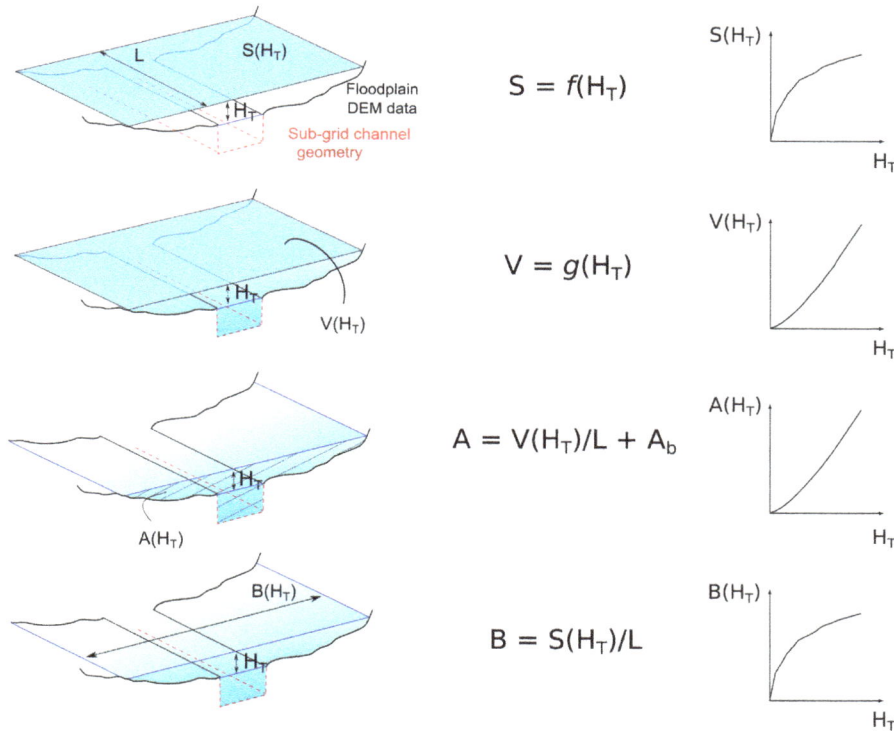

Figure 4. Representation of the reach-scale geometry derived from HAND and the DEM. $A(H_T)$ and $B(H_T)$ are derived from $V(H_T)$ and $S(H_T)$ respectively (Eqs. 1 and 2).

2.3 Boundary conditions

MHYST can work with either simulated or observed flows. In this paper, observed data from 12 measurement stations of the French HYDRO database (Leleu et al., 2014) were used to create an observed distributed streamflow map by interpolating flows based on drainage area (Eq. 11) for river pixels between outlets:

$$Q = Q_{up} + \frac{A_D - A_{D,up}}{A_{D,down} - A_{D,up}} \times \left(Q_{down} - Q_{up} \right), \quad (11)$$

where Q and A_D are the streamflow and drainage area of any river cell between two outlets, Q_{up}, Q_{down}, $A_{D,up}$ and

$A_{D,down}$ are the direct upstream and downstream outlet discharges and drainage areas. This way, streamflow is coherently interpolated over the network, and then averaged at the reach scale.

3 Material

3.1 Generic data

In this study, we used a 5 m resolution DEM with a vertical resolution of 0.01 m covering the Loing catchment (Fig. 5) from IGN (the French national institute for geographic information), which was filled and corrected to avoid depressions

Table 1. Names, units and interpretations of the variables used in the geometric and hydraulic equations of the MHYST model.

Variable	Unit	Interpretation
H_T	m	Threshold height
$V(H_T)$	m^3	Volume created by a height H_T over a reach
$S(H_T)$	m^2	Flooded area created by a height H_T over a reach
$A(H_T)$	m^2	Average cross section area created by a height H_T over a reach
$B(H_T)$	m	Average surface width created by a height H_T over a reach
$Q(H_T)$	$m^3\,s^{-1}$	Mean discharge created by a height H_T over a reach
D_e	$m^3\,s^{-1}$	Conveyance capacity
A_{ch}	m^2	Cross section area of the channel
A_{fp}	m^2	Cross section area of the floodplain
R_{ch}	m	Hydraulic radius of the channel
R_{fp}	m	Hydraulic radius of the floodplain
I_f	$m\,m^{-1}$	Slope of the channel
h_b	m	Bankfull water level of the channel
W_b	m	Bankfull width of the channel
A_b	m^2	Bankfull cross section area of the channel
Q_b	$m^3\,s^{-1}$	Bankfull discharge of the channel
A_D	m^2	Drainage area upstream a given cell
L	m	Target length of a reach (fixed)

Table 2. Names, units and interpretations of the free parameters of MHYST's structure.

Parameter	Unit	Interpretation
K_{ch}	$m^{1/3}\,s^{-1}$	Strickler roughness coefficient for the channel
K_{fp}	$m^{1/3}\,s^{-1}$	Strickler roughness coefficient for the floodplain

and to allow a strict coherence of flow directions, meaning that every pixel flows to the sea. Drainage directions and areas were derived from this DEM and used as model inputs along with elevations. The adaptations and modifications of the DEM were conducted using ESRI ArcGIS 10.3.

Daily observed discharges were obtained from the French HYDRO database (Leleu et al., 2014) and the stations were used to delineate the hydrological network over the catchment. Calibration data for the Loing catchment were obtained from the activation EMSN028 of the Copernicus Emergency Management Service (© 2016 European Union). The original Copernicus study covered a small part of the River Seine and half of the Loing catchment (Fig. 6). However, since the study area and the defined river network were smaller, we cropped the inundation extent to match the study area (Fig. 6). These calibration data are post-processed observed data, meaning that the original maps came from satellite observations but they were then modified to build a more homogeneous inundation extent, i.e. nearby areas whose elevations were below the observed flood level were added to the inundation extent and merged with all the others. The maximum flood extent was then validated by the European Service against reported flood damage and hydrological measurements (SERTIT, 2016).

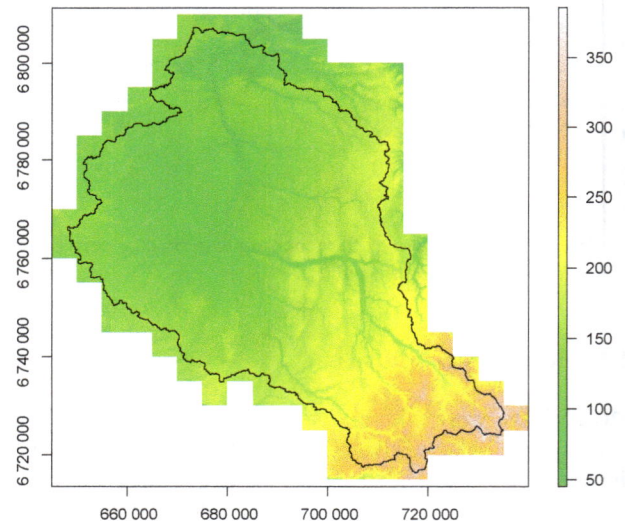

Figure 5. 5 m depressionless DEM used in this study. Elevations go from 45 to 390 m. Corrections have been applied so that each pixel flows to the sea.

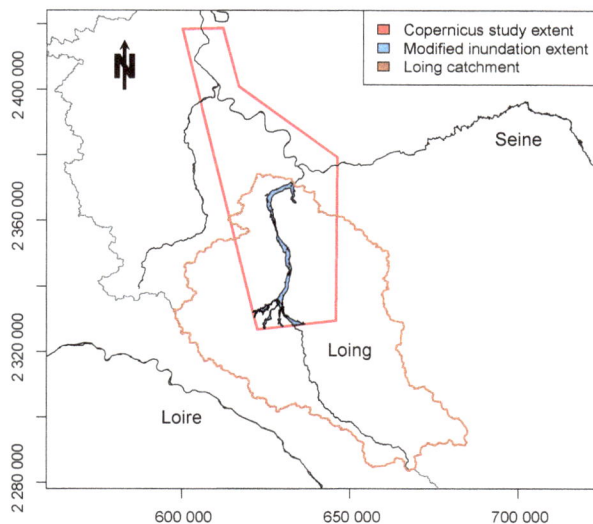

Figure 6. Maximum flood extent for the May–June 2016 event over the Loing catchment produced by the Copernicus Emergency Management Service.

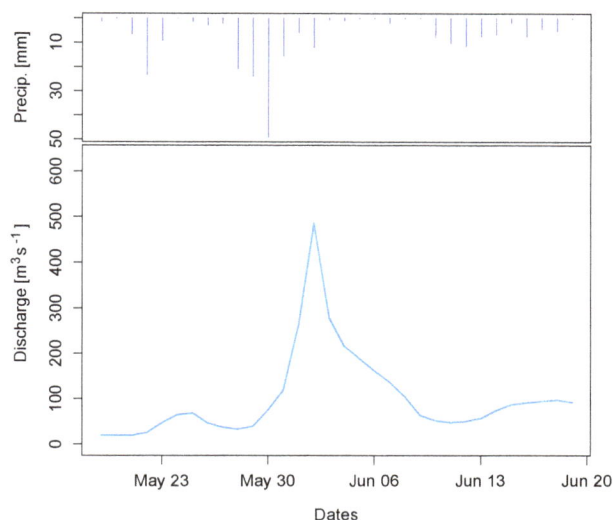

Figure 7. Daily hydrograph of the River Loing at Épisy (3900 km^2) during the event of June 2016. Overall precipitation reached 130 mm. The peak discharge was the largest ever observed on the catchment and reached about 500 m^3 s^{-1}.

3.2 Event of May–June 2016

Following an extremely wet month of May (namely the wettest on record for many stations), a heavy rainfall event started on 30 May 2016 over the centre of France, affecting the Upper and Middle Seine basin and the Middle Loire basin. This episode lasted until 6 June and, combined with highly saturated soils due to a series of preceding minor events, led to major flood inundations. Over this period, overall precipitation reached 180 mm in Paris and Orléans, while in some tributaries, such as the River Loing, peak flows largely exceeded those of the record 1910 flood event (Fig. 7). The flood resulted in 4 deaths, 24 people injured and EUR 1.4 billion worth of damage. A total of 1148 cities were declared to be in a state of natural disaster and insurance companies received about 182 000 claims (CCR, 2016).

Since calibration data were available for June 2016 event, we chose to use our model to simulate this episode and compare the results with observations. We conducted this study over the River Loing, tributary to the River Seine, with a catchment covering 3900 km^2, a mean elevation of 148 m and a mean slope of 0.03 m m^{-1}. This catchment was heavily impacted by the flood event and contains a significant proportion of the inundated area, making it a suitable area to carry out the study. Streamflow data were interpolated from measurements, so no hydrological model is involved in this paper.

4 Results

4.1 Calibration procedure

To assess the model's performance, we used several criteria based on the contingency table in Fig. 8. These scores are presented in detail by Jolliffe and Stephenson (2003) and are defined as a ratio between members of the table where n_1 is the number of hits, i.e. the number of flooded cells correctly forecast; n_4 is the number of pixels correctly forecast as dry; n_2 is the number of false alarms; and n_3 the number of observed flooded cells missed by the model. Table 3 summarises the formulas and the interpretations of each score used in this study.

The POD (probability of detection), which is also called Correct (Alfieri et al., 2014) or M1 (Teng et al., 2015), calculates the percentage of observed inundated pixels intersected by the simulation map. Its main drawback is that it does not take into account the false alarms and thus it can give good results for a clearly overestimating inundation extent. On the contrary, the FAR (false alarm ratio) or M2 (Teng et al., 2015) computes the proportion of cells wrongly flooded by the model. But similarly, if the model does not flood anything, the FAR can reach its optimal value. The critical success index (CSI), also known as Fit, F index or FAI (Alfieri et al., 2014; Bates and De Roo, 2000; Falter et al., 2015), is a criterion which tries to give an overall performance of the simulation by calculating the percentage of correctly flooded cells above the total number of flooded cells (observed and simulated). In this way, the score is penalised by the over- and underestimation. However, this criterion does not specify if the model is over- or underestimating the observed extent. This is why we also looked at the BIAS, which computes the

		Observed	
		Flood	Dry
Model	Flood	Hits (n_1)	False alarms (n_2)
	Dry	Misses (n_3)	Correct negative (n_4)

Figure 8. Contingency table gathering the different scenarios encountered during calibration (the numbers refer to pixels).

ratio between the number of simulated and observed flooded cells. If it is above 1, the model overestimates, and if it is below 1, it underestimates. However, a value of 1 does not equal a perfect simulation since there may be a balance between the misses and the false alarms.

These ratios are particularly reliable if they are used to compare simulations and exhaustive observations. This is almost the case with Copernicus calibration data, which represent a "maximum flood extent". However, MHYST outputs are dated, which is not the case for the observed map. This is why all daily simulated inundation extents were merged into one maximum simulated extent, meaning that we did not try to validate the temporal dynamic of the flood, but only aimed to assess its largest area. Thus, the preceding scores will only evaluate MHYST's ability to reproduce the maximum flood extent.

4.2 Parameterisation

MHYST has two free parameters (Table 2): K_{ch} (the Strickler roughness coefficient for the channel) and K_{fp} (the Strickler roughness coefficient for the floodplains). Preliminary studies showed that, for the Loing catchment, a length of 1000 m was a good trade-off between accuracy and computation time; consequently L was fixed at 1000 m in the rest of this study. K_{ch} and K_{fp} values were tested in the range [0.1; 30] in order to explore a wide range of possibilities (121 combinations were tested).

To help make a decision on the optimal parametrisation of the model, we used the following graphs, on which each (K_{ch}, K_{fp}) couple is characterised by one overall value:

- two contour plots (Fig. 9) showing the impact of K_{ch} and K_{fp} for the two main scores (BIAS and CSI);
- a Pareto plot (Fig. 10a) showing the role played by the K_{fp} parameter in balancing the POD and the FAR;

- a Pareto plot (Fig. 10b) showing that the CSI identifies the best compromises between the POD and the FAR.

Last, to be able to analyse the variability of results between reaches (we have a total of 90 reaches affected by the inundation), we also computed the CSI and BIAS reach by reach, and produced two cumulative distribution plots showing these results (Fig. 11). We found the following:

- The fit criteria are very sensitive to the K_{fp} value and much less to the K_{ch} value (Fig. 9): this should not be a surprise given that we deal with the maximum flood extents for calibration, where K_{ch} only plays a minor role. Remember also that (i) we are modelling a very extreme event ($T > 500$ years) with substantial overflowing, and (ii) that we are working with a channel geometry derived from hydraulic geometry relationships. All this contributes to making the estimation of the channel roughness coefficient more difficult.

- The CSI clearly shows an optimal zone around $K_{fp} = 5$ and $K_{ch} \leq 10\,\mathrm{m}^{1/3}\,\mathrm{s}^{-1}$. The best CSI values (greater than 0.66) correspond to combinations where $0.1 \leq K_{ch} \leq 10$ with $K_{fp} = 5$ or $1 \leq K_{ch} \leq 10$ with $K_{fp} = 4$. Given the equifinality, a good way to choose a combination in this range could be to use the most physical one, which, in this case, would be $K_{ch} = 10$ and $K_{fp} = 5\,\mathrm{m}^{1/3}\,\mathrm{s}^{-1}$. Indeed, over the catchment, floodplains mainly consist of 44 % non-irrigated arable land, 17 % broad-leaved forest and 10 % pastures with corresponding roughness coefficients reported in the literature of 8, 2 and $4\,\mathrm{m}^{1/3}\,\mathrm{s}^{-1}$, respectively (Grimaldi et al., 2010).

- Another way to confirm the validity of this choice ($K_{ch} = 10$ and $K_{fp} = 5\,\mathrm{m}^{1/3}\,\mathrm{s}^{-1}$) is to look at how this parametrisation behaves at the reach scale. Given a total of 90 reaches, we can compute the CSI and BIAS criteria for each of them and draw a distribution (Fig. 11): we observe that the "optimal" distribution is unbiased and that it represents a solution among the best available for each percentile, we can thus trust this parametrisation as a relatively "all-terrain" one for the Loing catchment.

Last, Fig. 10 provides a good illustration of how parameter sets interact with the FAR, POD and CSI criteria: choosing from the parameter sets with the best CSI makes it possible to find a compromise between a high POD and a low FAR.

4.3 Model behaviour

Figures 12 to 14 provide a further illustration with a colour-coded classification of each reach depending on its CSI and BIAS value. A total of 11 regions are highlighted and numbered because of their poor performance. The reasons of why MHYST was not able to reproduce the inundation extent in these regions are explained below.

Table 3. Table of forecast scores used to assess the performance of a flood simulation. All criteria are based on the contingency table (Fig. 8) and reflect one characteristic of the model. Taken together, they provide a comprehensive analysis of the model's behaviour.

Score	Ratio	Range	Perfect score	Characteristics
Bias (BIAS)	$\frac{n_1+n_2}{n_1+n_3}$	$[0, +\infty)$	1	Measures the overestimation (BIAS > 1) and underestimation (BIAS < 1) of the model.
False alarm ratio (FAR)	$\frac{n_2}{n_1+n_2}$	$[0, 1]$	0	Fraction of flooded pixels that were actually observed to be dry. Ignores misses.
Probability of detection (POD)	$\frac{n_1}{n_1+n_3}$	$[0, 1]$	1	Proportion of flooded cells intersected by the model. Ignores false alarms.
Critical success index (CSI)	$\frac{n_1}{n_1+n_2+n_3}$	$[0, 1]$	1	Counts the number of correct flooded cells, while penalising overestimation (false alarms) and underestimation (misses).

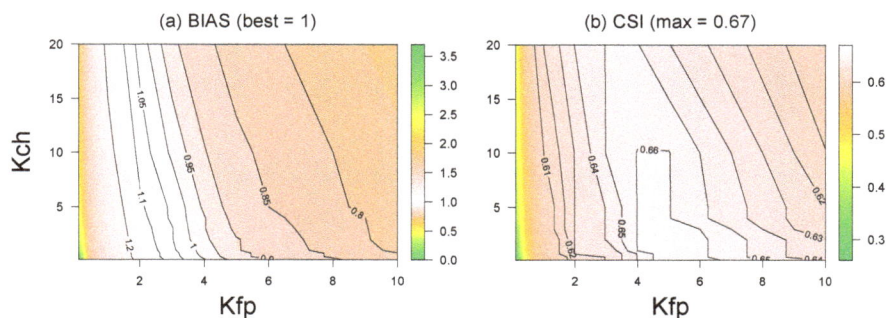

Figure 9. Forecast scores obtained by the model on the River Loing versus Copernicus data for all the parameter values tested, **(a)** BIAS contour lines and **(b)** CSI contour lines for various values of K_{ch} and K_{fp}.

1. For the downstream-most part of the Loing (Fig. 12), the reaches are red or orange because this area is only partially covered by the observation, which stops just after the confluence with the small tributary.

2. The small tributary (Fig. 12) is mainly red or orange for various reasons: downstream, at the confluence, the DEM is full of small high-elevation zones (not corrected in the DEM) which the model cannot reach, thus degrading the simulation. Along the tributary, the reason can be either the observed discharge values which seem small compared to the rest of the catchment or simply the effective geometry defined by the model, which does not correspond to the actual one. Finally, the upstream part of the tributary is not covered by the observation, which stops in the middle of what MHYST simulated. However, the study zone defined by the Copernicus Emergency Management Service goes further, so we cannot know whether it was not flooded or whether the service did not map this part because it was too insignificant.

3. The orange part in the middle of the BIAS map (Fig. 12) is due to the railway tracks which act like a wall in

the DEM, preventing the model from reaching the other side (from east to west), where a small tributary, which looks like a partly subterranean urban stream, overflowed in its open air part.

4. Finally, the red and orange zones in the south of the presented map (Fig. 12) correspond to a part of the river where the Loing man-made waterway plays a major role, running parallel with the main river. This configuration is difficult for MHYST because we only consider the main river, defined by the DEM, with an effective reach-scale geometry and we cannot take into account such specificities, which would require a 2-D hydraulic model.

5. The area identified (Fig. 13) shows a slight underestimation leading to a moderate CSI. This issue can be explained by a motorway which is represented in the DEM by a more elevated area. This motorway separates the reach into two parts linked by artificial openings made by the producers of the DEM. This and the Loing waterway and another road act as dikes that prevent the model from reaching a further part of the reach. The pa-

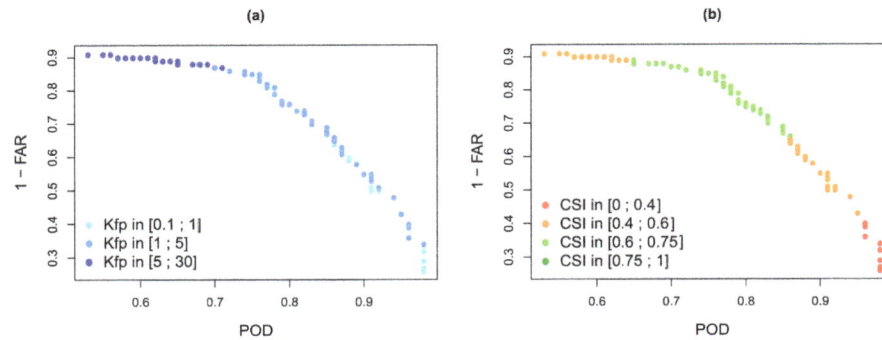

Figure 10. Pareto diagram for two forecast scores, POD and FAR. 1-FAR is used so that each criterion evolves in the same way, **(a)** distribution of K_{fp} values and **(b)** distribution of CSI values according to POD and 1-FAR.

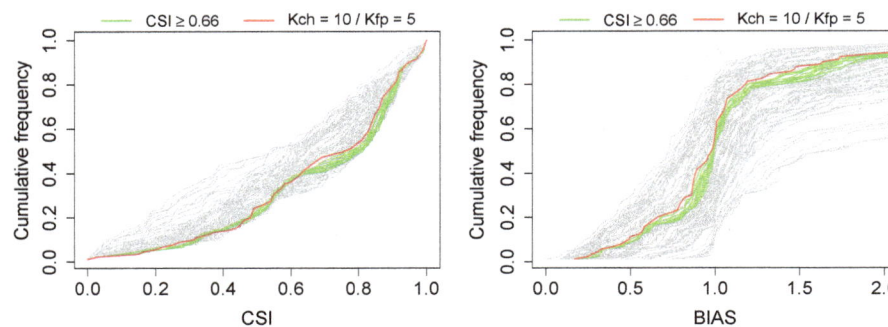

Figure 11. Cumulative frequency of CSI and BIAS values for all combinations of parameters and for the 90 affected reaches. Green lines correspond to the best combinations identified in Fig. 9 while the red line refers to the physical parametrisation. The other parameters are displayed in grey.

rameterisation of the model is not suitable to address this difficulty.

6. Similarly to the previous area (Fig. 13), a railway crosses the DEM from north to south with only one opening for the water. Given the parameterisation of the model, it is not possible to go over the railway to flood the missed area.

7. In that case (Fig. 13), the model clearly overestimates the flood. The water fills a depression which looks like a tributary but is only a thalweg. Once more, the parameterisation of the model does not provide an adequate representation of this reach.

8. In this area (Fig. 14), MHYST underestimates the inundation extent due to a road that works like a dike. However, with another parameterisation, the model would be able to provide enough water to go over the road.

9. In the western part of the upstream area (Fig. 14), MHYST overestimates the flood because it is a relatively flat zone. The exceeding water, still due to the parameterisation, is thus spread over the area.

10. This area (Fig. 14) is special because the overestimation of MHYST is due to a non-continuous observation map, creating large parts of reaches that are observed to be dry. However, since MHYST works at the reach scale, it necessarily floods the whole river reach. Moreover, one tributary, the Solin, is not defined in the hydrographic network used by the model, because no observed discharges were available, whereas it appears in the observed map, leading to an underestimation of the flooded area.

11. The most upstream part of the simulated area (Fig. 14) suffers from an excess of water and a non-continuous observation, leading to similar effects. Moreover, several elevated roads appear in the DEM and force the model to flood the area using artificial openings across the roads.

In order to complete our interpretation of MHYST behaviour, we conducted two sensitivity analyses, one with the Morris method (Morris, 1991) and the other with the Sobol method (Sobol, 2001, details can be found in Appendix A). We chose to assess the effect of six potential parameters, K_{ch}, K_{fp}, α, β, δ and ω, that may play a major role in the com-

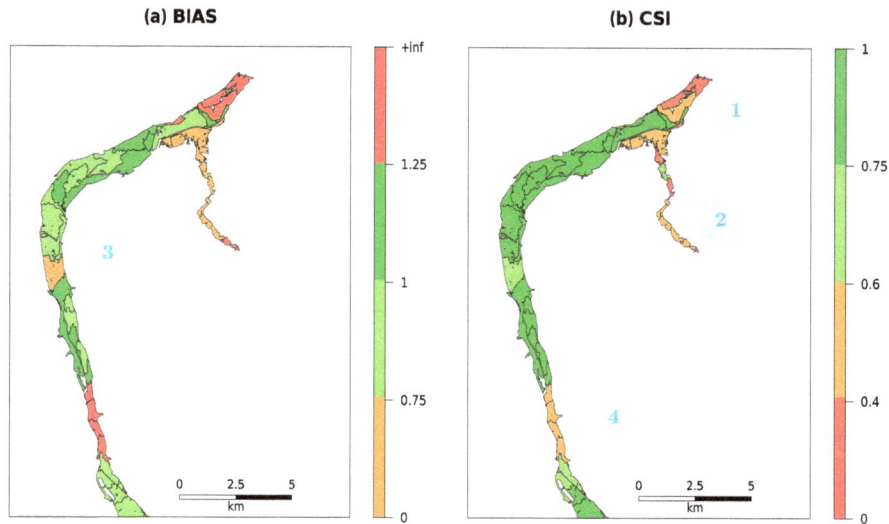

Figure 12. Reach-scale performance of **(a)** BIAS and **(b)** CSI for the physical combination of parameters, $K_{ch} = 10$ and $K_{fp} = 5\,\mathrm{m}^{1/3}\,\mathrm{s}^{-1}$, for the downstream part of the catchment. Criteria values have been categorised as follows: excellent (dark green), good (green), average (orange) and poor (red). The black lines delineate the reaches. Locations 1 to 4 correspond to areas where the model struggles to reproduce the observation (orange and/or red zones).

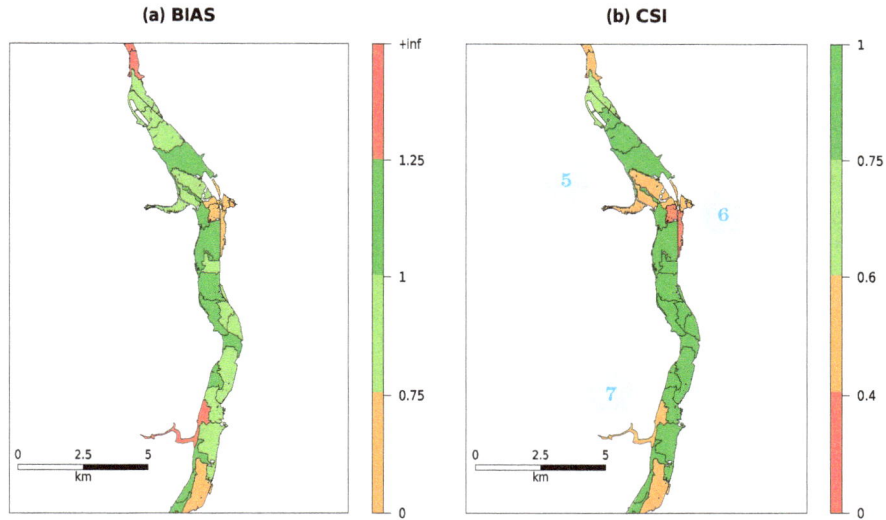

Figure 13. Reach-scale performance of **(a)** BIAS and **(b)** CSI for the physical combination of parameters, $K_{ch} = 10$ and $K_{fp} = 5\,\mathrm{m}^{1/3}\,\mathrm{s}^{-1}$, for the centre part of the catchment. Criteria values have been categorised as follows: excellent (dark green), good (green), average (orange) and poor (red). The black lines delineate the reaches. Locations 5 to 7 correspond to areas where the model struggles to reproduce the observation (orange and/or red zones).

putation of $H_T{-}Q$ relationships. In both analyses, we found that ω, which parameterises the regionalisation of bankfull heights, has the most substantial effect on the performance and that, surprisingly, K_{fp} has no influence at all. As a matter of fact, when we conducted the Sobol analysis with fixed hydraulic geometry parameters, we showed that K_{fp} is con-

siderably more influential than K_{ch}. We concluded that the previous results were due to the fact that these sensitivity analyses explore the parameter space in detail, and even with reasonable boundaries, they can reach values that may not be consistent with the characteristics of the catchment studied.

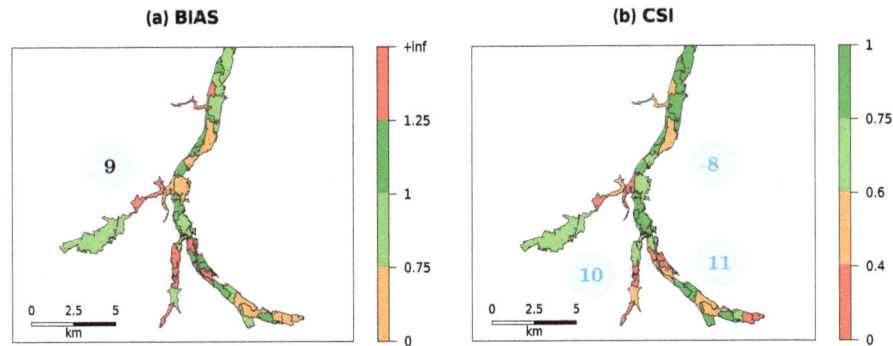

Figure 14. Reach-scale performance of **(a)** BIAS and **(b)** CSI for the physical combination of parameters, $K_{ch} = 10$ and $K_{fp} = 5\,m^{1/3}\,s^{-1}$, for the upstream part of the catchment. Criteria values have been categorised as follows: excellent (dark green), good (green), average (orange) and poor (red). The black lines delineate the reaches. Locations 8 to 11 correspond to areas where the model struggles to reproduce the observation (orange and/or red zones).

4.4   Influence of the DEM resolution

It is possible to assess the sensitivity to the DEM in two ways: first by aggregating our DEM from 5 m to various resolutions (10, 25, 50 and 100 m) and then by changing the source of the DEM. Figure 15 provides the CSI scores obtained by the model while changing the resolution. It shows that the resolution has relatively little effect on the optimal value, which varies between 0.65 and 0.69. However, the position of this optimal, i.e. the combination of parameters (K_{ch} and K_{fp}) leading to it, changes. We can also see that for some resolutions, such as 25 or 50 m, the equifinality zone is much smaller than the one for the 100 m resolution, for example. If we also look at the "physical" set of parameters we previously identified ($K_{ch} = 10$ and $K_{fp} = 5\,m^{1/3}\,s^{-1}$), we can see that the CSI reached by the model for this combination varies between the resolutions. Nevertheless, the result still seems satisfying, so it could be used as a "default" parameterisation, for instance for ungauged catchments. But this should be tested on other catchments with observed data to lead to a more comprehensive conclusion.

Before using the RGE 5 m DEM from IGN, we tried to use the 25 m EU-DEM from the European Environment Agency, and it showed poorer results, because it was not precise enough. Figure 16 shows the evolution of CSI for the same combinations of parameters as before. We see that the best combinations of parameters only lead to a 0.53 maximal CSI, which is more than 10 points below what we can obtain with the RGE DEM. There is also strictly no connection between the best values of BIAS and those of CSI, the latter being obtained for a clear overestimation of the flood extent (BIAS ~ 1.5). These results are due to the lack of precision of the EU-DEM, which does not distinguish the channel from the floodplain, leading to a 2 km wide channel in some parts of the river.

5   Conclusions and outlooks

The objective of this paper was to present and validate a simple hydraulic model for rapid inundation mapping in data-scarce areas. MHYST is based on DEM analyses and simple hydraulic equations, creating a reach-scale relation between the average discharge and the average "height above nearest drainage" which can then be used to simulate any event, past or future, as long as streamflow information (observed or simulated) is available. This model was calibrated against an observed exceptional flood which occurred in 2016 on the Loing River near Paris and showed results that are certainly not perfect, but from our point of view and for our objectives quite encouraging. Furthermore, we compared our methodology with the traditional HAND approach, using a single threshold height of 4 m (measured height at the outlet) for the whole catchment. The simple HAND model reached CSI = 0.49, BIAS = 1.55, POD = 0.84 and FAR = 0.46. It is clearly penalised by the overestimation (almost 50 % of false alarms), which is not surprising according to other studies (Nobre et al., 2016).

The simple structure of MHYST allows it to be used almost anywhere with few data and only two parameters. The model can, however, be used in first approximation, when a lack of time and data restrains the use of a more complex method.

For the sake of honesty, we would like to specify the theoretical limits of the MHYST approach:

– The model equations were solved by using the hypothesis of a reach-scale steady uniform flow (probably one of the most simplifying assumptions one can make). This simplification is probably too extreme for highly complex situations, especially in the presence of dikes and bridges. Indeed, on the one hand, the DEM resolution is too coarse to precisely take into account hydraulic structures, and on the other hand, the DEBORD

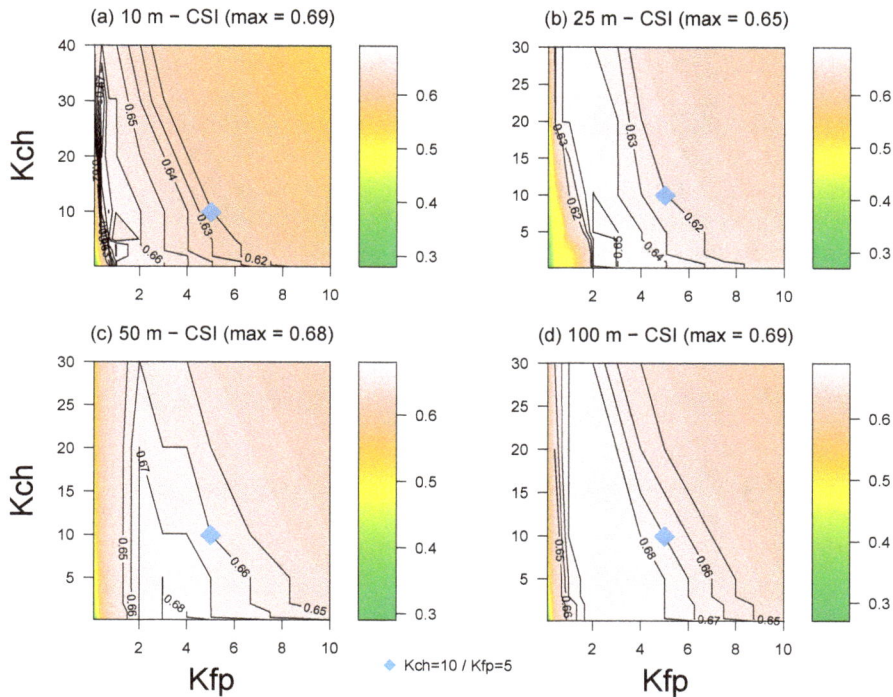

Figure 15. CSI scores obtained by the model on the River Loing versus Copernicus data for all the parameter values tested and for various resolutions of the DEM, aggregated from the 5 m resolution DEM: **(a)** 10 m, **(b)** 25 m, **(c)** 50 m and **(d)** 100 m.

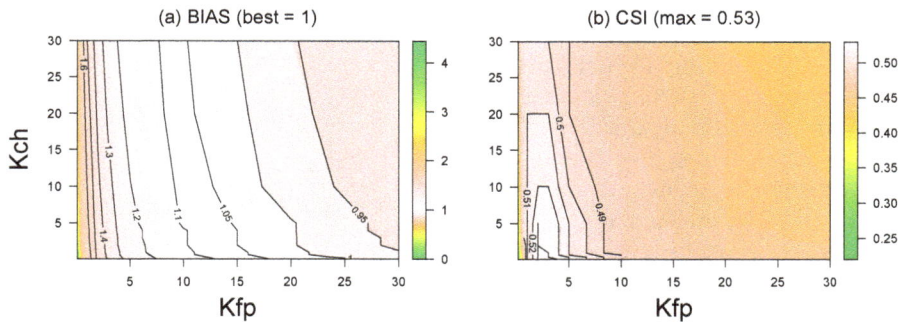

Figure 16. (a) BIAS and **(b)** CSI scores obtained by the model on the River Loing versus Copernicus data for all the parameter values tested and for another source of data: EU-DEM.

formulation is not sufficient to describe the interaction between the flow and these structures.

– The DEM is a critical part of the model, because geometrical relationships and variables are directly related to the shape and distribution of elevations. Another DEM was actually tested as model input and showed much poorer results.

– Moreover, since the channel geometry was unknown, hydraulic geometry equations were used to assess bankfull height and width, with fixed parameters from another study in the case of height, which may not be the

optimum for this catchment, adding its share of uncertainty.

– Finally, there is at this point no continuity equation between reaches, since the calculations were made for each reach separately. Uncertainties may therefore be higher in areas around connection points between reaches, especially if it is a confluence of rivers. One way to address this issue could be to add a continuity equation between the reaches, which might increase the overall coherence of the flood. However, at this point of the development of the model, we have not included this specificity.

Thus, the maps produced by MHYST should be seen as a maximum extent of the flood which can be used as a first and rapid estimation. To further test this approach, we consider that attention should first be given to the following: assessing the impact of the DEM choice, resolution and quality; testing the approach on a range of (less extreme) events and catchments, to better assess the range and stability of its parameters and performance; and improving the treatment of possible discontinuities between reaches.

Appendix A: Sensitivity analysis

In order to assess the sensitivity of the model to its main parameters (K_{fp}, K_{ch}, α, β, ω, δ), we conducted two sensitivity analyses, using different but complementary well-known methods: Morris (Morris, 1991) and Sobol (Sobol, 2001).

A1   Morris method

The Morris method (Morris, 1991) provides a qualification of the effect a parameter can have on the outputs. It is a OAT (one-at-a-time) methodology, which means that the effect of a parameter is measured by changing its value by adding $\pm\Delta$ without modifying the other parameters and by comparing the outputs. In order to provide a relevant analysis, we generated 160 sets of parameters, using the Latin hypercube sampling method, which acts as starting points from where the Morris method can assess the significance of parameters by changing their values one-at-a-time. Thus, more than 1000 simulations are needed to conduct the analysis. By using the 5 m resolution DEM we used in this paper, this study would take several days, if not weeks, to complete. But since we showed that the performance of MHYST did not really change with the resolution, we chose to use a coarser version of our DEM, which was aggregated at a 50 m resolution, by simply averaging the elevations, allowing us to complete this sensitivity analysis in only a few hours. For each permutation and for each parameter, D_i, the difference in CSI divided by the computing step, is calculated. The results in terms of means and standard deviations are presented in Fig. A1. The analysis shows that the model is very sensitive to changes of ω, the exponent in the calculation of the regionalised bankfull width (W_b). The most surprising part of the analysis is the fact that K_{fp} has little or no effect on the model, while K_{ch} has a moderate effect. This is contradicted by Fig. 9, which clearly shows that for a given value of K_{fp}, the CSI value varies only slightly for a K_{ch} between 0.1 and 20. K_{fp} is, contrary to what the Morris analysis shows, a significant parameter of the model, particularly in a major overflowing event such as the one studied here, where the channel only represents a fraction of the water.

The problem might be that despite the use of a Latin hypercube sampling method, the "good" values of the parameters never meet, i.e. when ω has a sensible value, K_{fp} has not and vice versa. And of course, if the ω value does not coherently represent the channel, the model is not able to conduct a correct simulation (i.e. little or no flooding), leading to little or no influence of the K_{fp} parameter.

Moreover, the issue with sensitivity analyses such as the Morris method is that the results can be very different depending on the catchment or the event modelled. Indeed, if the water is concentrated in the channel part for a very steep catchment, a very flat one will on the contrary rely on the floodplains, and so the parameterisation of the model will add more value to K_{ch} or K_{fp}. Thus, the conclusions one can

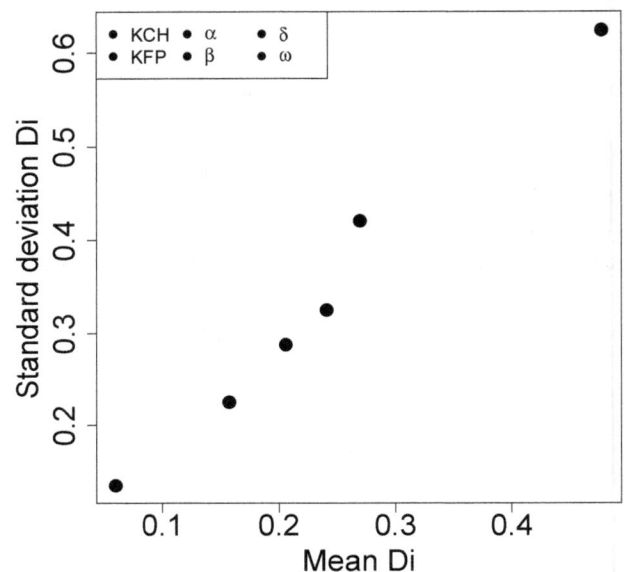

Figure A1. Results of the Morris method applied to MHYST with a 50 m resolution DEM on the Loing catchment for the six parameters (K_{fp}, K_{ch}, α, β, ω, δ).

make by interpreting one analysis of an example do not necessarily reflect the global behaviour of the model.

A2   Sobol method

The Sobol method (Sobol, 2001) is a variance-based sensitivity analysis which aims to compute the fraction of the variance that can be attributed to each parameter. For this study, 2×500 sets of parameters were randomly chosen with a Latin hypercube sampling method, thus creating two 500×6 matrices, $\mathbf{X}_A$ and $\mathbf{X}_B$. Each column of $\mathbf{X}_A$ has sequentially been substituted by a column of $\mathbf{X}_B$, corresponding to one of the six parameters, leading to six other matrices. In order to limit the computation time, the interaction of several parameters (i.e. substituting two or more columns of $\mathbf{X}_A$ by those of $\mathbf{X}_B$) has not been assessed. Indeed, MHYST has been launched with the 4000 sets of parameters, with a resolution of 50 m, which takes longer than the Morris method that only needed about a thousand simulations. The first-order Sobol indices S_i, which indicate the contribution of one parameter to the total variance, and the total-effect indices S_{-i}, which calculate the total contribution of one parameter to the variance, including the possible interactions between parameters, have been computed. Then, with a bootstrap re-sampling method, the distributions of S_i and S_{-i} have been assessed, allowing several characteristics such as the bias to be computed, the standard deviation and the confidence intervals.

The results of this analysis are presented in Table A1 for S_i and Table A2 for S_{-i}. The first-order indices confirm parts of what was concluded from the Morris analysis, interpreting ω as the most influential parameter, K_{ch} and α as moderately influential and K_{fp} as not influential, despite the observations we made in the article when we calibrated the parameters. The total-effect indices complete the analysis and confirm the conclusions we made with the Morris method, adding β to the list of influential parameters.

The distributions of S_i and S_{-i} show that the values calculated are not biased, but the 95 % confidence interval is rather large, which means that in some cases, the interpretation may differ. This might explain why when we set values for all downstream hydraulic geometry equations parameters (α, β, δ, ω) from regionalised studies or observations, K_{fp} has a greater influence which is not highlighted by the sensitivity analyses. These methodologies (Morris, Sobol) indeed explore the parameter space in detail, and even with reasonable boundaries, they can reach values that may not be consistent with the characteristics of the catchment studied. Another limitation is the fact that these analyses are only valid for this particular example (the Loing catchment and the event of May–June 2016). They should ideally be used with a larger set of catchments and events to be reliably trusted.

In order to understand why Morris and Sobol give, contrary to our initial expectation, so little importance to K_{fp}, we conducted a quick Sobol analysis with fixed hydraulic geometry parameters, i.e. we considered the α, β, δ and ω values used in the original study and only made K_{ch} and K_{fp} vary. This time, the results confirm what we observed: $S_{K_{ch}} = 0.15$ and $S_{K_{fp}} = 0.85$, which means that K_{fp} is a major parameter in our situation, and that K_{ch} has a smaller role.

The hydraulic geometry parameters are clearly important, but if they are fixed to legitimate values estimated by observations or tables of regionalised values, their impact becomes minor in front of the Strickler coefficients.

Table A1. Sobol first-order indices for the six parameters of MHYST. Confidence interval is denoted as conf. int. here.

Parameter	S_i value	Bias	Standard error	Min. conf. int.	Max. conf. int.
K_{ch}	0.121	0.004	0.193	−0.149	0.392
K_{fp}	0.043	−0.004	0.065	−0.071	0.156
α	0.158	0.013	0.205	−0.200	0.517
β	0.077	0.013	0.166	−0.187	0.341
δ	0.015	−0.0001	0.082	−0.116	0.146
ω	0.417	0.044	0.238	0.009	0.825

Table A2. Sobol total-effect index for the six parameters of MHYST. Confidence interval is denoted as conf. int. here.

Parameter	S_{-i} value	Bias	Standard error	Min. conf. int.	Max. conf. int.
K_{ch}	0.201	0.013	0.135	−0.007	0.410
K_{fp}	0.009	−0.00007	0.085	−0.139	0.157
α	0.238	−0.002	0.156	−0.038	0.514
β	0.167	0.001	0.128	−0.054	0.389
δ	0.047	−0.001	0.068	−0.060	0.156
ω	0.476	−0.003	0.22	0.120	0.832

Author contributions. The model presented in this paper was developed and analysed by CR during his PhD work. He also wrote the paper, which was corrected by VA and NLM.

Competing interests. The authors declare that they have no conflict of interest.

Acknowledgements. The first author was funded by a grant from the AXA Research Fund. Thanks are extended to Rafal Zielinksi, who helped us access data from the Copernicus Emergency Management Service. We would also like to thank the AXA Global P&C research team for their advice and our discussions on the development of simple conceptual inundation models. The MHYST model was developed using R (R Core Team, 2015) and GFortran, Gnu compiler collection (gcc) Version 4.9.2.

Edited by: Roger Moussa

References

Afshari, S., Tavakoly, A. A., Rajib, M. A., Zheng, X., Follum, M. L., Omranian, E., and Fekete, B. M.: Comparison of new generation low-complexity flood inundation mapping tools with a hydrodynamic model, J. Hydrol., 556, 539–556, https://doi.org/10.1016/j.jhydrol.2017.11.036, 2018.

Alfieri, L., Salamon, P., Bianchi, A., Neal, J., Bates, P., and Feyen, L.: Advances in pan-European flood hazard mapping, Hydrol. Process., 28, 4067–4077, https://doi.org/10.1002/hyp.9947, 2014.

Bates, P. D. and De Roo, A. P. J.: A simple raster-based model for flood inundation simulation, J. Hydrol., 236, 54–77, https://doi.org/10.1016/S0022-1694(00)00278-X, 2000.

Bates, P. D., Horritt, M. S., and Fewtrell, T. J.: A simple inertial formulation of the shallow water equations for efficient two-dimensional flood inundation modelling, J. Hydrol., 387, 33–45, https://doi.org/10.1016/j.jhydrol.2010.03.027, 2010.

Biancamaria, S., Bates, P. D., Boone, A., and Mognard, N. M.: Large-scale coupled hydrologic and hydraulic modelling of the Ob river in Siberia, J. Hydrol., 379, 136–150, https://doi.org/10.1016/j.jhydrol.2009.09.054, 2009.

Blackburn-Lynch, W., Agouridis, C. T., and Barton, C. D.: Development of Regional Curves for Hydrologic Landscape Regions (HLR) in the Contiguous United States, J. Am. Water Resour. As., 53, 903–928, https://doi.org/10.1111/1752-1688.12540, 2017.

CCR: Inondations de mai–juin 2016 en France – Modélisation de l'aléa et des dommages, Tech. rep., Service R&D modélisation – Direction des Réassurances & Fonds Publics, 2016.

Falter, D., Dung, N., Vorogushyn, S., Schröter, K., Hundecha, Y., Kreibich, H., Apel, H., Theisselmann, F., and Merz, B.: Continuous, large-scale simulation model for flood risk assessments: proof-of-concept: Large-scale flood risk assessment model, J. Flood Risk Manag., 9, 3–21, https://doi.org/10.1111/jfr3.12105, 2014.

Falter, D., Schröter, K., Dung, N. V., Vorogushyn, S., Kreibich, H., Hundecha, Y., Apel, H., and Merz, B.: Spatially coherent flood risk assessment based on long-term continuous simulation with a coupled model chain, J. Hydrol., 524, 182–193, https://doi.org/10.1016/j.jhydrol.2015.02.021, 2015.

Gouldby, B., Sayers, P., Mulet-Marti, J., Hassan, M. A. A. M., and Benwell, D.: A methodology for regional-scale flood risk assessment, Water Management, 161, 169–182, https://doi.org/10.1680/wama.2008.161.3.169, 2008.

Grimaldi, S., Petroselli, A., Alonso, G., and Nardi, F.: Flow time estimation with spatially variable hillslope velocity in ungauged basins, Adv. Water Res., 33, 1216–1223, https://doi.org/10.1016/j.advwatres.2010.06.003, 2010.

Horritt, M. S. and Bates, P. D.: Effects of spatial resolution on a raster based model of flood flow, J. Hydrol., 253, 239–249, https://doi.org/10.1016/S0022-1694(01)00490-5, 2001.

Horritt, M. S. and Bates, P. D.: Evaluation of 1D and 2D numerical models for predicting river flood inundation, J. Hydrol., 268, 87–99, https://doi.org/10.1016/S0022-1694(02)00121-X, 2002.

Hunter, N. M., Horritt, M. S., Bates, P. D., Wilson, M. D., and Werner, M. G. F.: An adaptive time step solution for raster-based storage cell modelling of floodplain inundation, Adv. Water Res., 28, 975–991, https://doi.org/10.1016/j.advwatres.2005.03.007, 2005.

Jafarzadegan, K. and Merwade, V.: A DEM-based approach for large-scale floodplain mapping in ungauged watersheds, J. Hydrol., 550, 650–662, https://doi.org/10.1016/j.jhydrol.2017.04.053, 2017.

Jolliffe, I. T. and Stephenson, D. B.: Forecast Verification: A Practitioner's Guide in Atmospheric Science, John Wiley & Sons, Chichester, 2003.

Le Bihan, G., Payrastre, O., Gaume, E., Moncoulon, D., and Pons, F.: The challenge of forecasting impacts of flash floods: test of a simplified hydraulic approach and validation based on insurance claim data, Hydrol. Earth Syst. Sci., 21, 5911–5928, https://doi.org/10.5194/hess-21-5911-2017, 2017.

Leleu, I., Tonnelier, I., Puechberty, R., Gouin, P., Viquendi, I., Cobos, L., Foray, A., Baillon, M., and Ndima, P.-O.: La refonte du système d'information national pour la gestion et la mise à disposition des données, La Houille Blanche, 1, 25–32, https://doi.org/10.1051/lhb/2014004, 2014.

Leopold, L. B. and Maddock, T.: The Hydraulic Geometry of Stream Channels and Some Physiographic Implications, Tech. Rep., 252, Washington, 1953.

McGrath, H., Bourgon, J.-F., Proulx-Bourque, J.-S., Nastev, M., and Abo El Ezz, A.: A comparison of simplified conceptual models for rapid web-based flood inundation mapping, Nat. Hazards, 93, 905–920, https://doi.org/10.1007/s11069-018-3331-y, 2018.

Morales-Hernández, M., Petaccia, G., Brufau, P., and García-Navarro, P.: Conservative 1D–2D coupled numerical strategies applied to river flooding: The Tiber (Rome), Appl. Math. Model., 40, 2087–2105, https://doi.org/10.1016/j.apm.2015.08.016, 2016.

Morris, M. D.: Factorial Sampling Plans for Preliminary Computational Experiments, Technometrics, 33, 161–174, https://doi.org/10.2307/1269043, 1991.

Moussa, R. and Cheviron, B.: Modeling of Floods – State of the Art and Research Challenges, in: Rivers – Physical, Fluvial and Environmental Processes, edited by: Rowiński, P. and Radecki-

Pawlik, A., 169–192, Springer International Publishing, Cham, https://doi.org/10.1007/978-3-319-17719-9_7, 2015.

Neal, J., Schumann, G., and Bates, P.: A subgrid channel model for simulating river hydraulics and floodplain inundation over large and data sparse areas, Water Resour. Res., 48, W11506, https://doi.org/10.1029/2012WR012514, 2012.

Nicollet, G. and Uan, M.: Écoulements permanents à surface libre en lits composés, La Houille Blanche, 1, 21–30, https://doi.org/10.1051/lhb/1979002, 1979.

Nobre, A. D., Cuartas, L. A., Hodnett, M., Rennó, C. D., Rodrigues, G., Silveira, A., Waterloo, M., and Saleska, S.: Height Above the Nearest Drainage – a hydrologically relevant new terrain model, J. Hydrol., 404, 13–29, https://doi.org/10.1016/j.jhydrol.2011.03.051, 2011.

Nobre, A. D., Cuartas, L. A., Momo, M. R., Severo, D. L., Pinheiro, A., and Nobre, C. A.: HAND contour: a new proxy predictor of inundation extent: Mapping Flood Hazard Potential Using Topography, Hydrol. Process., 30, 320–333, https://doi.org/10.1002/hyp.10581, 2016.

Pons, F., Delgado, J.-L., Guero, P., and Berthier, E.: EXZECO: A GIS and DEM based method for pre-determination of flood risk related to direct runoff and flash floods, in: 9th International Conference on Hydroinformatics, 7–11 September, Tianjin, China, 2010.

Rennó, C. D., Nobre, A. D., Cuartas, L. A., Soares, J. V., Hodnett, M. G., Tomasella, J., and Waterloo, M. J.: HAND, a new terrain descriptor using SRTM-DEM: Mapping terra-firme rain-forest environments in Amazonia, Remote Sens. Environ., 112, 3469–3481, https://doi.org/10.1016/j.rse.2008.03.018, 2008.

Schumann, G. J.-P., Neal, J. C., Voisin, N., Andreadis, K. M., Pappenberger, F., Phanthuwongpakdee, N., Hall, A. C., and Bates, P. D.: A first large-scale flood inundation forecasting model: Large-Scale Flood Inundation Forecasting, Water Resour. Res., 49, 6248–6257, https://doi.org/10.1002/wrcr.20521, 2013.

SERTIT: EMSN-028 Flood delineation and damage assessment, France, Technical Report, 2016.

Sobol, I. M.: Global sensitivity indices for nonlinear mathematical models and their Monte Carlo estimates, Math. Comput. Simulat., 55, 271–280, 2001.

Speckhann, G. A., Borges Chaffe, P. L., Fabris Goerl, R., Abreu, J. J. d., and Altamirano Flores, J. A.: Flood hazard mapping in Southern Brazil: a combination of flow frequency analysis and the HAND model, Hydrol. Sci. J., 63, 87–100, https://doi.org/10.1080/02626667.2017.1409896, 2018.

Teng, J., Vaze, J., Dutta, D., and Marvanek, S.: Rapid Inundation Modelling in Large Floodplains Using LiDAR DEM, Water Resour. Manage., 29, 2619–2636, https://doi.org/10.1007/s11269-015-0960-8, 2015.

Zheng, X., Tarboton, D. G., Maidment, D. R., Liu, Y. Y., and Passalacqua, P.: River channel geometry and rating curve estimation using height above the nearest drainage, J. Am. Water Resour. As., 54, 785–806, https://doi.org/10.1111/1752-1688.12661, 2018.

9

Dominant controls of transpiration along a hillslope transect inferred from ecohydrological measurements and thermodynamic limits

Maik Renner[1], **Sibylle K. Hassler**[2,6], **Theresa Blume**[2], **Markus Weiler**[3], **Anke Hildebrandt**[4,1], **Marcus Guderle**[4,1,7], **Stanislaus J. Schymanski**[5]**, and Axel Kleidon**[1]

[1]Max-Planck-Institut für Biogeochemie, Jena, Germany
[2]GFZ German Research Centre for Geosciences, Section Hydrology, Potsdam, Germany
[3]Universität Freiburg, Hydrologie, Freiburg, Germany
[4]Universität Jena, Ecological Modelling Group, Jena, Germany
[5]ETH Zürich, Department of Environmental Systems Science, Zurich, Switzerland
[6]Karlsruhe Institute of Technology, Institute of Water and River Basin Management, Karlsruhe, Germany
[7]Technische Universität München, Chair for Terrestrial Ecology, Department of Ecology and Ecosystemmanagement, Munich, Germany

Correspondence to: Maik Renner (mrenner@bgc-jena.mpg.de)

Abstract. We combine ecohydrological observations of sap flow and soil moisture with thermodynamically constrained estimates of atmospheric evaporative demand to infer the dominant controls of forest transpiration in complex terrain. We hypothesize that daily variations in transpiration are dominated by variations in atmospheric demand, while site-specific controls, including limiting soil moisture, act on longer timescales.

We test these hypotheses with data of a measurement setup consisting of five sites along a valley cross section in Luxembourg. Both hillslopes are covered by forest dominated by European beech (*Fagus sylvatica* L.). Two independent measurements are used to estimate stand transpiration: (i) sap flow and (ii) diurnal variations in soil moisture, which were used to estimate the daily root water uptake. Atmospheric evaporative demand is estimated through thermodynamically constrained evaporation, which only requires absorbed solar radiation and temperature as input data without any empirical parameters. Both transpiration estimates are strongly correlated to atmospheric demand at the daily timescale. We find that neither vapor pressure deficit nor wind speed add to the explained variance, supporting the idea that they are dependent variables on land–atmosphere exchange and the surface energy budget. Estimated stand transpiration was in a similar range at the north-facing and the south-facing hillslopes despite the different aspect and the largely different stand composition. We identified an inverse relationship between sap flux density and the site-average sapwood area per tree as estimated by the site forest inventories. This suggests that tree hydraulic adaptation can compensate for heterogeneous conditions. However, during dry summer periods differences in topographic factors and stand structure can cause spatially variable transpiration rates. We conclude that absorption of solar radiation at the surface forms a dominant control for turbulent heat and mass exchange and that vegetation across the hillslope adjusts to this constraint at the tree and stand level. These findings should help to improve the description of land-surface–atmosphere exchange at regional scales.

1 Introduction

Evapotranspiration E couples water and energy balances at the land surface and is constrained by both, the supply of water and the atmospheric demand for water. Total E is composed of evaporation from intercepted water on plants and the surface, soil evaporation, and the physiological process of plant transpiration (E_T) taking water from the soil, groundwater, and possibly bedrock (Shuttleworth, 1993; Miller et al., 2010; Schwinning, 2010). Transpiration is of key importance for the (local) climate by altering the surface energy balance (Oke, 1987), and for water resources where E_T is an important loss term of the water balance (Federer, 1973; Jasechko et al., 2013). The tight coupling to photosynthesis and thus primary productivity makes E_T also central for agriculture and forestry. There is thus a need to understand the spatial and temporal variation of E_T as a result of interacting climatic and biogeophysical processes. Especially in temperate forests these biogeophysical feedbacks are poorly understood, which strongly reduces our ability to assess their role in mitigating climate change (Bonan, 2008).

Forests are often found in complex topographical settings. Thus, we have to consider that the first-order physical controls on E are strongly altered by topography: hillslope angle and aspect systematically alter the amount of absorbed solar radiation and thus atmospheric demand (Baumgartner, 1960; Lee and Baumgartner, 1966; Ivanov et al., 2008), whereas slope and the topographic setting, such as contributing area, alter the lateral redistribution of water (Famiglietti et al., 1998; Bachmair and Weiler, 2011). Apparently, forests adjust to these conditions, but little is known how vegetation and site-scale transpiration in particular respond to these topographically altered boundary conditions of water supply and demand (Tromp-van Meerveld and McDonnell, 2006). For example Holst et al. (2010) found that detailed forest hydrological models yielded different trends of simulated mean stand transpiration when comparing European beech stands on different hillslope aspects. These differences add up when simulating runoff generation from these sites and emphasizes the difficulty in the assessment of forest water balance in complex terrain.

Although detailed models of various feedbacks between plant physiology and the environmental conditions are available (Monteith, 1965; Farquhar et al., 1980; Wang and Jarvis, 1990; Whitehead, 1998; Haas et al., 2013) their applicability is generally restricted by the need for detailed physiologic, atmospheric, and soil parameters. As an alternative to improve our understanding by increasingly detailed modeling, several authors proposed to deduce and predict E through physical and physiological constraints (Calder, 1998; West et al., 1999; Raupach, 2001; Zhang et al., 2008; Wang and Bras, 2011). Applications of these fundamental constraints do not only increase our understanding of the soil–plant atmosphere continuum, they may also lead to a few but independent predictors of E. For example, Kleidon and Renner (2013b) applied thermodynamic limits of convective heat exchange to the surface–atmosphere system. They found that under the assumption of maximum convective power and no

surface water limitation, atmospheric demand is only dependent on absorbed solar radiation and surface temperature, being consistent with the empirical formulations of potential evaporation of Makkink (1957) and the well-known equilibrium evaporation concept (Schmidt, 1915; Priestley and Taylor, 1972; de Bruin et al., 2016). Contrary to the classic formulation of Dalton (see Brutsaert (1982) for an overview), where vapor pressure deficit (VPD) and wind speed are used as forcing variables of potential evapotranspiration, Kleidon and Renner (2013b) argued that VPD and wind speed emerge from the land–atmosphere interaction. The practical implication of maximum convective power is that atmospheric demand can be estimated without these variables and empirical parameters.

In this study we aim to assess the dominant temporal and spatial controls of forest transpiration in complex terrain. Therefore, we use the parsimonious approach of maximum convective power to estimate potential evaporation. The approach of maximum convective power has so far only been applied for long-term annual estimates of E and large-scale geographic variations (Kleidon et al., 2014). Here we test this approach at the site level and for daily timescales in complex topographical terrain. We use data of a well-instrumented beech forest with measurement sites across a transect with a north-facing and a south-facing hillslope. The indirect assessment of E_T is based on ecohydrological measurements of sap flow and soil moisture. We present a systematic analysis to address the question of how much transpiration varies in complex terrain and what measurable site characteristics influence forest transpiration. In particular we hypothesize that potential evaporation representing the atmospheric demand for water can be estimated by thermodynamic limits of convection, which only relies on surface absorption of solar radiation and temperature. Specifically, we test how much of the daily variations of in situ transpiration observations can be explained by atmospheric demand and by how much they respond to atmospheric demand along the measurement transect. Therefore, we perform a linear regression and correlation analysis of the transpiration estimates to atmospheric demand comprising daily data of the growing season. Further, we evaluate how much additional variance can be explained by other time-varying parameters such as VPD, wind speed, and soil moisture. Differences in the strength of correlation between sites would imply that site-dependent, time-varying constraints on transpiration are relevant. In contrast, a high, consistent correlation would imply that atmospheric demand and thus energy limitation of transpiration is the dominant driver of day to day variability. The overall response of transpiration to atmospheric demand represented by the slope of the linear regression would then indicate the importance of site-dependent controls such as topographic, soil, and plant factors.

2 Methods

2.1 Site description

We analyze measurements at six different sites along a well-instrumented steep forested hillslope transect (north-facing vs. south-facing; see Fig. 1) in the Attert catchment in Luxembourg over the vegetation period of 2013. The hillslope transect is part of the Catchments As Organized Systems (CAOS) field observatory (Zehe et al., 2014) and is located in the western part of Luxembourg (5° E 48′13″, 49° N 49′34″) at about 460 m NN. The land cover of the transect is a mixed forest dominated by European beech (*Fagus sylvatica* L.). The north-facing slope has an inclination of $\approx 15°$ and is composed of a few dominant trees with large gaps and dense understorey mainly of young beech trees, whereas the south-facing slope is generally steeper ($\approx 22°$) and has no understorey and a denser canopy. Also tree species composition varies between slopes, with 97 % beech on the north-facing sites with single spruce trees and 90 % beech and 10 % oak on the south-facing sites. The valley site has 80 % beech with 10 % spruce and alder. Geologically, the site is situated in northeast–southwest-trending fold system of Schists of the Ardennes Massif. The shallow soils developed on periglacial slope deposits (Juilleret et al., 2011). The deposits are generally found at a depth of 70–90 cm.

Standard meteorological data, global radiation, air temperature, relative humidity, and wind speed were measured 2 m above ground at all sites. For the meteorological forcing we used the data from the open grassland site G1, which is located 240 m to the northwest of the forest site N1; see Fig. 1. Absorbed solar radiation $R_{sn} = (1 - \alpha)R_g$ in W m^{-2} is derived from global radiation R_g and an albedo estimate of $\alpha = 0.15$, which is representative for deciduous forests (Oke, 1987).

2.2 Estimation of atmospheric demand

Our aim is to estimate the potential evaporation from first principles and with few, independent input data. Therefore we make use of the concept of thermodynamic limits of convection, which was recently established by Kleidon and Renner (2013b) and used successfully to estimate the sensitivity of the hydrologic cycle to global warming (Kleidon and Renner, 2013a; Kleidon et al., 2015) and for grid-based global-scale predictions of annual average terrestrial evaporation (Kleidon et al., 2014). Here we only illustrate how the concept is used to estimate potential evaporation; for further details the reader is referred to the mentioned publications.

Convection can be thought of as a heat engine, which converts a temperature gradient into kinetic energy (Ozawa et al., 2003). To capture the fundamental trade-off of thermodynamic limits of convective exchange, we consider a simple two-box surface–atmosphere system in steady state, which is sketched in Fig. 2. We consider the steady-state energy balance of the surface $R_{sn} = J + R_{ln}$. The surface is heated by absorption of incoming solar radiation R_{sn}. The turbulent heat fluxes J and the net longwave exchange R_{ln} both cool the surface. The turbulent heat fluxes are composed of the sensible (H) and latent heat flux (λE). Longwave radiative exchange is represented by a simplified linearized radiation $R_{ln} = k_r(T_s - T_a)$, with T_s being the temperature of the surface and T_a the temperature of the atmosphere and k_r being a constant radiative exchange coefficient. The power of the convective heat engine G is fundamentally limited by the Carnot limit:

$$G = J \frac{T_s - T_a}{T_s}. \tag{1}$$

Different from the classic Carnot engines, the convective heat engine has flexible boundary conditions, namely, the temperature gradient $T_s - T_a$ responds to the strength of the convective heat fluxes J through its coupling to the surface energy balance. Rearranging the surface energy balance

$$T_s - T_a = \frac{R_{sn} - J}{k_r} \tag{2}$$

shows that a stronger convective heat flux at a given radiation reduces the temperature gradient. This effectively yields a thermodynamic limit of the maximal power of the convective heat engine. The limit is obtained by inserting this feedback into the Carnot limit Eq. (1) and solving for the maximum with respect to J. Hypothesizing that convection actually operates at this limit (subscript opt), we obtain a closure for the strength of convective fluxes:

$$J_{opt} = \frac{R_{sn}}{2}. \tag{3}$$

Once the convective fluxes are obtained we aim to estimate the partitioning into sensible and latent heat fluxes. Using the Bowen ratio B and bulk formulations for sensible (H) and latent heat fluxes (λE), we can write

$$B = \frac{H}{\lambda E} = \frac{\rho c_p k_h}{\rho \lambda k_e} \frac{T_s - T_a}{q_s - q_a} = \frac{c_p}{\lambda} \frac{T_s - T_a}{q_s - q_a} \tag{4}$$

where $c_p = 1.2 \, \text{J K}^{-1} \, \text{kg}^{-1}$ is the heat capacity of air, $\rho = 1 \, \text{kg m}^{-3}$ the air density, $\lambda = 2.5 \times 10^6 \, \text{J kg}^{-1}$ the latent heat of vaporization, q_s and q_a being the specific humidity (g kg^{-1}) at the surface and the atmosphere, respectively. If we assume that the air at the surface and in the atmosphere is saturated with water vapor and that the slope of the saturation vapor pressure curve ($s = \partial e_{sat} / \partial T$) is invariant within this range, the vertical humidity gradient can be written as $q_s - q_a = s(T_s - T_a) \cdot 0.622 / p_s$. e_{sat} is the temperature-dependent vapor pressure at saturation, p_s the surface pressure (hPa) and 0.622 being the ratio of the gas constants for air and water vapor. Introducing the psychrometric constant with $\gamma = c_p / \lambda \cdot p_s / 0.622 \approx 65 \, \text{Pa K}^{-1}$ and assuming that the

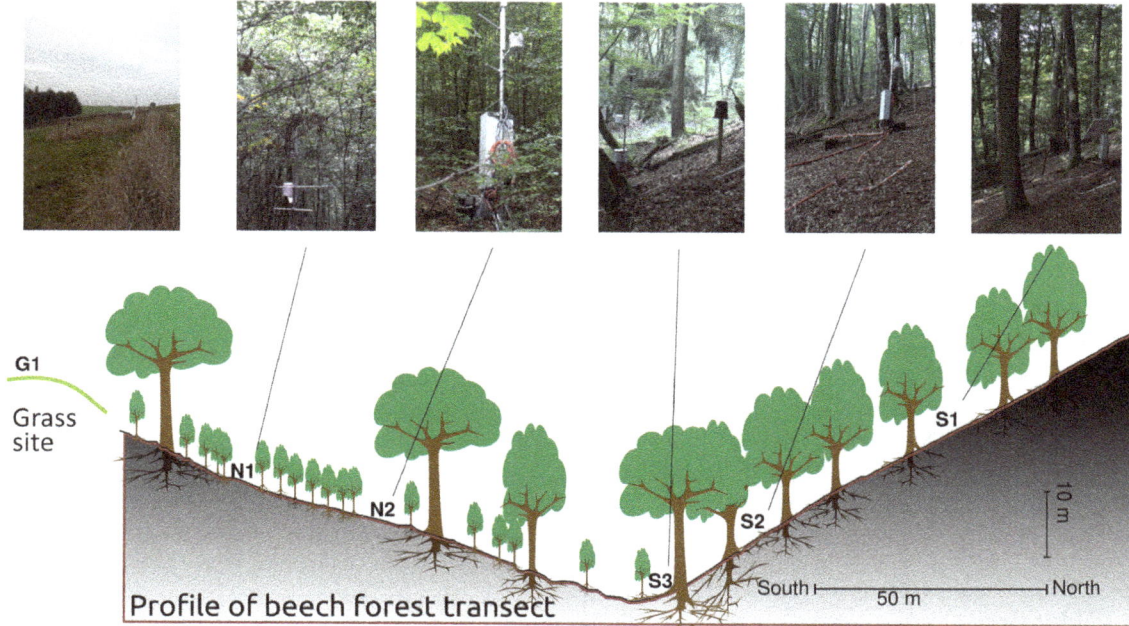

Figure 1. Measurement setup along a hillslope transect. Captital letters indicate aspect of the sites (N: north-facing, S: south-facing) and the numbering represents the position of the site on the hillslope: upslope (1), midslope (2), and downslope (3).

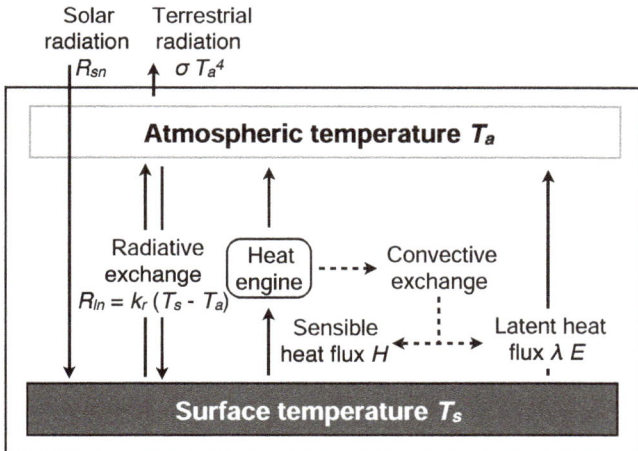

Figure 2. Land–atmosphere energy balance scheme for derivation of atmospheric demand adapted after Kleidon et al. (2014).

vertical exchange coefficients of air k_h and of vapor k_e are equal, we thus obtain the equilibrium Bowen ratio $B = \frac{\gamma}{s}$ (Schmidt, 1915; Stull, 1988). The equilibrium Bowen ratio depends only on temperature, because s is a nonlinear function of temperature and γ is approximately constant. We determine s by an empirical equation (Bohren and Albrecht, 1998): $s = s(T) = 6.11 \cdot 5417 \cdot T^{-2} \cdot e^{19.83 - 5417/T}$ with temperature T (K).

Combining the equilibrium Bowen ratio with the concept of maximum convective power we obtain an expression for the potential evaporation, herein referred to as atmospheric demand E_{opt} (Kleidon and Renner, 2013b; Kleidon et al., 2014):

$$E_{opt} = \frac{1}{\lambda} \frac{s}{s + \gamma} \frac{R_{sn}}{2}. \tag{5}$$

Hence, only absorbed solar radiation R_{sn} and temperature data are required to estimate atmospheric evaporative demand, when derived from thermodynamic limits using simplifying assumptions on longwave radiative exchange and the assumption for the equilibrium Bowen ratio.

We also compare our results with the well-established FAO Penman–Monteith grass reference evaporation equation (Allen et al., 1998):

$$E_{PM} = \frac{0.408\, s\, (R_n - G) + \gamma \frac{900}{T + 273} u \text{VPD}}{s + \gamma (1 + 0.34 u)}, \tag{6}$$

where R_n is net radiation, G the ground heat flux, T air temperature, u wind speed (both in 2 m height), and VPD is the saturation vapor pressure deficit in kilopascal (kPa). Equation (6) is a modified Penman–Monteith equation for a standardized grass surface at a daily resolution. As input variables we use the same variables as for E_{opt} (daily average air temperature, shortwave radiation) and in addition net radiation, ground heat flux, relative humidity, and wind speed. Because net radiation was not measured at our sites we used an empirical formulation for R_n (Allen et al., 1998, Eq. (39)). Therefore, data of daily minimum and maximum temperature as well as information on the day of year and latitude were required. The soil heat flux was estimated as $G = 0.1 \cdot R_n$.

2.3 Topographic effects on shortwave radiation

In complex terrain the incoming radiation is influenced by the slope and the aspect of the current position (Kondratyev and Fedorova, 1977). To account for these effects we use a topographic radiation correction method, which projects the extraterrestrial irradiance on inclined surfaces implemented in the software *r.sun* (Šúri and Hofierka, 2004), which is part of the open-source Geographic Information System (GIS) platform **GRASS GIS** (http://grass.osgeo.org). *r.sun* estimates potential clear-sky spatial global radiation $R_{g,pot}$ fields for each day of year. To account for cloudy conditions we use the open-field site G1 as a reference station and estimate a daily cloud factor $f_c = R_{g,G1} / R_{g,pot}$ using the closest grid point of the respective site. Then $R_{g,pot}$ is multiplied with f_c to obtain a radiation estimate for the sites in the forest $R_{g,c}$. As input for *r.sun* we use a 10 m resolution Digital Elevation Model (DEM) (with slope and aspect inputs from the DEM) and default parameters (Linke atmospheric turbidity coefficient = 2). We thus estimate global radiation on a tilted surface (aspect, slope) and also include topographic shading effects. The effect is largest when the sun angle is low (autumn–winter–spring) and decreases R_g and thus E_{opt} at the north-facing slope while increasing it on the south-facing slope; see Fig. S1 in the Supplement. These estimates are generally consistent with radiative observations in sloped terrain (Holst et al., 2005).

2.4 Biometric measurements

A forest inventory for all sites was done in March 2012. The circumference at breast height of all trees with circumference larger or equal to 4 cm was measured in a 20 m by 20 m plot for each site. Stem basal area was calculated and a total stand basal area was computed for each site.

Leaf area index (LAI) was measured with a LICOR LAI-2200 plant canopy analyzer at all forested sites in two campaigns. The summer campaign was carried out on 11 August 2012 and the winter campaign on 20 March 2014. Here we use the measurements taken with all rings of the LAI-2200. The LAI was averaged from 36 measurements points per site. The difference between summer and winter LAI should reveal the actual leaf area index without stems and topographic shading effects.

2.5 Sap flow measurements and upscaling stand transpiration

The five forested sites are instrumented with multiple sap flow sensors (four trees at each site, installed between mid-May to November). Heat pulse sensors (East 30 Sensors, Pullman, Washington 99163 USA) based on the heat ratio method (Burgess et al., 2001) have been used as they are less susceptible to natural heating gradients, and require less electrical power (Vandegehuchte and Steppe, 2013). The sensors measure the heat pulse velocity, which is corrected for wounding and sensor alignment; see Sect. A1 in the appendix. The heat pulse velocity is then converted into sap flux density, SFD ($cm^3 cm^{-2} h^{-1}$) (Burgess et al., 2001):

$$SFD = v_c \frac{\rho_b (c_w + m_c c_s)}{\rho_s c_s}. \tag{7}$$

Thus, to estimate the sap flux density, knowledge of xylem wood properties, namely, the basic density of wood ρ_b, the density of the sap ρ_s, the specific heat capacity of the wood matrix c_w, and of the sap c_s as well as the water content of the xylem is required. For sap we use the standard parameters for water at 20° C, with $\rho_s = 1 \, kg \, cm^3$, $c_s = 4182 \, J \, kg^{-1} \, K^{-1}$. The heat capacity of the woody matrix is generally given with $c_w = 1200 \, J \, kg^{-1} \, K^{-1}$ (Burgess et al., 2001). The basic density of sapwood ρ_b measured as dry weight over green volume and the moisture content of the xylem m_c are species-specific parameters. We used $\rho_b = 0.61 \, kg \, cm^{-3}$ and $m_c = 0.7$ for the xylem of European beech estimated by Glavac et al. (1990). They sampled about 260 trees at two different sites between 35 and 42 years old over the course of 1 year. For the thermal diffusivity of the xylem, we used a fixed value of $k_w = 2.5 \times 10^{-3} \, cm^2 \, s^{-1}$ (Burgess et al., 2001).

The raw measurements obtained every 30 min were quality controlled and suspect data were filtered before further analysis was performed. At three larger trees the outermost sensor readings were replaced by the second sensor depth reading because their annual mean was smaller than the inner sensors, which indicates a sensor misplacement into the bark of the tree. The sensors measure the heat pulse velocities at three different radial depths in the tree. The daily mean sap flux density per tree is obtained by an average of all readings per depth and day. Units of SFD were converted to $m^3 m^{-2} d^{-1}$. An arithmetic mean of the tree-average sap flux density was used to obtain the site-average sap flux density.

Measurements of sap flux density only provide a relative measure of the velocity of the ascending xylem sap. To obtain tree water fluxes a representative xylem area per sensor depth is assumed, which is then multiplied with the corresponding sap flux density (Burgess et al., 2001). Tall trees with a diffuse-porous xylem structure such as beech are known to have a large sapwood to basal area ratio (Köstner et al., 1998). Further, the radial profile of the sap flux density varies in these trees. Therefore, deep measurements are ideally required (Gebauer et al., 2008). In the absence of these deeper measurements, we follow the reasoning of Lüttschwager and Remus (2007) to derive uncertainty ranges for the inner conducting sapwood area; see appendix.

To upscale to the site level we use the inventory data, which provides the number of trees in the stand, diameter at breast height (DBH), and species information. Daily sap flux density per sensor depth was averaged for each species and tree status (dominant vs. suppressed) for each site. Missing sap flux density data were filled by linear regression with neighboring sites. The largest data gaps were filled at sites N2 and S2. Finally, the upscaled daily stand transpiration was obtained by summing up the product of sap flux density per depth D and sap wood area per depth $A_s(D)$ for all trees and dividing by the area of the inventory A_{stand}:

$$E_{sap} = \frac{1}{A_{stand}} \sum_{tree=1}^{n} \sum_{D=1}^{3} (SFD(D)A_s(D)). \quad (8)$$

2.6 Root water uptake estimation

As another means to estimate transpiration we use soil moisture measurements. Soil moisture sensors (Decagon 5TE soil moisture sensors, with an accuracy of $\pm 3\%$ volumetric water content and resolution of 0.08%) are installed at three profiles at each site at 10–30–50 cm depth and one deeper sensor (approx. 70 cm) at one of the profiles. There is a range of methods to estimate root water uptake from soil moisture observations (Shuttleworth, 1993; Cuenca et al., 1997; Wilson et al., 2001; Schwärzel et al., 2009; Breña Naranjo et al., 2011; Guderle and Hildebrandt, 2015). Here, we employ a simplified soil water budget method to estimate daily root water uptake E_{RWU} (units mm d^{-1}). The method is based the daytime reduction of measured soil moisture integrated over the soil profile. To avoid the influence of infiltration and other fluxes, which are not caused by root water uptake E_{RWU}, we filtered the days with significant nighttime soil moisture variations as well as rainy days. A representative site value of E_{RWU} was derived by averaging the estimates of all three soil profiles. Details of the method can be found in the Appendix A3.

2.7 Statistical analyses

To test if the transpiration estimates are driven by potential atmospheric demand, we use linear ordinary least squares (OLS) regression with the transpiration estimates as response variable and E_{opt} as independent variable. The slope of the regression with sap flux density is denoted by $b_{SFD} = \frac{dv_{sap}}{dE_{opt}}$ with units $[m^3\,m^{-2}\,d^{-1}]\,/\,[mm\,d^{-1}]$. For E_{sap} the slope is denoted $b_{sap} = \frac{dE_{sap}}{dE_{opt}}$ and for E_{RWU} the slope is denoted $b_{RWU} = \frac{dE_{RWU}}{dE_{opt}}$. The latter two regression slopes have nondimensional units $[mm\,d^{-1}]\,/\,[mm\,d^{-1}]$. To estimate the potential influence of other independent variables on transpiration, such as soil moisture or vapor pressure deficit, we use the residuals of the OLS regression as response variable of a further linear regression. The explanatory power of the other variable is reported by the adjusted explained variance (denoted by R_θ^2 for soil moisture effects), which is a diagnostic of the OLS regression.

Note that by using time series of daily data, which are shaped by the seasonal cycle, the assumption of independence of the predictor variables in the linear regression is not often justified. The effect of serial dependence generally does not bias the regression coefficients, but reduces the statistical significance of a regression. Therefore, we estimate the standard deviation of the regression coefficient (σ_{slope}) and its reported significance with a pre-whitening procedure of Newey and West (1994) provided by function *coeftest* of the R packages lmtest (Zeileis and Hothorn, 2002) and sandwich (Zeileis, 2004). All data analysis was done in R (R Core Team, 2016) and the R-package data.table (Dowle et al., 2014).

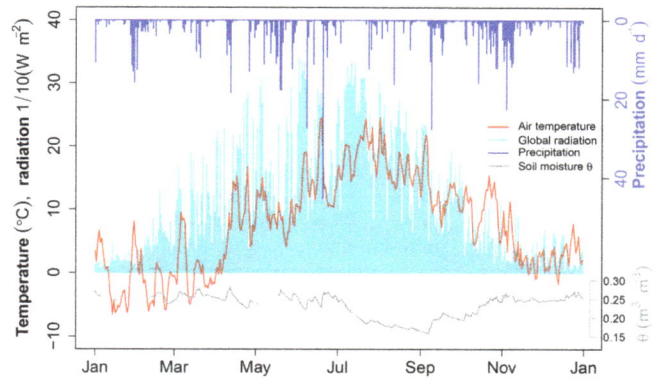

Figure 3. Meteorological forcing and average soil moisture at the grassland site (G1) for 2013. A few data gaps were filled with data from nearby grassland sites.

3 Results

3.1 Meteorological observations in 2013

Daily time series of temperature, global radiation, precipitation, and site-average soil moisture content for 2013 are shown in Fig. 3. The average annual (growing season, May–October) temperature was 7.9 (13.7) °C, with an average global radiation of 117 (169) W m^{-2}. The highest temperatures were observed in a short period in June and a longer period between mid-July and August. We observed a total annual precipitation of 866 mm (no catch correction applied), which was, however, unevenly distributed throughout the year. Most precipitation fell in June (110 mm), with low rainfall during July and August (34 and 38 mm, respectively) and more rainfall in the autumn months. Hence, the soil moisture recession lasted from July to September (bottom gray line in Fig. 3).

3.2 Site-scale transpiration estimates

Two independent site-scale transpiration estimates are evaluated in the following. The first estimate, E_{sap}, was obtained by upscaling sap flow measurements with the forest plot inventory data. The second estimate was derived by averaging E_{RWU} of the 2–3 soil profiles per site. Both transpiration

Table 1. Seasonal totals in mm of estimated stand transpiration E_{sap} and potential evaporation E_{opt} and E_{PM}. For the totals we used the period 12 May 2013–31 October 2013. For E_{sap} we present a range of estimates with minimal, linear decline and maximal estimate with respect to the depth of conducting inner sapwood (Eqn. A3, A4, and A5).

Site	E_{sap} min–lin–max	E_{opt}	E_{PM}
N1	106–128–156	240	351
N2	76–115–164	232	345
S3	127–176–244	296	387
S2	104–129–170	293	386
S1	96–122–160	295	387
G1		270	371

estimates are shown as daily time series for site N1 at the north-facing and S1 at the south-facing hillslope in Fig. 4. The north-facing sites have lower potential evaporative demand (estimated by topographically corrected incoming solar radiation) than the south-facing sites. Variations of daily soil water contents are quite similar between the sites. However, the north-facing site is on average slightly wetter ($\bar{\theta} = 0.21 \pm 0.04$) than the south-facing site ($\bar{\theta} = 0.18 \pm 0.03$). There is a large temporal variability in the atmospheric forcing represented by E_{opt}, which is also found for both transpiration estimates. The highest transpiration rates are seen in the dry and sunny period in July. Comparing the sites at the different hillslopes we find similar magnitudes of E_{sap}. However, E_{RWU} is found to be much higher than E_{sap} in spring and early summer at the north-facing sites.

Seasonal totals of E_{sap} and potential evaporation are reported in Table 1. The E_{sap} estimates for the upper north and south-facing hillslope sites are relatively similar. The highest E_{sap} is estimated at the valley bottom site S3, which amounts to a difference of about 50 mm for the seasonal total. Large uncertainty in the nominal value of E_{sap} is due to the assumed depth of the sapwood, which is estimated from data of the innermost sensor and allometric relationships of sapwood. Thereby the range between different conducting sapwood depths is largest at the valley site because there are more tall trees where this uncertainty matters.

3.3 Transpiration response to atmospheric demand

To test if the transpiration estimates are driven by potential atmospheric demand, we performed a linear regression of the site-average values with E_{opt} as independent variable. We found that E_{sap} was relatively low in May and early June after leaf out and after mid-October with leaf senescence. In order to avoid phenological effects in the statistical analysis, we restrict the following analyses to the vegetation period from 10 June to 20 October. Respective scatter plots are shown in Fig. 5 and a summary of the results is reported in Table 2.

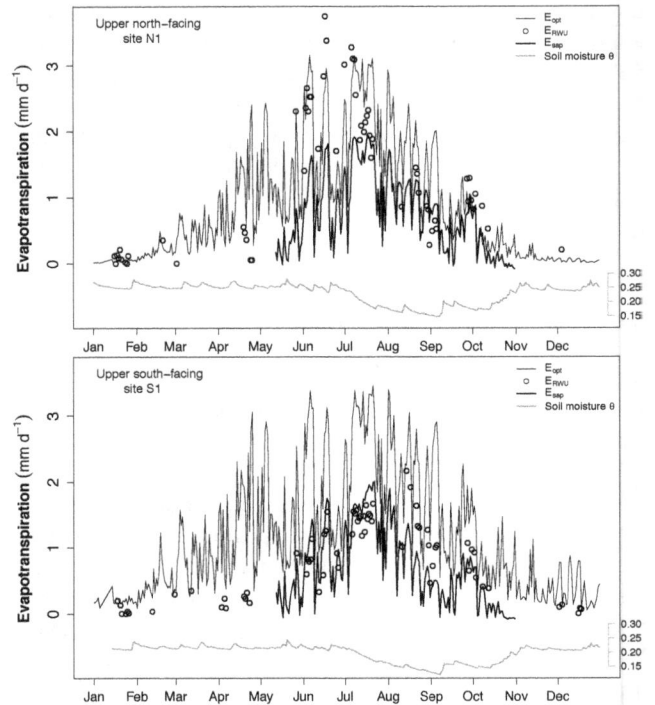

Figure 4. Time series of evapotranspiration estimates for north-facing hillslope site N1 in the left panel and south-facing site S1 in the right panel. E_{opt} is the potential evaporation derived from maximum power, E_{RWU} (blue points) and E_{sap} are transpiration estimates.

Figure 5 shows that both observations almost linearly increase with E_{opt}. For E_{sap} we find very high linear correlations with an average of $r^2_{E_{opt}} = 0.82$ (Table 2). The relatively large temporal variability in E_{sap} seen in Fig. 4 with a coefficient of variation CV = 0.68 across all forest sites is explained by the atmospheric demand E_{opt} (CV = 0.53). It is interesting to note that across all forest sites global radiation has only a slightly lower average $r^2 = 0.79$, whereas air temperature with $r^2 = 0.55$ and vapor pressure deficit with $r^2 = 0.66$ have less explanatory power. The correlation of site-average E_{RWU} to E_{opt} is much lower ($r^2 = 0.49$, averaged among all sites) than the correlation of E_{sap} to E_{opt}. All meteorological variables show lower correlations to E_{RWU} than to E_{sap}, which indicates a larger uncertainty of the soil moisture-derived transpiration estimate.

We also tested if the regression residuals show correlations with other daily observations. The reported adjusted squared correlation of a linear regression of the residual with VPD, wind speed, and soil moisture at the site level is reported in Table 2. Vapor pressure deficit does not explain residual variance of E_{sap} or E_{RWU}. Also wind speed does not show residual correlation for E_{sap} but higher positive residual correlation is found for E_{RWU} at the lower sites N2, S2, and significant at S3. Both of these variables are used as external

Table 2. Regression statistics for E_{sap} and E_{RWU} to E_{opt}, with n providing the number of observations (days). The slope and intercept are reported with the estimated standard deviation of the coefficients with $\pm\sigma$. Significance of the coefficients is indicated by stars: $p < .001$, ***; $p < .01$, **; $p < .05$, *. $r^2_{E_{opt}}$ and $r^2_{E_{PM}}$ are the linear squared correlation coefficients to E_{opt} and E_{PM}, respectively. The last three columns report the adjusted explained variance of a linear regression of the regression residuals for daily average vapor pressure deficit (R^2_{VPD}), daily average wind speed (R^2_u), and daily site-average volumetric water content (R^2_θ).

Variable	Site	n	Slope	Intercept	$r^2_{E_{opt}}$	$r^2_{E_{PM}}$	R^2_{VPD}	R^2_u	R^2_θ
E_{sap}	N1	130	0.61 ± 0.02***	-0.03 ± 0.05	0.88	0.89	0.01	0.00	0.02
	N2	129	0.57 ± 0.03***	-0.03 ± 0.05	0.88	0.89	0.01	0.01	-0.00
	S3	130	0.84 ± 0.08***	-0.36 ± 0.10***	0.83	0.87	-0.01	0.01	0.19**
	S2	109	0.62 ± 0.07***	-0.27 ± 0.09**	0.75	0.80	-0.01	0.01	0.25*
	S1	130	0.58 ± 0.07***	-0.23 ± 0.07**	0.78	0.83	-0.01	0.00	0.24**
E_{RWU}	N1	36	0.77 ± 0.18***	0.19 ± 0.27	0.52	0.50	0.04	0.03	0.64***
	N2	30	0.76 ± 0.19***	0.20 ± 0.40	0.67	0.69	-0.02	0.10	0.11
	S3	35	0.56 ± 0.11***	-0.21 ± 0.22	0.44	0.55	-0.03	0.32**	0.53***
	S2	30	0.52 ± 0.11***	0.11 ± 0.19	0.47	0.58	-0.02	0.12	0.45
	S1	39	0.44 ± 0.03***	0.13 ± 0.06*	0.63	0.60	-0.03	0.03	-0.02
	G1	28	0.75 ± 0.34*	0.32 ± 0.77	0.22	0.21	0.04	-0.04	0.63***

forcing in the Penman equation. Therefore, we calculated the squared correlation also for the FAO Penman–Monteith reference evaporation E_{PM}, Eq. (6). One can see from Table 2 that $r^2_{E_{PM}}$ is fairly similar, follows the same pattern, and is on average only slightly larger ($r^2_{E_{PM}} = 0.86$) than the correlation of E_{sap} to E_{opt}. Note that E_{PM} is on average 1.44 times larger than E_{opt} with a correlation between E_{PM} and E_{opt} of $r^2 = 0.98$.

In contrast to the meteorological variables, we found that the residuals are significantly correlated to the site-average soil moisture content at some sites. E_{sap} is significantly affected by soil moisture deficits at the south-facing sites. At these sites we find significant residual correlation of soil moisture $0.19 < R^2_\theta < 0.25$, which results in lower correlation to E_{opt} as compared to the two north-facing sites. Further, we find significant negative intercepts in the E_{sap}–E_{opt} relationship. Even more affected by soil water limitation is E_{RWU} with significant values of $R^2_\theta > 0.5$ at N1, S3, and G1. However, at S1 there is no residual correlation at the site average level, but this is probably an effect of deriving a representative site-average from E_{RWU} profile estimates, because one soil profile at S1 actually shows a significant residual correlation (Table S3 in the Supplement). The potential effect of soil moisture on transpiration is visualized by the size of the symbols in Fig. 5. Thereby, E_{RWU} tends to be above the regression line under moist and warm conditions in early summer particularly at sites N1 and G1.

Figure 5 and the regression statistics allow for a further comparison of E_{sap} and E_{RWU}. Generally there is a good agreement of both estimates with $0.5 < r^2 < 0.9$, but there are site-specific deviations between the site transpiration estimates. The upper south-facing sites S1 and S2 agree well in magnitude, while E_{sap} for site S3 appears to have an offset of about $0.5\,\mathrm{mm\,d^{-1}}$. This offset could be influenced by the

apparent site heterogeneity. The site is situated in the transition between the steep hillslope and the riparian zone of the nearby creek. The E_{sap} upscaling represents this transition zone by the 20 m by 20 m forest inventory size and assumes constant sapwood properties for all trees within this site. On the other hand, three soil moisture profiles may be too few in this heterogeneous zone, which complicates the estimation of a representative site-average E_{RWU} value. At the north-facing sites we find that E_{sap} is lower than the E_{RWU} estimates. Especially during early summer when soil moisture was still high, E_{RWU} was found to be almost twice as large as E_{sap} and of similar magnitude as E_{opt} at these sites. However, there was a better agreement in late summer, when soil moisture declined; see Fig. 4. This indicates that the root water uptake estimates tend to be higher under moist conditions, which is reflected by the positive residual correlation to soil moisture reported in Tables 2 and S3 in the Supplement.

3.3.1 Response at tree level

By using tree-averaged sap flow density measurements directly we can assess how strong single trees respond to E_{opt}. Results of a linear regression analysis of daily tree average SFD to E_{opt} are reported in Table 3 (see also Fig. S3 in the Supplement). SFD of every tree shows a strong linear relation to E_{opt}, with varying slopes and intercept terms. Because most of the variability is explained by E_{opt}, we can simply compare the slope of the regression b_{SFD}. First, we find that b_{SFD} is especially high at the tall beech trees at the north-facing sites (1.26–$1.64\,[\mathrm{m^3\,m^{-2}\,d^{-1}}]/[\mathrm{mm\,d^{-1}}]$), Table 3. These sampled tall trees at the north-facing sites are dominant trees without other tall trees in their vicinity. All other trees show much smaller values of b_{SFD} (0.26–$0.88\,[\mathrm{m^3\,m^{-2}\,d^{-1}}]/[\mathrm{mm\,d^{-1}}]$). Second, we find that b_{SFD} is on

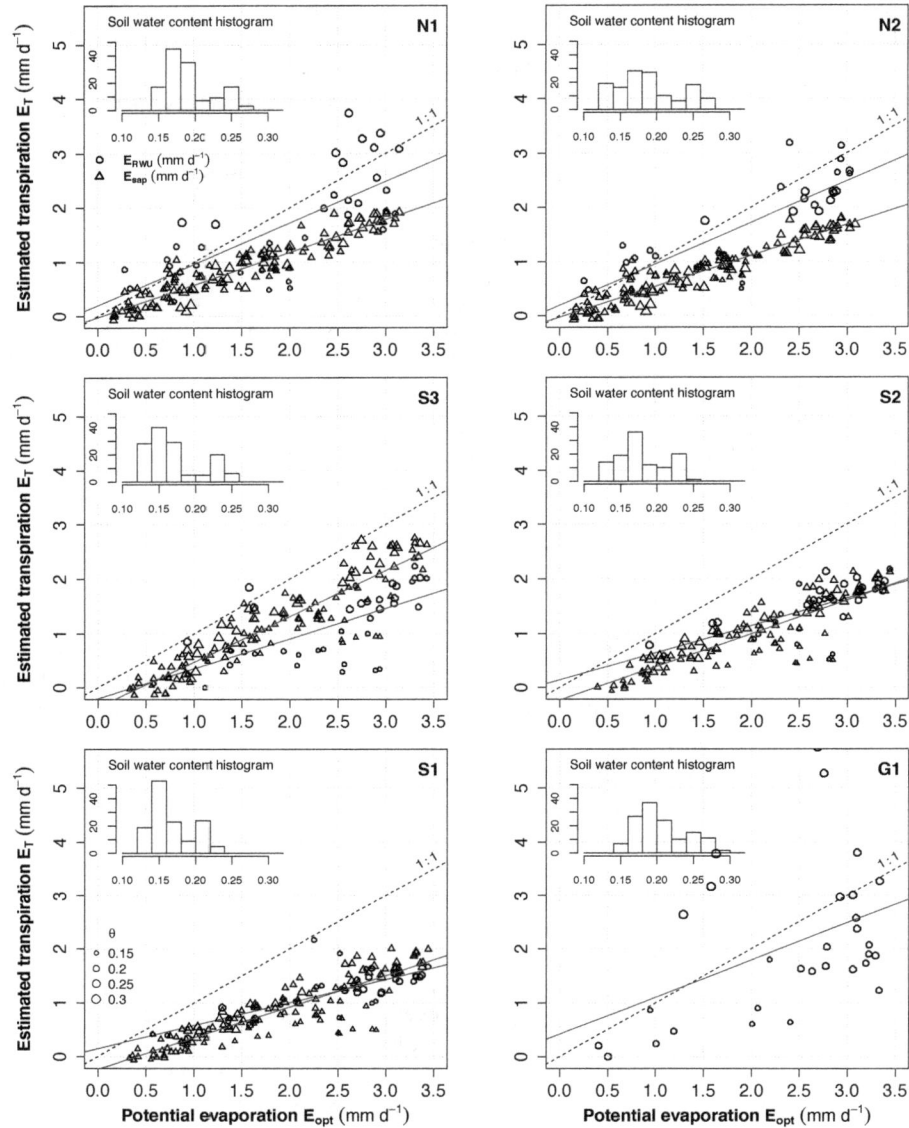

Figure 5. Site-average E_{RWU} (blue circles) and E_{sap} (red triangles) as function of E_{opt} during vegetation period (10 June–20 October) in 2013. Each panel presents data of one site (see Fig. 1 for site identifiers). The size of the symbols corresponds to the daily average soil moisture content. The histogram shows the distribution of site-average soil moisture. Solid lines depict the linear regression as tabulated in Table 2.

average larger at the north-facing than at the south-facing trees. Last, we also find that the oak tree at site S1 has the lowest b_{SFD}. For these reasons the upscaling of SFD to E_{sap} was done for each site separately and distinction was made between dominant and small trees as well as between species.

The residual regression analysis shows a similar pattern as the site-average values, with low influence of wind speed and VPD. Site-average soil moisture content (θ) shows significant effects at 9 out of 14 trees; see Table 3. Thereby all beech trees at the south-facing sites show an influence of θ on the sap flow residuals, with a maximum adjusted explained variance of $R_\theta^2 = 0.4$ at the upper south-facing site S1. In ad-

dition, the 3 of 4 small beech trees at the north-facing sites show significant influences of θ.

3.3.2 Soil profile response and root water uptake profiles

Generally, the measurement setup allows one to differentiate between three different levels of aggregation of root water uptake estimates: site-average, per profile, and per depth of the sensor.

Root water uptake per profile is obtained by summing up the estimates of each sensor. We find that 15 out of 16 pro-

Table 3. Regression statistics of daily SFD as average per tree and per site as extra row. Column DBH is the diameter at breast height in centimeters. n is the number of observations (days). The slope b_{SFD} and intercept of the linear univariate regression of SFD to E_{opt} with $\pm \sigma$ reporting the estimated standard deviation of the coefficients. Significance of the coefficients is indicated by stars: $p < .001$, ***; $p < .01$, **; $p < .05$, *. $r^2_{E_{opt}}$ and $r^2_{E_{PM}}$ are the linear squared correlation coefficients of SFD to E_{opt} and E_{PM}, respectively. The last three columns report the adjusted explained variance of a linear regression of the regression model residuals for the variables: vapor pressure deficit (R^2_{VPD}), wind speed (R^2_u), and site-average volumetric water content (R^2_θ).

Site	Tree	Species	DBH	n	b_{SFD}	Intercept	$r^2_{E_{opt}}$	$r^2_{E_{PM}}$	R^2_{VPD}	R^2_u	R^2_θ
N1	1	beech	66	130	1.30 ± 0.05***	0.09 ± 0.13	0.88	0.88	0.03	0.07**	0.11**
N1	2	beech	58	130	1.26 ± 0.04***	-0.06 ± 0.08	0.92	0.84	0.00	0.03*	−0.01
N1	3	beech	9	130	0.88 ± 0.05***	-0.08 ± 0.09	0.82	0.89	0.03	−0.00	0.08*
N1	4	beech	8	130	0.63 ± 0.08***	-0.22 ± 0.08**	0.79	0.88	−0.00	0.02	0.29***
N1				130	1.02 ± 0.03***	-0.07 ± 0.07	0.90	0.91	0.01	0.01	0.00
N2	1	beech	53	97	1.31 ± 0.08***	0.23 ± 0.16	0.83	0.83	−0.01	−0.00	0.03
N2	2	beech	49	97	1.64 ± 0.09***	-0.06 ± 0.18	0.87	0.88	0.00	−0.00	−0.01
N2	3	beech	10	97	0.68 ± 0.04***	-0.11 ± 0.05*	0.89	0.82	0.01	0.00	−0.00
N2	4	beech	8	97	0.51 ± 0.08***	-0.14 ± 0.08	0.80	0.78	−0.01	0.00	0.26**
N2				97	1.04 ± 0.06***	-0.02 ± 0.11	0.87	0.88	−0.00	−0.01	0.02
S1	1	beech	43	130	0.81 ± 0.08***	-0.23 ± 0.10*	0.77	0.80	−0.01	0.01	0.34**
S1	2	oak	40	130	0.26 ± 0.01***	-0.10 ± 0.02***	0.92	0.94	0.05*	0.01	−0.01
S1	3	beech	39	97	0.39 ± 0.12**	-0.14 ± 0.12	0.64	0.81	0.01	0.01	0.40***
S1	4	beech	46	130	0.40 ± 0.05***	-0.13 ± 0.06*	0.76	0.87	−0.01	0.01	0.40***
S1				130	0.49 ± 0.05***	-0.17 ± 0.06**	0.80	0.84	−0.01	0.00	0.30**
S3	1	beech	45	124	0.52 ± 0.06***	-0.22 ± 0.07***	0.83	0.86	−0.01	0.02	0.22***
S3	2	beech	39	124	0.86 ± 0.08***	-0.35 ± 0.10***	0.84	0.90	−0.00	−0.00	0.09*
S3				124	0.69 ± 0.07***	-0.28 ± 0.08***	0.84	0.87	−0.01	0.01	0.12*

files show significant slopes b_{RWU} (Table S3 in the Supplement), which reveals a significant linear relationship with E_{opt}. The profile-based results also highlight that there is considerable within-site variability of the response of E_{RWU} to E_{opt}, with significant slopes ranging between 0.15 and 1.17. Testing the regression residuals for an influence of soil moisture content we find 10 of 16 forest soil profiles with significant residual correlation R^2_θ. At the grass site two profiles showed even larger residual correlations to θ than to E_{opt} ($R^2_\theta = 0.71, 0.72$). Testing for additional correlation of VPD and wind speed on the regression residuals of E_{RWU} to E_{opt} we find that VPD has generally low additional value with only three profiles with significant influences. Wind speed shows significant residual correlation at 3 out of 16 profiles. Root water uptake at a specific sensor level ($E_{RWU,d,z}$) allows one to assess at which depths plant water extraction and/or capillary rise due to soil evaporation are effective. E_{RWU} is detected at almost all soil moisture sensors during the growing season in 2013 (Fig. S4 in the Supplement). However, E_{RWU} at the deepest sensor is much lower than what is observed at the top (0–20 cm) soil layer. Comparing the grass site (G1) with the forest sites we find that the grass site has much larger E_{RWU} from the top-layer (0–20 cm)

than from deeper layers. Further, we can see that the top-layer uptake decreases during the drier summer period July to September. The reduction is the strongest at the grass site, whereas the forest sites show a more evenly distributed uptake with soil depth.

3.4 Influence of topography and stand structure

The results show that most of the daily variability in both transpiration estimates is driven by atmospheric demand. The slope of the linear regression of E_{sap}, SFD, and E_{RWU} to E_{opt} provides a summary statistic of how strong a site or a tree responds to atmospheric demand. In the following we evaluate which location factors correlate with the slopes of the linear regressions. We concentrate on topographic and stand structural parameters.

The most obvious factor is the aspect of the sites. While b_{sap} does not show aspect-related differences, we find that b_{SFD} and b_{RWU} are higher at the north-facing sites and lower at the south-facing sites. This is found at the site-average level (Table 2), at the soil profile level (Table S3 in the Supplement) and at the tree level (Table 3). The differences are, however, only significant for b_{SFD} at the tree level being larger at the north-facing slope with average $b_{SFD} = 1.03$

Table 4. Site topographic characteristics with average inclination of each site is given in column "slope angle". Column, $R_{g,c}$ is the average global radiation $R_{g,c}$, and E_{opt} and E_{PM} are all averaged for the vegetation period 10 June–20 October. Values of site G1 are measured directly, values of the forested sites are estimated by the topographic correction of solar radiation.

Site	Slope angle (°)	$R_{g,c}$ (W m^{-2})	E_{opt} (mm d^{-1})	E_{PM} (mm d^{-1})
N1	14	153	1.46	2.19
N2	18	148	1.41	2.15
S3	30	192	1.82	2.44
S2	22	190	1.80	2.43
S1	24	191	1.81	2.44
G1	8	173	1.65	2.33

than at the south-facing slope $b_{SFD} = 0.63$, which is significant at the 0.05 level with a Student's t test.

Another topographic factor influencing the response to atmospheric demand is the inclination of the sites, where we find that the steeper the site the lower the slope of the linear regression of SFD to E_{opt} (cf. Tables 3 and 4). Both aspect and inclination affect the received solar radiation. According to the topographic correction of solar radiation, the north-facing sites received about 85 to 89 % of the solar radiation, which was observed at the grass site. In contrast the south-facing sites received about 108–110 %; see Table 4. This difference is confirmed by observations of air temperature within the forest, which reveal the same ranking across the sites as $R_{g,c}$.

Apart from the topographic data, the stand structure varies remarkably (see Table 5). At the two north-facing sites we find much understorey (> 98 % of trees smaller than DBH 15 cm) and a few tall trees, but no medium-sized trees between 15 and 50 cm DBH. The tall, dominant trees therefore have a well-lit canopy. The taller young trees form a secondary, lower but closed canopy layer. In contrast the south-facing sites S1 and S2 have about 20 medium sized trees per 20 by 20 m plot mainly between 16 and 65 cm DBH. These sites form a rather closed one-layer forest canopy. The valley site shows the largest tree size diversity with small, medium, and very tall trees. The stand composition differences of the sites result in quite different stand basal areas and thus total stand sapwood areas, which is the key factor in extrapolating SFD to site level. Although most trees are found at the north-facing sites, their total sapwood area is about half of the south-facing sites. The valley site (S3) has the largest sapwood area (Table 5). The LAI measurements also show the largest values for the valley site and slightly lower values for the south-facing site. Lower values are found at the north-facing sites. However, the differences in LAI between sites are comparably small when compared with sapwood area.

Plotting b_{SFD} as a function of the site-average sapwood area per tree (A_s / n_{tree}) (see Fig. 6), we find that b_{SFD} de-

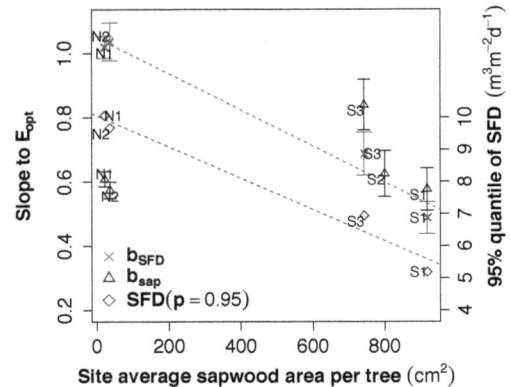

Figure 6. Sensitivity to E_{opt} as a function of the site-average sapwood area per tree computed by A_s / n_{tree}. On the left y axis: sensitivity of site-average sap flux density b_{SFD} (green crosses; see also Table 3) and the sensitivity of the upscaled transpiration estimate b_{sap} (red triangles). The vertical bars show the standard deviation of the sensitivity estimates tabulated in Tables 2 and 3. Right y axis: 95 % quantile of daily SFD maxima per site to estimate a robust maximum SFD. The dashed lines show linear regression model fits.

creases with A_s / n_{tree}. In addition, the maximum SFD estimated as the 95 % quantile of the daily maxima show a very similar decline. Thus, both measures show higher sap flux densities in the younger and smaller north-facing stands as compared to the older south-facing stands. In contrast the sensitivity of E_{sap} is rather constant $b_{sap} \approx 0.6$ [mm d^{-1}] / [mm d^{-1}] across sites at the hillslopes. Only at the valley site b_{sap} is significantly higher with ≈ 0.8 [mm d^{-1}] / [mm d^{-1}]. Although this site has a lower sapwood area per tree (A_s / n_{tree}) than the upper south-facing sites, the total sapwood area at S3 is larger, because of a few very tall trees; see Table 5.

4 Discussion

4.1 Using thermodynamic limits to estimate daily potential evaporation

Generally, there are two different physical limitations of atmospheric evaporative demand, which have been used for modeling E, (i) by energy limitation, and (ii) the mass exchange limitations, which requires that the vapor is removed before it might condense again (Brutsaert, 1982). The approach of maximum convective power applied here combines both of these physical constraints and emphasizes the strong link between the energy balance and the strength of convective motion.

The maximum power-derived estimate of potential evaporation E_{opt} is quite similar to the Priestley–Taylor formulation, but has two key differences. First, the Priestley–Taylor equation contains an additional empirical factor, which has been interpreted as an effect of large-scale advection or

Table 5. Observed biometric characteristics of the five forested sites along the transect. Column names: n_{tree} ...number of trees with circumference > 4 cm in a 20 by 20 m reference plot, diameter at breast height (DBH) distribution with minimum, first quartile, mean, third quartile, and maximum (cm), total stand basal area A_b (m^2 ha^{-1}), total stand sapwood area estimated from published allometric relationships A_s (m^2 ha^{-1}); see Sect. 2.4. Site leaf area index (LAI) was derived from the difference of summer LAI$_s$ and winter LAI$_w$.

| site | n_{tree} | DBH (cm) | | | | | A (m^2 ha^{-1}) | | LAI (m^2 m^{-2}) | | |
		min	Q25	mean	Q75	max	A_b	A_s	LAI$_s$	LAI$_w$	LAI
N1	346	< 1.3	2.1	3.8	4.4	65.4	25	19	7.5	2.3	5.2
N2	196	< 1.3	1.6	4.1	4.0	66.5	29	18	7.5	1.7	5.9
S3	28	1.8	12.7	30.2	40.7	80.9	75	52	8.1	1.7	6.5
S2	20	19.1	32.4	37.4	44.5	63.8	60	40	7.4	1.2	6.3
S1	17	16.4	31.8	41.1	49.8	58.9	61	39	7.1	0.7	6.4

boundary layer dynamics (Brutsaert, 1982). The second difference is the use of net shortwave radiation in our model, as opposed to net radiation in the Priestley–Taylor equation, the latter of which is not an independent forcing, as it depends on the surface temperature. Using shortwave radiation has important practical advantages because R_{sn} can be obtained from global radiation measurements and albedo information. Further topographic influences on incoming radiation can be directly computed by topographic radiation correction methods (Šúri and Hofierka, 2004). This also circumvents the necessity of collecting site-specific radiation data, which can be challenging above high vegetation like forests.

The derivation of E_{opt} is based on a representation of an atmosphere, which is in equilibrium with the underlying surface and the surface–atmosphere exchange of heat, moisture, and momentum is driven by locally absorbed solar radiation (Kleidon and Renner, 2013b). This implies that meteorological variables such as wind speed or vapor pressure deficit, which are commonly used in Penman formulations, cannot change independently, but are rather constrained by land–atmosphere interaction. In order to test if these assumptions are generally met, we tested for the effect of VPD on transpiration in our data set. The results showed that VPD and wind speed did not add consistently to the explained variance. In addition the correlation of the transpiration estimates to the FAO Penman–Monteith grass reference evaporation is on average slightly higher but follows the same patters across sites. Hence, the two additional input variables only slightly increased the predictability of atmospheric demand, which is consistent with the residual regression analysis above. Both of these results indicate that E_{opt}, which is a function of absorbed solar radiation and surface temperature only, captures the dominant drivers of daily sap velocity variations within this temperate forest without requiring further data on wind speed and VPD.

The dominance of absorbed solar radiation in explaining latent and sensible heat fluxes was also found by Best et al. (2015), who showed that simple linear regression models with solar radiation and temperature as input and thus without any information on wetness conditions had a similar performance than commonly used land-surface models at 20 diverse tower flux sites. However, an improved reproduction of sensible and latent heat fluxes was obtained by Best et al. (2015) when relative humidity was included into their regression model. It is likely that our potential evaporation estimate E_{opt} represents evaporation in a wet environment, whereas in drier environments, air humidity may become increasingly important. Interestingly, in contrast to our approach of eliminating air humidity while considering air temperature as an independent variable, Aminzadeh et al. (2016) estimated wet environment evaporation potential as a function of radiation and air humidity, while eliminating air temperature. Clearly, more investigations, encompassing larger and more diverse data sets, are needed to better understand general patterns of atmospheric control on transpiration. For example, it may well be that air humidity carries information about soil moisture and hence adds to explanatory power of transpiration models in water limited environments, while radiation represents the main control in energy-limited environments. In how far wind, air humidity, and temperature are affected by the land–atmosphere coupling and can hence be treated as internal variables, likely also depends on the scale of interest.

The derivation of E_{opt} shown in Sect. 2 was based on a range of simplifying assumptions to focus on the emerging maximum power limit in a coupled land–atmosphere system (Kleidon and Renner, 2013b; Kleidon et al., 2014). Most important features, which need to be addressed in future work are (i) changes in heat storage, (ii) a revision of the simplified scheme for longwave radiative exchange, and (iii) horizontal circulation patterns by large scale and mesoscale circulation. By assuming a steady state, heat storage effects were neglected in the derivation of the turbulent fluxes J_{opt}. However, heat storage in the surface–atmosphere system is an important mechanism to balance the seasonally and diurnally varying input of solar radiation. A comparison of $J_{opt} = R_{sn} / 2$ with the empirical net radiation estimate of Allen et al. (1998), which was used to estimate E_{PM} showed that J_{opt} is consistently lower during summer. Thus, to improve daily estimates based on thermodynamic limits, it is recommended to consider seasonal heat storage effects in the derivation. Heat storage is even more relevant at the diurnal timescale, which reduces the applicability of E_{opt} as given in Eq. (5). The relevance of considering heat storage is illustrated in Fig. 7, which shows a distinct hysteresis of both sap flow and air temperature with respect to global radiation on a sunny day. Both sap flow and temperature linearly increase with shortwave radiation in the morning hours, but remain high after midday until sunset. Thus, for the same amount of received global radiation there is a distinct difference between morning and afternoon transpiration. While moisture storage in soil and plants also affect the hysteresis, the dominant magnitude of the hysteresis is probably due to the lag of temperature to radiation leading to a VPD-radiation lag (Zhang et al., 2014). Hence, for predictions at the diurnal timescale, heat storage effects as well as boundary layer dynamics have to be accounted for.

Another limitation of the approach is the simple linearized scheme for longwave heat exchange. The description of radiative exchange affects the maximum power limit, because longwave radiation "competes" with the convective fluxes to cool the surface (Kleidon et al., 2015). Although the parameter describing the net longwave radiative exchange drops out in the maximization for the convective fluxes, the radiation scheme affects the partitioning of radiative and convective fluxes at the surface. Therefore, a more detailed description of radiation transfer processes will increase the ability to predict surface energy partitioning.

The third limitation is concerned with the spatial representativity of E_{opt}, but also other potential evaporation formulations and the role of horizontal heat and mass exchange. While the land surface is generally quite heterogeneous, the lower atmosphere is mostly well mixed (Claussen, 1991). By using the topographically corrected incoming radiation we effectively treated each site as an independent surface–atmosphere column. We thus neglected any lateral fluxes that could alter the estimated potential evaporation. The topographic correction increased the potential evaporation at the south-facing and decreased estimates for the north-facing sites. To evaluate this effect with respect to the transpiration estimates, we used the global radiation measured at the nearby grassland site as forcing for all forest sites. The correlation to the transpiration estimates was unaffected, but the topographic correction slightly amplified the aspect-related differences of the response to E_{opt} between sites (Fig. S1 in the Supplement). For a better understanding of such microclimatic effects spatial observations of the canopy surface energy balance are required to test more detailed models, which include horizontal exchange processes. Progress in this respect will be quite important for the parameterization of subgrid processes in numerical land-surface models.

4.2 Topographical and stand composition effects

Our results demonstrate a strong influence of daily variations in atmospheric demand on both transpiration estimates. This strong correlation at the daily timescale may allow us to separate the timescales of atmospheric demand and plant water

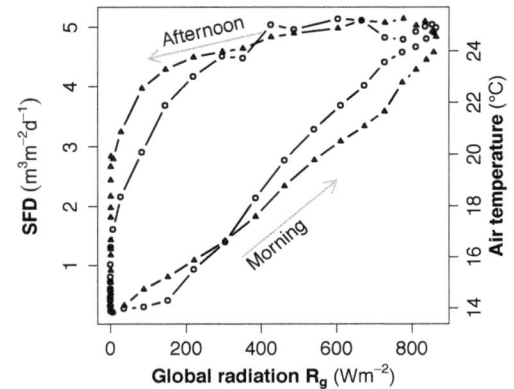

Figure 7. Hysteresis of sap flux density (left y axis) and air temperature (right y axis, red) plotted as a function of global radiation at the field site G1. Data are half-hourly values over one sunny summer day (7 July 2013). Sap flow data are taken from a beech (tree 1) at site S1.

limitations, which may become relevant at timescales larger than 1 day. Differences in the slope of the relationship to E_{opt} can then be compared across sites and be used to identify potential permanent site controls on transpiration.

Of the many factors that influence transpiration, stand composition and topography were the most important invariant controls during our measurement campaign. The measurement transect was placed on a valley cross section to primarily reveal the influence of hillslope aspect. Our results indicate only significant effects of aspect and hillslope angle for the sensitivity of SFD to E_{opt}, but no effects for the sensitivity of the upscaled transpiration estimate E_{sap}. E_{RWU} shows a weak relation to these topographic factors, which was, however, not significant because of the larger uncertainty of the soil moisture-based estimate.

The biometric measures of stand composition, however, also co-vary with aspect and hillslope inclination. Here we found a strongly negative linear relationship of b_{SFD} and the maximum sap flux densities to the site-average sapwood area per tree (Fig. 6). Such a decline in b_{SFD} and maximum SFD with sapwood area was found to be a universal tree hydraulic mechanism observed across different tree species (Andrade et al., 1998; Meinzer et al., 2001). Interestingly, Andrade et al. (1998) found that maximum sap velocity nonlinearly declined with sapwood area, with strongest effects at very small trees or branches with $A_s < 100\,\text{cm}^2$. Considering such an increase in b_{SFD} for the understorey vegetation at the north-facings sites in our study would increase their relevance in estimating stand transpiration.

In addition to the tree size differences, our sites show marked differences in canopy structure. The open canopy structure may explain the larger b_{SFD} of the dominant trees sampled at the north-facing sites, because tree growth could be enhanced by gaps in the canopy structure. For example Schweingruber et al. (2006) showed that annual tree ring width of beech trees increase quickly after neighboring trees have been cut. At the south-facing sites average tree size is high but there is a low number of stems. Here individual tree growth is limited by the neighboring trees with similar demands for light, water, nutrients, and space. Self-thinning or thinning by forest management has resulted in a more even-sized stand structure with old and large trees. Thus, the closed canopy structure, which is composed of well-established trees, may have lead to a more conservative water use per tree, which is reflected by low b_{SFD} and low maximum sap flux densities. These arguments are consistent with results of studies that show higher sap flux densities of thinned forest stands as compared to control stands (Morikawa et al., 1986; Bréda et al., 1995; Nahm et al., 2006).

The upscaling procedure integrates both the tree distribution within the stand and the sap flux density observations to yield the stand transpiration estimate E_{sap}. Apparently this has led to a compensation of the vastly different stand structure and sap flux densities across sites. Due to this tradeoff between SFD and the sapwood area similar transpiration rates are achieved at the hillslope sites. Therefore, forest transpiration may indeed be regarded as a conservative hydrological process (Roberts, 1983) even in this very complex terrain. Only for the valley site S3 significantly larger E_{sap} and b_{sap} was estimated. This site has comparably high maximal SFD and the largest total sapwood area, even though A_s / n_{tree} is lower than observed at the upper south-facing sites. The most likely reason for the comparably dense, but tall forest is the vicinity to the nearby stream, which reduces potential soil water limitations. A similar upslope–downslope effect was recently also established by Kume et al. (2015) for a mountainous stand of Japanese cypress with the lower site having taller trees with higher SFD but same tree density.

4.3 Response to soil moisture

Due to low rainfall amounts, soil moisture decreased from July to September at all sites and all depths. Thus the soil moisture reduction may have limited transpiration. We captured this effect of soil moisture limitation of transpiration through the regression of the residuals of the transpiration to E_{opt} relationships to site-average soil moisture conditions. Results showed that soil moisture explained a significant part of the residuals of E_{sap} at the south-facing sites (Table 2), which implies that site-scale transpiration was also water limited during the dry period in August. At the north-facing sites water limitation at the site scale was only apparent during a shorter period in September. However, we found significant residual correlation to soil moisture at three of four sampled small beech trees. This suggests that small trees are more susceptible to water shortage.

We argue that soil water limitation effects on transpiration could be topographically enhanced. First, by an aspect that affects the amount of received solar radiation and thus the atmospheric demand for water. A higher atmospheric demand increases evaporation from intercepted water and from the soil, which reduces the amount of precipitation entering the soil. Second, the hillslope inclination could have enhanced lateral runoff at the steeper south-facing sites, which reduces the soil water holding capacity (Bronstert and Plate, 1997). These topographical factors of soil moisture availability are also apparent in the tree species composition of our sites. While the north-facing sites are predominantly composed of beech and a few spruce trees, the south-facing sites have about 10 % oak trees, which are known to cope well at more dry sites (Zapater et al., 2011).

4.4 Limitations of transpiration estimates

4.4.1 Limitations of sap flow observations

Generally, sap flow observations are not limited by spatial heterogeneity and complex terrain, which would limit the applicability of micrometeorological measurements (Wilson et al., 2001). Installation is relatively simple and sensors are inexpensive. Despite these advantages, we experienced the following limitations:

(a) Deriving water fluxes requires extrapolation from the point measurement at some specific place within the stem to the entire tree. However, sapwood conductivity can have radial and circumferential differences and species-specific properties (Wullschleger and King, 2000; Saveyn et al., 2008). This can easily bias sap flow estimates (Köstner et al., 1998; Shinohara et al., 2013; Vandegehuchte and Steppe, 2013). An indication of this problem is that we found different b_{SFD} at the same tree installed the year before. Comparison with measurements taken at the same trees from the previous year showed differences in b_{SFD} ranging between -0.39 and 0.23 with an average of 0.04 $[m^3 m^{-2} d^{-1}] / [mm d^{-1}]$ (estimated for nine trees with more than 30 days of data each year).

(b) There is a sample bias towards larger trees as the method is more difficult with very small trees, which would require a different type of sensor, because the heat ratio method is designed only for lower sap velocities (Marshall, 1958; Burgess et al., 2001). This adds a sampling uncertainty in estimating site transpiration where much understorey exists. This is especially relevant for the north-facing sites, with a median DBH of 4 cm. This means that most trees were not sampled. The sampling rather reflected the trees that contributed most to the stand sapwood area.

(c) The inter-comparison of sap flux density measured in different trees is limited by the fact that xylem characteristics in the estimation of SFD are required (Burgess et al., 2001). Most important is the thermal diffusivity of sapwood, k_w as used in Eq. (A1). This conductivity is a function of wood density and wood water content m_c (Burgess et al., 2001; Vandegehuchte and Steppe, 2012a), both of which vary between species and trees (Gebauer et al., 2008). Xylem water content m_c, which in addition influences the apparent sap flux density through affecting the sapwood heat capacity (see Eq. 7), was shown to have seasonal changes in diffuse-porous species (Glavac et al., 1990; Hao et al., 2013). Glavac et al. (1990) found that m_c can reduce to about 25 % in sampled European beech trees during summer. Such a decline would thus reduce the apparent sap velocities. Therefore, it is recommended to use methods that take into account wood moisture content changes (Vandegehuchte and Steppe, 2012b). Here, in the absence of measurements of such wood properties we used the same parameters for all trees.

4.4.2 Limitations of root water uptake estimates from soil moisture variations

The advantage of using temporally highly resolved soil moisture readings is that it allows one to estimate root water uptake without further information on soil properties (Guderle and Hildebrandt, 2015). The accuracy of the method depends on various factors that can influence results:

(a) Data filtering: the method only applies under conditions with negligible soil water movement excluding events of infiltration, drainage, capillary rise, or hydraulic redistribution. These fluxes can have major influences on the observations of soil moisture and comprise the second term of the right-hand side of Eq. (A6). Thus, the estimates depend on the choice of suitable filter criteria. A very strict filter would reduce the number of estimates, whereas soft filter criteria may result in biased E_{RWU} estimates. Hence, seasonal or annual totals cannot be derived from this method alone. We use relatively strict filter criteria for nighttime fluxes of 0.1 Vol %, which is close to the sensor resolution. This filter criteria set the maximum accuracy per soil layer of 200 mm depth to 0.2 mm d^{-1}.

(b) Soil heterogeneities, dominant at the hillslopes, can induce large local variations in soil moisture and may lead to dissimilar / biased E_{RWU} compared to other methods (Wilson et al., 2001). Here, we found that the influence of E_{opt} and soil moisture content on E_{RWU} varied between soil profiles at a specific site (Table S3 and Fig. S5 in the Supplement). Large differences were observed at site N2, which results in a quite uncertain site-average estimate of E_{RWU}. Therefore, it is recommended to install several, representative measurement profiles when such a soil water budget method is used for transpiration estimation in heterogeneous terrain (Schwärzel et al., 2009).

(c) Deep root water uptake in response to drying topsoil may cause root water uptake below the deepest measurement depth in forest sites as observed e.g. by Teskey and Sheriff (1996); Wilson et al. (2001). Observations from the deepest sensor profile confirm root uptake at 60 cm depth (Fig. S4 in the Supplement), which also persists during the dry period. However, overall the contribution of deep root water uptake is assumed to be small, given the low observed diurnal variations.

4.4.3 Upscaling sap flow to site-scale transpiration

The estimates of site-scale transpiration based on up-scaled sap flow measurements were of similar magnitudes and correlated well with estimates derived from soil moisture variations. The seasonal estimates by E_{sap} are about 50 mm lower than other estimates for beech forests. For example Schipka et al. (2005) found 200–300 mm per year for European beech forests in Germany. Their sites, however, have been located in less steep terrain.

The comparison between E_{RWU} and E_{sap} also revealed striking differences, which could be an indication of the potential shortcomings of both methods, as discussed above. While the south-facing sites are in good agreement, E_{sap} at the north-facing sites seems to be quite low. First, this is due to the low basal area at the north-facing site. One reason could be that the assumed sapwood area of the few tall trees is much larger than that reported in the literature. Another possible reason could be that small trees (< 8 cm DBH), which were not sampled, had a significant contribution to stand transpiration. Also E_{RWU} might overestimate actual transpiration because soil evaporation would equally contribute to the diurnal signal in soil moisture. For example Bréda et al. (1993) also found consistently larger estimates of a soil water balance method than stand transpiration estimates, which was attributed to soil evaporation by the authors.

5 Summary and conclusions

We aimed to infer the dominant temporal and spatial controls on forest transpiration along a steep valley cross section through ecohydrological measurements of sap flow and soil moisture and their relation to atmospheric evaporative demand. The estimation of transpiration in space and time for forests in complex terrain is a challenge in its own right. Obtaining transpiration is only possible through indirect observations, whereby each method has its own limitations. Therefore, we used two independent observations to obtain site-scale estimates of transpiration along the hillslope transect. To estimate atmospheric demand, a formulation similar to the well-known Priestley–Taylor equilibrium evaporation concept was employed. The formulation is based on a simplified energy balance representation of the surface–atmospheric system and hypothesizing that convection operates at its upper thermodynamic limit. The formulation does not require empirical parameters and only requires data on the absorbed solar radiation and temperature. We find that at the daily timescale this approach explains most of variability in both transpiration estimates at the site and tree scale. This suggests that atmospheric demand is the dominant control on daily transpiration rates in this temperate forest. Although the well-established FAO Penman–Monteith reference evaporation yields slightly higher correlation and 20–30 % higher values, it requires additional data of net radiation, VPD and wind speed. Thereby both, VPD and wind speed did not add consistently to the explained variance and are also difficult to obtain above forests. While our results demonstrate that thermodynamic limits provide a first-order estimate for potential evaporation, we have to stress that the derivation is based on the simplest possible energy balance representation. Further refinements will probably improve the predictability of surface exchange fluxes.

Despite the prevailing topographic contrasts between the north-facing and the south-facing measurement sites, we find that up-scaled stand transpiration yields rather similar seasonal totals as well as a similar average response to atmospheric demand. This similarity is achieved through a compensation of the low sapwood area with high sap flux densities at the north facing sites, while at the south-facing sites a high sapwood area was accompanied with low sap flux densities. It appears that individual and stand average sap flux densities can vary strongly in heterogeneous terrain in order to compensate for tree size and stand structural differences through tree hydraulic mechanisms. The importance of these stand structural differences on stand transpiration thus masks the potential effects of topographical factors such as aspect and hillslope angle, which are cross correlated. However, during dry periods we find that topographic factors can enhance the response of transpiration to soil water limitation.

Despite unavoidable limitations in estimating stand transpiration and potential evaporation in complex terrain, we find that relating the employed ecohydrological observations to a thermodynamically constrained estimate of atmospheric demand enables important insights in the temporal drivers of transpiration and how they vary at the hillslope scale. First our results highlight the dominance of absorbed solar radiation as the main and independent driver of land–atmosphere exchange. Second, our results suggest an intriguing interplay of tree hydraulics and stand composition, which seemingly leads to transpiration rates close to its physical limits. We conclude that this approach should help us to better understand surface–atmosphere coupling in relation to thermodynamic constraints and how vegetation adapts to these.

Appendix A: Detailed description of transpiration estimates

A1　Measurement and correction of heat pulse velocity

We used sap flow sensors based on the heat ratio method (Burgess et al., 2001). Thereby three needles with equal distance are vertically inserted into the tree. The upper and lower needles measure the change in temperature after a short heat pulse was emitted by a heating element in the middle needle. The heat pulse velocity v_h is proportional to the logarithmic ratio of the temperature differences measured before and at t_1 usually 60 s after the heat pulse at the lower (ΔT_{dn}) and upper (ΔT_{up}) needles (Marshall, 1958; Burgess et al., 2001):

$$v_h = \frac{4k_w t_1 \log \frac{\Delta T_{up}}{\Delta T_{dn}} - x_2^2 + x_1^2}{2t_1(x_1 - x_2)}, \tag{A1}$$

whereby k_w is the thermal diffusivity of the xylem, and x_1 and x_2 are the vertical distances to the heater. The sensors have the standard distance of 0.6 cm. This distance can easily be slightly shifted during the installation into living trees, which causes a systematic bias. This sensor alignment bias can be corrected when sap flow is zero (Burgess et al., 2001). Setting $v_h = 0$ and rearranging Eq. (A1), we can estimate the distance of the upper needle x_1 while assuming $x_2 = 0.6$:

$$x_1 = \sqrt{x_2^2 - 4k_w t_1 \log \frac{\Delta T_{up}}{\Delta T_{dn}}}. \tag{A2}$$

We prefer Eq. A2 over the published correction in Burgess et al. (2001) because it allows to correct for both, positive and negative nighttime biases in heat velocities. We estimated the corrected distance x_1 by assuming zero flow during nighttime between 01:00 and 04:00 LT and the median of x_1 for the whole installation period. The installation of the sensor needles injures the surrounding xylem vessels and reduces the actual sap flow around the sensor. We applied a polynomial wounding correction $v_c = b\,V_h + c\,V_h^2 + d\,V_h^3$ with wounding correction parameters $b = 1.8558$, $c = -0.0018$, and $d = 0.003$, applicable for a sensor spacing of 0.6 cm and a drilling hole size of 2 mm, which is tabulated in Burgess et al. (2001).

A2　Upscaling of sap flow

The needles measure heat-pulse velocities at depths of 5, 18, and 30 mm within the stem. Following the user's manual we assigned for each sensor depth a representative radius of 15, 25, and 40 mm below the cambium radius r_x, which is obtained by assuming that the cambium takes 1 % of the total radius, thus $r_x = 0.99 \cdot \text{DBH} / 2$.

Because tall trees may have a wider active xylem depth than measured by the inserted sap flow needles (Gebauer et al., 2008), we derived three different estimates of the representative sapwood area of the innermost sensor, which provides the range and a best-guess estimate. A minimal estimate is obtained by assuming a representative annulus of 15 mm depth for the sensor and zero flow in the inner sapwood, which is calculated by $A_{S3,\text{min}}$. Assuming that sap flux density remains constant at the innermost sensor level throughout the inner sapwood area provides a maximal estimate of sap flux. Most realistic is the assumption of a linear decline reaching no flow at the estimated heartwood radius $A_{S3,\text{lin}}$ similar to Pausch et al. (2000):

$$A_{S3,\text{min}} = \pi \left((r_x - 25\,\text{mm})^2 - (r_x - 40\,\text{mm})^2 \right), \quad (A3)$$

$$A_{S3,\text{lin}} = \pi \left((r_x - 25\,\text{mm})^2 - \frac{1}{3} \left((r_x - 25\,\text{mm})^2 \right.\right.$$
$$\left.\left. + (r_x - 25\,\text{mm})\, r_h + r_h^2 \right) \right), \quad (A4)$$

$$A_{S3,\text{max}} = \pi \left((r_x - 25\,\text{mm})^2 - (r_x - r_h)^2 \right). \quad (A5)$$

Stem sapwood area was computed using published allometric power-law relationships of the form $A_s = a\,\text{DBH}^b$. For beech and alder we used the relationship published in Gebauer et al. (2008), for oak we used Schmidt (2007) and Alsheimer et al. (1998) for spruce. Then the radius of the heartwood r_h is obtained from sapwood area estimates. For small trees with radius smaller than the representative annulus depth we set the area to 0.

A3 Root water uptake estimation

The soil water continuity equation at a point in the soil may be written as Cuenca et al. (1997):

$$\frac{\partial \theta}{\partial t} = -\frac{\partial q_i}{\partial x_i} + S, \quad (A6)$$

where θ is the soil moisture content, q summarizes any soil water fluxes over an orthogonal coordinate system x_i with horizontal (x, y) and vertical directions (z). The sink term S is the root water uptake, which when integrated over the soil volume V_s yields the transpiration flux E_{RWU} (Cuenca et al., 1997):

$$E_{\text{RWU}} = \int_{V_s} S\, dV_s = \int_{V_s} \frac{\partial \theta}{\partial t} dV_s + \int_{V_s} \frac{\partial q_i}{\partial x_i} dV_s. \quad (A7)$$

Hence, to estimate E_{RWU} we need to resolve the two terms in Eq. (A7). The first term is the temporal evolution of soil moisture, which is in principle measured by the soil moisture sensors. The second term describes soil water fluxes within the soil such as downward fluxes during an infiltration event. The soil water fluxes q_i themselves depend on soil moisture, through its role in determining soil water potentials and unsaturated hydraulic conductivity. Soils along hillslopes have large spatial heterogeneity of their hydraulic properties and

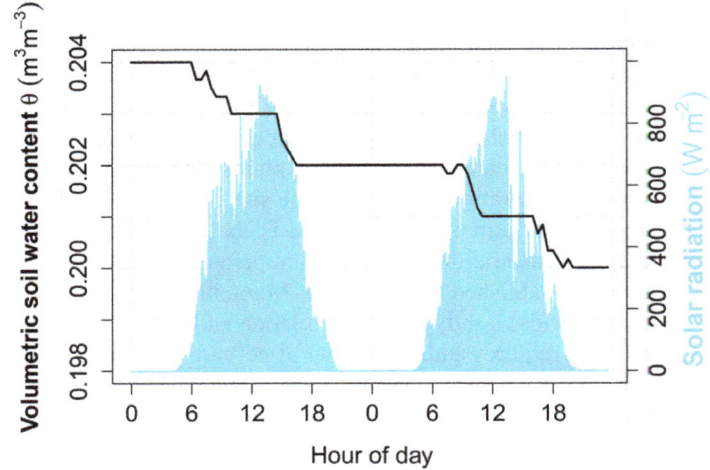

Figure A1. Observed diurnal decline in soil moisture over two sunny days in summer 2013 to illustrate the approach to estimate daily root water uptake from diurnal soil moisture variations. E_{RWU} is estimated at sensor level per day as a cumulative sum of differences from sunrise to sunset, in the absence of any nighttime variations or daytime increases in soil moisture.

pose many influences on the second term of Eq. (A7), which makes continuous estimation of root water uptake difficult.

There are, however, periods and locations where soil water fluxes q_i, such as drainage, capillary rise, etc., are of minor importance. Especially during dry conditions the reduction of soil moisture is dominated by root water uptake and soil evaporation during daytime, as illustrated in Fig. A1. Under these radiation-driven conditions, we observe rather constant nighttime moisture levels, which thus indicates that soil water fluxes are not active and nocturnal root–soil exchange is negligible. Thus, nighttime soil moisture dynamics are a practical filter to exclude days with dominant drainage or capillary rise fluxes. To estimate daily root water uptake we use soil moisture observations on a half-hourly basis. We first quantify the daytime change in soil moisture per sensor $\Delta\theta_{\text{d,s}}$ by cumulative sum of differences of soil moisture (at 30 min intervals) from astronomical sunrise to sunset:

$$\Delta\theta_{\text{d,s}} = \sum_{t=t_{\text{sunrise}}}^{t=t_{\text{sunset}}} (\theta_{t+1} - \theta_t). \quad (A8)$$

Assuming that the sensor observation is representative for the respective soil depth Δz of the sensor, we obtain a flux estimate $E_{\text{RWU},d,z} = -\Delta\theta_{\text{d,s}}\Delta z$ per soil layer. A constant soil depth of $\Delta z = 200$ mm per sensor level was assumed for all soil profiles. We then summed up the contributions

$E_{\text{RWU},d,z}$ of each soil layer to obtain a daily root water uptake $E_{\text{RWU},d}$ per profile. For most profiles this allowed us to estimate $E_{\text{RWU},d}$ down to a depth of 600 mm, and for three profiles down to 800 mm (see Table S1 for an overview of sensor placement). During sensor installation thick roots were mostly found until a depth of ≈ 300 mm with sporadic fine roots up to 800 mm depth.

The data were filtered for (i) precipitation (daily sum < 0.1 mm and rainfall of previous day < 1 mm), (ii) only negative daytime soil moisture changes $\Delta\theta_{\text{d,s}}$, and (iii) absolute cumulative nighttime changes in soil moisture $\left|\Delta\theta_{\text{n,s}}\right| < 0.1$ Vol %. Approximately 30 % of the actual data were retained, with details listed in Table S2 in the Supplement. Generally, diurnal variations of soil moisture are large in the upper soil depths compared to deeper layers. Thus, for estimating the total profile root water uptake, the upper sensors are very important. Unfortunately, we had to face sensor failure of top soil sensors at site N1, profile 2 and site S1 profiles 2 and 3; see Table S2 in the Supplement for an overview of available data. At these soil profiles we used the root water uptake estimates from the 2nd sensor level to fill the missing data. Two profiles, one at G1 and another at N2 were completely disregarded for site averaging due to sensor failures. The site-average was obtained from at least two profile-based E_{RWU} estimates. At the grass site this condition was relaxed because there were only a few days where E_{RWU} could be estimated from both profiles. At these few days the estimates of the two profiles were comparable.

Acknowledgements. This research contributes to the "Catchments As Organized Systems (CAOS)" research group (FOR 1598) funded by the German Science Foundation (DFG). AH was partially funded by grant SFB 1076 by the German Science Foundation. We thank Conrad Jackisch (KIT) for comments on an earlier draft of the paper. We thank all people involved in the field work. In particular the technical backbone Britta Kattenstroth (GFZ Potsdam), Tatiana Feskova (UFZ – Leipzig), Laurent Pfister and François Iffly (LIST, Luxembourg) and the landowners for giving access to their land. We acknowledge the encouraging and constructive comments of two reviewers, who helped to improve the manuscript.

Edited by: T. Bogaard

References

Allen, R., Pereira, L., Raes, D., and Smith, M.: Crop evapotranspiration-Guidelines for computing crop water requirements-FAO Irrigation and drainage paper 56, FAO, Rome, 300, 6541, 1998.

Alsheimer, M., Kästner, B., Falge, E., and Tenhunen, J. D.: Temporal and spatial variation in transpiration of Norway spruce stands within a forested catchment of the Fichtelgebirge, Germany, Ann. Sci. Forest Res., vol. 55, 103–123, EDP Sciences, 1998.

Aminzadeh, M., Roderick, M. L., and Or, D.: A generalized complementary relationship between actual and potential evaporation defined by a reference surface temperature, Water Resour. Res., 52, 385–406, doi:10.1002/2015WR017969, 2016.

Andrade, J. L., Meinzer, F. C., Goldstein, G., Holbrook, N. M., Cavelier, J., Jackson, P., and Silvera, K.: Regulation of water flux through trunks, branches, and leaves in trees of a lowland tropical forest, Oecologia, 115, 463–471, doi:10.1007/s004420050542, 1998.

Bachmair, S. and Weiler, M.: New Dimensions of Hillslope Hydrology, in: Forest Hydrology and Biogeochemistry, edited by: Levia, D. F., Carlyle-Moses, D., and Tanaka, T., vol. 216, 455–481, Springer Netherlands, Dordrecht, 2011.

Baumgartner, A.: Gelände und Sonnenstrahlung als Standortfaktor am Gr. Falkenstein (Bayerischer Wald), Forstwissenschaftliches Centralblatt, 79, 286–297, doi:10.1007/BF01815008, 1960.

Best, M. J., Abramowitz, G., Johnson, H. R., Pitman, A. J., Balsamo, G., Boone, A., Cuntz, M., Decharme, B., Dirmeyer, P. A., Dong, J., Ek, M., Guo, Z., Haverd, V., van den Hurk, B. J. J., Nearing, G. S., Pak, B., Peters-Lidard, C., Santanello, J. A., Stevens, L., and Vuichard, N.: The Plumbing of Land Surface Models: Benchmarking Model Performance, J. Hydrometeorol., 16, 1425–1442, doi:10.1175/JHM-D-14-0158.1, 2015.

Bohren, C. F. and Albrecht, B. A.: Atmospheric thermodynamics, Atmospheric thermodynamics, New York, Oxford, Oxford University Press, ISBN: 0195099044, 1, 1998.

Bonan, G. B.: Forests and Climate Change: Forcings, Feedbacks, and the Climate Benefits of Forests, Science, 320, 1444–1449, doi:10.1126/science.1155121, 2008.

Breña Naranjo, J. A., Weiler, M., and Stahl, K.: Sensitivity of a data-driven soil water balance model to estimate summer evapotranspiration along a forest chronosequence, Hydrol. Earth Syst. Sci., 15, 3461–3473, doi:10.5194/hess-15-3461-2011, 2011.

Bronstert, A. and Plate, E. J.: Modelling of runoff generation and soil moisture dynamics for hillslopes and micro-catchments, J. Hydrol., 198, 177–195, doi:10.1016/S0022-1694(96)03306-9, 1997.

Brutsaert, W.: Evaporation into the Atmosphere, Springer Netherlands, Dordrecht, 1982.

Bréda, N., Cochard, H., Dreyer, E., and Granier, A.: Water transfer in a mature oak stand (Quercuspetraea): seasonal evolution and effects of a severe drought, Canad. J. For. Res., 23, 1136–1143, doi:10.1139/x93-144, 1993.

Bréda, N., Granier, A., and Aussenac, G.: Effects of thinning on soil and tree water relations, transpiration and growth in an oak forest (Quercus petraea (Matt.) Liebl.), Tree Physiol., 15, 295–306, doi:10.1093/treephys/15.5.295, 1995.

Burgess, S. S. O., Adams, M. A., Turner, N. C., Beverly, C. R., Ong, C. K., Khan, A. A. H., and Bleby, T. M.: An improved heat pulse method to measure low and reverse rates of sap flow in woody plants, Tree Physiol., 21, 589–598, doi:10.1093/treephys/21.9.589, 2001.

Calder, I. R.: Water use by forests, limits and controls, Tree Physiol., 18, 625–631, doi:10.1093/treephys/18.8-9.625, 1998.

Claussen, M.: Estimation of areally-averaged surface fluxes, Bound.-Lay. Meteorol., 54, 387–410, doi:10.1007/BF00118868, 1991.

Cuenca, R. H., Stangel, D. E., and Kelly, S. F.: Soil water balance in a boreal forest, J. Geophys. Res.-Atmos., 102, 29355–29365, doi:10.1029/97JD02312, 1997.

de Bruin, H. A. R., Trigo, I. F., Bosveld, F. C., and Meirink, J. F.: A Thermodynamically Based Model for Actual Evapotranspiration of an Extensive Grass Field Close to FAO Reference, Suitable for Remote Sensing Application, J. Hydrometeorol., 17, 1373–1382, doi:10.1175/JHM-D-15-0006.1, 2016.

Dowle, M., Short, T., Lianoglou, S., and Srinivasan, A.: data. table: Extension of data. frame. R package version 1.9.4, 2014.

Famiglietti, J. S., Rudnicki, J. W., and Rodell, M.: Variability in surface moisture content along a hillslope transect: Rattlesnake Hill, Texas, J. Hydrol., 210, 259–281, doi:10.1016/S0022-1694(98)00187-5, 1998.

Farquhar, G. D., Caemmerer, S. v., and Berry, J. A.: A biochemical model of photosynthetic CO_2 assimilation in leaves of C3 species, Planta, 149, 78–90, doi:10.1007/BF00386231, 1980.

Federer, C. A.: Forest Transpiration Greatly Speeds Streamflow Recession, Water Resour. Res., 9, 1599–1604, 1973.

Gebauer, T., Horna, V., and Leuschner, C.: Variability in radial sap flux density patterns and sapwood area among seven co-occurring temperate broad-leaved tree species, Tree Physiol., 28, 1821–1830, doi:10.1093/treephys/28.12.1821, 2008.

Glavac, V., Koenies, H., and Ebben, U.: Auswirkung sommerlicher Trockenheit auf die Splintholz-Wassergehalte im Stammkörper der Buche (Fagus sylvatica L.), Holz als Roh- und Werkstoff, 48, 437–441, doi:10.1007/BF02627628, 1990.

Guderle, M. and Hildebrandt, A.: Using measured soil water contents to estimate evapotranspiration and root water uptake profiles – a comparative study, Hydrol. Earth Syst. Sci., 19, 409–425, doi:10.5194/hess-19-409-2015, 2015.

Haas, E., Klatt, S., Frählich, A., Kraft, P., Werner, C., Kiese, R., Grote, R., Breuer, L., and Butterbach-Bahl, K.: LandscapeD-NDC: a process model for simulation of biosphere-atmosphere-hydrosphere exchange processes at site and regional scale, Landsc. Ecol., 28, 615–636, doi:10.1007/s10980-012-9772-x, 2013.

Hao, G.-Y., Wheeler, J. K., Holbrook, N. M., and Goldstein, G.: Investigating xylem embolism formation, refilling and water storage in tree trunks using frequency domain reflectometry, J. Exp. Bot., 64, 2321–2332, doi:10.1093/jxb/ert090, 2013.

Holst, J., Grote, R., Offermann, C., Ferrio, J. P., Gessler, A., Mayer, H., and Rennenberg, H.: Water fluxes within beech stands in complex terrain, Int. J. Biometeorol., 54, 23–36, doi:10.1007/s00484-009-0248-x, 2010.

Holst, T., Rost, J., and Mayer, H.: Net radiation balance for two forested slopes on opposite sides of a valley, Int. J. Biometeorol., 49, 275–284, doi:10.1007/s00484-004-0251-1, 2005.

Ivanov, V. Y., Bras, R. L., and Vivoni, E. R.: Vegetation-hydrology dynamics in complex terrain of semiarid areas: 2. Energy-water controls of vegetation spatiotemporal dynamics and topographic niches of favorability, Water Resour. Res., 44, W03430, doi:10.1029/2006WR005595, 2008.

Jasechko, S., Sharp, Z. D., Gibson, J. J., Birks, S. J., Yi, Y., and Fawcett, P. J.: Terrestrial water fluxes dominated by transpiration, Nature, 496, 347–350, doi:10.1038/nature11983, 2013.

Juilleret, J., Iffly, J. F., Pfister, L., and Hissler, C.: Remarkable Pleistocene periglacial slope deposits in Luxembourg (Oesling): pedological implications and geosite potential, Bull. Soc. Nat. Luxembour., 112, 125–130, 2011.

Kleidon, A. and Renner, M.: A simple explanation for the sensitivity of the hydrologic cycle to surface temperature and solar radiation and its implications for global climate change, Earth Syst. Dynam., 4, 455–465, doi:10.5194/esd-4-455-2013, 2013a.

Kleidon, A. and Renner, M.: Thermodynamic limits of hydrologic cycling within the Earth system: concepts, estimates and implications, Hydrol. Earth Syst. Sci., 17, 2873–2892, doi:10.5194/hess-17-2873-2013, 2013b.

Kleidon, A., Renner, M., and Porada, P.: Estimates of the climatological land surface energy and water balance derived from maximum convective power, Hydrol. Earth Syst. Sci., 18, 2201–2218, doi:10.5194/hess-18-2201-2014, 2014.

Kleidon, A., Kravitz, B., and Renner, M.: The hydrological sensitivity to global warming and solar geoengineering derived from thermodynamic constraints, Geophys. Res. Lett., 42, 138–144, doi:10.1002/2014GL062589, 2015.

Kondratyev, K. Y. and Fedorova, M. P.: Radiation regime of inclined surfaces, Solar Energy, 1, 36–61, 1977.

Kume, T., Tsuruta, K., Komatsu, H., Shinohara, Y., Katayama, A., Ide, J., and Otsuki, K.: Differences in sap flux based stand transpiration between upper and lower slope positions in a Japanese cypress plantation watershed, Ecohydrology, doi:10.1002/eco.1709, 2015.

Köstner, B., Granier, A., and Cermák, J.: Sapflow measurements in forest stands: methods and uncertainties, Ann. For. Sci., 55, 13–27, doi:10.1051/forest:19980102, 1998.

Lee, R. and Baumgartner, A.: The Topography and Insolation Climate of a Mountainous Forest Area, For. Sci., 12, 258–267, 1966.

Lüttschwager, D. and Remus, R.: Radial distribution of sap flux density in trunks of a mature beech stand, Ann. For. Sci., 64, 431–438, doi:10.1051/forest:2007020, 2007.

Makkink, G.: Testing the Penman formula by means of lysimeters, J. Inst. Water Eng., 11, 277–288, 1957.

Marshall, D. C.: Measurement of Sap Flow in Conifers by Heat Transport. 1, Plant Physiol., 33, 385–396, 1958.

Meinzer, F. C., Goldstein, G., and Andrade, J. L.: Regulation of water flux through tropical forest canopy trees: Do universal rules apply?, Tree Physiol., 21, 19–26, doi:10.1093/treephys/21.1.19, 2001.

Miller, G. R., Chen, X., Rubin, Y., Ma, S., and Baldocchi, D. D.: Groundwater uptake by woody vegetation in a semiarid oak savanna: GROUNDWATER UPTAKE IN AN OAK SAVANNA, Water Resour. Res., 46, W10503, doi:10.1029/2009WR008902, 2010.

Monteith, J. L.: Evaporation and environment, Symposia of the Society for Experimental Biology, 19, 205–234, 1965.

Morikawa, Y., Hattori, S., and Kiyono, Y.: Transpiration of a 31-year-old Chamaecyparis obtusa Endl. stand before and after thinning, Tree Physiol., 2, 105–114, doi:10.1093/treephys/2.1-2-3.105, 1986.

Nahm, M., Holst, T., Matzarakis, A., Mayer, H., Rennenberg, H., and Geßler, A.: Soluble N compound profiles and concentrations in European beech (Fagus sylvatica L.) are influenced by local climate and thinning, Eur. J. For. Res., 125, 1–14, doi:10.1007/s10342-005-0103-5, 2006.

Newey, W. K. and West, K. D.: Automatic Lag Selection in Covariance Matrix Estimation, Rev. Econ. Stud., 61, 631–653, 1994.

Oke, T.: Boundary layer climates, Routledge, London and New York, 1987.

Ozawa, H., Ohmura, A., Lorenz, R. D., and Pujol, T.: The second law of thermodynamics and the global climate system: A review of the maximum entropy production principle, Rev. Geophys., 41, 1018, doi:10.1029/2002RG000113, 2003.

Pausch, R. C., Grote, E. E., and Dawson, T. E.: Estimating water use by sugar maple trees: considerations when using heat-pulse methods in trees with deep functional sapwood, Tree Physiol., 20, 217–227, 2000.

Priestley, C. and Taylor, R.: On the assessment of surface heat flux and evaporation using large-scale parameters, Mont. Eeather Rev., 100, 81–92, 1972.

R Core Team: R: A Language and Environment for Statistical Computing, R Foundation for Statistical Computing, Vienna, Austria, http://www.R-project.org/, 2016.

Raupach, M. R.: Combination theory and equilibrium evaporation, Quarterly J. Roy. Meteorol. Soc., 127, 1149–1181, doi:10.1002/qj.49712757402, 2001.

Roberts, J.: Forest transpiration: A conservative hydrological process?, J. Hydrol., 66, 133–141, doi:10.1016/0022-1694(83)90181-6, 1983.

Saveyn, A., Steppe, K., and Lemeur, R.: Spatial variability of xylem sap flow in mature beech (Fagus sylvatica) and its diurnal dynamics in relation to microclimate, Botany, 86, 1440–1448, doi:10.1139/B08-112, 2008.

Schipka, F., Heimann, J., and Leuschner, C.: Regional variation in canopy transpiration of Central European beech forests, Oecologia, 143, 260–270, doi:10.1007/s00442-004-1798-6, 2005.

Schmidt, M.: Canopy transpiration of beech forests in Northern Bavaria - Structure and function in pure and mixed stands with oak at colline and montane sites, Doctoral thesis, Bayreuth, 2007.

Schmidt, W.: Strahlung und Verdunstung an freien Wasserflächen; ein Beitrag zum Wärmehaushalt des Weltmeers und zum Wasserhaushalt der Erde, Ann. Calender Hydrographie und Maritimen Meteorologie, 43, 111–124, 1915.

Schweingruber, F. H., Bärner, A., and Schulze, E. D.: Atlas of Woody Plant Stems: Evolution, Structure, and Environmental Modifications, Springer Science & Business Media, 2006.

Schwinning, S.: The ecohydrology of roots in rocks, Ecohydrology, 3, 238–245, doi:10.1002/eco.134, 2010.

Schwärzel, K., Menzer, A., Clausnitzer, F., Spank, U., Häntzschel, J., Grünwald, T., Kästner, B., Bernhofer, C., and Feger, K.-H.: Soil water content measurements deliver reliable estimates of water fluxes: A comparative study in a beech and a spruce stand in the Tharandt forest (Saxony, Germany), Agr. For. Meteor., 149, 1994–2006, doi:10.1016/j.agrformet.2009.07.006, 2009.

Shinohara, Y., Tsuruta, K., Ogura, A., Noto, F., Komatsu, H., Otsuki, K., and Maruyama, T.: Azimuthal and radial variations in sap flux density and effects on stand-scale transpiration estimates in a Japanese cedar forest, Tree Physiol., 33, tpt029, doi:10.1093/treephys/tpt029, 2013.

Shuttleworth, W.: Evaporation. Handbook of Hydrology, DR Maidment, Ed, McGraw-Hill, New York, NY, USA, 1993.

Stull, R. B.: An introduction to boundary layer meteorology, vol. 13, Kluwer Academic Publishers, Dordrecht, The Netherlands, 1988.

Šúri, M. and Hofierka, J.: A New GIS-based Solar Radiation Model and Its Application to Photovoltaic Assessments, T. GIS, 8, 175–190, doi:10.1111/j.1467-9671.2004.00174.x, 2004.

Teskey, R. O. and Sheriff, D. W.: Water use by Pinus radiata trees in a plantation, Tree Physiol., 16, 273–279, doi:10.1093/treephys/16.1-2.273, 1996.

Tromp-van Meerveld, H. J. and McDonnell, J. J.: On the interrelations between topography, soil depth, soil moisture, transpiration rates and species distribution at the hillslope scale, Adv. Water Resour., 29, 293–310, doi:10.1016/j.advwatres.2005.02.016, 2006.

Vandegehuchte, M. W. and Steppe, K.: Improving sap flux density measurements by correctly determining thermal diffusivity, differentiating between bound and unbound water, Tree Physiol., 32, 930–942, doi:10.1093/treephys/tps034, 2012a.

Vandegehuchte, M. W. and Steppe, K.: Sapflow+: a four-needle heat-pulse sap flow sensor enabling nonempirical sap flux density and water content measurements, New Phytol., 196, 306–317, doi:10.1111/j.1469-8137.2012.04237.x, 2012b.

Vandegehuchte, M. W. and Steppe, K.: Sap-flux density measurement methods: working principles and applicability, Funct. Plant Biol., 40, 213–223, 2013.

Wang, J. and Bras, R. L.: A model of evapotranspiration based on the theory of maximum entropy production, Water Resour. Res., 47, W03521, doi:10.1029/2010WR009392, 2011.

Wang, Y. and Jarvis, P.: Description and validation of an array model - MAESTRO, Agr. For. Meteorol., 51, 257–280, doi:10.1016/0168-1923(90)90112-J, 1990.

West, G. B., Brown, J. H., and Enquist, B. J.: A general model for the structure and allometry of plant vascular systems, Nature, 400, 664–667, doi:10.1038/23251, 1999.

Whitehead, D.: Regulation of stomatal conductance and transpiration in forest canopies, Tree Physiol., 18, 633–644, doi:10.1093/treephys/18.8-9.633, 1998.

Wilson, K. B., Hanson, P. J., Mulholland, P. J., Baldocchi, D. D., and Wullschleger, S. D.: A comparison of methods for determining forest evapotranspiration and its components: sap-flow, soil water budget, eddy covariance and catchment water balance, Agr. For. Meteorol., 106, 153–168, 2001.

Wullschleger, S. D. and King, A. W.: Radial variation in sap velocity as a function of stem diameter and sapwood thickness in yellow-poplar trees, Tree Physiol., 20, 511–518, doi:10.1093/treephys/20.8.511, 2000.

Zapater, M., Hossann, C., Bréda, N., Bréchet, C., Bonal, D., and Granier, A.: Evidence of hydraulic lift in a young beech and oak mixed forest using 18O soil water labelling, Trees, 25, 885–894, doi:10.1007/s00468-011-0563-9, 2011.

Zehe, E., Ehret, U., Pfister, L., Blume, T., Schräder, B., Westhoff, M., Jackisch, C., Schymanski, S. J., Weiler, M., Schulz, K., Allroggen, N., Tronicke, J., van Schaik, L., Dietrich, P., Scherer, U., Eccard, J., Wulfmeyer, V., and Kleidon, A.: HESS Opinions: From response units to functional units: a thermodynamic reinterpretation of the HRU concept to link spatial organization and functioning of intermediate scale catchments, Hydrol. Earth Syst. Sci., 18, 4635–4655, doi:10.5194/hess-18-4635-2014, 2014.

Zeileis, A.: Econometric computing with HC and HAC covariance matrix estimators, J. Stat. Softw., 11, 1–17, 2004.

Zeileis, A. and Hothorn, T.: Diagnostic checking in regression relationships, R News, 2, 7–10, 2002.

Zhang, L., Potter, N., Hickel, K., Zhang, Y., and Shao, Q.: Water balance modeling over variable time scales based on the Budyko framework – Model development and testing, J. Hydrol., 360, 117–131, doi:10.1016/j.jhydrol.2008.07.021, 2008.

Zhang, Q., Manzoni, S., Katul, G., Porporato, A., and Yang, D.: The hysteretic evapotranspiration-Vapor pressure deficit relation: ET-VPD hysteresis, J. Geophys. Res.-Biogeosci., 119, 125–140, doi:10.1002/2013JG002484, 2014.

10

Higher-order statistical moments and a procedure that detects potentially anomalous years as two alternative methods describing alterations in continuous environmental data

I. Arismendi[1], **S. L. Johnson**[2], **and J. B. Dunham**[3]

[1]Department of Fisheries and Wildlife, Oregon State University, Corvallis, Oregon 97331, USA
[2]US Forest Service, Pacific Northwest Research Station, Corvallis, Oregon 97331, USA
[3]US Geological Survey, Forest and Rangeland Ecosystem Science Center, Corvallis, Oregon 97331, USA

Correspondence to: I. Arismendi (ivan.arismendi@oregonstate.edu)

Abstract. Statistics of central tendency and dispersion may not capture relevant or desired characteristics of the distribution of continuous phenomena and, thus, they may not adequately describe temporal patterns of change. Here, we present two methodological approaches that can help to identify temporal changes in environmental regimes. First, we use higher-order statistical moments (skewness and kurtosis) to examine potential changes of empirical distributions at decadal extents. Second, we adapt a statistical procedure combining a non-metric multidimensional scaling technique and higher density region plots to detect potentially anomalous years. We illustrate the use of these approaches by examining long-term stream temperature data from minimally and highly human-influenced streams. In particular, we contrast predictions about thermal regime responses to changing climates and human-related water uses. Using these methods, we effectively diagnose years with unusual thermal variability and patterns in variability through time, as well as spatial variability linked to regional and local factors that influence stream temperature. Our findings highlight the complexity of responses of thermal regimes of streams and reveal their differential vulnerability to climate warming and human-related water uses. The two approaches presented here can be applied with a variety of other continuous phenomena to address historical changes, extreme events, and their associated ecological responses.

1 Introduction

Environmental fluctuation is a fundamental feature that shapes ecological and evolutionary processes. Although empirical distributions of environmental data can be characterized in terms of the central tendency (or location), dispersion, and shape, most traditional statistical approaches are based on detecting changes in location and dispersion, and tend to oversimplify assumptions about temporal variation and shape. This issue is particularly troublesome for understanding the stationarity of temporally continuous phenomena and, thus, the detection of potential shifts in distributional properties beyond the location and dispersion. For instance, descriptors of location, such as mean, median or mode, may not be the most informative when extreme hydrological events are of primary attention (e.g., Chebana et al., 2012). In many regions, the future climate is expected to be characterized by increasing the frequency of extreme events (e.g., Jentsch et al., 2007; IPCC, 2012). Hence, the detection of changes in the shape of empirical distributions could be more informative than only using traditional descriptors of central tendency and dispersion (e.g., Shen et al., 2011; Donat and Alexander, 2012). More importantly, factors associated with changes in the shape of empirical distributions may have greater effects on species and ecosystems than do simple changes in location and dispersion (e.g., Colwell, 1974; Gaines and Denny, 1993; Thompson et al., 2013; Vasseur et al., 2014).

Here, we explore two approaches that identify and visualize temporal alterations in continuous environmental vari-

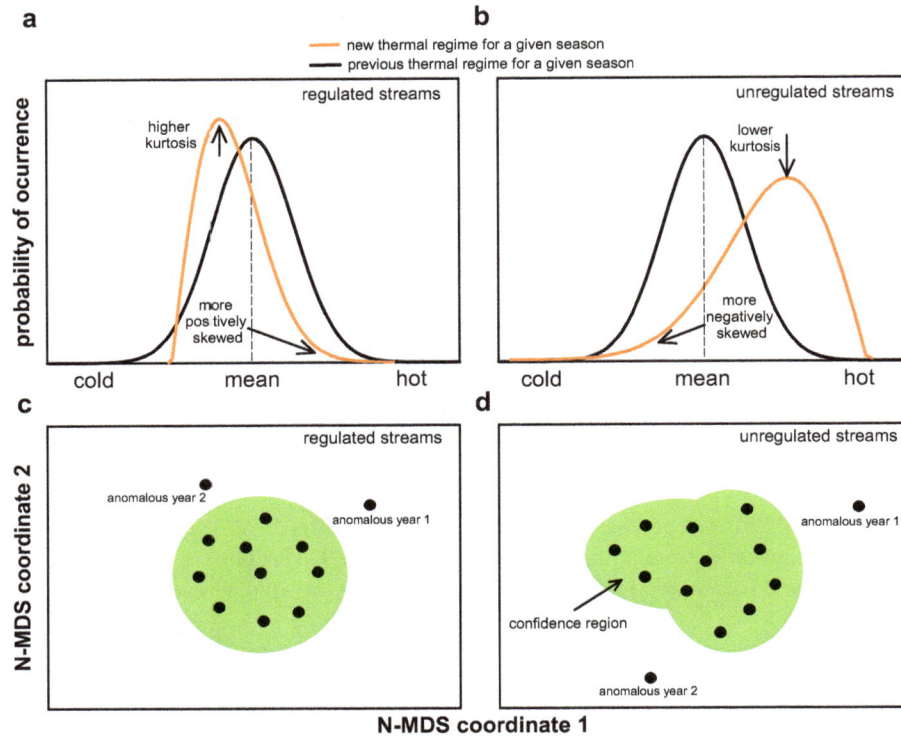

Figure 1. Conceptual diagram showing hypothesized shifts of distribution of water temperatures at both seasonal (upper panels) and annual (lower panels) scales in regulated (left panels) and unregulated (right panels) streams. In the upper panels examples of changes in skewness and kurtosis are shown for temperature distributions affected by stream regulation and a warming climate in a given season. For instance, in regulated streams the influence of the reservoir may reduce both extreme cold and warm temperatures confounding the effect from the climate (**a**) whereas less cold temperatures and an overall shift toward warming values may occur in unregulated streams (**b**). In the lower panels, we illustrate the use of N-MDS and HDR plots for detecting potentially anomalous years in regulated and unregulated streams (the shaded area represents a given coverage probability). Points located in the outer or the confidence region represent potentially anomalous years. For instance, in regulated streams individual years are more clustered because the reservoir may homogenize temperatures across years (**c**) whereas in unregulated streams individual years are less clustered due to more heterogeneous responses to the warming climate (**d**).

ables using thermal regimes of streams as an illustrative example. First, applying frequency analysis, we examine patterns of variability and long-term shifts in the shape of the empirical distribution of stream temperature using higher-order statistical moments (skewness and kurtosis) by season across decades. Second, we combine non-metric multidimensional scale ordination technique (N-MDS) and highest density region (HDR) plots to detect potentially anomalous years. To exemplify the utility of these approaches, we employ them to evaluate predictions about long-term responses of thermal regimes of streams to changing terrestrial climates and other human-related water uses (Fig. 1). Our main goal is to identify temporal changes of environmental regimes not captured by lower-order statistical moments. This is particularly relevant in streams because (1) global environmental change may affect water quality beyond the traditional lower-order statistical moments (e.g., Brock and Carpenter, 2012), and (2) ecosystems and organisms have been shown to be sensitive to such changes (e.g., Thompson et al., 2013; Vasseur et al., 2014).

1.1 Thermal regime of streams as an illustrative example

Temperature is a fundamental driver of ecosystem processes in freshwaters (Shelford, 1931; Fry, 1947; Magnuson et al., 1979; Vannote and Sweeney, 1980). Short-term (daily/weekly/monthly) descriptors of mean and maximum temperatures during summertime are frequently used for characterizations of thermal habitat availability and quality (McCullough et al., 2009), definitions of regulatory thresholds (Groom et al., 2011), and predictions about possible influences of climate change on streams (Mohseni et al., 2003; Mantua et al., 2010; Arismendi et al., 2013a, b). These simple descriptors can serve as useful first approximations but do not capture the full range of thermal conditions that the aquatic biota experience at daily, seasonal, or annual intervals (see Poole and Berman, 2001; Webb et al., 2008). Both human impacts and climate change have been shown to affect thermal regimes of streams at a variety of temporal scales (e.g., Steel and Lange, 2007; Arismendi et al., 2012, 2013a,

Table 1. Location and characteristics of unregulated ($n = 5$) and regulated ($n = 5$) streams at the gaging sites. Percent of daily gaps in the stream temperature time series from January 1979 to December 2009 used in this study.

River	Start of water regulation	Gage ID	ID	Lat. N	Long. W	Elevation (m)	Watershed area (km^2)	% of daily gaps
Fir Creek, OR	unregulated	14138870	site1	45.48	122.02	439	14.1	2.8%
SF Bull Run River, OR	unregulated	14139800	site2	45.45	122.11	302	39.9	2.0%
McRae Creek, OR	unregulated	TSMCRA	site3	44.26	22.17	840	5.9	3.5%
Lookout Creek, OR	unregulated	TSLOOK	site4	44.23	122.12	998	4.9	2.6%
Elk Creek, OR	unregulated	14338000	site5	42.68	122.74	455	334.1	5.2%
Clearwater River, ID	1971	13341050	site6	46.50	116.39	283	20658	4.0%
Bull Run River near Multnomah Falls, OR	1915*	14138850	site7	45.50	122.01	329	124.1	5.3%
NF Bull Run River, OR	1958	14138900	site8	45.49	122.04	323	21.6	2.6%
Rogue River near McLeod, OR	1977	14337600	site9	42.66	122.71	454	2,429	3.7%
Martis Creek near Truckee, CA	1971	10339400	site10	39.33	120.12	1747	103.4	6.5%

* regulation at times

b). For example, recent climate warming could lead to different responses of streams that may not be well described using average or maximum temperature values (Arismendi et al., 2012). Daily minimum stream temperatures in winter have warmed faster than daily maximum values during summer (Arismendi et al., 2013a; for air temperatures see Donat and Alexander, 2012). In human modified streams, seasonal shifts in stream temperatures and earlier warmer temperatures have been recorded following removal of riparian vegetation (Johnson and Jones, 2000). Simple threshold descriptors of central tendency (location) and dispersion cannot characterize these shifts.

Using higher-order statistical moments, we examine the question of whether the warming climate has led to shifts in the distribution of stream temperatures (Fig. 1a, b) or if all stream temperatures have warmed similarly and moved without any change in distribution or shape. In addition, we compare these potential shifts in the distribution of stream temperature between streams with unregulated and human-regulated streamflows. Using a technique that combines a non-metric multidimensional scaling procedure and higher density region plots, we address the question of whether potentially anomalous years are synoptically detected across streams types (regulated and unregulated) and examine if those potentially anomalous years represent the influence of regional climate or alternatively highlight the importance of local factors. Previous studies have shown that detecting changes in thermal regimes of streams is complex and the use of only traditional statistical approaches may oversimplify characterization of a variety of responses of ecological relevance (Arismendi et al., 2013a, b).

2 Material and methods

2.1 Study sites and time series

We selected long-term gage stations (US Geological Sur-

vey and US Forest Service) that monitored year-round daily stream temperature in Oregon, California, and Idaho ($n = 10$; Table 1). The sites were chosen based on (1) availability of continuous daily records for at least 31 years (1 January 1979–31 December 2009) and (2) complete information for time series of daily minimum (min), mean (mean), and maximum (max) stream temperature for at least 93 % of the period of record. Half of the sites ($n = 5$) were located in unregulated streams (sites 1–5) and the other half were in regulated streams (sites 6–10). Regulated streams were those with reservoirs constructed before 1978, whereas unregulated streams had no reservoirs upstream during the entire time period of the study (1979–2009). Time series were carefully inspected and the percentage of daily missing records of each time series was less than 7 % (Table 1). To ensure enough observations to adequately represent the tails of the respective distributions at a seasonal scale for analyses of higher-order statistical moments (i.e., winter: December–February; spring: March–May; summer: June–August; fall: September–November), we grouped and compared daily stream temperature data at each site among the three decades 1980–1989, 1990–1999, and 2000–2009. For the procedure that detects potentially anomalous years only (see below), we interpolated missing data following Arismendi et al. (2013a).

2.1.1 Higher-order statistical moments

To visualize and use a similar scale of stream temperatures across sites, we standardized time series of daily temperature values using a Z transformation as follows:

$$\mathrm{ST}_i = \frac{T_i - \mu}{\sigma},$$

where ST_i was the standardized temperature at day i, T_i was the actual temperature value at day i (°C), μ was the mean and σ was the standard deviation of the respective time series considering the entire time period.

Although common estimators of skewness and kurtosis are unbiased only for normal distributions, these moments can be useful to describe changes in the shape of the distribution of environmental variables over long-term periods (see Shen et al., 2011; Donat and Alexander, 2012). Skewness addresses the question of whether or not a certain variable is symmetrically distributed around its mean value. With respect to temperature, positive skewness of the distribution (or skewed right) indicates colder conditions are more common (Fig. 1a) whereas negative skewness (skewed left) represents increasing prevalence of warmer conditions (Fig. 1b). Therefore, increases in the skewness over time could occur with increases in warm conditions, decreases in cold conditions, or both.

Kurtosis describes the structure of the distribution between the center and the tails representing the dispersion around its "shoulders". In other words, as the probability mass decreases the shoulders of a distribution kurtosis it may increase in either the center, the tails, or both, resulting in a rise in the peakedness, the tail weight, or both, and thus the dispersion of the distribution around its shoulders increases. The reference standard is zero, a normal distribution with excess kurtosis equal to kurtosis minus three (mesokurtic). A sharp peak in a distribution that is more extreme than a normal distribution (excess kurtosis exceeding zero) is represented by less dispersion in the observations over the tails (leptokurtic). Distributions with higher kurtosis tend to have "tails" that are more accentuated. Therefore, observations are spread more evenly throughout the tails. A distribution with tails more flattened than the normal distribution (excess kurtosis below zero) is described by higher frequencies spread across the tails (platykurtic). With respect to temperature, a leptokurtic distribution may indicate that average conditions are much more frequent with a lower proportion of both extreme cold and warm values (Fig. 1a). A platykurtic distribution represents a more evenly distributed distribution across all values with a higher proportion of both extreme cold and warm values (Fig. 1b). Therefore, increases in the kurtosis over time would occur with decreases in extreme conditions, increases of average conditions, or both.

Time series of environmental data are generally large data sets that often have missing values and errors (see Table 1). Although the data we selected had no more than 7 % of missing values, we accounted for potential bias inherent to incomplete time series or small sample sizes by using sample skewness (adjusted Fisher–Pearson standardized moment coefficient) and sample excess kurtosis (Joanes and Gill, 1998).

The sample skewness and sample excess kurtosis are dimensionless and were estimated as follows:

$$\text{Skewness} = \frac{n}{(n-1)(n-2)} \sum_{i=1}^{n} \left(\frac{T_i - \mu}{\sigma} \right)^3,$$

$$\text{Kurtosis} = \left[\frac{n(n+1)}{(n-1)(n-2)(n-3)} \sum_{i=1}^{n} \left(\frac{T_i - \mu}{\sigma} \right)^4 \right] - \frac{3(n-1)^2}{(n-2)(n-3)},$$

where n represented the number of records of the time series, T_i was the temperature of the day i, μ and σ the mean and standard deviation of the time series.

To define the status of the skewness for the stream temperature distribution in a particular season and decade, we followed Bulmer (1979) in defining three categories as follows: "highly skewed" (if skewness was < -1 or > 1), "moderately skewed" (if skewness was between -1 and -0.5 or between 0.5 and 1), and "symmetric" (if skewness was between -0.5 and 0.5). We used similar procedures to define the status of excess kurtosis. We defined five categories that included "negative kurtosis or platykurtic" (if kurtosis was < -1), "moderately platykurtic" (if kurtosis was between -0.5 and -1), "positive kurtosis or leptokurtic" (if kurtosis was < 1), and "moderately leptokurtic" (if kurtosis was between 0.5 and 1). Finally, if kurtosis was between -0.5 and 0.5, we considered the distribution as "mesokurtic".

There are some caveats inherent to time series analyses of environmental data that should be considered. First, error terms for sequential time periods may be influenced by serial correlation affecting the independence of data. For hypothesis testing, when serial correlation occurs, the goodness of fit is inflated and the estimated standard error is smaller than the true standard error. Serial correlation often occurs on short-term scales (hourly, daily, weekly) in analyses of environmental water quality (Helsel and Hirsch, 1992). In this study, we reduced the potential for serial correlation by using higher-order statistical moments aggregated over longer time periods that allowed for a contrast among decades. Second, it is important to note that temporal changes in skewness and kurtosis could be influenced by several factors. Because skewness and kurtosis are ratios based on lower-order moments, their temporal changes may be the result of changes in only the lower-order moments, changes in the higher-order moments, or both. Thus, we recommend the use of higher-moment ratios in conjunction to the lower-order moments of central tendency and dispersion.

2.1.2 Statistical procedure to detect potentially anomalous years

We considered an entire year as one finite-dimensional observation (365 days of daily minimum stream temperature;

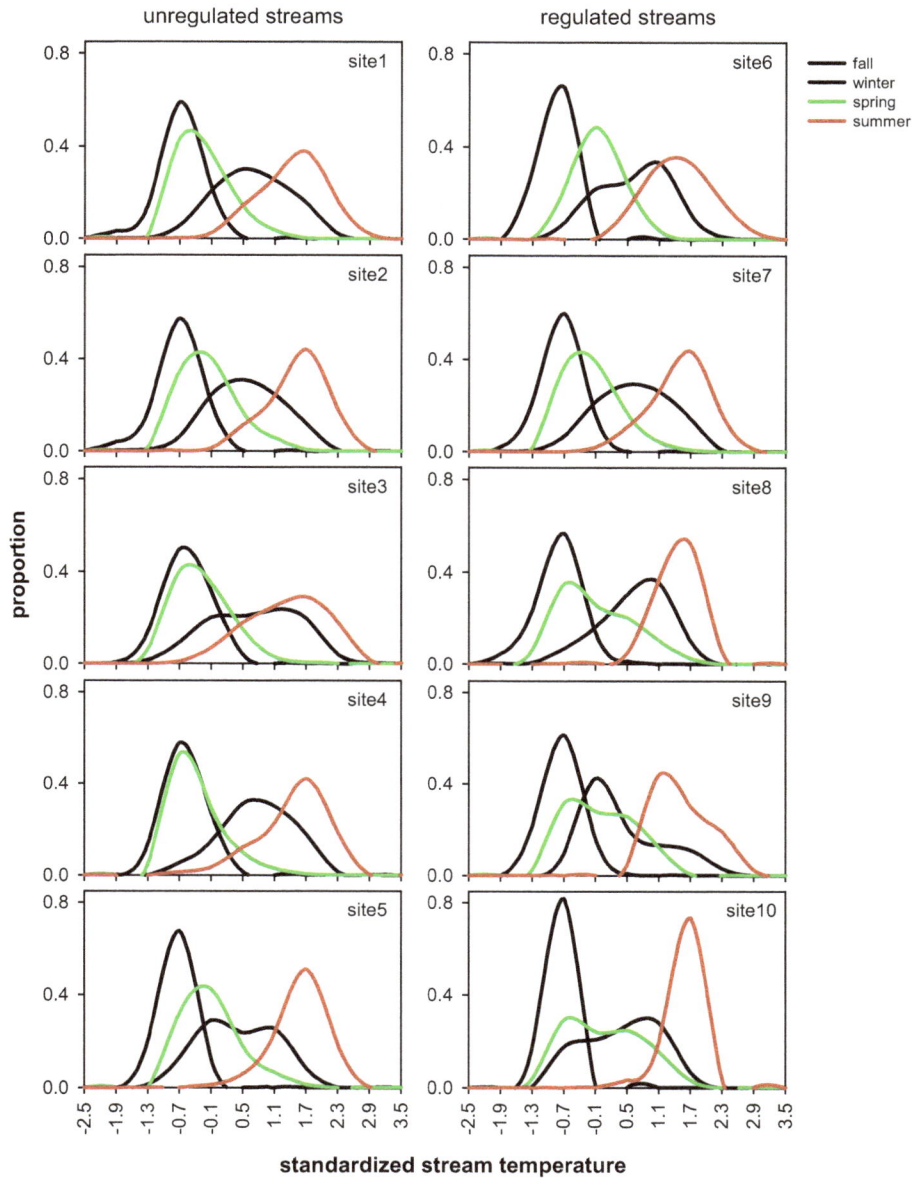

Figure 2. Density plots of standardized temperatures (1979–2009) by season (winter – blue line; spring – green line; summer – red line; fall – black line) in unregulated (left panel) and regulated (right panel) streams using time series of daily minimum.

see Sect. 2.1 above). Using N-MDS unconstrained ordination technique (Kruskal, 1964), we compared the similarity among years of the Euclidean distance of standardized temperatures for each day within a year across all years. The N-MDS analysis places each year in multivariate space in the most parsimonious arrangement (relative to each other) with no a priori hypotheses. Based on an iterative optimization procedure, we minimized a measure of disagreement or stress between their distances in 2-D using 999 random starts following the original MDSCAL algorithm (Kruskal, 1964; Clarke, 1993; Clarke and Gorley, 2006). The algorithm started with a random 2-D ordination of the years and

it regressed the inter-year 2-D distances to the actual multidimensional distances (365-D). The distance between the jth and the kth year of the random 2-D ordination is denoted as d_{jk} whereas the corresponding multidimensional distance is denoted as D_{jk}. The algorithm performed a non-parametric rank-order regression using all the jth and the kth pairs of values. The goodness of fit of the regression was estimated using the Kruskal stress as follows:

$$\text{Stress} = \sqrt{\frac{\sum_j \sum_k \left(d_{jk} - \hat{d}_{jk}\right)^2}{\sum_j \sum_k d_{jk}^2}},$$

where $\hat{d}_{jk}$ represented the predicted distance from the fitted regression between d_{jk} and D_{jk}. If $d_{jk} = \hat{d}_{jk}$ for all the distances, the stress is zero. The algorithm used a steepest descent numerical optimization method to evaluate the stress of the proposed ordination and it stops when the stress converges to a minimum. Clarke (1993) suggests the following benchmarks: stress < 0.05 – excellent ordination; stress < 0.1 – good ordination; stress < 0.2 – acceptable ordination; stress > 0.2 – poor ordination. The resulting coordinates 1 and 2 from the resulted optimized 2-D plot provided a collective index of how unique a given year was (Fig. 1c, d). In N-MDS the order of the axes was arbitrary and the coordinates represented no meaningful absolute scales for the axis. Fundamental to this method was the relative distances between points; those with greater proximity indicated a higher degree of similarity, whereas more dissimilar points were positioned further apart. We performed the N-MDS analyses using the software Primer version 6.1.15 (Clarke, 1993; Clarke and Gorley, 2006).

We created a bivariate highly dimensional region (HDR) boxplot using the two coordinates of each point (year) from the 2-D plot from the N-MDS ordination (Hyndman, 1996). The HDR plot has been typically produced using the two main principal component scores from a traditional principal component analysis (PCA; Hyndman, 1996; Chebana et al., 2012). However, in this study, we modified this procedure taking the advantage of the higher flexibility and lack of assumptions of the N-MDS analysis (Everitt, 1978; Kenkel and Orloci, 1986) to provide the two coordinates needed to create the HDR plot. In the HDR plot, there are regions defined based on a probability coverage (e.g., 50, 90, or 95 %) where all points (years) within the probability coverage region have higher density estimates than any of the points outside the region (Fig. 1c, d). The outer region of the probability coverage region (Fig. 1c, d) is bounded by points representing potentially anomalous years. We created the HDR plots using the package hdrcde (Hyndman et al., 2012) in R version 2.15.1 (R Development Core Team, 2012).

Similarly to the higher-order statistical moments, there are some caveats that should be considered when using the procedure that detects potentially anomalous years. First, it is important to note that this procedure identified years outside of a confidence region, in other words, those years that fall in the tails of the distribution. Because the confidence region represented an overall pattern extracted from the available data, it was constrained by the length of the time series. Thus, potentially anomalous years located outside of the confidence region may not necessarily represent true outliers. In addition, when the ordination is poor (stress > 0.2), interpreting the regularity/irregularity of the geometry of the confidence region should be done with caution. In our illustrative example, the regularity of the confidence region seen for regulated streams (Fig. 1c), when contrasted to unregulated sites, could be interpreted as influence of the reservoir

in dampening the inter-annual variability of downstream water temperature.

3 Results and discussion

Empirical distributions of stream temperature were distinctive among seasons, and seasons were relatively similar across sites (Fig. 2). Temperature distributions during winter had high overlap with those during spring. Winter had the narrowest range and, as would be expected, the highest frequency of observations occurring at colder standardized temperature categories (−1.3, −0.7). The second highest proportion of observations occurred in different seasons for regulated and unregulated sites: during spring in unregulated streams and during summer at four of the five regulated sites. This shift of frequency could be due to warming and release of the warmer water from the upstream reservoirs. Fall distributions showed the broadest range, with a similar proportion for a number of temperature values.

Changes in the shape of empirical distributions among seasons over decades were not immediately evident. However, the values of skewness or types of kurtosis captured these decadal changes in cases when lower-order statistical moments (average and standard deviation) did not show marked differences (e.g., site1 during fall and spring in Fig. 3; Tables 2 and 3; see also differences among decades at site1 during summer in Fig. S1 in the Supplement). The utility of combining skewness and kurtosis to detect changes in distributional shapes over time can be illustrated using site3 during winter and spring (Tables 2 and 3 and S1–S6). At this site, there was a shift across decades from symmetric towards a negatively skewed distribution in winter and from symmetric towards positively skewed in spring (Table 2), as well as from mesokurtic towards a leptokurtic distribution in both winter and spring (Table 3). Overall, in most unregulated sites, the type of kurtosis differed among decades during winter and summer (Tables 3 and S4–S6). Winter and summer frequently had negatively skewed distributions whereas spring generally had positively skewed distributions or those with little change across decades, except for site3 (Tables 2 and S1–S3).

Decadal changes in both skewness and kurtosis during winter and summer at unregulated sites suggest that the probability mass moved from its shoulders into warmer values at its center, but maintained the tail weight of the extreme cold temperature values (Figs. 3 and S1; Tables 2, 3 and S1–S6). However, in spring the probability mass diminished around its shoulders, likely due to decreases in the frequency of extreme cold temperature values. Hence, higher-order statistical moments may help in describing the complexity of temporal changes in stream temperature among seasons and highlight how shifts may occur at different portions of the distribution (e.g., extreme cold, average, or warm conditions) or among streams.

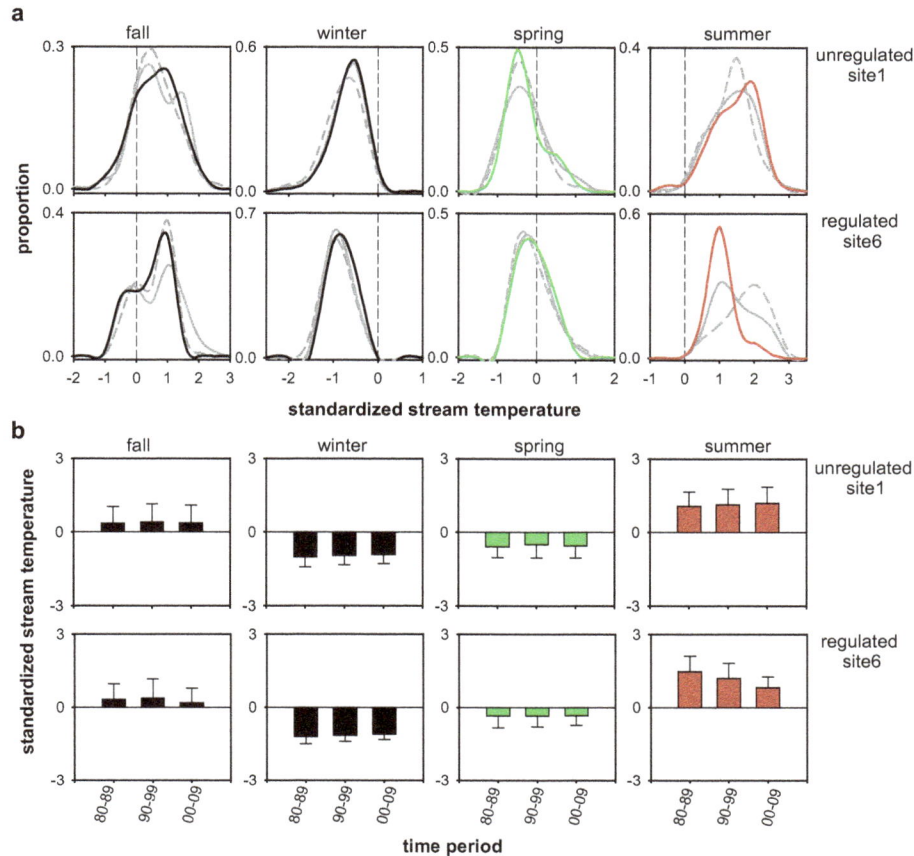

Figure 3. Examples of (**a**) density plots of standardized temperatures by decade (period 1980–1989 dashed line; period 1990–1999 grey line; period 2000–2009 solid color line) and season using time series of daily minimum in an unregulated (site1) and a regulated (site6) stream. In the lower panel (**b**) central tendency statistics (average ± SD) for each decade and season (winter – blue; spring – green; summer – red; fall – black) are also included. See results for all sites in Figs. S1 and S2.

In regulated sites, we observed shifts toward colder temperatures (e.g., site6 and site9 during summer and fall in Figs. 3 and S2), suggesting local influences of water regulation may dominate the impacts from warming climate. This is illustrated by the mixed patterns of skewness and kurtosis due to climate and water regulation, especially during spring, winter, and summer (Tables 2 and 3; Figs. 3 and S2). In particular, in spring, patterns of skewness in regulated sites were similar to unregulated sites, whereas patterns of kurtosis were in opposite directions (more platykurtic in regulated sites). This can be explained by the water discharged from reservoirs in spring that could be a mix of the cool inflows to the reservoir, the deep, colder water stored in the reservoir over the winter, and the accelerated warming of the exposed surface of the reservoir. Patterns of skewness and kurtosis seen in regulated sites also highlight the influences of site-dependent water management coupled with climatic influences. This is exemplified by the skewness of site7 and site8 compared to site9 and site10 in fall, winter, and spring (Table 2) and the high variability of the value of skewness among sites in summer.

Increased understanding of the shape of empirical distributions by season or by year will help researchers and resource managers evaluate potential impacts of shifting environmental regimes on organisms and processes across a range of disturbance types. Empirical distributions are a simple, but comprehensive way to examine high-frequency measurements that include the full range of values. Higher-order statistical moments provide useful information to characterize and compare environmental regimes and can show which seasons are most responsive to disturbances. The use of higher-order moments could help improve predictive models of climate change impacts in streams by incorporating full environmental regimes into scenarios rather than only using descriptors of central tendency and dispersion from summertime.

The technique for detection of potentially anomalous years used here was able to incorporate all daily data to provide a simple but comprehensive comparison of environmental regimes among years. We were able to characterize whole

Table 2. Magnitude and direction of the value of skewness in probability distributions of daily minimum stream temperature by season and decade at unregulated (sites 1–5) and regulated (sites 6–10) streams. Symmetric distributions are not shown. m: moderately skewed; h: highly skewed; (−): negatively skewed; (+): positively skewed (see Tables S1–S3 for more details).

Site type	Site ID	fall 1980–1989	fall 1990–1999	fall 2000–2009	winter 1980–1989	winter 1990–1999	winter 2000–2009	spring 1980–1989	spring 1990–1999	spring 2000–2009	summer 1980–1989	summer 1990–1999	summer 2000–2009
unregulated (1–5)	site1				m(−)	m(−)	m(−)	m(+)	m(+)	m(+)			
	site2					m(−)	m(−)	m(+)	m(+)	m(+)	m(−)		m(−)
	site3						m(−)		m(+)	h(+)			m(−)
	site4							h(+)	m(+)	h(+)	m(−)	m(−)	m(−)
	site5							m(+)	h(+)	m(+)	m(−)	m(−)	m(−)
regulated (6–10)	site6							m(+)					m(+)
	site7						m(−)	m(+)	m(+)	m(+)			m(−)
	site8	m(−)	m(−)					m(+)		m(+)			h(−)
	site9	m(+)	m(+)	m(+)	m(+)	m(+)							m(+)
	site10					m(+)					h(−)	m(−)	

Table 3. Types of kurtosis of probability distributions of daily minimum stream temperature by season and decade at unregulated and regulated sites. ↔↔: platykurtic; ↔: moderately platykurtic; ↕↕: leptokurtic, and ↕: moderately leptokurtic. Mesokurtic distributions are not shown (see Tables S4–S6 for more details).

Site type	Site ID	fall 1980–1989	fall 1990–1999	fall 2000–2009	winter 1980–1989	winter 1990–1999	winter 2000–2009	spring 1980–1989	spring 1990–1999	spring 2000–2009	summer 1980–1989	summer 1990–1999	summer 2000–2009
unregulated (1-5)	site1			↔	↕	↕	↕	↕↕					
	site2			↔			↕	↕↕				↔	↕
	site3	↔	↔	↔			↕↕			↕	↔		
	site4			↔				↕↕		↕↕			↕
	site5	↔	↔	↔	↔			↕	↕	↕↕		↕	
regulated (6-10)	site6	↔	↔	↔								↔	↕
	site7	↔		↔				↕↕					
	site8	↕		↔	↕		↕		↔			↕	↕↕
	site9			↔	↕			↔	↔	↔	↔	↔	
	site10	↔↔	↔↔	↔↔	↔	↕↕		↔↔	↔	↔	↕↕	↕	↕

year responses and identify where regional climatic or hydrologic trends dominated versus where local influences distinctively influenced stream temperature. For example, year 1992 was identified as potentially anomalous at three unregulated sites (or four at 90 % CI) and at two regulated sites (or four at 90 % CI), and identified that across the region, the majority of stream temperatures were being influenced (Figs. 4 and 5; Table S7). Stream temperatures in years 1987 and 2008 were less synchronous across the region, but regulated and unregulated sites located in the same watershed (site2, site7, and site8 in Tables 1 and S7; Figs. 4 and 5) shared similar potentially anomalous years. We also observed site-specific anomalous years, suggesting that more local conditions of watersheds influenced stream temperature (e.g., Arismendi et al., 2012). Indeed, sites located close to one another (site3 and site4 in Tables 1 and S7; Fig. 4) did not necessarily share all potentially anomalous years, suggesting that local drivers were more influential than regional climate forces during those years. Hence, the procedure for detection of potentially anomalous years used here may be useful to evaluate and contrast the vulnerability of streams to regional or local climate changes by characterizing years with anomalous conditions.

The technique that detects potentially anomalous years identified years with differences in either magnitude or timing of events (Figs. 4 and 5) and mapped these differences

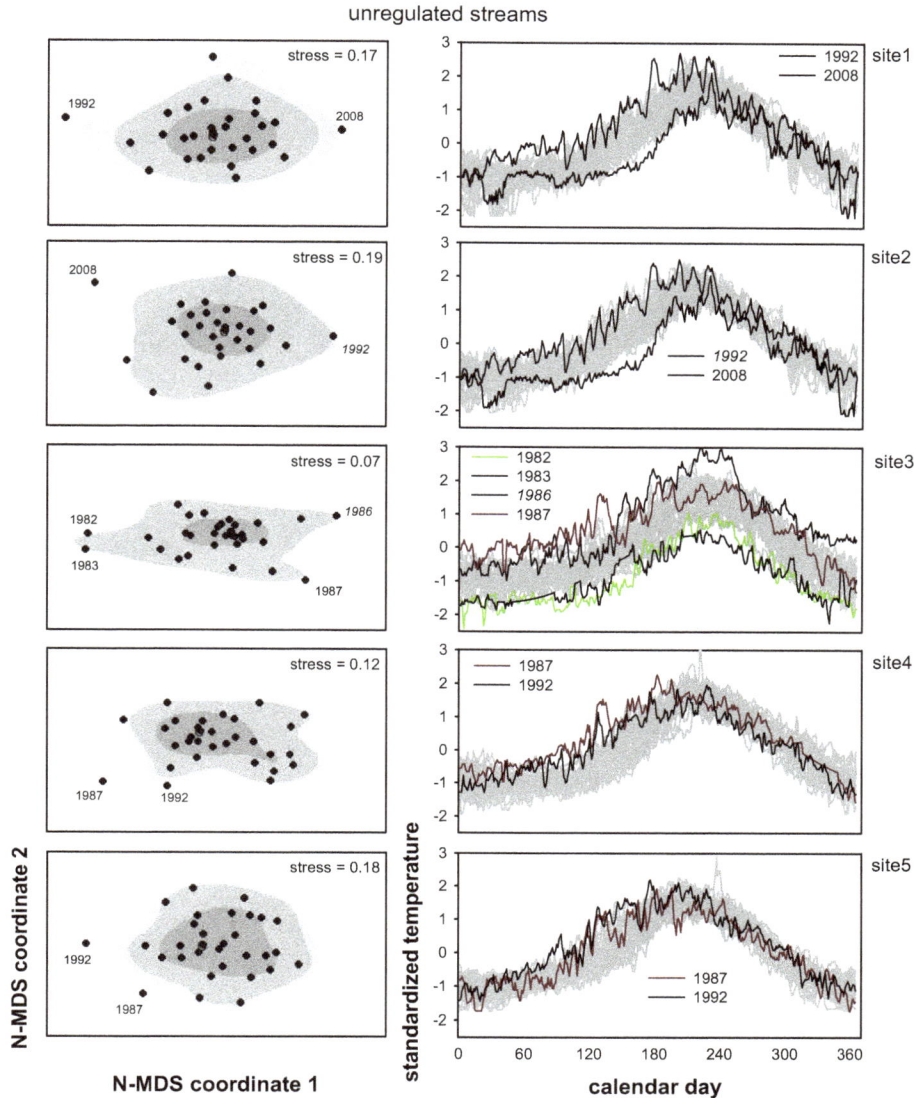

Figure 4. Bivariate HDR boxplots (left panel) and standardized daily temperature distribution (right panel) in unregulated streams using annual time series of daily minimum. The dark and light grey regions show the 50, 90, and 95 % coverage probabilities. The symbols outside the grey regions and darker lines represent potentially anomalous years. Examples of years between 90 and 95 % of the coverage probability were italicized.

within the ordination plot. For example, years 1992 and 1987 were potentially anomalous likely due the magnitude of warming throughout the year. At other sites, such as site3, site4 and site5 (Fig. 4), the potentially anomalous years were most likely due to increased temperatures in seasons other than summertime, and not related to higher summertime temperatures. Years 1992 and 2008 plotted at the opposite extremes of the ordination plot for site1, site2 and site7 (Figs. 4 and 5); see also years 1982–1983 and 1986–1987 for site3. These years contained warm and cold conditions, respectively, and likely influenced the shape of the confidence region (Figs. 4 and 5; Table S7). Interestingly, we observed that the confidence region for unregulated sites (Fig. 4) appeared

to be more irregularly shaped than regulated sites (Fig. 5), which suggests that stream regulation may tightly cluster and homogenize temperature values across years (e.g., Fig. 1c, d). Further attention on the interpretation of the geometry of a confidence region may be useful to contrast purely climatic from human influences on streams.

There are some considerations when detecting potential changes in continuous environmental phenomena that are inherent to time series analysis, including the length, timing, and quality of the time series as well as the type of the driver that is investigated as responsible for such change. Of-

regulated streams

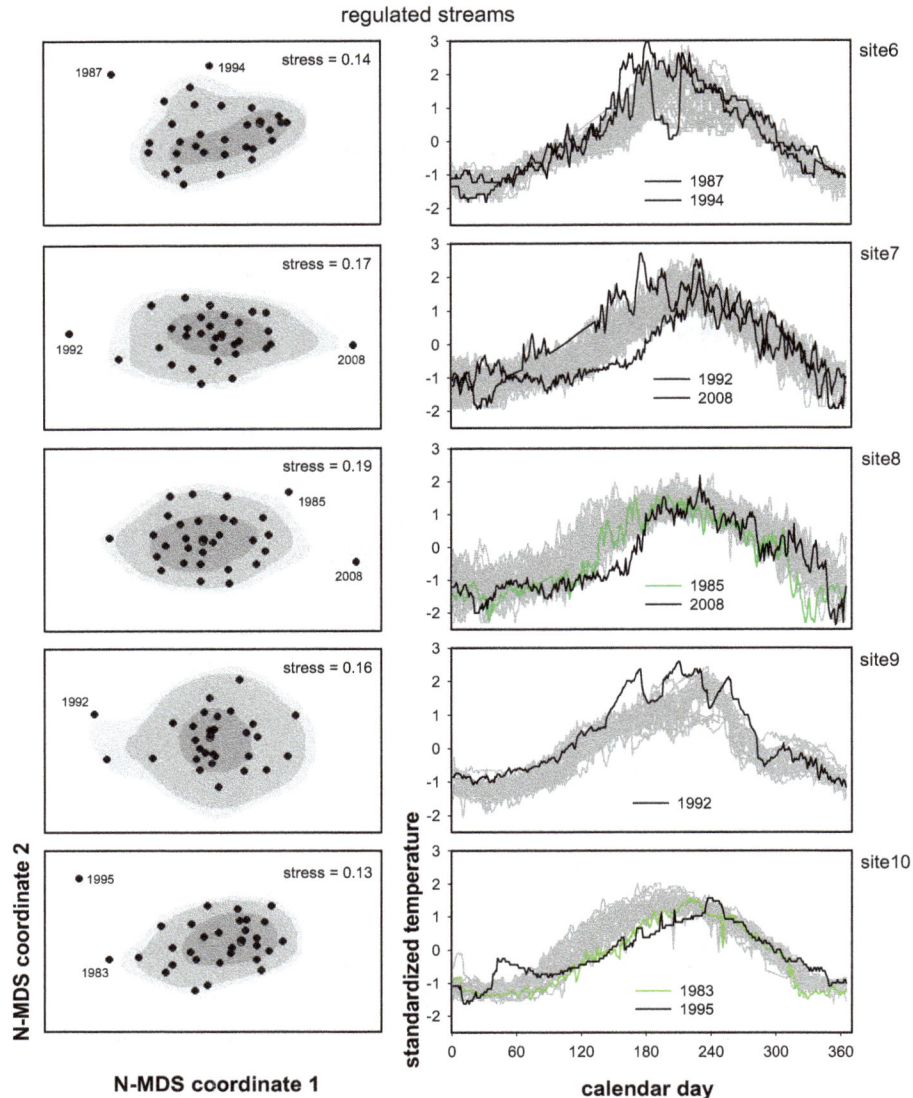

Figure 5. Bivariate HDR boxplots (left panel) and standardized daily temperature distribution (right panel) in regulated streams using annual time series of daily minimum. The dark and light grey regions show the 50, 90, and 95 % coverage probabilities. The symbols outside the grey regions and darker lines represent potentially anomalous years. Examples of years between 90 and 95 % of the coverage probability were italicized.

ten, the detection of shifts in time series of environmental data is affected by the amount of censored data that limits the length and timing of the time series (e.g., Arismendi et al., 2012). There are uncertainties regarding the importance of regional drivers and the representativeness of sites (e.g., complex mountain terrain) and periods of record (e.g., ENSO, and PDO climatic oscillations). Lastly, the type of climatic influences may affect the magnitude and duration of the responses resulting in short-term abrupt shifts (e.g., extreme climatic events), persistent long-term shifts (e.g., climate change), or a more complex combination of them (e.g., regime shifts; Brock and Carpenter, 2012).

4   Summary and conclusions

Here we show the utility of using higher-order statistical moments and a procedure that detects potentially anomalous years as complementary approaches to identify temporal changes in environmental regimes and evaluate whether these changes are consistent across years and sites. Stream ecosystems are exposed to multiple climatic and non-climatic forces which may differentially affect their hydrological regimes (e.g., temperature and streamflow). In particular, we show that potential timing and magnitude of responses of stream temperature to recent climate warming and other human-related impacts may vary among seasons, years, and across sites. Statistics of central tendency and dispersion may or may not distinguish between thermal regimes or characterize changes to thermal regimes, which could be relevant to understanding their ecological and management implications. In addition, when only single metrics are used to describe environmental regimes, they have to be selected carefully. Often, selection involves simplification resulting in the compression or loss of information (e.g., Arismendi et al., 2013a). By examining the whole empirical distribution and multiple moments, we can provide a better characterization of shifts over time or following disturbances than simple thresholds or descriptors.

In conclusion, our two approaches complement traditional summary statistics by helping to characterize continuous environmental regimes across seasons and years, which we illustrate using stream temperatures in unregulated and regulated sites as an example. Although we did not include a broad range of stream types, they were sufficiently different to demonstrate the utility of the two approaches. These two approaches are transferable to many types of continuous environmental variables and regions and suitable for examining seasonal and annual responses as well as climate or human-related influences (e.g., for streamflow see Chebana et al., 2012; for air temperature see Shen et al., 2011). These analyses will be useful to characterize the strength of the resilience of regimes and to identify how regimes of continuous phenomena have changed in the past and may respond in the future.

Acknowledgements. Brooke Penaluna and two anonymous reviewers provided comments that improved the manuscript. Vicente Monleon revised statistical concepts. Part of the data was provided by the HJ Andrews Experimental Forest research program, funded by the National Science Foundation's Long-Term Ecological Research Program (DEB 08-23380), US Forest Service Pacific Northwest Research Station, and Oregon State University. Financial support for I. Arismendi was provided by US Geological Survey, the US Forest Service Pacific Northwest Research Station and Oregon State University through joint venture agreement 10-JV-11261991-055. Use of firm or trade names is for reader information only and does not imply endorsement of any product or service by the US Government.

Edited by: S. Archfield

References

Arismendi, I., Johnson, S. L., Dunham, J. B., Haggerty, R., and Hockman-Wert, D.: The paradox of cooling streams in a warming world: Regional climate trends do not parallel variable local trends in stream temperature in the Pacific continental United States, Geophys. Res. Lett., 39, L10401, doi:10.1029/2012GL051448, 2012.

Arismendi, I., Johnson, S. L., Dunham J. B., and Haggerty, R.: Descriptors of natural thermal regimes in streams and their responsiveness to change in the Pacific Northwest of North America, Freshwater Biol., 58, 880–894, 2013a.

Arismendi, I., Safeeq, M., Johnson, S. L., Dunham, J. B., and Haggerty, R.: Increasing synchrony of high temperature and low flow in western North American streams: double trouble for coldwater biota? Hydrobiologia, 712, 61–70, 2013b.

Brock, W. A. and Carpenter, S. R.: Early warnings of regime shift when the ecosystem structure is unknown., PLoS ONE, 7, e45586, doi:10.1371/journal.pone.0045586, 2012.

Bulmer, M. G.: Principles of Statistics, Dover Publications Inc., New York, USA, 252 pp., 1979.

Chebana, F., Dabo-Niang, S., and Ouarda, T. B. M. J.: Exploratory functional flood frequency analysis and outlier detection, Water Resour. Res., 48, W04514, doi:10.1029/2011WR011040, 2012.

Clarke, K. R.: Nonparametric multivariate analyses of changes in community structure, Aust. J. Ecol., 18, 117–143, 1993.

Clarke, K. R. and Gorley, R. N.: PRIMER v6: User Manual/Tutorial, PRIMER-E, Plymouth, UK, 2006.

Colwell, R. K.: Predictability, constancy, and contingency of periodic phenomena, Ecology, 55, 1148–1153, 1974.

Donat, M. G. and Alexander, L. V.: The shifting probability distribution of global daytime and night-time temperatures, Geophys. Res. Lett., 39, L14707, doi:10.1029/2012GL052459, 2012.

Everitt, B.: Graphical techniques for multivariate data, North-Holland, New York, USA, 117 pp., 1978.

Fry, F. E. J.: Effects of the environment on animal activity, University of Toronto Studies, Biological Series 55, Publication of the Ontario Fisheries Research Laboratory, 68, 1–62, 1947.

Gaines, S. D. and Denny, M. W.: The largest, smallest, highest, lowest, longest, and shortest: extremes in ecology, Ecology, 74, 1677–1692, 1993.

Groom, J. D., Dent, L., Madsen, L. J., and Fleuret, J.: Response of western Oregon (USA) stream temperatures to contemporary forest management, Forest Ecol. Manag., 262, 1618–1629, 2011.

Helsel, D. R and Hirsch, R. M.: Statistical methods in water resources, Elsevier, the Netherlands, 522 pp., 1992.

Hyndman, R. J.: Computing and graphing highest density regions, Am. Stat., 50, 120–126, 1996.

Hyndman, R. J., Einbeck, J., and Wand, M.: Package "hdrcde": highest density regions and conditional density estimation, available at: http://cran.r-project.org/web/packages/hdrcde/hdrcde.pdf (last access: 4 November 2014), 2012.

IPCC: Managing the risks of extreme events and disasters to advance climate change adaptation, in: A Special Report of Working Groups I and II of the Intergovernmental Panel on Climate Change, edited by: Field, C. B., Barros, V., Stocker, T. F., Qin, D., Dokken, D. J., Ebi, K. L., Mastrandrea, M. D., Mach, K. J., Plattner, G. K., Allen, S. K., Tignor, M., and Midgley, P. M., Cambridge University Press, Cambridge, United Kingdom and New York, NY, USA, 1–19, 2012.

Jentsch, A., Kreyling, J., and Beierkuhnlein, C.: A new generation of climate change experiments: events, not trends, Front. Ecol. Environ., 5, 365–374, 2007.

Joanes, D. N. and Gill, C. A.: Comparing measures of sample skewness and kurtosis, J. Roy. Stat. Soc.: Am. Stat., 47, 183–189, 1998.

Johnson, S. L. and Jones, J. A.: Stream temperature response to forest harvest and debris flows in western Cascades, Oregon, Can. J. Fish. Aquat. Sc., 57, 30–39, 2000.

Kenkel, N. C. and Orloci, L.: Applying metric and nonmetric multidimensional scaling to ecological studies: some new results, Ecology, 67, 919–928, 1986.

Kruskal, J. B.: Non-metric multidimensional scaling: a numerical method, Psychometrika, 29, 115–129, 1964.

Magnuson, J. J, Crowder, L. B, and Medvick, P. A.: Temperature as an ecological resource, Am. Zool., 19, 331–343, 1979.

Mantua, N., Tohver, I., and Hamlet, A.: Climate change impacts on streamflow extremes and summertime stream temperature and their possible consequences for freshwater salmon habitat in Washington State, Clim. Change, 102, 187–223, 2010.

McCullough, D. A, Bartholow, J. M., Jager, H.I., Beschta, R. L., Cheslak, E. F., Deas, M. L., Ebersole, J. L., Foott, J. S., Johnson, S. L., Marine, K. R., Mesa, M. G., Petersen, J. H., Souchon, Y., Tiffan, K. F., and Wurtsbaugh, W. A.: Research in Thermal Biology: Burning Questions for Coldwater Stream Fishes, Rev. Fish. Sc., 17, 90–115, 2009.

Mohseni, O., Stefan, H. G., and Eaton, J. G.: Global warming and potential changes in fish habitat in US streams, Clim. Change, 59, 389–409, 2003.

Poole, G. C. and Berman, C. H.: An ecological perspective on in-stream temperature: natural heat dynamics and mechanisms of human-caused thermal degradation, Environ. Manag., 27, 787–802, 2001.

Shelford, V. E.: Some concepts of bioecology, Ecology, 123, 455–467, 1931.

Shen, S. S. P., Gurung, A. B., Oh, H., Shu, T., and Easterling, D. R.: The twentieth century contiguous US temperature changes indicated by daily data and higher statistical moments, Climatic Change, 109, 287–317, 2011.

Steel, E. A. and Lange, I. A.: Using wavelet analysis to detect changes in water temperature regimes at multiple scales: effects of multi-purpose dams in the Willamette River basin, Riv. Res. Appl., 23, 351–359, 2007.

Thompson, R. M., Beardall, J., Beringer, J., Grace, M., and Sardina, P.: Means and extremes: building variability into community-level climate change experiments, Ecol. Let., 16, 799–806, 2013.

Vannote, R. L. and Sweeney, B. W.: Geographic analysis of thermal equilibria: a conceptual model for evaluating the effects of natural and modified thermal regimes on aquatic insect communities, Am. Nat., 115, 667–695, 1980.

Vasseur, D. A., DeLong, J. P., Gilbert, B., Greig, H. S., Harley, C. D. G., McCann, K. S., Savage, V., Tunney, T. D., and O'Connor, M. I..: Increased temperature variation poses a greater risk to species than climate warming, Proc. Roy. Soc. B, 281, 20132612, 2014.

Webb, B. W., Hannah, D. M., Moore, R. D., Brown, L. E., and Nobilis, F..: Recent advances in stream and river temperature research, Hydrol. Process., 22, 902–918, 2008.

11

Geostatistical upscaling of rain gauge data to support uncertainty analysis of lumped urban hydrological models

Manoranjan Muthusamy[1], **Alma Schellart**[1], **Simon Tait**[1], **and Gerard B. M. Heuvelink**[2]

[1]Department of Civil and Structural Engineering, University of Sheffield, Sheffield, S1 3JD, UK
[2]Soil Geography and Landscape group, Wageningen University, Wageningen, 6700, the Netherlands

Correspondence to: Manoranjan Muthusamy (m.muthusamy@sheffield.ac.uk)

Abstract. In this study we develop a method to estimate the spatially averaged rainfall intensity together with associated level of uncertainty using geostatistical upscaling. Rainfall data collected from a cluster of eight paired rain gauges in a 400 m × 200 m urban catchment are used in combination with spatial stochastic simulation to obtain optimal predictions of the spatially averaged rainfall intensity at any point in time within the urban catchment. The uncertainty in the prediction of catchment average rainfall intensity is obtained for multiple combinations of intensity ranges and temporal averaging intervals. The two main challenges addressed in this study are scarcity of rainfall measurement locations and non-normality of rainfall data, both of which need to be considered when adopting a geostatistical approach. Scarcity of measurement points is dealt with by pooling sample variograms of repeated rainfall measurements with similar characteristics. Normality of rainfall data is achieved through the use of normal score transformation. Geostatistical models in the form of variograms are derived for transformed rainfall intensity. Next spatial stochastic simulation which is robust to nonlinear data transformation is applied to produce realisations of rainfall fields. These realisations in transformed space are first back-transformed and next spatially aggregated to derive a random sample of the spatially averaged rainfall intensity. Results show that the prediction uncertainty comes mainly from two sources: spatial variability of rainfall and measurement error. At smaller temporal averaging intervals both these effects are high, resulting in a relatively high uncertainty in prediction. With longer temporal averaging intervals the uncertainty becomes lower due to stronger spatial correlation of rainfall data and relatively smaller measurement error. Results also show that the measurement error in-

creases with decreasing rainfall intensity resulting in a higher uncertainty at lower intensities. Results from this study can be used for uncertainty analyses of hydrologic and hydrodynamic modelling of similar-sized urban catchments as it provides information on uncertainty associated with rainfall estimation, which is arguably the most important input in these models. This will help to better interpret model results and avoid false calibration and force-fitting of model parameters.

1 Introduction

Being the process driving runoff, rainfall is arguably the most important input parameter in any hydrological modelling study. But it is a challenging task to accurately measure rainfall due to its highly variable nature over time and space, especially in small urban catchments. Despite recent advances in radar technologies rain gauge measurements are still considered to be the most accurate way of measuring rainfall, especially at short temporal averaging intervals (< 1 h), which are of most interest in urban hydrology studies (Ochoa-Rodriguez et al., 2015). However, many commonly used urban hydrological models (e.g. SWMM, HBV) are lump catchment models (LCMs) where time series of areal average rainfall intensity (AARI) are needed as model input. Therefore, point observations of rainfall need to be scaled up using spatial aggregation in order to be fed in to a LCM. There are a number of interpolation methods available for spatial aggregation and used in the various LCMs to scale up point rainfall data. The simplest method is to take the arithmetic average (Chow, 1964) of the point observations within the catchment. But this method does not account for the spatial correlation structure of the rainfall and the spatial organisation of the rain gauge locations. Another commonly used method in hydrological modelling is the nearest neighbour interpolation (Chow, 1964; Nalder and Wein, 1998) which leads to Thiessen polygons. In this method the nearest observation is given a weight of one and other observations are given zero weights during interpolation, thereby ignoring spatial variability of rainfall to a certain extent. There are also other methods, with varying complexity levels, including inverse distance weighting (Dirks et al., 1998), polynomial interpolation (Tabios III and Salas, 1985) and moving window regression (Lloyd, 2005). The predictive performance of the above methods are found to be case dependent and no single method has been shown to be optimal for all catchments and rainfall conditions (Ly et al., 2013). One common drawback with all the above methods is that they do not provide any information on the uncertainty of the predictions of AARI as all the methods are deterministic. The uncertainty in prediction of AARI mainly comes from two sources; uncertainty due to measurement errors and uncertainty associated with spatial variability of rainfall. The characteristics of measurement errors can vary depending on the rain gauge type. For example, errors associated with commonly used tipping bucket rain gauges range from errors due to wind, wetting, evaporation, and splashing (Fankhauser, 1998; Sevruk and Hamon, 1984) to errors due to its sampling mechanism (Habib et al., 2001). In addition to measurement errors and since rainfall can vary over space significantly, any spatial aggregation method for scaling up the point rainfall measurements incorporates more uncertainty (Villarini et al., 2008). The magnitude of the uncertainty depends on many factors including rain gauge density and location, rainfall variability, catchment size, topography, and the spatial interpolation technique used. Quantification of the level of uncertainty is essential for robust interpretation of hydrological model outputs. For instance, the absence of information on uncertainty can lead to force fitting of hydrological model parameters to compensate for the uncertainty in rainfall input data (Schuurmans and Bierkens, 2007).

Geostatistical methods such as kriging present a solution to this problem by providing a measure of prediction error. In addition to this capability, these statistical methods also take into account the spatial dependence structure of the measured rainfall data (Ly et al., 2013; Mair and Fares, 2011). Although these features make geostatistical methods more attractive than deterministic methods, they are rarely used in LCMs due to their inherent complexity and heavy data requirements. Since they are statistical methods encompassing multiple parameters the amount of spatial data required for model inference is higher compared to deterministic methods. In addition the underlying assumption of geostatistical approaches typically requires data to be normally distributed (Isaaks and Srivastava, 1989). In general, catchments, especially those at small urban scales, do not contain as many measurement locations as required by geostatisti-

cal methods. Furthermore, rainfall intensity data are almost never normally distributed, especially at smaller averaging intervals (< 1 h) (Glasbey and Nevison, 1997). Despite these challenges geostatistical methods can provide information on uncertainty associated with predicted AARI. This capability can be utilised in uncertainty propagation analysis in hydrological models. In literature, geostatistical methods have been used to analyse the spatial correlation structure of rainfall at various spatial scales (Berne et al., 2004; Ciach and Krajewski, 2006; Emmanuel et al., 2012; Jaffrain and Berne, 2012), however its application to support uncertainty analyses of upscaling rainfall data has not been explored.

In this paper we present a geostatistical approach to derive AARI and the level of uncertainty associated with it from observations obtained from multiple "paired" rain gauges located in a small urban catchment. The proposed approach presents solutions to the above-described challenges of geostatistical methods. First, it uses pooling of sample variograms of rainfall measurements at different times but with similar characteristics to increase the number of paired observations used to fit variogram models. Second, a data transformation method is employed to transform the rainfall data to obtain a normally distributed data set. The level of uncertainty in the prediction of AARI is then quantified for different combinations of temporal averaging intervals and intensity ranges for the studied urban catchment. We focused on a small urban catchment with a spatial extent of less than a kilometre given the findings of recent research on the high significance of unmeasured spatial rainfall variability at such spatial scales, especially for urban hydrological and hydrodynamic modelling applications (Gires et al., 2012, 2014; Ochoa-Rodriguez et al., 2015).

2 Data collection

2.1 Location and rain gauge network design

The study area is located in Bradford, a city in West Yorkshire, England. Bradford has a maritime climate, with an average yearly rainfall of 873 mm recorded from 1981–2010 (MetOffice UK, 2016). The rain gauge network, used in this study was located at the premises of Bradford University (Fig. 1) and rainfall data were collected from paired tipping bucket rain gauges placed at eight locations covering an area of 400 m × 200 m. Data used in this study were collected from April to August 2012 and from April to August 2013. These stations were located on selected roofs of the university buildings, thereby providing controlled, secure and obstruction-free measurement locations. Each station consists of two tipping bucket type rain gauges mounted 1 m apart. On each roof the paired gauges were placed such that the height of the nearest obstruction is less than two times the distance between the gauges and the obstruction. The rim of each rain gauge was set up around 0.5 m above the surround-

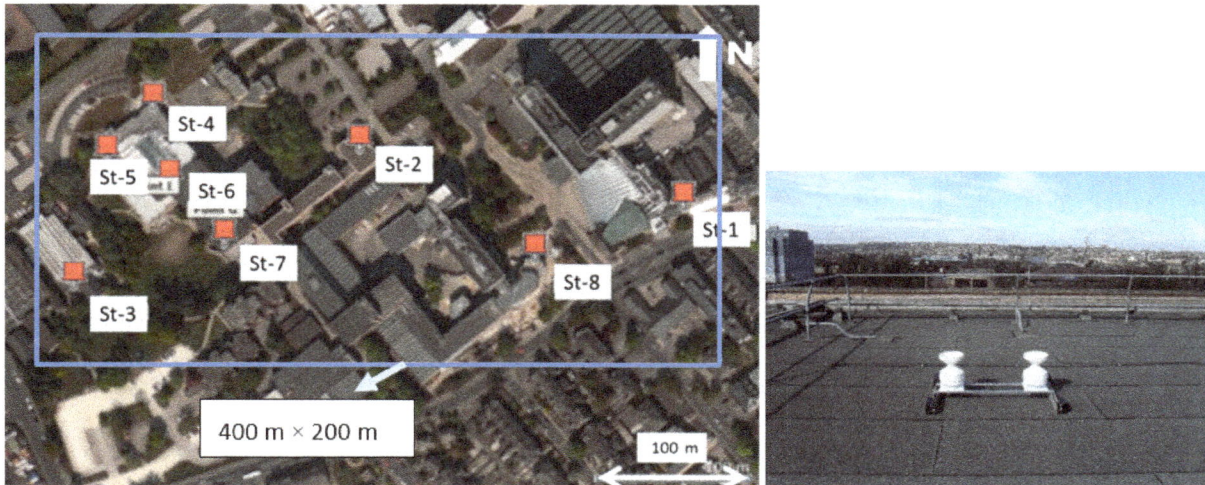

Figure 1. Left: aerial view of rain gauge network covering an area of 400 m × 200 m at Bradford University, UK. Right: a photograph of paired rain gauges at station 6.

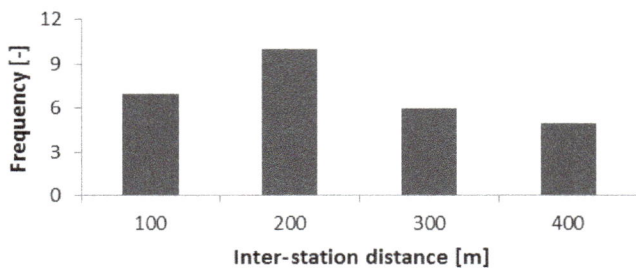

Figure 2. Histogram with class interval width of 100 m showing frequency distribution of inter-station distances (m).

ing ground level following UK standard practice (MetOffice UK, 2016). An example of the measurement setup (station 6) is also shown in Fig. 1. A histogram of the inter-station distances of the rain gauge network is presented in Fig. 2. Lag distances covered in this network are distributed between 21 m (stations 4–5) and 399 m (stations 1–3).

All rain gauges are ARG100 tipping bucket type with an orifice diameter of 254 mm and a resolution of 0.2 mm. Dynamic calibration was carried out for each individual gauge before deployment and visual checks were carried out every 4–5 weeks during the measurement period to ensure that the instruments were free of dirt and debris. Data loggers were reset every 4–5 weeks during data collection to avoid any significant time drift. Measurements (number of tips) were taken every minute and recorded on TinyTag data loggers mounted in each rain gauge.

Quality control procedures were performed prior to statistical analysis, taking advantage of the paired gauge setup to detect gross measurement errors. The paired gauge design provides efficient quality control of the rain gauge data records as it helps to identify the instances when one of the gauges fails, and to clearly identify periods of missing or incorrect data (Ciach and Krajewski, 2006). During the dynamic calibration of all rain gauges in the laboratory before deployment, it was identified that the highest and lowest values of the calibration factors for the tipping bucket size are 0.196 and 0.204 mm. The gauges were recalibrated in the laboratory after the first period of measurement and it was found that the largest change in calibration factor for any gauge was a maximum of 4 % of the original calibration factor. Therefore a maximum difference of 4 % in volume per tip was assumed to be caused by inherent instrument error. It was therefore decided that this is the maximum acceptable difference between any pair of gauges. Sets of cumulative rainfall data corresponding to specific events from the paired gauges were checked against each other and if the (absolute) difference in cumulative rainfall was greater than 4 %, that complete set was identified as unreliable and removed from further analysis.

2.2 Characteristics of the data

The total average network rainfall depth for the summer seasons of 2012 and 2013 are 538 and 207 mm, respectively. Figure 3 shows time series of daily rainfall averaged over the network for 2012 and 2013. There is a significant difference in cumulative rainfall between 2012 and 2013. This is because 2012 was the wettest year recorded in 100 years in the UK (MetOffice UK, 2016) and 558 mm of rainfall during 2012 summer was unusually high. An average rainfall of only 360 mm was recorded during April to September over the 1981–2010 period at the nearest operational rain gauge station at Bingley, which is around 8 km from the study site with a similar ground elevation (MetOffice UK, 2016).

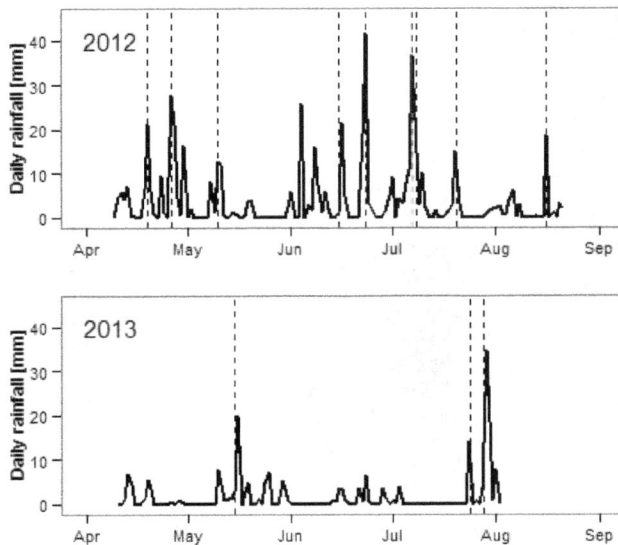

Figure 3. Time series of network average daily rainfall in the two seasons of 2012 and 2013 with vertical dashed lines indicating the events presented in Table 1.

The data set for 2012 and 2013 contains 13 events yielding more than 10 mm network average rainfall depth each and lasting for more than 20 min. A summary of these events is presented in Table 1. Note that this event separation is only used for the presentation of results in Sect. 4.2. Hence it does not leave out any data from the development and calibration of the geostatistical model as presented in Sect. 3. Table 1 shows that the total event duration ranges from 1.5 to 11.4 h while the event network average rainfall intensity varies from 1.79 to 7.96 mm h^{-1}. Table 1 also includes summary statistics of peaks of events (temporal averaging interval of 5 min) for the eight stations within the network. Although the spatial extent of the area is only 400 m × 200 m, it is clear that there is a considerable difference in rainfall intensity measurements indicated by the standard deviation and range of peaks observed in the individual events. The maximum standard deviation between peaks of individual events is 9.27 mm h^{-1} for event 8, which is around 12.5 % of the mean network peak intensity of 74.4 mm h^{-1}. This variation provides evidence of the potential importance of analysing uncertainty in the estimation of AARI even in such a small urban catchment.

3   Methodology

Figure 4 summarises the procedure of geostatistical upscaling of the rainfall data adapted in this study in a step-by-step instruction followed by the detail descriptions of each step. This complete procedure was repeated for temporal averaging intervals of 2, 5, 15, and 30 min in order to investigate the effect of temporal aggregation on the prediction of AARI.

The entire 10 months of collected data were used for the development and calibration of the geostatistical model.

3.1   Step 1: pooling of sample variograms

The rain gauge network contains eight measurement locations. These eight measurement locations give 28 spatial pairs at a given time instant which yields too few spatial lags than would normally be used in geostatistical modelling. For example, Webster and Oliver (2007) recommend around 100 measurement points to calibrate a geostatistical model. The procedure adapted in this study increases the number of pairs by pooling sample variograms for time instants with similar rainfall characteristics. With n measurement locations and measurements taken at t time instants, the pooling over t time instants creates $t \times 1/2 \times n \times (n-1)$ spatial pairs. Although this procedure increases the number of spatial pairs by a factor t, the spatial separation distances for which information is available will be limited to the original configuration of the n measurement locations.

The underlying assumption of this pooling procedure is that the spatial variability over the pooled time instants is the same. Therefore it is important to pool sample variograms of rainfall measurements with similar rainfall characteristics. Since the spatial rainfall variability is often intensity dependent (Ciach and Krajewski, 2006), the characteristics of a less intense rainfall event may not be the same as that of a high-intensity rainfall event. Hence to make the assumption of consistency of spatial variability, the range of rainfall intensity over the pooled time instants should be reasonably small. On the other hand, one should also make sure that there are enough time instants within a pooled subset to meet the data requirement to calibrate the geostatistical model. Based on the above two criteria, three rainfall intensity classes were selected. The maximum threshold value was limited to 10 mm h^{-1} to have enough time instants for the highest range (i.e. > 10 mm h^{-1}) in order to produce stable variograms even at 30 min temporal averaging interval. It was then decided to divide the 0–10 mm h^{-1} class to two equal subclasses (i.e. < 5 and 5–10 mm h^{-1}). This resulted in three subclasses, which is a reasonable number given the size of the data set and computational demand. The number of time instants (t) within each rainfall intensity class is presented for three temporal averaging intervals in Fig. 5. The natural characteristic of rainfall data results in the dominance of lower intensity rainfall (0.1–5.0 mm h^{-1}) over the recording period. In addition, the number of time instants t obviously reduces with increasing temporal averaging intervals due to the aggregation process. As a consequence there are only seven time instants for the intensity range > 10 mm h^{-1} at the 30 min temporal averaging interval. This limits the maximum temporal averaging interval to 30 min for our analyses. For a catchment of this size (400 m × 200 m) it is very unlikely to have a response time of more than 30 min. Hence, from a hydrological point of view consideration of temporal

Table 1. Summary of events which yielded more than 10 mm rainfall and lasted for more than 20 min with summary statistics of event peaks (derived at 5 min temporal averaging interval) from all stations.

Event ID	Date	Network average duration	Network average intensity	Network average rainfall	Summary statistics of peaks between different stations (mm h^{-1})			
		(h)	(mm h^{-1})	(mm)	Mean	SD	Max	Min
1	18/04/2012	6.33	2.20	13.9	5.10	0.550	6.02	4.74
2	25/04/2012	6.42	2.55	16.3	7.05	0.751	8.32	5.92
3	09/05/2012	8.92	1.79	16.0	5.10	0.537	5.97	4.74
4	14/06/2012	6.83	1.99	13.6	5.25	0.636	6.04	4.74
5	22/06/2012	11.4	2.39	27.3	12.7	1.72	15.4	9.67
6	06/07/2012	4.42	5.31	23.4	38.5	4.52	42.9	30.5
7	06/07/2012	3.25	3.23	10.5	7.20	0.679	8.46	5.93
8	07/07/2012	1.50	7.84	11.8	74.4	9.27	86.5	61.9
9	19/07/2012	3.08	3.35	10.3	12.7	2.01	14.5	9.74
10	15/08/2012	2.00	7.96	15.9	43.0	3.69	47.8	37.5
11	14/05/2013	7.92	2.14	17.0	8.08	1.20	9.55	6.09
12	23/07/2013	1.75	6.51	11.4	37.7	2.09	42.6	35.7
13	27/07/2013	8.17	4.34	35.5	26.6	1.23	27.5	23.8

averaging intervals longer than 30 min would not be sensible. Note that although there are only seven time instants, the pooling procedure will produce 196 ($= 7 \times 28$) points to calculate and calibrate the geostatistical model for that intensity class.

3.2 Step 2: standardisation of rainfall intensities

Having chosen the rainfall intensity classes to create pooled time instants, there can still be inconsistency in spatial variability between time instants within a class and therefore assuming a single geostatistical model for the whole subset may not be realistic. To reduce this effect to a certain extent, all observations within an intensity class were standardised using the mean and standard deviation of each time instant as follows:

$$\widetilde{r}_{ix} = \frac{r_{ix} - m_i}{\mathrm{SD}_i}, \tag{1}$$

where $i = 1 \ldots t$, $x = 1 \ldots n$; $\widetilde{r}_{ix}$ is standardised rainfall intensity at a time instant i and location x; r_{ix} is rainfall intensity at time instant i and location x; and m_i, SD_i are mean and standard deviation of rainfall intensities at time instant i, respectively. Further steps were carried out on the standardised rainfall intensity.

3.3 Step 3: normal transformation of data

The upper part of Fig. 6 shows the distribution of standardised rainfall intensity for a temporal averaging interval of 5 min derived using Eq. (1). From the figure it is clear that the data are not normally distributed. Distributions for other temporal averaging intervals (i.e. 2, 15, and 30 min) show a similar behaviour. But the geostatistical upscaling method to

be used is based on the normal distribution. This requires the rainfall data to be normally distributed prior to the calibration of the geostatistical model. The normal score transformation (NST, also known as normal quantile transformation; Van der Waerden, 1952) is a widely used method to transform a variable distribution to the Gaussian distribution. It has widely been applied in many hydrological applications (Bogner et al., 2012; Montanari and Brath, 2004; Todini, 2008; Weerts et al., 2011). The concept of NST is to match the p quantile of the data distribution with the p quantile of the standard normal distribution. Consider a standardised rainfall intensity $\widetilde{r}$ with cumulative distribution $F_{\widetilde{R}}(\widetilde{r})$. It is transformed to a r_N value with a Gaussian cumulative distribution $F_{R_N}(r_N)$ as follows:

$$r_N = F_{R_N}^{-1}\left(F_{\widetilde{R}}(\widetilde{r})\right). \tag{2}$$

Detailed description of NST including the steps involved can be found in Bogner et al. (2012), Van der Waerden (1952) and Weerts et al. (2011). The lower part of Fig. 6 shows the transformed standardised intensity for the temporal averaging interval of 5 min.

3.4 Step 4: calibration of geostatistical model

A geostatistical model of (normalised) rainfall intensity r_N (derived from Sect. 3.3) at any location x can be written as

$$r_N(x) = p(x) + \varepsilon(x), \tag{3}$$

where $p(x)$ is the trend (explanatory part) and $\varepsilon(x)$ is the stochastic residual (unexplanatory part). Considering the availability of data, small catchment size, and scope of this study, it was assumed that the trend is constant and does not

1. Pooling of sample variograms of time instants using predefined range of rainfall intensities, r

2. Standardisation of rainfall intensities, $\tilde{r} = S(r)$

3. Normal score transformation of standardised intensities,

$r_N = NST(\tilde{r})$

4. Calibration of geostatistical model for r_N in the form of a variogram

5. Spatial stochastic simulation producing a large number of realisations of r_N

6. Back-transformation of all realisations using NST^{-1}

7. Spatial aggregation of each of the back-transformed simulations

8. Estimation of the mean prediction (mean of the aggregates) and standard deviation (standard deviation of the aggregates)

9. Inverse standardisation of mean prediction (=AARI) and standard deviation (uncertainty measure) using S^{-1}

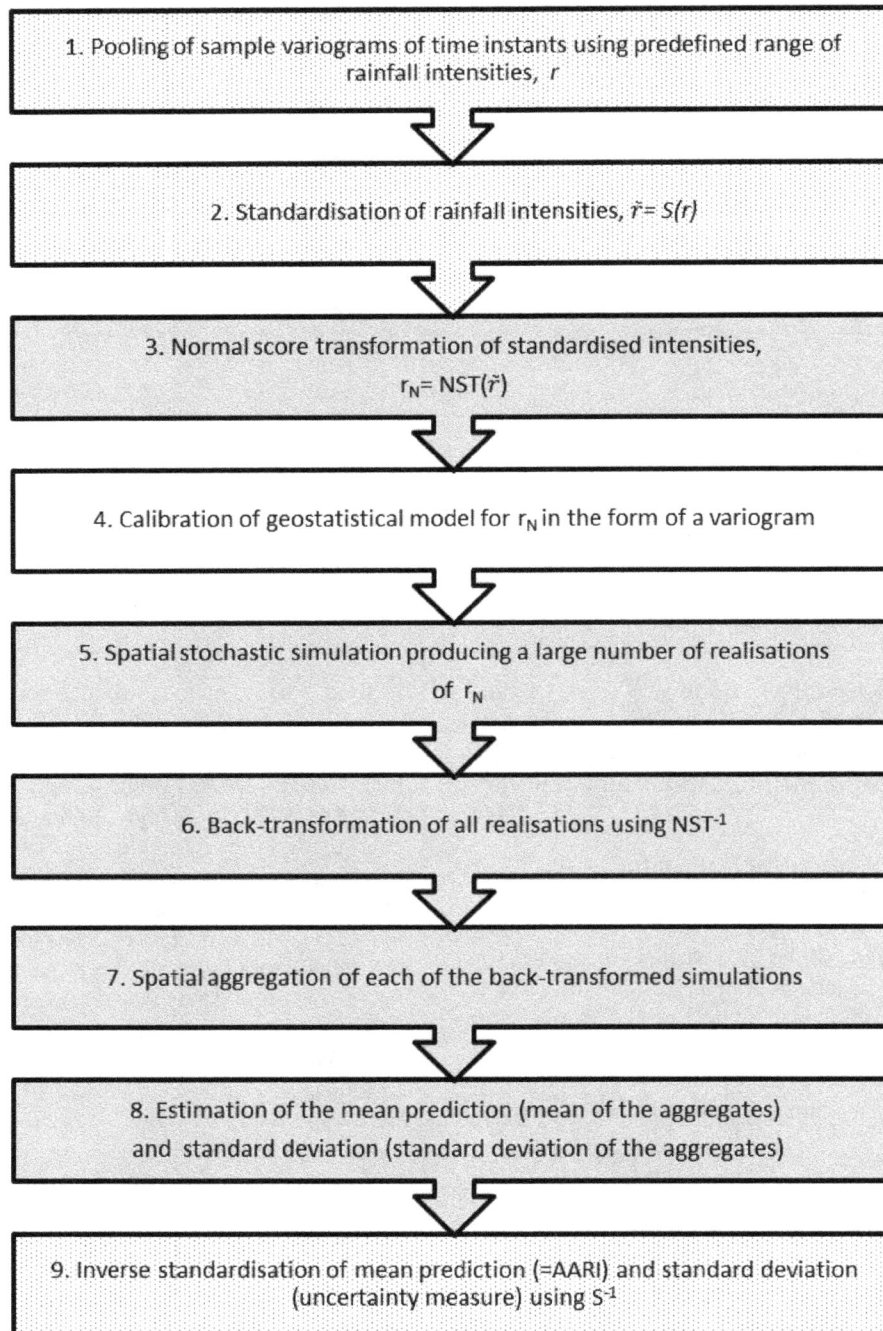

Figure 4. Step-by-step procedure developed in this study to predict AARI and associated level of uncertainty. Boxes highlighted in dots indicate the steps to resolve the problem of scarcity in measurement locations, grey boxes show the steps introduced to address non-normality of rainfall data.

depend on explanatory variables (e.g. topography of the area, wind direction). The stochastic term ε is spatially correlated and characterised by a variogram model. A variogram model typically consists of three parameters; nugget, sill, and range (Isaaks and Srivastava, 1989). The nugget is the value of the semi-variance at near-zero distance. It is often greater than zero because of random measurement error and micro-scale spatial variation. The range is the distance beyond which the data are no longer spatially correlated. The sill is the maximum variogram value and equal to the variance of the variable of interest (Isaaks and Srivastava, 1989)

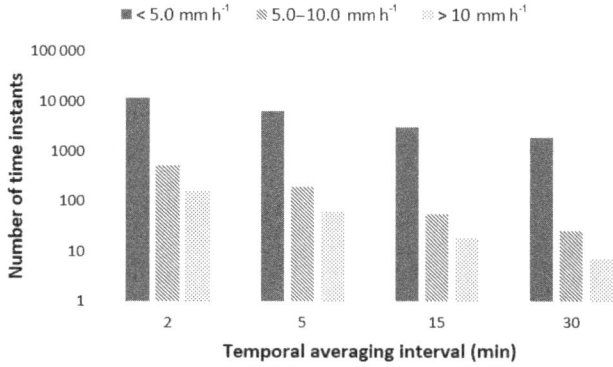

Figure 5. Number of time instants for each temporal averaging interval and rainfall intensity class combination.

3.5   Step 5: spatial stochastic simulation

The assumption of a constant trend makes that the spatial interpolation can be solved using an ordinary kriging system (Isaaks and Srivastava, 1989):

$$\sum_{y=1}^{n} w_y \gamma_{xy} - \mu = \gamma_{xz} \ \forall x = 1, \ldots, n \tag{4}$$

$$\sum_{x=1}^{n} w_x = 1, \tag{5}$$

where w_x, $x = 1, \ldots, n$ are ordinary kriging weights, γ_{xy} is the semivariance between rainfall intensities at locations x and y, γ_{xz} is the semivariance between rainfall intensities at location y and prediction location z, and μ is a Lagrange parameter. Once the ordinary kriging weights are calculated using Eqs. (4) and (5), point rainfall intensities can be predicted using point kriging at any given point by taking the weighted average of the observed rainfall intensities, using the w_x as weights. In this case we need a change of support from point to block as our intention is to predict the average rainfall intensity over the catchment. This is usually done by predicting at all points inside the catchment and integrating these over the catchment. This procedure is known as block kriging (Isaaks and Srivastava, 1989), which also has provisions for calculating the prediction error variance of the catchment average. But the procedure of NST as explained in Sect. 3.3 also involves back-transformation of kriging predictions to the original domain at the end (step 6). Since this transformation is typically non-linear, the back-transform of the spatial average of the transformed variable that is obtained from block kriging is not the same as the spatial average of the back-transformed variable; we need the latter and not the former. In principle, we could predict at all points within the block, back-transform all and next calculate the spatial average, but standard block kriging software implementations do not support this and neither is it possible to compute the associated prediction error variance. Hence block kriging cannot

be applied. The alternative used in this study is to apply a computationally more demanding spatial stochastic simulation approach, which involves generation of a larger number of realisations and spatial averaging of these realisations. Unlike kriging, spatial stochastic simulation does not aim to minimise the prediction error variance but focuses on the reproduction of the statistics such as the histogram and variogram model (Goovaerts, 2000). The output from spatial stochastic simulation is a set of alternative rainfall realisations ("possible realities"). The mean of a large set of realisations approximates the kriging prediction, while their standard deviation approximates the kriging standard deviation. We used the sequential Gaussian simulation algorithm which involves the following steps (Goovaerts, 2000):

i. Define a prediction grid (a 25 m × 25 m regular grid in this case).

ii. Visit a randomly selected grid cell that has not been visited before and predict the transformed rainfall intensity at the grid cell centre using ordinary kriging; this yields a kriging prediction and a kriging standard deviation.

iii. Use a pseudo-random number generator to sample from a normal distribution mean equal to the kriging prediction and standard deviation equal to the kriging standard deviation and assign this value to the grid cell centre.

iv. Add the simulated value to the conditioning data set; in other words treat the simulated value as if it were another observation.

v. Go back to step (ii) and repeat the procedure until there are no more unvisited grid cells left.

The five steps above produce a single realisation. This must be repeated as many times as the number of realisations required (500 in this study). It must also be repeated for each time instant, which explains that the computational burden can be high. Implementation of these steps with the gstat package in R (Pebesma, 2004) is straightforward.

The grid size and number of simulations (i.e., the sample size) were selected considering the spatial resolution of available measurements and computational demand. It was observed that neither a finer grid nor more simulations improved the results significantly. Increasing the resolution to 10 m × 10 m only reduces the standard deviation of the prediction by less than 5 % in most cases while making the computational time six times higher (a summary on computation power is presented as Supplement).

3.6   Steps 6–9: calculation of AARI and associated uncertainty

Once the realisations have been prepared these are back-transformed by applying the inverse of Eq. (2) to all grid cells (step 6). Some values derived from spatial stochastic simulation were outside the transformed data range. Hence during

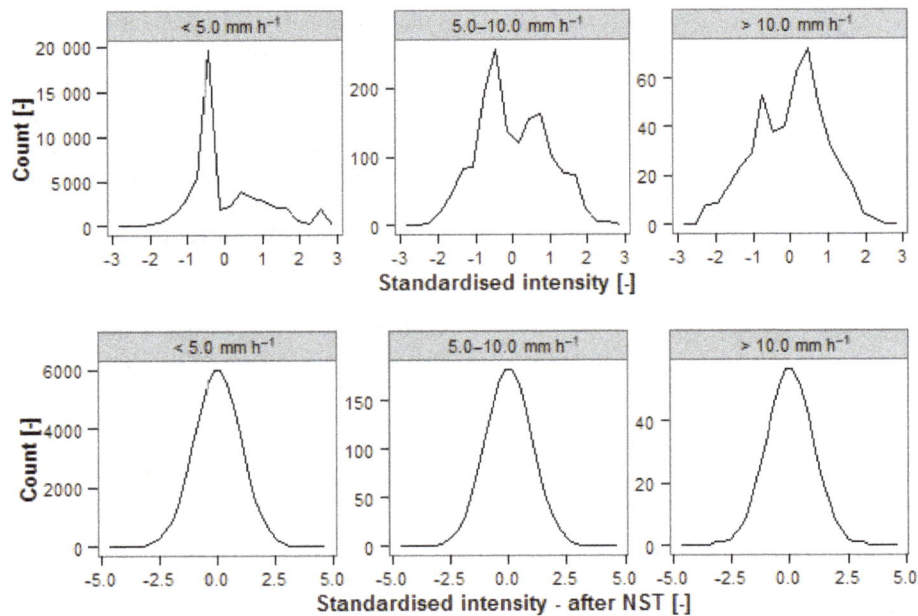

Figure 6. Distribution of standardised rainfall intensity for different rainfall intensity classes at a temporal averaging interval of 5 min before (upper part) and after (lower part) normal score transformation (NST).

back transformation (step 6) of these values linear extrapolation was used. These linear models were derived using a selected number of head and tail portion of normal Q–Q plot. This is one of the simplest and most commonly used solutions for NST back-transformation (Bogner et al., 2012; Weerts et al., 2011). Considering the scope of this study and the relatively small number of data which had to be extrapolated, other extrapolation methods were not explored. After step 6, the back-transformed realisations are spatially averaged one by one (step 7). This yields as many spatially averages as the number of realisations that had been generated in step 5. This set of values is a simple random sample from the probability distribution of the catchment average rainfall. Thus, the sample mean and standard deviation provide estimates of the mean and standard deviation of the distribution, respectively (step 8). Finally, by doing the inverse standardisation of the mean and standard deviation of the distribution to account for step 2, the AARI and associated uncertainty measure (standard deviation) were derived (step 9).

4 Results and discussion

4.1 Calibration of the geostatistical model of rainfall

As explained in Sect. 3.4, the geostatistical model of transformed rainfall data were calibrated using variograms for three different intensity ranges. This procedure was repeated for temporal averaging intervals of 2, 5, 15, and 30 min. Ex-

ponential models were fitted to empirical variograms. The resulting variograms are presented in Fig. 7.

The variograms illustrate two properties of the collected rainfall measurements: spatial variability of rainfall and measurement error. One of the main parameters which characterises these properties is the nugget. Theoretically at zero lag distance the variance should be zero. However, most of the variograms exhibit a positive nugget effect (generally presented as nugget-to-sill ratio) at zero lag distance. This nugget effect can be due to two reasons: random measurement error and microscale spatial variability of rainfall. Unfortunately we cannot quantify these causes individually using the variograms. But there is a consistent pattern of nugget against both rainfall intensity class and temporal averaging interval which helps to interpret the variograms.

Considering the behaviour of nugget-to-sill ratio against rainfall intensity class, it can be observed that the smaller the intensity the higher the nugget-to-sill ratio, regardless of temporal averaging interval. For example, at 2 min averaging interval the nugget-to-sill ratio increases from zero to almost one (nugget variogram) as the rainfall intensity class changes from > 10 to < 5 mm h^{-1}. The pure nugget variogram at < 5 mm h^{-1} means that either there is no spatial correlation at the regarded distance, or the spatial correlation of the field cannot be detected by the measurements because of the measurement error. Looking at the behaviour of nugget-to-sill ratio against temporal averaging interval, Fig. 7 shows that the smaller the averaging interval the higher the nugget-to-sill ratio, regardless of rainfall intensity class. For exam-

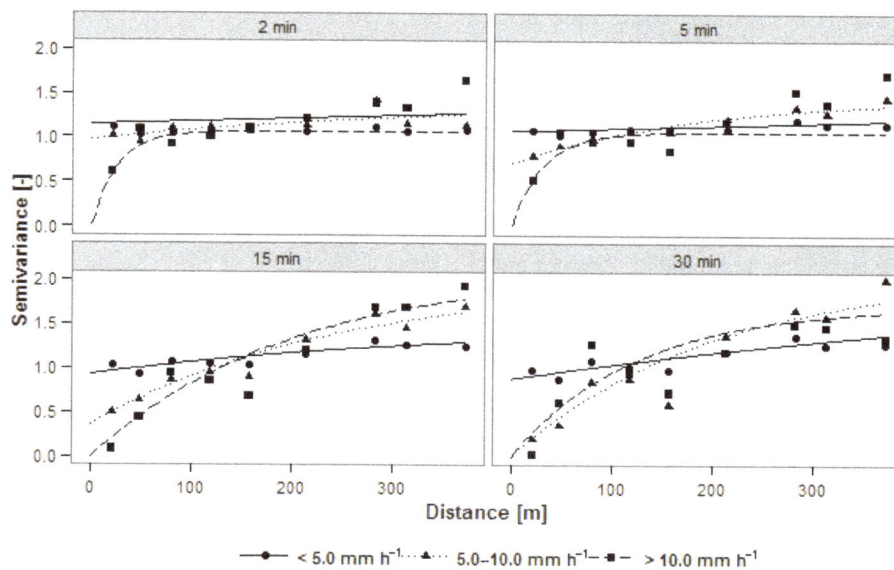

Figure 7. Calculated variograms for each intensity class within each temporal averaging interval.

ple, for rainfall intensity class 5.0–10.0 mm h^{-1} the nugget-to-sill ratio decreases from almost one to zero as the temporal averaging interval increases from 2 to 30 min. Overall these observations show that the combined effect of random measurement error and microscale special variability of rainfall characterised by nugget-to-sill ratio decreases with increasing (a) rainfall intensity class and (b) averaging interval.

Regarding the behaviour of the nugget-to-sill ratio against averaging interval, it is expected that with the averaging interval the (microscale) spatial correlation of rainfall would increase, which partly explains the observed pattern. The increase in spatial correlation of rainfall intensity with increasing temporal averaging interval agrees with other similar studies (e.g. Ciach and Krajewski, 2006; Fiener and Auerswald, 2009; Krajewski et al., 2003; Peleg et al., 2013; Villarini et al., 2008). For example, Krajewski et al. (2003) observed in their study on analysis of spatial correlation structure of small-scale rainfall in central Oklahoma a similar behaviour using correlogram functions for different temporal averaging intervals. But commenting on the decreasing trend of the nugget-to-sill ratio against intensity class, it cannot be attributed to improvement in microscale spatial correlation as it is neither natural nor proven. In fact, in Fig. 7 the behaviour of spatial correlation against rainfall intensity class does not show a distinctive trend except at the origin, i.e. the nugget effect. The absence of any consistent trend of spatial variability against intensity class was also observed in Ciach and Krajewski (2006). Meanwhile this decreasing trend of nugget-to-sill ratio against rainfall intensity corresponds well with measurement errors of tipping bucket type rain gauges caused by its sampling mechanism (hereafter referred to as TB error). This is due to the rain gauges' inability to capture small temporal variability of the rainfall time series. The

behaviour of TB error against rainfall intensity as seen from Fig. 7 complements results from previous studies (Habib et al., 2001; Villarini et al., 2008). These studies also show that the TB error decreases with temporal averaging interval. Habib et al. (2001) found similar behaviour of TB error with increasing intensity (0–100 mm h^{-1}) and also with increasing averaging interval (1, 5, and 15 min). Although the bucket size used in their study (0.254 mm) is slightly different from our rain gauge bucket size of 0.2 mm, the characteristic of the TB error against rainfall intensity for different averaging interval is consistent in both cases. In summary, the behaviour of nugget-to-sill ratio of the variograms against temporal averaging interval can be explained by the combined effect of microscale spatial variability of rainfall and TB error, while the behaviour of nugget-to-sill ratio against intensity range can mainly be attributed to the latter.

In addition to the nugget-to-sill ratio, another parameter that characterises the variograms is the range, i.e. the distance up to which there is spatial correlation. At lower temporal averaging intervals (≤ 5 min) the variograms for all rainfall intensity classes reach the variogram range very quickly (< 100 m). But at averaging intervals ≥ 15 min, the range has not been reached even at a maximum separation distance, showing the improvement in spatial correlation. High spatial variability of rainfall at shorter temporal averaging interval (≤ 5 min) is an important observation in the context of urban drainage runoff modelling, as the time step used in such models is generally around 2 min for small catchments.

The fact that the data set covers only 10 months of data from 2 years with varying climatology is something that needs to be acknowledged. However, for previous studies using such a dense network the duration of data collection is similar (e.g. 15 months – Ciach and Krajewski, 2006;

16 months – Jaffrain and Berne, 2012). These time periods are reflection of the practical and funding issues to maintain such dense networks operating accurately for extended periods. The characteristics of our data are comparable with Ciach and Krajewski (2006) and Fiener and Auerswald (2009) as these studies also used rainfall data from warm months to investigate the spatial correlation structure. Despite the fact that the data cover only 10 months all derived variogram models are stable and reliable. Webster and Oliver (2007) suggested around 100 samples to reliably estimate a variogram model. Even in the case of 30 min temporal averaging interval and $> 10\,mm\,h^{-1}$ (where we had the fewest number of observations) we had a total of 196 spatial lags to calculate the variogram. Furthermore, we demonstrated that all derived variogram models are stable and reliable by examining sub-sets of the data. We randomly selected 80 % of the data from each intensity class and reproduced the variograms to compare them with the variograms presented in Fig. 7. We had to limit the subclass percentage to 80 % to give enough time instants to reproduce variograms for all subclasses. We repeated this procedure a few times. Comparing these variograms with Fig. 7 shows that these variograms are very similar. One set of the variograms computed from 80 % of the data are presented in the Supplement. This analysis supports our claim that the variograms shown in Fig. 7 are stable and an adequate representation of the rainfall spatial variation for each intensity class and temporal averaging interval.

One of the assumptions we made during the pooling procedure is that the spatial variability is reasonably consistent within a pooled intensity class. We acknowledge that with narrower intervals the assumption of consistency in spatial variability would be more realistic. But with the available data we had to find a compromise with the number of time instants. We believe that using three intensity subclasses is a reasonable compromise. Further we also introduced step 2 (Sect. 3.2) which standardises the rainfall for each time instant within a subset. Although variograms are derived only for the whole subset, step 2 (before geostatistical upscaling) and step 9 (after geostatistical upscaling) ensure that the probabilistic model is adjusted for each time instant separately. Effectively, we assume the same correlogram for time instants of the same subclass, not the same variogram. Although this does not justify the assumption of similar spatial correlation structure within the pooled classes, it at least relaxes the assumption of the same variogram within subclasses. To compare the behaviour of variogram models for a narrower intensity interval, we produced variograms for narrower intensity classes ranging from 0 to $14\,mm\,h^{-1}$ for the 5 min averaging interval. The highest intensity class is limited to ≥ 12 to $< 14\,mm\,h^{-1}$ as for further narrower ranges (i.e ≥ 14 to $< 16\,mm\,h^{-1}$ and so on) there are not enough sample points to produce a meaningful variogram. Narrower intensity classes means that the assumption of similar spatial variability within a pooled subset is more realistic. Compar-

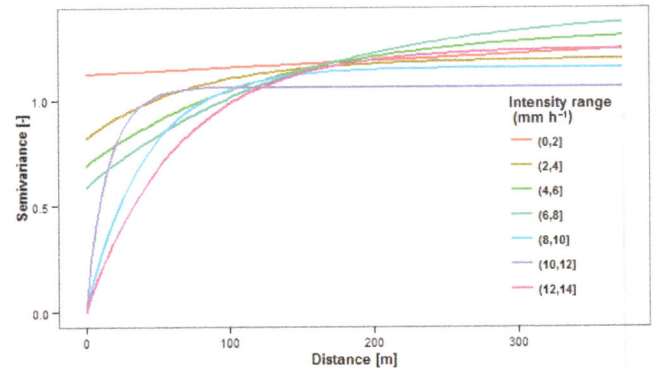

Figure 8. Calculated variograms for a narrower range of intensity at 5 min averaging interval.

ing Figs. 7 and 8, we conclude that the variograms shown in Fig. 7 are accurate representations of the average spatial variability conditions for corresponding intensity classes.

4.2 Geostatistical upscaling of rainfall data

Having calculated all variograms, the next step is to apply spatial stochastic simulation for the time instants of interest followed by steps 6 to 9 in Fig. 4 to calculate the AARI together with associated uncertainty. This procedure was carried out for all events presented in Table 1. The following sections present and discuss the predicted AARI and associated uncertainty levels derived from step 9.

4.2.1 Prediction error vs. AARI

The scatter plot in Fig. 9 shows the coefficient of variation of the prediction error (CV; see Eq. 6) plotted against predicted AARI at 5 min averaging interval for all time instants of all events presented in Table 1:

$$CV = \frac{AARI\ prediction\ error\ standard\ deviation}{Predicted\ AARI} \times 100\%. \quad (6)$$

The uncertainty level in the prediction of AARI represented by the CV is due to the combined effect of both spatial variability of rainfall and TB error in the rainfall data. It can be seen here that there is a clear trend of decreasing CV with increasing AARI. The CV values are as high as 80 % when the AARI is smaller than $1\,mm\,h^{-1}$ and they get reduced to less than 10 % when AARI is larger than $10\,mm\,h^1$. In a previous study by Pedersen et al. (2010) using rainfall measurements from similar tipping bucket type rain gauges, they also found that the uncertainty in prediction of mean rainfall depth decreases with increasing mean rainfall depth, but due to the limited information in their results they could not analyse this observation in detail. But here it is clear that this observation corresponds well with what we already observed in variograms in Fig. 7. These variograms show higher nugget-to-sill ratio at lower intensity due to high TB error consequently

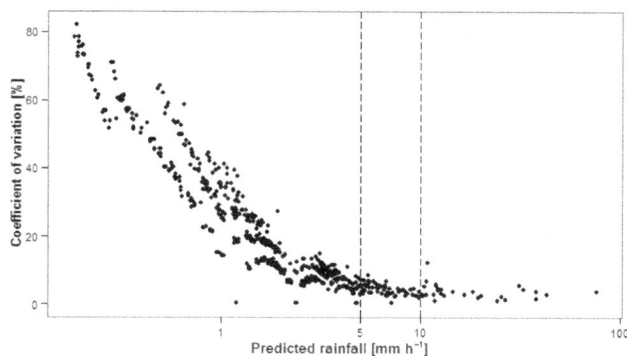

Figure 9. AARI prediction error CV (%) values against predicted AARI for averaging interval of 5 min.

causing higher uncertainty in the prediction of AARI. At intensity class $0–5\,\mathrm{mm\,h^{-1}}$ the nugget-to-sill ratio was almost one (nugget variogram) and as a result the derived CV values are significantly higher than other two intensity classes. It is interesting to note that, in the range of $1–10\,\mathrm{mm\,h^{-1}}$, there are few points that are separated from the larger cluster with almost zero CV. It shows a consistent rainfall measurement over the area at these time instants, which results in a very small CV in the predicted AARI.

The above discussion is based on results from 5 min temporal averaging interval. The following section discusses the effect of temporal averaging interval on prediction error. Further, although CV in Fig. 9 gets as high as 80 %, the corresponding AARI is less than $1\,\mathrm{mm\,h^{-1}}$, thus the prediction error has a very less significance in urban hydrology. Hence we also analysed the prediction error associated with rainfall events' peaks in the last section.

4.2.2 Prediction error vs. temporal averaging interval

Having analysed the behaviour of the prediction error CV against predicted AARI, this section presents the effect of temporal averaging interval on the prediction error of AARI. Figure 10 shows the kriging predictions with 95 % prediction intervals derived from the prediction standard deviation for temporal averaging intervals of 2, 5, 15, and 30 min for event 11. Event 11 has average conditions in terms of event duration and peak intensity. Prediction errors of other events against the temporal averaging interval follow the same pattern of behaviour.

While short time intervals are of greater interest in urban hydrology, they also lead to large uncertainties. Figure 10 shows the smaller the temporal averaging interval, the larger the prediction interval and the larger the level of uncertainty. This is due to the combined effect of higher spatial variability and larger TB error at lower temporal averaging interval as seen from Fig. 7. When the averaging interval is larger than 15 min the prediction interval width becomes negligible. But temporal scales of interest in urban hydrology of a similar-

sized catchment can be as low as 2 min where there is still considerable uncertainty. The 95 % prediction interval shows around ± 13 % of error in rainfall intensity corresponding to a prediction of peak rainfall of $47\,\mathrm{mm\,h^{-1}}$ at 2 min averaging interval. While temporal aggregation decreases uncertainty, it obviously leads to a significant reduction of the predicted peaks of AARI. For example, the peak of event 11 gets reduced to around $20\,\mathrm{mm\,h^{-1}}$ from around $50\,\mathrm{mm\,h^{-1}}$ when averaging interval increases from 2 to 30 min. Hence a careful trade-off between temporal resolution and accuracy in rainfall prediction is needed to decide the most appropriate time step for averaging point rainfall data for urban hydrologic applications.

The decreasing trend of uncertainty in the prediction of AARI with increasing temporal averaging interval agrees with a previous study by Villarini et al. (2008). Although the spatial extent of their study is much larger ($360\,\mathrm{km^2}$), their results also show that the spatial sampling uncertainties tend to decrease with increasing temporal averaging interval due to improvement in measurement accuracy and improved spatial correlation.

4.2.3 Prediction error vs. peak rainfall intensity

In addition to rainfall event durations, rainfall event peaks are also of significant interest in urban hydrology as most of the hydraulic structures in urban drainage systems are designed based on peak discharge which is often derived from peak rainfall. Hence it is important to consider the uncertainty in prediction of peaks of AARI. Figure 11 presents predicted peaks of AARI for all 13 events presented in Table 1, together with labels indicating corresponding CV (%) values. The peak intensities range from 6 to $92\,\mathrm{mm\,h^{-1}}$ at 2 min averaging interval and this range narrows down to 3–$21\,\mathrm{mm\,h^{-1}}$ at averaging interval of 30 min as a result of temporal aggregation. As expected, temporal aggregation from 2 to 30 min also results in the reduction of CV. The highest CV at 2 min averaging intervals is 13 % for event 4 and reduces to 1.7 % at 30 min averaging interval. But it can also be noted that events 5, 6, 8, and 11 show their highest CV at 5 min averaging interval and not at 2 min averaging interval. Tracking back these events, they indeed show more spatial variation over 5 min period compared to 2 min period around the peak.

As discussed in Sect. 4.2.1, CV decreases with increasing predicted rainfall peaks and this effect is dominant when the averaging interval is at the lowest, i.e. 2 min. This is when the TB error is at its highest. When the temporal averaging interval is 30 min where the TB error is at its lowest, the difference between CV for lower ($<10\,\mathrm{mm\,h^{-1}}$) and higher ($>10\,\mathrm{mm\,h^{-1}}$) intensity becomes smaller. At 30 min averaging interval the mean CV below and above $10\,\mathrm{mm\,h^{-1}}$ are 1.7 and 1.2 % respectively, but they increase to 6.6 and 3.5 % at 2 min averaging interval. The maximum CV at 2 min averaging interval are 13 and 6.8 % for lower ($<10\,\mathrm{mm\,h^{-1}}$) and

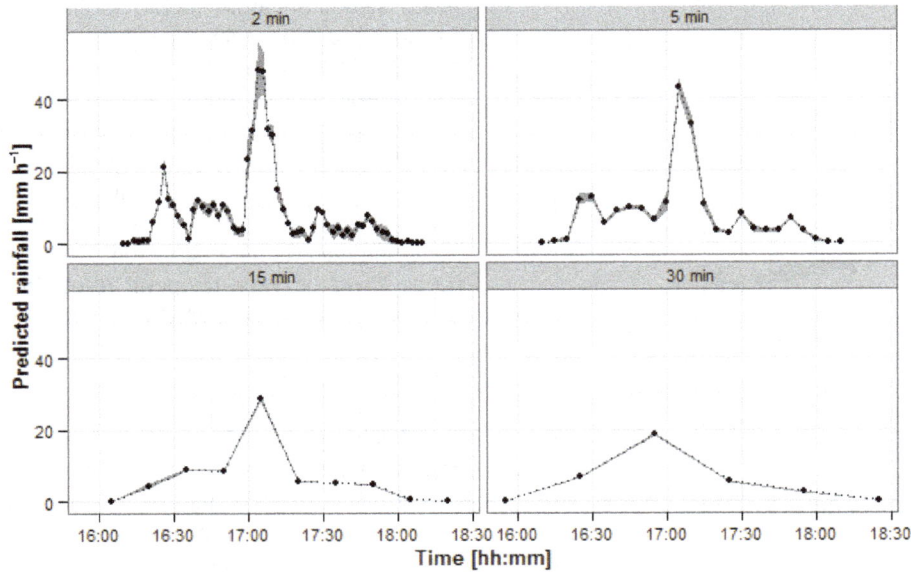

Figure 10. Predictions of AARI (indicated by points) together with 95 % prediction intervals (indicated by grey ribbon) for rainfall event 11 for different averaging intervals.

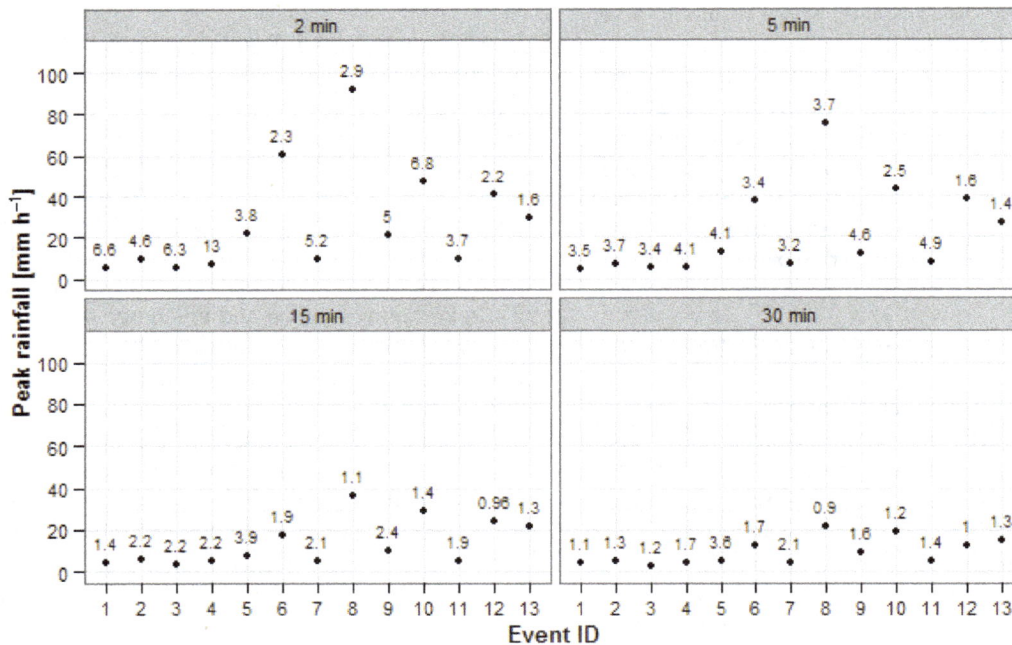

Figure 11. Predictions of event peaks of AARI (indicated by points) together with labels indicating corresponding CV (%) values.

higher ($> 10 \, \text{mm} \, \text{h}^{-1}$) rainfall intensity respectively. Even though these values are significantly less than what we observed from Fig. 9 when the rainfall intensity is less than $1 \, \text{mm} \, \text{h}^{-1}$, they are still high considering the required accuracy defined in standard guidelines of urban hydrological modelling practice. For example, the current urban drainage verification guideline (WaPUG, 2012) in the UK defines a maximum allowable deviation of 25 to −15 % in peak runoff

demanding more accurate prediction of rainfall which is the main driver of the runoff process in urban areas. A 13 % uncertainty in rainfall will result in a similar level of uncertainty in runoff prediction for a completely impervious surface according to the well-established rational formula (Viessman Jr. and Lewis, 1995) which is still widely used for estimating design discharge in small urban catchments.

5 Conclusions

Geostatistical methods have been used to analyse the spatial correlation structure of rainfall at various spatial scales, but its application to estimate the level of uncertainty in rainfall upscaling has not been fully explored mainly due to its inherent complexity and demanding data requirements. In this study we presented a method to overcome these challenges and predict AARI together with associated uncertainty using geostatistical upscaling. We used a spatial stochastic simulation approach to address the combination of change of support (from point to catchment) and non-normality of rainfall observations for prediction of AARI and the associated uncertainty. We addressed the issue of scarcity in measurement points by using repetitive rainfall measurements (pooling) to increase the number of spatial samples used for variogram estimation. The methods were illustrated with rainfall data collected from a cluster of eight paired rain gauges in a 400 m × 200 m urban catchment in Bradford, UK. The spatial lag ranges from 21 to 399 m. As far we are aware these are the smallest lag ranges in which spatial variability in rainfall is examined in an urban area using point rainfall measurements. We defined intensity classes and derived different geostatistical models (variograms) for each intensity class separately. We also used different temporal averaging intervals, ranging from 2 to 30 min, which are of interest in urban hydrology. To the best of our knowledge this is the first such attempt to assign geostatistical models for a combination of intensity class and temporal averaging interval. Finally, we quantified the level of uncertainty in the prediction of AARI for these different combinations of temporal averaging intervals and rainfall intensity ranges.

A summary of the significant findings is listed below:

- Several studies (e.g. Berne et al., 2004; Gebremichael and Krajewski, 2004; Krajewski et al., 2003) used a single geostatistical model in the form of variogram/correlogram for the entire range of rainfall intensity. The current study shows that for small time and space scales the use of a single geostatistical model based on a single variogram is not appropriate and a distinction between rainfall intensity classes and length of temporal averaging intervals should be made.

- The level of uncertainty in the prediction of AARI using point measurement data essentially comes from two sources: spatial variability of the rainfall and measurement error. The significance and characteristics of the measurement error observed here mainly corresponds to sampling related error of tipping bucket type rain gauges (TB error) and may vary for other types of rain gauges.

- TB error decreases with increasing rainfall intensity. As a result of that, the prediction error decreases with increasing AARI. At 5 min averaging interval the CV values are as high as 80 % when the AARI is smaller than 1 mm h^{-1} and they get reduced to less than 10 % when AARI is larger than 10 mm h^{-1}.

- At smaller temporal averaging intervals, the effect of both spatial variability and TB error is high, resulting in higher uncertainty levels in the prediction of AARI. With increasing temporal averaging interval the uncertainty becomes smaller as the spatial correlation increases and the TB error reduces. At 2 min temporal averaging interval the average CV in the prediction of peak AARI is 6.6 % and the maximum CV is 13 % and they are reduced to 1.5 and 3.6 % respectively at 30 min averaging interval.

- TB error at averaging intervals of less than 5 min, especially at low-intensity rainfall measurements, is as significant as spatial variability. Hence proper attention to TB error should be given in any application of these measurements, especially in urban hydrology, where averaging intervals are often as small as 2 min.

Although the spatial stochastic simulation method used in this study needs more computational power (a summary on computation power is presented in the Supplement) than block kriging, it is a robust approach and allows data transformation during spatial interpolation and aggregation. Such data transformation is important because rainfall data are not normally distributed for small temporal averaging intervals. The pooling procedure used in this study helps provide a solution to meet the data requirements for geostatistical methods as it extends the available information for variogram estimation. Commenting on the minimum number of measurement points needed to employ this method is difficult, because like any other geostatistical interpolation method, the efficiency of this method also heavily depends on reliable estimation of the geostatistical model (variogram). Hence, it basically comes down to the question of whether or not a given measurement network can produce a meaningful variogram. As mentioned, Webster and Oliver (2007) advised that around 100 measurement points are needed to adequately estimate a geostatistical model. But there is no single universal rule to define the minimum number of bins and the number of samples for each bin to produce a reliable variogram. Further, since pooling sample variograms of repeated measurements would produce a multiplication of spatial lags, the size of the available data set would also play a role in deciding the minimum number of measurement points.

An urban catchment of this size needs rainfall data at a temporal and spatial resolution which is higher than the resolution of most commonly available radar data (1000 m, 5 min). In addition the level of uncertainty in radar measurements would be much higher than that of point measurements, especially at a small averaging interval (< 5 min, Seo and Krajewski, 2010; Villarini et al., 2008), which are often of interest in urban hydrology. Hence, experimental rain gauge data similar to the ones used in this study are crucial for similar studies focused on small urban catchments.

Results from this study can be used for uncertainty analyses of hydrologic and hydrodynamic modelling of similar-sized urban catchments in similar climates as it provides information on uncertainty associated with rainfall estimation which is arguably the most important input in these models. This information will help to differentiate input uncertainty from total uncertainty thereby helping to understand other sources of uncertainty due to model parameter and model structure. This estimate of the relative importance of uncertainty sources can help to avoid false calibration and force fitting of model parameters (Vrugt et al., 2008). This study can also help to judge optimal temporal averaging interval for rainfall estimation of hydrologic and hydrodynamic modelling especially for small urban catchments.

Competing interests. The authors declare that they have no conflict of interest.

Acknowledgements. This research was done as part of the Marie Curie ITN – Quantifying Uncertainty in Integrated Catchment Studies (QUICS) project. This project has received funding from the European Union's Seventh Framework Programme for research, technological development and demonstration under grant agreement no. 607000.

Edited by: P. Molnar

References

Berne, A., Delrieu, G., Creutin, J. D., and Obled, C.: Temporal and spatial resolution of rainfall measurements required for urban hydrology, J. Hydrol., 299, 166–179, doi:10.1016/j.jhydrol.2004.08.002, 2004.

Bogner, K., Pappenberger, F., and Cloke, H. L.: Technical Note: The normal quantile transformation and its application in a flood forecasting system, Hydrol. Earth Syst. Sci., 16, 1085–1094, doi:10.5194/hess-16-1085-2012, 2012.

Chow, V. T.: Handbook of Applied Hydrology. A compendium of water-resources technology, McGraw-Hill, New York, 1964.

Ciach, G. J. and Krajewski, W. F.: Analysis and modeling of spatial correlation structure in small-scale rainfall in Central Oklahoma, Adv. Water Resour., 29, 1450–1463, doi:10.1016/j.advwatres.2005.11.003, 2006.

Dirks, K. N., Hay, J. E., Stow, C. D., and Harris, D.: High-resolution studies of rainfall on Norfolk Island Part II: Interpolation of rainfall data, J. Hydrol., 208, 187–193, doi:10.1016/S0022-1694(98)00155-3, 1998.

Emmanuel, I., Andrieu, H., Leblois, E., and Flahaut, B.: Temporal and spatial variability of rainfall at the urban hydrological scale, J. Hydrol., 430–431, 162–172, doi:10.1016/j.jhydrol.2012.02.013, 2012.

Fankhauser, R.: Influence of systematic errors from tipping bucket rain gauges on recorded rainfall data, Water Sci. Technol., 37, 121–129, doi:10.1016/S0273-1223(98)00324-2, 1998.

Fiener, P. and Auerswald, K.: Spatial variability of rainfall on a sub-kilometre scale, Earth Surf. Processes, 34, 848–859, doi:10.1002/esp.1779, 2009.

Gebremichael, M. and Krajewski, W. F.: Assessment of the statistical characterization of small-scale rainfall variability from radar: Analysis of TRMM ground validation datasets, J. Appl. Meteorol., 43, 1180–1199, doi:10.1175/1520-0450(2004)043<1180:AOTSCO>2.0.CO;2, 2004.

Gires, A., Onof, C., Maksimovic, C., Schertzer, D., Tchiguirinskaia, I., and Simoes, N.: Quantifying the impact of small scale unmeasured rainfall variability on urban runoff through multifractal downscaling: A case study, J. Hydrol., 442–443, 117–128, doi:10.1016/j.jhydrol.2012.04.005, 2012.

Gires, A., Tchiguirinskaia, I., Schertzer, D., Schellart, A., Berne, A., and Lovejoy, S.: Influence of small scale rainfall variability on standard comparison tools between radar and rain gauge data, Atmos. Res., 138, 125–138, doi:10.1016/j.atmosres.2013.11.008, 2014.

Glasbey, C. A. and Nevison, I. M.: Modelling Longitudinal and Spatially Correlated Data, edited by: Gregoire, T. G., Brillinger, D. R., Diggle, P. J., Russek-Cohen, E., Warren, W. G., and Wolfinger, R. D., Springer, New York, 233–242, 1997.

Goovaerts, P.: Estimation or simulation of soil properties? An optimization problem with conflicting criteria, Geoderma, 97, 165–186, doi:10.1016/S0016-7061(00)00037-9, 2000.

Habib, E., Krajewski, W. F., and Kruger, A.: Sampling Errors of Tipping-Bucket Rain Gauge Measurements, J. Hydrol. Eng., 6, 159–166, doi:10.1061/(ASCE)1084-0699(2001)6:2(159), 2001.

Isaaks, E. H. and Srivastava, R. M.: An Introduction to Applied Geostatistics, Oxford University Press, New York, USA, 1989.

Jaffrain, J. and Berne, A.: Quantification of the small-scale spatial structure of the raindrop size distribution from a network of disdrometers, J. Appl. Meteorol. Clim., 51, 941–953, doi:10.1175/JAMC-D-11-0136.1, 2012.

Krajewski, W. F., Ciach, G. J., and Habib, E.: An analysis of small-scale rainfall variability in different climatic regimes, Hydrolog. Sci. J., 48, 151–162, doi:10.1623/hysj.48.2.151.44694, 2003.

Lloyd, C. D.: Assessing the effect of integrating elevation data into the estimation of monthly precipitation in Great Britain, J. Hydrol., 308, 128–150, doi:10.1016/j.jhydrol.2004.10.026, 2005.

Ly, S., Charles, C., and Degré, A.: Different methods for spatial interpolation of rainfall data for operational hydrology and hydrological modeling at watershed scale?: a review, Biotechnol. Agron. Soc., 17, 392–406, 2013.

Mair, A. and Fares, A.: Comparison of Rainfall Interpolation Methods in a Mountainous Region of a Tropical Island, J. Hydrol. Eng., 16, 371–383, doi:10.1061/(ASCE)HE.1943-5584.0000330, 2011.

MetOffice UK: MetOffice, UK, available at: http://www.metoffice.gov.uk/, last access: December 2016.

Montanari, A. and Brath, A.: A stochastic approach for assessing the uncertainty of rainfall-runoff simulations, Water Resour. Res., 40, 1–11, doi:10.1029/2003WR002540, 2004.

Nalder, I. A. and Wein, R. W.: Spatial interpolation of climatic Normals: test of a new method in the Canadian boreal forest, Agr. Forest Meteorol., 92, 211–225, doi:10.1016/S0168-1923(98)00102-6, 1998.

Ochoa-Rodriguez, S., Wang, L. P., Gires, A., Pina, R. D., Reinoso-Rondinel, R., Bruni, G., Ichiba, A., Gaitan, S., Cristiano,

E., van Assel, J., Kroll, S., Murlà-Tuyls, D., Tisserand, B., Schertzer, D., Tchiguirinskaia, I., Onof, C., Willems, P., and ten Veldhuis, M. C.: Impact of spatial and temporal resolution of rainfall inputs on urban hydrodynamic modelling outputs: A multi-catchment investigation, J. Hydrol., 531, 389–407, doi:10.1016/j.jhydrol.2015.05.035, 2015.

Pebesma, E. J.: Multivariable geostatistics in S: The gstat package, Comput. Geosci., 30, 683–691, doi:10.1016/j.cageo.2004.03.012, 2004.

Pedersen, L., Jensen, N. E., Christensen, L. E., and Madsen, H.: Quantification of the spatial variability of rainfall based on a dense network of rain gauges, Atmos. Res., 95, 441–454, doi:10.1016/j.atmosres.2009.11.007, 2010.

Peleg, N., Ben-Asher, M., and Morin, E.: Radar subpixel-scale rainfall variability and uncertainty: lessons learned from observations of a dense rain-gauge network, Hydrol. Earth Syst. Sci., 17, 2195–2208, doi:10.5194/hess-17-2195-2013, 2013.

Schuurmans, J. M. and Bierkens, M. F. P.: Effect of spatial distribution of daily rainfall on interior catchment response of a distributed hydrological model, Hydrol. Earth Syst. Sci., 11, 677–693, doi:10.5194/hess-11-677-2007, 2007.

Seo, B.-C. and Krajewski, W. F.: Scale Dependence of Radar Rainfall Uncertainty: Initial Evaluation of NEXRAD's New Super-Resolution Data for Hydrologic Applications, J. Hydrometeorol., 11, 1191–1198, doi:10.1175/2010JHM1265.1, 2010.

Sevruk, B. and Hamon, W. R.: International comparison of national precipitation gauges with a reference pit gauge, Instruments and Observing Methods, Report No. 17, WMO/TD-No. 38, World Meteorological Organization, Geneva, Switzerland, 1984.

Tabios III, G. Q. and Salas, J. D.: A comparative analysis of techniques for spatial interpolation of precipitation, J. Am. Water Resour. As., 21, 365–380, 1985.

Todini, E.: A model conditional processor to assess predictive uncertainty in flood forecasting, Int. J. River Basin Manag., 6, 123–137, doi:10.1080/15715124.2008.9635342, 2008.

Viessman Jr., W. and Lewis, G. L.: Introduction to hydrology, in: Introduction to hydrology, Harper Collins, p. 311, 1995.

Villarini, G., Mandapaka, P. V., Krajewski, W. F., and Moore, R. J.: Rainfall and sampling uncertainties: A rain gauge perspective, J. Geophys. Res.-Atmos., 113, 1–12, doi:10.1029/2007JD009214, 2008.

Van der Waerden, B. L.: Order tests for two-sample problem and their power, Indag. Math., 14, 453–458, 1952.

Vrugt, J. A., ter Braak, C. J. F., Clark, M. P., Hyman, J. M., and Robinson, B. A.: Treatment of input uncertainty in hydrologic modeling: Doing hydrology backward with Markov chain Monte Carlo simulation, Water Resour. Res., 44, W00B09, doi:10.1029/2007WR006720, 2008.

WaPUG: Code of Practice For The Hydraulic Modelling of Sewer Systems, Wastewater Planning Users Group, UK, 2002.

Webster, R. and Oliver, M. A.: Geostatistics for environmental scientists, Second Edn., John Wiley & Sons, Ltd, West Sussex, England, 2007.

Weerts, A. H., Winsemius, H. C., and Verkade, J. S.: Estimation of predictive hydrological uncertainty using quantile regression: examples from the National Flood Forecasting System (England and Wales), Hydrol. Earth Syst. Sci., 15, 255–265, doi:10.5194/hess-15-255-2011, 2011.

PERMISSIONS

LIST OF CONTRIBUTORS

R. G. Anderson
USDA, Agricultural Research Service, San Joaquin Valley Agricultural Sciences Center, Water Management Research Unit, Parlier, California, USA
USDA, Agricultural Research Service, U.S. Salinity Laboratory, Contaminant Fate and Transport Unit, Riverside, California, USA

D. Wang and J. E. Ayars
USDA, Agricultural Research Service, San Joaquin Valley Agricultural Sciences Center, Water Management Research Unit, Parlier, California, USA

R. Tirado-Corbalá
USDA, Agricultural Research Service, San Joaquin Valley Agricultural Sciences Center, Water Management Research Unit, Parlier, California, USA
Crops and Agro-Environmental Science Department, University of Puerto Rico, Mayagüez, Puerto Rico, USA

H. Zhang
USDA, Agricultural Research Service, San Joaquin Valley Agricultural Sciences Center, Water Management Research Unit, Parlier, California, USA
USDA, Agricultural Research Service, Water Management Research Unit, Fort Collins, Colorado, USA

Jutta Metzger, Manuela Nied, Ulrich Corsmeier and Christoph Kottmeier
Institute of Meteorology and Climate Research, Karlsruhe Institute of Technology (KIT), 76021 Karlsruhe, Germany

Jörg Kleffmann
Physikalische und Theoretische Chemie, Fakultät für Mathematik und Naturwissenschaften, Bergische Universität Wuppertal, 42097 Wuppertal, Germany

M. F. Müller and S. E. Thompson
Department of Civil and Environmental Engineering, Davis Hall, University of California, Berkeley CA, USA

Tobias Mosthaf and András Bárdossy
Institute for Modeling Hydraulic and Environmental Systems, Universität Stuttgart, Stuttgart, Germany

M. F. Rios Gaona and R. Uijlenhoet
Hydrology and Quantitative Water Management Group, Department of Environmental Sciences, Wageningen University, 6708 PB Wageningen, the Netherlands

H. Leijnse
R&D Observations and Data Technology, Royal Netherlands Meteorological Institute, 3731 GA De Bilt, the Netherlands

A. Overeem
Hydrology and Quantitative Water Management Group, Department of Environmental Sciences, Wageningen University, 6708 PB Wageningen, the Netherlands
R&D Observations and Data Technology, Royal Netherlands Meteorological Institute, 3731 GA De Bilt, the Netherlands

Sanjib Sharma and Alfonso Mejia
Department of Civil and Environmental Engineering, The Pennsylvania State University, University Park, PA, USA

Ridwan Siddique
Northeast Climate Science Center, University of Massachusetts, Amherst, MA, USA

Seann Reed and Peter Ahnert
National Weather Service, Middle Atlantic River Forecast Center, State College, PA, USA

Pablo Mendoza
Advanced Mining Technology Center (AMTC), Universidad de Chile, Santiago, Chile

Feinan Xu
Key Laboratory of Remote Sensing of Gansu Province, Heihe Remote Sensing Experimental Research Station, Northwest Institute of Eco-Environment and Resources, Chinese Academy of Sciences, Lanzhou, 730000, China
University of Chinese Academy of Sciences, Beijing, 100049, China

Weizhen Wang, Jiemin Wang, Yuan Qi and Yueru Wu
Key Laboratory of Remote Sensing of Gansu Province, Heihe Remote Sensing Experimental Research Station, Northwest Institute of Eco-Environment and Resources, Chinese Academy of Sciences, Lanzhou, 730000, China

Ziwei Xu
State Key Laboratory of Earth Surface Processes and Resource Ecology, Faculty of Geographical Science, Beijing Normal University, Beijing, 100875, China

Cédric Rebolho and Vazken Andréassian
Irstea, UR HYCAR, 1 Rue Pierre-Gilles de Gennes, 92160 Antony, France

Nicolas Le Moine
Sorbonne Universités, UPMC Univ Paris 06, CNRS, EPHE, UMR 7619 Metis, 4 Place Jussieu, 75005 Paris, France

Maik Renner and Axel Kleidon
Max-Planck-Institut für Biogeochemie, Jena, Germany

Theresa Blume
GFZ German Research Centre for Geosciences, Section Hydrology, Potsdam, Germany

Markus Weiler
Universität Freiburg, Hydrologie, Freiburg, Germany

Sibylle K. Hassler
GFZ German Research Centre for Geosciences, Section Hydrology, Potsdam, Germany
Karlsruhe Institute of Technology, Institute of Water and River Basin Management, Karlsruhe, Germany

Anke Hildebrandt
Max-Planck-Institut für Biogeochemie, Jena, Germany
Universität Jena, Ecological Modelling Group, Jena, Germany

Marcus Guderle
Max-Planck-Institut für Biogeochemie, Jena, Germany
Universität Jena, Ecological Modelling Group, Jena, Germany

Technische Universität München, Chair for Terrestrial Ecology, Department of Ecology and Ecosystemmanagement, Munich, Germany

Stanislaus J. Schymanski
ETH Zürich, Department of Environmental Systems Science, Zurich, Switzerland

I. Arismendi
Department of Fisheries and Wildlife, Oregon State University, Corvallis, Oregon 97331, USA

S. L. Johnson
US Forest Service, Pacific Northwest Research Station, Corvallis, Oregon 97331, USA

J. B. Dunham
US Geological Survey, Forest and Rangeland Ecosystem Science Center, Corvallis, Oregon 97331, USA

Manoranjan Muthusamy, Alma Schellart and Simon Tait
Department of Civil and Structural Engineering, University of Sheffield, Sheffield, S1 3JD, UK

Gerard B. M. Heuvelink
Soil Geography and Landscape group, Wageningen University, Wageningen, 6700, the Netherlands

Index

Radiology: A Case-Based Guide

Edited by Tyler Collins

hayle
medical

New York

Hayle Medical,
750 Third Avenue, 9th Floor,
New York, NY 10017, USA

Visit us on the World Wide Web at:
www.haylemedical.com

ISBN: 978-1-63241-580-6

Cataloging-in-Publication Data

Radiology : a case-based guide / edited by Tyler Collins.
p. cm.
Includes bibliographical references and index.
ISBN 978-1-63241-580-6
1. Medical radiology. 2. Medical radiology--Case studies.
3. Radiography, Medical. I. Collins, Tyler.
R895 .R33 2019
616.075 7--dc23

Table of Contents

Preface

Radiology is a specialization in medicine, which uses imaging techniques to diagnose and treat diseases. Imaging modalities used in radiology are projection radiography, computed tomography, magnetic resonance imaging, fluoroscopy, nuclear medicine, etc. A subspecialty of radiology is interventional radiology in which minimally invasive procedures are carried out using image guidance. Examples of these are angiography and angioplasty. Such procedures are used in the case of renal artery stenosis, peripheral vascular disease, inferior vena cava filter placement, hepatic intervention, biliary stent, etc. This book provides comprehensive insights into the field of diagnostic radiology. It traces the progress of this field and highlights some of its key aspects and applications. The extensive content of this book provides the readers with a thorough understanding of the subject.

This book is the end result of constructive efforts and intensive research done by experts in this field. The aim of this book is to enlighten the readers with recent information in this area of research. The information provided in this profound book would serve as a valuable reference to students and researchers in this field.

At the end, I would like to thank all the authors for devoting their precious time and providing their valuable contribution to this book. I would also like to express my gratitude to my fellow colleagues who encouraged me throughout the process.

Editor

Cervical Spinal Meningeal Melanocytoma Presenting as Intracranial Superficial Siderosis

Savitha Srirama Jayamma,[1] Seema Sud,[1] TBS Buxi,[1] VS Madan,[2] Ashish Goyal,[2] and Shashi Dhawan[3]

[1]Department of CT and MRI, Sir Ganga Ram Hospital, New Delhi 110060, India
[2]Department of Neurosurgery, Sir Ganga Ram Hospital, New Delhi 110060, India
[3]Department of Pathology, Sir Ganga Ram Hospital, New Delhi 110060, India

Correspondence should be addressed to Savitha Srirama Jayamma; drsavithasrija@gmail.com

Academic Editor: Atsushi Komemushi

Meningeal melanocytoma is a rare pigmented tumor of the leptomeningeal melanocytes. This rare entity results in diagnostic difficulty in imaging unless clinical and histopathology correlation is performed. In this case report, we describe a case of meningeal melanocytoma of the cervical region presenting with superficial siderosis. Extensive neuroradiological examination is necessary to locate the source of the bleeding in such patients. Usually, the patient will be cured by the complete surgical excision of the lesion.

1. Introduction

Meningeal melanocytoma is a rare tumour which arises from leptomeningeal melanocytes. They are considered to be the benign end of the neoplastic lesions of the leptomeningeal melanocytes. The preoperative diagnosis of the meningeal melanocytoma is often a diagnostic challenge as the clinical and neurological features are often nonspecific. These tumors have better prognosis than their malignant counterparts [1]. They can present as diffuse disseminations within the subarachnoid spaces, space occupying solid masses within the central nervous system. They present as slowly growing mass lesions with focal neurological deficits due to the mass effect on the adjacent tissues [2]. Here, the authors describe a case of meningeal melanocytoma of the cervical spine presenting as superficial siderosis of the central nervous system that was treated successfully with complete excision of the lesion by cervical laminectomy from C3 to C5 level. Local recurrence and leptomeningeal spread of these tumors secondary to the malignant transformation are well reported in the literature [3].

2. Case Report

A 62-year-old hypertensive, obese, diabetic, male patient with short neck presented with episodic falls, difficulty in walking, and brief loss of consciousness to our hospital. Weakness of both hands and progressive weakness of all four limbs were present since 1 week. The patient was diagnosed as complex partial seizures clinically. There was no history of trauma at the presentation or in the past. The patient had a history of subarachnoid haemorrhage 6 months back for which conservative management was done elsewhere. Magnetic resonance (MR) imaging of the brain showed mild effacement of the sulcal spaces on T1 weighted images (T1WI) and T2 weighted images (T2WI) (Figures 1(a) and 1(b)). There was a positive phase shift and blooming along the sulcal spaces on phase contrast and maximum intensity projection (MIP) susceptibility weighted images (arrow heads in Figures 1(c) and 1(d)) suggestive of superficial siderosis. There was no evidence to suggest the possible source of haemorrhage in the MR images of the brain. The remainder of general physical and systemic clinical examination was

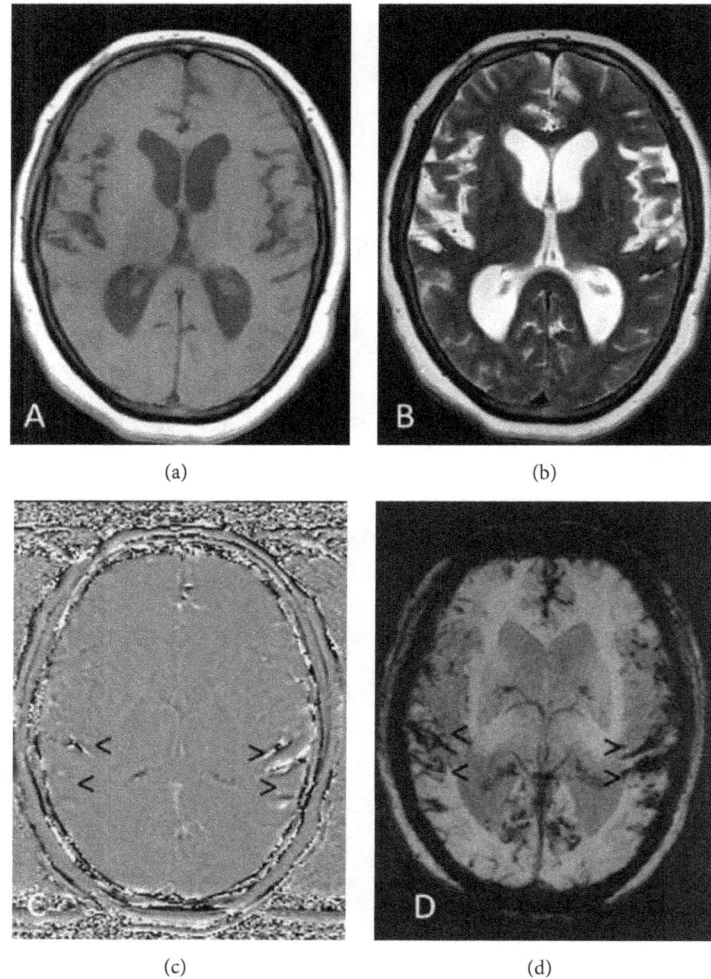

FIGURE 1: Axial sections of the brain: (a) T1 weighted and (b) T2 weighted images showing mild effacement of the sulci. Positive phase shift and blooming along the sulcal spaces (arrow heads) on (c) phase and (d) MIP (susceptibility weighted) images suggestive of superficial siderosis.

also unremarkable. Screening of the spine was done to locate the possible source of the bleeding leading to superficial siderosis.

Cervical spinal magnetic resonance imaging (MRI) revealed an intradural extramedullary mass occupying the anterior intradural space which displayed hypointense signal on T2WI and hyperintense signal on T1WI (Figures 2(a), 2(b), 3(a), and 3(b)). Contrast enhancement of the mass was not evidently revealed by visual assessment on postcontrast fat saturated T1 weighted images (Figure 2(c)) due to strong T1 hyperintensity of the mass on unenhanced images. Mild peripheral heterogeneous enhancement of the tumor was verified by the subtracted images (Figure 2(d)). The lesion was compressing and displacing the spinal cord posteriorly (arrow in Figures 3(a), 3(b), and 3(c)). The PET CT images showed moderately FDG avid extramedullary intradural mass (arrow in Figure 4) and no other foci of FDG avidity were noted elsewhere in the body. The patient underwent C3–C5 laminectomy and excision of the intradural

extramedullary mass lesion. The Lesion was placed anterior to the cord and was firm, smooth surfaced, and blackish in colour. Retrospectively, examinations of the skin and the fundus of the eye did not reveal any melanotic lesions. Hence the lesion was treated as primary cervical spinal melanocytoma which was confirmed on histopathological examination which revealed diffusely pigmented tumor with peritheliomatous arrangement which obscured the cytological details (H and E 40x) (Figure 5(a)). After bleaching, tumor had pleomorphic round to oval nuclei with inconspicuous nucleoli and moderate amount of cytoplasm (H and E 40x) (Figure 5(b)). Tumor on higher magnification had a mild nuclear pleomorphism, moderate to abundant cytoplasm with few containing blackish brown pigment. No significant increase in mitosis (<1 mitosis per 10 high power fields) was seen (H and E 200x) (Figure 5(c)). The proliferation index Ki-67 was low (<1%) which is diagnostic of melanocytomas. The patient improved symptomatically in the postoperative period.

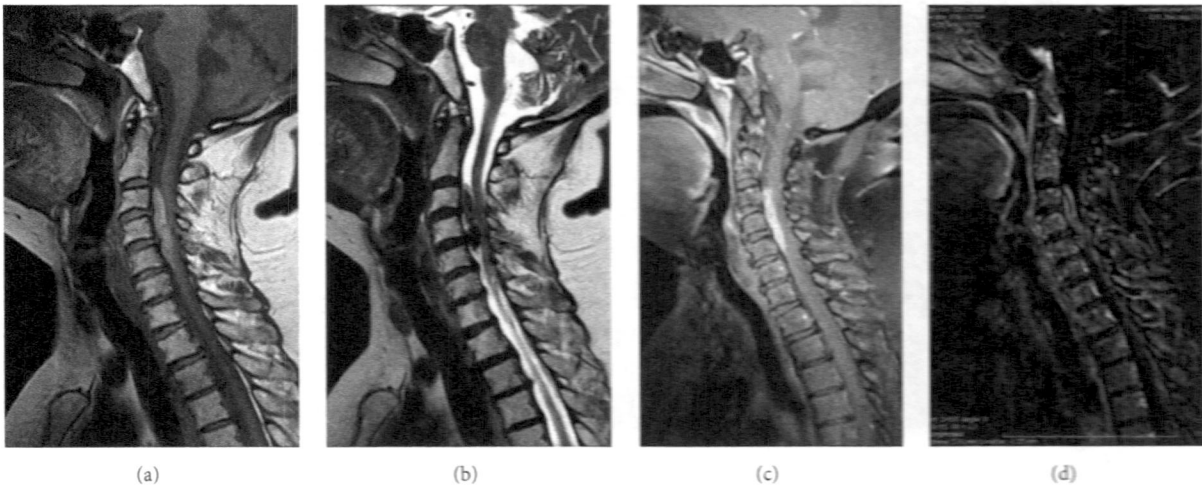

(a) (b) (c) (d)

FIGURE 2: Sagittal sections of the cervical vertebral column: (a) T2 weighted and (b) T1 weighted images showing an intradural extramedullary mass which appears hypointense on T2 weighted image and hyperintense on T1 weighted image. (c) Postcontrast fat saturated T1 weighted and (d) subtracted contrast image showing mild peripheral heterogeneous enhancement of the tumor since the true enhancement pattern was obscured by the strong T1 hyperintensity of the mass on unenhanced images.

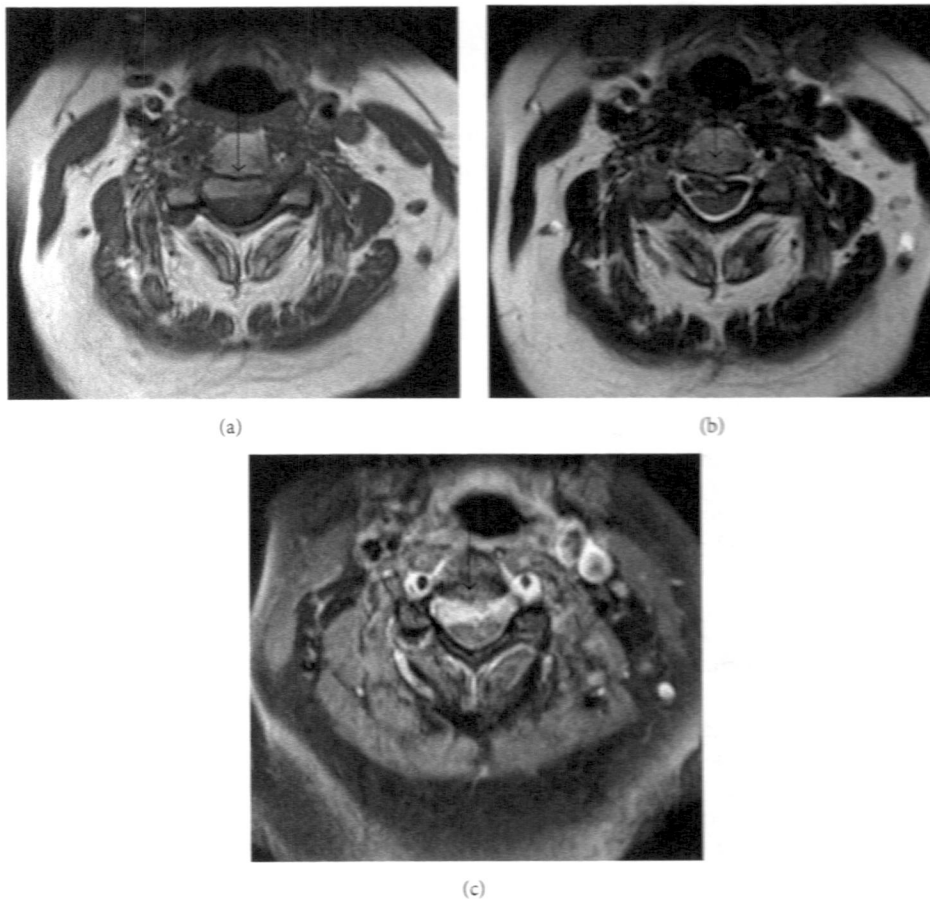

(a) (b)

(c)

FIGURE 3: Axial images of the cervical vertebral column: (a) T1 weighted and (b) T2 weighted images showing an intradural extramedullary mass (arrow) occupying the anterior intradural space, compressing and displacing the spinal cord posteriorly. (c) On postcontrast fat saturated T1 weighted images the lesion shows peripheral heterogeneous enhancement.

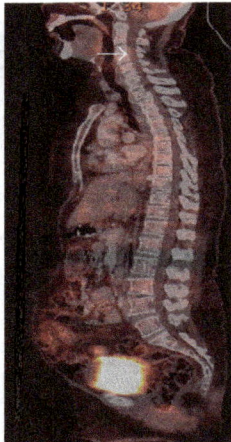

FIGURE 4: PET image showing moderately FDG avid intradural mass lesion (arrow).

3. Discussion

Melanocytes are cells of neural crest origin and are normally found in the leptomeningeal layers. Melanocytes may cause primary melanocytic neoplasms which can be classified as melanocytoma (benign end of the spectrum), intermediate grade melanocytic neoplasms, and primary malignant melanoma. These neoplasms can be differentiated histologically based on the identification of mitotic activity, cytological atypia, necrosis, and invasion of the adjacent structures [2]. Melanocytoma can occur along the neural axis most commonly in posterior fossa, adjacent to the cranial nerve nuclei, and Meckel's cave and in the foramen magnum. Within the spine, it present as intradural extramedullary masses, mostly found in the upper cervical region, as the melanocytes are most concentrated in this region [4]. They tend to present in fourth and fifth decades of life. Spinal melanocytoma present with weakness, sensory deficits, and progressive pain. Subarachnoid haemorrhage can be a rare presentation of these tumors [1]. Meningeal melanocytoma was introduced as a primary melanocytic lesion from the leptomeninges with benign clinical and pathologic features by Limas and Tio in 1972 [5]. There are few case reports of meningeal melanocytoma from the spinal column [6, 7] and cerebellopontine region [8] presenting as superficial siderosis of the central nervous system in the literature.

On computed tomography, these lesions present as well-defined, isodense to hyperdense, homogenous, and contrast enhancing mass lesions. The MRI appearance of meningeal melanocytoma is variable, depending on the amount of melanin content present. Signal characteristics include isointense or hyperintense on T1 weighted images and isointense or hypointense on T2 weighted images which give a heterogeneous enhancement on postcontrast images and show blooming of low signal on gradient images [9]. The contrast enhanced T1 weighted images and subtracted contrast enhanced images should be obtained as these tumors show T1W hyperintensity.

Complete excision is the treatment of choice; however this is often not possible as intraoperative haemorrhage may be severe. Furthermore, local recurrence has been reported even after gross total removal. Risk of tumour recurrence is noted even after complete excision; hence, adjuvant radiation therapy is advised in cases of both complete and incomplete resection [10].

Melanocytoma should be considered in the differential diagnosis of the pigmented tumors which include meningeal meningioma, melanotic schwannoma, and metastatic and primary malignant melanoma [11, 12]. Radiologically, the differential diagnosis of these primary pigmented lesions is difficult due to their similar appearances on CT and MR imaging, thus necessitating further diagnostic confirmation. Tumor calcification and hyperostosis of adjacent bones are visualised in melanocytomas; however, a lack of these signs obviously does not exclude the presence of a meningioma. Small schwannomas usually show uniform contrast enhancement, whereas larger schwannomas may be heterogeneous [2]. Histopathological examination is the most crucial method of differentiating primary melanocytoma from other pigmented lesions of the leptomeninges.

Due to abundant melanin production, meningeal melanocytomas appear dark brown to black on gross examination. They feature pigmented spindle cells growing in tight nests. Cytologically, tumor cells are bland with less pleomorphism and indistinct nucleoli. These must be distinguished from intermediate grade melanocytic neoplasms and malignant melanoma owing to the same cell of origin. Intermediate grade melanocytic neoplasms should be considered when bland cytology with some CNS invasion, mitotic activity (0–3 mitoses per 10 high power fields), and Ki-67 labelling index ranging from 1–4 (% of nuclei) is seen. Features in favour of malignant melanoma include high mitotic activity (2–15 mitoses per 10 high power fields), central nervous system invasion, tumor necrosis, Ki-67 labelling index ranging from 2–15 (% of nuclei), and cellular and nuclear atypia [2]. Melanocytomas are typically immunoreactive for S-100 protein and HMB-45. A positive immunostain for epithelial membrane antibody can confirm the diagnosis of melanotic meningioma. Melanotic schwannomas show considerable histologic overlap with melanocytomas and both tumors are immunoreactive for S-100 protein, therefore making the distinction between these tumors difficult. However, immunohistochemical staining for HMB-45 to detect the melanocytic origin can help to reach an accurate diagnosis as in our case [2]. The histologic differentiation between malignant melanoma and melanocytoma is difficult. The lack of mitotic activity, presence of large nucleoli, the lack of nuclear pleomorphism, hyperchromicity, and slow growth are the pointers toward the diagnosis of melanocytoma [1]. Although extremely rare, meningeal melanocytoma may be associated with histological benign leptomeningeal spread and aggressive clinical course in spite of absence of malignant transformation.

In summary, although the imaging might not differentiate the pigmented lesions very well, it helps in early identification of the complications like diffuse benign spread and recurrence. Making a correct differential diagnosis and being

(a)

(b)

(c)

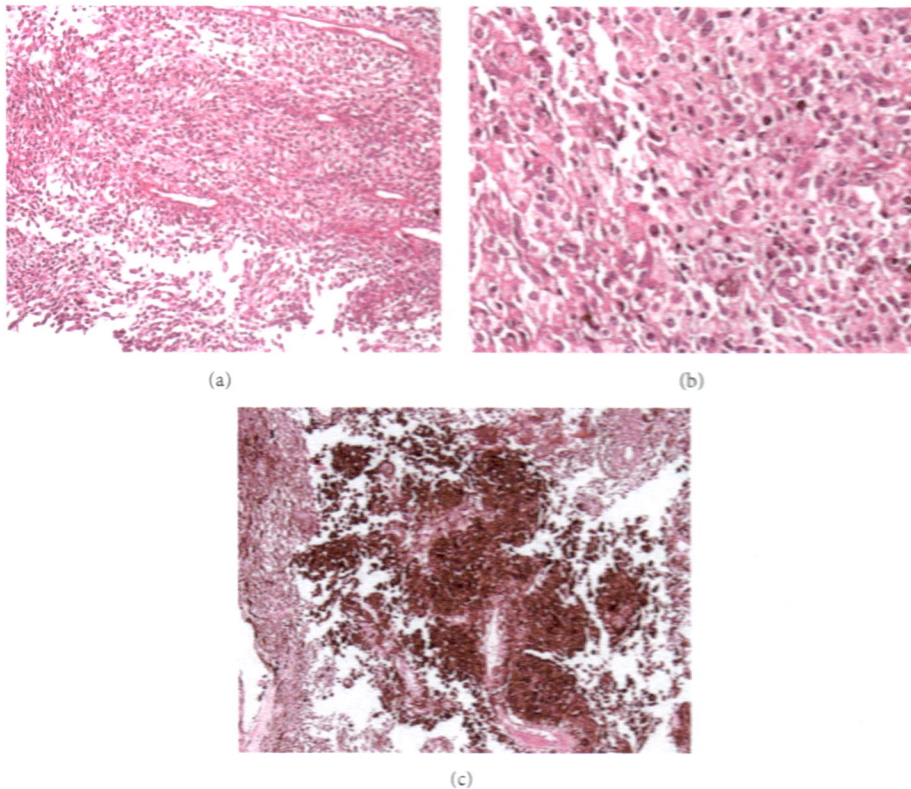

FIGURE 5: Histopathological examination. (a) Diffusely pigmented tumor with peritheliomatous arrangement which obscured the cytological details (H and E 40x). (b) After bleaching, tumor had pleomorphic round to oval nuclei with inconspicuous nucleoli and moderate amount of cytoplasm (H and E 40x). (c) Tumor on higher magnification had mild nuclear pleomorphism, moderate to abundant cytoplasm with few containing blackish brown pigment. No significant increase in mitosis was seen (H and E 200x).

alert to the unusual presentation of meningeal melanocytoma as superficial siderosis will result in appropriate therapy as well as identification of the complications in the follow-up.

References

[1] T. J. Painter, G. Chaljub, R. Sethi, H. Singh, and B. Gelman, "Intracranial and intraspinal meningeal melanocytosis," *American Journal of Neuroradiology*, vol. 21, no. 7, pp. 1349–1353, 2000.

[2] D. J. Brat, "Melanocytic neoplasms of the central nervous system," in *Practical Surgical Neuropathology: A Diagnostic Approach*, A. Perry and D. J. Brat, Eds., pp. 353–359, Churchill Livingstone, Philadelphia, Pa, USA, 2010.

[3] F. Roser, M. Nakamura, A. Brandis, V. Hans, P. Vorkapic, and M. Samii, "Transition from meningeal melanocytoma to primary cerebral melanoma. Case report," *Journal of Neurosurgery*, vol. 101, no. 3, pp. 528–531, 2004.

[4] G. Chacko and V. Rajshekhar, "Thoracic intramedullary melanocytoma with long-term follow-up: case report," *Journal of Neurosurgery: Spine*, vol. 9, no. 6, pp. 589–592, 2008.

[5] C. Limas and F. O. Tio, "Meningeal melanocytoma ('melanotic meningioma'). Its melanocytic origin as revealed by electron microscopy," *Cancer*, vol. 30, no. 5, pp. 1286–1294, 1972.

[6] S. Matsumoto, Y. Kang, S. Sato et al., "Spinal meningeal melanocytoma presenting with superficial siderosis of the central nervous system. Case report and review of the literature," *Journal of Neurosurgery*, vol. 88, no. 5, pp. 890–894, 1998.

[7] A. Das, P. Ratnagopal, K. Puvanendran, and J. G. C. Teo, "Spinal meningeal melanocytoma with hydrocephalus and intracranial superficial siderosis," *Internal Medicine Journal*, vol. 31, no. 9, pp. 562–564, 2001.

[8] G. Vreto, A. Rroji, A. Xhumari, L. Leka, M. Rakacolli, and M. Petrela, "Meningeal melanocytoma of the cerebellopontine angle as the unusual cause of superficial siderosis," *Neuroradiology*, vol. 53, no. 11, pp. 927–930, 2011.

[9] C. J. Chen, Y. I. Hsu, Y. S. Ho, Y. H. Hsu, L. J. Wang, and Y. C. Wong, "Intracranial meningeal melanocytoma. CT and MRI," *Neuroradiology*, vol. 39, no. 11, pp. 811–814, 1997.

[10] O. H. Kim, S. J. Kim, H. J. Choo et al., "Spinal meningeal melanocytoma with benign histology showing leptomeningeal spread: case report," *Korean Journal of Radiology*, vol. 14, no. 3, pp. 470–476, 2013.

[11] N. S. Litofsky, C.-S. Zee, R. E. Breeze, P. T. Chandrasoma, M. E. Leavens, and K. R. Winston, "Meningeal melanocytoma: diagnostic criteria for a rare lesion," *Neurosurgery*, vol. 31, no. 5, pp. 945–948, 1992.

[12] M. Tatagiba, D.-K. Boker, A. Brandis et al., "Meningeal melanocytoma of the C8 nerve root: case report," *Neurosurgery*, vol. 31, no. 5, pp. 958–961, 1992.

2

Segmental Arterial Mediolysis of Omental Arteries with Haemoperitoneum: Case Report with Embolization of the Left Omental Artery and Brief Review of Literature

Gernot Rott (iD) **and Frieder Boecker** (iD)

Radiological Department, Bethesda-Hospital Duisburg, Heerstr. 219, 47053 Duisburg, Germany

Correspondence should be addressed to Gernot Rott; info@myom-therapie.de

Academic Editor: Bruce J. Barron

Segmental arterial mediolysis of an omental artery is an exceptionally rare condition. A 69-year-old man presented with haemoperitoneum six days after being hospitalized due to pneumogenic sepsis. Computed tomography of the abdomen showed a short segment dilatation of an omental artery in the left upper abdomen, compatible with segmental arterial mediolysis. Angiographic examination revealed alterations of omental branches of the right gastroepiploic artery and an aneurysm of the left omental artery, both characteristic of segmental arterial mediolysis. Embolization of the left omental artery with use of N-butyl-2-cyanoacrylate was performed. The postinterventional course was uneventful with increase of haemoglobin levels and without symptoms of omental infarction. Transcatheter embolization in the setting of haemoperitoneum due to segmental arterial mediolysis of an omental branch is technically feasible and a valuable alternative to emergency operation.

1. Introduction

Segmental arterial mediolysis (SAM) is an uncommon, nonatherosclerotic, noninflammatory arteriopathy of unknown pathogenesis characterized by lysis of the outer arterial media resulting in dissecting hematomas and/or dissecting aneurysms of primarily medium- to large-sized splanchnic arteries. The diagnosis of SAM historically has been confirmed by histopathologic examination of the arteries involved; in modern practice, however, it is diagnosed with imaging [1–4]. The diagnosis is made radiographically by demonstrating the typical radiographic features of focal aneurysms, beading, and narrowing of affected splanchnic and renal vessels, in patients without clinical and laboratory findings indicating a vasculitis or fibromuscular dysplasia as primary diagnosis.

Recently SAM has been of interest particularly in radiological circles as the potential for investigating radiological therapeutic modalities. In the acute stage of SAM complicated by haemorrhage related to ruptured aneurysms, interventional treatment with transcatheter embolization or placement of covered stents can be the treatment of first choice. Most publications on the subject are case reports and small case series of embolization of large or middle calibre splanchnic arteries, respectively, the splenic, hepatic, gastroduodenal, gastroepiploic, or pancreaticoduodenal arteries [2, 5].

Very few cases of embolization of omental arteries due to bleeding have been published, however almost exclusively in cases of so-called idiopathic rupture of omental arteries.

To our knowledge, this is the first description of a successful embolization of an omental artery (omental branch) in the setting of acute bleeding attributable to SAM.

2. Case Report

A 69-year-old man presented with dyspnoea, relapsing dizziness, falls, and systemic inflammatory response syndrome in our institution. Notable lab values were a white blood count of 17.7 /μl (normal range 4.0-10.0 /μl), an elevated C-reactive protein of 9.6 ng/ml (normal < 0.5 ng/ml), a haemoglobin level of 13.5 g/dl (normal range 13.5-17.5 g/dl),

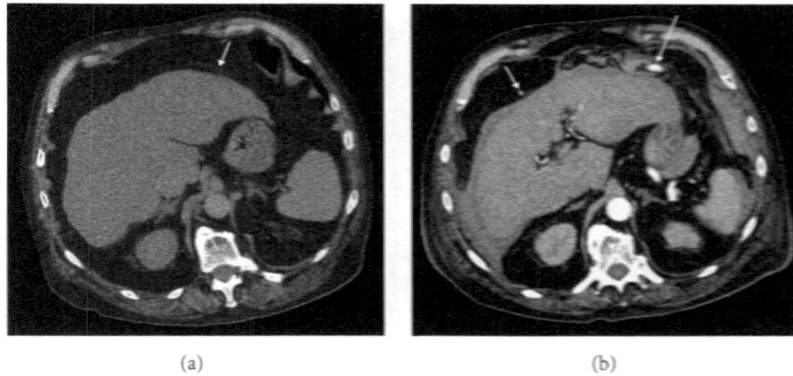

(a) (b)

FIGURE 1: Computed tomography (CT) images. CT on admission (a) demonstrating liver cirrhosis with the greater omentum in the upper abdomen and tiny omental vessels (arrow) in the left upper abdomen. Arterial phase of contrast-enhanced computed tomography image six days later (b) with haemoperitoneum, segmental dilatated omental artery in the left (arrow), and spot of enhanced tiny omental artery in the right upper abdomen (small arrow).

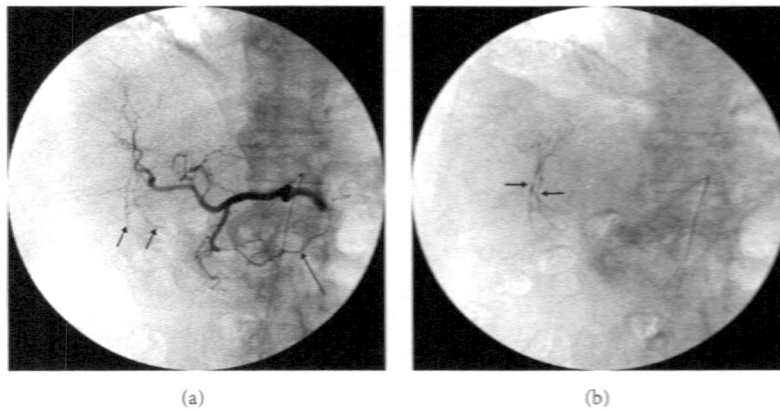

(a) (b)

FIGURE 2: Digital subtraction angiograms. Common hepatic arteriogram (a) with the right gastroepiploic artery forming the omental arch (arrow) and two small right omental branches with multisegmental dilatation (small arrows), better visible in the late phase (b).

and an elevated international normalized ratio (INR) of 1.41 without anticoagulant medication.

Initial workup included computed tomography (CT) of the chest to rule out pulmonary embolism, which revealed right lower lobe pneumonia. In doing so, scans of the upper parts of the abdomen demonstrated liver cirrhosis without ascites or additional pathologies and with the greater omentum positioned almost entirely in the upper abdomen (Figure 1(a)). Intravenous antibiotic therapy was started for diagnosis of pneumogenic sepsis.

After six days of hospitalization the patient developed mild abdominal symptoms and his haemoglobin level decreased from 13.5 to 9.9 g/dl while INR increased from 1.41 to 1.51. An abdominal CT showed a moderate sized haemoperitoneum, particularly in the upper abdomen, left anterior perihepatic space, and surrounding a significantly enlarged segment of an omental artery in the left upper abdomen. Spots of enhanced tiny vessels also were visible in the right upper abdomen ventral to the liver (Figure 1(b)). SAM of the left omental artery (LOA) was suspected. As the patient remained haemodynamically stable with a borderline

coagulopathic status, a noninterventional therapeutic approach was initially agreed upon. Fresh frozen plasma and erythrocyte concentrates were administered. Despite this therapy the haemoglobin levels further decreased to 7.8 g/dl during the next three days, so that an abdominal control CT was performed. This demonstrated slight progression of the haemoperitoneum. After multidisciplinary discussion the radiological department was asked to perform catheter angiography, if possible with transcatheter embolization.

Digital subtraction angiography (DSA) with selective superior mesenteric, common hepatic and splenic arteriogram was performed. These revealed an anatomic variant with absence of the left gastroepiploic artery (left gastroomental artery), the gastroepiploic arch being exclusively formed by the right gastroepiploic artery (right gastroomental artery), and consequently a separate origin of the LOA (left omental branch) from an inferior pole splenic artery (Figures 2 and 3). Additionally, two tiny omental branches of the right gastroepiploic artery with small multisegmental dilatation and a typical "wind-sock" formed aneurysm of the LOA were visible (Figures 2 and 3). Although no active bleeding was

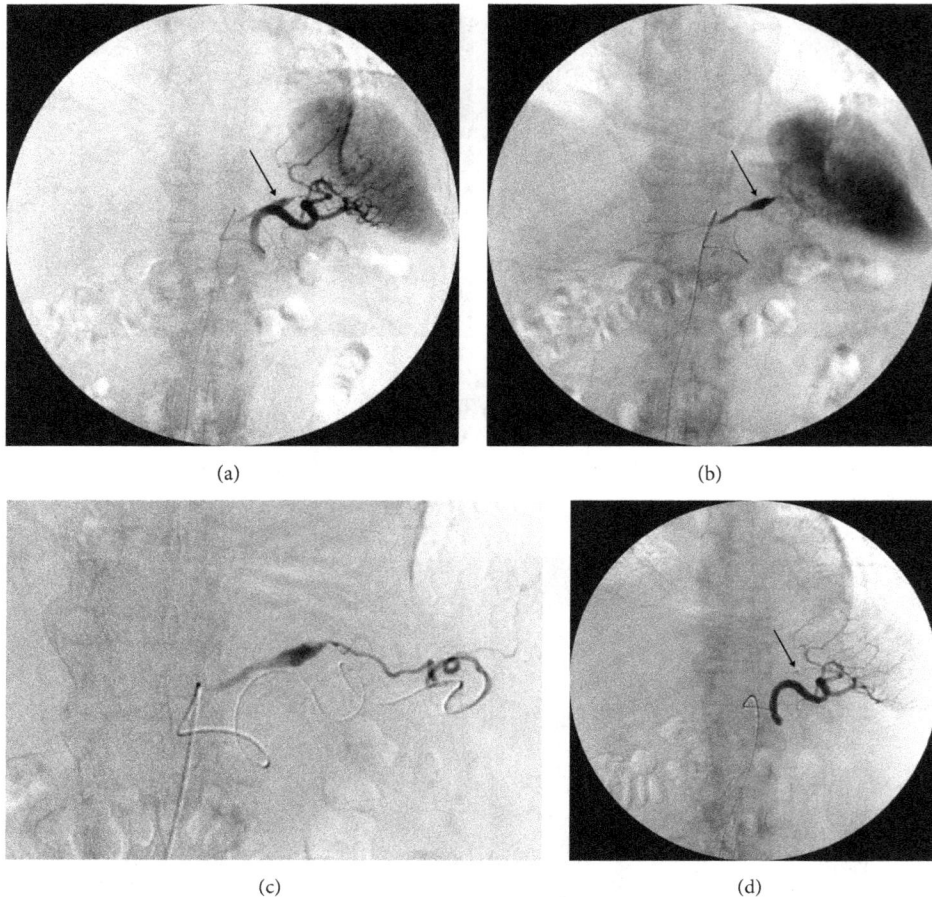

FIGURE 3: Digital subtraction angiograms. Splenic arteriogram (a) with a markedly dilatated segment, respectively, spindle-shaped aneurysm (arrow) of the left omental artery (LOA), better visible in the late phase (b) and with separate origin of the LOA from an inferior splenic pole artery. Superselective arteriogram of the LOA (c) before embolization (noteworthy, the tortuous microcatheter in the substantially elongated splenic artery). Postembolization arteriogram of the splenic artery demonstrating NBCA-casting and stasis of the LOA (arrow) with preserved splenic arteries (d).

detected, the aneurysm of the LOA was considered to be the obvious cause for the haemoperitoneum.

Superselective catheterization of the LOA through splenic artery and lower pole splenic artery with a microcatheter (2.7-F Progreat, Terumo, Tokyo, Japan) was technically challenging due to vessel tortuosity but succeeded, however, only just a few centimeters proximal to the aneurysm. The decision was made to embolize the LOA with N-butyl-2-cyanoacrylate (NBCA). Approximately 1 ml of a 1:3 mixture of NBCA (Histoacryl; B. Braun, Melsungen, Germany) with iodized oil (Lipiodol ultrafluid; Guerbet, Villepinte, France) was injected in the usual manner. The final DSA control confirmed the complete embolization of the aneurysm of the LOA with preservation of the splenic vessels (Figure 3).

Postinterventional course was uneventful with no signs of omental infarction and with increase of haemoglobin levels up to normal levels.

One month after embolization and after therapy of his protracted pneumonia the patient was transferred to another hospital for early rehabilitation in a satisfactory general condition.

3. Discussion

Omental haemorrhage with or without haemoperitoneum is caused by trauma, aneurysm, vasculitis, neoplasm, torsion of the greater omentum, peritonitis, anticoagulant therapy or coagulopathy, and omental pregnancy or with no recognizable cause is referred to as idiopathic. The condition will generally be treated conservatively, in particular by correction of coagulopathy or surgically by means of emergency operation.

With regard to splanchnic aneurysms, the omental artery aneurysm is considered to be the most uncommon [6]. Omental artery aneurysms usually are recognized only after aneurysm rupture and surgery [7, 8].

Radiological literature covering the topic "embolization of omental arteries" is mostly exclusively limited on chemoembolization of hepatocellular carcinomas with additional blood supply from parasitized omental arteries, in which case the corresponding omental artery often is hypertrophied and easier to be catheterized than in healthy subjects [9–11]. In a complete different context of tumour disease Parmar et al. described a case of embolization of omental branches for treatment of disseminated peritoneal leiomyomatosis [12].

Apart from this, there are very limited reports of transcatheter embolization in the event of omental bleeding. We were able to find only four documented cases, most of them relating to "idiopathic" or "spontaneous" omental bleeding without suspected SAM [13–16]. Of these, at least two case descriptions are of embolization of the left gastroepiploic artery and not an omental artery [14, 15]. Takahashi embolized the left gastroepiploic artery, because the culprit vessel of the omentum was too thin to be selected, as did Matsumoto. Only the publication of Yasuoka et al. describes a case of haemoperitoneum due to SAM with embolization of "the greater omental artery branched off from the intrasplenic artery," that is to say, a similar case to ours; however embolization "did not realize hemostasis" and patient was operated on with partial resection of the greater omentum the same day [13].

In addition we found one case report with haemoperitoneum "as a result of segmental mediolytic arteritis of an omental artery" diagnosed after primary surgery with omental segment resection [17].

In this respect, our case report is the first one of embolization of an omental artery for bleeding due to SAM with both technical and clinical success.

One reason for this might be that omental branches usually are small and branch at an acute angle from the gastroepiploic artery [18] and so superselective catheterization of them might be technically challenging and not always be feasible. Even in cases of hepatocellular carcinomas with additional blood supply from parasitized omental arteries these can be catheterized only in 64-73%, depending on the calibre of the used microcatheter [10, 19].

Blood supply of the omentum according to standard anatomical textbooks is provided by the right and left gastroepiploic artery, both deriving from the celiac trunk. However, anatomical and radiological studies demonstrate a wide range of varieties in this respect and the impossibility of predicting any standard pattern of vascularization [20].

The anatomical variant of our patient with absence of the left gastroepiploic artery and an omental branch directly from the splenic artery has been described in the radiological literature [9, 21] and corresponds to a type 5 of the omental vascular arcade as described by Alday and Goldsmith in their classic paper [22], where "the terminal branch of the splenic artery is not part of the gastroepiploic arch but instead connects directly into the left omental artery."

In our case with the LOA as a direct branch of the splenic artery on one hand this could have made superselective catheterization of the vessel easier, although the pronounced tortuosity of the splenic artery on the other hand was not a favourable prerequisite for this at all.

4. Conclusion

In cases of spontaneous haemoperitoneum SAM should be considered as a cause. The diagnosis can be confirmed and the culprit artery be identified with CT. In the event of

an omental artery being the culprit vessel and under the provision of a hemodynamically stable patient transcatheter embolization of the omental branch or, if this is not feasible, of the corresponding gastroepiploic artery should be considered at least as a valuable alternative to a surgical procedure.

References

[1] S. P. Kalva, B. Somarouthu, M. R. Jaff, and S. Wicky, "Segmental arterial mediolysis: Clinical and imaging features at presentation and during follow-up," *Journal of Vascular and Interventional Radiology*, vol. 22, no. 10, pp. 1380–1387, 2011.

[2] M. Shenouda, C. Riga, Y. Naji, and S. Renton, "Segmental arterial mediolysis: A systematic review of 85 cases," *Annals of Vascular Surgery*, vol. 28, no. 1, pp. 269–277, 2014.

[3] S. G. Naidu, C. O. Menias, R. Oklu et al., "Segmental arterial mediolysis: abdominal imaging of and disease course in 111 patients," *American Journal of Roentgenology*, vol. 210, no. 4, pp. 899–905, 2018.

[4] A. K. Pillai, S. I. Iqbal, R. W. Liu, N. Rachamreddy, and S. P. Kalva, "Segmental arterial mediolysis," *CardioVascular and Interventional Radiology*, vol. 37, no. 3, pp. 604–612, 2014.

[5] A. Vance, A. Graif, M. McGarry et al., "Segmental arterial mediolysis (SAM): clinical presentation, angiographic manifestations and endovascular management," *Journal of Vascular and Interventional Radiology*, vol. 27, no. 3, p. S117, 2016.

[6] S. Jacobs, S. Houthoofd, I. Fourneau, and K. Daenens, "A ruptured omental aneurysm, a rare cause of intraabdominal bleeding," *Annals of Vascular Surgery*, vol. 28, no. 2, pp. 491–e11, 2014.

[7] R. Fernandes, B. Dhillon, and B. Andrews, "Haemoperitoneum secondary to rupture of an omental artery aneurysm," *BMJ Case Reports*, vol. 2009, 2009.

[8] D. J. Park, K. Oh, S. J. Kim et al., "True Aneurysm rupture of omental artery leading to hemoperitoneum and shock in a CAPD patient," *Nephrology Dialysis Transplantation* , vol. 20, no. 10, pp. 2292-2292, 2005.

[9] J. W. Choi, H.-C. Kim, J. W. Chung et al., "Chemoembolization via branches from the splenic artery in patients with hepatocellular carcinoma," *CardioVascular and Interventional Radiology*, vol. 35, no. 1, pp. 90–96, 2012.

[10] S. Gao, R.-J. Yang, and J.-H. Dong, "Hepatocellular carcinoma with blood supply from parasitized Omental artery: Angiographic appearance and chemoembolization," *Chinese Journal of Cancer Research*, vol. 24, no. 3, pp. 207–212, 2012.

[11] J. Y. Won, D. Y. Lee, J. T. Lee et al., "Supplemental transcatheter arterial chemoembolization through a collateral omental artery: Treatment for hepatocellular carcinoma," *CardioVascular and Interventional Radiology*, vol. 26, no. 2, pp. 136–140, 2003.

[12] J. Parmar, C. Mohan, D. Hans, and M. Vora, "A diagnostic dilemma of recurrent disseminated peritoneal leiomyomatosis with hypertrophied omental vessels: imaging and embolization of omental branches with positive outcome," *Case Reports in Obstetrics and Gynecology*, vol. 2017, Article ID 8427240, 6 pages, 2017.

[13] R. Yasuoka, S. Nishino, S. Ogino et al., "A case of the greater omental hemorrhage due to segmental arterial mediolysis," *The Japanese Journal of Gastroenterological Surgery*, vol. 41, no. 1, pp. 46–51, 2008.

[14] M. Takahashi, Y. Matsuoka, T. Yasutake, H. Abe, K. Sugiyama, and K. Oyama, "Spontaneous rupture of the omental artery treated by transcatheter arterial embolization," *Case Reports in Radiology*, vol. 2012, Article ID 273027, 4 pages, 2012.

[15] T. Matsumoto, T. Yamagami, H. Morishita et al., "Transcatheter arterial embolization for spontaneous rupture of the omental artery," *CardioVascular and Interventional Radiology*, vol. 34, no. 2, pp. S142–S145, 2011.

[16] R. Tsuchiya, S. Takahashi, T. Takaoka et al., "A case of idiopathic omental bleeding treated successfully with transarterial embolization," *Nippon Shokakibyo Gakkai zasshi The Japanese journal of gastro-enterology*, vol. 106, no. 4, pp. 554–559, 2009 (Japanese).

[17] D. M. Heritz, J. Butany, K. W. Johnston, and K. W. Sniderman, "Intraabdominal hemorrhage as a result of segmental mediolytic arteritis of an omental artery: Case report," *Journal of Vascular Surgery*, vol. 12, no. 5, pp. 561–565, 1990.

[18] H. C. Kim and J. W. Chung, "Embolization of liver tumors: anatomy," in *Interventional Oncology: Principles and Practice of Image-Guided Cancer Therapy*, J. F. H. Geschwind and M. C. Soulen, Eds., p. 117, Cambridge University Press, Cambridge, UK, 2nd edition, 2016.

[19] S. Miyayama, O. Matsui, Y. Akakura et al., "Hepatocellular carcinoma with blood supply from omental branches: Treatment with transcatheter arterial embolization," *Journal of Vascular and Interventional Radiology*, vol. 12, no. 11, pp. 1285–1290, 2001.

[20] D. Liebermann-Meffert, H. White, and E. Vaubel, "Anatomy and functional anatomy: vessels and innervation," in *The Greater Omentum: Anatomy, Physiology, Pathology, Surgery with an Historical Survey*, pp. 30–36, Springer, Berlin, Germany, 1983.

[21] L. Hannoun, C. Le Breton, V. Bors, C. Helenon, J. M. Bigot, and R. Parc, "Radiological anatomy of the right gastroepiploic artery," *Surgical and Radiologic Anatomy*, vol. 5, no. 4, pp. 265–271, 1984.

[22] E. S. Alday and H. S. Goldsmith, "Surgical technique for omental lengthening based on arterial anatomy," *Surgery, Gynecology & Obstetrics*, vol. 135, no. 1, pp. 103–107, 1972.

3

Radiographic Thrombus within the External Jugular Vein: Report of a Rare Case and Review of the Literature

Sayyad Yaseen Zia,[1] Richard L. Bakst,[1] Qiusheng Si,[2] Mike Yao,[3] and Peter M. Som[4]

[1]Department of Radiation Oncology, Icahn School of Medicine at Mount Sinai Hospital, New York, NY, USA
[2]Department of Pathology, Icahn School of Medicine at Mount Sinai Hospital, New York, NY, USA
[3]Department of Otolaryngology, Icahn School of Medicine at Mount Sinai Hospital, New York, NY, USA
[4]Department of Radiology, Icahn School of Medicine at Mount Sinai Hospital, New York, NY, USA

Correspondence should be addressed to Sayyad Yaseen Zia; sayyad.zia@mountsinai.org

Academic Editor: Samer Ezziddin

We are reporting a case of a 91-year-old male with a primary malignancy of the right parotid gland with radiographic thrombus extension within the right external jugular vein. He was treated with palliative radiation therapy to the right parotid mass with a marked clinical response. The rarity of this occurrence as documented in the review of the literature provides for uncertainty with regard to proper management. Radiographic evidence of thrombus in the absence of clinical manifestations, the role of anticoagulation, and the proper radiation target delineation were all challenges encountered in the care of this patient. Our case represents a rare occurrence with unique radiologic findings that has implications for management.

1. Introduction

We are reporting a case of a 91-year-old male with a primary malignancy of the right parotid gland with thrombus extension within the right external jugular vein. He was treated with palliative radiotherapy to the right parotid mass with a marked clinical response. Tumor thrombus within neck veins is an uncommon event and in the head and neck it has most often been reported in association with thyroid malignancies [1–6]. Thrombus associated with malignancy may manifest clinically but in our case was detected radiographically. Here we report a case of a parotid malignancy with direct intravenous thrombus extending down the right external jugular vein to the level of the subclavian vein documented with computed tomography (CT) and positron emission tomography (PET) and review the literature on this rare radiographic finding.

2. A Case Report

A 91-year-old male with no smoking, drinking, or prior malignancy history was initially referred to the Department of Radiation Oncology for palliative irradiation to a growing painless right parotid mass that had been present for one year.

One year prior to presenting to our clinic, the patient had noticed a right neck mass. CT of the neck demonstrated an unencapsulated, poorly defined hypodense mass measuring $2.3 \times 1.5 \times 2.1$ cm in the superficial portion of the right parotid gland with fine needle aspiration (FNA) suggesting Warthin tumor. Further management was complicated by a stroke that left him wheelchair bound with aphasia. CT of the neck one year later noted the mass to have increased in size to $8.8 \times 6 \times 4$ cm with no evidence of associated thrombus.

Upon presentation at our institution, physical examination demonstrated significant enlargement in the size of the mass with no evidence of swelling, palpable veins, or tenderness in the neck. FNA of the mass showed pleomorphic malignant cells suggestive of a high grade carcinoma (Figure 1). CT of the neck showed a large mass in the right parotid gland measuring $8.5 \times 5.4 \times 8.3$ cm with some invasion of the adjacent masseter and sternocleidomastoid muscles. There was invasion of the right external jugular vein with thrombus within the vein extending to the level of

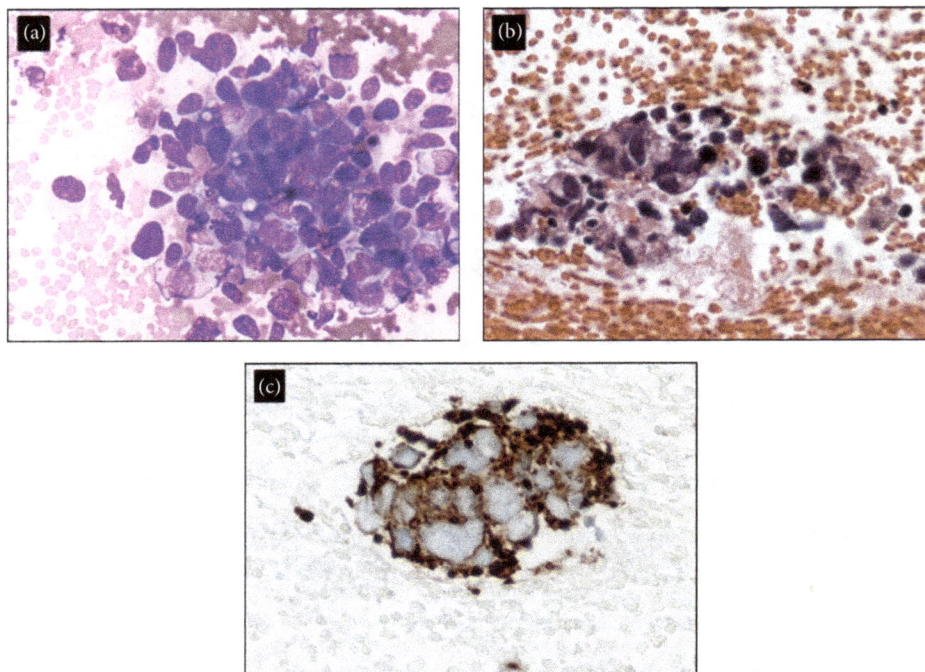

FIGURE 1: Parotid tumor pathology. Smear and cell block section from FNA biopsy showing pleomorphic malignant cells ((a) Diff-Quick stain, ×600; (b) cell block, ×600). The tumor cells are positive for Cam 5.2 ((c) ×600) and negative for AE1/AE3, p63, and CK5/6. Melan A, androgen receptor, mammaglobin, S-100, CD45, CD3, CD20, and TTF1. The cytological features with immunostaining results support a high grade carcinoma.

the right subclavian vein (Figures 2(a) and 2(b)). There was no gross evidence for invasion of the bone. PET demonstrated a large heterogeneous mass abutting the mandibular ramus measuring up to 8.6 cm with an SUV max of 14 involving the right parotid gland. Arising from the posterior region of this mass was a tubular structure with FDG avidity and internal thrombosis that appeared to form a collateral with the right brachiocephalic vein (Figures 3(a) and 3(b)). FNA of the tubular structure was negative for malignant cells with lymphocytes present. The patient was not started on any anticoagulation for the thrombus secondary to his recent stroke. A goal of care discussion deemed the patient to be medically inoperable due to his poor performance status, advanced age, and comorbidities. The patient was initiated and completed a course of radiation therapy to the parotid mass to a total dose of 6000 cGy with significant tumor shrinkage noted during treatment. On one month follow-up the patient experienced no acute side effects of radiation therapy with tumor shrinkage to 1/3 of presenting volume.

3. Discussion

Thrombus associated with malignancy can result from either tumor compression of a vein leading to stasis or direct extension of the primary tumor. Clinically this is important to distinguish when determining either the appropriate extent of surgical resection or radiation target volume delineation. Direct tumor extension into an adjacent vein from a head and neck malignancy is rare and was first reported by Kaufmann

in 1879. Since then, less than 20 cases have been reported associated with thyroid cancer [3, 5, 7, 8]. The invasion of the internal jugular vein was from either direct tumor extension from the thyroid gland or extension from a metastatic node [3]. The presence of a tumor thrombus in the internal jugular vein from a thyroid cancer was first documented with CT in 1991 [4]. Since then there was a report of a renal cell carcinoma metastatic to the parotid gland with tumor thrombus in the adjacent veins. There are also reports of deep lobe parotid tumors causing compression thrombosis of the internal jugular vein and an associated report of a patient with an acute parotitis and a thrombosis of the internal jugular vein [9–11]. However, we could not identify a previous report of a primary parotid malignancy with direct tumor thrombus into an adjacent vein. In our case, the thrombus was in the external jugular vein extending down to the subclavian vein. There are other head and neck tumors that have been reported to invade or grow within the great vessels; among these are the paragangliomas [12, 13].

There are reports of increased FDG avidity in an intra-venous tumor thrombus from a thyroid carcinoma [6]. However, there is also a reported increased intravenous FDG avidity in jugular vein thrombosis without a tumor thrombus [14]. Thus increased avidity alone within a vein is not conclusive of an intravenous tumor thrombus. The tumor thrombus is the intraluminal extension of the primary tumor, around which fibrin is deposited, protecting the tumor and allowing further growth into the vessel lumen [4].

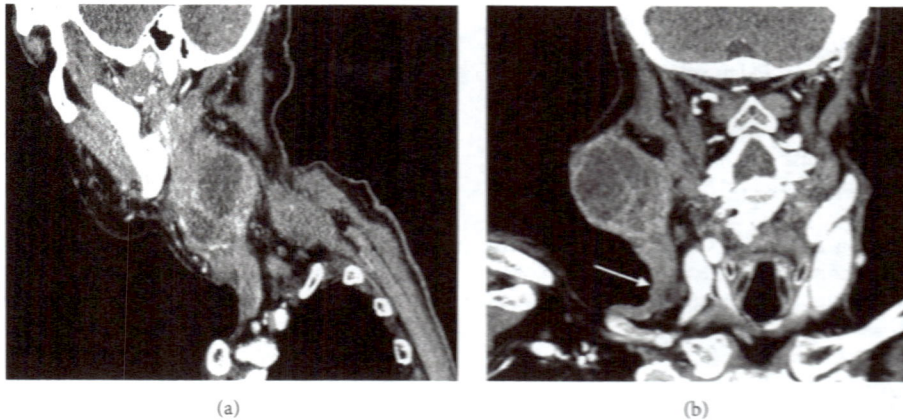

(a)　　　　　　　　　　　　　　　　(b)

FIGURE 2: CT neck: sagittal and coronal images with contrast demonstrating parotid tumor with external jugular vein invasion with extension to the level of the right subclavian vein.

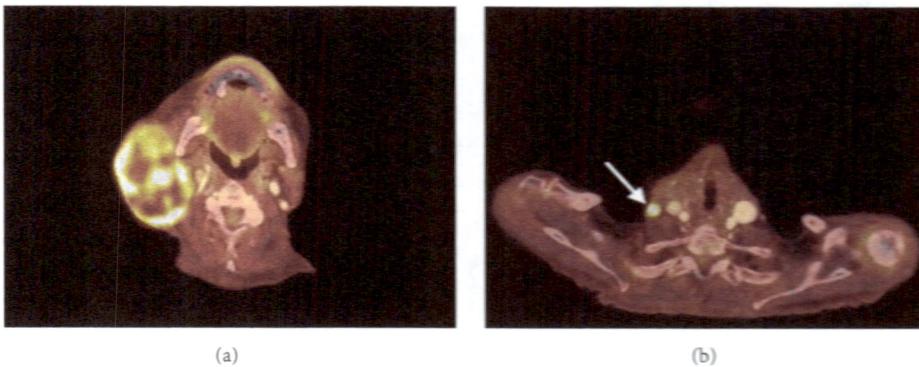

(a)　　　　　　　　　　　　　　　　(b)

FIGURE 3: PET: axial images demonstrating FDG avidity in right parotid mass and extension to right subclavian vein.

Our case illustrates several interesting radiological findings that have implications for management. The first being the extensive nature of the thrombus associated with the tumor extending down the length of the right external jugular vein to the level of the right subclavian vein. Although in our case anticoagulation was not given, the decision to give anticoagulation should be carefully considered particularly with malignancy related hypercoagulable states. The long time course from initial detection to initiation of therapy adds ambiguity to the nature of the thrombus. During that time, either tumor extended into the vessel or prolonged stasis allowed blood thrombus formation.

The uncertainty of the nature of the thrombus with regard to whether it represented tumor or simple blood thrombus presents further therapeutic challenges, and the fact that it showed FDG avidity on PET adds another layer of opacity. As noted previously, there have been reports in the literature of increased avidity not being conclusive of tumor in a thrombus, as was in our case. Caution should be exercised when making treatment decisions solely based on increased FDG avidity. If doubt exists about the nature of the thrombus and the site in question is readily accessible, pathologically sampling could greatly aid in guiding the correct therapeutic intervention. In our case, FNA demonstrated lymphocytes,

which were more consistent with blood thrombus than tumor extension. If a FNA is not feasible, a magnetic resonance imaging (MRI) scan may be useful in distinguishing between tumor and blood thrombus. Lastly, correct target delineation corresponding to the known tumor is critical in delivering safe and effective radiation therapy. If the thrombus had been treated as if it contained tumor, a far larger area would have been irradiated with minimal to no increase in tumor control or palliation of symptoms. Furthermore, increased amounts of normal tissue would have been irradiated leading to increased acute and late radiation toxicities. Our case represents a rare case with unique radiologic findings that has implications for management.

References

[1] A. Alzaraa, J. Stone, G. Williams, I. Ahmed, and M. Quraishi, "Direct spread of thyroid follicular carcinoma to the parotid gland and the internal jugular vein: a case report," *Journal of Medical Case Reports*, vol. 2, article 297, 2008.

[2] M. Gross, Y. Mintz, B. Maly, R. Pinchas, and M. Muggia-Sullam, "Internal jugular vein tumor thrombus associated with thyroid carcinoma," *Annals of Otology, Rhinology and Laryngology*, vol. 113, no. 9, pp. 738–740, 2004.

[3] E. Koike, H. Yamashita, S. Watanabe, H. Yamashita, and S. Noguchi, "Brachiocephalic vein thrombus of papillary thyroid cancer: report of a case," *Surgery Today*, vol. 32, no. 1, pp. 59–62, 2002.

[4] S. Thomas, S. Sawhney, and B. M. L. Kapur, "Case report: bilateral massive internal jugular vein thrombosis in carcinoma of the thyroid: CT evaluation," *Clinical Radiology*, vol. 43, no. 6, pp. 433–434, 1991.

[5] O. Wiseman, P. G. Preston, and J. M. F. Clarke, "Presentation of thyroid carcinoma as a thrombosed external jugular vein, with intraluminal tumour thrombus in the great veins," *European Journal of Surgical Oncology*, vol. 26, no. 8, pp. 816–817, 2000.

[6] M. Tripathi, R. Sharma, A. Jaimini et al., "Metastatic follicular carcinoma of the thyroid with tumor thrombus in the superior vena cava and right brachiocephalic and internal jugular veins: FDG-PET/CT findings," *Clinical Nuclear Medicine*, vol. 33, no. 6, pp. 426–428, 2008.

[7] J.-L. Leong, H. W. Yuen, V. A. LiVolsi et al., "Insular carcinoma of the thyroid with jugular vein invasion," *Head and Neck*, vol. 26, no. 7, pp. 642–646, 2004.

[8] Y. Onaran, T. Terzioğlu, H. Oğuz, Y. Kapran, and S. Tezelman, "Great cervical vein invasion of thyroid carcinoma," *Thyroid*, vol. 8, no. 1, pp. 59–61, 1998.

[9] P. Maralani, S. Mohan, C. H. Rassekh, and L. A. Loevner, "Salivary neoplasms presenting with radiologic venous invasion: an imaging pearl to diagnosing metastatic renal cell carcinoma," *ORL*, vol. 76, no. 2, pp. 105–109, 2014.

[10] E. Hadjihannas, K. W. Kesse, and A. P. d'E Meredith, "Thrombosis of internal jugular vein associated with acute parotitis," *Journal of Laryngology and Otology*, vol. 114, no. 9, pp. 721–723, 2000.

[11] J. K. Fortson, M. Rosenthal, J. S. Lin, and G. Lawrence, "Sigmoid sinus thrombosis associated with a parapharyngeal deep lobe parotid gland tumor causing seizure," *Otolaryngology—Head and Neck Surgery*, vol. 143, no. 2, pp. 313–314, 2010.

[12] H. F. Biller, W. Lawson, P. Som, and R. Rosenfeld, "Glomus vagale tumors," *Annals of Otology, Rhinology and Laryngology*, vol. 98, no. 1, pp. 21–26, 1989.

[13] D. Arsene, C. Ardeleanu, and L. Dănăilă, "An intrajugular paraganglioma. Unusual presentation of a classical tumor," *Romanian Journal of Morphology and Embryology*, vol. 48, no. 2, pp. 189–193, 2007.

[14] M. Kikuchi, E. Yamamoto, Y. Shiomi et al., "Internal and external jugular vein thrombosis with marked accumulation of FDG," *British Journal of Radiology*, vol. 77, no. 922, pp. 888–890, 2004.

4

Multisystem Radiologic Manifestations of Erdheim-Chester Disease

Umairullah Lodhi, Uzair Sarmast, Saadullah Khan, and Kavitha Yaddanapudi

Department of Radiology, Stony Brook University Hospital, 101 Nicolls Road, Stony Brook, NY 11794, USA

Correspondence should be addressed to Umairullah Lodhi; umairullah.lodhi@stonybrookmedicine.edu

Academic Editor: Turab Chakera

Erdheim-Chester Disease is a rare form of multiorgan non-Langerhans' cell histiocytosis that affects individuals between the ages of 50 and 70 with an equal distribution among males and females. It is associated with significant morbidity and mortality that is mostly due to infiltration of critical organs. Some of the sites that Erdheim-Chester Disease affects include the skeletal system, central nervous system, cardiovascular system, lungs, kidneys (retroperitoneum), and skin. The most common presenting symptom of Erdheim-Chester Disease is bone pain although a large majority of patients are diagnosed incidentally during a workup for a different disease process. Diagnosing Erdheim-Chester Disease is challenging due its rarity and mimicry to other infiltrative processes. Therefore, a multimodality diagnostic approach is employed with imaging being at the forefront. As of date, a comprehensive radiologic review of the manifestations of Erdheim-Chester Disease has rarely been reported. Here we present radiologic findings of an individual suffering from Erdheim-Chester Disease.

1. Introduction

Erdheim-Chester Disease (ECD) is a rare form of non-Langerhan cell histiocytosis that was first described by Chester in 1930 [1]. Since then about 550 cases have been reported in the medical literature [2]. The characteristic features of this disease are related to the multiorgan tissue infiltration of lipid-laden macrophages, multinucleated giant cells, and inflammatory cells composed of lymphocytes and histiocytes [3]. Clinical manifestations of ECD vary among individuals, ranging from an indolent focal disease to a life threatening organ failure [4]. Therefore, prompt diagnosis of this disease is paramount for a favorable outcome. ECD primarily affects adults who are between their 5th and 7th decades of life [5]; however cases have been reported in patients between 7 and 84 years [6]. The etiology of this disease is largely unknown [7] and the presenting symptoms are also often nonspecific [6] which add hindrance in accurately diagnosing this disease in a timely fashion.

Therefore, diagnosis of ECD relies largely on radiologic evidence leading to histologic confirmation. Previously, findings have been described on radiographs, 99mTc bone scintigraphs, computed tomography (CT), and magnetic resonance (MR) imaging scans that could clue one into diagnosing ECD. For example, on conventional radiographs of the long bones, bilateral cortical sclerosis involving the diametaphyseal regions is commonly seen in ECD [8]. Abnormally strong uptake of radioactive tracer at the distal ends of the long bones as observed on 99mTc bone scintigraphs is also noted. Either of these findings could lead to tissue sampling of the lesions for histologic analysis. The histological diagnosis is met when typical ECD histiocytes are found in the examined lesion while testing positive for CD68, CD163, and Factor XIIIa and negative for CD1a on immunohistochemical staining [2, 9]. As of date, a comprehensive review of the radiologic manifestation of ECD with pathognomonic features such as "hairy kidney" and "coated aorta" especially when it occurs in the same individual has rarely been reported. These findings are described and illustrated along with various other radiologic manifestations with the goal of helping physicians in accurately diagnosing this disease.

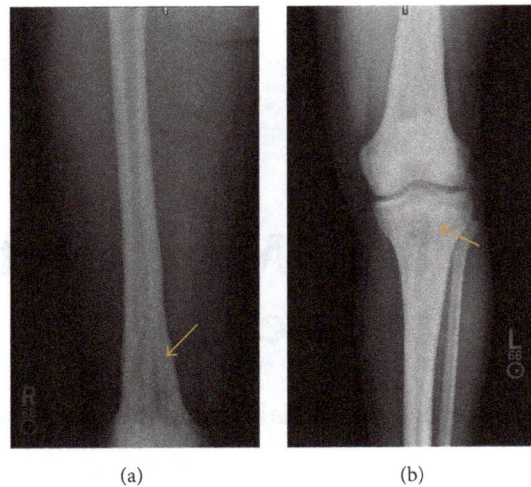

(a)   (b)

FIGURE 1: A 35-year-old male with frontal radiographs of the distal right femur and left knee: there is permeative mixed sclerosis and lucency (arrow) in the distal femoral and proximal tibial shafts.

2. Case Report

This patient is a 35-year-old male with no significant past medical history who presented to the emergency department with symptoms of "redness and swelling in his eyes associated with purulent discharge" that had failed to improve on outpatient antibiotic regimen. On further history, patient revealed that for about a year and a half, he had been experiencing grittiness and a "bulging" feeling in his eyes that he thought were due to seasonal allergies as well as loss of balance when standing or walking. His ophthalmologist had ordered an orbital CT scan that revealed retroorbital soft tissue masses. An outpatient biopsy of these masses was nondiagnostic and the patient was subsequently admitted for further evaluation. On physical exam, the patient had bilateral erythematous conjunctiva associated with exophthalmoses without any evidence of lid lag or thyromegaly. Cardiovascular and pulmonary examinations were normal. There was no hepatosplenomegaly or palpable lymphadenopathy. Laboratory values on admission were notable for leukocytosis of 21×10^3/uL (nL $\leq 10 \times 10^3$/uL), normal erythrocyte sedimentation rate of 10 mm/hr, elevated C-reactive protein of 2.7 mg/dL (nL < 1 mg/dL), normal thyroid stimulating hormone level of 3.13 u/IU/mL, and a normal lactate dehydrogenase level of 121. He tested negative for HIV. Flow cytometry studies performed on peripheral blood did not reveal any abnormalities.

The patient was initially started on intravenous antibiotics for suspected preseptal cellulitis of the orbit. He subsequently underwent a CT scan of the abdomen and pelvis with oral contrast (OMNIPAQUE 300 milliliters), a radiographic bone survey, and a contrast (MAGNEVIST (Gadopentetate Dimeglumine) 15 milliliters) enhanced MR of the brain and orbits. The radiographic metastatic survey yielded diffuse and bilateral appendicular permeative lucencies mixed with sclerosis (Figures 1–3). The CT of the orbits revealed bilateral soft tissue mass-like lesions in the retrobulbar orbits involving

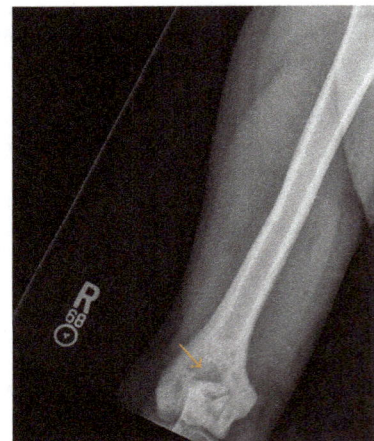

FIGURE 2: A 35-year-old male with frontal radiograph of the distal right humerus: there is mixed sclerosis and lucency in the distal metaphysis and epiphysis (arrow). Similar findings were present on the contralateral side.

intraconal and extraconal compartments (Figure 4). On MR imaging, these masses were heterogeneously hyperintense on T2-weighted imaging, homogeneously hypointense on T1-weighted imaging, with homogeneous enhancement on postcontrast imaging (Figures 5–7). An expansile mass-like lesion in the pons was also noted which was heterogeneously hyperintense on T2-weighed imaging and demonstrated heterogeneous enhancement on contrast administration (Figures 8 and 9). The contrast-enhanced CT of the abdomen demonstrated a rind of soft tissue covering the lateral and posterior margins of the aorta extending from below the level of the renal arteries down to the aortic bifurcation and proximal right common iliac artery (Figure 10). This classically gave an appearance of a "coated aorta." Additionally, a ring of enhancing soft tissue surrounding the bilateral

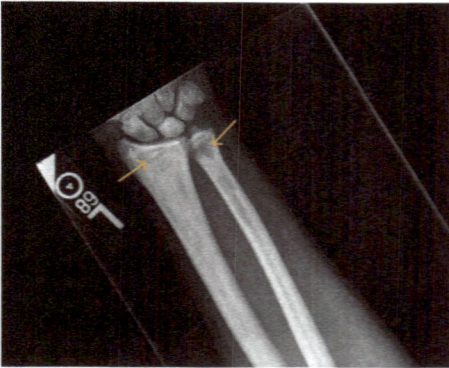

FIGURE 3: A 35-year-old male with frontal radiograph of the distal right forearm: there is permeative mixed sclerosis and lucency in the distal radial and ulnar metaphysis and epiphysis (arrows). Similar findings were present on the contralateral side.

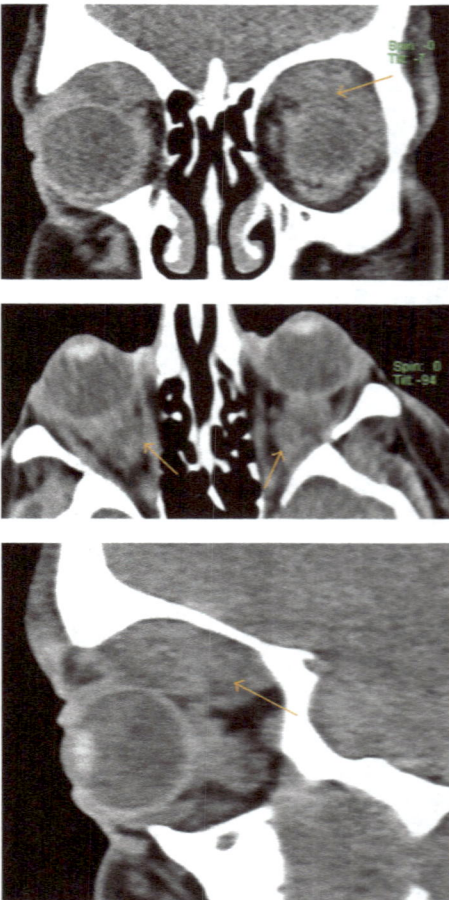

FIGURE 4: A 35-year-old male with unenhanced axial, coronal, and sagittal reformatted CT images of the orbit: mass-like soft tissue lesions in the intra- and extraconal retrobulbar space (arrow).

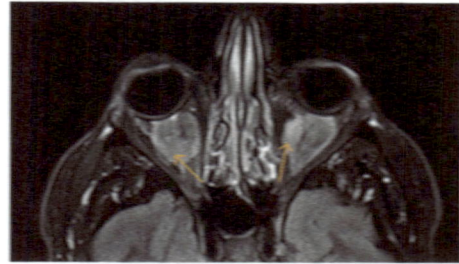

FIGURE 5: A 35-year-old male with axial T2-weighted FLAIR image showing heterogeneously hyperintense soft tissue lesions (arrow) in the intra- and extraconal retrobulbar space.

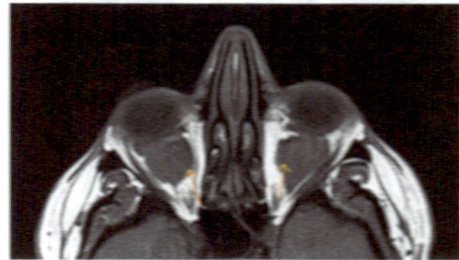

FIGURE 6: A 35-year-old male with unenhanced axial T1 image showing bulky, homogeneously hypointense soft tissue lesions (arrows) in the intra- and extraconal retrobulbar space.

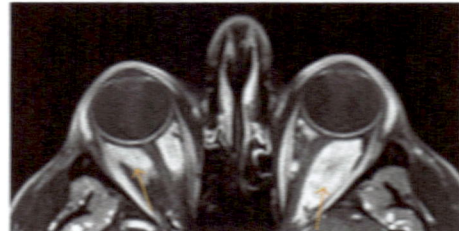

FIGURE 7: A 35-year-old male with contrast-enhanced, fat-suppressed axial T1 image showing homogeneous enhancement of soft tissue lesions (arrows) in the retrobulbar space which encases the optic nerves bilaterally. The globes are unremarkable. The optic nerves are normal in signal and without enhancement.

kidneys, extending towards the renal sinus and constricting the proximal ureters and renal pelvises, was also seen on the CT abdomen giving the classic appearance of a "hairy kidney" (Figure 10). Lastly, as part of the workup of ECD, a cardiac MR was performed that revealed a small pericardial effusion with a heterogeneously enhancing soft tissue mass measuring approximately 3.6 × 3.4 cm abutting the right atrium (Figures 11 and 12). The patient subsequently underwent a CT guided bone biopsy of a right distal metaphyseal lesion which yielded sclerotic lamellar bone with marrow fibrosis and histiocytic infiltration consistent with ECD.

3. Discussion

ECD is a rare multiorgan disease that almost invariably requires tissue sampling for definitive diagnosis. The goal of this report is to highlight those characteristic radiologic findings that are pathognomonic for ECD. Tissue biopsy, though, is still mandatory for definitive diagnosis. Reportedly, as

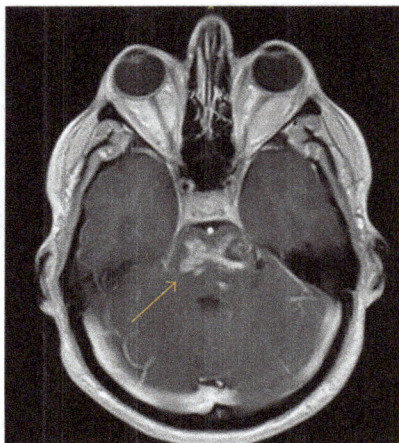

FIGURE 8: A 35-year-old male with contrast-enhanced axial T1 image showing an expansile heterogeneous patchy enhancement within the pons (arrow).

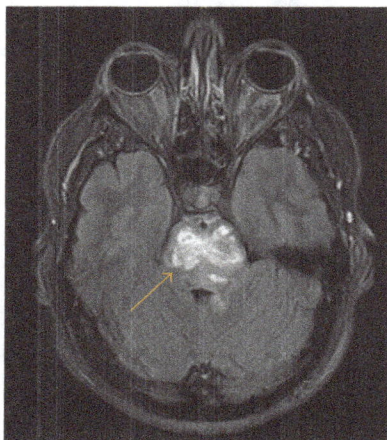

FIGURE 9: A 35-year-old male with axial T2-weighted FLAI image showing heterogeneous, predominantly hyperintense signal in the pons (arrow).

many as 96% of individuals diagnosed with ECD will have skeletal involvement; however only 50% are symptomatic [10]. The most frequently affected bones are the femur, tibia, and fibula and less frequently the ulna, radius, and humerus with pain usually manifesting around the knees and ankles. The hallmark feature, as was the case in our patient, is symmetric osteosclerosis in the diametaphyseal regions of the long bones with relative sparing of the axial skeleton and epiphyseal regions [11]. The radiographic osseous differential diagnosis in this case would include and is not limited to bony lymphoma, Paget's disease, osteomyelitis, and metastases.

Additionally, ECD can also involve the central nervous system including meninges, the orbits, and facial bones and manifest in a wide array of symptoms such as diabetes insipidus, exophthalmos, cerebellar ataxia, panhypopituitarism, and papilledema [12]. In our patient, the presenting symptom was of "bulging eyes" that was misinterpreted

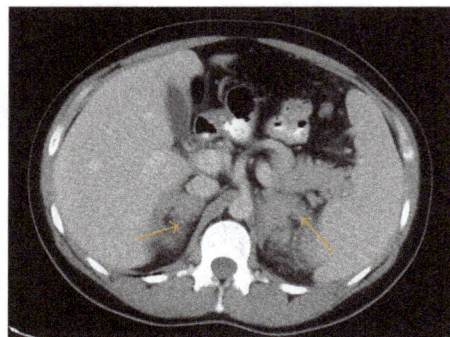

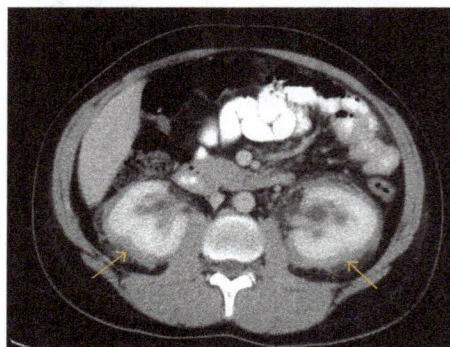

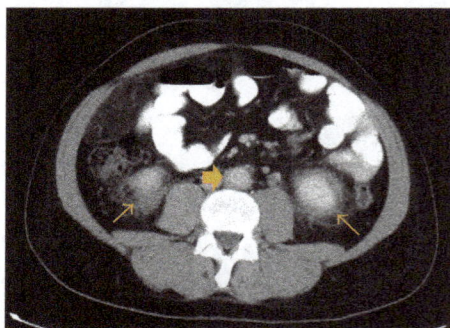

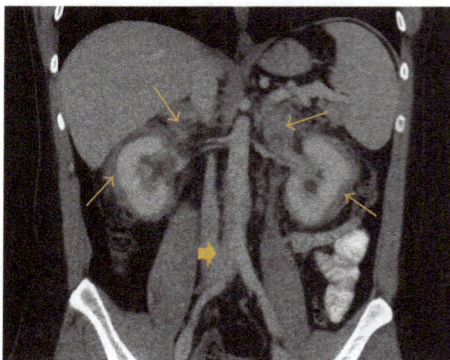

FIGURE 10: A 35-year-old male with axial and coronal reformatted contrast-enhanced CT images of the abdomen and pelvis showing enhancing soft tissue infiltration of the adrenal glands and kidneys (hairy kidney appearance) bilaterally (thin arrows). This is causing constriction of the renal pelvis and proximal ureters resulting in hydronephrosis. There is also retroperitoneal soft tissue infiltration (thick arrow) covering the posterolateral aorta (coated aorta appearance) which extends from just below level of renal arteries to aortic bifurcation and proximal right common iliac artery.

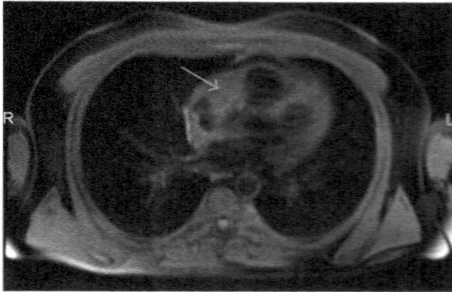

FIGURE 11: A 35-year-old male with fat-saturated T1-weighted image showing a lesion abutting the right atrium that is isointense to the surrounding muscle.

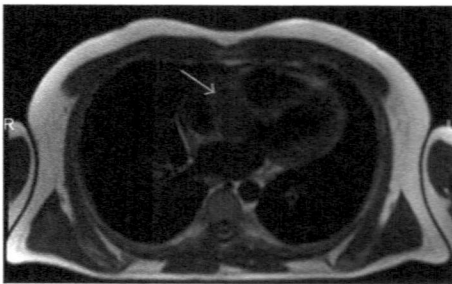

FIGURE 12: A 35-year-old male with non-fat-saturated T2-weighted image showing a lesion (arrow) abutting the right atrium that is hyperintense to the surrounding muscle.

around the ureters. This classically is known as the "hairy kidney" appearance which is seen on iodinated CT contrast as the infiltrative process extends into the perirenal fat giving an appearance of an irregular renal border [15]. There was also evidence of cardiac involvement in our patient in the form of a heterogeneous soft tissue mass abutting the right atrium. Cardiac involvement of ECD confers a poor prognosis related to a suboptimal response to treatment [16]. Reportedly, 60% of the 75% suffering from cardiac involvement die from cardiac complications [17] such as congestive heart failure, valvular disease, myocardial infarction, thromboembolism, and cardiac remodeling. Pericardial involvement usually manifests at first in the form of an effusion which is followed by myocardial involvement particularly of the right atrium [18]. The only finding that our patient did not have on presentation was diabetes insipidus that according to Cavalli et al. is one of the more common manifestations of ECD [13].

ECD continues to be a rare form of infiltrative disease but one that is associated with high morbidity and mortality. Timely diagnosis with immediate treatment of this disease remains paramount in achieving some success in countering the progression of this disease. Since most patients will present with nonspecific symptoms of bone pain or eye redness and may subsequently get imaged, one needs to be familiar with the key radiologic findings that may point one into considering ECD in the differential.

to be preseptal cellulitis not responding to antibiotics. CT of the orbits revealed soft tissue mass-like lesions in the retrobulbar orbits that on MRI were shown to be heterogeneously hyperintense on T2-weighted imaging, homogeneously hypointense on T1-weighted imaging, and homogeneously enhancing on postcontrast imaging. All of these findings are consistent with the diagnosis of ECD. In some cases, mass effect of the retroorbital lesions may result in thickening and tortuosity of the optic nerves; however that was not the case in our patient. The differential again is wide and includes but is not limited to Wegener's granulomatosis, Graves' disease, Langerhans cell histiocytosis, lymphoma, sarcoidosis, and Sjogren's disease [12]. There was also evidence of cerebellar involvement in our patient in the form of an expansile pontine lesion which supported his developing symptoms of ataxia. Reportedly, neurological involvement is a prominent feature of ECD, occurring in approximately 51% of the patients over the course of the disease and 23% at disease onset, and an independent predictor of death [13].

Two-thirds of patients with ECD also have evidence of retroperitoneal involvement [14] which is usually an incidental finding. In our patient, there was involvement of the abdominal aorta extending from below the level of the renal arteries down to the aortic bifurcation and proximal right common iliac artery giving the classic appearance of a "coated aorta" [2]. There was also renal involvement resulting in progressive renal failure as the infiltrative tissue compressed on the renal pelvis and subsequently fibrosed

References

[1] W. Chester, "Über Lipoidgranulomatose," Virchows Archiv für Pathologische Anatomie und Physiologie und für Klinische Medizin, vol. 279, no. 2, pp. 561–602, 1930.

[2] E. L. Diamond, L. Dagna, D. M. Hyman et al., "Consensus guidelines for the diagnosis and clinical management of Erdheim-Chester disease," Blood, vol. 124, no. 4, pp. 483–492, 2014.

[3] H. L. Jaffe, Metabolic, Degenerative and Inflammatory Diseases of Bones and Joints, Urban and Schwarzenberg, Munich, Germany, 1970.

[4] R. Carpinteri, I. Patelli, F. F. Casanueva, and A. Giustina, "Pituitary tumours: inflammatory and granulomatous expansive lesions of the pituitary," Best Practice & Research Clinical Endocrinology & Metabolism, vol. 23, no. 5, pp. 639–650, 2009.

[5] E. R. Volpicelli, L. Doyle, J. P. Annes et al., "Erdheim-Chester disease presenting with cutaneous involvement: a case report and literature review," Journal of Cutaneous Pathology, vol. 38, no. 3, pp. 280–285, 2011.

[6] C. Veyssier-Belot, P. Cacoub, D. Caparros-Lefebvre et al., "Erdheim-Chester disease. Clinical and radiologic characteristics of 59 cases," Medicine, vol. 75, no. 3, pp. 157–169, 1996.

[7] J. Haroche, Z. Amoura, S. G. Trad et al., "Variability in the efficacy of interferon-α in Erdheim-Chester disease by patient and site of involvement: results in eight patients," Arthritis and Rheumatism, vol. 54, no. 10, pp. 3330–3336, 2006.

[8] J. E. Sanchez, C. Mora, M. Macia, and J. F. Navarro, "Erdheim-Chester disease as cause of end-stage renal failure: a case report and review of the literature," International Urology and Nephrology, vol. 42, no. 4, pp. 1107–1112, 2010.

[9] L. Arnaud, I. Pierre, C. Beigelman-Aubry et al., "Pulmonary involvement in Erdheim-Chester disease: a single-center study of thirty-four patients and a review of the literature," *Arthritis and Rheumatism*, vol. 62, no. 11, pp. 3504–3512, 2010.

[10] J. Haroche, L. Arnaud, and Z. Amoura, "Erdheim-Chester disease," *Current Opinion in Rheumatology*, vol. 24, no. 1, pp. 53–59, 2012.

[11] D. Murray, M. Marshall, E. England, J. Mander, and T. M. H. Chakera, "Erdheim-chester disease," *Clinical Radiology*, vol. 56, no. 6, pp. 481–484, 2001.

[12] A. Drier, J. Haroche, J. Savatovsky et al., "Cerebral, facial, and orbital involvement in Erdheim-Chester disease: CT and MR imaging findings," *Radiology*, vol. 255, no. 2, pp. 586–594, 2010.

[13] G. Cavalli, B. Guglielmi, A. Berti, C. Campochiaro, M. G. Sabbadini, and L. Dagna, "The multifaceted clinical presentations and manifestations of Erdheim-Chester disease: comprehensive review of the literature and of 10 new cases," *Annals of the Rheumatic Diseases*, vol. 72, no. 10, pp. 1691–1695, 2013.

[14] L. Arnaud, B. Hervier, A. Neel et al., "CNS involvement and treatment with interferon-α are independent prognostic factors in Erdheim-Chester disease: a multicenter survival analysis of 53 patients," *Blood*, vol. 117, no. 10, pp. 2778–2782, 2011.

[15] E. Dion, C. Graef, J. Haroche et al., "Imaging of thoracoabdominal involvement in Erdheim-Chester disease," *American Journal of Roentgenology*, vol. 183, no. 5, pp. 1253–1260, 2004.

[16] J. Haroche, Z. Amoura, F. Charlotte et al., "Imatinib mesylate for platelet-derived growth factor receptor-beta positive Erdheim-Chester histiocytosis," *Blood*, vol. 111, no. 11, pp. 5413–5415, 2008.

[17] M. S. Alharthi, A. Calleja, P. Panse et al., "Multimodality imaging showing complete cardiovascular involvement by Erdheim-Chester disease," *European Journal of Echocardiography*, vol. 11, article E25, 2010.

[18] J. Haroche, P. Cluzel, D. Toledano et al., "Images in cardiovascular medicine. Cardiac involvement in erdheim-chester disease: magnetic resonance and computed tomographic scan imaging in a monocentric series of 37 patients," *Circulation*, vol. 119, no. 25, pp. e597–e598, 2009.

5

Giant Cell Tumor within the Proximal Tibia after ACL Reconstruction

Takashi Takahashi,[1] **Lauren MacCormick,**[2] **Jutta Ellermann,**[1]
Denis Clohisy,[2] **and Shelly Marette**[1]

[1]*Department of Radiology, University of Minnesota Medical Center, 420 Delaware Street SE, MMC 292, Minneapolis,
MN 55455, USA*
[2]*Department of Orthopaedic Surgery, University of Minnesota Medical Center, 2450 Riverside Avenue, Suite R200,
Minneapolis, MN 55454, USA*

Correspondence should be addressed to Takashi Takahashi; takas005@umn.edu

Academic Editor: Toshihiro Akisue

26-year-old female with prior anterior cruciate ligament reconstruction developed an enlarging lytic bone lesion around the tibial screw with sequential imaging over the course of one year demonstrating progression of this finding, which was confirmed histologically to be a giant cell tumor of bone. The lesion originated around the postoperative bed, making the diagnosis challenging during the early course of the presentation. The case demonstrates giant cell tumor which originated in the metaphysis and subsequently grew to involve the epiphysis; therefore, early course of the disease not involving the epiphysis should not exclude this diagnosis.

1. Introduction

There have been only two previous reported cases of giant cell tumor around the fixation screws after bone-patella-bone anterior cruciate ligament reconstruction. This is the third reportable case where sequential images demonstrate development of a new giant cell tumor within the proximal tibia around the tibial screw. Retrospectively, image findings demonstrate classic features of giant cell tumor of the bone. However, the postoperative timing as well as the initial location of the lesion in the metaphysis with subsequent extension in the epiphysis made the diagnosis challenging for experienced musculoskeletal radiologists, as well as orthopedic surgeons in the early disease course. The case demonstrates and confirms the origin of the tumor in the metaphyseal region as previously suggested.

2. Case Report

A 26-year-old female sustained an injury to her right knee at a dance tryout. Radiographs and MRI were obtained for the evaluation of the acute injury (Figures 1 and 2). Imaging demonstrated a complete tear of her anterior cruciate ligament, terminal sulcus impaction injury, and a corner fracture of the posterior aspect of the lateral tibial plateau.

The patient subsequently underwent an ACL reconstruction utilizing bone-patellar tendon-bone autograft at an outside institution with an uneventful recovery. The initial postoperative radiographs from the outside institution were unremarkable (Figure 3).

Approximately 6 months after the ACL reconstruction surgery, the patient had reinjured her right knee from a slip and fall on the ice. Radiographs were obtained at this time (Figure 4), with the initial interpretation by an outside institution, read as being unremarkable. The patient's symptoms persisted; therefore, A MRI was obtained which demonstrated a new lesion in the anteromedial aspect of the tibia, adjacent to the tibial tunnel (Figure 5). No specific differential diagnosis was given by the outside radiologist. Additional radiographs were obtained approximately 10 months after the ACL reconstruction due to the patient's persistent symptoms. Radiographs from this time (Figure 6) demonstrated slight

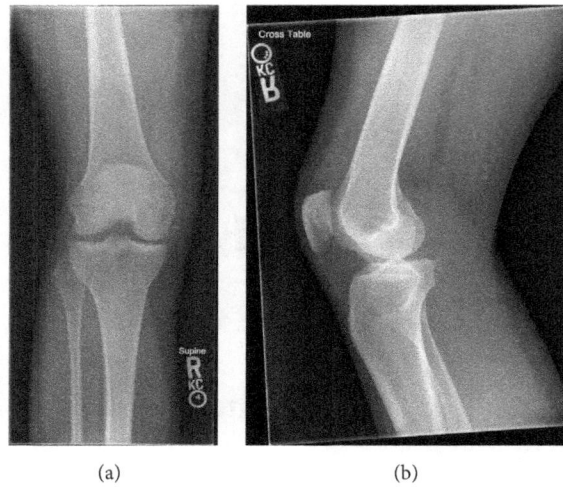

(a)   (b)

FIGURE 1: AP (a) and lateral (b) view radiographs of right knee for the evaluation of acute injury. Lateral view demonstrates moderate joint effusion, deepening of the terminal sulcus, and irregularity of the posterior aspect of the lateral tibial plateau.

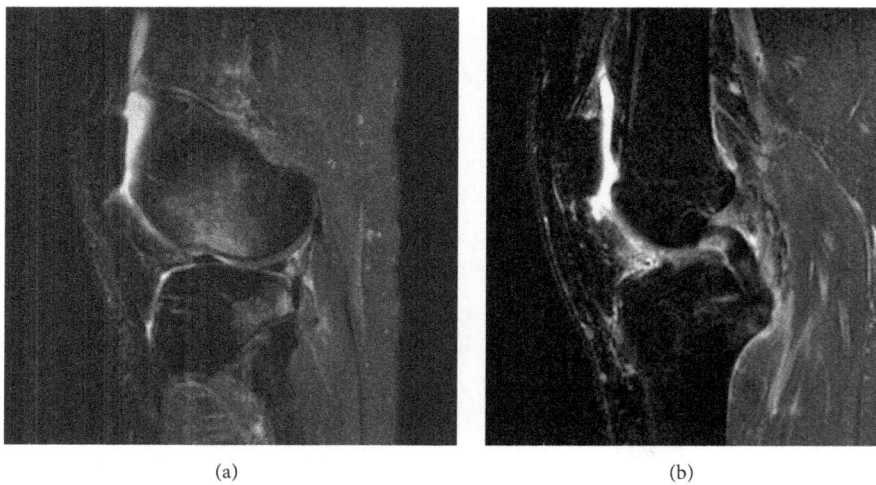

(a)   (b)

FIGURE 2: T2 weighted fat suppressed sagittal plane MRI images for evaluation of acute knee injury. Classic marrow contusion pattern at the lateral terminal sulcus and posterior aspect of the lateral tibial plateau (a) and complete rupture of anterior cruciate ligament (b).

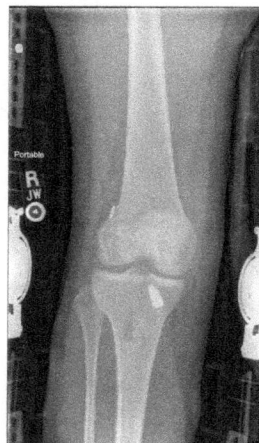

FIGURE 3: Immediate ACL reconstruction postoperative radiographs in AP view.

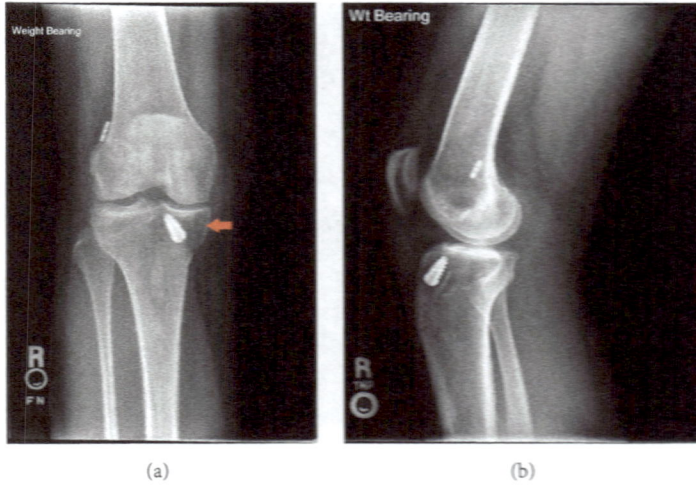

FIGURE 4: AP (a) and lateral (b) weight bearing radiographs of right knee obtained at the time of knee reinjury, approximately 6 months after the ACL reconstruction surgery. Retrospectively, new lucent lesion adjacent to the tibial tunnel is present (arrow).

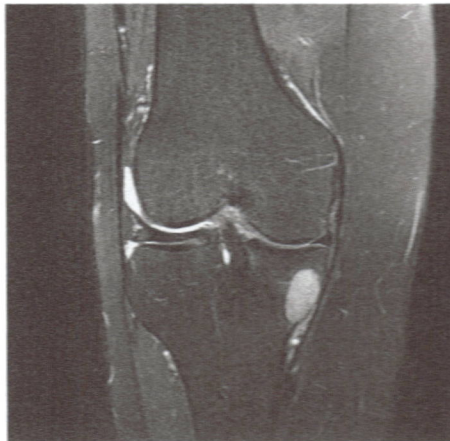

FIGURE 5: T2 weighted fat suppressed MRI in coronal plane shows well circumscribed hyperintense lesion in the medial tibial plateau without surrounding marrow edema. No communication of the lesion with tibial tunnel was present.

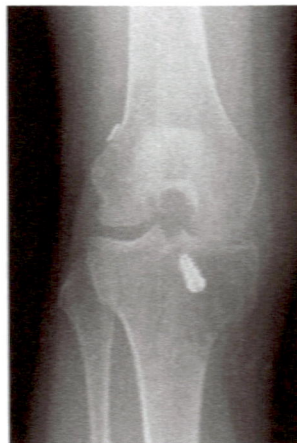

FIGURE 6: AP radiograph of the right knee shows significant progression of the tibial plateau osteolytic lesion in the medial tibial plateau, which now extends to the epiphysis.

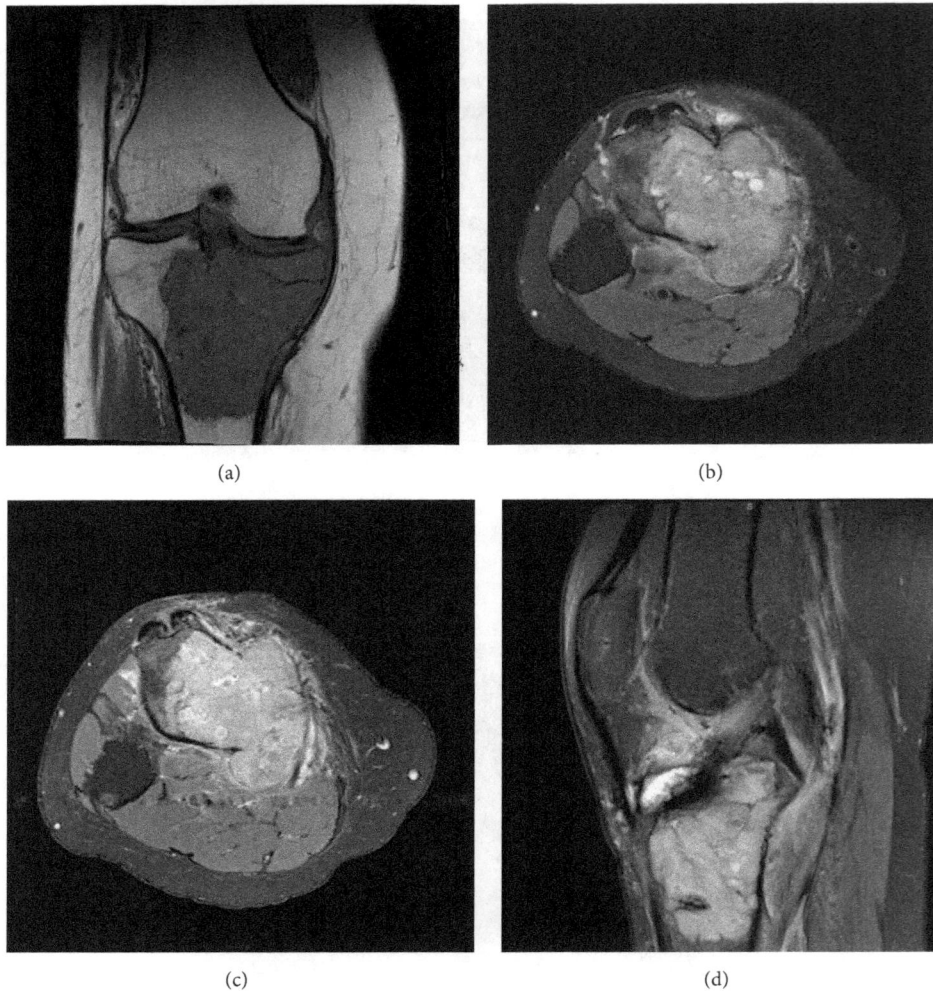

FIGURE 7: Coronal T1 weighted (a), axial fat suppressed proton density (b), and axial fat suppressed T1 postcontrast (c) images show enhancing bony lesion involving the tibial plateau with cortical breakthrough posteriorly. Selective lateral fat suppressed proton density (d) shows no evidence of other lesions in the distal femur and patella.

interval increase in the size of the medial tibial lesion. At this time, the patient was treated with steroid injection for her symptoms by the outside institution.

The patient was referred to our tertiary referral center for further evaluation of the bone lesion and associated tenderness over the anteromedial aspect of the patient's knee in the area of tibial tunnel. Review of outside films by a musculoskeletal radiologist and an orthopedic surgeon in our center included a differential diagnosis of infection, benign fibrous lesion such as fibrous dysplasia, and nonossifying fibroma. At this point, the patient was lost for further follow-up at our institution due to combination of patient's status of serving in the military and living out of state. Approximately 1 year and 7 months after the ACL reconstruction, the patient was referred back to our tertiary referral center for further assessment and treatment of the medial tibial bone lesion. Radiographs and MRI from the outside institution (Figures 6 and 7) showed significant progression of the lesion. A biopsy was recommended at this time. We discussed with the

patient that the most likely diagnosis was a giant cell tumor, but it was also possible that this was a malignancy due to the aggressive nature of the tumor over the course of one year. We counseled the patient on our plan to proceed with treatment of the tumor if histology confirmed a giant cell tumor, or we would stage and plan surgical intervention at a later date if histology was consistent with a malignancy. Intraoperatively, frozen section of the lesion was consistent with giant cell tumor. At that time, we proceeded with curettage and cementation of the tumor as well as removal of prior ACL screw. Due to extension of the tumor into the medial articular surface, posterior cortex, and insertion site of the medial collateral ligament (MCL), prophylactic fixation of the right tibia and repair of the MCL with soft tissue mobilization and transfer were also performed (Figure 8). Hematoxylin and eosin stain histological slide demonstrates characteristic giant cells (Figure 11).

Approximately 3 months after the giant tumor curettage and cementation, MRI and CT were obtained as a baseline.

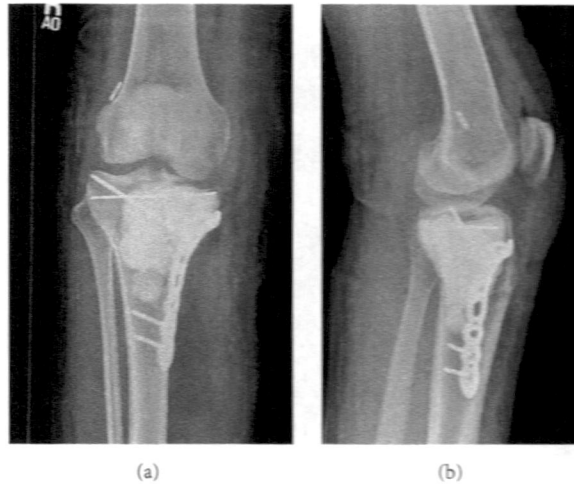

FIGURE 8: AP and lateral view radiographs immediately after curettage and packing of the proximal tibial giant cell tumor along with medial collateral ligament (MCL) repair; prophylactic fixation of the right tibia.

These studies demonstrated a new area of uncemented intramedullary cavity anterior to the cement (Figure 9). Given the short interval since prior MRI (Figure 7), findings are clinically thought to represent postoperative inflammatory change. Subsequent follow-up radiograph, which is approximately 4 months after the curettage and packing surgery, shows stable surgical change (Figure 10). The patient continues to show clinical improvement in her symptoms and continues with clinical and imaging surveillance.

3. Discussion

As the lesion progressed, it became apparent that the radiographic and MRI features of the lesion are suggestive of giant cell tumor of bone. Our case demonstrates serial imaging of histologically proven new development of giant cell tumor of bone within the previous ACL reconstruction surgical bed. To our knowledge, there have been two previous case reports of giant cell tumor which have occurred in the proximal tibia and distal femur around the fixation screws after bone-patella-bone anterior cruciate ligament reconstruction [1, 2].

The exact site of origin of GCT has been controversial. Murphey et al. suggest that GCTB likely arise from the metaphyseal side of the epiphyseal plate [3]. The present case also demonstrates new development and progression of the pathologically proven giant cell tumor consistent with this hypothesis.

World Health Organization (WHO) classifies giant cell tumor of bone (GCTB) as a benign, locally aggressive tumor. It is generally a benign, common bone tumor accounting for approximately 20% of benign osseous neoplasms, approximately 10% of primary osseous neoplasms, and 5% of all bone tumors [3, 4]. There is slight female gender predominance in its benign form, and it is typically seen between 20 and 40 years of age, with 80% of cases occurring between 20 and 50 years of age. The lesion is usually solitary. GCBT are most commonly seen about the knee, with the distal femur

being more frequently involved than the proximal tibia. Clinically, affected patients often present with pain secondary to underlying bone destruction, which can predispose to pathologic fractures. Multifocal GCTB is rare, accounting for less than 1% of all cases. In 5–10% of cases, GCTB may be classified as malignant, which includes benign metastasizing giant cell tumor, sarcoma developing following radiation or other intervention of a preexisting giant cell tumor, primary transformation of preexisting giant cell tumor, and osteoclastic sarcoma.

Conventional radiographs often have classic findings and can be highly suggestive of the diagnosis of GCBT. These findings include eccentric, lytic lesion centered in the metaepiphysis extending up to the subchondral bone plate without internal mineralization in a patient with closed physis [3]. The margin of the lesion is typically nonsclerotic. Pereira et al. analyzed 30 patients diagnosed with giant cell tumor of bone for their MRI findings [4]. In this study, MRI features suggesting giant cell tumor typically included low T2 signal due to the fibrous component along with the deposition of hemosiderin within the tumor. Following the administration of intravenous contrast, typically there is heterogeneous enhancement pattern. Secondary aneurysmal bone cysts are relatively common within a GCT lesion of bone. Over 40% of cases had evidence of extraosseous involvement on MRI, among which four cases were radiographically occult.

Imaging differential diagnosis includes primary aneurysmal bone cyst (ABC) and chondroblastoma. Intravenous contrast administration on MRI is helpful in distinguishing GCTB with secondary ABC from primary ABC as the presence of enhancing soft tissue component is typically present in GCTB but not in primary ABC. Presence of extensive surrounding reactive edema within the marrow and soft tissues, sclerotic margin, and presence of chondroid matrix are helpful features distinguishing chondroblastoma from GCT. Additional differential diagnoses include metastasis, plasmacytoma, or multiple myeloma, which should be

(a)

(b)

(c)

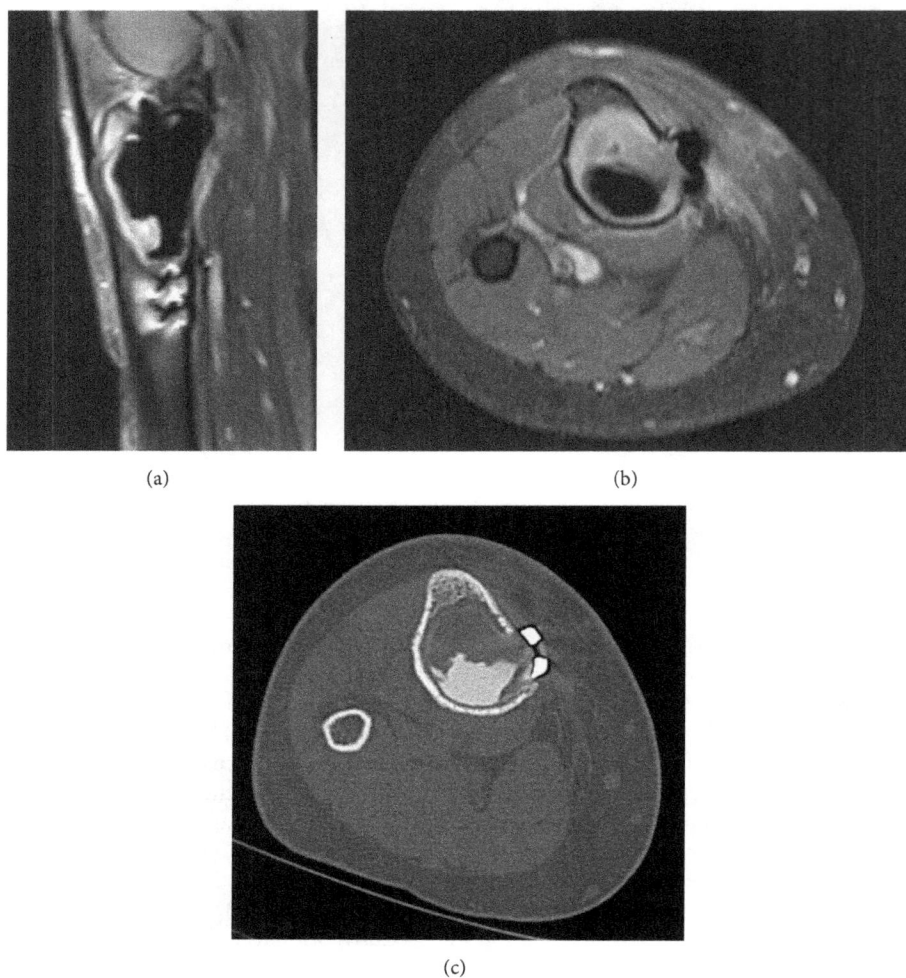

FIGURE 9: Sagittal T2 weighted fat suppressed image (a) and axial postcontrast enhanced T1 weighted image (b) of the right knee 3 months after the giant cell tumor curettage and packing demonstrate area of T2 hyperintensity and associated enhancement anterior to the surgical curettage cavity. CT scan image (c) from the same date also shows that there is corresponding uncemented area.

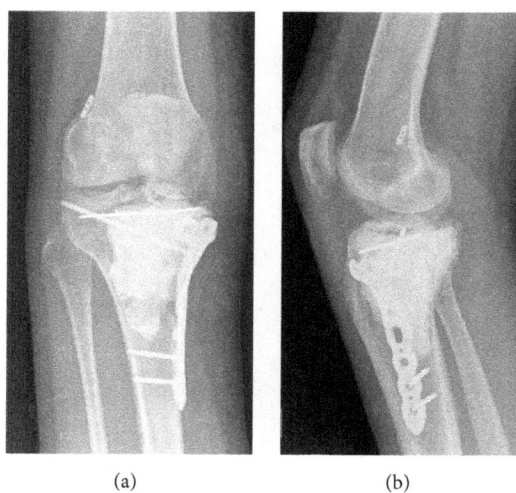

(a)

(b)

FIGURE 10: AP and lateral standing radiographs 4 months after giant cell tumor curettage and packing demonstrate stable postoperative change.

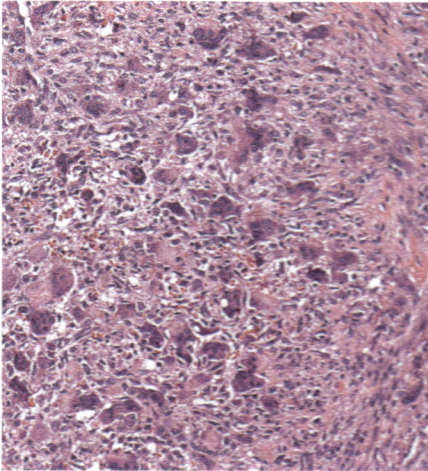

FIGURE 11: Hematoxylin and eosin stain histological slide.

included based on patient's age, multifocality, and clinical history of known primary neoplasm. Telangiectatic osteosarcoma, giant cell-rich osteosarcoma, and fibroblastic osteosarcoma should be included in differential considerations when aggressive giant cell tumor of bone is suspected, as these subtypes of osteosarcoma do not produce osteoid matrix [5]. Furthermore, additional differential diagnosis should include clear cell chondrosarcoma and joint centric processes such as pigmented villonodular synovitis and geodes.

In the past, GCTB has been treated primarily by the means of surgical resection with curettage and placement of cement. Aggressive giant cell tumor may require wide excision and reconstruction utilizing modular endoprosthesis. Even with aggressive surgical treatment, the local recurrence rate was noted to range from 15 to 50% [4–6].

References

[1] S. E. Fitzsimmons, N. Chinitz, and J. Glashow, "Giant cell tumor at tibial screw site after anterior cruciate ligament reconstruction," *The American Journal of Orthopedics*, vol. 39, no. 6, pp. E54–E56, 2010.

[2] D. Dowen, R. Kakkar, P. Dildey, and C. Gerrand, "Pain and fracture after anterior cruciate ligament reconstruction caused by giant cell tumour of the distal femur," *BMJ Case Reports*, 2013.

[3] M. D. Murphey, G. C. Nomikos, D. J. Flemming, F. H. Gannon, H. T. Temple, and M. J. Kransdorf, "From the archives of the AFIP. Imaging of giant cell tumor and giant cell reparative granuloma of bone: radiologic-pathologic correlation," *Radiographics*, vol. 21, no. 5, pp. 1283–1309, 2001.

[4] H. M. Pereira, E. Marchiori, and A. Severo, "Magnetic resonance imaging aspects of giant-cell tumours of bone," *Journal of Medical Imaging and Radiation Oncology*, vol. 58, no. 6, pp. 674–678, 2014.

[5] C. J. Chakarun, D. M. Forrester, C. J. Gottsegen, D. B. Patel, E. A. White, and G. R. Matcuk, "Giant cell tumor of bone: review, mimics, and new developments in treatment," *Radiographics*, vol. 33, no. 1, pp. 197–211, 2013.

[6] H. Wang, N. Wan, and Y. Hu, "Giant cell tumour of bone: a new evaluating system is necessary," *International Orthopaedics*, vol. 36, no. 12, pp. 2521–2527, 2012.

6

Double Meniscal Ossicle, the First Description: CT and MRI Findings—Different Etiologies

Puneeth Kumar, Amit Kumar Dey, Kartik Mittal, Rajaram Sharma, and Priya Hira

Department of Radiology, Seth G. S. Medical College and KEM Hospital, Mumbai 400012, India

Correspondence should be addressed to Puneeth Kumar; puneethkumar1234567@gmail.com

Academic Editor: Aruna Vade

We present a case of 2 ossicles in the medial meniscus with emphasis on MRI and CT findings. Meniscal ossicle is a rare entity and is quite uncommon on the medial side. By showing the typical signal characteristics and intrameniscal location, MRI can be helpful in distinguishing this from other more clinically significant abnormalities. It should be kept as differential from synovial chondromatosis or sesamoid bones like fabella as management is different for all of these entities.

1. Introduction

Ossicles within the meniscus of the knee are reported as a rare finding [1]. The ossicle could be described as corticocancellous bones with central fatty marrow completely surrounded by the meniscal fibrocartilage. They are usually symptomatic and discovered on knee radiographs [1–4]. Incidentally, they can occur in asymptomatic person. Radiological differentiation can be made from osteochondral loose bodies and chondrocalcinosis by its ossified appearance and its typical location within the meniscus. Correct diagnosis is required so that unnecessary surgery is avoided and protracted search of free fragment is not carried out [5, 6]. We present a case of 2 ossicles in the medial meniscus with emphasis on MRI and CT findings.

2. Case Report

A 21-year young male student presented with chronic pain in the right knee joint for 1-2 years. There was no recent history of trauma. There was not any other relevant past history. On clinical examination, there was mild swelling without any restriction of movements. Posterior drawer sign and the posterior tibial sag sign were negative. There was not any history of sudden violent trauma like dashboard injury. So PCL avulsion was ruled out. The radiographs were

unavailable but were reported to be normal. Patient was sent for further investigations in the form of MRI. MRI showed two small lesions isointense to bone marrow in relation to the posterior horn of the medial meniscus, with a hypointense rim suggestive of meniscal ossicle (Figures 1(a), 1(b), 1(c), and 1(d)). It was confirmed on plain CT axial, coronal, and sagittal bone window images which showed well defined lamellated bone density lesions, two in number in intra-articular region on medial aspect of right knee (Figures 2(a), 2(b), and 2(c)). Patient was put on analgesics and advised to take rest but did not improve. Arthroscopic findings of the patient confirmed our findings and avulsion of the PCL was ruled out. On entering the joint, a bulge in posterior horn of medial meniscus was seen and the rest of the joint appeared normal. PCL was normal with no signs of avulsion. So he was operated and on follow-up pain has subsided and patient is doing well. The surface of the excised bony fragment was noncystic, hard, and not irregular as usually found in the case of the ossicle.

3. Discussion

In 1934, the first case of meniscal ossicle was reported by Burrows [7]. To the best of our knowledge, in these 70 odd years, it has been reported 41 times [8]. It is not clear whether this is because it is an underdiagnosed/underreported condition or because it is actually an uncommon occurrence. Different

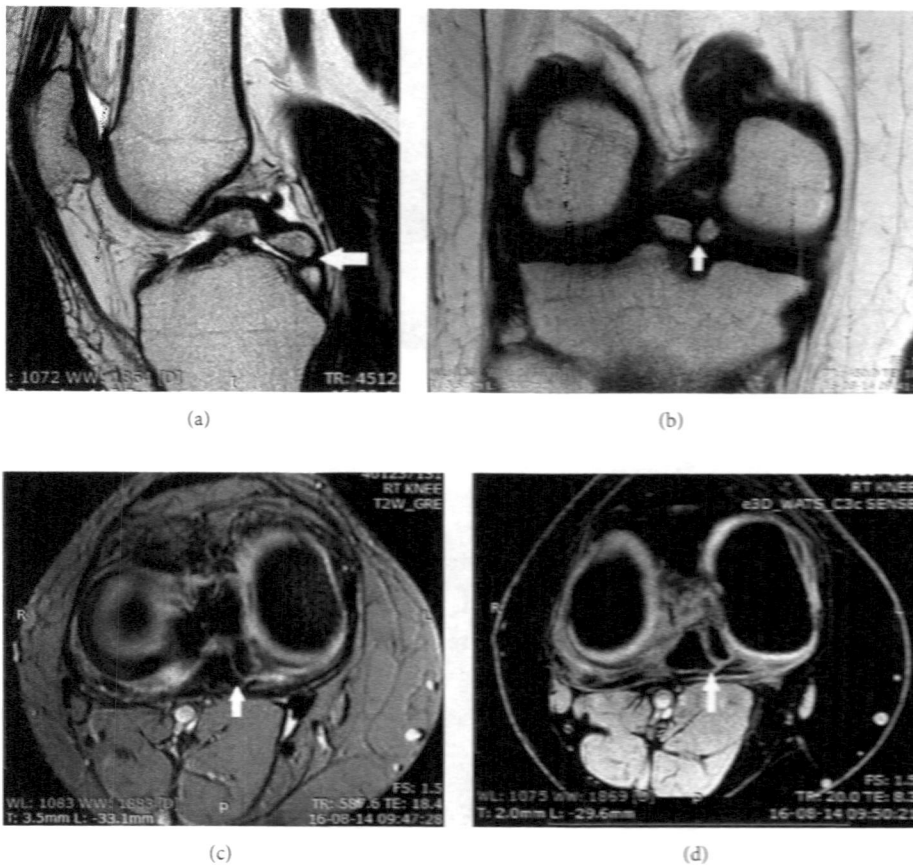

FIGURE 1: A 21-year young male student presented with chronic pain in the right knee joint for 1-2 years subsequently diagnosed as double meniscal ossicle of the knee. (a) Sagittal T1W image shows two small lesions (solid arrow), isointense to bone marrow in relation to the posterior horn of the medial meniscus, with a hypointense rim. (b) Coronal T1 W images confirm the isointensity of the lesions (solid arrow) to the bone marrow in relation to posterior aspect of medial meniscus. (c) Axial high-resolution T2 GRE image confirms the relationship of the meniscal ossicle (solid arrow) with the posterior horn of the medial meniscus. (d) Axial high-resolution T2 DESS image confirms the relationship of the meniscal ossicle (solid arrow) with the posterior horn of the medial meniscus.

theories are proposed for the etiology of the meniscal ossicle. Firstly, some consider it to be a degenerative phenomenon where areas of mucoid degeneration are replaced by bone [9]. Secondly some suggest it as posttraumatic sequelae with development of heterotopic ossification [2, 5–7]. Third theory proposes it to be a vestigial structure based on its presence in animal species like domestic cats, rodents, and Bengal tigers [1, 5]. The last theory suggests meniscal ossicles as bone fragments coming from the tibia at meniscal root insertion sites [8, 10]. The normal contour of the adjoining bone on MRI however, as in this case, argues against the last theory. In short, there is no definite consensus on the etiology of meniscal ossicles. Most patients complain of intermittent pain; however, since many patients also have other associated abnormalities, the relationship between the ossicles and pain is not definite [8]. A locking sensation is usually not experienced as would be expected with a free intra-articular body [8]. Most cases describe meniscal ossicles in

the posterior horn of medial meniscus and very rarely in the anterior horn of the lateral meniscus [8]. It is important to differentiate meniscal ossicle from osteochondral loose bodies and chondrocalcinosis. Osteochondral loose bodies can easily be differentiated from ossicle because of defect in the articular cartilage of distal femur [11]. Loose body is frequently found lying in the superolateral part of the anterior compartment of the knee and is composed predominantly of calcified cartilage and subchondral bone [11]. When there is confusion to differentiate loose bodies from meniscal ossicle on plain radiographs, a meniscal ossicle can be differentiated by its MRI characteristics which include an intrameniscal location, internal signal intensity of marrow, and a surrounding rim of low signal intensity corresponding to cortex [12]. Chondrocalcinosis may create calcific density within the body of meniscus; it is typically punctuated and linearly arranged; it will not have well defined cancellous bone of meniscal ossicle [11].

(a)

(b)

(c)

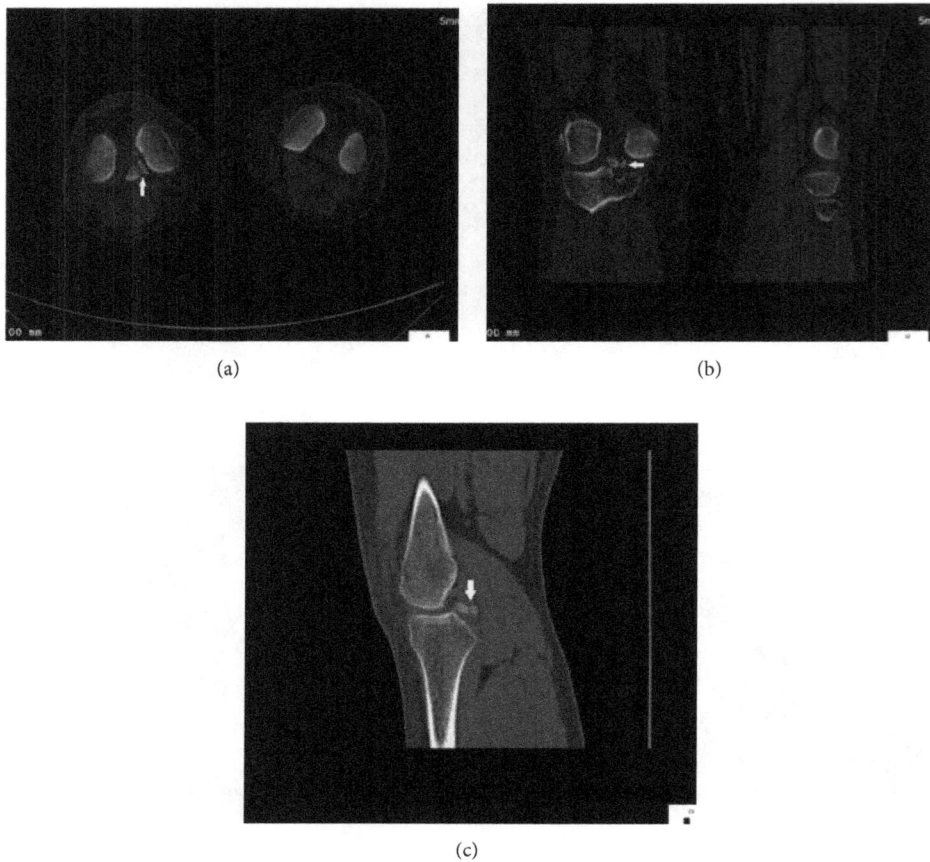

FIGURE 2: A 21-year young male student presented with chronic pain in the right knee joint for 1-2 years subsequently diagnosed as double meniscal ossicle of the knee. (a) Plain CT axial bone window images show well defined lamellated bone density lesions, two in number (solid arrow) in intra-articular region on medial aspect of right knee. (b) Plain CT coronal bone window images show well defined lamellated bone density lesions, two in number (solid arrow) in intra-articular region on medial aspect of right knee. (c) Plain CT sagittal bone window images show well defined lamellated bone density lesions, two in number (solid arrow) in intra-articular region on medial aspect of right knee.

4. Conclusion

Meniscal ossicle is a rare entity and is quite uncommon on the medial side. By showing the typical signal characteristics and intrameniscal location, MRI can be helpful in distinguishing this from other more clinically significant abnormalities. It should be kept as a differential from synovial chondromatosis or sesamoid bones like fabella as management is different for all of these entities.

Disclosure

Each of the authors was involved in preparation of the paper. The paper has been read and approved by all the authors and represents honest work.

References

[1] B. Conforty and M. Lotem, "Ossicles in human menisci: report of two cases," *Clinical Orthopaedics and Related Research*, vol. 144, pp. 272–275, 1979.

[2] R. S. Glass, W. M. Barnes, D. U. Kells, S. Thomas, and C. Campbell, "Ossicles of knee menisci: report of seven cases," *Clinical Orthopaedics and Related Research*, vol. 111, pp. 163–171, 1975.

[3] P. Symeonides and G. Ioannides, "Ossicles in the knee menisci. Report of three cases," *The Journal of Bone & Joint Surgery—American Volume*, vol. 54, no. 6, pp. 1288–1288, 1972.

[4] J. Kossoff, A. Naimark, and M. Corbett, "Case report 85," *Skeletal Radiology*, vol. 4, no. 1, pp. 45–46, 1979.

[5] R. W. Jones and R. E. Roberts, "Calcification, decalcification, and ossification," *British Journal of Surgery*, vol. 21, no. 83, pp. 461–499, 1934.

[6] J. B. Weaver, "Calcifications and ossification of the menisci," *The Journal of Bone & Joint Surgery*, vol. 24, pp. 873–882, 1942.

[7] H. J. Burrows, "Two cases of ossification in the internal semilunar cartilage," *British Journal of Surgery*, vol. 21, no. 83, pp. 404–410, 1934.

[8] I. Van Breuseghem, E. Geusens, S. Pans, and P. Brys, "The meniscal ossicle revisited," *JBR-BTR*, vol. 86, no. 5, pp. 276–277, 2003.

[9] H. A. Harris, "Calcification and ossification in the semilunar cartilages," *The Lancet*, vol. 223, no. 5778, pp. 1114–1116, 1934.

[10] E. E. Berg, "The meniscal ossicle: the consequence of a meniscal avulsion," *Arthroscopy*, vol. 7, no. 2, pp. 241–243, 1991.

[11] R. M. Bernsetin, H. E. Olsson, R. M. Spitzer, K. E. Robinson, and M. W. Korn, "Ossicle of the meniscus," *American Journal of Roentgenology*, vol. 127, no. 5, pp. 785–788, 1976.

[12] P. Schnarkowski, P. F. J. Tirman, K. D. Fuchigami, J. V. Crues, M. G. Butler, and H. K. Genant, "Meniscal ossicle: radiographic and MR imaging findings," *Radiology*, vol. 196, no. 1, pp. 47–50, 1995.

7

Urachal Adenocarcinoma: A Case Report with Key Imaging Findings and Radiologic-Pathologic Correlation

Willian Schmitt ⓘ,[1] Marta Baptista,[1] Marco Ferreira,[2] António Gomes,[3] and Ana Germano[1]

[1]*Department of Radiology, Hospital Prof. Doutor Fernando Fonseca, Amadora, Portugal*
[2]*Department of Pathology, Hospital Prof. Doutor Fernando Fonseca, Amadora, Portugal*
[3]*Department of Surgery, Hospital Prof. Doutor Fernando Fonseca, Amadora, Portugal*

Correspondence should be addressed to Willian Schmitt; schmitt.wr@gmail.com

Academic Editor: Vincent Low

Urachal pathologies are rare and can mimic numerous abdominal and pelvic diseases. Differential diagnosis of urachal anomalies can be narrowed down by proper assessment of lesion location, morphology, imaging findings, patient demographics, and clinical history. We report a case of a 60-year-old male, with a history of unintentional weight loss without associated symptoms, who was diagnosed with locally invasive urachal adenocarcinoma. With this article, we pretend to emphasize urachal adenocarcinoma clinical features along with its key imaging findings with radiologic-pathologic correlation.

1. Introduction

Urachal cancer (UrC) is an uncommon neoplasm, representing 0.5–2% of all bladder cancer. Although transitional epithelium usually lines the urachus, most urachal tumors are adenocarcinomas (90%) and represent 20%–40% of all primary bladder adenocarcinomas [1]. Despite being such rare entities, the number of publications regarding UrC has increased from 3 (between 1980 and 2005) to 10 cases per year on the last decade [2].

2. Case Presentation

We describe the case of a 60-year-old male who presented with a four-month history of 15 kg unintentional weight loss, without associated gastrointestinal or urogenital symptoms. On physical examination a visible, nontender, and nonmobile infraumbilical mass was noted. The patient denied significant past medical or family history. Blood work showed mildly elevated white cell count 14.4×10^9 (normal range: $4–11 \times 10^9$) and CRP 9.1 (normal range < 0.30 mg/dL).

An abdominal ultrasound was performed to determine the mass origin. It revealed a midline mass with a central gas-filled cavity contacting the superior bladder wall and extending to the anterior abdominal wall (Figure 1).

Additional evaluation with contrast-enhanced abdominopelvic computed tomography (CT) confirmed the presence of a median infraumbilical large intra-abdominal mass, measuring $13 \times 8 \times 14$ cm. Inferiorly it contacted the bladder dome, and anteriorly there was infiltration of the entire thickness of the anterior abdominal wall, invading the umbilicus (Figure 2).

Posteriorly there was no fat interface between the tumor and the transverse colon. Two fistulous tracts were perceived, connecting the mass with this bowel segment. The remaining study was unremarkable, with no sign of distant metastasis.

The morphology of the mass, as well as it aggressive behavior, with abdominal wall, colon, and bladder invasion, suggested a neoplastic lesion. Its location favored a urachal carcinoma as the main diagnosis. Differential diagnosis included a colon carcinoma with abdominal wall and bladder extension, a sarcoma, and an infected urachal remnant.

A colonoscopic study was requested, but it was not completed due to fixation of a colonic loop at 30 cm. After the procedure, discharge of fecal material through skin fistula was noted.

Urgent laparotomy was performed, with total resection of the tumor, with an en bloc resection of the anterior abdominal wall, transverse colon, and superior bladder wall. A temporary colostomy was performed.

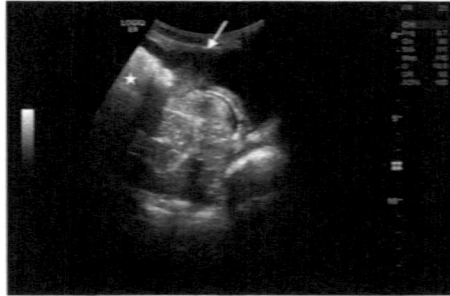

FIGURE 1: Sagittal US image of the lower abdomen obtained with a convex probe (1–6 Mhz). Large mass filled with gas (star), contacting the superior bladder wall (arrow), with focal thickening. Presence of free abdominal fluid is noted.

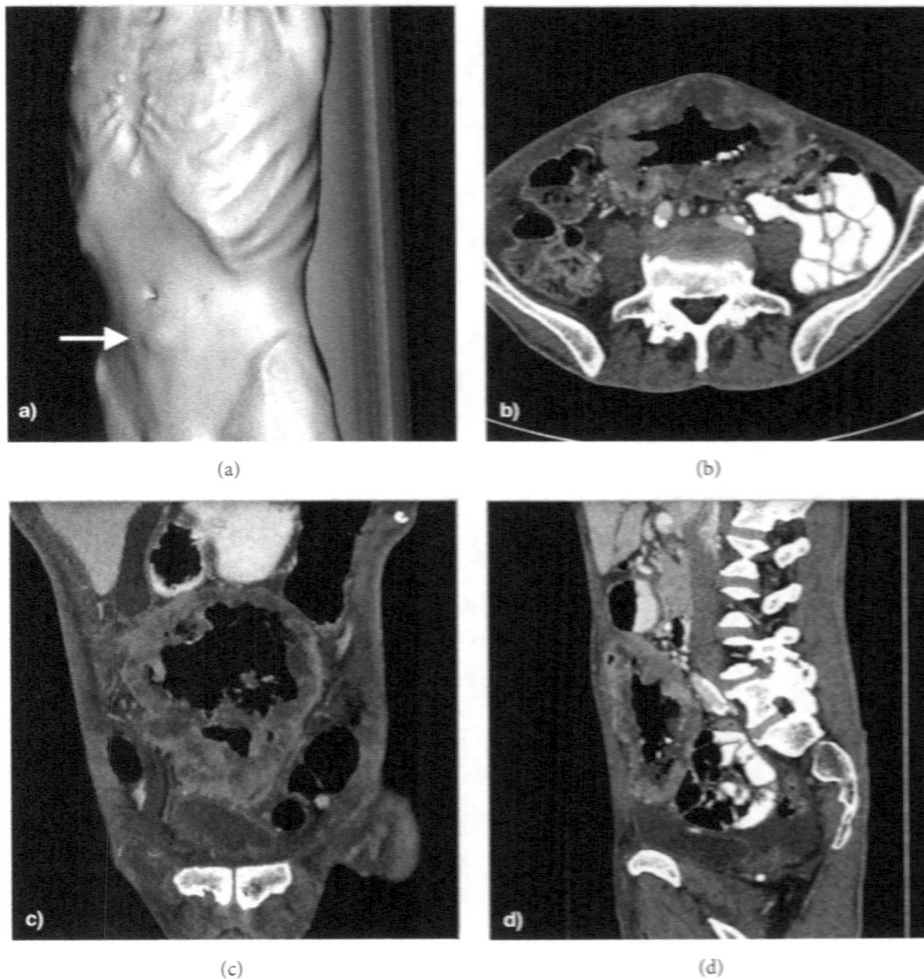

(a)

(b)

(c)

(d)

FIGURE 2: (a) 3D CT reconstruction showing an emaciated patient with a visible infraumbilical mass (arrow). Axial (b), coronal (c), and sagittal (d) contrast-enhanced CT images depicting a large and heterogeneous median abdominal mass, extending from the anterosuperior aspect of the bladder toward the umbilicus.

Specimen's pathology revealed a poorly differentiated adenocarcinoma with signet ring cell component and mucinous features. It infiltrated the skin, as well as the colonic and bladder walls, with mucosa ulceration (Figure 3).

Immunohistochemical positivity for 34-beta-E12 and beta-catenin (membrane) was seen (Figure 4), along with focal positivity for CDX2 CK7, and CK20, and uroplakins were nonreactive.

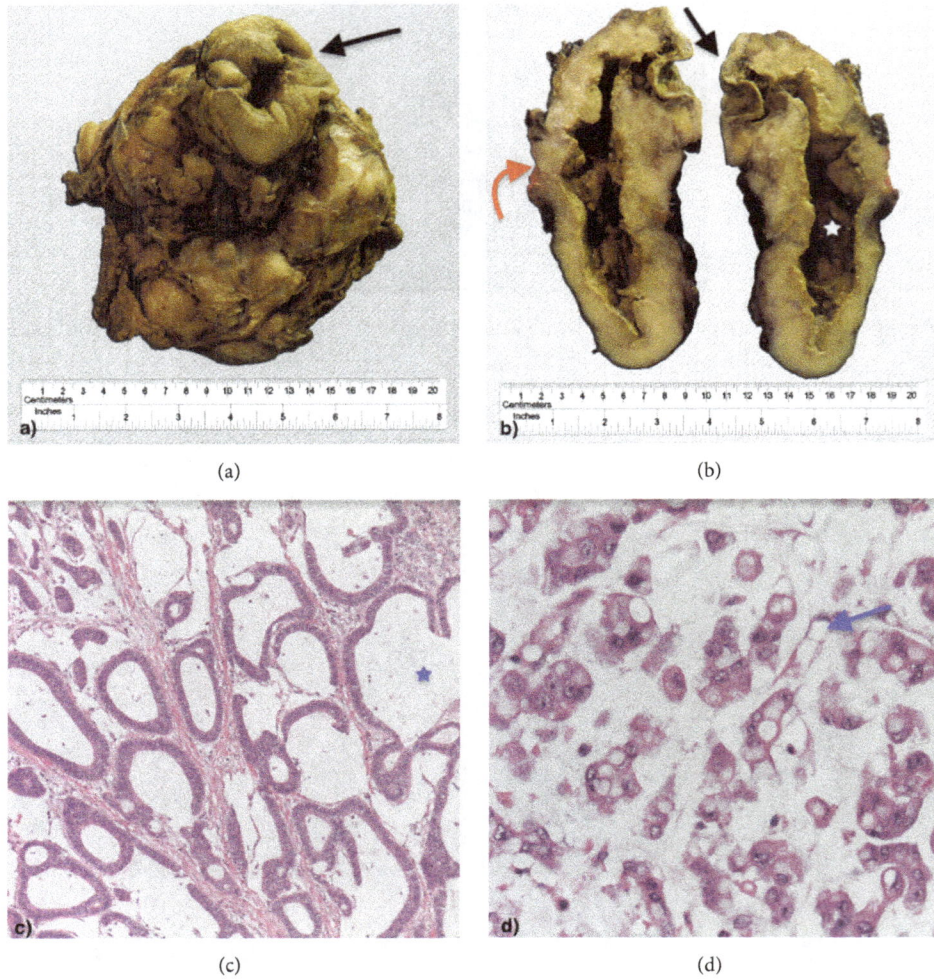

FIGURE 3: Gross specimen of the tumor ((a) anterior view; (b) cross section), showing a multinodular mass with fistulization to the umbilical skin (black arrows) and involvement of the bladder wall (red arrow) and a central necrotic cavity (star). Histologic image of the resected tumor ((c) H&E 100x; (d) H&E 400x) showing areas with a mucinous pattern (blue star) and a signet ring cell morphology (blue arrow).

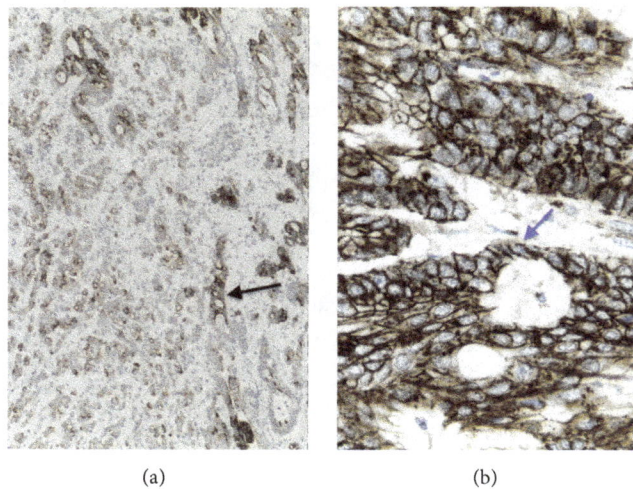

FIGURE 4: (a) Histologic image of the resected tumor (cytokeratin 34-beta-E12, 100x). Most of the neoplastic cells expressed high molecular-weight cytokeratins (black arrow). (b) Neoplastic cells showing diffuse membranous and cytoplasmic expression of beta-catenin (blue arrow). Note that there is no nuclei staining (beta-catenin, 400x).

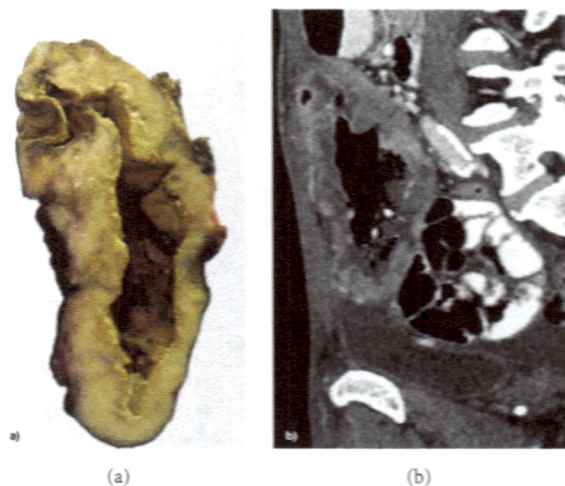

(a)   (b)

FIGURE 5: Radiologic-pathologic correlation of the tumor ((a) sagittal CT image; (b) cross section of the gross specimen), showing a large and heterogeneous mass, with thickened wall and central necrotic cavity.

This morphological and immunohistochemical features favored the diagnosis of a urachal adenocarcinoma (Figure 5).

The postoperative period was complicated with surgical wound infection, and the patient was discharged from hospital after 25 days, with wound care instructions and primary care physician referral.

3. Discussion

The urachus is an embryological remnant that originates from the involution of the allantois and cloaca, extending from the umbilicus to the bladder dome. During the gestational development, it involutes and becomes the median umbilical ligament, with obliteration of its lumen. Most of the urachus pathologies are found incidentally, and, with the increasing use of cross-sectional imaging, they have become more frequently diagnosed [3].

UrC accounts for 0.01% of all cancers in adults [1]. It was originally described by Hue and Jacquin in 1963 and it has a male predilection, with a mean age of presentation of 60 years (ranging from 40 to 70) [4, 5].

Symptoms do not usually accompany early UrC, which causes patients to present at advanced stages, with local invasion or distant metastases. The most frequently reported symptom is hematuria, followed by abdominal pain, dysuria, and mucosuria [2]. Less frequent clinical presentations include other urogenital related manifestations and nonspecific symptoms such as nausea, diarrhea, or weight loss. In contrast with other bladder cancers, UrC often presents with a palpable lower abdominal mass that along with weight loss was the presentation in our case.

Cystoscopy remains one of the most valuable diagnostic tools for UrC management. It allows a direct visualization of the tumor and it can also determine if the lesion is covered with normal mucosa, as seen in early stages, or if it is a broad-based ulcerated mass [2].

Imaging plays also an invaluable role at UrC workup. On Ultrasound (US), it is commonly recognized as a midline soft tissue mass or a fluid-filled cavity with mixed echogenicity and calcifications. CT scan is often used for local staging and evaluation of distant metastasis. It is usually depicted as a midline mass, superior to the bladder dome and adjacent to the abdominal wall. In the majority of cases, the tumor is mixed solid and cystic, the latter representing its mucin composition. Peripheral calcifications are often seen and are considered pathognomonic for urachal adenocarcinoma. On MRI, focal areas of high signal in T2 indicate the presence of mucinous component. CT and MRI are useful in demonstrating intra- or extravesical extension of the tumor.

The diagnosis of a UrC in biopsy and transurethral resection specimens remains a challenge. The criteria for pathologic diagnosis include the following: (1) location of the tumor in the dome/anterior wall; (2) epicenter of carcinoma in the bladder wall; (3) absence of widespread atypical intestinal metaplasia and cystitis/glandularis beyond the dome/anterior wall; (4) absence of urothelial neoplasia in the bladder; (5) absence of a known primary tumor elsewhere [6, 7].

The most frequent histologic subtype is adenocarcinoma with enteric features with or without mucin production. Some have a signet ring component, as seen in our case, and others have the morphology of colloid carcinomas [5].

Among the used immunomarkers, β-catenin and CK7 are the most important in establishing the distinction of urachal from colorectal adenocarcinoma. Diffuse nuclear b-catenin and CK7 help to differentiate urachal adenocarcinoma of enteric subtype (nuclear b-catenin −, CK7 +/−), as seen in our case, from colonic adenocarcinoma (diffuse nuclear b-catenin +, CK7 −) [6].

The main differential diagnosis includes benign urachal tumors, nonurachal carcinomas of the bladder, and metastasis from different organs (prostate, colon, rectum, and female genital tract). Infection of a urachal anomaly can also mimic a UrC, and the diagnosis is often challenging at imaging. In our case, the presence of a midline cavity, anterosuperior to the bladder dome, filled with low attenuation fluid, suggested adenocarcinoma of the urachus as the main diagnosis.

Although several staging systems have been proposed for UrC (Sheldon et al. [8], Ashley et al. [5], and Pinthus et al. [9]), their prognostic relevance has still to be validated in larger series. It is becoming apparent that staging for UrC is more consistent when utterly dichotomized to those confined to urachus, bladder, and perivesical tissue (surgical specimen) and those with spread to the peritoneum and other organs [6, 10]. Mean survival for locally advanced or metastatic disease is 12–24 months, with a 5-year survival rate of 43% [5].

Currently, surgery is the treatment of choice for UrC. The most significant predictor of prognosis is surgical margin status. Because UrC can present with metachronous or synchronous tumors along the urachal tract, the standard surgical approach includes excision of the urachus, umbilicus, and partial/radical cystectomy, combined with pelvic lymphadenectomy [5, 8, 11].

The effective role of adjuvant or neoadjuvant chemotherapy for UrC is not yet established [1, 12]. The fact that most of the available data results from case reports with several chemotherapy regimens and low case numbers does not provide the required statistical significance. Available data suggest that 5-fluorouracil (FU) based chemotherapies are superior to cisplatin-based regimens regarding radiologic tumor response, whereas the combination of 5-FU with cisplatin provides the most beneficial response in metastatic UrC [2, 6]. A fairly recent study described the combination of cytoreductive surgery and hyperthermic intraperitoneal chemotherapy (HIPEC) to be effective in prolonging survival in UrC patients with peritoneal metastases [13].

Abbreviations

UrC: Urachal cancer
US: Ultrasound
CT: Computed tomography
MRI: Magnetic resonance imaging
FU: Fluorouracil
HIPEC: Hyperthermic intraperitoneal chemotherapy.

Additional Points

Key Message. UrC is a rare malignant tumor and constitutes a clinical challenge. Its diagnosis, management, and treatment decisions rely on a multidisciplinary approach.

References

[1] B. Bao, M. Hatem, and J. K. Wong, "Urachal adenocarcinoma: a rare case report," *Radiology Case Reports*, vol. 12, no. 1, pp. 65–69, 2017.

[2] T. Szarvas, O. Módos, C. Niedworok et al., "Clinical, prognostic, and therapeutic aspects of urachal carcinoma—a comprehensive review with meta-analysis of 1,010 cases," *Urologic Oncology: Seminars and Original Investigations*, vol. 34, no. 9, pp. 388–398, 2016.

[3] C. Parada Villavicencio, S. Z. Adam, P. Nikolaidis, V. Yaghmai, and F. H. Miller, "Imaging of the urachus: Anomalies, complications, and mimics," *RadioGraphics*, vol. 36, no. 7, pp. 2049–2063, 2016.

[4] V. Monteiro and T. M. Cunha, "Urachal Carcinoma: Imaging Findings," *Acta Radiologica Short Reports*, vol. 1, no. 1, pp. 1–3, 2012.

[5] R. A. Ashley, B. A. Inman, T. J. Sebo et al., "Urachal carcinoma: clinicopathologic features and long-term outcomes of an aggressive malignancy," *Cancer*, vol. 107, no. 4, pp. 712–720, 2006.

[6] G. P. Paner, A. Lopez-Beltran, D. Sirohi, and M. B. Amin, "Updates in the pathologic diagnosis and classification of epithelial neoplasms of urachal origin," *Advances in Anatomic Pathology*, vol. 23, no. 2, pp. 71–83, 2016.

[7] A. Gopalan, D. S. Sharp, S. W. Fine et al., "Urachal carcinoma: a clinicopathologic analysis of 24 cases with outcome correlation," *The American Journal of Surgical Pathology*, vol. 33, no. 5, pp. 659–668, 2009.

[8] C. A. Sheldon, R. V. Clayman, R. Gonzalez, R. D. Williams, and E. E. Fraley, "Malignant urachal lesions," *The Journal of Urology*, vol. 131, no. 1, pp. 1–8, 1984.

[9] J. H. Pinthus, R. Haddad, J. Trachtenberg et al., "Population based survival data on urachal tumors," *The Journal of Urology*, vol. 175, no. 6, pp. 2042–2047, 2006.

[10] H. W. Herr, B. H. Bochner, D. Sharp, G. Dalbagni, and V. E. Reuter, "Urachal carcinoma: contemporary surgical outcomes," *The Journal of Urology*, vol. 178, no. 1, pp. 74–78, 2007.

[11] I. M. Koster, P. Cleyndert, and R. W. M. Giard, "Best cases from the AFIP: Urachal carcinoma," *RadioGraphics*, vol. 29, no. 3, pp. 939–942, 2009.

[12] J. R. Molina, J. F. Quevedo, A. F. Furth, R. L. Richardson, H. Zincke, and P. A. Burch, "Predictors of survival from urachal cancer: a Mayo Clinic study of 49 cases," *Cancer*, vol. 110, no. 11, pp. 2434–2440, 2007.

[13] F. Mercier, G. Passot, L. Villeneuve et al., "Peritoneal Carcinomatosis of Urachus Origin Treated by Cytoreductive Surgery and Hyperthermic Intraperitoneal Chemotherapy (HIPEC): An International Registry of 36 Patients," *Annals of Surgical Oncology*.

8

Port Placement via the Anterior Jugular Venous System: Case Report, Anatomic Considerations, and Literature Review

Gernot Rott and Frieder Boecker

Radiological Department, Bethesda-Hospital Duisburg, Heerstr. 219, 47053 Duisburg, Germany

Correspondence should be addressed to Gernot Rott; info@myom-therapie.de

Academic Editor: Dimitrios Tsetis

We report on a patient who was referred for port implantation with a two-chamber pacemaker aggregate on the right and total occlusion of the central veins on the left side. Venous access for port implantation was performed via left side puncture of the horizontal segment of the anterior jugular vein system (AJVS) and insertion of the port catheter using a crossover technique from the left to the right venous system via the jugular venous arch (JVA). The clinical significance of the AJVS and the JVA for central venous access and port implantation is emphasised and the corresponding literature is reviewed.

1. Introduction

Venous access for chest port implantation may be compromised for a variety of reasons and finding a suitable alternative can be difficult. If a conventional approach for this purpose is not an option, collateral veins may offer an attractive alternative. A crossover route utilising the anterior jugular venous system (AJVS) and the jugular venous arch (JVA) may be considered. We present, to the best of our knowledge, a unique case where anatomical constraints led us to utilise this crossover technique for port placement.

The placement of a chest collateral port in crossover technique through the anterior jugular vein system can be quite simple and the most appropriate method in the mentioned or a comparable scenario.

2. Case Report

A 70-year-old woman was admitted to our hospital for staging and therapy of a recently diagnosed adenocarcinoma of the esophagogastric junction. The significant past medical history of the patient included implantation of a two-chamber pacemaker on the right side and subtotal sternum resection with a pectoral muscular flap graft coverage for treatment of sternum osteomyelitis. Contrast enhanced computed tomography (CT) of the thorax showed chronic occlusion of the

left-sided central veins with the presence of a cervical collateral venous flow from the left to the right side reaching the nonoccluded right-sided central veins, specifically the confluence to the right innominate vein (IV) and, as part of the collateral circulation, a transverse connecting, in parts quite tortuous, and u-shape midline jugular vein measuring about 3 mm in diameter (Figure 1). Staging examinations revealed a T3 cancer and our interdisciplinary tumor board recommended the implementation of neoadjuvant chemotherapy.

The patient was presented to our radiological department for port implantation. We decided to opt for venous access on the left upper body to avoid problems with the pacemaker aggregate on the right upper body. The plan was to recanalise the left IV. If this proved unsuccessful, alternative access via puncture of a left-sided collateral vein to reach the contralateral central veins would be instigated.

The procedure was performed under local anesthesia in our angiographic suite. Phlebography of the left arm demonstrated the venous anatomy of the shoulder region, as described above in the previously performed CT (Figure 2). After the usual preparations, the left subclavian vein (SV) was punctured under fluoroscopy and guidance of contrast venography, and a 4-French catheter was introduced. Recanalisation of the totally occluded left IV was attempted using different types of guidewires, however without success due to high tissue resistance and absence of a funnel-shaped

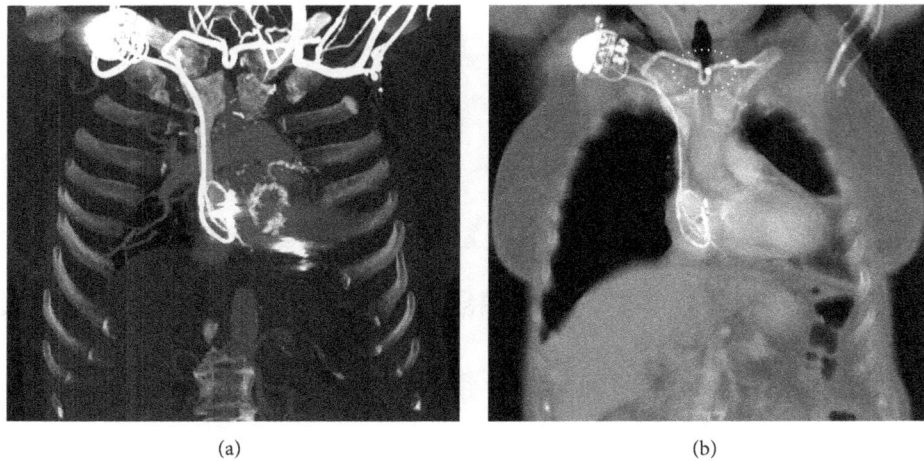

(a)   (b)

FIGURE 1: CT with administration of contrast media through a left peripheral vein. (a) Coronal maximum intensity projection image shows chronic occlusion of the left-sided central veins with cervical collateral veins from the left to the right side and here reaching the nonoccluded right-sided central veins; slight metal artifacts by the right-sided pacemaker. (b) Coronal multiplanar reconstruction image showing a partially quite tortuous u-shape horizontal-transverse midline vein in the jugulum (dotted ellipse), the AJVS with the JVA.

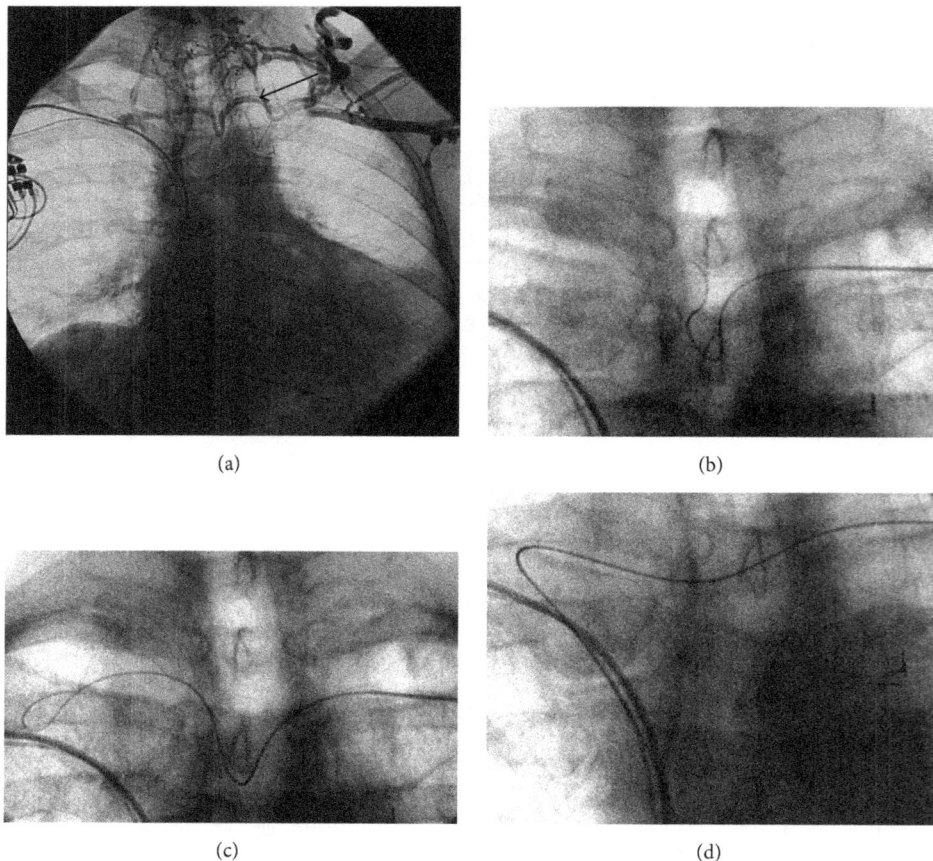

(a)   (b)

(c)   (d)

FIGURE 2: Phlebography correlation of the left arm demonstrates the same collateral circulation as in the CT, fully comparable to Figure 1(a) (a). Note arrow indicating the exact location of the following vascular access. (b–d) Fluoroscopy with last-image hold images showing a guidewire stepwise navigated through the AJVS crossover to the right side and from here through the right innominate vein along the pacemaker leads into the inferior vena cava.

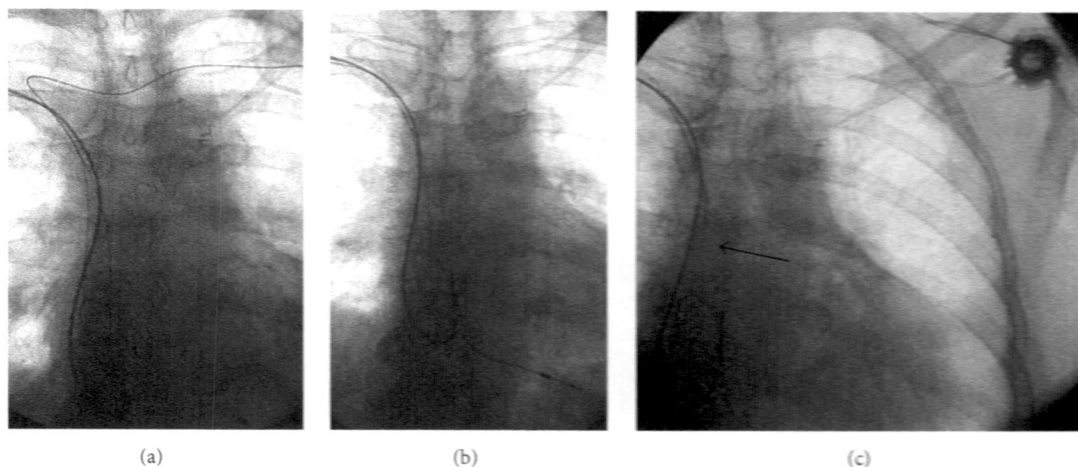

FIGURE 3: Fluoroscopy-guided images showing guidewire (a) and catheter (b) placement into the inferior vena cava and final positioning of the venous port system (c).

stump in the area of the chronic occlusion. After another review of the CT series, a left-sided thyrocervical transverse collateral vein, localised relatively superficially and directly above the sternal end of the left clavicle, was selected. This vessel was easily punctured with an 18-gauge needle using solely the bony landmark of the left clavicle. A 0.035-inch angled tip hydrophilic guidewire was stepwise navigated through the tortuous midline u-shaped segment of the collateral vein and crossed over to the right side and from here at an acute angle via the right IV, along the pacemaker leads into the inferior vena cava (Figure 2). The 6-French catheter of the port set (Titanium Low-Profile Implantable Port, Bard Access System Inc., Salt Lake City, USA) was then inserted over the wire just as easily and without complication or kinking up to the cavoatrial junction. A port pocket was prepared below the lateral end of the left clavicle, and the port catheter was tunnelled from the puncture side to the pocket completing the procedure as usual (Figure 3). The port could be used for the entire chemotherapy without any problems.

3. Discussion

In our institution, both radiologists and surgeons utilise the anterior chest wall for the standard approach for port placement. As insertion of arm ports does not necessarily show always the best results [1, 2], we consider peripherally placed central venous access ports only to be an alternative for patients in whom a chest-placed device is inappropriate or undesired.

For patients with implantable cardiac devices, the general recommendation is to place the port on the contralateral side to avoid any damage to pacing or defibrillator leads during port placement [3, 4]. Furthermore, ipsilateral central venous access along with a pacemaker is a compounding risk variable for central venous occlusion. Although ports are much smaller and less bulky than a pacemaker, positioning a port chamber directly next to a pacemaker aggregate or pacemaker lead in our experience is not a good solution. In such a situation, the options for port-chamber placement

are clearly limited and the port chamber may cause an uncomfortable result for the patient due to its proximity to the clavicle.

In our patient, we considered a right-sided approach via the internal jugular vein (IJV) with tunnelling the catheter over a long distance to the other side, but we refrained from this because of the extended presternal scars after sternal osteomyelitis and subtotal sternal resection.

In this clinical case, we proposed port placement at the left upper body to avoid problems with the pacemaker aggregate on the right side.

Translumbar port placement had been discussed with the patient as an alternative, in the event of failed thoracic placement.

4. Anatomical Considerations

A closer look at the anatomical conditions of our patient reveals that the supraclavicular punctured collateral vein segment is the horizontal lateral segment of the AJVS, as has been described and defined under functional and clinical aspects by Chasen and Charnsangavej [5] and Schummer et al. [6]. The midline u-shape thyrocervical collateral vein in fact is the JVA (Terminologia Anatomica: arcus venosus jugularis).

The JVA is an infrequently found transverse connecting trunk extending across the midline between the two anterior jugular veins (AJV) of either side and lying in the suprasternal space between superficial and pretracheal layers of the cervical fascia. The JVA serves as a natural crossover collateral and may become prominent in cases of deep venous outflow obstruction. It is the midline part of the AJVS, typically in u-shape or v-shape configuration. Apart from textbooks of surgery in the context of, for example, thyroid surgery or tracheostomy, it is mentioned in the literature mostly in relation to malposition of central vein catheters or unintended crossover placement of central lines [7, 8].

The AJVS is an important collateral venous network across the midline of the superoanterior aspect of the thorax

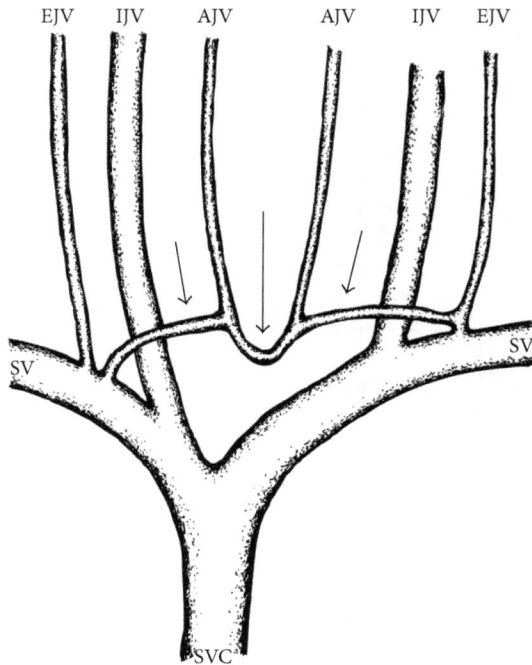

FIGURE 4: Schematic drawing illustrating the segmental anatomy of a typically and fully developed anterior jugular venous system (AJVS) consisting of three segments (three arrows) with the jugular venous arch (JVA) as the transverse midline segment (large arrow). AJV: anterior jugular vein; EJV: external jugular vein; IJV: internal jugular vein; SV: subclavian vein; SVC: superior vena cava.

and, if fully developed, is composed of three segments: the JVA as the transverse midline segment [5] and the two as infrequently found horizontal lateral segments connecting the JVA to the subclavian vein (SV), external jugular vein (EJV), or more rarely IJV. It is worth emphasising that "anterior jugular venous system" is to be understood as a clinical and not as an anatomical term. In the anatomical images or the textbooks of anatomy, the horizontal lateral segment of the AJVS often is unmarked or rather regarded as the termination of the AJV. Corresponding to its variability, the AJVS shows a wide array of formation, course, communication, and termination [9].

The segmental anatomy of a fully developed AJVS is illustrated in the schematic drawing (Figure 4).

Schummer et al. [6] stated that correct placement of central venous catheters through the AJVS may be possible.

5. Literature Review

The literature covering collateral vein access for port insertion is rather limited. Teichgräber et al. [10] reported the placement of a port catheter through collateral veins in a patient with central venous occlusion through ipsilateral cervical collateral veins. Very few reports exist of the placement of a port through internal mammary veins [11, 12]. Walser [13] described and illustrated a case, in which a "port catheter was negotiated into the superior vena cava by way

of a large, crossed-cervical venous collateral", but without exact anatomical description of the veins and illustrating a catheter with probably just a subcutaneous "cross-cervical" route and then passing through a large mediastinal vein. Even intercostal veins may serve as transcollateral venous access for port placement; however, this procedure requires a thoracotomy [14]. To the best of our knowledge, the only published case roughly comparable to ours is the one of Marcy et al. [15]. Here, a description of a port implantation in a crossover technique, utilising puncture of a rather normal-sized left EJV, and subsequently through an anastomosis of the left and right AJV, naming it "anterior jugular arch anastomosis" and with greater enlargement of this than in our case, is presented. Marcy titled the procedure "external jugular vein cross-over." In contrast to this, the access to the EJV was not an option in our patient, as her left EJV drained into the SV that in the further course was occluded and had no connection with the AJVS.

The important role of the AJVS as a collateral also becomes apparent in the context presented by Yamada et al. [16], where a case in which a malfunctioning implantable cardioverter-defibrillator lead was exchanged "via an angulated and tortuous collateral vein" was published. Here, lead placement is obviously via a crossover technique through the AJVS, accessed by subclavicular puncture of the occluded SV. Brieda et al. [17] reported on a patient, who was referred for upgrading of a dual-chamber implantable cardioverter-defibrillator to a biventricular cardioverter-defibrillator in the presence of a suboccluded left SV, "using a collateral vein that drained into the contralateral subclavian vein." Again in this case, the collateral vein is not named. A look at the attached venogram clearly demonstrates an approach via a collateral vein of the SV and again a crossover technique via the AJVS.

6. Conclusion

We describe a case of placement of a port catheter by direct puncture of the horizontal lateral segment of the AJVS and crossover through the JVA that was technically possible without the use of special equipment, not necessitating greater effort or significantly higher costs than an ordinary port implantation, and feasible despite a tortuous and relatively narrow diameter JVA.

In the unusual constellation of patients with central venous occlusion on one side and requiring ipsilateral port implantation, closer consideration of a potentially fully developed and enlarged AJVS is warranted, as this vessel has been proven clinically to be a major cervical crossover collateral vein.

For clinicians, the AJVS can play an important role as a collateral for the insertion of port catheters, pacemaker leads, or other types of central devices.

References

[1] A. Akahane, M. Sone, S. Ehara, K. Kato, R. Tanaka, and T. Nakasato, "Subclavian vein versus arm vein for totally implantable central venous port for patients with head and neck cancer: a retrospective comparative analysis," *CardioVascular and Interventional Radiology*, vol. 34, no. 6, pp. 1222–1229, 2011.

[2] P. Y. Marcy, A. Lacout, J. Thariat, A. Figl, and J. Merckx, "Ultrasound-guided arm ports: indications, techniques, and management," *Journal of the Association for Vascular Access*, vol. 20, no. 1, pp. 26–31, 2015.

[3] R. A. Gottlieb and P. K. Mehta, *Cardio-Oncology: Principles, Prevention and Management*, Academic Press, London, UK, 1st edition, 2017.

[4] C. L. Pacana and J.-B. Durand, "The risk of central venous catheter placement ipsilateral to the permanent pacemaker," *The Journal of the Association for Vascular Access*, vol. 14, no. 1, pp. 28–32, 2009.

[5] M. H. Chasen and C. Charnsangavej, "Venous chest anatomy: clinical implications," *European Journal of Radiology*, vol. 27, no. 1, pp. 2–14, 1998.

[6] W. Schummer, C. Schummer, D. Bredle, and R. Fröber, "The anterior jugular venous system: variability and clinical impact," *Anesthesia and Analgesia*, vol. 99, no. 6, pp. 1625–1629, 2004.

[7] P. Lalwani, S. Aggarwal, R. Uppal, and Somchandra, "A case of malpositioned catheter via supraclavicular approach for subclavian vein cannulation: a rare technique revisited," *Journal of Anaesthesiology Clinical Pharmacology*, vol. 32, no. 1, pp. 120–122, 2016.

[8] T.-E. Jung and D. Jee, "Misplaced central venous catheter in the jugular venous arch exposed during dissection before sternotomy," *Journal of Clinical Anesthesia*, vol. 20, no. 7, pp. 542–545, 2008.

[9] D. Premavathy and P. Seppan, "Anatomical variation in the anterior jugular veins and its clinical implications—a case report," *Journal of the Anatomical Society of India*, vol. 64, supplement 1, pp. S40–S43, 2015.

[10] U. K.-M. Teichgräber, F. Streitparth, B. Gebauer, and T. Benter, "Placement of a port catheter through collateral veins in a patient with central venous occlusion," *CardioVascular and Interventional Radiology*, vol. 33, no. 2, pp. 417–420, 2010.

[11] T. Urbania, D. M. Hershock, and T. W. I. Clark, "Percutaneous placement of an implantable arm port through an internal mammary venous pathway: case report," *Journal of Vascular Access*, vol. 4, no. 4, pp. 154–157, 2003.

[12] E. Jaime-Solis, M. Anaya-Ortega, and J. Moctezuma-Espinosa, "The internal mammary vein: an alternate route for central venous access with an implantable port," *Journal of Pediatric Surgery*, vol. 29, no. 10, pp. 1328–1330, 1994.

[13] E. M. Walser, "Venous access ports: indications, implantation technique, follow-up, and complications," *CardioVascular and Interventional Radiology*, vol. 35, no. 4, pp. 751–764, 2012.

[14] U. Tannuri, A. C. A. Tannuri, and J. G. Maksoud, "The second and third right posterior intercostal veins: an alternate route for central venous access with an implantable port in children," *Journal of Pediatric Surgery*, vol. 40, no. 11, pp. e27–e30, 2005.

[15] P. Y. Marcy, M. El Hajjam, A. Lacout, C. Nöel, J. J. Simon, and A. Figl, "External jugular vein cross-over as a new technique for percutaneous central venous port access in case of left central venous occlusion," *Journal of Vascular Access*, vol. 14, no. 4, pp. 388–391, 2013.

[16] T. Yamada, P. J. Robertson, and G. N. Kay, "Successful ICD lead implantation via an angulated and tortuous collateral vein after subclavian vein occlusion," *Europace*, vol. 13, no. 2, pp. 286–287, 2011.

[17] M. Brieda, L. De Mattia, E. Dametto, F. Del Bianco, and G. Nicolosi, "Placement of a coronary sinus pacing lead from a sub-occluded left subclavian vein using a collateral vein to the right subclavian vein," *Indian Pacing and Electrophysiology Journal*, vol. 11, no. 6, pp. 176–179, 2011.

9

Mass-Like Ground-Glass Opacities in Sarcoidosis: A Rare Presentation Not Previously Described

Marie Tominna ⓘ **and Sayf Al-Katib** ⓘ

Beaumont Hospital, Oakland University William Beaumont School of Medicine,
Department of Diagnostic Radiology and Molecular Imaging, 3601 W 13 Mile Rd, Royal Oak, MI 48073, USA

Correspondence should be addressed to Marie Tominna; marie.tominna@beaumont.edu

Academic Editor: Atsushi Komemushi

Various typical and atypical imaging findings for pulmonary sarcoidosis have been described in the literature. Ground-glass opacities are one of the atypical manifestations, reported as diffuse or patchy ill-defined opacities frequently associated with additional findings and interstitial nodules. We performed a literature review to determine if our case had previously been described. The literature describes cases of mass-like consolidations, but there are no reports of mass-like ground-glass opacities. The appearance of the ground-glass opacities in our case is unique, appearing as discrete well-defined mass-like ground-glass opacities in a peribronchovascular distribution without additional parenchymal findings typically seen in sarcoidosis.

1. Introduction

Sarcoidosis can present with a myriad of imaging findings in the chest. Typical and atypical manifestations have been described within the literature. Finding ground-glass opacities within the lungs on imaging is one of the atypical imaging features. These ground-glass opacities are typically diffuse or patchy and ill-defined, and frequently associated with additional imaging findings such as interstitial nodules [1, 2]. Ground-glass attenuation correlates with granulomatous lesions in alveolar septa and around small vessels on histopathology, rather than an alveolitis picture [3]. For the first time, we will report a case of discrete and well-defined mass-like ground-glass opacities as a presentation of sarcoidosis.

2. Case Presentation

An African American woman in her mid-30s presented to the emergency room with neurologic-type symptoms. Initial work-up included a chest X-ray which demonstrated multiple bilateral pulmonary masses and right paratracheal lymphadenopathy (Figure 1). Further evaluation with chest CT (Figures 2 and 3(a)) revealed discrete, sharply demarcated, mass-like ground-glass opacities involving all lobes bilaterally. The CT confirmed the right paratracheal lymphadenopathy and also showed bilateral hilar, para-aortic, subcarinal, and prevascular lymphadenopathy (Figure 3(b)). No interstitial opacities or perilymphatic nodules were identified otherwise. The differential diagnosis based on the imaging findings included lymphoma, vasculitis, and atypical pulmonary infection.

The patient did not report any significant respiratory symptoms. An extensive work-up was performed as the clinical differential included inflammatory, infectious, vasculitic, embolic, and neoplastic etiologies. Rheumatologic, serologic, and infectious work-up were negative. Her angiotensin converting enzyme (ACE) was elevated at 57 (reference range 8–52 U/L).

She went on to have bronchoscopy. Cultures from bronchoalveolar lavage were negative for bacterial, fungal, mycobacterial, and viral etiologies. There was no evidence of malignancy. Fine-needle aspiration of an enlarged right paratracheal lymph node and biopsy of the right lower lobe (Figure 4) both revealed noncaseating granulomas, consistent with sarcoidosis.

Given the histopathologic findings with the elevated ACE level, the diagnosis was consistent with sarcoidosis and she was started on treatment with corticosteroids.

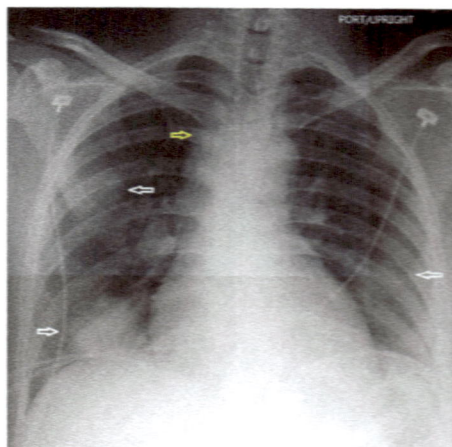

FIGURE 1: Chest X-ray: multiple bilateral pulmonary masses (white arrows) with right paratracheal lymphadenopathy (yellow arrow).

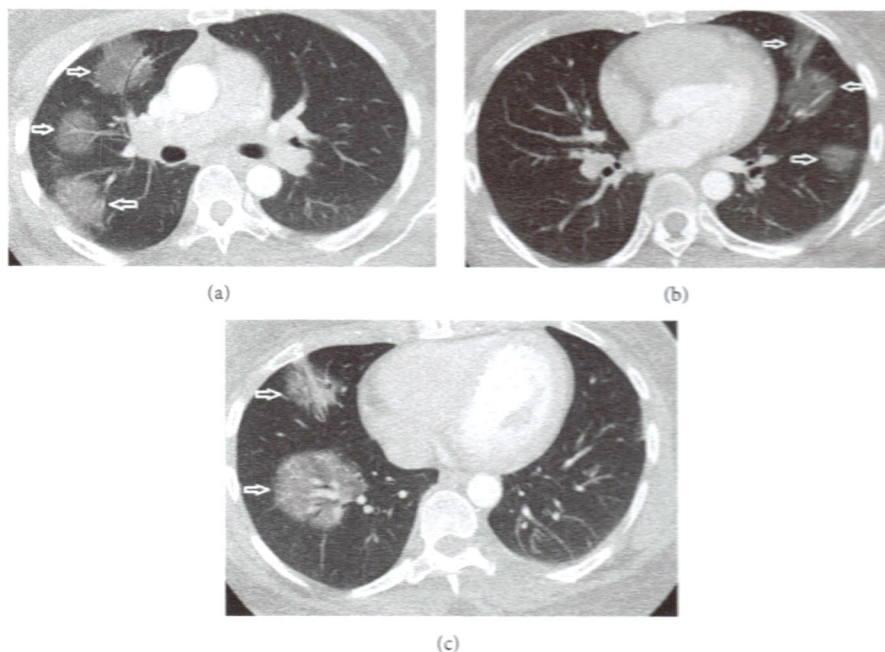

(a)

(b)

(c)

FIGURE 2: Axial contrast-enhanced CT obtained at different levels (a, b, c), demonstrating bilateral mass-like ground-glass opacities in a peribronchovascular distribution (arrows) with air-bronchograms and vessels seen coursing through.

3. Discussion

To the best of our knowledge, this is the first case reporting mass-like ground-glass opacities in a peribronchovascular distribution as a presentation of sarcoidosis. Typical manifestations of thoracic sarcoidosis include bilateral symmetric hilar lymphadenopathy and interstitial lung disease [1]. Atypical manifestations are seen in approximately 25–30% of cases and have been described as mass-like airspace consolidations, miliary opacities, fibrocystic changes, airway involvement, and pleural involvement [4, 5]. Ground-glass opacities are also one of the atypical manifestations in sarcoidosis but

have been typically described as diffuse or patchy ill-defined opacities with a background of interstitial nodules. While our case did show some typical findings including right paratracheal and bilateral hilar lymphadenopathy, the discrete mass-like ground-glass opacities have not been described previously. Additionally, parenchymal findings in sarcoidosis are primarily seen within the upper and middle lobes, and in our case all lobes were affected [6]. It is known that up to half of cases of sarcoidosis may be asymptomatic with incidental findings on imaging [1]. While our patient did not have pulmonary-type symptoms, during the patient's hospital course she had elevated troponins which led to a cardiac

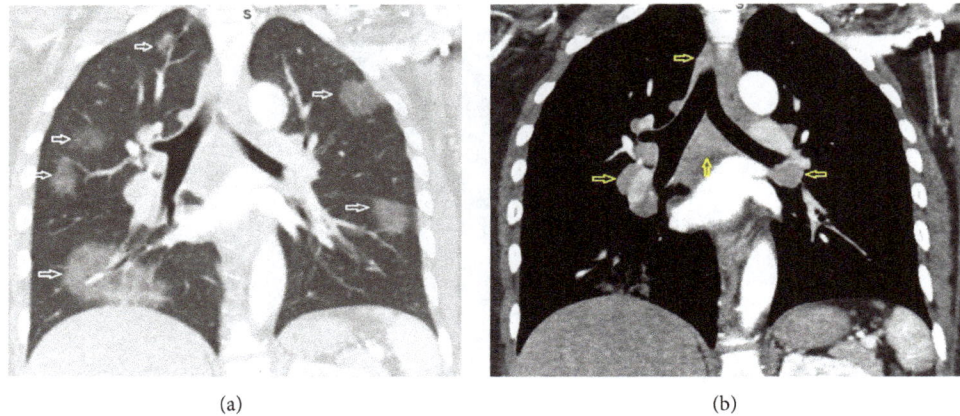

(a)   (b)

FIGURE 3: Coronal reformatted image from contrast-enhanced CT scan, with lung window (a), redemonstrating the findings in Figure 2 (white arrows), and with soft tissue window (b) demonstrating lymphadenopathy in the right paratracheal, bilateral hilar, and subcarinal regions (yellow arrows).

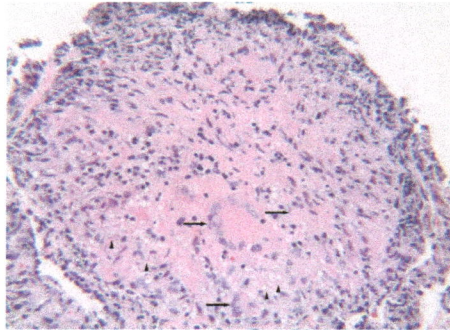

FIGURE 4: Hematoxylin-eosin stain image from right lower lobe biopsy shows a non-necrotizing granuloma with epithelioid histiocytes (arrowheads) and giant cells (arrows).

MRI suggesting preclinical sarcoidosis involvement, and her neurologic symptoms were attributed to neurosarcoidosis. Most patients do not require treatment. Indications for treatment include symptoms, worsening organ damage, or declining pulmonary function. Typically, corticosteroids are commonly used for the initial treatment [7]. A majority of patients will have spontaneous remission, but some may develop a chronic or progressive course, which may ultimately lead to lung transplantation in cases of end-stage disease.

When presented with the thoracic findings of mass-like ground-glass opacities in a peribronchovascular distribution and lymphadenopathy, one may include sarcoidosis in the differential diagnosis if provided with the appropriate clinical picture.

References

[1] E. Criado, M. Sánchez, J. Ramírez et al., "Pulmonary sarcoidosis: Typical and atypical manifestations at high-resolution CT with pathologic correlation," *RadioGraphics*, vol. 30, no. 6, pp. 1567–1586, 2010.

[2] C. Ma, Y. Zhao, and T. Wu, "Predominant diffuse ground glass opacity in both lung fields: A case of sarcoidosis with atypical CT findings," *Respiratory Medicine Case Reports*, vol. 17, pp. 61–63, 2016.

[3] K. Nishimura, H. Itoh, M. Kitaichi, S. Nagai, and T. Izumi, "Pulmonary sarcoidosis: Correlation of CT and histopathologic findings," *Radiology*, vol. 189, no. 1, pp. 105–109, 1993.

[4] D. Cozzi, E. Bargagli, A. G. Calabrò et al., "Atypical HRCT manifestations of pulmonary sarcoidosis," *La Radiologia Medica*, vol. 123, no. 3, pp. 174–184, 2018.

[5] H. J. Park, I. Jung, M. H. Chung et al., "Typical and atypical manifestations of intrathoracic sarcoidosis," *Korean Journal of Radiology*, vol. 10, no. 6, pp. 623–631, 2009.

[6] H. Nunes, Y. Uzunhan, T. Gille, C. Lamberto, D. Valeyre, and P.-Y. Brillet, "Imaging of sarcoidosis of the airways and lung parenchyma and correlation with lung function," *European Respiratory Journal*, vol. 40, no. 3, pp. 750–765, 2012.

[7] V. Ramachandraiah, W. Aronow, and D. Chandy, "Pulmonary sarcoidosis: an update," *Postgraduate Medical Journal*, vol. 129, no. 1, pp. 149–158, 2017.

10

Retroodontoid Pseudotumor Related to Development of Myelopathy Secondary to Atlantoaxial Instability on Os Odontoideum

M. Hamard ⓘ, **S. P. Martin** ⓘ, **and S. Boudabbous**

Department of Imaging and Medical Information Sciences, Division of Radiology, Geneva University Hospitals, Geneva, Switzerland

Correspondence should be addressed to M. Hamard; marion.hamard@hcuge.ch

Academic Editor: Atsushi Komemushi

Retroodontoid pseudotumor (ROP) is a nonneoplasic lesion of unknown etiology, commonly associated with inflammatory conditions, and the term of pannus is usually used. Less frequently, ROP formation can develop with other noninflammatory entities, with atlantoaxial instability as most accepted pathophysiological mechanism for posttraumatic or degenerative ROP. As it can clinically and radiologically mimic a malignant tumor, it is paramount for the radiologist to know this entity. Magnetic resonance imaging is the modality of choice to reveal the possible severe complication of ROP in the form of a compressive myelopathy of the upper cervical cord. The purpose of the surgical treatment is the regression or complete disappearance of ROP, with posterior decompression by laminectomy and posterior C1-C2 or occipitocervical fixation. We present the case of an elderly patient with retroodontoid soft tissue mass secondary to a chronic atlantoaxial instability on os odontoideum, an extremely rare cause of ROP. The patient developed a posttraumatic cervical myelopathy related to the decompensation of this C1-C2 instability responsible for the formation of a compressive ROP. We will overview the retroodontoid pseudotumor and its differential diagnosis.

1. Introduction

Retroodontoid pseudotumor (ROP) is an entity that can mimic malignant tumors and is from uncertain etiology. Some consider it as a low-grade fibrosarcoma. ROP are mainly related to infectious processes and less frequently to inflammatory disorders. ROP might have neurological complications due to mass effect on the spinal cord.

2. Case Description

A 77-year-old female patient was admitted in our institution following a ground-level fall due to relatively sudden grade 3-4 right hemiparesis with lower limb predominance.

This patient was not known for any systemic disease, no rheumatoid arthritis, or other joint-related generalized disease.

An initial enhanced CT was performed for the suspicion of an ischemic stroke. The exam revealed a smooth and well-corticated bone ossicle measuring 14 mm and located superiorly to the odontoid process corresponding to an os odontoideum (Figures 1(a)–1(c)). The ossicle was associated with an atlantoaxial subluxation and with the posterior wall of 14 mm on spinal canal (Figures 1(e) and 1(f), white lines) that has increased since a previous CT 8 years ago. The late enhanced phase showed an intracanal hyperattenuated but no enhancing pseudomass situated just posterior to the ossicle (Figure 1(d)).

A complementary cervical enhanced MRI with administration of Gadolinium confirmed a well-corticated ossicle and demonstrated a tissular retroodontoid process (Figures 2(a) and 2(b)). The tissue component showed a low signal on T1- and T2-weighted images and no enhancement (Figure 2(g)), compatible with a ROP. The main diagnosis was a noninflammatory ROP developed on atlantoaxial instability, secondary to an os odontoideum. The main differential diagnosis was pseudoarthrosis of an old fracture of the dens of axis. Inflammatory arthritis such as gout, rheumatoid, or psoriatic arthritis was suggested as differential diagnosis, but less likely because of the negative history of those diseases.

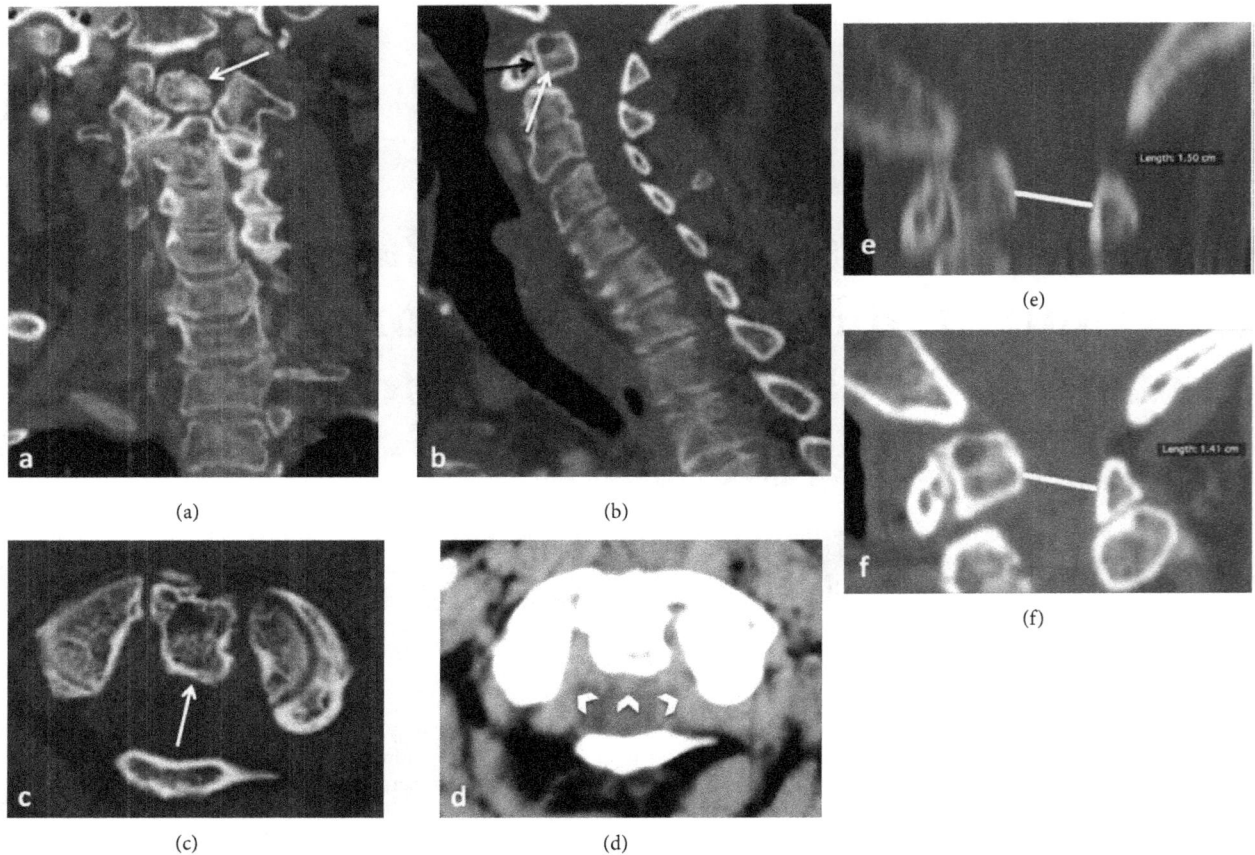

(a)

(b)

(c)

(d)

(e)

(f)

FIGURE 1: Angiography brain CT obtained to investigate the sudden right hemiparesis of the patient, showing a well-corticated bone fragment located superiorly of odontoid process in coronal reconstruction (a, white arrow) and in axial plane (c, white arrow) in bone window. This fragment seems an os odontoideum, with a pseudoarthrosis of a fracture of the dens as main differential diagnosis. In sagittal plane with bone window, it is associated with an atlantoaxial subluxation at this level (anterior arch of C1 above C2 body), with a narrowing of the space available for spinal cord between the previous 2006 CT exam (e, white line measuring 1.5 cm) and recent 2014 CT exam (f, white line measuring 1.4 cm) in bone window. There is no enlargement of the space between anterior surface of os odontoideum and the anterior arch of C1 on sagittal plane (b, black arrow). We can already see a pseudomass hyperattenuated but without enhancement in the late phase, posteriorly of os odontoideum in the cervical spinal canal in axial plane and soft tissue window (d, white arrowheads), compatible with a retroodontoid pseudotumor in ROP.

There was no enlargement of space between this os odontoideum and the anterior arch of C1 (Figure 2(e)). A subcentimetric geode, in low signal on T1-weighted images and enhancement after contrast administration, was seen in the posterior dens basis of C2 (Figures 2(b) and 2(c)). The pseudotumor indenting into cervicomedullary cord and resulting in cord compression is shown (Figure 2(a)), with cervical myelopathy seen in high signal on T2-weighted images in sagittal plane (Figure 2(d)).

Given the severity of the radiological findings and the clinical impact due to spinal cord compression and the life-threatening risk, patient was treated with cervical posterior screw fixation and a decompressive laminectomy at C1-C2 level (Figures 3(a)–3(d)). Unenhanced CT after posterior C1–C2 shows posterior cervical fixation and spinal canal decompression by laminectomy. The intracanal ROP is not removed but regressed in size in each postoperative CT compared to preoperative CT (Figures 4(a)–4(d)).

The patient showed a progressive improvement of her neurological recovery, with complete neurological recovery 6 months after surgery. In the postoperative follow-up, entire cervical spine showed marked degenerative changes, best viewed on postoperative cervical spine X-rays (Figures 5(a) and 5(b)) as disco-uncarthrosis (white arrowheads), interarticular posterior arthrosis (black arrowheads), and anterior marginal osteophytosis (black-framed white arrowheads).

3. Discussion

The etiology of retroodontoid pseudotumor (ROP), which can mimic malignant tumors, is unknown. Some authors believe that this entity is a low-grade fibrosarcoma [1]. When pseudotumors are related to an infectious process, the more frequent encountered organisms are mycobacteria, mycoplasma, Epstein-Barr virus, actinomycetes, and nocardiae. The noninfectious inflammatory ROP, also named

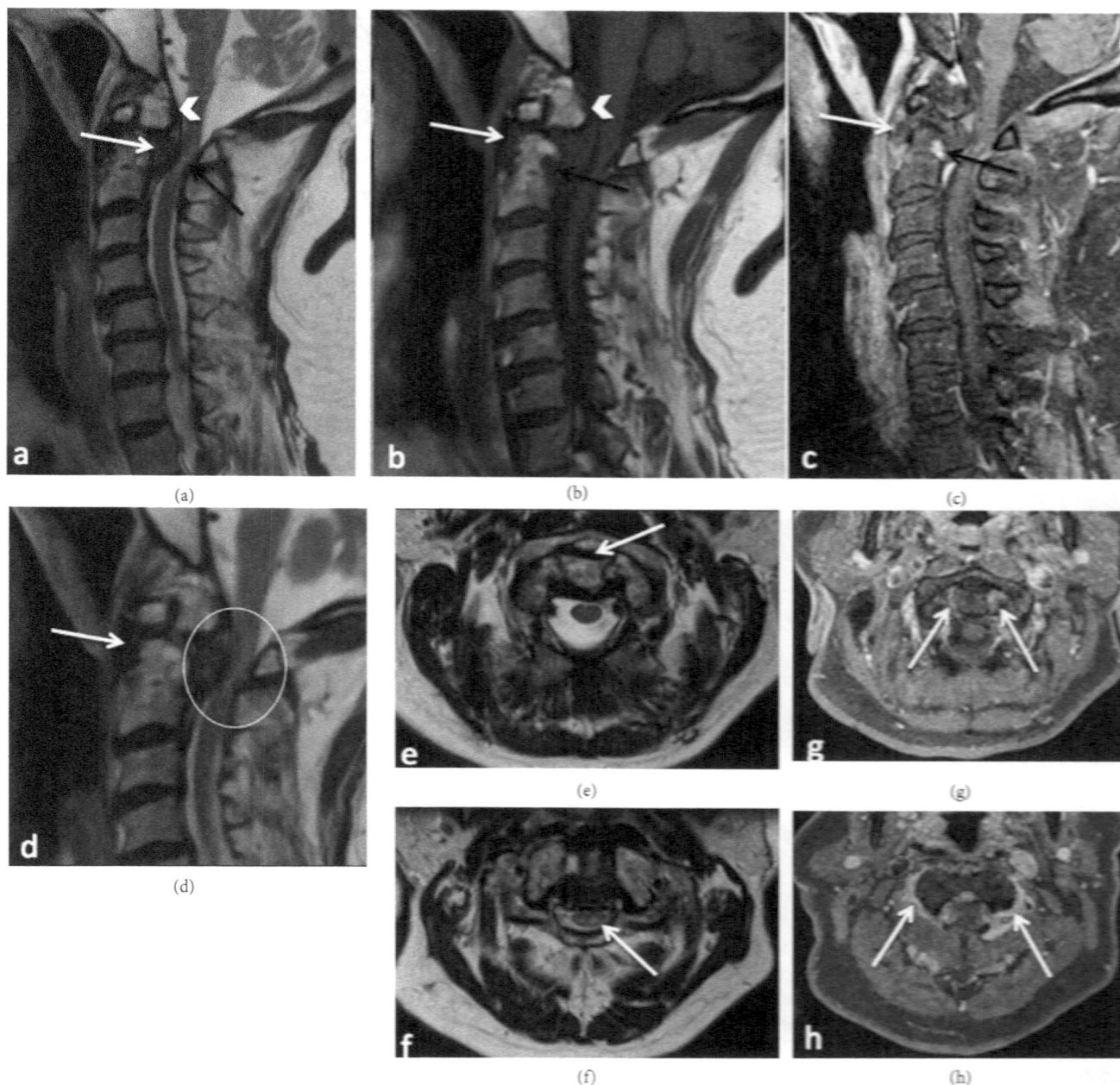

FIGURE 2: Complementary cervical enhanced MRI with Gadolinium administration, acquired to investigate the cervical spinal cord compression, showing the ROP as low signal intensity on both T1w and T2w images in sagittal planes surrounding the body of C2 and this ROP (a,b,d, white arrows). It appears without enhancement on T1w fat sat images after gadolinium administration in sagittal (c, white arrow) and axial (g, white arrows) planes. We can see some enhancement on axial T1w fat sat after contrast administration around interfacetar articulations of C1-C2 (h, white arrows), probably from degenerative origin. A sub centimetric geode in the basis of C2 shows low signal on T1w image in sagittal plane (b, black arrow) with an enhancement after contrast administration in axial plane (c, black arrow). The T2w image in sagittal plane reveals an area of high intensity in the intramedullary regions of C1 and C2 (d, white circle), due to compression by the ROP, but without diastasis between anterior arch of C1 and the dens (e, white arrow).

pannus or phantom tumors [2], is a rare nonneoplasic reactive inflammatory granulation tissue at the craniovertebral junction typically arising from the synovium around the dens and usually associated with rheumatoid arthritis [3–5], psoriatic arthritis [6], or gout [7]. Less frequently, the noninflammatory and noninfectious entities with predisposing atlantoaxial instability (AAI) can be associated with a ROP, as os odontoideum, posttraumatic pseudoarthrosis of the odontoid fracture, and degenerative retroodontoid cysts [8]. Other entities can mimic ROP as amyloidosis [9–11],

(a)   (b)   (c)

(d)

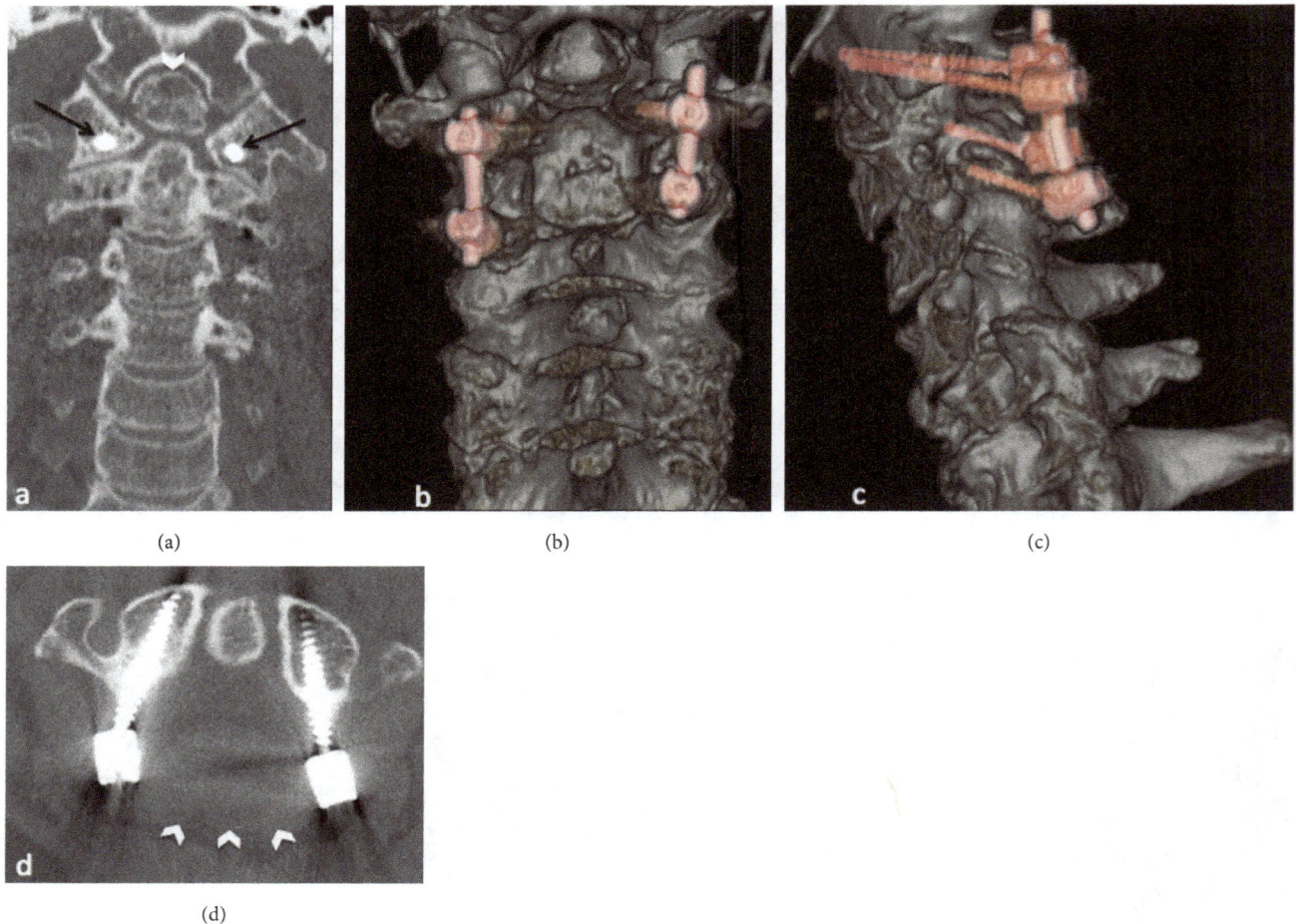

FIGURE 3: Unenhanced CT after posterior C1–C2 fixation. We see in coronal plane with bone window the transarticular C1-C2 screws (a, black arrows) and on volume rendering in coronal and sagittal reconstructions (b,c, the screws are highlighting). Spinal canal decompression by laminectomy is also made, seen in axial plane with bone window (d, white arrowheads) showing the absence of posterior arch of C1 (in comparison to image c of Figure 1). The intracanal ROP is not removed but regressed in size compared to preoperative CT.

pigmented villonodular synovitis [12], epidural lipomatosis [13], and idiopathic skeletal hyperostosis [14, 15] (Table 1).

In 2004, Goel A. et al. identified that ROP indicated atlantoaxial instability (AAI) [2]. The chronic mechanical stress exerted by the AAI induces repeated tears and transverse ligamentous hypertrophy with formation of reactive fibrous granulation tissue. Patients are usually asymptomatic or can report chronic cervical pain. In more advanced and severe stage, ROP could cause cervical spinal cord compression with neurological symptoms. Occasionally, an added event can decompensate the clinical condition, like ground-level fall in our case. Pathophysiologically, noninflammatory ROP consists of fibrocartilaginous tissue between the odontoid process and the anterior arch of the C1 or between the dens and the transversal ligament [2, 16]. Thus, despite the lack of histological findings, the clinical and radiological features of our case are consistent with a noninflammatory cause. Tanaka et al. have classified ROP into 3 types, according to the etiology and MRI findings. Type 1 is the classical and common ROP, caused by AAI, and is mainly associated with rheumatoid arthritis. Type 2 shows similar MRI findings to type 1, but it is associated with spondylosis or ankylosis of the cervical segment and ossification of the anterior longitudinal ligament. In type 3, ROP is sustained by a C2-C3 disc herniation, which penetrates the posterior longitudinal ligament and migrates upward to the retroodontoid space [17].

X-rays analysis allows estimating of the atlantoaxial instability, usually with dynamic study in flexion and extension. The radiographic characteristics as ossification of the anterior longitudinal ligament and ankylosis of the adjacent segments seem to be risk factors for the formation of a ROP [18]. Distance from the posterior border of the odontoid process to the anterior border of the posterior arch of C1 can also be assessed in order to estimate the space available for spinal cord (SAC). A typical MRI finding of noninflammatory ROP is a hypo- or isointense area on T1-weighted images and an area of low or mixed intensity on T2-weighted images. ROP typically demonstrates no enhancement after gadolinium administration [19].

TABLE 1: Differential diagnosis of ROP.

	Clinical history	Symptoms	X-ray	CT scans	MRI	Diagnostic	Treatment and follow-up
ROP from atlantoaxial instability	Elderly patients. Long term neck injury with atlantoaxial subluxation. Fracture of the odontoid process. Os odontoideum.	Usually asymptomatic. Chronic cervical pain.	Dynamic study of atlantoaxial instability. Ossification of anterior longitudinal ligament. Ankylosis.	Degenerative changes: (i) Disco-uncarthrosis. (ii) Interarticular posterior arthrosis. (iii) Osteophytosis (iv) Atlantoaxial subluxation (v) Space available for spinal cord (SAC) Ankylosis.	ROP is hypo- to iso intense on T1w and predominantly hypo intense or mixed intensity on T2w.	Fibrous or cartilaginous nature. Degenerated ligaments.	Posterior cervical fixation results in immediate postoperative neurological improvement.
Rhumatoid or psoriasitic arthritis (RA)	History of rhumatoid or psoriasitic arthritis.	Long standing and progressive neurologic deficits. Types 1 and 3 disappears faster than type 2.	Dynamic X-rays: severe atlantoaxial subluxation during flexion.	Erosive alterations.	Pannus predominates anteriorly to the dens, surrounds the eroded odontoid process. 3 types exists: (1) pannus in low signal in T2w and high signal in T1w with enhancement post contrast administration. (2) pseudotumour in low signal in T2w and T1w. (3) mixed with combination of high and low signal in T2w.		Spontaneous resolution or diminution of retrodental pannus after posterior atlantoaxial stabilisation.
Retro-odontoid cysts	Elderly patients with degenerative spine disease.				Cystic and hypertrophic degeneration of the transverse ligament of axis. We can differentiate this pattern from synovial cyst with echo-gradient T2-weighted images on MRI showing hemosiderin deposits and no or thin enhancement.	Transverse ligament develops granuloma formation and angiogenesis, with chronic recurrent micro hemorrhages in case of rupture, leading to cyst formation.	Excision of the cystic lesion by a trans condylar approach.

TABLE 1: Continued.

	Clinical history	Symptoms	X-ray	CT scans	MRI	Diagnostic	Treatment and follow-up
Gouty deposits	History of goutt (acute peripheral arthritis) Hyperuricemia.	Lombalgy > cervicalgy. All spine components can be affected. Long standing and progressive neurologic deficits.	Cord compression on myelogram.	Erosive lesion. Tohpus is hyperdense on CT, with positive diagnostic to crystal of uric acid on Dual Energy CT.	Extradural intraspinal cervical fibrous tophus, in hypo to intermediate on T1w, hypo- or hyper signal on T2w with diffuse enhancement. Ventral and lateral position to the cord. Ligament softening. Subluxation C1-C2.	Histological results showing crystals of uric acid, multinucleate giant cell and histiocyte proliferation. Macroscopically, whitish and chalky material.	Progressive clinical improvement.
Amyloidosis	History of long-term hemodialysis, multiple myeloma, chronic inflammatory diseases, chronic infection.	Joints pains that mimic rheumatologic disorders.	Proeminent erosive alterations.	Dorsal vertebral > lombar spine > cervical spine.	Soft tissue intra neural mass with in low intensity T1w images and hypo- or mixed intensity in T2w images. Paraspinal extension. Severe medullary cord compression is possible.	Extracellular deposition of amyloid, with normal serum amyloid associated protein. Tissue biopsy: green birefringent fibrils by polarization microscopy after staining with Congo red.	Pronostic depends of the type of amyloidosis: primary solitary amyloidosis has best recovery and of lack of recurrence with a complete resolution of amyloidoma after surgical treatment.
Pigmented villonodular synovitis	Uncommon Young adults Knee > hip > ankle > shoulder.	Long standing and progressive neurologic deficits.	Mass epidural between the posterior longitudinal ligament and the vertebral bodies on cervical myelogram.	No erosion.	ROP hypo intense T1.		No data.
Epidural lipomatosis	Diffuse overgrowth of nonencapsulated adipose tissue in the epidural space: (i) Severe obese patients. (ii) Longstanding high-dose > low dose steroid therapy (iatrogenic Cushing's syndrome).	Long standing and progressive neurologic deficits.		Fatty epidural mass with mean density of -100 to -50 HU.	Massive diffuse epidural fat compressing the entire spinal cord in hyper signal T1/T2, hypo signal in STIR. Ventral displacement of the cord. Ischemic myelopathy with central gray matter hyper signal T2w in thoracic region.		No data.

TABLE 1: Continued.

	Clinical history	Symptoms	X-ray	CT scans	MRI	Diagnostic	Treatment and follow-up
Forestier's disease	Older men. Diffuse idiopathic skeletal hyperostosis (DISH).	Progressive neurologic symptoms resulting from anterior cervicomedullary junction.	Massive anterior longitudinal ligament calcification with bridge on the anterior border of the thoracic and subaxial cervical spine.	Massive anterior longitudinal ligament (ALL) calcification with bridge on the anterior border of the thoracic and subaxial cervical spine.	Calcification of ALL shows hypo signal in all sequences T1/T2/STIR and no enhancement. Permits to differenciate with shiny corner in ankylosing spondylitis.	Hypertrophic degenerative cartilage.	Transoral resection of the ligamentous mass followed by occipitocervical fusion. Early postoperative neurological improvement.

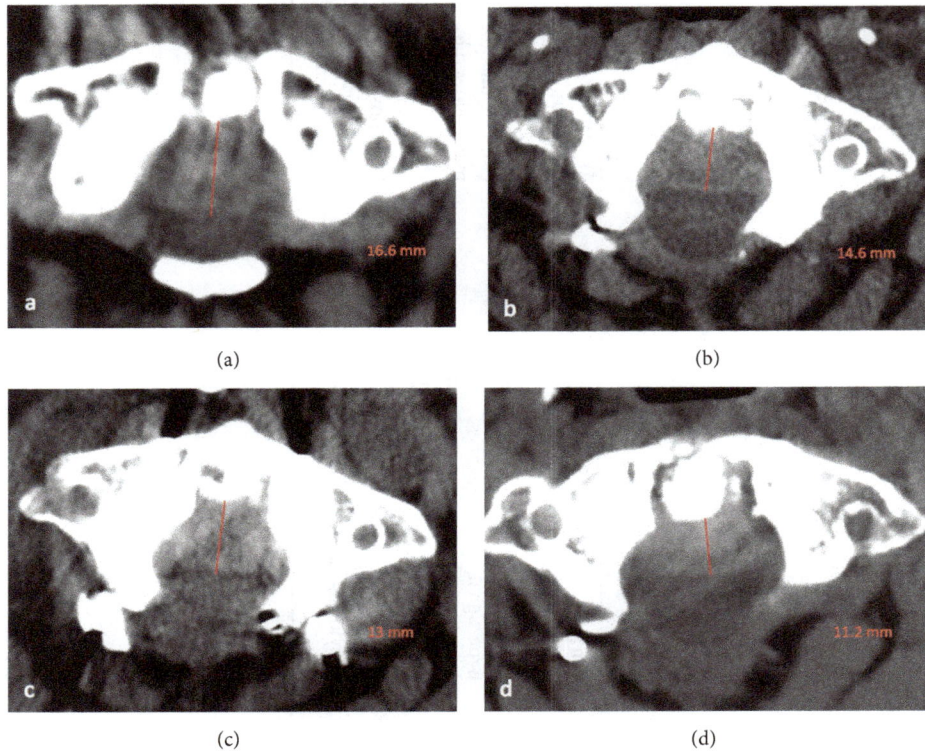

FIGURE 4: Comparison of the size of ROP on preoperative CT and postoperative CT with soft tissue window and in axial plane. We see the regression of size of the ROP between preoperative CT (a) with thickness of 16.6 mm under the os odontoideum and the postoperative CT just after surgical management (b) with thickness of 14.6 mm, at 4 months with thickness of 13 mm and at 6 months with thickness of 11.2 mm after surgery.

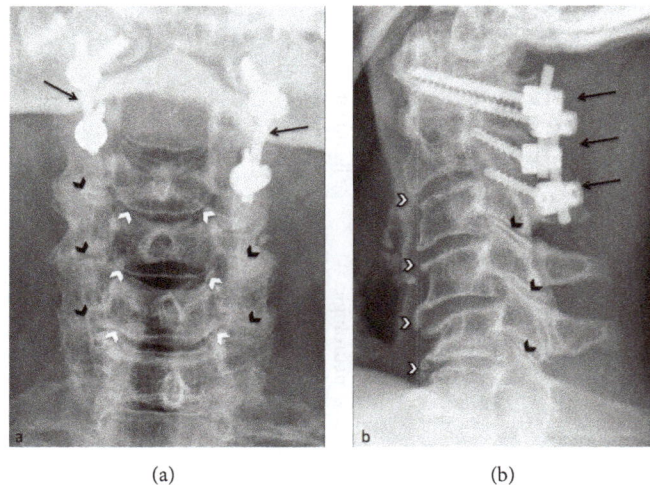

FIGURE 5: Postoperative X-rays of the cervical spine are also acquired for the follow-up. We see the posterior C1–C2 fixation with transarticular screws (a,b, black arrows). These X-rays show better degenerative changes of the cervical spine; disco-uncarthrosis (a, white arrowheads), interarticular posterior staggered arthrosis (a, b, black arrowheads), and anterior marginal osteophytosis (b, black-framed white arrowheads).

The surgical management of ROP aims to restore the AII and to relieve the mechanical stress at the C1-C2 junction, with spontaneous regression of ROP. Many cases report the complete disappearance of this degenerative and noninflammatory ROP following posterior C1–C2 fixation [3, 16, 20–22]. In our case, the patient showed a progressive regression of the size of the ROP on three postoperative CT scans (Figure 4) and improvement of her neurological recovery during, with complete recovery 6 months after surgery. The most used and recommended surgical procedures, with or without atlantoaxial subluxation, are the combination of spinal canal decompression by laminectomy and a posterior C1–C2 or occipitocervical fixation with transarticular screws, without removal of the intracanal ROP [2]. Tanaka et al. proposed the C1-C2 fixation with laminectomy as surgical treatment of types 1 and 2 [17].

In this case, os odontoideum is the cause for the AII with ROP. This bone ossicle is situated superiorly to the dens of axis and differs morphologically from persistent ossiculum terminale or from odontoid fracture. According to the congenital hypothesis, it corresponds to a congenital anatomic variant, due to a failure of fusion of dens center with the body of C2. The other current hypothesis supports a traumatic origin, corresponding to an odontoid process fracture type II of Anderson and Alonso classification, sustained by history of traumatism in many patients [23]. In our case, the traumatic hypothesis is more likely because of the many degenerative changes associated. On cervical spine CT, the gap between the os odontoideum and the odontoid process usually extends above the level of the superior axis facets, as in our case. Even though extremely rare, os odontoideum could be associated with atlantoaxial instability and lead to ROP [24]. There are two main types. First, the orthotopic type shows a normal position, merged with the anterior arch of C1 and a wide gap between the body of C2 and os odontoideum, as in our case. Second, the dystopic os odontoideum is cranially displaced usually fused to the basion of the clivus. This smooth small ossicle is of variable sizes, around half the size of the normal dens and associated with hypertrophied and rounded anterior arch of C1. The development of the ROP in the C2 fracture is most commonly seen in elderly population after a low-energy traumatism, with high mortality. Pseudoarthrosis of dens fracture is then usually seen in patients with history of cervical spine traumatism and osteoporosis [25].

4. Conclusion

We present an extremely rare case of ROP, confirmed by surgery, secondary to chronic atlantoaxial instability on os odontoideum. Our case, being the fifth reported in the literature, gives us the opportunity to remind the reader of the recent knowledge about this anatomical variant suggesting a traumatic etiology rather than a congenital source.

References

[1] L. Das Narla, B. Newman, S. S. Spottswood, S. Narla, and R. Kolli, "Inflammatory pseudotumor," *RadioGraphics*, vol. 23, no. 3, pp. 719–729, 2003.

[2] A. Goel, U. Phalke, F. Cacciola, and D. Muzumdar, "Atlantoaxial instability and retroodontoid mass - Two case reports," *Neurologia medico-chirurgica*, vol. 44, no. 11, pp. 603–606, 2004.

[3] E.-M. Larsson, S. Holtas, and S. Zygmunt, "Pre- and postoperative MR imaging of the craniocervical junction in rheumatoid arthritis," *American Journal of Roentgenology*, vol. 152, no. 3, pp. 561–566, 1989.

[4] I. Yonezawa, T. Okuda, J. Won et al., "Retrodental mass in rheumatoid arthritis," *Journal of Spinal Disorders & Techniques*, vol. 26, no. 2, pp. E65–E69, 2013.

[5] D. Grob, R. Würsch, W. Grauer, J. Sturzenegger, and J. Dvorak, "Atlantoaxial fusion and retrodental pannus in rheumatoid arthritis," *The Spine Journal*, vol. 22, no. 14, pp. 1580–1584, 1997.

[6] K. Lu and T.-C. Lee, "Spontaneous regression of periodontoid pannus mass in psoriatic atlantoaxial subluxation case report," *The Spine Journal*, vol. 24, no. 6, pp. 578–581, 1999.

[7] B. J. Leaney and J. M. Calvert, "Tophaceous gout producing spinal cord compression," *Journal of Neurosurgery*, vol. 58, no. 4, pp. 580–582, 1983.

[8] M. Takeuchi, M. Yasuda, E. Takahashi, M. Funai, M. Joko, and M. Takayasu, "A large retro-odontoid cystic mass caused by transverse ligament degeneration with atlantoaxial subluxation leading to granuloma formation and chronic recurrent microbleeding case report," *The Spine Journal*, vol. 11, no. 12, pp. 1152–1156, 2011.

[9] B. C. Werner, F. H. Shen, and A. L. Shimer, "Primary cervical amyloidoma: A case report and review of the literature," *The Spine Journal*, vol. 13, no. 10, p. -e7, 2013.

[10] E. Nitta, K. Sakajiri, and A. Kawashima, "Cervical epidural β2-microglobulin amyloidoma presenting with acute paraplegia 5 months after introduction of hemodialysis," *Journal of Clinical Neurology*, vol. 55, no. 9, pp. 645–650, 2015.

[11] G. Moonis, E. R. Savolaine, S. A. Anvar, and A. Khan, "MRI findings of isolated beta-2 microglobulin amyloidosis presenting as a cervical spine mass: Case report and review of literature," *Clinical Imaging*, vol. 23, no. 1, pp. 11–14, 1999.

[12] G. M. Kleinman, T. F. Dagi, and C. E. Poletti, "Villonodular synovitis in the spinal canal. Case report," *Journal of Neurosurgery*, vol. 52, no. 6, pp. 846–848, 1980.

[13] A. Muñoz, J. A. Barkovich, F. Mateos, and R. Simón, "Symptomatic epidural lipomatosis of the spinal cord in a child: MR demonstration of spinal cord injury," *Pediatric Radiology*, vol. 32, no. 12, pp. 865–868, 2002.

[14] N. P. Patel and N. M. Wright, "Forestier disease associated with a retroodontoid mass causing cervicomedullary compression," *Journal of Neurosurgery*, vol. 96(2 Suppl.), pp. 190–196, 2002.

[15] R. Mader, J. Verlaan, I. Eshed et al., "Diffuse idiopathic skeletal hyperostosis (DISH): where we are now and where to go next," *RMD Open*, vol. 3, no. 1, Article ID e000472, 2017.

[16] I. Yamaguchi, S. Shibuya, N. Arima, S. Oka, Y. Kanda, and T. Yamamoto, "Remarkable reduction or disappearance of retroodontoid pseudotumors after occipitocervical fusion: Report of three cases," *Journal of Neurosurgery: Spine*, vol. 5, no. 2, pp. 156–160, 2006.

[17] S. Tanaka, M. Nakada, Y. Hayashi et al., "Retro-odontoid pseudotumor without atlantoaxial subluxation," *Journal of Clinical Neuroscience*, vol. 17, no. 5, pp. 649–652, 2010.

[18] H. Chikuda, A. Seichi, K. Takeshita et al., "Radiographic analysis of the cervical spine in patients with retro-odontoid pseudotumors," *The Spine Journal*, vol. 34, no. 3, pp. E110–E114, 2009.

[19] G. Sze, M. N. Brant-Zawadzki, C. R. Wilson, D. Norman, and T. H. Newton, "Pseudotumor of the craniovertebral junction associated with chronic subluxation: MR imaging studies," *Radiology*, vol. 161, no. 2, pp. 391–394, 1986.

[20] G. M. Barbagallo, F. Certo, M. Visocchi, S. Palmucci, G. Sci-
acca, and V. Albanese, "Disappearance of degenerative, noninflammatory, retro-odontoid pseudotumor following posterior C1–C2 fixation: case series and review of the literature,"
European Spine Journal, vol. 22, no. S6, pp. 879–888, 2013.

[21] A. Lagares, I. Arrese, B. Pascual, P. A. Gòmez, A. Ramos, and
R. D. Lobato, "Pannus resolution after occipitocervical fusion
in a non-rheumatoid atlanto-axial instability," *European Spine
Journal*, vol. 15, no. 3, pp. 366–369, 2006.

[22] A. Shah, S. Jain, A. Kaswa, and A. Goel, "Immediate Postoperative Disappearance of Retro-Odontoid "Pseudotumor","" *World
Neurosurgery*, vol. 91, pp. 419–423, 2016.

[23] B. Arvin, M.-P. Fournier-Gosselin, and M. G. Fehlings, "Os
Odontoideum: etiology and surgical management," *Neurosurgery*, vol. 66, supplement 3, pp. A22–A31, 2010.

[24] B.-Y. Jun, "Complete reduction of retro-odontoid soft tissue
mass in os odontoideum following the posterior C1-C2 tranar-
ticular screw fixation," *The Spine Journal*, vol. 24, no. 18, pp.
1961–1964, 1999.

[25] M. S. Shinseki, N. L. Zusman, J. Hiratzka, L. M. Marshall, and
J. U. Yoo, "Association between advanced degenerative changes
of the atlanto-dens joint and presence of dens fracture," *Journal
of Bone and Joint Surgery - American Volume*, vol. 96, no. 9, pp.
712–717, 2014.

11

Progressive Diffuse Osteonecrosis in a Patient with Secondary Hemophagocytic Lymphohistiocytosis

Takashi Takahashi and Jeffrey Rykken

Department of Radiology, University of Minnesota Medical Center, Minneapolis, MN, USA

Correspondence should be addressed to Takashi Takahashi; takas005@umn.edu

Academic Editor: Salah D. Qanadli

This is a case report with serial imaging showing progression of diffuse osteonecrosis in a patient after a diagnosis of secondary hemophagocytic lymphohistiocytosis (HLH). While bone marrow involvement in HLH has been long noted at histological evaluation and is itself one of the diagnosis criteria, to the best of our knowledge, there has been no previous publication addressing osseous image findings in a patient with HLH.

1. Introduction

Hemophagocytic lymphohistiocytosis (HLH) also known as hemophagocytic syndrome is a rare immune disorder of macrophages [1]. The secondary (acquired) form affects older children and adults and is most commonly associated with infection, underlying malignancy, and prolonged immunosuppression such as in patients with organ or stem cell transplantation.

Although HLH primary affects the pediatric population, it can be diagnosed in patients of all ages [2]. In earlier reports, the reported incidence HLH was believed to be low, 1.2 children per million per year [3]. More recently, this is believed to be an underestimate, and a tertiary care pediatric hospital is expected to see 1 case per 3000 inpatient admissions [4]. The incidence of HLH among the adult population is less studied with one study specifically targeting previously published cases across the world that had merely 2197 reported cases clearly meeting the criteria based on a PubMed search [5]. Regardless of the actual incidence, it still is a relatively rare diagnosis. Despite its rarity, HLH has been receiving increasing attention in the field of medicine with over 1600 articles published between 2006 and 2014, whereas there are only 354 articles preceding 2006 based on a PubMed search. Given the relatively high mortality rate, early diagnosis with appropriate management is crucial.

Diagnosis of HLH is based on the combination of clinical and laboratory findings [6]. Although there are several well documented nonspecific imaging findings among HLH patients, these findings are limited to the brain and abdomen/pelvis [1]. To our surprise, despite the histological finding of hemophagocytosis in bone marrow being one of one of the diagnostic criteria for this disease entity, to the best of our knowledge, there has been no previous publication addressing the osseous imaging findings or complications in a patient with HLH.

This is a case report of a patient who had been diagnosed with secondary hemophagocytic lymphohistiocytosis and received high dose corticosteroid, IVIG, and cyclosporine as treatment with serial imaging showing progressive osteonecrosis in bone marrow involving the spine, pelvis, hips, and shoulders. Osteonecrosis of the pelvic bone was confirmed by bone marrow biopsy and involvement of the other bones assumed based on the imaging appearance.

2. Case Presentation

32-year-old male with past medical history significant for type 1 diabetes, hypertension, hypothyroidism, and peripheral vascular disease initially presented to an outside institution emergency department with a one-day history of bloody diarrhea, 2-3 days of nausea/vomiting, jaundice, and altered mental status. Two weeks prior to the admission, the patient and his family members had an upper respiratory infection with symptoms/signs including low grade fever, cough, and congestion. These symptoms/signs had been gradually improving; however, 2-3 days prior to presentation in the outside emergency department, the patient started to have new symptoms of nausea and vomiting, fever, and jaundice with new onset of watery stool, which developed into bloody diarrhea one day prior to initial presentation. Upon admission at the outside institution, the patient was diagnosed with hemolytic anemia for which he was initiated on steroid therapy and received a single dose of IVIG treatment.

Initial CT of the abdomen and pelvis was obtained on the day of admission at the outside institution (Figure 1). At this time, CT appearance of the bone marrow was essentially normal. Following the admission, the patient rapidly developed acute renal failure, respiratory failure, altered mental status, and thrombocytopenia. Initial iliac bone biopsies/aspirations at the outside institution showed evidence of hemophagocytosis, and the possibility of HLH had been raised at this point.

For further specialized care, the patient was transferred to our tertiary institution where he had a prolonged hospitalization of over 2 months. During this hospitalization, the patient had an additional iliac bone biopsy/aspiration, which demonstrated extensive bone necrosis and again the morphologic features of hemophagocytosis. These additional biopsies were primarily performed in response to image progression of changes in the iliac bone to exclude underlying malignancy or infection, as these can be seen as a cause of secondary HLH. However, for lack of culture growth other than on an initial outside sputum culture, infectious etiologies were excluded early in this hospitalization and the patient was not placed on any antibiotics throughout the rest of the admission. No evidence of malignancy was found by laboratory, iliac biopsy, or other imaging tests. Other etiologies such as autoimmune disorder were also ruled out after a thorough clinical workup involving multiple medicine subspecialties including rheumatology, hematology, and nephrology evaluation. Laboratory evaluation included negative antiphospholipid and anti-dsDNA antibodies.

Based on clinical and laboratory criteria including bone marrow biopsy with "morphologic features of hemophagocytosis," elevated IL-2R receptors, absent NK cell function, pancytopenia, elevated ferritin, fevers, splenomegaly, and hypertriglyceridemia, the diagnosis of HLH was clinically established. Following the HLH-2004 treatment guidelines, the patient was treated with cyclosporine and corticosteroids. Etoposide was considered per the HLH-2004 guidelines; however, it was not administered as it can cause pancytopenia in actively bleeding patients.

During this hospitalization, the patient had multiple noncontrast enhanced imaging studies as acute renal failure prevented the patient from receiving contrast. These imaging studies were primarily performed for evaluation of the cause of the secondary HLH and multiorgan failure. Noncontrast enhanced MRI approximately 8 weeks after the initial presentation showed fluid collections anterior to the sacroiliac joints

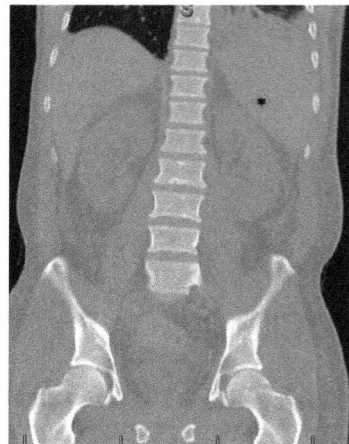

FIGURE 1: Coronal reconstruction CT in bone window at the time of admission to the outside institution shows essentially normal bone marrow other than some Schmorl's nodes. Note additional findings of splenomegaly and left basilar consolidation/atelectasis. Focus of hypoattenuation in the spleen (*) was interpreted as a focus of splenic infarction. Also note that there is no evidence of osteonecrosis in either hip.

and along the right piriformis muscle, which was interpreted as concerning for abscesses by the radiologist (Figure 2). However, based on the clinical assessment, a source of active infection seemed unlikely, and this was deemed to represent inflammatory change possibly related to adjacent bone marrow necrosis. The patient remained off antibiotics throughout this hospitalization. Three days prior to discharge (11 weeks after initial presentation), a noncontrast enhanced CT of chest, abdomen, and pelvis was obtained (Figure 3).

Approximately three months after the initial presentation (2 weeks after discharge from our institution) the patient returned to the emergency department for decreased urine output and was readmitted for acute renal failure. Although the fluid collections noted in the bilateral iliacus muscles remained relatively stable in size, given their persistence, during this second admission, the infectious disease service requested drain placement, and the patient was initiated on ertapenem therapy. The abscess culture from the right iliacus muscle fluid collection grew *Bacteroides fragilis* only in broth; nonetheless, the full course of ertapenem was completed as an outpatient.

Approximately 6 months after initial presentation, the patient presented again to the emergency department with new onset of back pain and fever. MRI of the lumbar spine (Figure 4) was obtained to evaluate for discitis/osteomyelitis given the patient's prior history of possible abscess. Three days later, a CT scan of the abdomen and pelvis with contrast was also obtained (Figure 5). These images showed continued progression of osteonecrosis with worsening pelvic bone osteonecrosis and new imaging findings of diffuse lumbar spine osteonecrosis. A dedicated CT of the lumbar spine without contrast obtained approximately 8 months after the initial presentation showed further progression of the osteonecrosis (Figure 6). Due to increasing left shoulder

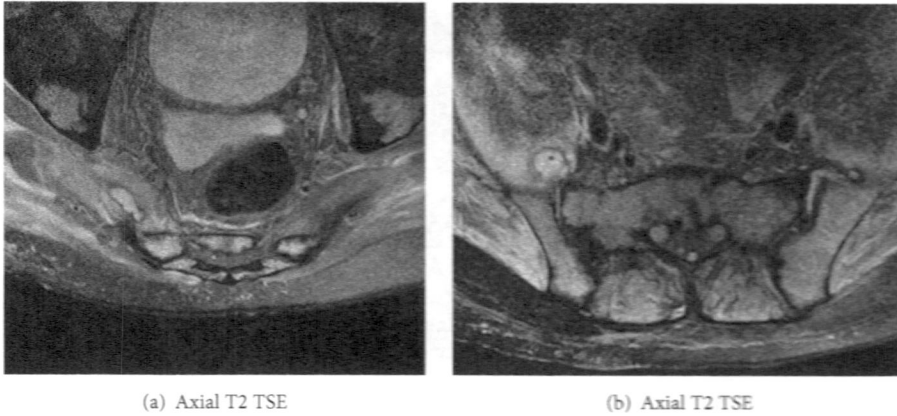

(a) Axial T2 TSE   (b) Axial T2 TSE

FIGURE 2: Axial T2 weighted images following a lumbosacral plexus protocol show right psoas fluid collections (a) and a fluid collection in the right iliopsoas muscle anterior to the sacroiliac joint (b). The smaller contralateral collection anterior to the left sacroiliac joint is not shown. Note diffuse muscle edema asymmetrically affecting the gluteal and iliacus musculature.

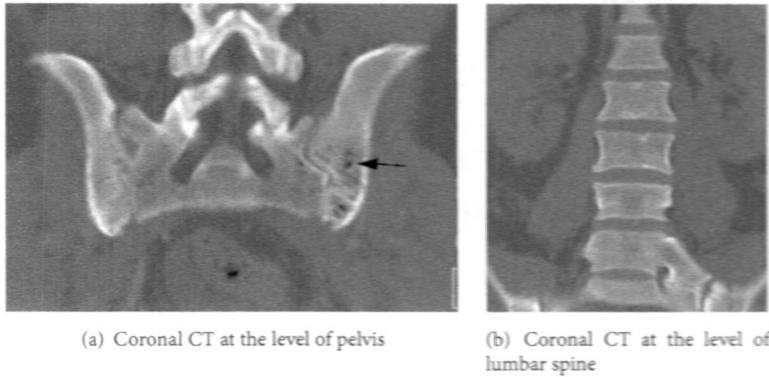

(a) Coronal CT at the level of pelvis   (b) Coronal CT at the level of lumbar spine

FIGURE 3: Coronal noncontrast CT three days prior to discharge/approximately 11 weeks after initial presentation shows (a) new foci of gas (arrow) within the sacrum and iliac bones, left greater than right. Findings in keeping with the preceding iliac bone marrow biopsy showing extensive bone necrosis. (b) Spine at this point still remains normal in CT appearance.

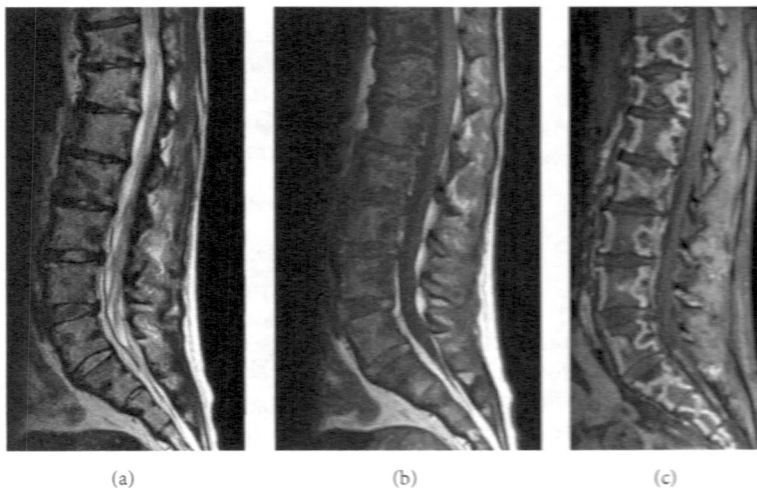

(a)   (b)   (c)

FIGURE 4: Lumbar spine MRI without and with contrast obtained 6 months after the initial presentation. (a) Sagittal T2 weighted and T1 weighted images without contrast and (c) sagittal postcontrast T1-weighted fat suppressed image. These images show marrow signal alteration with associated band-like areas of enhancement surrounding areas of nonenhancement with associated T1 and T2 hypointensity, consistent with diffuse osteonecrosis.

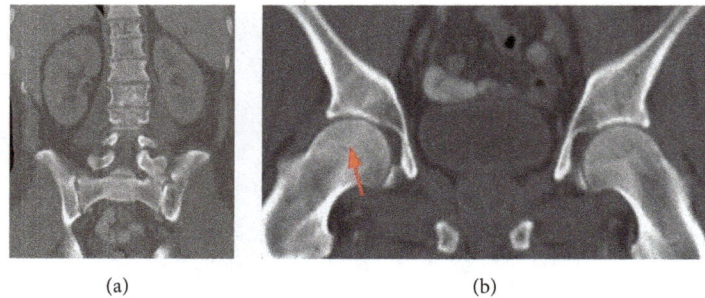

(a)   (b)

FIGURE 5: Coronal intravenous contrast enhanced CT images obtained 3 days after Figure 4. These images show (a) progression of osteonecrosis now involving the lumbar spine. (b) There is also subtle evidence of osteonecrosis of the right hip (arrow).

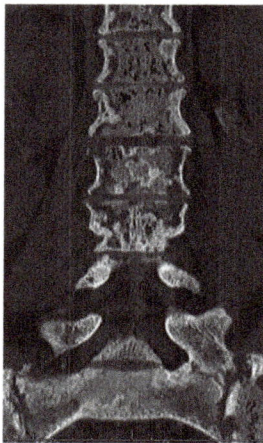

FIGURE 6: CT lumbar spine without contrast approximately 8 months since the initial presentation shows continued progression of diffuse osteonecrosis of the visualized marrow. The bone marrow is abnormal with multiple foci of gas.

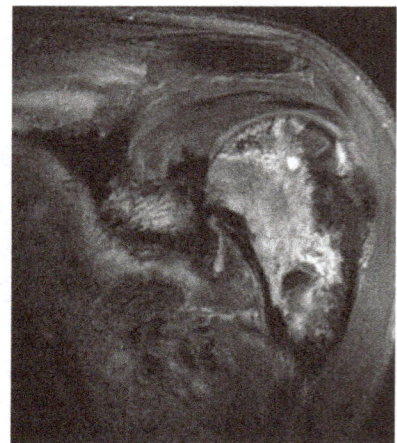

FIGURE 7: Coronal T2 weighted fat suppressed image of the left shoulder obtained approximately 1 year and 9 months since the initial presentation shows image findings consistent with avascular necrosis of the left humeral head. In this image, the classic double line sign with inner T2 hyperintensity and outer T2 hypointensity for osteonecrosis is well depicted.

pain, MRI of the shoulder without contrast was obtained approximately 1 year after the initial presentation (Figure 7), which demonstrated imaging evidence of avascular necrosis. Approximately 14 months after the initial presentation, the patient's abdominal CT also showed further progression of osteonecrosis involving both femoral heads (Figure 8). Furthermore, the patient also eventually developed right shoulder osteonecrosis.

3. Discussion

Hemophagocytic lymphohistiocytosis (HLH) also known as hemophagocytic syndrome is a rare immune disorder of macrophages, characterized by the overactivation of the mononuclear phagocytic system [1, 7]. HLH is a potentially life-threatening, hyperinflammatory syndrome caused by a lack of normal downregulation of activated macrophages and lymphocytes [8, 9].

Diagnosis of HLH requires either a molecular diagnosis or fulfilment of the combination of clinical and laboratory findings, that is, Henter's criteria. In Henter's criteria, none of the clinical or laboratory findings are specific by themselves.

Therefore, five out of eight criteria must be fulfilled for the diagnosis as shown below.

Henter's Criteria for HLH Diagnosis 2004 Version. Five out of 8 criteria must be satisfied for the diagnosis:

Fever.

Splenomegaly.

Cytopenia (affecting ≥ 2 of 3 lineages in the peripheral blood).

Hypertriglyceridemia and/or hypofibrinogenemia.

Hemophagocytosis in bone marrow, spleen, or lymph nodes.

No evidence of malignancy.

Low or absent NK cell activity.

Ferritin level ≥ 500 μg/L.

Soluble CD 25/soluble IL-2 receptor ≥ 2400 U/mL.

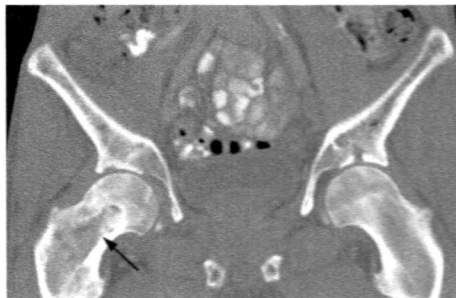

FIGURE 8: Coronal CT with oral contrast shows progression of bilateral femoral head osteonecrosis. There is also diffuse marrow signal abnormality of the visualized pelvis and right proximal femur (arrow), also believed to be consistent with osteonecrosis by imaging.

HLH is divided into two forms: primary (familial) and secondary (acquired) [1, 6–8, 10]. The primary form (25% of cases) is an autosomal recessive condition associated with several genes involved in regulating the immune system, including proteins in the targeted cell death pathway. Infants and young children are most frequently affected, but the disorder is increasingly recognized in older patients [2].

The secondary (acquired) form affects older children and adults and is most commonly associated with infection, underlying malignancy, and prolonged immunosuppression, such as in patients with organ or stem cell transplantation.

HLH is a multisystemic disorder and particularly known to involve bone marrow, spleen, liver, and lymph nodes. Regardless of the form, the phenotype of the disease is the same [11]. The natural course of the disease varies depending on the cause and treatment. For example, a retrospective study showed 0% survival among patients who had EBV as a cause of HLH who had only received conservative therapy, whereas idiopathic HLH had a 100% survival rate in the same study, even with conservative therapy alone [5]. Overall the prognosis of HLH is poor in adults with a mortality rate of 41% in one study [5]. Relapse of HLH after a good therapeutic response is not uncommon; however, the actual rate and risk factors are not known. This patient has not fully recovered to his preadmission state with development of chronic renal failure, but to this date he has not had any recurrent episodes or clinical flare-ups related to HLH.

Regarding the imaging findings, several articles have been published depicting HLH associated nonspecific abnormal brain findings including diffuse cerebral and cerebellar volume loss, white matter T2 hyperintensity and calcifications, leptomeningeal and perivascular enhancement, nodular or ring enhancing parenchymal lesion, and findings of posterior reversible leukoencephalopathy [12–15]. Nonspecific pulmonary findings including atelectasis, interstitial opacities, consolidation, and pleural effusion may be seen [15]. Nonspecific abdominal findings include hepatomegaly, heterogeneous liver echogenicity, splenomegaly, marked gallbladder wall thickening, increased echogenicity of the porta hepatitis, nephromegaly, ascites, and abdominal and inguinal adenopathy [15, 16].

Despite the well documented underlying histologic involvement of bone marrow in HLH, the associated imaging changes of bone in HLH have not yet been well described [14].

In this case, we have shown progressive diffuse osteonecrosis that started in the pelvis, followed by the spine, and eventually involved the appendicular skeleton, including the hips and shoulder of a patient who had been newly diagnosed with the secondary form of HLH. Although the etiologies of the osteonecrosis are numerous, we hypothesize two potential etiologies in this clinical setting: steroid therapy and the underlying condition of HLH itself.

There is a well-documented association of steroid use and osteonecrosis, which depends on duration, daily dose, and cumulative dose of the steroid; however, the threshold is not yet known [17, 18]. As this patient had received steroids as a component of his treatment beginning on the day of admission at the outside hospital, it is not possible to totally exclude the possibility of steroids as a cause of diffuse osteonecrosis.

However, one of the bone marrow biopsies which was obtained merely 11 days after the initiation of steroid therapy had already been histologically reported as extensive osteonecrosis in the iliac bone, at which time the cumulative dose of steroid was calculated to be 6 grams of methylprednisolone. Although some studies have shown steroid induced osteonecrosis occurring as early as 36 days after the initiation of therapy, the median time between commencing steroid medication and developing osteonecrosis of the femoral head has been noted to be 18 months [18–20]. Therefore, we would consider this a rather atypically rapid manifestation and progression of osteonecrosis in this patient if steroid therapy is deemed to be the sole cause of osteonecrosis. Similarly, the extent of osteonecrosis was thought to be rather atypically diffuse for just steroid related osteonecrosis, though no literature addressing this can be found. Hence, we hypothesize the rapid progressive osteonecrosis in this newly diagnosed HLH patient is likely related at least partly to the HLH itself rather than solely from steroid use.

A limitation of this case report is osteonecrosis in the spine and appendicular skeleton diagnosed based on the classic imaging appearance without histologic confirmation other than iliac bone biopsies showing evidence of osteonecrosis. Furthermore, given that this is a single case report, the incidence of imaging findings of osteonecrosis among HLH patients is uncertain; we especially have seen a few cases with relatively normal appearing bone marrow in patients who had previously been diagnosed with HLH.

References

[1] J. Zaveri, Q. La, G. Yarmish, and J. Neuman, "More than just langerhans cell histiocytosis: a radiologic review of histiocytic disorders," *Radiographics*, vol. 34, no. 7, pp. 2008–2024, 2014.

[2] K. Nagafuji, A. Nonami, T. Kumano et al., "Perforin gene mutations in adult-onset hemophagocytic lymphohistiocytosis," *Haematologica*, vol. 92, no. 7, pp. 978–981, 2007.

[3] J.-I. Henter, G. Elinder, O. Söder, and A. Ost, "Incidence in Sweden and clinical features of familial hemophagocytic lymphohistiocytosis," *Acta Paediatrica Scandinavica*, vol. 80, no. 4, pp. 428–435, 1991.

[4] M. B. Jordan, C. E. Allen, S. Weitzman, A. H. Filipovich, and K. L. McClain, "How I treat hemophagocytic lymphohistiocytosis," *Blood*, vol. 118, no. 15, pp. 4041–4052, 2011.

[5] M. Ramos-Casals, P. Brito-Zerón, A. López-Guillermo, M. A. Khamashta, and X. Bosch, "Adult haemophagocytic syndrome," *The Lancet*, vol. 383, no. 9927, pp. 1503–1516, 2014.

[6] J.-I. Henter, A. Horne, M. Aricó et al., "HLH-2004: diagnostic and therapeutic guidelines for hemophagocytic lymphohistiocytosis," *Pediatric Blood and Cancer*, vol. 48, no. 2, pp. 124–131, 2007.

[7] Z. K. Otrock and C. S. Eby, "Clinical characteristics, prognostic factors, and outcomes of adult patients with hemophagocytic lymphohistiocytosis," *American Journal of Hematology*, vol. 90, no. 3, pp. 220–224, 2015.

[8] G. E. Janka and K. Lehmberg, "Hemophagocytic syndromes— an update," *Blood Reviews*, vol. 28, no. 4, pp. 135–142, 2014.

[9] A. Filipovich, K. McClain, and A. Grom, "Histiocytic disorders: recent insights into pathophysiology and practical guidelines," *Biology of Blood and Marrow Transplantation*, vol. 16, pp. S82–S89, 2010.

[10] S. Chandrakasan and A. H. Filipovich, "Hemophagocytic lymphohistiocytosis: advances in pathophysiology, diagnosis, and treatment," *Journal of Pediatrics*, vol. 163, no. 5, pp. 1253–1259, 2013.

[11] C. Larroche, "Hemophagocytic lymphohistiocytosis in adults: diagnosis and treatment," *Joint Bone Spine*, vol. 79, no. 4, pp. 356–361, 2012.

[12] T. W. Chung, "CNS involvement in hemophagocytic lymphohistiocytosis: CT and MR findings," *Korean Journal of Radiology*, vol. 81, pp. 78–81, 2007.

[13] K. P. N. Forbes, D. A. Collie, and A. Parker, "CNS involvement of virus-associated hemophagocytic syndrome. MR imaging appearance," *American Journal of Neuroradiology*, vol. 21, no. 7, pp. 1248–1250, 2000.

[14] B. Ozgen, K. Karli-Oguz, B. Sarikaya, B. Tavil, and A. Gurgey, "Diffusion-weighted cranial MR imaging findings in a patient with hemophagocytic syndrome," *American Journal of Neuroradiology*, vol. 27, no. 6, pp. 1312–1314, 2006.

[15] N. E. Fitzgerald and K. L. McClain, "Imaging characteristics of hemophagocytic lymphohistiocytosis," *Pediatric Radiology*, vol. 33, no. 6, pp. 392–401, 2003.

[16] M. H. Schmidt, L. Sung, and B. M. Shuckett, "Hemophagocytic lymphohistiocytosis in children: abdominal US findings within 1 week of presentation," *Radiology*, vol. 230, no. 3, pp. 685–689, 2004.

[17] C. Powell, C. Chang, S. M. Naguwa, G. Cheema, and M. E. Gershwin, "Steroid induced osteonecrosis: an analysis of steroid dosing risk," *Autoimmunity Reviews*, vol. 9, no. 11, pp. 721–743, 2010.

[18] M. F. Dilisio, "Osteonecrosis following short-term, low-dose oral corticosteroids: a population-based study of 24 million patients," *Orthopedics*, vol. 37, no. 7, pp. e631–e636, 2014.

[19] D. Liu, A. Ahmet, L. Ward et al., "A practical guide to the monitoring and management of the complications of systemic corticosteroid therapy," *Allergy, Asthma & Clinical Immunology*, vol. 9, article 30, 2013.

[20] R. S. Weinstein, "Glucocorticoid-induced osteonecrosis," *Endocrine*, vol. 41, no. 2, pp. 183–190, 2012.

12

Thymic Epidermoid Cyst: Clinical and Imaging Manifestations of This Rare Anterior Mediastinal Mass

Jawad M. Qureshi,[1] Brian Pagano,[2] Jeffrey Mueller,[1] Lana Schumacher,[3] Claudia Velosa,[4] and Matthew S. Hartman[1]

[1]*Department of Radiology, Allegheny Health Network, Pittsburgh, PA, USA*
[2]*Temple University School of Medicine, Philadelphia, PA, USA*
[3]*Department of Surgery, Allegheny Health Network, Pittsburgh, PA, USA*
[4]*Department of Pathology, Allegheny Health Network, Pittsburgh, PA, USA*

Correspondence should be addressed to Jawad M. Qureshi; jqureshi@wpahs.org

Academic Editor: Amit Agrawal

Thymic epidermoid cysts are an extremely rare entity. These arise from epidermal cells that migrate to the thymus. The radiologic diagnosis of this rare lesion is challenging. We describe a case of an otherwise healthy 35-year-old woman who presented with an acute onset of chest pain and shortness of breath. She was found to have an anterior mediastinal mass. The imaging findings were, however, not characteristic for any single diagnostic entity. Since the imaging was inconclusive, surgical resection was performed for definitive diagnosis. The mass was found to be a thymic epidermoid cyst. This case underlines the significance for radiologists to be aware that epidermoid cysts can occur in the thymus and should be considered in the differential diagnosis for a heterogeneous anterior mediastinal mass.

1. Introduction

Thymic epidermoid cysts are extremely rare with only three cases described in literature. These are benign and carry an overall good prognosis. However, the location is atypical and imaging findings are nonspecific. Despite their benign nature, surgical resection is required to exclude malignancy and attain a definitive tissue diagnosis.

2. Case History

A 35-year-old lady presented with chest pain and shortness of breath, one week after undergoing a C-section. She was evaluated with a chest computed tomography angiography (CTA) to evaluate possible pulmonary emboli. The chest CTA was negative for pulmonary emboli but incidentally demonstrated a homogenous 5 cm mass in the anterior mediastinum (Figures 1 and 2). The patient was scheduled for a positron emission tomography CT (PET CT) which showed no significant FDG activity in the mass (Figure 5). Follow-up magnetic resonance imaging (MRI) of the chest demonstrated a nonenhancing, heterogeneous anterior mediastinal mass with cystic components and no macroscopic fat (Figures 3 and 4).

CT-guided needle biopsy was performed for definitive diagnosis. This showed benign squamous and fibroconnective tissue and was inconclusive. She subsequently underwent a right robotic assisted thoracoscopy for resection of the mass. A 9.5 cm × 7.0 cm × 3.0 cm soft, round mass with a red, glistening capsule was resected following careful dissection from the adhering mediastinal structures. The surgical specimen was submitted for pathological analysis (Figure 6). The final histopathology of the surgical specimen showed a benign epidermoid cyst, with abundant internal keratin debris, that was attached to benign thymic tissue (Figures 7 and 8). The patient had a good outcome and is currently asymptomatic.

3. Discussion

Thymic epidermoid cysts are an extremely rare entity. To the best of our knowledge, this is the 4th reported case. Rare cases

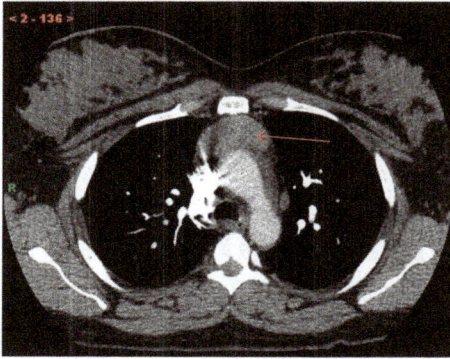

FIGURE 1: Axial CTA through the level of the aortic arch shows a homogenous anterior mediastinal mass (red arrow).

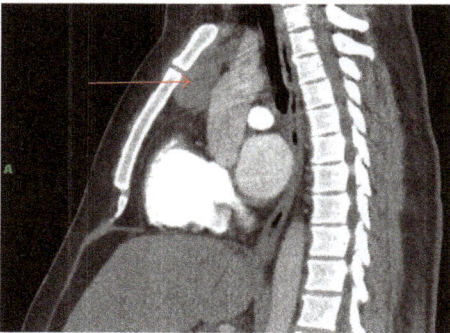

FIGURE 2: Sagittal reconstruction shows the same mass in the vertical plane (red arrow).

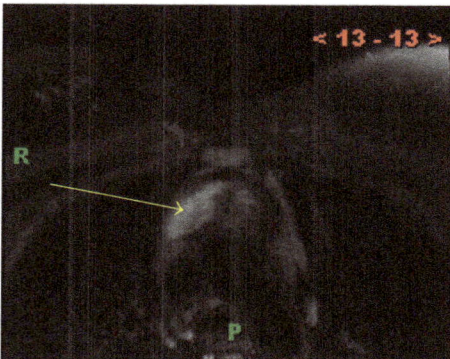

FIGURE 3: (MR T2) axial T2 fat saturated image showing a hyperintense heterogenous mass suggestive of cystic components (yellow arrow).

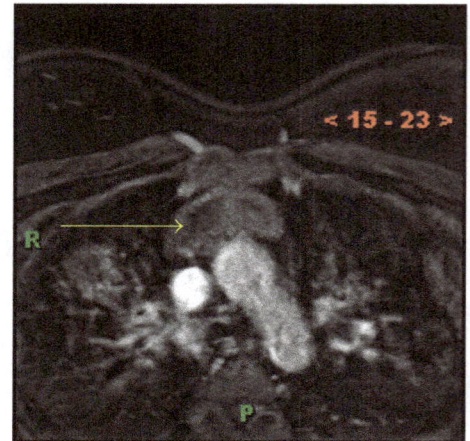

FIGURE 4: (MR T1 postcontrast) heterogeneous anterior mediastinal mass (yellow arrow) without obvious enhancement.

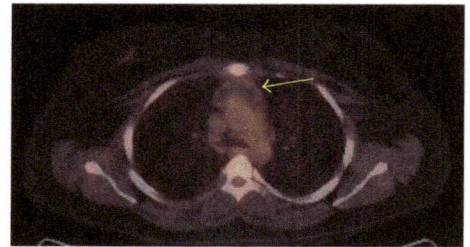

FIGURE 5: (PET) no significant metabolic activity in the anterior mediastinal mass (yellow arrow).

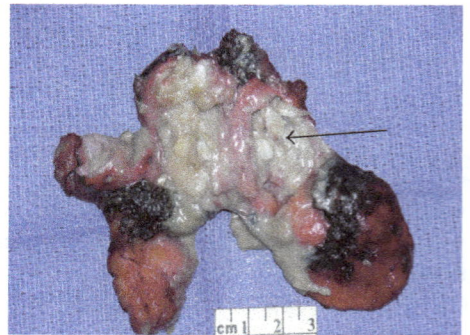

FIGURE 6: Gross specimen with keratinaceous debris within the cyst (black arrow).

of epidermoid cysts have been reported within the spleen, kidney, and the GI and GU tracts [1].

The exact etiology of thymic epidermoid cysts remains unknown. Developmentally, the thymus forms primarily from epithelial cells derived from the endoderm with a mesenchymal thymic remnant. Epidermoid cysts are sequestration cysts that form by proliferation of epidermal cells that arise from the ectoderm within an unusual location within the thymus [2]. Acquired epidermoid cysts in the thymus are

hypothesized to result from epidermal tissue migration into the anterior mediastinum and their subsequent proliferation within the thymus. Congenital epidermoid cysts may potentially form in the thymus, as in other locations. However, no confirmed case has been reported in the literature to date [3].

There has been a case report of an acquired, posttraumatic thymic epidermoid cyst. This was thought to result from the introduction of epidermoid cells into the thymus following trauma [4]. Epidermoid tissue may be introduced in the thymus following surgery as well. One case report has suggested an association between Gardner's syndrome and

TABLE 1: Differential diagnosis of an anterior mediastinal mass.

	CT	PET/CT	MRI	Enhancement pattern
Thymic epidermoid cyst	Heterogeneous	Not FDG avid	Heterogenous	Possible restricted diffusion on DWI
Thymic hyperplasia	Homogenous, soft tissue attenuation	Mildly FDG avid; difficult to exclude malignancy due to physiologic uptake in the thymus	Homogenous	Loss of signal on out of phase
Teratoma	Heterogenous, may contain fat/calcification	Not FDG avid	Heterogenous, may contain fat and calcification	Heterogenous
Thymic neoplasm	Focal mass; possible metastases, local invasion, and/or lymphadenopathy	Mildly FDG avid; difficult to exclude malignancy due to physiologic uptake in the thymus	Focal mass; possible metastases, local invasion, and/or lymphadenopathy	T1 isointense to muscle/normal thymus; heterogenous on T2WI
Lymphoma	Enlarged, heterogeneously enhancing mass	FDG avid	Heterogenous, could be necrotic	Shows enhancement

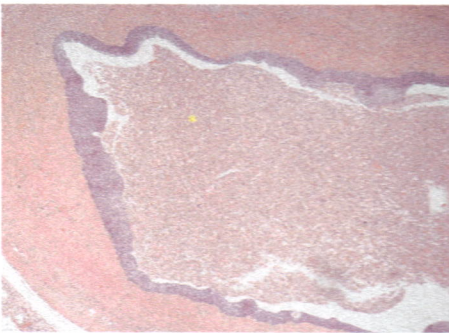

FIGURE 7: H&E 10x image showing abundant keratin debris (*) within the cyst.

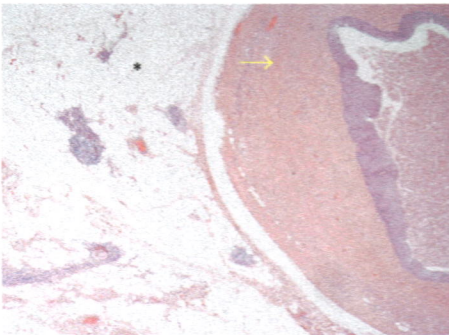

FIGURE 8: H&E 2.5x image showing the epidermoid cyst wall (yellow arrow) and the normal thymus tissue (*).

the development of thymic epidermoid cysts [5]. However, our patient did not have gastrointestinal or other abdominal signs or symptoms.

The clinical presentation is variable. In the past, epidermoid cysts in the thymus have been diagnosed in asymptotic patients. They have also been found during the workup for chest pain, dyspnea, fever, or hemoptysis. A history of chest trauma or recent surgery may be present. There was no history of trauma in our patient. She first noticed her symptoms towards the end of pregnancy.

The imaging findings are nonspecific, which may be misleading for the radiologists. The differential diagnosis includes thymic hyperplasia, teratoma, thymic neoplasm, or lymphoma. Due to its rarity, an epidermoid cyst of the thymus is not typically part of the differential diagnosis (Table 1). In our patient, the chest CTA demonstrated nonspecific findings of a homogenous anterior mediastinal mass. The subsequent PET CT showed no significant metabolic activity. In retrospect, in our patient the MRI was the most suggestive imaging test for an epidermoid cyst. The MRI showed a heterogenous mass with cystic components. Typically, thymic hyperplasia has a homogenous mass which may lose signal on out-of-phase MR imaging. Although this mass has cystic components, typical thymic cysts are not heterogeneous. Lymphoma can be heterogenous and necrotic but usually enhances and is metabolically active on PET-CT. Teratomas are commonly heterogenous but usually contain macroscopic fat and/or calcifications. There was no obvious macroscopic fat on either CT or MRI in our patient. Epidermoid tumors can show restricted diffusion in other parts of the body, particularly in the brain. Unfortunately, diffusion weighted imaging was not utilized as it is not part of the usual anterior mediastinal mass protocol. A more definitive diagnosis could not be made based on these nonspecific findings on imaging. Therefore, the patient required surgical resection to reach a diagnosis.

Pathology showed that the cystic mass was attached to the normal thymus. The epithelial lining of the cyst consisted of stratified squamous epithelium with abundant keratin debris present within the cyst. No other epithelial or mesenchymal components were identified. These findings were consistent with a thymic epidermoid cyst.

4. Conclusion

Epidermoid cysts can occur anywhere in the body. However, the thymus is a very unusual location. This is only the fourth reported case, to the best of our knowledge. These may be found incidentally on asymptomatic patients or on routine workup for chest pain and shortness of breath. A history of trauma or surgery may be associated. Imaging findings are nonspecific. That being said, surgical resection may eventually be necessary to reach a definitive diagnosis. While extremely rare, this entity should be considered in the radiologic differential diagnostic of a heterogeneous anterior mediastinal mass.

References

[1] M. R. Sahoo, M. S. Gowda, and S. S. Behera, "Unusual site and uncommon presentation of epidermoid cyst: a rare case report and review of literature," *BMJ Case Reports*, 2013.

[2] T. Boehm, "Thymus development and function," *Current Opinion in Immunology*, vol. 20, no. 2, pp. 178–184, 2008.

[3] B. L. Pear, "Epidermoid and dermoid sequestration cysts," *The American Journal of Roentgenology, Radium Therapy, and Nuclear Medicine*, vol. 110, no. 1, pp. 148–155, 1970.

[4] F. Monaco, M. Barone, and M. Monaco, "Intrathymic epidermoid cyst: a very rare condition," *Asian Cardiovascular & Thoracic Annals*, vol. 23, no. 3, pp. 323–324, 2015.

[5] J. Delamarre, J. L. Dupas, J. F. Muir, B. Deschepper, H. Sevestre, and J. P. Capron, "Gardner's syndrome and epidermoid cyst of the thymus," *Gastroenterologie Clinique et Biologique*, vol. 11, no. 5, pp. 421–423, 1987.

13

Magnetic Resonance Imaging-Guided Focused Ultrasound Surgery for the Treatment of Symptomatic Uterine Fibroids

Laura Geraci,[1] **Alessandro Napoli,**[2] **Carlo Catalano,**[2] **Massimo Midiri,**[1] **and Cesare Gagliardo**[1]

[1]*Section of Radiological Sciences, Department of Biopathology and Medical Biotechnologies, University of Palermo, Palermo, Italy*
[2]*Radiology Section, Department of Radiological, Oncological and Anatomopathological Sciences,*
 "Sapienza" University of Rome, Rome, Italy

Correspondence should be addressed to Cesare Gagliardo; cesare.gagliardo@unipa.it

Academic Editor: Dimitrios Tsetis

Uterine fibroids, the most common benign tumor in women of childbearing age, may cause symptoms including pelvic pain, menorrhagia, dysmenorrhea, pressure, urinary symptoms, and infertility. Various approaches are available to treat symptomatic uterine fibroids. Magnetic Resonance-guided Focused Ultrasound Surgery (MRgFUS) represents a recently introduced noninvasive safe and effective technique that can be performed without general anesthesia, in an outpatient setting. We review the principles of MRgFUS, describing patient selection criteria for the treatments performed at our center and we present a series of five selected patients with symptomatic uterine fibroids treated with this not yet widely known technique, showing its efficacy in symptom improvement and fibroid volume reduction.

1. Introduction

Uterine fibroids (or leiomyomas) are the most common benign tumor of the genital tract in women of reproductive age. According to recent longitudinal studies, the lifetime risk of fibroids in a woman over the age of 45 years is more than 60%, with incidence higher in blacks than in whites [1].

Although they often are asymptomatic, nearly half of women with fibroids have debilitating symptoms, such as menorrhagia, dysmenorrhea, anemia, pelvic pressure or pain, urinary symptoms, constipation, acute pain from degeneration or torsion of a pedunculated fibroid, dyspareunia, infertility, or miscarriage [2]. Table 1 shows a summary of uterine fibroids characteristics.

Several approaches are now available for the treatment of uterine fibroids, including pharmacologic options, such as hormonal therapies and Gonadotropin-Releasing Hormone (GnRH) agonists, and surgical approaches, such as hysterectomy and myomectomy (see Table 2). The new treatment modalities include the following: hysteroscopic resection for submucous fibroids, laparoscopic and vaginal myomectomy, uterine artery embolization (UAE), myolysis by heat, cold coagulation, and laser, laparoscopic uterine artery occlusion, temporary transvaginal uterine artery occlusion, and Magnetic Resonance-guided Focused Ultrasound Surgery (MRgFUS). Factors such as the importance of uterine preservation for patient's desire to become pregnant in the future, symptoms severity, and tumor characteristics may affect the choice of the best possible approach [2, 3].

MRgFUS is a noninvasive thermoablative technique that combines the anatomic detail and thermal monitoring capabilities of Magnetic Resonance Imaging (MRI) with the therapeutic potential of High-Intensity Focused Ultrasound (HI-FU). MRI offers excellent three-dimensional anatomic resolution and real-time thermal monitoring, measuring tissue temperature with an accuracy of ±2°C [4]. HI-FU waves can pass through the anterior abdominal wall and aim at the targeted volume, where tissue temperature increases rapidly up to 60°C or higher, inducing a thermal lesion by protein denaturation and resulting coagulative necrosis, while the

TABLE 1: Summary table for uterine fibroids.

Etiology	Unknown
Incidence	>60% over the age of 45 years
Age predilection	>30 years
Risk factors	Age, black race, early age at menarche, familial predisposition, overweight, polycystic ovary syndrome, diabetes, hypertension, nulliparity
Symptoms	Menorrhagia, dysmenorrhea, anemia, pelvic pressure or pain, urinary symptoms, constipation, backache or leg pains, dyspareunia, infertility, or miscarriage
Treatment	Treatment is required in up to 25% of women. Treatment options include medications, such as gonadotropin-releasing hormone (Gn-RH) agonists, hysterectomy, myomectomy, myolysis, uterine artery embolization, MR-guided Focused Ultrasound Surgery (MRgFUS)
Prognosis	Benign tumor, excellent prognosis. In general, they begin to shrink after menopause, and they can grow quickly during pregnancy. They may also bleed into themselves, degenerate, become cystic, calcify, or undergo sarcomatous degeneration (<1% of cases)
Findings on MR imaging	Well-defined uterine mass with uniformly low signal intensity as compared to the myometrium on T2-w images and iso-hypointense to the myometrium on T1-w images that enhances homogeneously when gadolinium is administered intravenously. Degenerated fibroids show complex appearance with high or heterogeneous signal on T2-w and postcontrast images

skin and overlying tissue layers outside the ablated area remain unaffected [5].

Compared to other treatment options, MRgFUS represents a safe, effective, and noninvasive approach that may be alternatively used as a fertility-preserving technique in selected cases [4].

MRgFUS was approved by the European Community (CE) in 2002 and by the US Food and Drug Administration (FDA) in 2004.

This paper provides an overview of our initial clinical experience with MRgFUS, including a brief description of the treatment system, selection criteria, and procedure workflow, followed by the presentation of a case series of five patients with symptomatic uterine fibroids treated with this technique. These cases have distinctive features that offer good points for discussion.

2. Clinical Series

Patients with symptomatic fibroids included in this report were screened by means of a medical examination by a general practitioner or a gynecologist and by a pelvic ultrasounds examination and MRI in order to determine patients' clinical and technical suitability for MRgFUS treatment. We included patient with a definitive diagnosis of uterine fibroid(s) as the cause for their symptoms, uterus size less than 24 cm without the cervix, the presence of less than 6 clinically significant fibroids, absence of contraindications to MRI examination, no massive abdominal scarring that cannot be avoided by manipulations or covered by a US-blocking scar patch (made of a polyethylene film mixed with air bubbles), no evidence of high grade Squamous Intraepithelial Lesion (SIL), and no ongoing pregnancy or unstable medical conditions. All the patients included in this report have read and signed an informed consent which included the use of collected anonymized data for scientific publications.

2.1. MRI Selection Criteria. Patients underwent a pelvic MRI screening examination (Signa HDxt 1.5T scanner from GE Medical Systems; Milwaukee, WI, USA) that included the following sequences acquired with 4 mm thick consecutive slices: axial, sagittal, and coronal T2-weighted (T2-w) Fast Recalled Fast Spin Echo (FRFSE), axial T1-weighted (T1-w) Fast Spin Echo (FSE) with and without fat saturation, and, after intravenous (i.v.) administration of 0.1 mmol/kg of gadobenate dimeglumine (MultiHance®, Gd-BOPTA; Bracco Imaging SpA, Milano, Italy), sagittal and coronal T1-w FSE and axial T1-w FSE with fat saturation.

The purpose of the MRI screening examination was to check the accessibility, viability, and texture of the fibroids. MR images were also useful to obtain information on size, location, number, signal intensity on T2-w images, and postcontrast enhancement of uterine fibroid(s) as well as abdominal scars in the intended beam path, presence of adenomyosis, and existence of other uterine disorders or any other pathology outside the uterus. Fibroids suitable for MRgFUS were selected in according with guidelines outlined by Yoon et al. [6], Lénárd et al. [7], and more recently by Mindjuk et al. [8].

We included patients with less than 6 clinically significant submucosal, intramural, and/or subserosal fibroids greater than 3 cm (since the smaller sonication spot with our MRgFUS system is 2.5 cm; see asterisks in Figure 1(a)) and smaller than 10 cm (to avoid longer treatment time; Figure 1(b)). Patients with pedunculated fibroids on a small and narrow stalk (less than 50% of fibroid diameter) were excluded as they may potentially disconnect from the uterus after the treatment and require further surgical procedures to remove them from the pelvic cavity (asterisk in Figure 1(c)). Fibroids with a significantly calcified envelope were excluded as well as cellular fibroids showing a bright signal on T2-w images (relative to the uterus wall; see Figure 1(d)) and nonenhancing fibroids on postcontrast T1-w images (Figure 1(e)).

TABLE 2: Comparison of hysterectomy, myomectomy, uterine artery embolization (UAE), and MRgFUS.

Procedure	Hysterectomy	Myomectomy	Uterine artery embolization (UAE)	MRgFUS
Description	Surgical removal of the uterus with or without the cervix. There are several different surgical approaches: vaginal hysterectomy (performed through an incision in the vagina), abdominal hysterectomy (through a horizontal incision on the lower abdomen), and laparoscopic hysterectomy (through four tiny incisions on the abdomen).	Surgical removal of one or more fibroids from within the uterus. It can be performed through several different ways: abdominal myomectomy, laparoscopic myomectomy, and hysteroscopic myomectomy (only for women with submucosal fibroids).	UAE involves blocking, with small particles injected through a catheter, the blood vessels that supply the fibroids, causing them to shrink.	High intensity focused ultrasound waves heat and destroy fibroid tissue. The MRI allows guiding treatment and monitoring tissue temperature in realtime.
Return to normal activities	7 to 56 days.	1 to 44 days.	3 to 10 days,	1 day.
Hospital days	1 to 5 days.	0 to 3 days.	0 to 1 day,	Outpatient procedure; no hospital stay.
Procedure time	1.5 to 3 hours.	1 to 3 hours.	30 minutes to 1.5 hours,	1.5 to 4 hours.
Advantages	Fibroids will not recur because the uterus is removed. The ovaries may be removed or spared.	Only the fibroids are removed; reproductive potential is spared.	Most fibroids can be treated. Incision is small and uterus is retained. Hospital stay is short (1 day) and in some cases may be performed as an outpatient procedure. Recurrence of treated fibroids is very rare. Return to normal activity within 10 days.	Day care procedure requiring no hospitalization, no incisions, no ionizing radiation, no general anesthesia. Severe complications virtually absent. Return to daily activities from the next day of treatment. Fertility is preserved.
Disadvantages/risks	Reproductive potential is lost. Side effects may include early menopause and a reduction in libido. Removal of the ovaries in a premenopausal woman can lead to hot flashes, vaginal dryness, and osteoporosis. Possible surgical risks include bleeding, infections, adhesions, injury to the intestines, or bladder.	Fibroids can regrow and/or new fibroids can develop resulting in recurrent symptoms and additional procedures. The younger the woman is and the more the fibroids are present at the time of myomectomy, the more likely she is to develop fibroids in the future. Possible surgical risks include bleeding, adhesions, and infections.	Low risk of menopause and blockage of blood supply to ovaries. Possible surgical risks include bleeding, uterine infection, blood clots, and injury of the ovaries and to the uterus, potentially leading to a hysterectomy.	Not all type of fibroids can be treated. Fibroids may recur with time. It is a safe procedure with minimal risk; infrequent complications are abdominal pain/cramping, back or leg pain, urinary tract infection, vaginal discharge, skin injury (burns), and transient nerve damage.
Future fertility	Reproductive potential is lost.	Possibility of pregnancy after adequate healing time. A cesarean section may be required for delivery.	Unpredictable effect on fertility.	Fertility is preserved.

Fibroids with a significant portion (more than 50%) of their volume deeper than 14 cm from the skin line (i.e., the maximum penetration achievable with the HI-FU transducer we used) required mitigation techniques (thinner acoustic coupling gel pad or rectum filling with ultrasound gel) to reduce the distance between the targeted fibroid and the transducer. Special attention has been given to the following elements:

(i) *Proximity to Sacrum.* The center of the targeted fibroid should be more than 4 cm from any bone surface. Bone can be indirectly heated by the HI-FU far field energy, which could potentially cause secondary nerve heating, resulting in pain and frequent interruptions of the treatment sessions that can result in risk of nerve reversible or permanent injury. Mitigation techniques to avoid sacrum proximity may

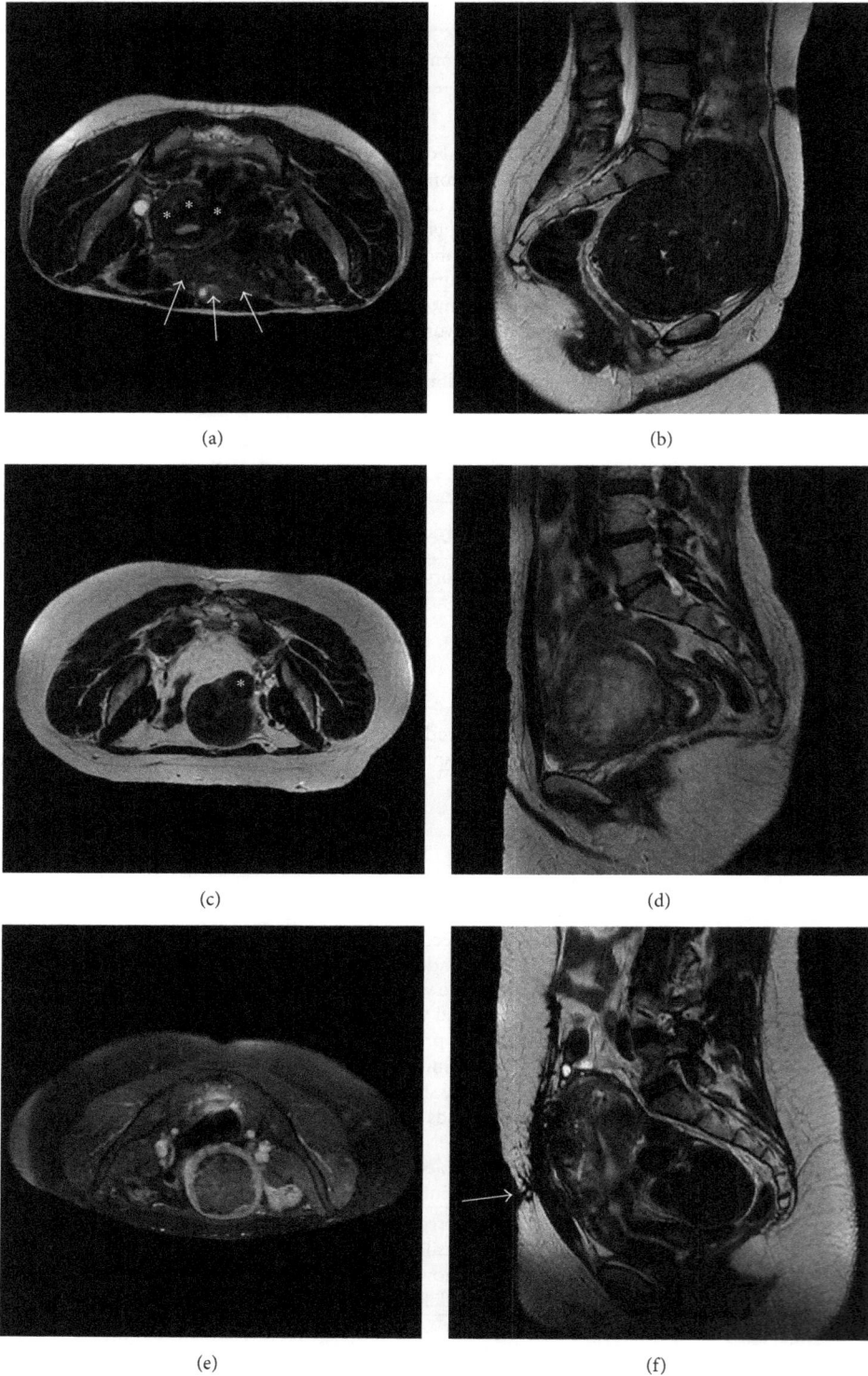

FIGURE 1: Illustrative pelvic MRI scans of excluded patients: (a) axial T2-w showing three small fibroids (asterisks) and many bowel loops (white arrows) that are interposed between the skin surface and the hypothetic target; (b) sagittal T2-w of a large fibroid which almost occupies the whole pelvis and is dangerously close to sacrum bone and nerves (this patient performed the screening MRI in supine feet first position since she reported some discomfort in maintaining the prone position); (c) axial T2-w showing a small and pedunculated subserosal fibroid (asterisk); (d) sagittal T2-w of a "bright" untreatable cellular fibroid; (e) axial T1-w with fat saturation acquired after intravenous injection of paramagnetic contrast medium showing a nonenhancing fibroid; (f) sagittal T2-w of a patients with a bulky scar in abdominal skin (white arrow).

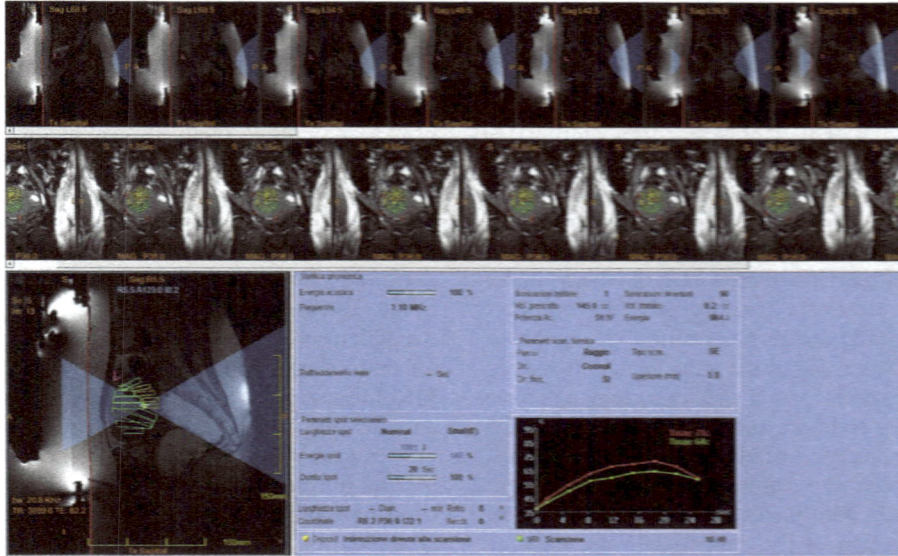

FIGURE 2: Screenshot from the ExAblate workstation showing the planned target of patient in case 3 after the first sonication performed. In bottom left, the HI-FU beam representation is shown in light blue, the red line indicates the skin-gel pad interface, and critical structures are secured by the use of specific low-energy density region (LEDR) and no-pass regions markers (bowel in pink, pubic bone in light blue). The target volume (region of treatment, ROT) has been split into multiple subvolumes (green and yellow voxels) each of which will be ablated by a specific sonication. Patients' movements during treatment are monitored by reviewing fiducials (red crosshairs) placed by the treating physician on distinct anatomic structures which can be monitored during real-time MR imaging. Real-time thermal mapping after the first sonication is shown in the bottom right graph (maximum temperature achieved in the focal spot is 71˚C).

include HI-FU transducer or beam tilting and rectal filling.

(ii) *Accessibility of the Fibroid(s)*. Bowel (white arrows in Figure 1(a)) and scars (white arrow in Figure 1(f)) may represent an obstacle to the ultrasound beam path. As bowel may contain air or energy absorbing particles, only patients with bowel that can be shifted away from the beam path (by bladder or rectal filling) or by beam angulation may be treated. Extensive abdominal scars may absorb the ultrasound energy and cause pain or result in a skin burn. Therefore, patients with abdominal scars that cannot be avoided using bladder filling and/or beam angulation or that cannot be fully covered by an ultrasound (US) blocking scar patch may not be suitable for MRgFUS.

(iii) *Malignancy*. Patients should not have any imaging finding suggestive of uterine, ovarian, or cervical malignancy on MRI screening evaluation.

2.2. Treatment: Planning and Execution. On the day of the treatment, patients are shaved from the umbilicus to the pubis; an i.v. line and a urinary catheter are inserted. The procedure is performed with the patient lying prone (feet toward the MR gantry) on a dedicated MR detachable table which houses the MRgFUS system. In our practice we used an InSightec ExAblate 2100 (InSightec Ltd., Haifa, Israel) MRg-FUS system connected to a GE HDxt 1.5T MR scanner (GE Medical Systems; Milwaukee, WI, USA). The abdomen of the patients lies in a water bath of deionized degassed water, in

contact with an acoustic coupling gel pad located above the ultrasound transducer.

A light i.v. conscious sedation is administrated by an anesthesiologist to relieve anxiety, prevent movement, and minimize discomfort during the procedure. Patients received 1 gr paracetamol (15 minutes before laying on the MR table) and 0.03 mg/Kg midazolam (at the beginning of the treatment). Additional midazolam boluses (0.01–0.03 mg/Kg) were administered to keep the appropriate sedation level.

As a safety measure, a "stop sonication button" is placed in patient's hand and one more button is available on the workstation that the operating physician uses during treatment. These buttons are capable of the immediate suspension of energy delivery anytime during the procedure.

Preprocedural MRI is essential to evaluate the correct patient positioning and the adequate transducer-to-fibroid alignment. To avoid extensive abdominal scars, bowel, or ovaries, the transducer can be moved and tilted. The HI-FU beam path can be changed by acting on electronic steering through the dedicated software. Moreover, the physician may accordingly modify the position of the uterus by bladder and/or rectal filling (using saline and US gel, resp.).

Before starting a treatment, the skin-gel pad interface needs to be defined in the dedicated software (red line in Figure 2). This is followed by marking out the critical structures using specific low-energy density region (LEDR) and no-pass region markers to curtail or prevent the beam path to pass through sensitive organs (see Figure 2: bowel in pink; pubic bone in light blue; sacrum in orange). For added safety, patients' movements during treatment are monitored by reviewing fiducials placed by the treating physician on distinct anatomic structures which can be monitored during real-time MR imaging acquired in treatment (red crosshairs in Figure 2). The target volume, referred to as the region of treatment (ROT), is then defined with its safety margins. A few low-energy sonications (usually up to 3 or 4) are performed for final targeting calibration. Clinical treatment will be performed with multiple therapeutic sonications until a sufficient fibroid volume will be covered (in Figure 2, the ROT has been split into multiple subvolumes each of which will be ablated by a specific sonication). Real-time MR thermometry will reveal any potentially dangerous heating or unwanted exposure. Before the beginning of the treatment, patients are explained what they may feel and if and when they will have to stop the sonication: mild warming, transient pain, and uterine cramping are usually reported by patients during sonications, while acute pain, signs of neurological involvement of the lower limbs, and skin burning sensations require an instantaneous interruption of the energy delivery. Patients are also asked to relay all relevant sensations after each sonication and they are constantly monitored and assessed to check for unusual symptoms.

In our practice, the average treatment time (from the first low-energy sonication to the last therapeutic one) was about 150 minutes.

After the treatment, postcontrast T1-w images are acquired to assess the area of treated tissue (calculated as the nonenhancing area in terms of nonperfused volume, NPV) and to exclude any procedure-related complications.

Even if procedure-related complications are mostly minor and attributable to the effects from focused ultrasound along the whole beam trajectory (near field, targeted site, and far field), the operating physician must be steadily vigilant during the treatment. The most common near field side effect is the risk of skin burns due to bad skin-to-gel pad coupling (scars or skin folds must be carefully avoided and/or settled) or to targeting a site too close to the skin surface. An accurate preparation of the patient (that include proper skin shaving and cleansing and the use of scar patches) usually helps preventing thermal injuries in the near field. In some patients, abdominal wall subcutaneous and muscular edema may occur, but this usually gradually resolves spontaneously [9]. A proper planning and real-time MR imaging prevent target exposures of adjacent organs. Injuries may be also avoided by filling or emptying the bladder via a catheter and/or filling the rectum via a rectal tube for displacing structures such as bowel loops or ovaries out of the HI-FU beam path or to displace the area of treatment with respect to a skin scar; in some cases, the use of bowel mitigation techniques has been also described to treat patients with small fibroids [8]. Far field side effects and risks are attributed to bone and nerves heating due to HI-FU absorbance. During the planning stage of the procedure, these structures are carefully marked (see Figure 2) and, for safety reasons, the minimal distance between the posterior treatment area and bone surface should be no less than 4 cm [10–12]. Potential risk of deep venous thrombosis from lying in a supine position for up to 3-4 hours is substantially mitigated by the use of compression stockings.

We calculated the NPV with an independent workstation (Apple iMac 27″ late 2013) using the region of interest tool of the Horos software (⟨https://www.horosproject.org/⟩), a free, open source medical image viewer software based upon OsiriX and other open source medical imaging libraries available under the GNU Lesser General Public License, Version 3 (LGPL-3.0).

An MRI examination follow-up was scheduled at 3, 6, and 12 months after treatment to evaluate the amount of reduction in fibroid volume and to assess the presence of residual viable tissue. The Symptom Severity Score (SSS) was assessed in all the patients who have signed the additional informed consent for research purposes at the time of enrolment and on each MRI follow-up, by the use of the Uterine Fibroids Symptom and Quality of Life (UFS-QoL) questionnaire [13].

We present a selection of five patients with symptomatic uterine fibroids treated with MRgFUS at our center. These cases have distinctive features that offer good points for discussion.

3. Case 1

A 43-year-old nulliparous female presented with heavy menstrual bleeding, anemia, pelvic pain, and infertility. The screening MRI showed multiple fibroids (at least six); the three largest ranged from 3.5 cm to 5.7 cm (total volume 107.18 cc) and were hypointense on T2-w images (relative to the uterine wall) and viable on postcontrast T1-w images that were considered clinically significant for patient symptoms (SSS = 61).

In this case, the rectum was filled with about 250 cc of US gel to displace the uterus anteriorly and bowel out of the treatment path. T2-w images were acquired for treatment planning. Three ROT were then defined on the targeted fibroids. Treatment duration was 3 h 55 min (calculated from the first to the last sonication); 102 sonications were performed with an HI-FU energy range of 1600 to 4500 J. The treatment protocol used included small, large, and elongated sonication spots depending on the size of the targeted fibroid. Temperature achieved was in the range of 65–90°C. Contrast-enhanced images were acquired at the end of the treatment and showed a total nonperfused volume (NPV) of 80.45 cc, achieving an NPV ratio (NPV ratio = nonperfused volume/perfused volume expressed as percentage) of 75%.

This case illustrates that smaller fibroids located in the near field and not directly treated may show as nonperfused fibroids after treatment (white arrows in Figure 3(a)). A possible hypothesis to explain this finding may lie in damage to feeding vessels shared between targeted fibroid and nontargeted fibroids [14] or, more likely, to a transient vasospasm of the feeding vessels of nontargeted fibroids, since, in a subsequent MRI follow-up, the untargeted fibroids showed back a normal vascularization (black arrows in Figure 3(b)). This finding is not common but should be taken into account when treating patients with similar conditions as it could mislead the physician. At 3, 6, and 12 months of follow-up, her SSS score has reduced to 23, 22, and 18 points, respectively, and the reduction in total tumor size, as a percentage of initial tumor volume, was 42%, 46%, and 50%, respectively.

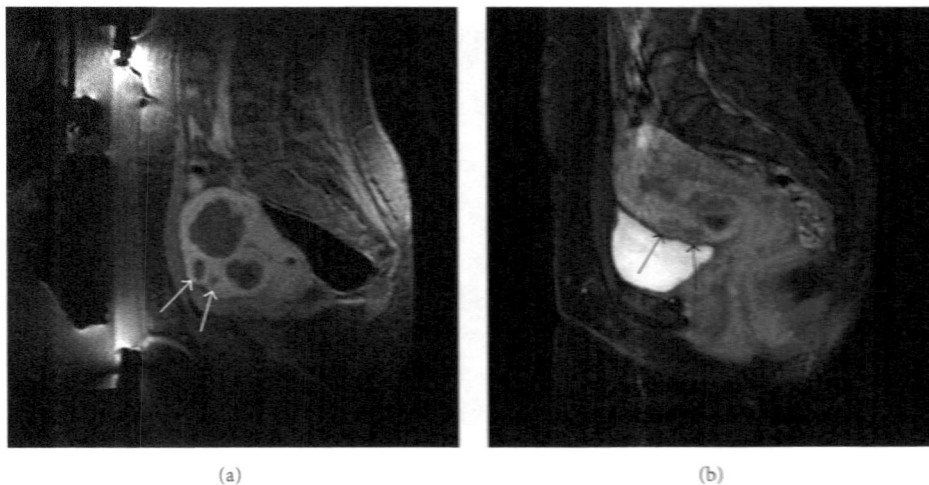

(a)   (b)

FIGURE 3: Case 1: (a) sagittal T1-w with fat saturation acquired after intravenous injection of paramagnetic contrast medium showing the two small fibroids (white arrows) located in the near field that were not directly treated but that became nonperfused too after the treatment; two bigger fibroids treated are visible too; (b) a follow-up MRI (T1-w with fat saturation acquired after intravenous injection of paramagnetic contrast medium) showing a normal vascularization of the two small fibroids.

4. Case 2

A 33-year-old female presented with menometrorrhagia and infertility (SSS = 67). The screening MRI showed multiple small fibroids, at least 11, of which there was only one treatable (size: 5 cm; volume: 51.50 cc) intramural fibroid, homogenously hypointense on T2-w images (asterisk in Figure 4), involving the right side of the fundus of the uterus with an heterogeneous enhancement on postcontrast T1-w images. An abdominal scar and some bowel loops were preventing a direct targeting of the region of treatment, but some expedients have allowed us to effectively treat this patient. The abdominal scar was covered by an energy-blocking scar patch to avoid the risk of any skin burns. Bowel obstructing the beam path was mitigated using a custom sliced 45 mm thick cylindrical gel pad (a circular segment equal to about one-sixth of the diameter was sliced off from the cranial part; see dashed line in Figure 4); furthermore, rectum was filled with about 300 cc of US gel and bladder with about 350 cc of saline. Bowel loops were able to be displaced superiorly and out of the beam pathway (curved arrow in Figure 4). The treatment was performed through the full bladder (B in Figure 4) demonstrating how even challenging cases can be treated if the physician succeeds to find a way to expose the ROT. In this patient, we defined only one ROT, around the biggest fibroid, because the other smaller fibroids were not accessible and smaller than 3 cm.

The fibroid was treated in 1 h 40 min, using 43 sonications (energy range: 1100–2250 J; temperature range: 59–102°C). The treatment protocol used included large and elongated spots. Posttreatment, a nonperfused volume of 46 cc and an NPV ratio of 89% were calculated. At 6 months of follow-up, her SSS score has reduced to 35 points and the shrinkage of the treated fibroid was 46% compared to initial fibroid volume.

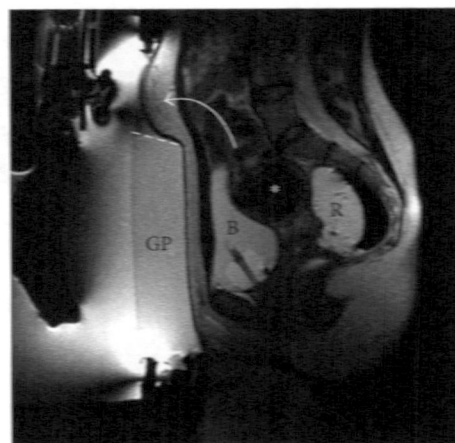

FIGURE 4: Case 2: sagittal T2-w intraoperative scan showing the targeted fibroid (asterisk) and full bladder (B) with urinary catheter present. Bowels obstructing the beam path were mitigated using a custom sliced (dashed line) gel pad (GP) and rectal filling with ultrasound gel (R). Curved arrow mimics the dislocation of the bowel loops. This treatment was successfully performed through the full bladder.

5. Case 3

A 45-year-old female presented with dysmenorrhea, heavy bleeding, anemia, and pelvic pain (SSS scored 66 points). MRI of the pelvis showed two fibroids, both hypointense on T2-w images; the largest one (diameter 7.3 cm; volume 164.07 cc) was intramural, involving the right side of the fundus of the uterus and the other one was a very small (diameter 2.5 cm; volume 5.73 cc) intracavitary submucosal fibroid (white arrow in Figure 5(a)). Postgadolinium contrast images

(a)

(b)

(c)

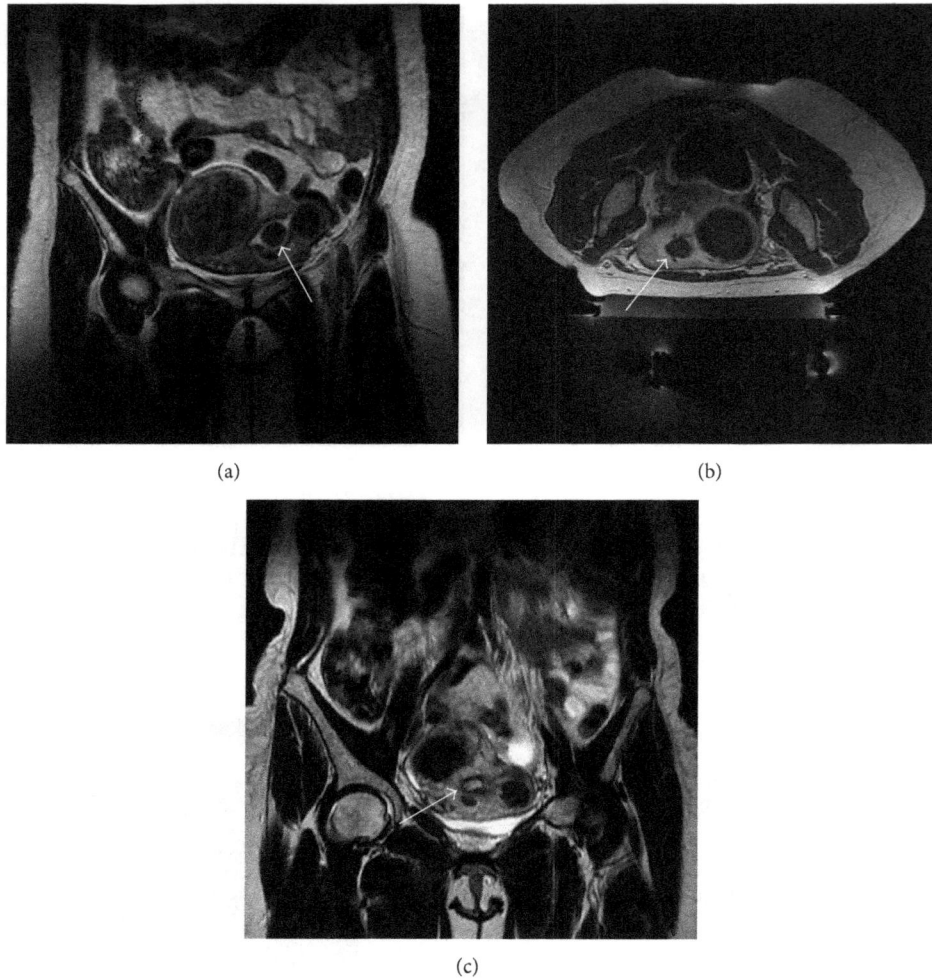

FIGURE 5: Case 3: (a) screening coronal T2-w showing the bigger intramural fibroid on the right wall of the uterus, the small intracavitary submucosal fibroid on the mid left (white arrow), and one more small intramural fibroid on the left wall of the uterus; (b) axial T1-w acquired after intravenous injection of paramagnetic contrast medium acquired at the end of the treatment showing treated fibroids as nonenhancing round lesions (white arrow on the small intracavitary submucosal fibroid); (c) in a follow-up MRI, the small intracavitary submucosal fibroid was no more appreciable; the white arrow shows the "empty" uterine cavity after spontaneous vaginal expulsion.

revealed homogeneous enhancement of both fibroids. Rectum was filled with 250 cc of US gel bringing the uterus closer to the anterior abdominal wall.

Treatment duration was 2 h 15 min, using 82 sonications with an energy range of 1700–5100 J. Temperature achieved was in the range of 60–100°C.

Posttreatment MR images revealed a nonperfused volume of 125.85 cc (NPV ratio of 77%) in the intramural fibroid and a nonperfused volume of 6.01 cc (NPV ratio > 100%) in the intracavitary fibroid.

This case shows that small intracavitary fibroids may be treated safely; ablating an intracavitary fibroid (white arrow in Figure 5(b)) and its stalks will cause its disconnection from uterine wall, followed by spontaneous vaginal expulsion without any complications (see the 3-month follow-up MRI shown in Figure 5(c)).

At 3, 6, and 12 months of follow-up, her SSS score has reduced to 26, 23, and 19 points, respectively, and the shrinkage of the treated intramural fibroid was 40%, 42%, and 49%, respectively, compared to initial fibroid volume.

6. Cases 4 and 5

Both 38-year-old (case 4) and 44-year-old (case 5) multiparous females presented with heavy and prolonged menstrual flow, pelvic pain, and pressure. Their SSS were 68 and 65, respectively.

The patient in case 4 had a screening MRI demonstrating a single fibroid (3.6 cm in size with volume of 15.19 cc; Figure 6(a)), involving the fundus of the uterus; patient in case 5 showed two fibroids, of which only the larger anterior one (diameter 4 cm; volume 21.45 cc; Figure 7(a)) was

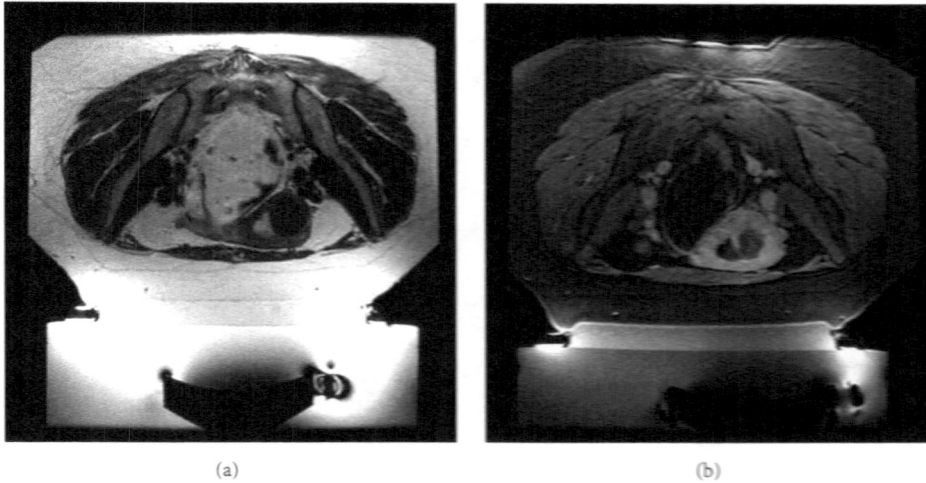

FIGURE 6: Case 4: example of a treatment with a low ablation rate (NPV = 57%); (a) screening axial T2-w scan; (b) sagittal T1-w with fat saturation acquired after intravenous injection of paramagnetic contrast medium acquired at the end of the treatment.

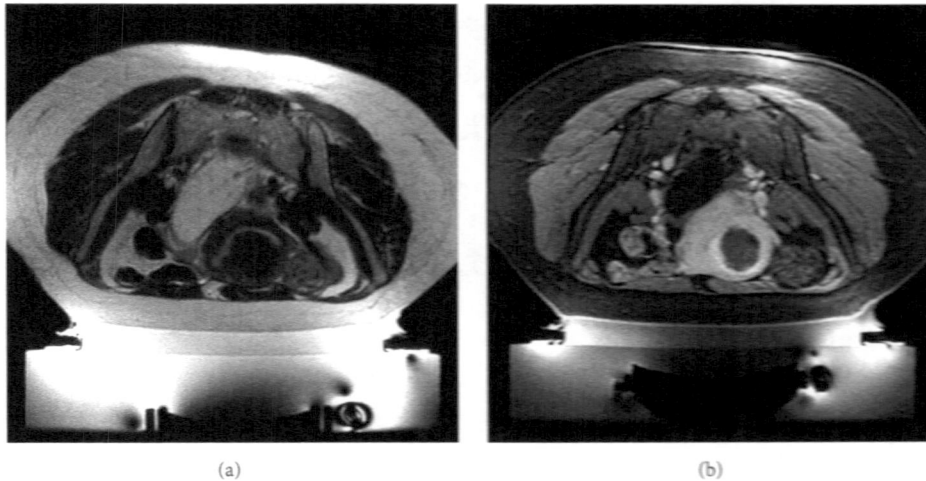

FIGURE 7: Case 5: example of a treatment with a very high ablation rate (NPV = 99%); (a) screening axial T2-w scan; (b) sagittal T1-w with fat saturation acquired after intravenous injection of paramagnetic contrast medium acquired at the end of the treatment.

treatable because the smaller one was posterior and not fully accessible. In both cases, suitable fibroids were homogenously hypointense on T2-w images and showed homogenous enhancement on postcontrast T1-w images. Patient in case 4 had an abdominal transverse scar that was covered by a scar patch. This patient was treated after filling the rectum with 250 cc of US gel, while patient in case 5 was treated with a partially full bladder to displace bowels superiorly.

Patient in case 4 was treated in 1 h 55 min, using 36 sonications with an energy range of 1259–2770 J using large and elongated spots. Temperature achieved was in the range of 39–96°C. Posttreatment images showed a nonperfused volume of 8.62 cc and NPV ratio of 57% (Figure 6(b)). At 3, 6, and 12 months of follow-up, there was a significant decrease in her SSS score to 21, 19, and 12 points, respectively, and fibroid size had been reduced from baseline by 27%, 62%, and 81%, respectively.

Patient in case 5 was treated in 1 h 15 min, using 31 sonications with an energy range of 1051–446 J using small, large, and elongated spots. Temperature achieved was in the range of 45–97°C. Posttreatment, a nonperfused volume of 21.30 cc and NPV ratio of 99% were achieved (Figure 7(b)). At 3, 6, and 12 months of follow-up, her SSS has reduced to 58, 55, and 52 points, respectively, and the shrinkage of the treated fibroid from baseline was 46%, 55%, and 58%, respectively.

Despite similar symptoms and SSSs, patient in case 4 resulted in only a 57% NPV ratio with a significant decrease in her SSS (from 68 to 21 at the 3-month follow-up) if compared to patient in case 5 that resulted in a 99% NPV ratio with only a slight reduction in her SSS (from 65 to 52 even one year after the treatment). We presented these two cases together to highlight how there is not always a clear correlation between reduction in symptoms and reduction in tumor size even in patients with similar symptoms and SSS.

7. Discussion

Several studies have been published focusing on patient symptomatic relief and fibroid shrinkage.

In our experience, all the treatments were successfully completed without complications and the patients were discharged 30 min later after the anesthesiologist confirmed a full recovery from the conscious sedation protocol used. The following day patients returned to their normal daily activities and job outside their home when applicable.

Although this is not always true, in our practice, we have often observed that patients with larger treated volumes have reported an higher degree of symptom improvement and required no alternative treatments. This trend has been already reported in the English literature [15]. Thus, it is fair to always aim at the highest nonperfused volume (at least 60% for a successful treatment), as there is a close relationship between the NPV and clinical outcomes [16]. However, our report suggests that the extent of change in both volume and symptom score varies largely among the patients and there is not always a close correlation between the volume of ablated tissue and the clinical outcome. Nevertheless, there are several predictors of success of an MRgFUS treatment, namely, low signal intensity on T2-weighted images before treatment and nonperfused tissue volume over at least 20% [7].

MRgFUS for uterine fibroids offers several advantages over other fibroid therapies including the following: it is an outpatient procedure, requiring no hospitalization, has a very low complication rate, utilizes no ionizing radiation, and allows the patient to return to her normal daily activities the day after treatment. Furthermore, patients need only conscious sedation allowing patients to communicate with the treating physician and halt treatment if needed.

MRgFUS has a good safety profile. Careful supervision by the operating physician and detailed preparation of the patient are important in minimizing uncommon severe complications as bowel perforation, skin burns, and deep vein thrombosis. It can be carried out even in anemic patients and those who are at risk for surgical operation. It represents an effective and almost noninvasive approach that may be used as a fertility-preserving technique [4]. A number of reports of successful treatment have been published [17–19] showing the feasibility of pregnancy following MRgFUS.

8. Conclusion

We have presented a selection of five cases of adult females with symptomatic uterine fibroids treated with MRgFUS. The cases we presented had some distinctive features of interest ranging from technical measures which enabled establishing a safe treatment pathway to uncommon posttreatment findings and demonstration of variable clinical response that does not always correlate with the percentage of tumor ablation calculated. As for all newly introduced therapeutic procedures, many points will be clarified only with further investigations, additional research, and multicenter studies.

MRgFUS represents an effective and almost noninvasive approach that may be used as a fertility-preserving technique for the treatment of symptomatic uterine fibroids. It is a disruptive technology with a remarkable potential in terms of improved outcomes, safety and efficacy, faster recovery, and sustained symptomatic relief and, in the near future, it could cause a sea change in current clinical surgical care. However, this will really happen only when health regulatory systems begin to include this procedure among those reimbursable to extend this treatment option to undeserved patients.

Nowadays, minimally invasive or noninvasive surgical applications are steadily increasing thanks to the continuous technological developments that characterize this particular historical period. Increasingly, imaging-based methods today are used not only for a preliminary assessment or a conventional preoperative planning of many surgical procedures but also for live monitoring too.

MRgFUS is one of the emerging technologies offering treating providers image guidance and thermal monitoring. MRgFUS is a noninvasive safe and effective treatment option for uterine fibroids [20]. The feasibility and the safety of MRgFUS have been tested in a growing number of clinical studies and applications such as painful bone metastases [21], musculoskeletal diseases [22], and various other benign and malignant tumors, like breast cancer [23], bone tumors [24], prostate [25], liver [26], and pancreatic cancer [26, 27] and, more recently, it has been used for noninvasive transcranial applications too [28–32]. MRgFUS is evolving as an alternative or complementary therapy to current treatment options such as surgery, radiotherapy, other thermal ablation procedures, gene therapy, and chemotherapy.

Acknowledgments

The research described in this paper was supported by Project "Programma Operativo Nazionale (PON) 2007/2013" no. PON a3_00011 (Project Leader: Professor C. Catalano) funded by MIUR (Italian Ministry of Education, University and Research). Dr. A. Napoli has been a researcher within the Projects "PON 2007/2013" no. PON01_01059 and no. PON a3_00011 (Project Leader: Professor C. Catalano) funded by MIUR. Dr. C. Gagliardo has been a researcher within the Project "PON 2007/2013" no. PON01_01059 (Project Leader: Professor C. Catalano) funded by MIUR.

References

[1] S. Okolo, "Incidence, aetiology and epidemiology of uterine fibroids," *Best Practice & Research: Clinical Obstetrics & Gynaecology*, vol. 22, no. 4, pp. 571–588, 2008.

[2] B. S. Levy, "Modern management of uterine fibroids," *Acta Obstetricia et Gynecologica Scandinavica*, vol. 87, no. 8, pp. 812–823, 2008.

[3] O. Istre, "Management of symptomatic fibroids: conservative surgical treatment modalities other than abdominal or laparoscopic myomectomy," *Best Practice and Research: Clinical Obstetrics and Gynaecology*, vol. 22, no. 4, pp. 735–747, 2008.

[4] A. Napoli, M. Anzidei, F. Ciolina et al., "MR-guided high-intensity focused ultrasound: current status of an emerging technology," *CardioVascular and Interventional Radiology*, vol. 36, no. 5, pp. 1190–1203, 2013.

[5] J. W. Jenne, T. Preusser, and M. Günther, "High-intensity focused ultrasound: principles, therapy guidance, simulations and applications," *Zeitschrift fur Medizinische Physik*, vol. 22, no. 4, pp. 311–322, 2012.

[6] S.-W. Yoon, C. Lee, S. H. Cha et al., "Patient selection guidelines in MR-guided focused ultrasound surgery of uterine fibroids: A pictorial guide to relevant findings in screening pelvic MRI," *European Radiology*, vol. 18, no. 12, pp. 2997–3006, 2008.

[7] Z. M. Lénárd, N. J. McDannold, F. M. Fennessy et al., "Uterine leiomyomas: MR imaging-guided focused ultrasound surgery-imaging predictors of success," *Radiology*, vol. 249, no. 1, pp. 187–194, 2008.

[8] I. Mindjuk, C. G. Trumm, P. Herzog, R. Stahl, and M. Matzko, "MRI predictors of clinical success in MR-guided focused ultrasound (MRgFUS) treatments of uterine fibroids: results from a single centre," *European Radiology*, vol. 25, no. 5, pp. 1317–1328, 2015.

[9] G. K. Hesley, K. R. Gorny, and D. A. Woodrum, "MR-guided focused ultrasound for the treatment of uterine fibroids," *CardioVascular and Interventional Radiology*, vol. 36, no. 1, pp. 5–13, 2013.

[10] J. Hindley, W. M. Gedroyc, L. Regan et al., "MRI guidance of focused ultrasound therapy of uterine fibroids: early results," *American Journal of Roentgenology*, vol. 183, no. 6, pp. 1713–1719, 2004.

[11] E. A. Stewart, J. Rabinovici, C. M. C. Tempany et al., "Clinical outcomes of focused ultrasound surgery for the treatment of uterine fibroids," *Fertility and Sterility*, vol. 85, no. 1, pp. 22–29, 2006.

[12] A. Roberts, "Magnetic resonance-guided focused ultrasound for uterine fibroids," *Seminars in Interventional Radiology*, vol. 25, no. 4, pp. 394–405, 2008.

[13] J. B. Spies, K. Coyne, N. Guaou Guaou, D. Boyle, K. Skyrnarz-Murphy, and S. M. Gonzalves, "The UFS-QOL, a new disease-specific symptom and health-related quality of life questionnaire for leiomyomata," *Obstetrics and Gynecology*, vol. 99, no. 2, pp. 290–300, 2002.

[14] N. McDannold, C. M. Tempany, F. M. Fennessy et al., "Uterine leiomyomas: MR imaging- based thermometry and thermal dosimetry during focused ultrasound thermal ablation," *Radiology*, vol. 240, no. 1, pp. 263–272, 2006.

[15] E. A. Stewart, B. Gostout, J. Rabinovici, H. S. Kim, L. Regan, and C. M. C. Tempany, "Sustained relief of leiomyoma symptoms by using focused ultrasound surgery," *Obstetrics and Gynecology*, vol. 110, no. 2, Part 1, pp. 279–287, 2007.

[16] C. M. Tempany, "From the RSNA refresher courses: Image-guided thermal therapy of uterine fibroids," *Radiographics*, vol. 27, no. 6, pp. 1819–1826, 2007.

[17] L. P. Gavrilova-Jordan, C. H. Rose, K. D. Traynor, B. C. Brost, and B. S. Gostout, "Successful term pregnancy following MR-guided focused ultrasound treatment of uterine leiomyoma," *Journal of Perinatology*, vol. 27, no. 1, pp. 59–61, 2007.

[18] Y. Morita, N. Ito, and H. Ohashi, "Pregnancy following MR-guided focused ultrasound surgery for a uterine fibroid," *International Journal of Gynecology and Obstetrics*, vol. 99, no. 1, pp. 56–57, 2007.

[19] J. Rabinovici, M. David, H. Fukunishi, Y. Morita, B. S. Gostout, and E. A. Stewart, "Pregnancy outcome after magnetic resonance-guided focused ultrasound surgery (MRgFUS) for conservative treatment of uterine fibroids," *Fertility and Sterility*, vol. 93, no. 1, pp. 199–209, 2010.

[20] F. Ciolina, L. Manganaro, R. Scipione, and A. Napoli, "Alternatives to surgery for the treatment of myomas," *Minerva Ginecologica*, vol. 68, no. 3, pp. 364–379, 2016.

[21] M. D. Hurwitz, P. Ghanouni, S. V. Kanaev et al., "Magnetic resonance-guided focused ultrasound for patients with painful bone metastases: phase III trial results," *JNCI Journal of the National Cancer Institute*, vol. 106, no. 5, 2014.

[22] A. Bazzocchi, A. Napoli, B. Sacconi et al., "MRI-guided focused ultrasound surgery in musculoskeletal diseases: the hot topics," *British Journal of Radiology*, vol. 89, no. 1057, Article ID 20150358, 2015.

[23] B. Cavallo Marincola, F. Pediconi, M. Anzidei et al., "High-intensity focused ultrasound in breast pathology: non-invasive treatment of benign and malignant lesions," *Expert Review of Medical Devices*, vol. 12, no. 2, pp. 191–199, 2015.

[24] A. Napoli, M. Mastantuono, B. C. Marincola et al., "Osteoid osteoma: MR-guided focused ultrasound for entirely noninvasive treatment," *Radiology*, vol. 267, no. 2, pp. 514–521, 2013.

[25] A. Napoli, M. Anzidei, C. de Nunzio et al., "Real-time magnetic resonance-guided high-intensity focused ultrasound focal therapy for localised prostate cancer: preliminary experience," *European Urology*, vol. 63, no. 2, pp. 395–398, 2013.

[26] M. Anzidei, A. Napoli, F. Sandolo et al., "Magnetic resonance-guided focused ultrasound ablation in abdominal moving organs: a feasibility study in selected cases of pancreatic and liver cancer," *Cardiovascular and Interventional Radiology*, vol. 37, no. 6, pp. 1611–1617, 2014.

[27] M. Anzidei, B. C. Marincola, M. Bezzi et al., "Magnetic resonance-guided high-intensity focused ultrasound treatment of locally advanced pancreatic adenocarcinoma: preliminary experience for pain palliation and local tumor control," *Investigative Radiology*, vol. 49, no. 12, pp. 759–765, 2014.

[28] D. Coluccia, J. Fandino, L. Schwyzer et al., "First noninvasive thermal ablation of a brain tumor with MR-guided focused ultrasound," *Journal of Therapeutic Ultrasound*, vol. 16, no. 2, article 17, 2014.

[29] W. J. Elias, D. Huss, T. Voss et al., "A pilot study of focused ultrasound thalamotomy for essential tremor," *The New England Journal of Medicine*, vol. 369, no. 7, pp. 640–648, 2013.

[30] D. Jeanmonod, B. Werner, A. Morel et al., "Transcranial magnetic resonance imaging-guided focused ultrasound: noninvasive central lateral thalamotomy for chronic neuropathic pain," *Neurosurgical Focus*, vol. 32, no. 1, article E1, 2012.

[31] D. G. Iacopino, R. Giugno, A. Maugeri et al., "Is there still a role for lesioning in functional neurosurgery: the Italian experience of delivering focused ultrasound high energy through a 1,5 tesla MR apparatus," *Journal of Neurosurgical Sciences*, in Press.

[32] C. Gagliardo, L. Geraci, A. Napoli et al., "Non-invasive transcranial magnetic resonance imaging-guided focused ultrasounds surgery to treat neurologic disorders," *Recenti Progressi in Medicina*, vol. 107, pp. 1–4, 2016.

14

Late Type 3b Endoleak with an Endurant Endograft

Mehmet Barburoglu,[1] **Bulent Acunas,**[1] **Yilmaz Onal,**[1] **Murat Ugurlucan,**[2]
Omer Ali Sayin,[2] **and Ufuk Alpagut**[2]

[1]*Istanbul Medical Faculty, Istanbul University, Department of Radiology, Millet Caddesi, Capa, Fatih, 34390 Istanbul, Turkey*
[2]*Istanbul Medical Faculty, Istanbul University, Department of Cardiovascular Surgery, Istanbul, Turkey*

Correspondence should be addressed to Mehmet Barburoglu; barburmehmet@gmail.com

Academic Editor: Daniel P. Link

Endovascular stent grafting with different commercially available stent graft systems is widely applied for the treatment of abdominal aortic aneurysms with high success rates in the current era. Various types of endoleaks are potential complications of the procedure. They usually occur in the early period. In this report, we present type 3b endoleak occurring 14 months after a successful endovascular abdominal aortic aneurysm repair with a Medtronic Endurant stent graft.

1. Introduction

Endovascular abdominal aortic aneurysm repair (EVAR) is a safe and efficient method, which has been widely used for treatment at abdominal aortic aneurysms (AAA). New generation endografts make it possible to treat more complex AAA with less perioperative mortality and systemic complications [1–4]. However, endoleak is still a common problem of EVAR that is characterized by persistent blood flow within the aneurysm sac [5, 6]. White et al. classified endoleaks into four types and type 3b endoleak originates from a defect in the graft fabric [7].

We report a case of late type 3b endoleak from a tear in the main body of an Endurant (Medtronic Endovascular, Santa Rosa, Calif., USA) endograft 14 months after deployment and treatment with coil embolization. Consent from the patient and his family for possible academic activities including publication acts regarding his medical history was obtained. To the best of our knowledge, this is the first report of a late type 3b endoleak with an Endurant endograft.

2. Case Report

An 81-year-old man underwent EVAR for the treatment of a 70 mm AAA with deployment of 36 × 16 × 166 mm main body Endurant bifurcated endograft, 16 × 20 × 120 mm contralateral limb, and extensions of 16 × 16 × 95 mm to the right and to 20 × 20 × 80 mm to the left. His medical history included heart failure with an ejection fraction of 25%, hypertension, chronic obstructive pulmonary disease, and chronic renal failure requiring hemodialysis. He has been followed up regularly with Doppler ultrasonography and computerized tomography (CT) angiography at 3, 6, and 12 months showed successful exclusion of the aneurysm sac with no leak. Bilateral internal iliac arteries were patent (Figures 1(a) and 1(b)).

About 14 months after EVAR, the patient was admitted to the emergency clinic with sudden onset back pain. Clinical examination revealed a pulsatile mass at the epigastrium. Computerized tomography showed enlarged AAA bounded with hematoma and an endoleak originating from distal main body of the EVAR graft (Figure 2(a)).

The patient was taken to the angiography unit in the emergency night conditions. Several angiography runs were performed to determine the location of the endoleak; type 1a endoleak was excluded. The pigtail catheter was pulled into the main body. Angiogram showed endoleak at the level of the bifurcation endograft (Figure 2(b)). This time a type 3a endoleak (disconnection of main graft components) was suspected and 16 × 20 × 120 mm stent graft was placed from the contralateral limb. Control angiogram showed

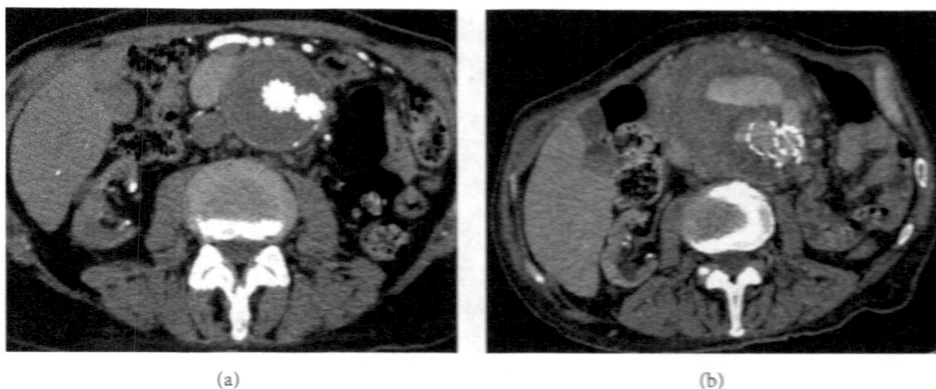

FIGURE 1: (a) 12-month follow-up CT scan indicating complete exclusion of the aneurysm with sac shrinkage after the initial procedure. (b) CT angiography showing enlarged AAA with hematoma and endoleak originating from distal main body of the EVAR graft.

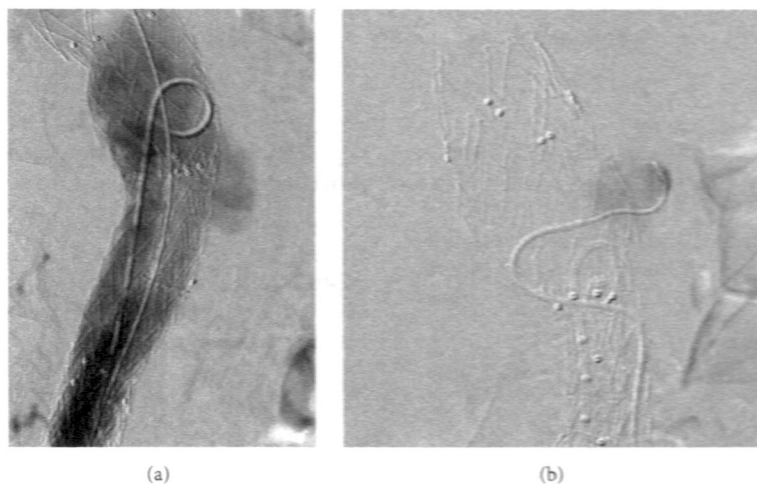

FIGURE 2: (a) Angiogram showed endoleak at the level of the bifurcation endograft. (b) Direct catheterization of the aneurysm sac through the tear of the graft.

persistent endoleak; however, this time a type 3b endoleak was suspected.

We discussed the treatment options with the cardiovascular surgeons and we decided to treat the patient endovascularly because of the high mortality risk of the surgery. Endovascular options for the treatment included placement of a new aorto-uni-iliac or bi-iliac stent graft or coil embolization of the sac through translumbar or transarterial approach. We decided to deploy an aorto-uni-iliac stent graft, as there was not enough distance to deploy an aorto-bi-iliac graft between the bifurcation of the previous endograft and ostia of the renal arteries. In addition, we lacked enough and appropriate sized coils for the embolization of the sac in the night conditions. We deployed an Endurant uni-iliac stent graft with a body diameter of 36 mm (36 × 14 × 105 mm, Endurant, Medtronic Endovascular, Santa Rosa, Calif., USA). The control angiogram showed persistent endoleak despite additional ballooning. The antegrade flow was still present into the right limb decoding incomplete apposition of the latter stent graft to the previous one. The patient was clinically stable and there

was no marked filling into the aneurysm sac so we decided to end the procedure and evaluate the filling of the right limb with a postoperative CT examination.

The second day CT angiography showed inaccurate apposition of the new stent graft into the previous stent graft because of the inadequate diameter of the aorto-uni-iliac stent graft; 36 mm is the largest diameter in the market. The gap between the stent grafts allowed persistent blood flow into the right limb and into the aneurysm sac through the fabric tear.

We decided to coil-embolize the fabric tear and aneurysm sac through the gap between the two-stent grafts after providing the appropriate coils sizes. The gap was catheterized with 5 F vertebral catheter rather than a 2.7 F microcatheter inserted into the aneurysm sac through the fabric tear (Figure 2(b)). We coiled the aneurismal sac with detachable coils until the angiography showed the complete disappearance of type 3b endoleak (Figure 3(a)). Control CT examination denoted no contrast filling into the sac; there was peripheral high-density area in the sac, which could be related to fresh

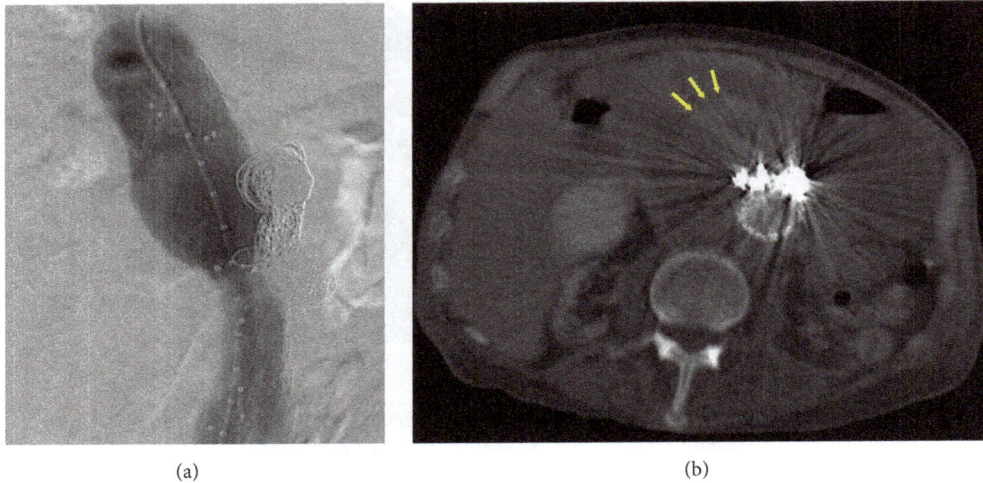

(a)   (b)

FIGURE 3: (a) Angiogram showed the complete disappearance of type 3b endoleak. (b) CT angiography showing peripheral high-density area in the sac, which could be related to fresh clot or residual contrast media during coils embolization.

clot or residual contrast media during coils embolization (Figure 3(b)). Doppler ultrasonography also confirmed no filling in the sac.

The procedure was terminated and the patient was transferred to the intensive care unit. Unfortunately, the patient was lost due to severe heart failure 72 hours after the second procedure.

3. Discussion

Type 3 endoleak is a rare complication but potentially has a high risk of aneurismal rupture and always warrants urgent intervention [8]. Type 3 endoleak divides into two types: type 3a originates from disconnection of the graft compounds and type 3b originates from fabric tear [7]. In our patient, the endoleak was type 3b confirmed with direct sac catheterization through the defect with a microcatheter. We did not know the reason for the fabric defect after 14 months after the deployment. In the literature, there are reports about type 3b endoleaks with other aortic stent graft devices [9–11] but there was only one case report about early type 3b endoleak associated with Endurant endograft [12]. Our case is the first case of a late type 3b endoleak reported with an Endurant stent graft. Medtronic was informed of the case and provided no explanation.

Type 3 endoleak can be treated endovascularly or surgically. Endovascular options include repairing the defect with aortic cuff extension or placement of a new aorto-bi-iliac graft or placement of an aorto-uni-iliac graft with cross-femoral bypass. Our patient came with ruptured aneurysm and he was unstable clinically. Then we evaluated his CT angiogram; we suggested a type 3a endoleak from contralateral leg disconnection. Later type 3b endoleak from a fabric tear was proved.

We tried to close the defect with an aorto-uni-iliac graft but there was bad apposition between the new and the previous grafts despite several balloon dilatations. Therefore, we closed the endoleak with coiling the aneurismal sac

through the defect. However, the patient died, 3 days after the procedure.

In conclusion, type 3b endoleak can occur on new generation endografts and it is associated with high-risk aneurysm rupture morbidity and mortality. It is difficult to diagnose the type of the endoleak only with CT angiogram without catheter angiography.

References

[1] I. E. Steingruber, B. Neuhauser, R. Seiler et al., "Technical and clinical success of infrarenal endovascular abdominal aortic aneurysm repair: a 10-year single-center experience," *European Journal of Radiology*, vol. 59, no. 3, pp. 384–392, 2006.

[2] M. Prinssen, E. L. G. Verhoeven, J. Buth et al., "A randomized trial comparing conventional and endovascular repair of abdominal aortic aneurysms," *The New England Journal of Medicine*, vol. 351, no. 16, pp. 1607–1618, 2004.

[3] J. N. Albertini, T. Perdikides, C. V. Soong et al., "Endovascular repair of abdominal aortic aneurysms in patients with severe angulation of the proximal neck using a flexible stent-graft: European Multicenter Experience," *Journal of Cardiovascular Surgery*, vol. 47, no. 3, pp. 245–250, 2006.

[4] EVAR Trial Participants, "Endovascular aneurysm repair versus open repair in patients with abdominal aortic aneurym (EVAR trial 1): randomised controlled trial," *The Lancet*, vol. 365, no. 9478, pp. 2179–2186, 2005.

[5] M. Chane and R. R. Heuser, "Review of interventional repair for abdominal aortic aneurysm," *Journal of Interventional Cardiology*, vol. 19, no. 6, pp. 530–538, 2006.

[6] F. Pozzi Mucelli, M. Doddi, S. Bruni, R. Adovasio, F. Pancrazio, and M. Cova, "Endovascular treatment of endoleaks after endovascular abdominal aortic aneurysm repair: personal experience," *Radiologia Medica*, vol. 112, no. 3, pp. 409–419, 2007.

[7] G. H. White, W. Yu, J. May, X. Chaufour, and M. S. Stephen, "Endoleak as a complication of endoluminal grafting of abdominal aortic aneurysms: classification, incidence, diagnosis, and management," *Journal of Endovascular Surgery*, vol. 4, no. 2, pp. 152–168, 1997.

[8] G. H. White, J. May, R. C. Waugh, X. Chaufour, and W. Yu, "Type III and type IV endoleak: toward a complete definition of blood flow in the sac after endoluminal AAA repair," *Journal of Endovascular Surgery*, vol. 5, no. 4, pp. 305–309, 1998.

[9] H. Banno, H. Morimae, T. Ihara, M. Kobayashi, K. Yamamoto, and K. Komori, "Late type III endoleak from fabric tears of a zenith stent graft: report of a case," *Surgery Today*, vol. 42, no. 12, pp. 1206–1209, 2012.

[10] A. Wanhainen, R. Nyman, M.-O. Eriksson, and M. Björck, "First report of a late type III endoleak from fabric tears of a Zenith stent graft," *Journal of Vascular Surgery*, vol. 48, no. 3, pp. 723–726, 2008.

[11] A. Teutelink, M. J. van der Laan, R. Milner, and J. D. Blankensteijn, "Fabric tears as a new cause of type III endoleak with Ancure endograft," *Journal of Vascular Surgery*, vol. 38, no. 4, pp. 843–846, 2003.

[12] I. Abouliatim, D. Gouicem, H. Kobeiter, M. Majeski, and J.-P. Becquemin, "Early type III endoleak with an Endurant endograft," *Journal of Vascular Surgery*, vol. 52, no. 6, pp. 1665–1667, 2010.

15

Solitary Spinal Epidural Metastasis from Prostatic Small Cell Carcinoma

Kyung Ryeol Lee[1] and Young Hee Maeng[2]

[1]*Department of Radiology, Jeju National University Hospital, Aran 13gil 15 (Ara-1-Dong), Jeju-si,*
Jeju Special Self-Governing Province 63241, Republic of Korea
[2]*Department of Pathology, Jeju National University Hospital, Aran 13gil 15 (Ara-1-Dong), Jeju-si,*
Jeju Special Self-Governing Province 63241, Republic of Korea

Correspondence should be addressed to Kyung Ryeol Lee; wel977@naver.com

Academic Editor: Kazushige Hayakawa

Solitary, spinal epidural metastasis (SEM) that is not related to vertebral metastasis is very rare. And solitary SEM from prostatic cancer is rarely found in previously published reports. However, it is clinically significant due to the possibility of neurologic dysfunction, and it can be assessed by MRI. In this report, we show a case of solitary SEM arising from prostatic small cell carcinoma detected by MRI.

1. Introduction

Metastasis to the spinal nervous system can be classified as intramedullary, epidural, and leptomeningeal metastases. Most of the epidural metastases in the spine are associated with vertebral metastases [1]. Solitary spinal epidural metastasis (SEM) which is not related to vertebral metastasis is very rare. However, the diagnosis of solitary SEM is very important because it can result in neurological complications. MRI is the imaging modality of choice for the assessment of solitary SEM [2]. We report a rare case of solitary SEM from prostatic small cell carcinoma detected by MRI.

2. Case Presentation

A 68-year-old man diagnosed two years ago with adenocarcinoma of the prostate came to our hospital because of low back pain. At that time, he had already received hormone therapy for two years. His low back pain started three years ago, although it became severely aggravated two weeks ago before visiting the hospital. His past medical history included hypertension, although he had no history of another cancer except for prostate cancer. On physical examination, hypoesthesia

was detected in the L5 dermatome. His laboratory findings were within a normal range including his PSA level.

As his attending urologist suspected that his chronic low back pain was due to a degenerative disease such as spinal stenosis, he recommended MR imaging which showed a space-occupying lesion in the posterior epidural space at the L4/5 level. On T1-weighted and T2-weighted images, the lesion showed slightly high signal intensity compared with that of the skeletal muscle (Figure 1). The mass had an irregular shape and resulted in a thecal sac and left L5 nerve root compression (Figure 2). The lesion did not develop as an outgrowth and with invasion of vertebral body metastasis to the epidural space. In addition, gadolinium-enhanced, fat-suppressed, T1-weighted images revealed intense and heterogeneous enhancement (Figure 3). Because the patient had prostate cancer, we suspected the possibility of epidural metastasis. However, the differential diagnosis included other space-occupying lesions such as a sequestrated lumbar disc, complicated synovial cyst, hematoma, and malignant lymphoma. In other imaging studies, including a PET/CT and bone scintigraphy, we could not find any other systemic metastasis (Figures 4 and 5), and, therefore, a consulted neurosurgeon planned and performed excision of the epidural mass lesion.

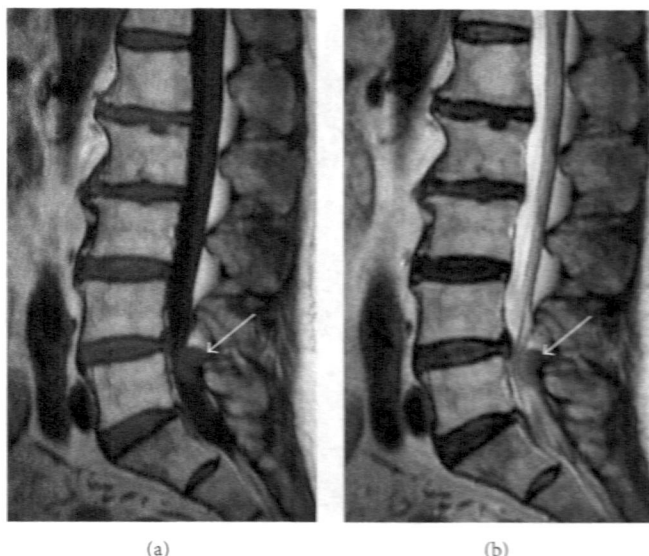

(a)   (b)

FIGURE 1: Sagittal T1-weighted MR image (a) and sagittal T2-weighted MR image reveal solitary spinal epidural mass (arrow) showing slightly high signal intensity as compared with signal intensity of skeletal muscle.

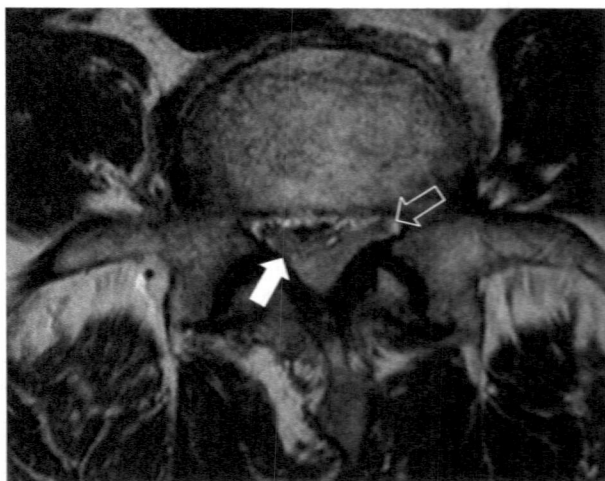

FIGURE 2: Axial T2-weighted image in upper body level of L5 reveals compression of thecal sac (solid arrow) and left S1 nerve root (open arrow).

Eventually, the mass was removed by en bloc resection. According to the intraoperative findings, the tumor was dissected well in the dura. On the gross pathology examination, the excised tumor, measuring $2.0 \times 1.5 \times 0.8$ cm, had a grayish white color. On the microscopic pathology examination, the tumor was proven to be a small cell carcinoma and immunohistochemically the tumor was positive for CD56, synaptophysin, and CK, but with negative PSA (Figure 6). The pathologist indicated that the immunohistochemical findings suggested the possibility of neuroendocrine carcinoma of the small cell type and that the primary site of the tumor cells (adenocarcinoma) could not be identified. Therefore, we cannot be sure that the tumor was a solitary spinal epidural metastasis (SEM) of the prostate cancer.

However, four months following excision of the solitary SEM, the patient underwent transurethral resection of the prostate (TURP) at an outside hospital due to his obstruction symptoms and nocturia, and the resected prostate cancer was histologically determined to be small cell carcinoma. Therefore, we concluded that the patient's prostate cancer was mixed type and that the spinal epidural mass was a solitary SEM.

The patient underwent L4-L5 postoperative radiotherapy with 30 Gy in 10 fractions for two weeks, after which palliative chemotherapy and radiation therapy for residual prostate cancer followed. Following excision of the solitary SEM of the L4/5 level, the patient's symptoms were slightly relieved. However, progression of the residual prostate cancer and lung metastasis were subsequently detected and the patient died 11 months after excision of the solitary SEM.

3. Discussion

Metastasis to the spinal nervous system includes intramedullary metastases, epidural metastases, and leptomeningeal metastases [3]. These spinal nervous system metastases are medical emergencies resulting in neurological dysfunction for which a rapid diagnosis and therapeutic intervention are required [1]. Usually, spinal epidural metastasis (SEM) arises from spinal vertebral body metastases, and the vertebral body is primarily involved because of its highly vascular red marrow. As a tumor grows, bone resorption due to osteoclast activating factors and prostaglandins, as well as bone destruction, occurs. Subsequently, SEM develops as an outgrowth and invasion of vertebral body metastasis to the epidural space [1, 3].

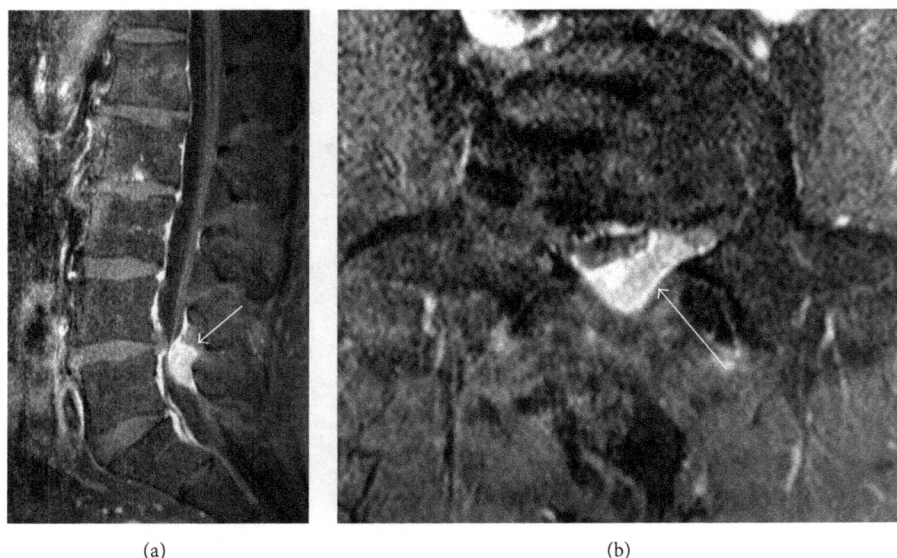

(a)   (b)

FIGURE 3: Sagittal (a) and axial (b) gadolinium-enhanced T1-weighted MR images with fat saturation show intense enhancement of solitary spinal epidural mass (arrow).

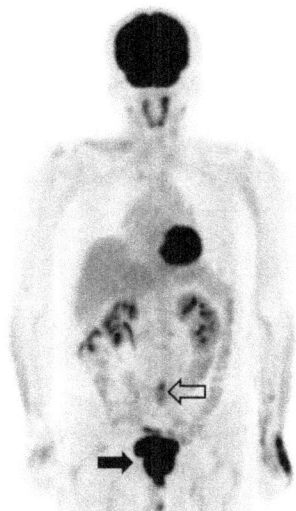

FIGURE 4: Coronal FDG PET image performed after surgery shows no abnormal uptake suggesting malignancy or metastatic tumor in the body except intense uptake due to prostate cancer and urine in the pelvis (solid arrow) and shows focal increased uptake due to postoperative change in L4/5 level (open arrow).

Lung, breast, and prostate cancer each account for 15–20% of all cases of SEM, and non-Hodgkin's lymphoma, renal cell cancer, and multiple myeloma account for 5–10% of these cases, and the remainder of the cases of SEM are colorectal cancers, sarcomas, and unknown primary tumors. This tendency is closely related to the tendency of the type of tumor to metastasize to bone and the spine [1].

Solitary SEM that is not related to vertebral metastasis is very rare. We have no concept regarding the incidence or the type of the primary tumor of solitary SEM based on

our literature review. There have been three reports regarding solitary SEM cases. Madden et al. reported thoracic SEM of Merkel cell carcinoma in an immunocompromised patient. In this patient, SEM represented a bilobed, epidural mass extending from T6 to T8 with extension into the left T6/T7 neural foramen, and caudal extension into the left T8/T9 neural foramen as seen on precontrast CT images. As this patient had remaining metal fragments in his right eye following previous surgery, MRI could not be performed. The patient presented with mid-thoracic back pain radiating around the trunk, and this symptom was relieved after his surgery [4]. Gupta et al. published a report regarding SEM from lung carcinoma. In this patient, SEM presented with focal, enhancing, epidural, and soft-tissue thickening in the right side of the T12-L1 level, as seen on MR imaging. The authors did not mention the signal intensities of the lesion seen on T1- and T2-weighted images. This patient complained of back pain, and her symptom was relieved after she underwent radiotherapy [5]. Brown et al. reported SEM in an endometrial carcinoma patient. In this patient, SEM appeared as a peripheral enhancing epidural mass located at the L5 level on MR images. The authors also did not mention the signal intensities of the lesion, as seen on T1- and T2-weighted images. The patient complained of back pain and paresthesia radiating in her lower legs, and she underwent surgical resection and postoperative radiotherapy [6]. Based on these case reviews, our reported case is very important because it reveals both the signal intensities on T1- and T2-weighted images and the enhancement character of solitary SEM in a prostate cancer patient.

The imaging differential diagnosis of epidural, space-occupying lesions includes both tumor lesions and nontumorous lesions mimicking tumors. Tumor lesions include metastasis, schwannoma, and malignant lymphoma, while tumor-mimicking lesions include sequestrated intervertebral

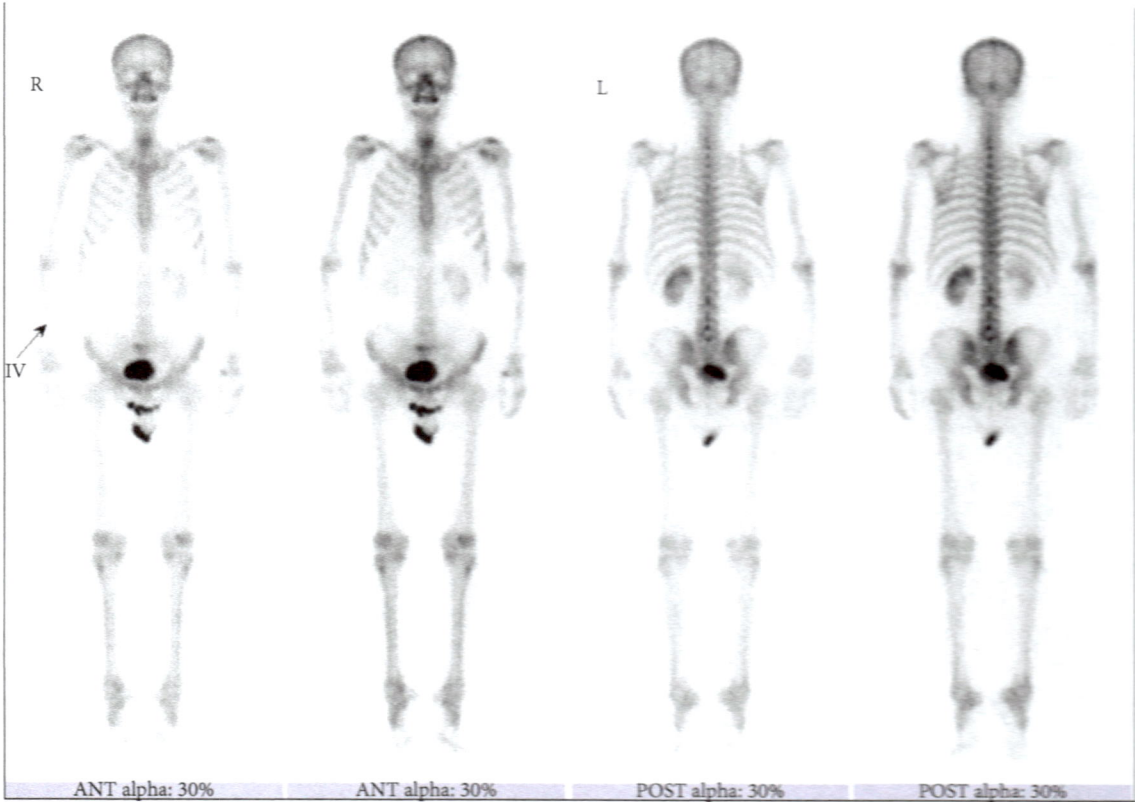

FIGURE 5: Bone scintigraphy performed after surgery shows no intense uptake suggesting bone metastasis.

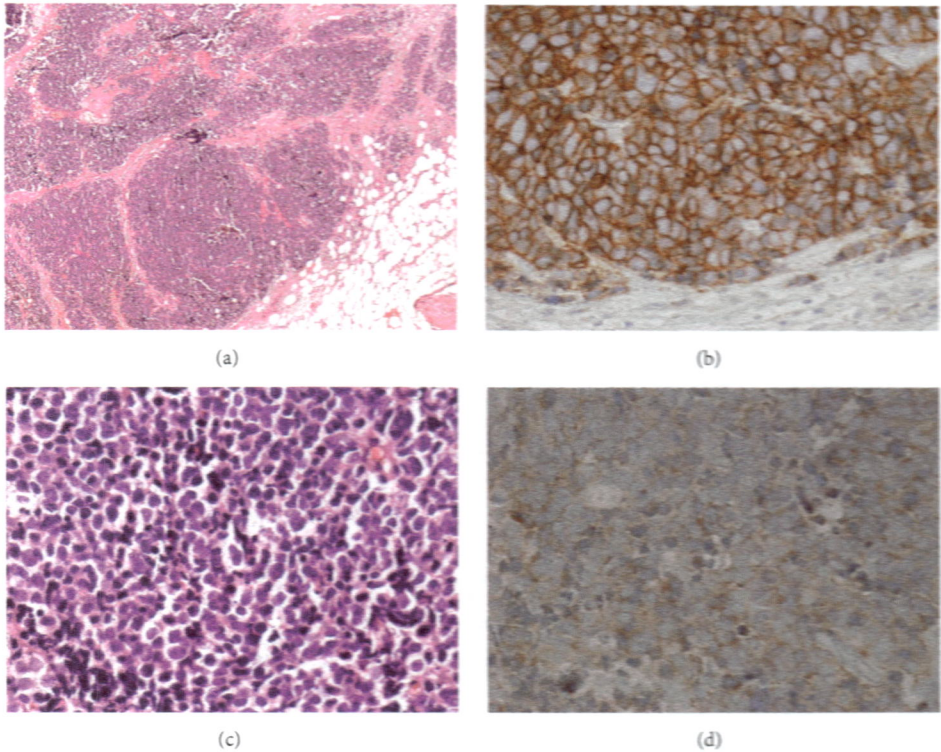

FIGURE 6: Photomicrographs showing diffuse infiltration of small anaplastic cells ((a) H&E, ×40; (b) H&E, ×400) and immunohistochemical reactivity of tumor cells for CD56 ((c) ×400) and synaptophysin ((d) ×400).

TABLE 1: The differential diagnosis according to MRI findings of the epidural space occupying lesion.

	T1WI	T2WI	Gd enhancement
Metastatic tumor	Low	High	$-\sim+$
Schwannoma	Low~iso	High	++
Lymphoma	Low~iso	High or low	+
Sequestrated intervertebral disc	Low	High	+ (rim enhancement)
Hemorrhagic synovial cyst	High	High or low	$-/+$
Hematoma	Variable	Variable	$-$
Abscess	Low~iso	High	+ (rim enhancement)

disc, hemorrhagic synovial cyst, hematoma, and abscess [7]. When we discover an epidural, space-occupying lesion on MRI, it is very difficult to distinguish whether it is a tumor or tumor-mimicking lesion. The signal intensities on T1- and T2-weighted images and the enhancement pattern help us to differentially diagnose epidural, space-occupying lesions. However, there is limited suspicion regarding the specific disease. Furthermore, reported cases of sequestrated disc and hemorrhagic synovial cyst showed a mass effect and intense enhancement and mimicked tumor occurring in the epidural space of the spine [7, 8]. Therefore, based on our literature review, it is important to detect a lesion in the epidural space using MRI before neurologic complications occur and to surgically excise the lesion in order to obtain an accurate diagnosis and initiate appropriate treatment. The signal intensities on T1- and T2-weighted images and the enhancement pattern of epidural, space-occupying lesions are described in Table 1 [7, 9, 10].

Our reported case was solitary SEM of prostate cancer of the small cell type. The majority of prostate cancers are acinar adenocarcinomas. Histological variants of prostate cancers can be defined as acinar adenocarcinoma and nonacinar carcinoma variants. The nonacinar carcinoma variants account for approximately 5–10% of the carcinomas that occur in the prostate. These include sarcomatoid carcinoma, ductal adenocarcinoma, urothelial carcinoma, basal cell carcinoma, and small cell carcinoma, in other words, neuroendocrine tumors. Small cell carcinoma developing in the prostate is a rare and very aggressive tumor, which frequently presents with disseminated disease. The incidence of prostate small cell carcinoma ranges from 0.3% to 1.0% in all prostate cancers. Prostatic small cell carcinoma shows different clinical features from those of prostatic acinar adenocarcinoma. Distinguishing clinical features include a lower percentage of men who present with an elevated serum PSA level at advanced stages of prostate cancer, poor hormonal responsiveness, and a short patient survival time. Most patients are 65–72 years old, and the most frequent presenting symptoms arise from bladder outlet obstruction and disseminated disease. One-third to two-thirds of patients with prostatic small cell carcinoma show an elevated serum PSA level, which could be due to an admixed adenocarcinomatous component [11–13].

Approximately half of the cases of prostatic small cell carcinomas are mixed tumors with conventional prostate cancer. The grade as well as extent of the acinar adeno-carcinoma component in mixed small cell adenocarcinoma in the prostate is variable. In most patients, if adenocarcinoma is diagnosed earlier than small cell carcinoma, the histologic grade of the adenocarcinoma portion is low-grade. In metastatic lesions derived from primary, mixed, small-cell carcinoma-adenocarcinoma, the small cell carcinoma element is sometimes seen, although both components can be found [11].

The diagnosis of solitary SEM begins with clinical suspicion in a known cancer patient complaining of new onset pain. And the solitary SEM is assessed by imaging modalities including plain radiographs, myelography, CT/CT myelography, and MRI. MRI is the method of choice for the diagnosis of solitary SEM. It is most useful and efficient that when a physician performs MRI, MRI of the whole spine should be performed. CT myelography remains an alternative method when MRI is unavailable or cannot be used because of uncontrolled pain, patient size, implanted metallic objects, the inability to lay flat during the exam time, or severe claustrophobia. The recommended treatment varies and includes surgical treatment, radiation therapy and chemotherapy, corticosteroids, or bisphosphonate [1, 3].

References

[1] J. S. Cole and R. A. Patchell, "Metastatic epidural spinal cord compression," *The Lancet Neurology*, vol. 7, no. 5, pp. 459–466, 2008.

[2] J. K. Kim, T. J. Learch, P. M. Colletti, J. W. Lee, S. D. Tran, and M. R. Terk, "Diagnosis of vertebral metastasis, epidural metastasis, and malignant spinal cord compression: are T1-weighted sagittal images sufficient?" *Magnetic Resonance Imaging*, vol. 18, no. 7, pp. 819–824, 2000.

[3] M. Mut, D. Schiff, and M. E. Shaffrey, "Metastasis to nervous system: spinal epidural and intramedullary metastases," *Journal of Neuro-Oncology*, vol. 75, no. 1, pp. 43–56, 2005.

[4] N. A. Madden, P. A. Thomas, P. L. Johnson, K. K. Anderson, and P. M. Arnold, "Thoracic spinal metastasis of Merkel cell carcinoma in an immunocompromised patient: case report," *Evidence-Based Spine-Care Journal*, vol. 4, no. 1, pp. 54–58, 2013.

[5] M. Gupta, P. Choudhary, A. Jain, and A. Pruthi, "Solitary spinal epidural metastasis from lung carcinoma," *Indian Journal of Nuclear Medicine*, vol. 29, no. 1, pp. 38–39, 2014.

[6] J. V. Brown, J. M. Stallman, H. Wong, C. M. Duma, and B. H. Goldstein, "Spinal epidural metastasis in an endometrial carcinoma patient," *Gynecologic Oncology Case Reports*, vol. 2, no. 1, pp. 20–22, 2012.

[7] H. Matsui, S. Imagama, Z. Ito et al., "Chronic spontaneous lumbar epidural hematoma simulating extradural spinal tumor: a case report," *Nagoya Journal of Medical Science*, vol. 76, no. 1-2, pp. 195–201, 2014.

[8] K. Li, Z. Li, W. Geng, C. Wang, and J. Ma, "Postdural disc herniation at L5/S1 level mimicking an extradural spinal tumor," *European Spine Journal*, 2015.

[9] G. Cannarsa, S. W. Clark, N. Chalouhi, M. Zanaty, and J. Heller, "Hemorrhagic lumbar synovial cyst: case report and literature review," *Nagoya Journal of Medical Science*, vol. 77, no. 3, pp. 481–492, 2015.

[10] C. Y. Chen, Y. L. Chuang, M. S. Yao, W. T. Chiu, C. L. Chen, and W. P. Chan, "Posterior epidural migration of a sequestrated lumbar disk fragment: MR imaging findings," *American Journal of Neuroradiology*, vol. 27, no. 7, pp. 1592–1594, 2006.

[11] P. A. Humphrey, "Histological variants of prostatic carcinoma and their significance," *Histopathology*, vol. 60, no. 1, pp. 59–74, 2012.

[12] S. W. Fine, "Variants and unusual patterns of prostate cancer: clinicopathologic and differential diagnostic considerations," *Advances in Anatomic Pathology*, vol. 19, no. 4, pp. 204–216, 2012.

[13] D. J. Grignon, "Unusual subtypes of prostate cancer," *Modern Pathology*, vol. 17, no. 3, pp. 316–327, 2004.

16

The Clinical Impact of Accurate Cystine Calculi Characterization Using Dual-Energy Computed Tomography

William E. Haley,[1] El-Sayed H. Ibrahim,[1,2] Mingliang Qu,[3] Joseph G. Cernigliaro,[1] David S. Goldfarb,[4] and Cynthia H. McCollough[3]

[1]Mayo Clinic, Jacksonville, FL 32224, USA
[2]University of Michigan, Ann Arbor, MI 48109, USA
[3]Mayo Clinic, Rochester, MN 55905, USA
[4]New York Harbor VA Healthcare System, Brooklyn, NY 11209, USA

Correspondence should be addressed to El-Sayed H. Ibrahim; elsayei@umich.edu

Academic Editor: Stefania Rizzo

Dual-energy computed tomography (DECT) has recently been suggested as the imaging modality of choice for kidney stones due to its ability to provide information on stone composition. Standard postprocessing of the dual-energy images accurately identifies uric acid stones, but not other types. Cystine stones can be identified from DECT images when analyzed with advanced postprocessing. This case report describes clinical implications of accurate diagnosis of cystine stones using DECT.

1. Introduction

Cystinuria is the most common of the inherited kidney stone diseases, accounting for about 1-2% and 25% of adult and pediatric patients, respectively, with kidney stones [1]. It is an autosomal recessive disorder caused by mutations in the SLC3A1 and SLC7A9 genes, leading to an increased excretion of the amino acid cystine, which is poorly soluble in urine, resulting in the formation of recurrent kidney stones [2]. The divergent treatment strategies for cystine and calcium stones reflect the importance of identifying the stone composition accurately [3]. Dual-energy computed tomography (DECT) is a relatively new imaging modality that has proven successful in differentiating between uric acid (UA) and non-UA stones with near 100% specificity for stones greater than 3 mm in size [4, 5]. In contrast to conventional CT, DECT utilizes X-ray tube and detector structures, which are set at different tube potentials, to simultaneously acquire two data sets that allow differentiation of stone materials based on the attenuation ratio between the two peak X-ray energies [5, 6]. This capability potentially obviates the need for stone analysis to guide treatment. Nevertheless, with the

current standard processing algorithm of DECT images, all non-UA stones are characterized as a single group without being further separated and are not differentiated from the more common calcium stones and other less common types including cystine [5]. We report a case of a patient with a long history of presumed calcium stones, in whom advanced processing of the DECT images correctly identified cystine stones, leading to change in treatment and resulting in improvement in stone-related outcomes.

2. Case Presentation

A 65-year-old man presented for evaluation with recent onset of intermittent left and right flank pain and nausea. History was notable for numerous stones over 35 years often requiring urological procedures, including extracorporeal shock wave lithotripsy (ESWL), ureteroscopy, and percutaneous nephrolithotomy. Increasing stone burden was documented on serial X-ray studies. Other medical histories included pulmonary sarcoidosis, with no recent hypercalcemia. The patient had no family history of kidney stones.

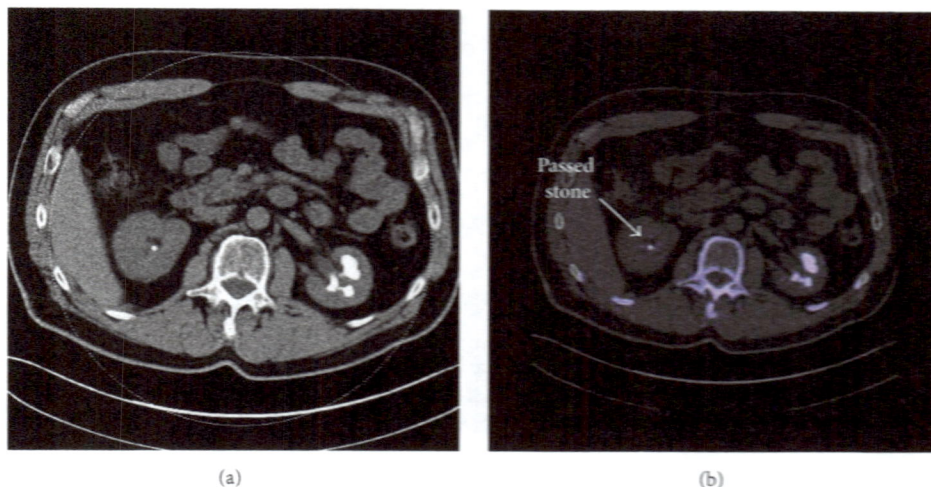

(a)   (b)

FIGURE 1: Conventional CT (a) and DECT (b) images showing a large staghorn stone and a small stone in the left and right kidneys, respectively. The standard DECT postprocessing algorithm colors all non-uric-acid stones blue. A follow-up scan confirmed the location of the passed stone.

Body mass index was $28 \, \text{kg/m}^2$; physical examination was unremarkable. Laboratory testing revealed normal serum creatinine, calcium, phosphorus, and electrolytes. Urinalysis showed 21 WBC/hpf, 6 RBC/hpf, and pH 6. 24 h urine panel included volume 2750 mL, calcium 206 mg/24 h, sodium 336 mEq/24 h, citrate 976 mg/24 h, uric acid 853 mg/24 h, oxalate 44 mg/24 h, phosphorus 1568 mg/24 h, and pH 6.2, with elevated supersaturation of calcium phosphate (brushite and apatite).

A DECT scan was performed using a Somatom Definition Flash dual-source dual-energy CT scanner (Siemens Healthcare, Forchheim, Germany), which utilizes 2 independent X-ray tube and detector structures and includes a tin filter for better separation between different X-ray beam spectra [7]. A dedicated renal stone imaging protocol was used, without intravenous or oral contrast. The DECT peak tube potentials (kVp) were set to 80 kV and 140 kV. Images were obtained helically from the diaphragm to the pubic symphysis with 0.75 mm slice thickness. Data were reconstructed on a multimodality workstation (Syngo Kidney Stone, Siemens Healthcare, Forchheim, Germany), which color-codes the UA and non-UA stones in red and blue, respectively, based on a 3-material decomposition algorithm (water, calcium, and uric acid). Images were reconstructed with a 0.75 mm slice thickness and 0.5 mm interval using a D30f convolution kernel for enhanced resolution. These images revealed bilateral renal calculi: a large staghorn on the left and two smaller stones on the right. There were no obstructing stones. All stones were characterized as non-UA (color-coded in blue), as shown in Figure 1. Images were postprocessed using an advanced algorithm previously demonstrated to differentiate different types of non-UA stones. The postprocessing involved manually determining a region of interest, followed by software segmentation from surrounding tissues using a predefined CT number threshold and extraction of a 3D volume of interest containing the stone. The CT number was calculated

at 80 kV and divided by the number at 140 kV for each voxel and the average dual-energy ratio (DER) value of all voxels within the segmented stone was used for characterizing the composition [8, 9]. The DER measurements were calculated for the large left kidney stone (DER (mean ± SD) = 1.25 ± 0.22), and the right-side stones (stone #1 DER = 1.24 ± 0.36 and stone #2 DER = 1.22 ± 0.34) (Figure 2). These DER values lie in the range associated with cystine [8, 9].

Analysis performed on a spontaneously passed stone by infrared spectroscopy revealed 100% cystine. Cystine excretion in 24 h urine was measured: 3.5 mmol (875 mg)/24 h. The patient was treated with tiopronin and potassium citrate. He was instructed to increase fluid intake, decrease dietary sodium, and moderate protein intake. On this regimen, follow-up for 2 years has been remarkable for stability of residual stones with no new stone growth and no urological procedures required.

3. Discussion

In this case report we demonstrate that cystine stones can be accurately identified from DECT images when analyzed with advanced postprocessing, and we describe clinical implications including significant impact on prescribing appropriate treatment that led to improved outcomes. This observation has significant clinical impact because the medical treatment of cystine stones requires a different approach from calcium stones [3]. Indeed, the prescribed regimen aimed at prevention of cystine stone growth proved to be highly successful in this case. Prior to that, the patient had suffered from a long history of presumed calcium-based stones, including frequent attacks and urological procedures, including ESWL, ureteroscopy, and percutaneous nephrolithotomy. This patient's history of sarcoidosis, with its association with calcium stones, clouded the clinical picture. On presentation, he was found to have increasing stone burden on serial X-ray

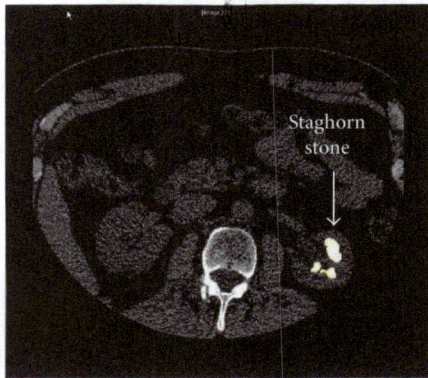

(a)

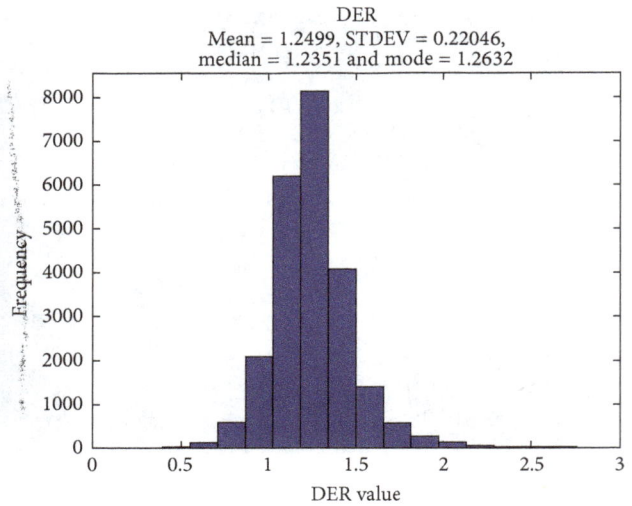

(b)

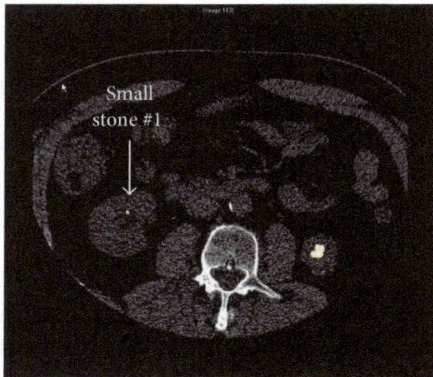

(c)

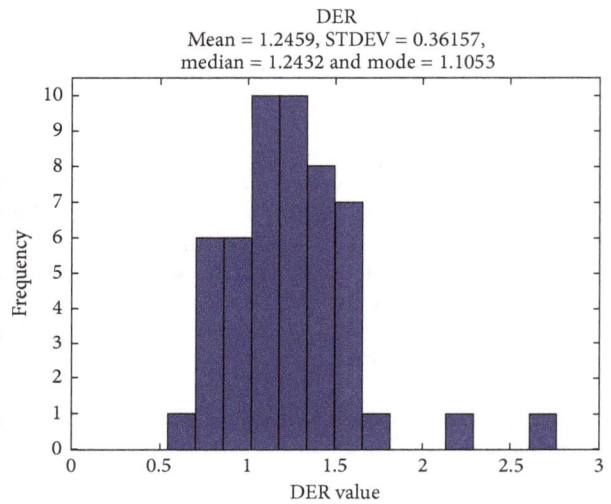

(d)

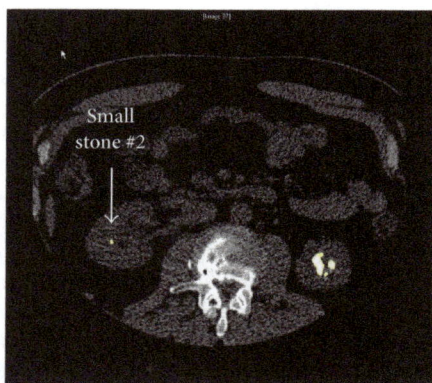

(e)

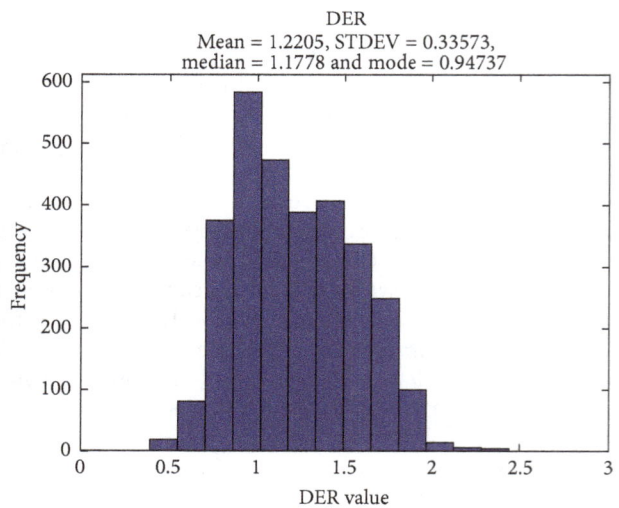

(f)

FIGURE 2: DECT scan with advanced postprocessing, which identified the stone material as cystine, in (a) left-side staghorn stone appears in this slice, and in (b), (c) two small right-side stones appear in these slices labeled small stone #1 and small stone #2, assigned a unique color (yellow). Histograms of the dual-energy ratio (DER) distribution showing mean ± SD = 1.25 ± 0.22 of staghorn stone (d), small stone #1 in (e) = 1.24 ± 0.36, and small stone #2 in (f) = 1.22 ± 0.34, which are in the range associated with cystine.

studies including the development of a symptomatic staghorn calculus.

DECT has been successfully used in recent years to distinguish UA from non-UA stones [5]. It allows for immediate identification of UA stones, raising the prospect of medical dissolution as an option, in addition to or in place of urological procedures. Whereas non-UA stones comprise about 90% of stones [5], the standard commercially available DECT 3-material decomposition image processing algorithm cannot differentiate between different non-UA stone types with sufficient reliability. Consequently, all non-UA stones are grouped together and assigned the same color. Distinguishing the more common non-UA stones, calcium oxalate and calcium phosphate, from less common stone types including cystine is important to assure the best outcome, as illustrated in this case report. Cystine stones, like uric acid stones, are often less radiopaque than calcium stones and may not be seen on plain films of the abdomen. Cystine stones are detectable but indistinguishable from other composition types on conventional noncontrast CT as well as standard DECT (Figure 1).

Although cystinuria meets the criteria of a rare disease (less than 200,00 cases in the United States), it is not as uncommon as one might expect from that designation, given that kidney stones affect about 10% of the adult US population and 1-2% of those may have cystinuria; the true prevalence may be even higher as some proportion of affected persons never have the diagnosis made. The present case highlights the potential for misdiagnosis of rare stones and may help explain the discrepancy in case-finding compared to estimated prevalence noted by the Rare Kidney Stone Consortium (http://www.rarekidneystones.org/). A number of factors contribute to the underrecognition of this disease. The presence of pathognomonic hexagonal cystine crystals on urinalysis is noted in only a fraction of cases. The diagnosis of cystinuria may be made based on measuring 24 h urine cystine excretion; however, this can be problematic [2, 10]. Nitroprusside screening of urine, performed by many laboratories specializing in assessment of urinary risk factors, is inexpensive and sensitive, but many patients never have urine sent to such specialty labs. Stone analysis by infrared spectroscopy or X-ray diffraction confirming cystine is diagnostic; however, such analysis is often not available to assist with diagnosis. Family history is usually not helpful, unless siblings are known to be affected. Absent the latter, a nitroprusside screening test, or the pathognomonic crystals on urinalysis, there would be little suspicion for this diagnosis.

The natural history of cystine stone disease is variably expressed. Most affected individuals present in childhood and 75% by teen years marked by multiple stone recurrences; however, a surprising number manifest in 40–80 year olds [11]. A significant proportion of adults develop renal pathological changes and loss of kidney function relating to crystalline nephropathy, in addition to recurrent obstructive uropathy and repeated urological interventions [2, 12]. Therefore, making the correct diagnosis has great clinical importance, reducing urological procedures and potential complications through medical management. Hydration, dietary modification, and urinary alkalinization have been successfully used for prevention of new cystine stones [3].

The restriction of dietary protein and sodium reduces urinary excretion of cystine, alkalinizing the urine improves cystine solubility, and cystine-binding thiol drugs, tiopronin or d-penicillamine, form soluble drug-cysteine complexes [10].

CT radiation dose deserves comment in the context of this report. Thomas reported a mean effective dose 2.7 mSv among 40 patients using a similar low-dose unenhanced DECT protocol, equipment, and settings [8]. With this technology, radiation dose is increased in obese patients. 100 kV can be used for the low energy setting to enhance image quality [5]. Experts have urged caution in the use of effective dose in estimating risks of ionizing radiation, especially in diagnostic X-ray studies. Low dose stone studies fall within the range of natural background levels of radiation found in our environment and any adverse health effect at such low dose levels of radiation used in medical imaging is either too small to be demonstrated or is nonexistent [13, 14].

Important questions remain. The sensitivity and specificity of DECT for the diagnosis of cystine calculi need further study. The detection limit of DECT technology is not known. 3D imaging may be utilized to quantify cystine burden and track response to treatment. Mixed stones are common; to what extent DECT with its ability to separate 5 types of kidney stones (uric acid, cystine, struvite, calcium oxalate, and calcium phosphate) [5, 8] fits into the management of patients in general practice is yet to be established.

Disclaimer

This paper's contents are solely the responsibility of the authors and do not necessarily represent the official views of the NIH.

Acknowledgments

This work was supported by the Rare Kidney Stone Consortium, a part of NIH Rare Diseases Clinical Research Network 1U54DK083908-01 from the NIDDK and the Office of Rare Diseases Research, part of the National Center for Advancing Translational Sciences, and by the Mayo Clinic O'Brien Urology Research Center U54DK10022 from the NIDDK.

References

[1] V. O. Edvardsson, D. S. Goldfarb, J. C. Lieske et al., "Hereditary causes of kidney stones and chronic kidney disease," *Pediatric Nephrology*, vol. 28, no. 10, pp. 1923–1942, 2013.

[2] N. Sumorok and D. S. Goldfarb, "Update on cystinuria," *Current Opinion in Nephrology and Hypertension*, vol. 22, no. 4, pp. 427–431, 2013.

[3] J. Chillarón, M. Font-Llitjós, J. Fort et al., "Pathophysiology and treatment of cystinuria," *Nature Reviews Nephrology*, vol. 6, no. 7, pp. 424–434, 2010.

[4] A. N. Primak, J. G. Fletcher, T. J. Vrtiska et al., "Noninvasive differentiation of uric acid versus non-uric acid kidney stones using dual-energy CT," *Academic Radiology*, vol. 14, no. 12, pp. 1441–1447, 2007.

[5] R. Hartman, A. Kawashima, N. Takahashi et al., "Applications of dual-energy CT in urologic imaging: an update," *Radiologic Clinics of North America*, vol. 50, no. 2, pp. 191–205, 2012.

[6] M. N. Ferrandino, S. A. Pierre, W. N. Simmons, E. K. Paulson, D. M. Albala, and G. M. Preminger, "Dual-energy computed tomography with advanced postimage acquisition data processing: improved determination of urinary stone composition," *Journal of Endourology*, vol. 24, no. 3, pp. 347–354, 2010.

[7] A. N. Primak, J. C. Giraldo, C. D. Eusemann et al., "Dual-source dual-energy CT with additional tin filtration: dose and image quality evaluation in phantoms and in vivo," *American Journal of Roentgenology*, vol. 195, no. 5, pp. 1164–1174, 2010.

[8] C. Thomas, M. Heuschmid, D. Schilling et al., "Urinary calculi composed of uric acid, cystine, and mineral salts: differentiation with dual-energy CT at a radiation dose comparable to that of intravenous pyelography," *Radiology*, vol. 257, no. 2, pp. 402–409, 2010.

[9] G. Hidas, R. Eliahou, M. Duvdevani et al., "Determination of renal stone composition with dual-energy CT: in vivo analysis and comparison with x-ray diffraction," *Radiology*, vol. 257, no. 2, pp. 394–401, 2010.

[10] D. S. Goldfarb, F. L. Coe, and J. R. Asplin, "Urinary cystine excretion and capacity in patients with cystinuria," *Kidney International*, vol. 69, no. 6, pp. 1041–1047, 2006.

[11] E. H. Lambert, J. R. Asplin, S. D. Herrell, and N. L. Miller, "Analysis of 24-hour urine parameters as it relates to age of onset of cystine stone formation," *Journal of Endourology*, vol. 24, no. 7, pp. 1179–1182, 2010.

[12] D. G. Assimos, S. W. Leslie, C. Ng, S. B. Streem, and L. J. Hart, "The impact of cystinuria on renal function," *Journal of Urology*, vol. 168, no. 1, pp. 27–30, 2002.

[13] W. R. Hendee and M. K. O'Connor, "Radiation risks of medical imaging: separating fact from fantasy," *Radiology*, vol. 264, no. 2, pp. 312–321, 2012.

[14] C. H. McCollough, J. T. Bushberg, J. G. Fletcher, and L. Eckel, "Answers to common questions about the use and safety of CT scans," *Mayo Clinic Proceedings*, vol. 90, no. 10, pp. 1380–1392, 2015.

17

Direct Percutaneous Embolization of Peristomal Ileostomy Varices in an Emergency Setting

William Ryan (ID),[1] **Farouk Dako,**[2] **Gary Cohen,**[2] **David Pryluck,**[2] **Joseph Panaro,**[2] **Emily Cuthbertson,**[2] **and Dmitry Niman**[2]

[1]*Lewis Katz School of Medicine at Temple University, USA*
[2]*Temple University Hospital, USA*

Correspondence should be addressed to William Ryan; williamjryan.ii@gmail.com

Academic Editor: Daniel P. Link

Patients with liver disease and portal hypertension who have had surgical formation of an abdominal stoma are at risk of developing peristomal varices. These varices have a predilection for bleeding. Ideally, portal decompression via TIPS procedure is performed, with or without direct embolization of the bleeding varix. When TIPS is not an appropriate option due to significant liver disease and hepatic encephalopathy there are other approaches to treat peristomal variceal hemorrhage. We report the embolization of such a varix via direct percutaneous puncture under ultrasound guidance when portal decompression was not an appropriate option.

1. Intro

Ectopic varices are defined as varices found outside of the esophageal or gastric mucosae [1]. Similar to esophageal and gastric varices, they are commonly related to portal hypertension and can cause clinically significant hemorrhage. Patients with portal hypertension who undergo surgical enterostomy formation have a tendency to form peristomal ectopic varices, and these varices are prone to bleeding [2, 3]. Mortality from these peristomal variceal bleeds is estimated to be 3-4% per episode; thus many of these patients undergo rapid intervention aimed at either embolization, portal venous decompression, or a combination of the two [1]. Because these cases are not especially common, comprising only about 5% of variceal bleeding, no consensus has been reached on optimal interventional approach [4]. A review of the literature illustrates that the percutaneous transhepatic approach, transjugular transhepatic approach, balloon-occluded retrograde transvenous occlusion technique, and direct varix puncture have all shown varying degrees of success in controlling peristomal variceal hemorrhage, with the latter being the least reported upon. We present a case of peristomal variceal hemorrhage that was successfully treated via direct percutaneous access of an engorged peristomal varix under ultrasound guidance and subsequent embolization with coils and Gelfoam to the feeding portal venous flow.

2. Case

A 59-year-old woman presented in July 2017 with extensive bleeding from her ileostomy site. Her history included locally advanced bladder cancer for which she had undergone pelvic exenteration and ileal conduit formation in November 2015. At that time, she had a known primary lung adenocarcinoma as well, but had no known liver metastases or other liver disease. Intravenous contrast-enhanced CT of the abdomen and pelvis performed in January 2016 raised the possibility of cirrhosis; however this was not biopsy-proven. In April 2016, she began to notice intermittent bleeding from her stoma which was initially thought to be mechanical tissue breakdown from the stomal flange. Concern for hepatic encephalopathy was raised when she had her first episode of confusion in December 2016. At that time CT of the abdomen and pelvis demonstrated strong radiographic suspicion for cirrhosis together with prominent vessels surrounding the urinary diversion site suspicious for portal hypertension.

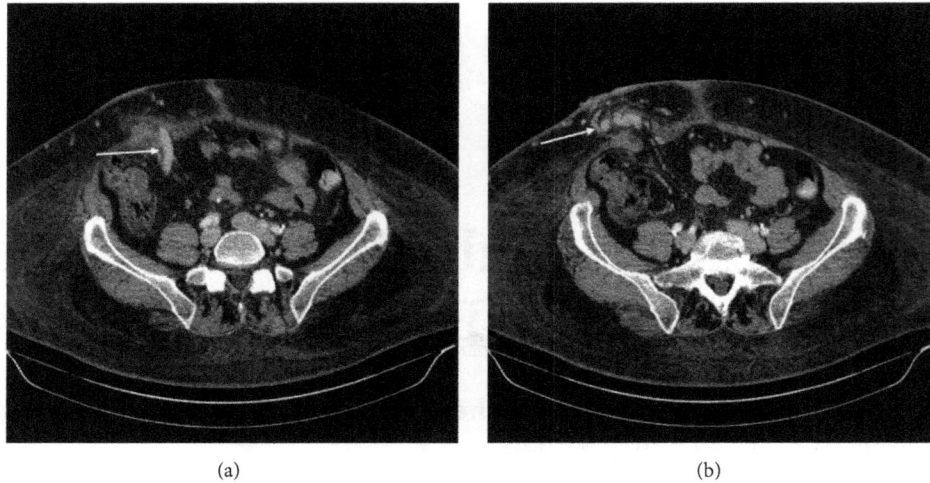

(a)   (b)

FIGURE 1: IV contrast-enhanced CT images of the abdomen and pelvic. (a) Large draining varix from the portal venous system (arrow). (b) Multiple varices within the right lower quadrant stoma (arrow).

Despite not having a tissue biopsy, she was diagnosed clinically with cryptogenic cirrhosis in May 2017 during a hospitalization for fatigue, anasarca, and altered mental status. An upper endoscopy performed in June 2017 demonstrated portal hypertensive gastropathy but no esophageal varices.

Upon presentation to the Emergency Department in July 2017 she had significant hemorrhage from her stoma resulting in hemodynamic instability. She was anemic with a hemoglobin of 8.3 g/dL that improved to 9.4 g/dL after blood transfusion, but gradually fell to 8.2 g/dL by the time of the procedure. Her MELD score was retrospectively calculated to be 19 at presentation, with an INR of 1.5 and total bilirubin of 4.3 mg/dl. She was emergently taken to the interventional suite for embolization with or without portal venous decompression via portosystemic shunt formation. A review of intravenous contrast-enhanced CT imaging showed extensive venous varices around the stoma involving the abdominal wall with a large draining varix arising from the portal system, likely the inferior mesenteric vein [Figure 1]. Also visualized was a variceal connection to the right common femoral vein. The portal and mesenteric veins were noted to be patent. Multiple approaches were considered for this patient. The transjugular intrahepatic portosystemic shunt (TIPS) and transjugular transhepatic approach with portosystemic shunt creation offered the benefit of portal decompression; however, the patient's recurrent hepatic encephalopathy was felt to be a relative contraindication. Transjugular transhepatic approach without formation of a permanent portosystemic shunt was also considered, since it would eliminate the risk of progressive hepatic encephalopathy. Percutaneous transhepatic approach would also eliminate the risk of progressive hepatic encephalopathy but was believed to pose increased risks of hepatic injury and bleeding. Transsplenic venous access to the portal venous system was considered as a viable, albeit technically challenging, option. The superficial nature of the abdominal wall stomal varix presented a less challenging and seemingly more time-efficient approach for access and was chosen as the target.

Using a micropuncture kit, the peristomal varix was directly accessed under ultrasound guidance and a micropuncture sheath was placed. Venography was performed and showed a large variceal collateral conglomerate around the stoma with variceal anastomosis with the right common femoral vein [Figure 2]. A wire was advanced and a 5F sheath was secured over the wire. A Kumpe catheter was introduced and advanced into the distal intra-abdominal aspect of the large draining varix. Catheter position was confirmed with repeat venography, and multiple coils were deployed [Figure 3]. This was followed by Gelfoam embolization. Postembolization venography showed sluggish flow in the draining varix with multiple filling defects within the visualized collaterals consistent with embolization [Figure 3]. The coils remained well-situated after placement and there was no evidence of migration. To ensure that there was no filling from the systemic venous system, the right superficial femoral vein was then accessed with a micropuncture kit and a femoral-iliac venogram and IVC venogram were both performed. These demonstrated brisk flow from the right common femoral vein through the iliac system and into the IVC. There was no filling of the stomal variceal collaterals visualized [Figure 3]. Hemostasis was thereby achieved, and the patient became hemodynamically stable shortly thereafter. At 6-month follow-up time no further imaging had been performed and the patient had not had any further episodes of hemorrhage from the ileostomy site

3. Discussion

Ectopic varices likely account for up to 5% of all variceal bleeding [4]. While the exact prevalence of peristomal variceal bleeding is unknown, one review indicated that of 169 ectopic variceal bleeds, 17% (26/169) of them were peristomal in nature [1]. An episode of peristomal variceal bleeding in the setting of portal hypertension carries an estimated mortality rate of 3-4% per episode, thus understanding different mechanisms of treatment is important [1].

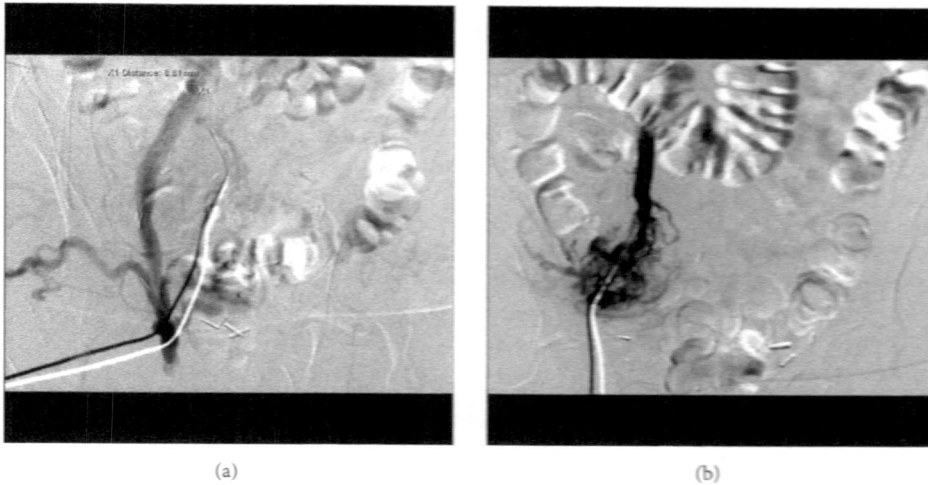

FIGURE 2: Fluoroscopic images of right lower quadrant. (a) Contrast injection of the peristomal varix and opacification of the right iliac vein. (b) Opacification of the peristomal varices and the right iliac vein.

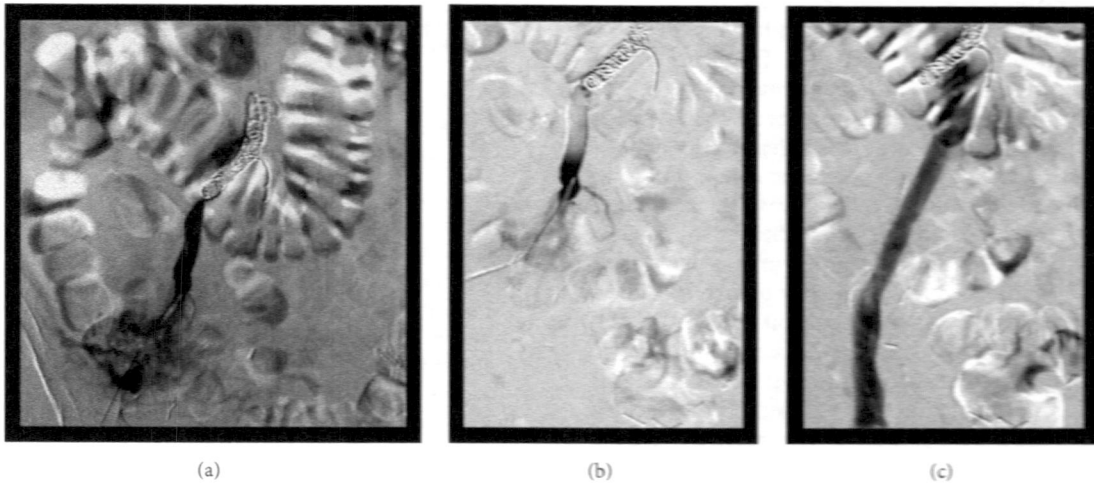

FIGURE 3: Fluoroscopic images of the right lower quadrant. (a) Venography after embolization with coils demonstrates hemostasis. (b) Subsequent venography after embolization with glue demonstrated sluggish flow in the portosystemic shunt with multiple filling defects within the visualized collaterals consistent with embolization. (c) Injection of the right femoral vein demonstrated flow from the right common femoral vein through the iliac system and IVC without opacification of the peristomal varices.

The creation of an abdominal wall stoma on its own does not necessarily pose a significant risk of peristomal varix formation. In a review of 244 patients without prior liver disease it was found that only 1.5% of patients with ileal conduit formation went on to develop peristomal varices [5]. The problem ensues when liver disease and portal hypertension are present concurrently with an abdominal wall stoma. Estimates from 27-53% of patients with chronic liver disease and ileostomy/colostomy creation will go on to develop peristomal varices [2, 3]. Hemorrhage requiring blood transfusion may occur in up to 70% of these patients [3]. The specific etiology and severity of liver disease and reason for small bowel/colonic stomal formation may mediate the risk of varix development. For example, patients with primary sclerosing cholangitis (PSC) have an up to 90% risk of underlying ulcerative colitis (UC) [6]. The progressive hepatic fibrosis, driven by PSC, and the high likelihood of colonic resection with stoma formation, driven by UC, put these patients at a significant risk for developing peristomal varices [3].

When a peristomal varix begins to bleed, there are multiple approaches to achieve hemostasis, none of which have been deemed "first line". Because portal hypertension is the driving force responsible for the bleeding, it is wise to consider decompressing the portal venous system using TIPS, assuming that pre- and postsinusoidal causes of portal hypertension have been ruled out. There have been numerous reports on the successful use of TIPS in this setting [7]. TIPS may offer the best chance to prevent future rebleeding when compared to more conservative therapies such as suture ligation, sclerotherapy, and intravascular coil or glue embolization [8]. However, the rates of rebleeding may still be as high as 25% [9]. A common reason to forgo performing TIPS, as was the case with the current report, is the presence of decompensated liver failure and hepatic encephalopathy. It is well known that TIPS formation increases the likelihood and severity of hepatic encephalopathy, and it is estimated that 30% of patients who receive TIPS for stomal variceal bleed will develop hepatic encephalopathy [9]. Shunt creation also poses the risk of interfering with future liver transplantation, assuming that the patient is an appropriate candidate. For example, improper placement of TIPS stent may impair a surgeon's ability to connect a new liver to the portal and hepatic veins. Alternatively, there have been reports of using a transjugular transhepatic approach without creation of a permanent shunt [10].

Another described successful approach to treat symptomatic peristomal varices has been through a percutaneous transhepatic route [10–12]. This gains direct access into the portal system through the liver parenchyma without creating portosystemic shunting. Although this approach has been successfully used, it carries a significant risk of bleeding, particularly because of the underlying liver disease and commonly elevated INR [13]. The percutaneous transsplenic approach offers a large target in the splenic vein, especially when engorged from portal hypertension. However the tortuous nature of the splenic vein and the risk of damaging surrounding structures make this approach technically difficult and would likely increase periprocedural bleeding risk. It is likely for these reasons that the transsplenic approach has not been reported upon in this emergency setting. A final, more recently described approach is balloon-occluded retrograde transvenous occlusion (BRTO) [12–14]. This adopts the technique commonly used to treat gastric varices and applies it to the portosystemic shunt at the stoma.

Direct percutaneous puncture of the intra-abdominal varix with subsequent embolization has been reported as safe and effective [15–18]. As was in our case, the direct percutaneous puncture of the varix can be easily performed under ultrasound guidance. Once accessed, the landing zones for subsequent embolic material are almost immediately available, which theoretically decreases the time to hemostasis. Again, it is important to consider the individual patient's candidacy and need for future liver transplant and to assure that intervention will not impede this candidacy. Despite this particular patient's poor candidacy for liver transplantation, it is worth noting that the described direct percutaneous approach has virtually no risk of interfering with future liver transplantation. In our report, coils were used followed by Gelfoam to embolize the portal venous inflow. Coils and Gelfoam were chosen due to availability and operator comfort. Other reports confirm that coils alone or in conjunction with Gelfoam are both safe and effective, although the risk of coil migration must be taken seriously when choosing size and landing zone [16, 17]. Other embolic materials, such as Histoacryl glue and iodized oil, have been used with success as well [15]. At this time, embolic material should be chosen based on operator comfort as there is no consensus on a superior material in this clinical setting. As detailed above, embolization of the feeding vessels does not treat the underlying portal hypertension. Thus, this approach is of limited value in isolation in patients who are good candidates for TIPS procedure. However, when a patient has significant relative contraindications for TIPS, percutaneous embolization via direct varix sac puncture is a safe and feasible option.

References

[1] I. D. Norton, J. C. Andrews, and P. S. Kamath, "Management of ectopic varices," *Hepatology*, vol. 28, no. 4, pp. 1154–1158, 1998.

[2] C. Fucini, B. G. Wolff, and R. R. Dozois, "Bleeding from peristomal varices: perspectives on prevention and treatment," *Diseases of the Colon & Rectum*, vol. 34, no. 12, pp. 1073–1078, 1991.

[3] R. H. Wiesner, N. F. LaRusso, R. R. Dozois, and S. J. Beaver, "Peristomal varices after proctocolectomy in patients with primary sclerosing cholangitis," *Gastroenterology*, vol. 90, no. 2, pp. 316–322, 1986.

[4] M. Kinkhabwala, A. Mousavi, S. Iyer, and R. Adamsons, "Bleeding ileal varicosity demonstrated by transhepatic portography," *American Journal of Roentgenology*, vol. 129, no. 3, pp. 514–516, 1977.

[5] L. B. Harbach, R. L. Hall, A. T. Cockett et al., "Ileal loop cutaneous urinary diversion: a critical review." *The Journal of Urology*, vol. 105, no. 4, pp. 511–514, 1971.

[6] B. Y. Tung, T. Brentnall, K. V. Kowdley et al., "Diagnosis and prevalence of ulcerative colitis in patients with sclerosing cholangitis," *Hepatology*, vol. 24, no. 169, 1996.

[7] A. R. Deipolyi, S. P. Kalva, R. Oklu, T. G. Walker, S. Wicky, and S. Ganguli, "Reduction in portal venous pressure by transjugular intrahepatic portosystemic shunt for treatment of hemorrhagic stomal varices," *American Journal of Roentgenology*, vol. 203, no. 3, pp. 668–673, 2014.

[8] M. O. Pennick and D. Y. Artioukh, "Management of parastomal varices: Who re-bleeds and who does not? A systematic review of the literature," *Techniques in Coloproctology*, vol. 17, no. 2, pp. 163–170, 2013.

[9] N. Kochar, D. Tripathi, N. C. Mcavoy, H. I. Reland, D. N. Redhead, and P. C. Hayes, "Bleeding ectopic varices in cirrhosis: The role of transjugular intrahepatic portosystemic stent shunts," *Alimentary Pharmacology & Therapeutics*, vol. 28, no. 3, pp. 294–303, 2008.

[10] D. B. Lashley, R. R. Saxon, E. F. Fuchs, D. H. Chin, and B. A. Lowe, "Bleeding ileal conduit stomal varices: diagnosis and management using transjugular transhepatic angiography and embolization," *Urology*, vol. 50, no. 4, pp. 612–614, 1997.

[11] M. J. Maciel, O. Pereira, J. Motta Leal Filho et al., "Peristomal variceal bleeding treated by coil embolization using a percutaneous transhepatic approach," *World Journal of Clinical Cases*, vol. 4, no. 1, pp. 25–29, 2016.

[12] D.-H. Yao, X.-F. Luo, B. Zhou, and X. Li, "Ileal conduit stomal variceal bleeding managed by endovascular embolization," *World Journal of Gastroenterology*, vol. 19, no. 44, pp. 8156–8159, 2013.

[13] K. Kishimoto, A. Hara, T. Arita et al., "Stomal varices: Treatment by percutaneous transhepatic coil embolization," *CardioVascular and Interventional Radiology*, vol. 22, no. 6, pp. 523–525, 1999.

Direct Percutaneous Embolization of Peristomal Ileostomy Varices in an Emergency...

95

[14] S. Minami, K. Okada, M. Matsuo, Y. Kamohara, I. Sakamoto, and T. Kanematsu, "Treatment of bleeding stomal varices by balloon-occluded retrograde transvenous obliteration," *Journal of Gastroenterology*, vol. 42, no. 1, pp. 91–95, 2007.

[15] F. Thouveny, C. Aubé, A. Konaté, J. Lebigot, A. Bouvier, and F. Oberti, "Direct Percutaneous Approach for Endoluminal Glue Embolization of Stomal Varices," *Journal of Vascular and Interventional Radiology*, vol. 19, no. 5, pp. 774–777, 2008.

[16] S. G. Naidu, E. P. Castle, J. S. Kriegshauser, and E. A. Huettl, "Direct percutaneous embolization of bleeding stomal varices," *CardioVascular and Interventional Radiology*, vol. 33, no. 1, pp. 201–204, 2010.

[17] R. Arulraj, K. S. Mangat, and D. Tripathi, "Embolization of bleeding stomal varices by direct percutaneous approach," *CardioVascular and Interventional Radiology*, vol. 34, no. 2, pp. S210–S213, 2011.

[18] A. C. H. Kwok, F. Wang, R. Maher et al., "The Role of Minimally Invasive Percutaneous Embolisation Technique in the Management of Bleeding Stomal Varices," *Journal of Gastrointestinal Surgery*, vol. 17, no. 7, pp. 1327–1330, 2013.

18

A Thymic Hyperplasia Case without Suppressing on Chemical Shift Magnetic Resonance Imaging

Tuan Phung⬥,[1] Thach Nguyen,[2] Dung Tran,[3] Nga Phan,[4] and Hung Nguyen[3]

[1]*Radiology Department, Military Hospital 103, Vietnam Military Medical University, Hanoi, Vietnam*
[2]*Anesthesiology Department, National Institute of Burns, Vietnam Military Medical University, Hanoi, Vietnam*
[3]*Histopathology Department, Military Hospital 103, Vietnam Military Medical University, Hanoi, Vietnam*
[4]*Neurology Department, Military Hospital 103, Vietnam Military Medical University, Hanoi, Vietnam*

Correspondence should be addressed to Tuan Phung; phunganhtuanbv103@gmail.com

Academic Editor: Roberto Iezzi

A 22-year-old woman with myasthenia gravis (MG) presented with ptosis and mild muscle weakness symptoms for one year. Computed tomography (CT) presented a diffuse bilobulate enlargement gland with a high density of soft tissue. Magnetic resonance imaging (MRI) showed the gland with no suppression on the opposed-phase chemical shift. After the thymic tumor diagnosis, she underwent thoracoscopic surgery for tumor resection. The postoperative histopathological finding was thymic lymphoid hyperplasia. This case suggests chemical shift MRI is not enough in distinguishing, and supplementary examination is essential to avoid unnecessary thymic biopsy and surgery.

1. Introduction

Thymic lymphoid hyperplasia (TLH) is very common in patients with myasthenia gravis. Contrary to true thymic hyperplasia, thymic lymphoid hyperplasia has diverse types and shapes. It can exhibit a normal shape and size, diffuse enlargement of both lobes, or a focal soft tissue mass [1]. Therefore, distinguishing thymic hyperplasia from thymic tumors is difficult. Inaoka et al.'s and Popa et al.'s studies showed that chemical shift magnetic resonance imaging is valuable in distinguishing thymic hyperplasia from thymic tumors [2, 3]. On the opposed-phase image, hyperplasia presents a signal intensity decrease, whereas the thymic tumor does not. We reported a thymic hyperplasia case that did not present the signal intensity decrease on chemical shift magnetic resonance imaging.

2. Case Report

A 22-year-old woman had ptosis and mild muscle weakness symptoms for one year. The symptoms were mild in the morning, severe in the evening, worse on exertion, and improved with rest. She had not diplopia, dyspnea,

or dysphagia symptoms. Prostigmine and repetitive nerve stimulation tests were positive. She was diagnosed with myasthenia gravis and treated with corticosteroids and mytelase and her symptoms got better. After CT and MRI examination, she was diagnosed with a thymic tumor and underwent thoracoscopic surgery for tumor resection. The postoperative histopathological finding was thymic lymphoid hyperplasia.

CT examination was performed using a 2-section CT system (Siemens, Somatom Spirit, Germany) in a single-breath hold at end inspiration. Technical parameters included 120 kVp, 180 mAs, pitch of 1, section thickness of 5 mm, contiguous section interval, and 512 × 512 matrix without contrast agent intravenous injection. Observation was performed on soft tissue window W350, L100 HU.

MRI examination was obtained using a 1.5 T MRI unit (Intera Achieva, Philips Healthcare, Netherlands). She underwent transverse gradient-echo T1-weighted non-dual-echo in-phase and opposed-phase imaging, using an anterior-to-posterior phase-encoding direction, in separate breath holds. Imaging parameters included 350 mm field of view, 256 × 256 image matrix, 5 mm section thickness, 151 ms time repetition (TR), and in-phase and opposed-phase time echo (TE) of 4.6 and 2.3 ms. She also underwent axial T1-weighted and

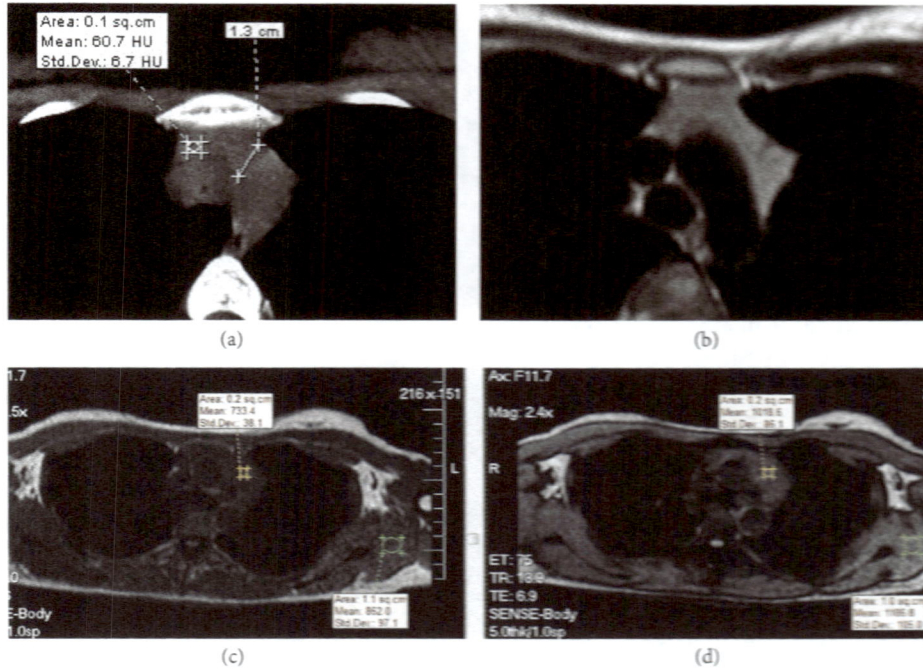

FIGURE 1: (a) CT image, (b) T2-weighted MRI, and (c) in-phase and (d) opposed-phase images presenting the gland without adipose tissue with CSR 1.0.

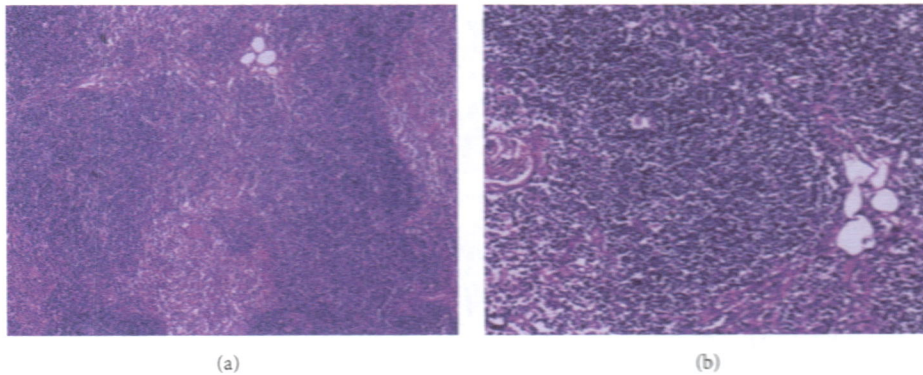

FIGURE 2: Microscopic images of thymic hyperplasia: great lymphoid follicles ((a) hematoxylin and eosin stain (H&E), magnification ×100), Hassall's corpuscles, and the rarity of adipose tissue ((b) hematoxylin and eosin stain (H&E), magnification ×200).

T2-weighted imaging without and with fat suppression and black blood technique with cardiac gate. Imaging parameters included TR 1000, 2000 ms, TE 10, 60 ms, and section thickness of 6 mm.

CT findings showed the relative homogeneous bilobulate gland with a high density of soft tissue (60.7 HU). The left lobe with a thickness of 13 mm was greater than the right lobe. The gland had straight margins. MRI findings presented the gland with an increased homogenous intensity on both T1-weighted and T2-weighted imaging. It was greater than in the wall muscle but lower than in the adipose tissue. Chemical shift magnetic resonance images demonstrated no decrease in signal intensity of the gland on the opposed-phase images relative to the in-phase images, suggesting the absence of fat component. The mean chemical shift ratio (CSR) was 1.0 (Figure 1).

On gross examination, the resected thymus weighed 55 g, and the external surface was smooth and lobulated, triangle shaped, and yellowish, with a size of 25 × 20 × 10 mm. A microscopic examination revealed numerous reactive lymphoid follicles with large germinal centers in the thymic medulla. Lobules were separated by a thin fibrous septum and had a little fat tissue. In addition, there were an increased number of Hassall's corpuscles (Figure 2).

3. Discussion

Myasthenia gravis is an acquired autoimmune disease caused by an immune response in which IgG autoantibodies are produced against the acetylcholine receptors of the neuromuscular junction postsynaptic membrane. The thymic gland plays a very important role in the pathogenesis of MG. Above 90% of myasthenia gravis cases had abnormal thymus including 70% thymic lymphoid hyperplasia and 20% thymoma [4]. Differentiation between thymic lymphoid hyperplasia and thymoma is essential for surgical management. Thymectomy is strongly recommended in all thymoma cases. Conversely, the surgical indication for hyperplasia cases should be only considered with less effective conservation treatments [5]. CT is the most common diagnostic tool for distinguishing based on morphological assessment. On the CT, thymic hyperplasia manifests diffuse large gland image in two lobes while thymoma is in the form of localized soft tissue mass. However, Nicolaou' et al. study showed 45% hyperplasia cases with normal form, 35% two-lobe diffuse enlargement cases, and 20% soft tissue mass form cases [1]. Conversely, thymoma could show a diffuse enlargement gland. Therefore, chemical shift MRI is useful for distinguishing in atypical cases. It is able to detect microscopic fatty infiltration within the normal or hyperplastic thymus, which would be indistinct at CT, by showing homogeneous signal decrease on opposed-phase images relative to in-phase images, whereas signal loss is absent in thymoma that does not include fat [5]. However, fat infiltration in thymic gland occurs together with age. By evaluation of normal fatty replacement of the thymus, Inaoka et al. concluded that the CSR value should not be used in children under 16 years of age [6]. In another study, the author found that the CSR values of the tumor group and the hyperplasia group were 1.026 ± 0.039 and 0.614 ± 0.13, respectively [2]. Popa et al.'s study showed the tumor group CSR value and the hyperplasia group CSR value of 1.0398 ± 0.0244 and 0.4964 ± 0.1841, respectively [3]. These studies showed an accuracy of 100% for CSR, with no overlap in the range between the hyperplasia and tumor groups.

Besides CSR, another index also used to quantify fat tissue is signal intensity index (SII). The SII is often used when chemical shift imaging is obtained by a dual-echo technique. According to Priola et al., reference tissue was not only unnecessary but also incorrect because the tissue may contain a determined amount of fat [7]. The study of the author showed that SII had sensitivity (Se) and specificity (Sp) of 100% at cutoff point 8.92% and CSR had respective Se of 100% and Sp of 96.7% at cutoff point 0.849. No overlap was found for SII values between the two groups while CSR values overlapped in some cases [8]. However, the difference between two indexes was negligible. Furthermore, no other study has been published about the matter. In our study, due to using a non-dual-echo technique, we applied CSR to quantify fat tissue, which confirmed that this was a thymic hyperplasia case without fat.

Ackman et al. presented a pathologically proven case of normal thymus in a 21-year-old woman that demonstrated no fat replacement on the opposed-phase chemical shift MRI (CSR = 1.1) [9]. Priola et al. also reported a true hyperplasia case in a 60-year-old female being treated with corticosteroids without fat infiltration on chemical shift MRI (SII = −7.57%) [10]. Therefore, the soft tissue mass without fat in the position of thymus gland with the intermediate signal intensity on T1-weighted and T2-weighted MRI was not enough to determine the tumor, especially in young women. Seo et al.'s study about lipid-poor adrenal adenoma also showed the same results [11]. The author found that adrenal adenoma with the density on CT ≤ 20 HU (lipid-rich) had the sensitivity of adrenal-spleen ratio (ASR) on MRI of 100%, while adrenal adenoma with the density on CT > 30 HU (lipid-poor) had the sensitivity of ASR on MRI 61.5%.

With thymic hyperplasia cases without fat as our case, Priola's studies have suggested that diffusion-weighted MRI could be valuable because of its capability to reflect cell density and cellular architecture and to detect malignant tissues by demonstrating restricted diffusion and low (ADC) values. He reported two thymic hyperplasia cases without fat that was not detected on chemical shift imaging but was found on diffusion-weighted MRI due to the high ADC values $1.97 \times 10^{-3} \ mm^2/s$ and $2.47 \times 10^{-3} \ mm^2/s$ [5, 10]. His other study showed that diffusion-weighted MRI could be applied to distinguish thymic tumors from nonthymic tumors at cutoff ADC $1.625 \times 10^{-3} \ mm^2/sec$ with Se 96.8% and Sp 79.2% [12]. However, until now, this has been the only study using diffusion-weighted MRI to distinguish thymic tumors from normal and hyperplasia thymus. Studies of Razek et al. [13, 14], Usuda et al. [15], Seki et al. [16], and Gümüştaş et al. [17] always use diffusion-weighted MRI to distinguish benign from malignant tumors. Therefore, the matter should be further studied.

In thymic hyperplasia, histopathological characteristics revealed numerous reactive lymphoid follicles with prominent germinal centers in the thymic medulla. In adults, both normal thymus and thymic hyperplasia contain a great amount of fat tissue. However, in this case, postoperative histopathological findings showed only a few fat cells, which was not sufficient to detect the decrease of signal intensity on chemical shift MRI (Figure 2).

In conclusion, the normal and hyperplasia thymus glands present great fat infiltration. Conversely, thymoma does not show adipose tissue. Chemical shift magnetic resonance imaging is helpful in differentiating thymic lymphoid hyperplasia from thymic neoplasm. However, in a few cases, especially in young women, the chemical shift magnetic resonance imaging is not enough in distinguishing. Diffusion-weighted MRI and supplementary examination in these cases are essential to eliminate unnecessary thymic biopsy and thymectomy.

References

[1] S. Nicolaou, N. L. Müller, D. K. B. Li, and J. J. F. Oger, "Thymus in myasthenia gravis: Comparison of CT and pathologic findings and clinical outcome after thymectomy," *Radiology*, vol. 201, no. 2, pp. 471–474, 1996.

[2] T. Inaoka, K. Takahashi, M. Mineta et al., "Thymic hyperplasia and thymus gland tumors: Differentiation with chemical shift MR imaging," *Radiology*, vol. 243, no. 3, pp. 869–876, 2007.

[3] G. A. Popa, E. M. Preda, C. Scheau, C. Vilciu, and I. G. Lupescu, "Updates in MRI characterization of the thymus in myasthenic patients," *Journal of Medicine and Life*, vol. 5, no. 2, pp. 206–210, 2012.

[4] H. Onodera, "The role of the thymus in the pathogenesis of myasthenia gravis," *The Tohoku Journal of Experimental Medicine*, vol. 207, no. 2, pp. 87–98, 2005.

[5] A. M. Priola and S. M. Priola, "Imaging of thymus in myasthenia gravis: From thymic hyperplasia to thymic tumor," *Clinical Radiology*, vol. 69, no. 5, pp. e230–e245, 2014.

[6] T. Inaoka, K. Takahashi, K. Iwata et al., "Evaluation of normal fatty replacement of the thymus with chemical-shift MR imaging for identification of the normal thymus," *Journal of Magnetic Resonance Imaging*, vol. 22, no. 3, pp. 341–346, 2005.

[7] A. M. Priola, D. Gned, A. Veltri, and S. M. Priola, "Chemical shift and diffusion-weighted magnetic resonance imaging of the anterior mediastinum in oncology: Current clinical applications in qualitative and quantitative assessment," *Critical Review in Oncology/Hematology*, vol. 98, pp. 335–357, 2016.

[8] A. M. Priola, S. M. Priola, G. Ciccone et al., "Differentiation of rebound and lymphoid Thymic hyperplasia from anterior mediastinal tumors with dual-echo chemical-shift MR imaging in adulthood: Reliability of the chemical-shift ratio and signal intensity index," *Radiology*, vol. 274, no. 1, pp. 238–249, 2015.

[9] J. B. Ackman, M. Mino-Kenudson, and C. R. Morse, "Nonsuppressing normal thymus on chemical shift magnetic resonance imaging in a young woman," *Journal of Thoracic Imaging*, vol. 27, no. 6, pp. W196–W198, 2012.

[10] A. M. Priola, D. Gned, V. Marci, A. Veltri, and S. M. Priola, "Diffusion-weighted MRI in a case of nonsuppressing rebound thymic hyperplasia on chemical-shift MRI," *Japanese Journal of Radiology*, vol. 33, no. 3, pp. 158–163, 2015.

[11] J. M. Seo, B. K. Park, S. Y. Park, and C. K. Kim, "Characterization of lipid-poor adrenal adenoma: Chemical-shift MRI and washout CT," *American Journal of Roentgenology*, vol. 202, no. 5, pp. 1043–1050, 2014.

[12] A. M. Priola, S. M. Priola, M. T. Giraudo et al., "Chemical-shift and diffusion-weighted magnetic resonance imaging of thymus in myasthenia gravis: Usefulness of quantitative assessment," *Investigative Radiology*, vol. 50, no. 4, pp. 228–238, 2015.

[13] A. A. Abdel Razek, "Characterization of thymic tumors with diffusion weighted MR imaging," *European Society of Radiology*, 2010.

[14] A. A. K. A. Razek, M. Khairy, and N. Nada, "Diffusion-weighted MR imaging in thymic epithelial tumors: Correlation with world health organization classification and clinical staging," *Radiology*, vol. 273, no. 1, pp. 268–275, 2014.

[15] K. Usuda, S. Maeda, N. Motono et al., "Diffusion weighted imaging can distinguish benign from malignant mediastinal tumors and mass lesions: Comparison with positron emission tomography," *Asian Pacific Journal of Cancer Prevention*, vol. 16, no. 15, pp. 6469–6475, 2015.

[16] S. Seki, H. Koyama, Y. Ohno et al., "Diffusion-weighted MR imaging vs. multi-detector row CT: direct comparison of capability for assessment of management needs for anterior mediastinal solitary tumors," *European Journal of Radiology*, vol. 83, no. 5, pp. 835–842, 2014.

[17] S. Gümüştaş, N. Inan, H. T. Sarisoy et al., "Malignant versus benign mediastinal lesions: Quantitative assessment with diffusion weighted MR imaging," *European Radiology*, vol. 21, no. 11, pp. 2255–2260, 2011.

19

Müllerian Remnant Cyst as a Cause of Acute Abdomen in a Female Patient with Müllerian Agenesis: Radiologic and Pathologic Findings

Mujtaba Mohammed,[1] Mary Allen-Proctor,[2] and Andrij Wojtowycz[1]

[1]Department of Radiology, SUNY Upstate University Hospital, Syracuse, NY 13210, USA
[2]Department of Pathology, SUNY Upstate University Hospital, Syracuse, NY 13210, USA

Correspondence should be addressed to Mujtaba Mohammed; mohammem@upstate.edu

Academic Editor: Yoshito Tsushima

We report a case of a 17-year-old female with Müllerian agenesis who presented with right sided abdominal pain clinically suspicious for acute appendicitis. Multimodality imaging workup revealed a heterogeneous cystic right upper quadrant mass with surrounding fluid and inflammatory changes. Surgical resection of this mass was performed and a histopathologic diagnosis of a hemorrhagic Müllerian remnant cyst was made, which to the best of our knowledge has never been described in a patient with Müllerian agenesis.

1. Introduction

Müllerian agenesis, also referred to as Mayer-Rokitansky-Kuster-Hauser (MRKH) syndrome, is a rare congenital abnormality that occurs in females and primarily affects the reproductive system. It is characterized by agenesis or underdevelopment of the uterus and vagina with normal development of the ovaries and secondary sexual characteristics [1]. Abnormalities of the reproductive system increase their predisposition to unusual causes of abdominal pain such as hematocolpos/hematometra. There is also increased incidence of ovarian maldescent in females with Müllerian duct anomalies who may present with atypical abdominal pain when ovaries are located within the abdomen [2]. We describe an unusual case of acute abdomen in a patient with Müllerian agenesis.

2. Case Presentation

A 17-year-old female presented with acute right sided abdominal pain that had been progressively getting worse for the previous 3 days. Her past medical history was significant for Müllerian agenesis and lack of a uterus. Her past surgical history was significant for small bowel atresia and resection of a portion of her small bowel during infancy. Physical examination revealed tenderness to palpation in the right lower quadrant and the periumbilical region. Laboratory tests were significant for leukocytosis with a left shift and an elevation of C-reactive protein.

CT scan of the abdomen and pelvis demonstrated a heterogeneous structure measuring 3.8 × 2.2 × 2.6 cm (transverse × anteroposterior × craniocaudal) inferior to the right hepatic lobe and posterolateral to the ascending colon with blood supply from a branch arising off the right renal artery and a draining vessel into the inferior vena cava (Figure 1). A low density cystic component was seen inferiorly within the mass. There was a small amount of surrounding free fluid and fat stranding. The uterus was absent and both ovaries were present in the pelvis. Ultrasound exam (Figure 2) of the right upper quadrant revealed a heterogeneous mass with no internal blood flow and confirmed the arterial and venous connections noted on the CT scan. MRI exam of the abdomen (Figure 3) showed a mass with intermediate to low signal intensity on T1 and a low signal intensity on T2 with a cystic component inferiorly. Extensive surrounding edema and rim enhancement were also present.

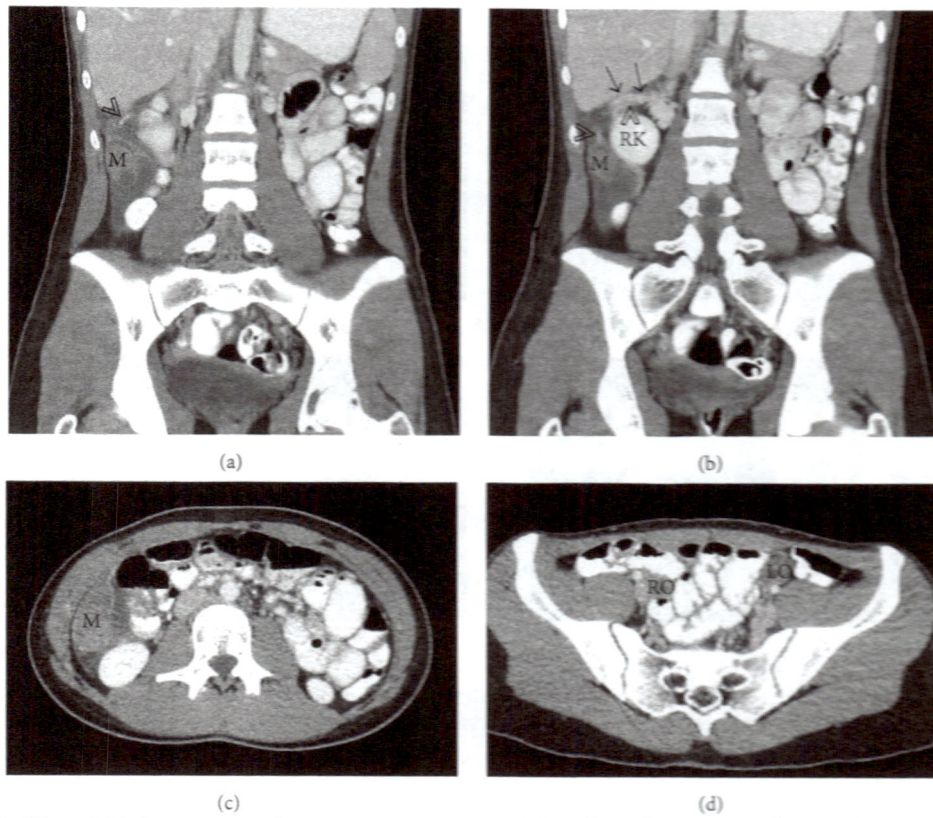

FIGURE 1: Images (a), (b), and (c) demonstrate a heterogeneous structure (M) in the right upper quadrant with its supplying artery (arrowheads) and venous drainage (arrows) adjacent to the right kidney (RK). Axial CT image (d) shows both the right (RO) and left (LO) ovaries in the pelvis.

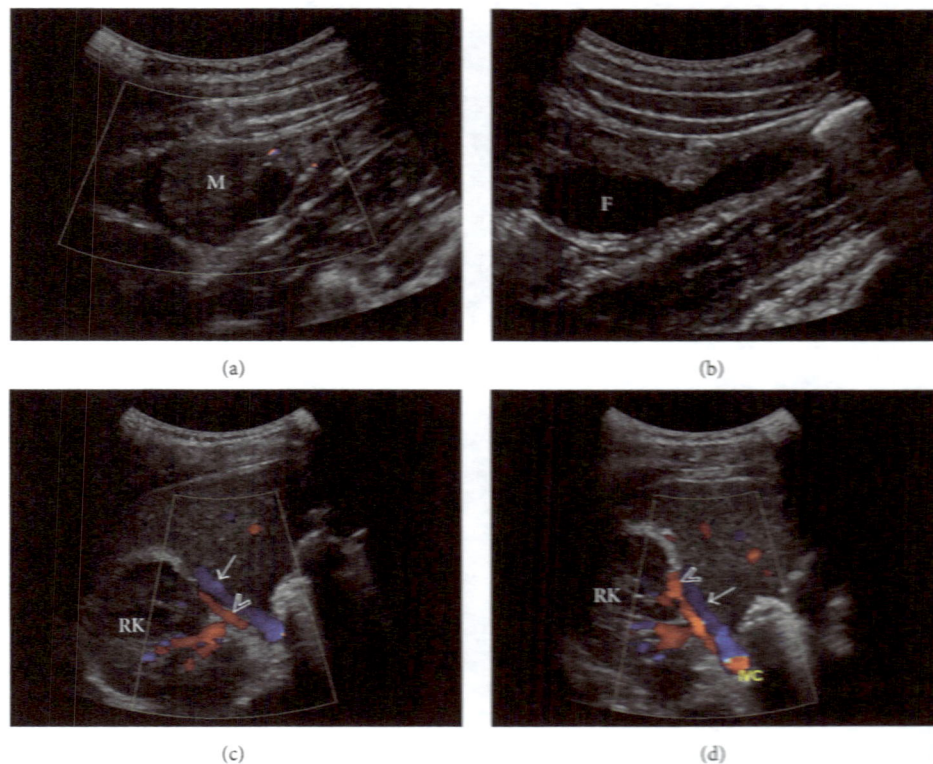

FIGURE 2: Image (a) is a sagittal sonographic image of the right upper quadrant demonstrating a structure (M) with no internal flow on color Doppler. Fluid (F) is seen just below the mass on image (b). Axial sonographic images (c) and (d) demonstrate the supplying artery (arrowhead) and the venous drainage (arrow) adjacent to the right kidney (RK).

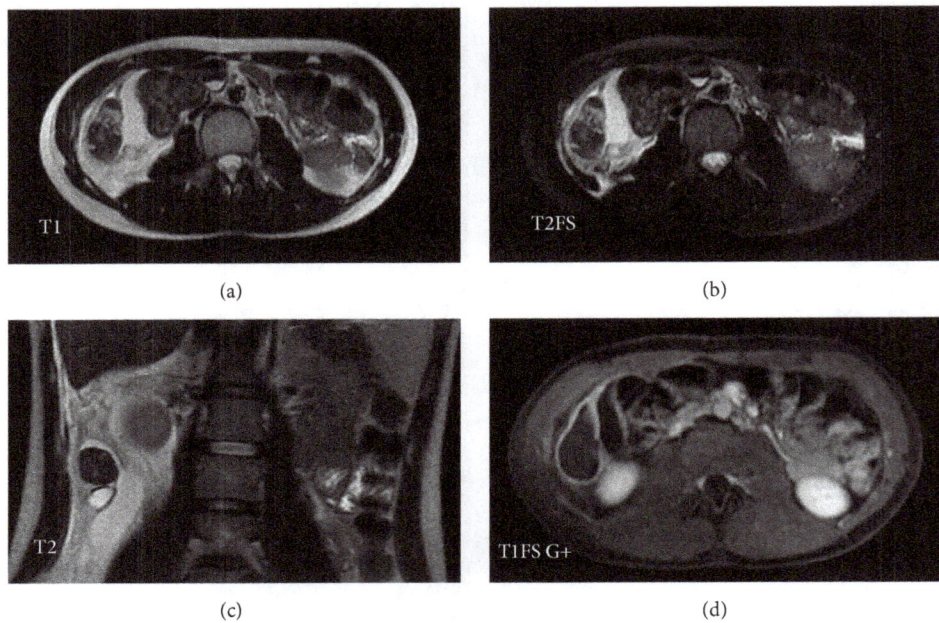

(a)

(b)

(c)

(d)

FIGURE 3: Axial MRI images (a) and (b) demonstrate a mostly hypointense structure on T1 and T2FS sequences in the right upper quadrant with surrounding fluid and edema. On the coronal T2 image (c), a cystic component is seen inferior to this mostly T2 hypointense mass. Axial post contrast image (d) shows peripheral enhancement.

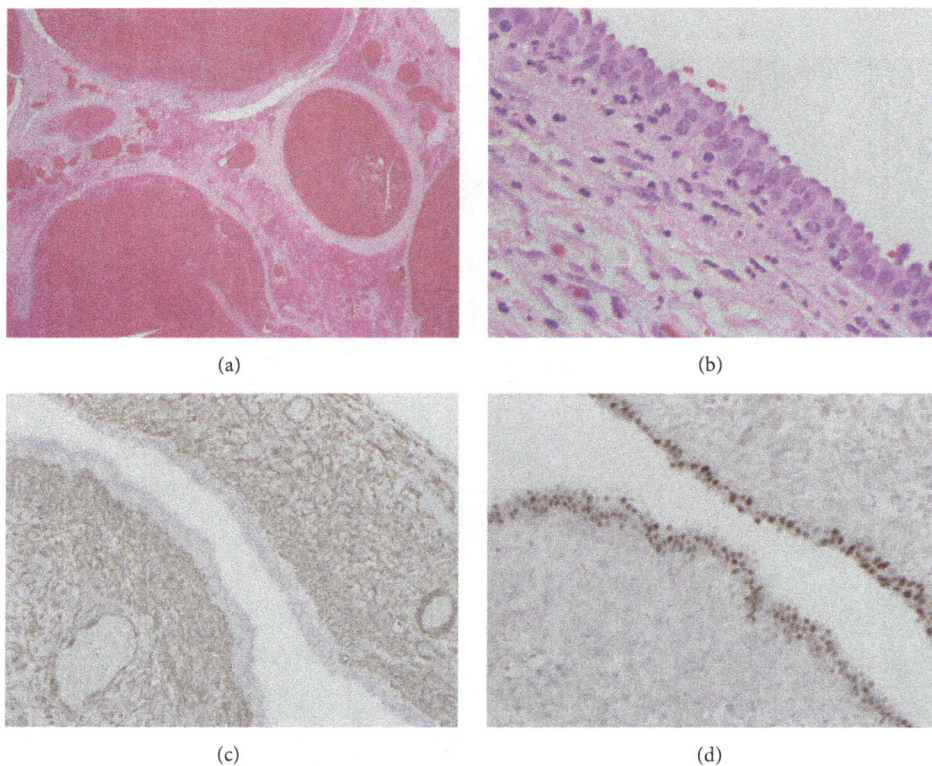

(a)

(b)

(c)

(d)

FIGURE 4: Histopathological images of the resected specimen. Image (a) demonstrates extensive hemorrhagic soft tissue (low power, H&E). Image (b) shows cyst wall lined by cuboidal epithelium (high power, H&E). Image (c) demonstrates smooth muscle in the cyst wall after immunohistochemical staining with smooth muscle actin (medium power, SMA). Image (d) demonstrates positive nuclear immunostaining with PAX-8 (medium power, PAX-8).

The patient underwent laparoscopic surgery for a presumptive diagnosis of a possible torsed supernumerary ovary or acute appendicitis with perforation. The resected gross specimen consisted predominantly of yellow lobulated adipose tissue and a portion of hemorrhagic tubulocystic structure. Microscopic examination demonstrated mature adipose tissue and abundant hemorrhagic tissues (Figure 4(a)). The wall of the cystic structure was lined by cuboidal epithelium (Figure 4(b)). Immunohistochemistry for PAX-8 and smooth muscle actin (SMA) showed the presence of Müllerian type epithelium and smooth muscle within the cyst (Figures 4(c) and 4(d)). The hemorrhagic soft tissue showed no evidence of follicles. The histopathological diagnosis was consistent with Müllerian derived remnant cyst with extensive hemorrhage.

3. Discussion

Müllerian cysts in females are usually seen within the pelvis and typically mimic cysts of ovarian origin [3]. Müllerian cysts of the upper abdomen are an extremely rare occurrence in females [4] and have never previously been reported in a patient with Müllerian agenesis. Histologically, the cysts have a smooth muscle wall and are lined with either cuboidal or columnar epithelial cells. The histogenesis of the development of these cysts is unclear with several proposed theories of origin including caudal growth of the developing mesonephric duct, ectopic ovarian tissue, a secondary Müllerian system, and endometriosis [3, 5]. The differential considerations in this adolescent female presenting with abdominal pain would most commonly include appendiceal or ovarian pathology as well as other unusual causes such as hematocolpos/hematometra and ovarian maldescent, especially if there is known history of Müllerian agenesis. Imaging findings of an upper abdominal heterogeneous cystic mass with distinct arterial supply and venous drainage, in a patient with Müllerian agenesis, should raise the possibility of a Müllerian remnant cyst.

References

[1] K. Morcel and L. Camborieux, "Programme de Recherches sur les Aplasies Müllériennes (PRAM), Guerrier D. Mayer-Rokitansky-Küster-Hauser (MRKH) syndrome," *Orphanet Journal of Rare Diseases*, vol. 2, article 13, 2007.

[2] J. W. Allen, S. Cardall, M. Kittijarukhajorn, and C. L. Siegel, "Incidence of ovarian maldescent in women with müllerian duct anomalies: evaluation by MRI," *American Journal of Roentgenology*, vol. 198, no. 4, pp. W381–W385, 2012.

[3] H. Shayan, D. Owen, and G. Warnock, "Surgical images: soft tissue. Müllerian cyst of the upper abdomen: a lesion mimicking pancreatic cystadenoma," *Canadian Journal of Surgery*, vol. 47, no. 5, pp. 369–371, 2004.

[4] L. Steinberg, D. Rothman, and N. W. Drey, "Mullerian cyst of the retroperitoneum," *American Journal of Obstetrics & Gynecology*, vol. 107, no. 6, pp. 963–964, 1970.

[5] J. Lee, S.-Y. Song, C. S. Park, and B. Kim, "Mullerian cysts of the mesentery and retroperitoneum: a case report and literature review," *Pathology International*, vol. 48, no. 11, pp. 902–906, 1998.

20

An Extrafollicular Adenomatoid Odontogenic Tumor Mimicking a Periapical Cyst

Farzaneh Mosavat ⓘ,[1] Roxana Rashtchian,[1] Negar Zeini ⓘ,[1] Daryoush Goodarzi Pour,[1] Shabnam Mohammed Charlie,[1] and Nazanin Mahdavi[2]

[1]Oral and Maxillofacial Radiology Department, School of Dentistry, Tehran University of Medical Sciences, Tehran, Iran
[2]Oral and Maxillofacial Pathology Department, School of Dentistry, Tehran University of Medical Sciences, Tehran, Iran

Correspondence should be addressed to Negar Zeini; Negarzeini@yahoo.com

Academic Editor: Soon Thye Lim

Adenomatoid odontogenic tumor (AOT) is a rare noninvasive odontogenic tumor that occurs mostly in the second decade of life. Based on its tooth association, AOT can be classified into three categories of follicular, extrafollicular, and peripheral types; the follicular classification is considered as the most common type of AOT. This study reported a large extrafollicular case of AOT in a 40-year-old female. She was asymptomatic and tumor was detected accidentally by her dental practitioner. Since the panoramic radiograph showed a well-defined unilocular radiolucent lesion, we observed radiopaque spots within the lesion by using cone beam computed tomography. The extrafollicular type can mimic a periapical radiolucent lesion.

1. Introduction

Adenomatoid odontogenic tumor (AOT) is a slow-growing, well-defined tumor accounting for 3–7% of all odontogenic tumors [1]. Some authors consider AOTs to be benign and noninvasive neoplasms; however others describe them as developmental hamartomas odontogenic growths [2]. Although the AOT is considered as a low occurrence tumor in the literature, Philipsen et al. reported that AOT ranks fourth among the odontogenic tumors. The increasing number of reports in literature on AOT shows that the tumor develops more frequently than expected [3–5]. Depending on its location and tooth association, AOT can be divided into three classifications of follicular, extrafollicular, and peripheral type. About 70% of AOTs were identified as follicular, which is associated with an impacted permanent or supernumerary tooth; radiographic examination showed a well-circumscribed, unilocular radiolucent lesion which is diagnosed earlier in life than extrafollicular type (mean age of 17 years) [6–8].

The extrafollicular type is a central lesion that is not related to the embedded teeth, and the peripheral type is attached to the gingival structures [9]. Internal radiopaque focus was considered as one of the significant features of AOT, which can help its differential diagnosis from other bone cystic lesions [10]. Philipsen and Reichart showed that nearly two-thirds of AOTs had radiopaque spots inside the lesion [11]. The differential diagnosis of AOT from other lesions similar to AOT (e.g., dentigerous cyst, keratocyst odontogenic tumors, unicystic ameloblastoma, and calcifying cystic odontogenic tumors) in radiographic findings may be difficult. The ability of radiographic modality on showing the radiopaque foci within a lesion is essential for the diagnosis of AOT [7]. In the case of small opacification or superimposed area in the anterior region, CBCT is beneficial modality in demonstrating the detailed internal structures of lesions including radiopaque calcified spots [10].

2. Case Report

A 40-year-old female patient visited the Department of Oral and Maxillofacial Radiology of Tehran Dental School.

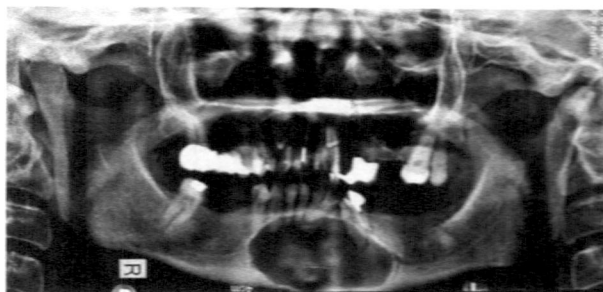

FIGURE 1: Panoramic radiograph shows a single large radiolucent lesion with well-defined border.

She was asymptomatic and the lesion was detected incidentally at routine radiography by her dental practitioner.

Intraorally, the patient had mild bony hard swelling in the anterior region of the mandible. The overlying mucosa was normal, and there was no sign of acute dentoalveolar or mucosal infection in the mandible region. The anterior mandibular teeth were displaced without mobility. The panoramic radiograph revealed a well-defined unilocular radiolucency with corticated rim, which extended from right to left mental foramens. Because of the lesion, the roots of the left lateral mandibular incisor and canine were deviated and resorbed (Figures 1 and 2(c)). The shadow of cervical spine was superimposed over the central part of the lesion (Figure 2(b)). Axial slice showed expansion of buccal and lingual cortical plates in the anterior mandible with perforation along the outer cortical plate at the left side (Figure 2(a)). Differential diagnosis included calcifying odontogenic cyst, central giant cell granuloma, AOT, and ameloblastoma. The lesion was completely enucleated. Microscopically, epithelial cells were arranged as spindle shaped cells in sheets and trabecular pattern and can form duct-like and rosette-like structures in a scant hyalinized stroma (Figure 3(d)).

On gross examination the lesion appears as an elliptical tissue, measuring about 3.5 × 2.7 cm in size (Figures 3(a) and 3(b)). Cut section of the mass revealed multiple cystic spaces and solid area. Small calcifications foci are scattered throughout the tumor. Small islands of tumoral cells have infiltrated the fibrous capsule (Figure 3(c)). Thus, the final diagnosis was given as extrafollicular AOT.

3. Discussion

AOT is a rare odontogenic tumor [12]. The prevalence of AOT is less than odontoma, cementoma, myxoma, and ameloblastoma [13]. AOT is a noninvasive, benign lesion representing 2–7% of all odontogenic tumors [14]. AOT usually appears in the age group of 5–50 years; two-thirds of the cases are diagnosed in the second decade of life, with an average age of 16 years. There is a predilection of AOT in females (female to male ratio = 1.9:1). At least 75% of lesions occur in the anterior maxilla, followed by the anterior mandible, and radiopacities were developed inside 77% of radiolucent lesions [2, 15]. As mentioned above, this tumor has two variants, that is, central and peripheral type (3% of all

cases) [2, 16]. The peripheral type can be similar to a gingival fibroma or epulis [17]. Central tumor may have two types: (1) follicular type is associated with an impacted tooth (73% of all cases) and is often detected in mean age of 17 years and (2) extrafollicular type is often detected in mean age of 24 years (24% of all cases) [2, 3]. The extrafollicular type may appear as a periapical radiolucent lesion mimicking periapical cyst or intrabony defect [18, 19].

Radiographically, central AOT presents as well-defined, almost always unilocular radiolucency [20]. Expansion of the cortical plate can be presented. As a result of tumor expansion, adjacent teeth may be displaced. Tooth displacement is more common than root resorption [21]. This case had unusual radiographic features; it was huge extrafollicular AOT without any radiopaque foci in panoramic radiograph mimicking a periapical lesion. Although AOT occurs most often in second decade, the patient was a 40-year-old female. Late diagnosis of the present case could be due to slow growth and lack of interaction with tooth eruption. The most common site of extrafollicular AOT is anterior region of maxilla (incisor to canine). Our case was observed in the anterior region of mandible, which is the second common site [22]. It has reported that only 28% of AOT lesions occurred in the mandibular incisor area [23]. Generally in patient with AOT lesion, the lamina dura is commonly intact and periodontal ligament is normal. But, in our case, lamina dura cannot be radiographically detected and there was significant root resorption of the involved teeth. The lack of intact periodontal ligament and lamina dura in the involved teeth makes a more likely diagnosis of radicular cyst [18]. Since root resorption rarely occurred in AOT lesion, we detected displacement of the adjacent teeth (especially at the right side) and root resorption of the involved teeth [2]. The size of the current lesion was 3.5 × 2.7 cm; this was is consistent with the size of tumor used in the previous study, which was 1.5–3 cm in diameter [24]. Yilmaz et al. described an AOT causing painless swelling in the anterior mandible which was bony hard with no previous history of trauma, tenderness, discharge, or any other symptoms. These findings were consistent with that of our case [9]. CBCT has the superiority over panoramic radiograph in providing information on the detailed internal structure of the lesion; this can be ascribed to the small calcified area in the lesion. CBCT is the preferred option due to elimination of superimposition and high contrast resolution for

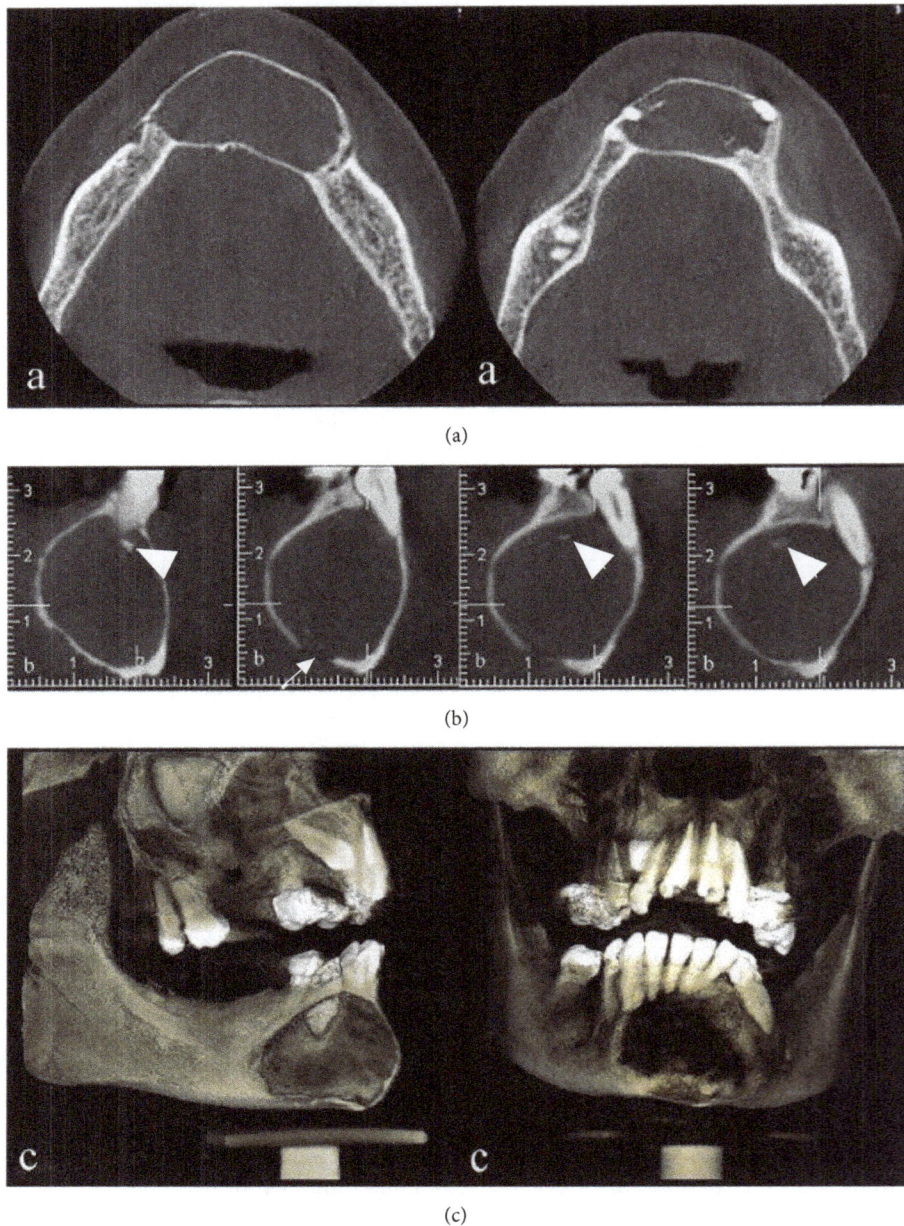

(a)

(b)

(c)

Figure 2: (a) Axial sections show that mental foramen is not involved but has close contact with border of the lesion at the left side. (b) Cross-sectional CBCT images reveal radiopaque spots inside the lesion indicated by white arrows in the image. (c) Three-dimensional volumetric surface rendering.

mineralized tissue such as bones and calcified foci. Therefore, every single detail of a lesion is well depicted on CBCT images.

In summary, some clinical and radiographic features including age and radiolucent appearance of the lesion in a panoramic radiograph did not resemble AOT. However, CBCT assessment, due to its ability to provide more information from the internal structure of the lesion, suggests a differential diagnosis of AOT. Conservation surgical excision, with reoccurrence rate of 0.2%, is today's standard treatment.

Some authors have reported that even incompletely removed lesion does not recur [17].

4. Conclusion

The present case was described as an extrafollicular AOT mimicking a periapical lesion in a panoramic radiograph. In the case of small opacification or superimposed area in the anterior region, CBCT is beneficial modality in

(a)

(b)

(c)

(d)

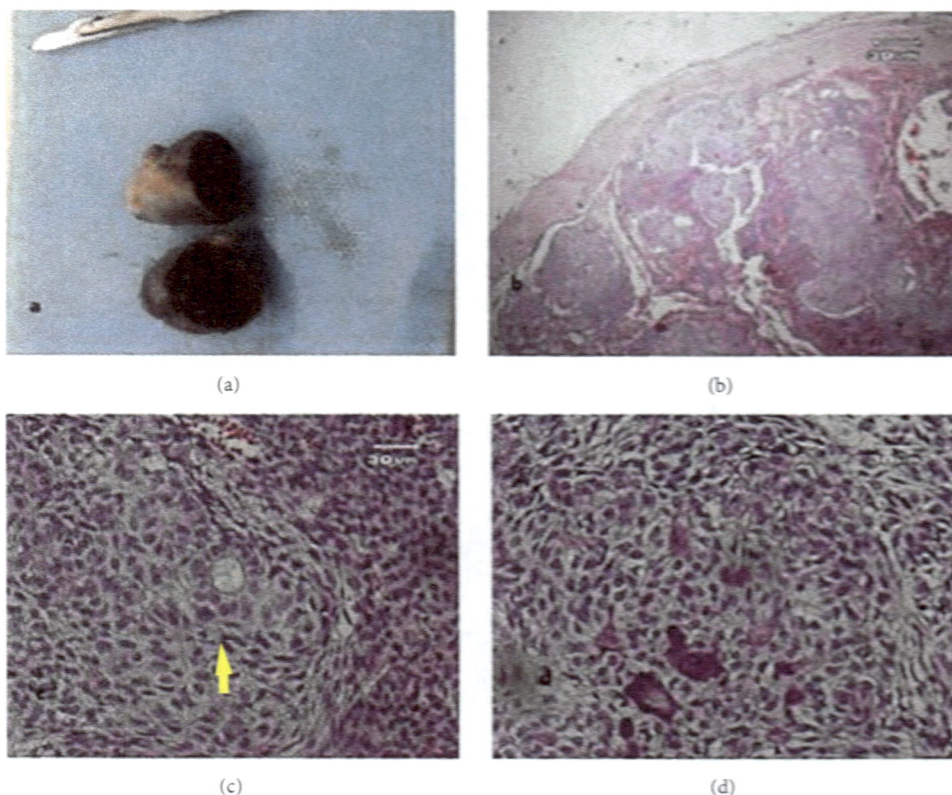

FIGURE 3: (a) On gross examination the lesion appears as an elliptical tissue, with 3.5 × 2.7 cm diameter. Cut section reveals a solid mass with multiple cystic spaces. (b) Low power view demonstrating a thick capsule surrounding the tumor (×40). (c) Duct-like structures which are the characteristic feature of AOT indicated by yellow arrow (400). (d) Spindle shaped cells that form whorled masses and rosette-like structures are noticeable (×400).

demonstrating the detailed internal structures of lesions including radiopaque calcified spots.

References

[1] B. W. Neville, D. D. Damm, C. M. Allen, and J. E. Bouquet, *Oral and Maxillofacial Pathology*, WB Saunders, Philadelphia, PA, USA, 2nd edition, 2002.

[2] White SC PM, *Oral Radiology Principles and Interpretation*, Elsevier, Pennsylvania, Mosby, 7th edition, 2014.

[3] H. P. Philipsen, P. A. Reichart, K. H. Zhang, H. Nikai, and Q. X. Yu, "Adenomatoid odontogenic tumor: Biologic profile based on 499 cases," *Journal of Oral Pathology & Medicine*, vol. 20, no. 4, pp. 149–158, 1991.

[4] H. P. Philipsen and H. Birn, "The adenomatoid odontogenic tumour, ameloblastic adenomatoid tumour or adeno-ameloblastoma," *APMIS-Acta Pathologica, Microbiologica et Immunologica Scandinavica*, vol. 75, no. 3, pp. 375–398, 1969.

[5] P. A. Reichart and H. P. Philipsen, *Odontogenic Tumors and Allied Lesions*, Quintessence Publ, London, UK, 2004.

[6] G. M. Rick, "Adenomatoid odontogenic tumor," *Oral and Maxillofacial Surgery Clinics of North America*, vol. 16, no. 3, pp. 333–354, 2004.

[7] H. P. Philipsen, P. A. Reichart, C. H. Siar, N. g. KH, S. H. Lau, X. Zhang et al., "An updated clinical and epidemiological profile of the Adenomatoid odontogenictumor: a collaborative retrospective study," *Journal of Oral Pathology & Medicine*, vol. 36, pp. 383–393, 2007.

[8] J. Chindasombatjaroen, S. Poomsawat, N. Kakimoto, and H. Shimamoto, "Calcifying cystic odontogenic tumor and adenomatoid odontogenic tumor: Radiographic evaluation," *Oral Surgery, Oral Medicine, Oral Pathology, Oral Radiology, and Endodontology*, vol. 114, no. 6, pp. 796–803, 2012.

[9] N. Yilmaz, A. Acikgoz, N. Celebi, A. Z. Zengin, and O. Gunhan, "Extrafollicular Adenomatoid odontogenic tumor of the mandible: report of a case," *European Journal of Dentistry*, vol. 3, pp. 71–74, 2009.

[10] M. Jiang, M. You, H. Wang, and L. Xu, "Characteristic features of the adenomatoid odontogenic tumour on cone beam CT," *Dentomaxillofacial Radiology*, vol. 43, no. 6, Article ID 20140016, 2014.

[11] H. P. Philipsen and P. A. Reichart, "Adenomatoid odontogenic tumour: facts and figures," *Oral Oncology*, vol. 35, no. 2, pp. 125–131, 1999.

[12] J. G. Handschel, R. A. Depprich, A. C. Zimmermann, S. Braunstein, and N. R. Kübler, "Adenomatoid odontogenic tumor of

the mandible: review of the literature and report of a rare case," *Head & Face Medicine*, vol. 1, article no. 3, 2005.

[13] C. Anand Kumar, R. eddy J, and S. Gupta, "A usual site of AOT presenting as periapical cyst: a rare case report," *JIAOMR*, vol. 22, pp. 39–41, 2010.

[14] W. G. Seo, C. H. Kim, H. S. Park, J. Jang, and W. Chung, "Adenomatoid odontogenic tumor associated with an unerupted mandibular lateral incisor: a case report," *Journal of the Korean Association of Oral and Maxillofacial Surgeons*, vol. 41, no. 6, p. 342, 2015.

[15] T. Becker, A. Buchner, and I. Kaffe, "Critical evaluation of the radiological and clinical features of adenomatoid odontogenic tumour," *Dentomaxillofacial Radiology*, vol. 41, no. 7, pp. 533–540, 2012.

[16] K. Krishnamurthy, R. S. Balaji, S. Devadiga, and R. G. R. Prasad, "Adenomatoid odontogenic tumor in the maxillary antrum: A rare case entity," *Journal of Pharmacy and Bioallied Sciences*, vol. 6, no. 1, pp. S196–S199, 2014.

[17] E. Dayi, G. Gürbüz, O. M. Bilge, and M. A. Çiftcioğlu, "Adenomatoid odontogenic tumour (adenoameloblastoma): case report and review of the literature," *Australian Dental Journal*, vol. 42, no. 5, pp. 315–318, 1997.

[18] H. P. Philipsen, T. Srisuwan, and P. A. Reichart, "Adenomatoid odontogenic tumor mimicking a periapical (radicular) cyst: A case report," *Oral Surgery, Oral Medicine, Oral Pathology, Oral Radiology, and Endodontology*, vol. 94, no. 2, pp. 246–248, 2002.

[19] N. M. Blumenthal and R. Mostofi, "Repair of an intrabony defect from an adenomatoid odontogenic tumor," *Journal of Periodontology*, vol. 71, no. 10, pp. 1637–1640, 2000.

[20] H. L. Shafer, *Textbook of Oral Pathology*, Elsevier Publication, Mosby, Pennsylvania, 5th edition, 2006.

[21] B. W. Neville, "Update on current trends in oral and maxillofacial pathology," *Head & Neck Pathology*, vol. 1, no. 1, pp. 75–80, 2007.

[22] H. P. Philipsen and P. A. Reichart, "Adenomatoid odontogenic tumor: facts and figures," *Oral Oncol*, vol. 35, pp. 1–7, 1998.

[23] J. E. Leon, G. M. Mata, E. R. Fregnani et al., "Clinicopathological and immunohistochemical study of 39 cases of adenomatoid odontogenic tumour: a multicentric study," *Oral Oncology*, vol. 41, no. 8, pp. 835–842, 2005.

[24] G. Prasad, P. Nair, S. Thomas, H. Gharote, N. Singh, and A. Bhambal, "Extrafollicular adenomatoid odontogenic tumour," *BMJ Case Reports*, 2011.

21

Multiple Brain Abscesses due to *Streptococcus anginosus*: Prediction of Mortality by an Imaging Severity Index Score

K. O. Kragha

Department of Radiology, University of Louisville, 530 S. S. Jackson Street, CCB C07, Louisville, KY 40202, USA

Correspondence should be addressed to K. O. Kragha; kokragha@msn.com

Academic Editor: Amit Agrawal

An elderly patient with altered mental status, brain abscesses, ventriculitis, and empyemas died of septic shock and brain abscesses secondary to *Streptococcus anginosus* despite aggressive treatment. An imaging severity index score with a better prognostic value than the Glasgow coma scale predicted mortality in this patient.

1. Introduction

Brain abscess is a local infection within the brain parenchyma that starts as cerebritis and then becomes collection of pus with well vascularized capsule. There are four routes of brain infection: contiguous infection, hematogenous spread to the brain from distant infection, direct implant (trauma or neurosurgery), and peripheral nerves. Brain contiguous infection is usually by odontogenic abscesses, sinusitis, and otomastoiditis. Hematogenous brain abscess is usually in the distribution of the middle cerebral artery, near the gray-white matter junction, and multiple. Brain abscess has an estimated incidence of 0.13–0.9 per 100,000 person-years and mortality of up to 20–70%, mean age of 35–37 years, and a male to female ratio of 1.3 : 1 to 3.0 : 1. Brain abscess may cause long-term neurological deficits or be life threatening. Brain abscess is usually caused by multiple bacterial, fungi, or parasites with *Streptococcus* which is the most common organism. Other less common microbes are *Listeria monocytogenes*, staphylococci, Gram-negative bacilli, or *Haemophilus influenza*. Up to about 30% of brain abscesses are cryptogenic. Brain abscess mortality has decreased over the past few decades due to improved diagnosis and improved antibiotic therapy (cf. Table 1) [1–7].

This case presents the clinical, imaging, and imaging severity index score features and treatment of multiple intracranial abscesses due to *Streptococcus anginosus* that are important for its management and prediction of the outcome.

2. Case Report

This is a 61-year-old female who was admitted with altered mental status. The patient had 1-day history of confusion and hemoptysis that began about 4 days prior to admission. Patient was very agitated, unable to communicate, kept her eyes clenched throughout the entire examination, and would not allow examiners to open her eyes. They were pried open briefly and pupils appeared to be reactive. However, pupillary reaction could not be assessed secondary to her agitation and keeping her eyes clenched tight. The patient moved all her extremities well and tried to climb out of bed. Her right patellar and biceps reflexes were 2/4; reflexes on the left were diminished.

CT and MRI with IV contrast showed 5 abscesses and periabscess edema (cf. Table 2, Figures 1-2). CT understated the size of the abscesses probably because of bone beaming hardening artifact. However, the size of the edema surrounding the abscesses on CT and MRI is similar. On MRI, there was pus in the right lateral ventricle and 4th ventricle and posterior falx and medial left occipital lobe empyemas. There was edema and mass effect with a right to left midline shift of about 9 mm.

Patient underwent neurosurgery to drain the abscesses twice, first right cerebral abscesses, and then left cerebral abscesses. Patient had dural sinus thrombosis and was started on full dose heparin; she improved after surgery and

TABLE 1: Imaging characteristics of abscess [1–20].

Stage	Description	Histology	MRI	MR spectroscopy	CT	CT perfusion
I	Early cerebritis, intermediate cerebritis	Early infection, inflammation. Poorly demarcated, toxic changes, and perivascular infiltrates	T1 isointensity to hypointensity. T2 hyperintensity without a defined margin; ill-defined nodular enhancement		Ill-defined low attenuation, variable contrast enhancement (no enhancement, nodular, or ring enhancement)	
II	Late cerebritis	Reticular matrix (collagen precursor), developing necrotic center	Increasing edema, development of rim of granulation tissue. T1 hypointensity. T2 isointensity to hyperintensity with surrounding edema; incomplete to complete zone of enhancement		Poorly defined low attenuation edema; thick ring or nodular enhancement	
III	Early capsule	Neovascularity, necrotic center, and periphery reticular matrix	Peripheral zone of enhancement is thinner, more uniform in contour, and relative to mesial thinning. No persistent central enhancement		Core is round or ovoid low attenuation, sometimes faint surrounding capsule ring. Ring enhancement corresponds to granulation tissue of capsule; medial or ventricular thinner than lateral wall due to differences in capsule blood supply	
			Loss of capsular hypointensity on T2-weighted images, reduction in size of central necrotic cavity. Enhancement may persist for several months, but progressively decreases on serial examination			
IV	Late capsule	Collagen capsule, necrotic center, and gliosis around capsule	Increased signal intensity of DWI, decreased signal intensity on ADC map	Metabolites identified in abscess: lactate, succinate, acetate, amino acids (valine, leucine, and isoleucine), and aspartate	Same as III above	Increase in cerebral blood flow, cerebral blood volume seen in 24 hours, and peak in 48–72 hours
			Increased fractional anisotropy and reduced mean diffusivity	Late capsule: necrotic center lack normal brain metabolites of NAA, choline, and creatine. Elevated cytosolic amino acids (valine, leucine, and isoleucine), lactate, acetate, and succinate		

TABLE 2: Brain abscesses by MRI and CT with IV contrast done on the same date.

Location	Transverse diameter (cm)		Anterior posterior diameter (cm)		Craniocaudal diameter (cm)		Periabscess edema (cm)	
	MRI	CT	MRI	CT	MRI	CT	MRI	CT
Right temporal lobe	2.6	2.3	2.3	1.8	3.5	N/A	1.3	1.3
Right temporal lobe	3.3	2.6	2.1	1.3	3.4	N/A	0.6	0.8
Right occipital lobe	2.7	2.6	2.9	2.4	3.0	N/A	0.8	0.8
Left occipital lobe	2.5	2.4	3.3	2.5	2.0	N/A	0.6	0.6
Left parietal lobe	3.1	2.4	3.3	2.5	3.2	N/A	1.1	1.3

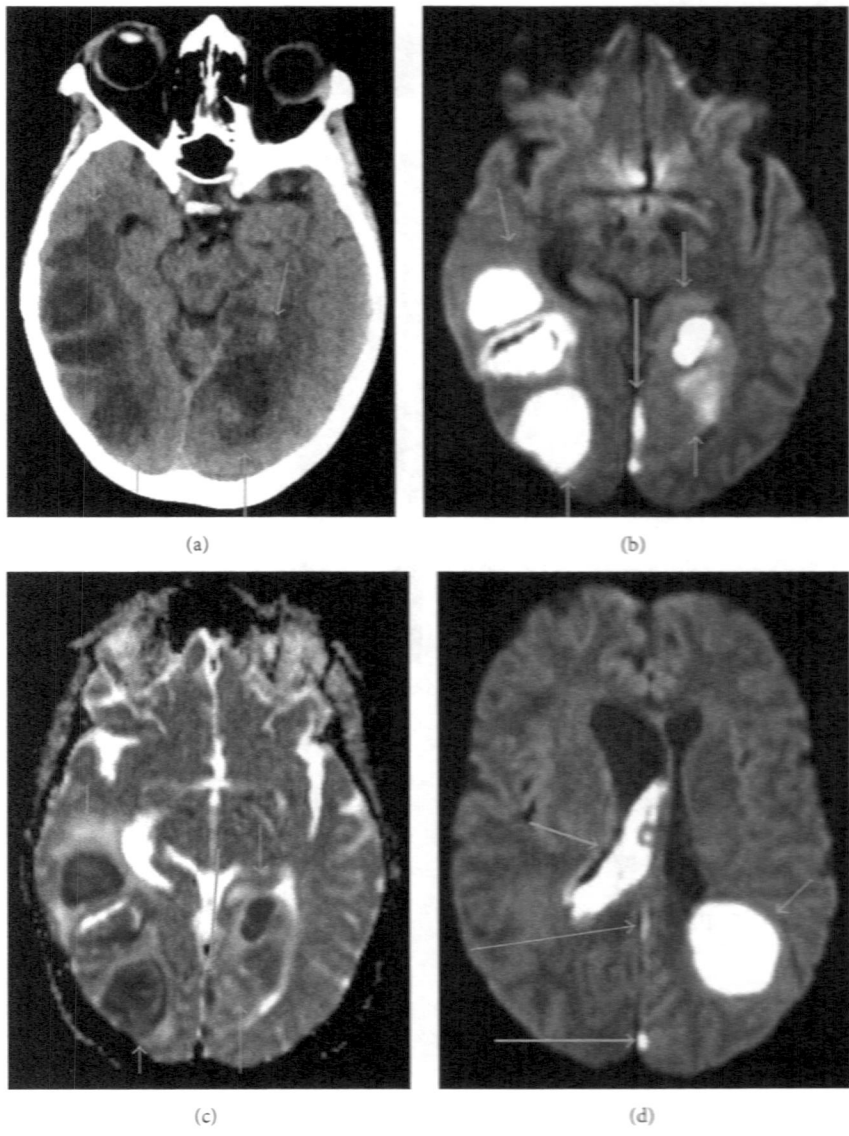

FIGURE 1: (a) Noncontrast axial CT image, (b) axial diffusion weighted images, (c) axial adjusted diffusion coefficient map that shows bilateral temporal-occipital lobe abscesses (thin arrows) and empyema in the medial left occipital lobe (thick arrow), and (d) axial diffusion weighted image that shows ventriculitis in the right lateral ventricle (short thick arrow), abscess in the left parietal lobe (short thin arrow), and empyema in the posterior falx (long thin arrow) and medial left occipital lobe (long thick arrow).

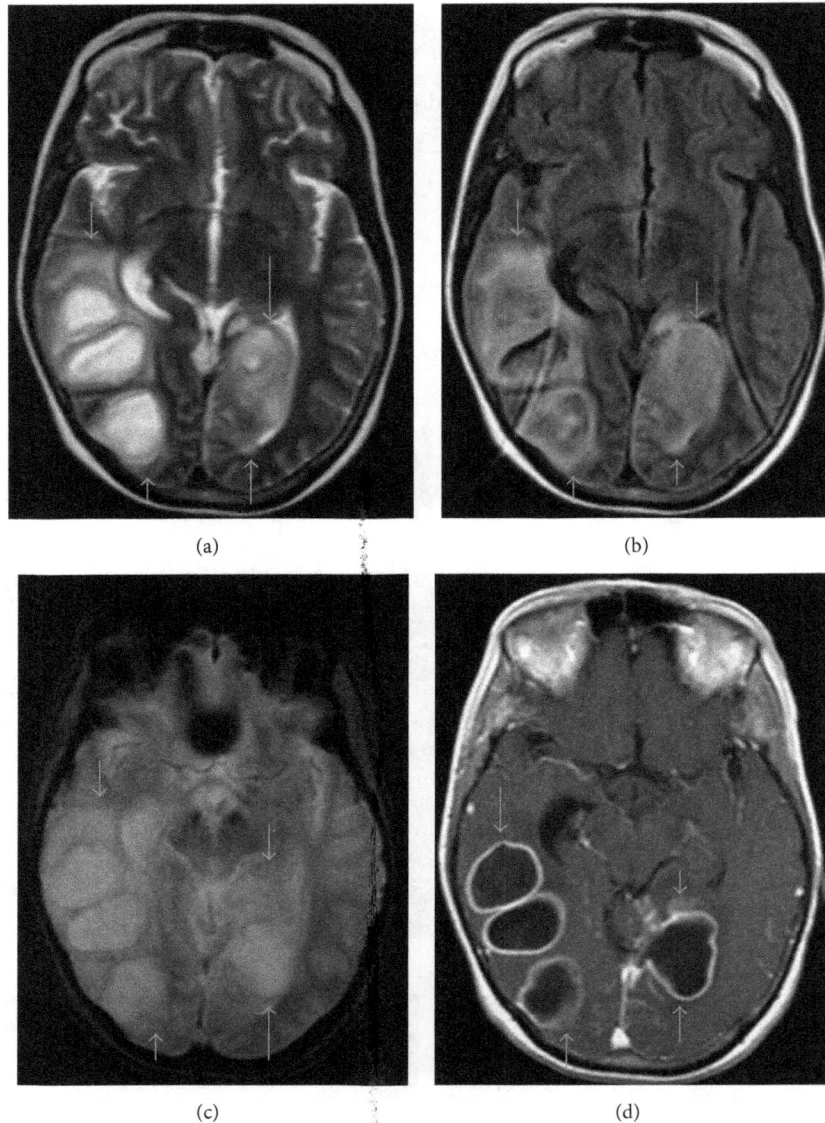

FIGURE 2: (a) Axial T2 weighted image, (b) axial FLAIR weighted image, (c) axial GRE weighted image, and (d) axial T1 weighted postcontrast image show bilateral temporal-occipital lobe abscesses (arrows).

treatment. She became hypotensive likely secondary to reinfection and developed large retroperitoneal hematoma which further exacerbated her shock. Poor prognosis was discussed with family and their decision to withdraw artificial means of survival was made. The patient died due to septic shock and brain abscesses secondary to *Streptococcus anginosus* despite aggressive antibiotic and surgical treatment.

3. Discussion

Brain abscess goes through four stages: early cerebritis, late cerebritis, early capsule, and late capsule. Cerebritis is acute inflammation consisting of neutrophils, macrophages, lymphocytes, and plasma cells with or without parenchymal necrosis. Abscess capsule is a vascular tissue lining, collagen, reactive gliosis, firm, and fibrous and contains lipid-laden macrophages, capillary proliferation, and chronic inflammatory cells [1].

Most abscesses are successfully treated with antibiotics. Surgical drainage may be required depending on the patient's clinical condition, the location of the abscess, and risk of intraventricular rupture (which carries a high mortality). Nearly 90% of abscesses are pyogenic and the mortality rate may be as high as 14% despite antibiotic treatment [5]. Brain abscess may successfully be treated by antibiotic treatment and surgical procedures, either aspiration or excision [6]. This is 61-year-old female with brain abscess that was promptly diagnosed and treated medically and surgically but died about 5 weeks after admission. According to Seydoux and Francioli [8], the mean delay between occurrence of first symptoms and hospitalization is significantly shorter for patients with poor outcome (death or severe sequelae) than for patients who recovered (fully or with moderate sequelae). Moreover, severely impaired mental status and neurological impairment at admission were associated with a poor outcome in terms of both mortality and sequelae. In all cases with fatal outcome or severe sequelae, the diagnosis was made and treatment was initiated within 24 hours of admission.

3.1. Presentation. This patient presented with altered mental status, confusion and agitation, and neurologic deficit. Symptoms of brain abscess depend on location, mass effect, and complications. Headache, fever, focal neurologic deficit, nausea, vomiting, neck stiffness, lethargy, hemiparesis, or seizures may be presenting symptoms of brain abscess. Laboratory tests may add little to the diagnosis of brain abscess [2, 5].

3.2. Imaging. Most abscesses are in the frontal and temporal lobes, normally at the gray-white matter junction. The MRI features of abscess are related to free radicals by phagocytic macrophages in the abscess capsule wall. Ring enhancement is not specific for pyogenic abscess and may be seen in other conditions such as nonpyogenic abscess, high grade neoplasm, primary central nervous system lymphoma, metastasis, infarct, hematoma, thrombosed giant aneurysm, radiation necrosis, demyelinating disease, and *Toxoplasmosis gondii* infection. Imaging characteristics of abscess are (1) 2–7 mm continuous smooth thin rim of enhancement, (2) T2 hyperintense rim, and (3) thinning along the medial wall. Abscess demonstrates high signal intensity on diffusion weighted image and low signal intensity of adjusted diffusion coefficient map. The sensitivity of diffusion weighted imaging in differentiation brain abscess from other brain lesions is 72–95% and 96–100%, respectively. A higher temperature around the abscess capsule than in the brain tissue more distant to the capsule, indicative of inflammation has been demonstrated by a thermosensitive MRI protocol [1, 2, 5, 8–16]. In this patient, there were abscesses in both temporal lobes, both occipital lobes, and left parietal lobe (cf. Table 2).

The necrotic centers of bacterial abscess lack the normal brain metabolites of N-acetylaspartate (NAA), choline, and creatine. The typical resonances within the cavity of untreated pyogenic abscess cavity are elevated cytosolic amino acids (0.9 ppm) and lactate (1.3 ppm) with or without acetate (1.9 ppm) and succinate (2.4 ppm). Although lactate and lipid may be found in brain tumor and brain abscess, cytosolic amino acids (valine, leucine, and isoleucine) are only present in pyogenic brain abscess. Lactate, acetate, and succinate are the byproducts of glycolysis and fermentation by the causative bacteria whereas the amino acids are the results of proteolysis by polymorphonucleocytes in pus. The sensitivity and specificity of MR spectroscopy in distinguishing pyogenic brain abscess from other brain lesion are 72–96% and 30–100%, respectively. [1, 2, 5, 17–19].

Brain perfusion imaging demonstrates that the mean relative cerebral blood volume of brain tumor is significantly higher than that of brain abscess due to higher vascularity and blood brain barrier breakdown in the wall of tumor relative to the collagen wall of abscess. Diffusion tensor imaging shows reduced diffusivity and increased fractional anisotropy due to increased regulation of neuroinflammatory adhesion molecules causing a structure orientation of inflammatory cells in abscess cavity. The sensitivity of increased fractional anisotropy and reduced diffusivity in predicting pyogenic abscess is 100% and 75%, respectively [2].

3.3. Treatment. The size, location, number of abscess, and type of abscess influence the choice of treatment of brain abscess. Brain abscesses of about 2-3 cm are usually treated medically whilst brain abscesses larger than 3 cm or associated with mass effect are usually treated with stereotactic aspiration or excision combined with intravenous antibiotics. Most brain abscesses may be successfully treated with intravenous antibiotics alone for 6–8 weeks in about 10% of brain abscesses. Surgical aspiration/drainage and excision of brain abscess may be needed in about 75% and 15%, respectively, depending on the location of the brain abscess; risk of intraventricular rupture (which carries a high mortality), patient's clinical condition, when the abscess increases in size, and neurological problems develop; the abscess fails to decrease in size after about 3-4 weeks of antibiotic treatment, or recurrent large and superficial abscess despite multiple drainages. A routine follow-up CT or MRI every 2 weeks if new neurological signs develop may be required [2–5]. Nearly 90% of abscesses are pyogenic and the mortality rate may be as high despite antibiotic treatment [5]. The patient died due to septic shock despite aggressive antibiotic and surgical treatment.

3.4. Complications. Brain abscess complications include local mass effect with intracranial herniation, hydrocephalus, meningitis, extra-axial fluid collection, sinus thrombosis, infarct, cranial nerve involvement, ventriculitis, and extra-axial fluid collections (empyema and hygroma). This patient had mass effect, sinus thrombosis, empyemas, and ventriculitis. Intraventricular rupture with ventriculitis is a devastating complication associated with high mortality and appears as ventricular debris and abnormal ependymal enhancement. Empyemas have thicker enhancing rim than that of sterile effusions and nonspecific internal septations [2, 5, 20].

3.5. Prognosis. This is a 61-year-old female with brain abscess that was promptly diagnosed and treated medically and surgically but died about 5 weeks after admission. For brain abscess, the mean delay between occurrence of first symptoms and hospitalization is significantly shorter for patients with poor outcome (death or severe sequelae) than for patients who recovered (fully or with moderate sequelae). Severely impaired mental status and neurological impairment at admission are associated with a poor outcome in terms of both mortality and sequelae. In fatal outcome or severe sequelae, the diagnosis is made and treatment was initiated within 24 hours of admission [8]. Prognosis of brain abscess may be worse in multiple abscesses, intraventricular rupture of abscesses, and deeply located abscesses. The prognosis of brain abscess appears to be mainly determined by the rapidity of progression of the disease before hospitalization and the patient's mental status on admission [3, 8]. This patient had ventriculitis and poor mental status at presentation and

TABLE 3: Image severity index score [3].

Parameters	Points
Number (single, multilocular, or attached to each other like a bunch of grapes = 1; score increased for every additional abscess)	
Solitary	1
Multiple	2–6
Location (superficial: cerebral, cerebellar hemispheres; deep: basal ganglia, thalamus, corpus callosum, brain stem, vermis, and within ventricles; extensive or combined: both superficial and deep)	
Superficial	1
Deep	2
Combined	3
Diameter (large diameter of abscess in transverse plane on CT and MRI; largest diameter of multilocular abscess is the largest diameter of the abscess taken as a whole)	
<2 cm	1
2–4 cm	2
>4 cm	3
Perilesional edema (surrounding edema observed as high signal intensity on T2 weighted MRI image or hypodensity on CT image. For multiple abscesses, highest for edema was used)	
Minimal (maximum thickness < radius of abscess)	1
Moderate (maximum thickness between the radius and diameter of abscess)	2
Large (maximum thickness > diameter of abscess)	3
Midline shift (in mm)	
Mild (<5 mm)	1
Moderate (5–10 mm)	2
Severe (>10 mm)	3

expired about 5 weeks after admission despite aggressive medical and surgical interventions.

3.6. Predictive Value of Imaging Severity Index Score. To the best of my knowledge, the only imaging severity index score (ISI) is that proposed by Demir et al. [3] (Table 3) which had a better prognostic value than the Glasgow coma scale. Using an ISI score cut-off of 8, only 3 out of 41 patients with image severity index score of 8 or less demonstrated any adverse events whereas 38 of 55 patients having image severity index score of 9 or higher had disability. The image severity index score of all their patients that died was 10 or higher. This patient's image severity index score was 12; the patient died about 5 weeks after admission despite aggressive antibiotic and surgical treatments. Image severity index score which can be used by clinicians and radiologists is simple and reproducible and uses only five imaging parameters: number, location, size, amount of surrounding edema, and extent of midline shift. There is a recommendation for modification of the image severity index score to specifically take into account empyema and ventriculitis because these complications are associated with worse prognosis (Table 4).

3.7. Streptococcus anginosus. S. anginosus (milleri) group streptococci are the most common microbes associated with bacterial intracerebral abscesses. The *Streptococcus anginosus* group is made up of three species, *S. anginosus, S. constellatus,* and *S. intermedius,* and is part of viridans group of

streptococci. *S. anginosus* group is normal flora in human oropharyngeal, gastrointestinal, and genitourinary tracts but are strongly associated with abscess formation or pyogenic invasive infections including multiple types of infection: head and neck (periodontitis, odontogenic abscesses, sinusitis, and pharyngitis), central nervous system (brain and spinal cord abscesses), pulmonary (lung parenchyma and pleural abscesses, and pneumonia), cardiovascular (endocarditis and bacteremia), and abdominal cavity (hepatic abscesses, gastrointestinal tract infections, peritoneal abscesses, cholangitis, and appendicitis), soft tissue, bone, and skin (cellulitis, osteomyelitis, subcutaneous and muscle abscesses, and wound infections). The ability of *S. anginosus* group bacteria to form abscesses may be due to its ability to inhibit lysis within neutrophils after phagocytosis. Although greater than 90% of the anginosus group streptococci are susceptible to penicillin, abscesses caused by these microbes may require surgical treatment [7, 21–25].

This patient had multiple abscesses in both temporal lobes, both occipital lobes, and left parietal lobe due to *Streptococcus anginosus* (cf. Table 2). In a retrospective analysis of 49 cases of brain abscesses, (i) *S. anginosus* (milleri) group was the most commonly isolated microorganism in brain abscesses (11 of the 49 cases, 22%), and (ii) multiple organisms were isolated from 8 specimens out of 42 cases (19%) with majority containing Gram-positive cocci and Gram-negative rods [26]. *Streptococcus anginosus* may cause hypothyroidism, meningitis, cerebral venous system

TABLE 4: Image severity index score [3] and proposed modifications in italic letters.

Parameters	Points
Number (single, multilocular, or attached to each other like a bunch of grapes = 1; score increased for every additional abscess)	
Solitary	1
Multiple	2–6
Location (superficial: cerebral, cerebellar hemispheres; deep: basal ganglia, thalamus, corpus callosum, brain stem, vermis, and within ventricles; extensive or combined: both superficial and deep)	
Superficial	1
Deep	2
Empyema	3
Ventriculitis	4
Combined	5
Diameter (large diameter of abscess in transverse plane on CT and MRI; largest diameter of multilocular abscess is the largest diameter of the abscess taken as a whole)	
<2 cm	1
2–4 cm	2
>4 cm	3
Perilesional edema (surrounding edema observed as high signal intensity on T2 weighted MRI image or hypodensity on CT image. For multiple abscesses, highest for edema was used)	
Minimal (maximum thickness < radius of abscess)	1
Moderate (maximum thickness between the radius and diameter of abscess)	2
Large (maximum thickness > diameter of abscess)	3
Midline shift (in mm)	
Mild (<5 mm)	1
Moderate (5–10 mm)	2
Severe (>10 mm)	3

thrombophlebitis, intracranial arteritis, inflammation and thrombosis of the cavernous sinus, and inflammation of the carotid sheath causing monoplegia and hemiplegia [27]. This patient had dural sinus thrombosis, was started on full dose heparin, became hypotensive likely secondary to reinfection, and developed large retroperitoneal hematoma which further exacerbated her shock. Direct specimen culture is the gold standard method of identifying the microorganisms causing *Streptococcus anginosus* brain abscess. A negative blood culture does not exclude the presence of live bacteria; blood cultures are negative in 24–20% of intracerebral abscesses [28, 29]. Patient underwent neurosurgery to drain the abscesses and *Streptococcus anginosus* was cultured from the purulent material evacuated from the abscess.

References

[1] J. B. Nguyen, B. R. Black, M. M. Leimkuehler, V. Halder, J. V. Nguyen, and N. Ahktar, "Intracranial pyogenic abscess: imaging diagnosis utilizing recent advances in computed tomography and magnetic resonance imaging," *Critical Reviews in Computed Tomography*, vol. 45, no. 3, pp. 181–224, 2004.

[2] T. J. Rath, M. Hughes, M. Arabi, and G. V. Shah, "Imaging of cerebritis, encephalitis, and brain abscess," *Neuroimaging Clinics of North America*, vol. 22, no. 4, pp. 585–607, 2012.

[3] M. K. Demir, T. Hakan, G. Kilicoglu et al., "Bacterial brain abscesses: prognostic value of an imaging severity index," *Clinical Radiology*, vol. 62, no. 6, pp. 564–572, 2007.

[4] S. Fuentes, P. Bouillot, J. Regis, G. Lena, and M. Choux, "Management of brain stem abscess," *British Journal of Neurosurgery*, vol. 15, no. 1, pp. 57–62, 2001.

[5] D. C. Hughes, A. Raghavan, S. R. Mordekar, P. D. Griffiths, and D. J. A. Connolly, "Role of imaging in the diagnosis of acute bacterial meningitis and its complications," *Postgraduate Medical Journal*, vol. 86, no. 1018, pp. 478–485, 2010.

[6] M. Yamamoto, T. Fukushima, K. Hirakawa, H. Kimura, and M. Tomonaga, "Treatment of bacterial brain abscess by repeated aspiration—follow up by serial computed tomography," *Neurologia Medico-Chirurgica*, vol. 40, no. 2, pp. 98–105, 2000.

[7] C. A. Petti, K. E. Simmon, J. Bender et al., "Culture-negative intracerebral abscesses in children and adolescents from *Streptococcus anginosus* group infection: a case series," *Clinical Infectious Diseases*, vol. 46, no. 10, pp. 1578–1580, 2008.

[8] C. Seydoux and P. Francioli, "Bacterial brain abscesses: factors influencing mortality and sequelae," *Clinical Infectious Diseases*, vol. 15, no. 3, pp. 394–401, 1992.

[9] T. Ebisu, C. Tanaka, M. Umeda et al., "Discrimination of brain abscess from necrotic or cystic tumors by diffusion-weighted

echo planar imaging," *Magnetic Resonance Imaging*, vol. 14, no. 9, pp. 1113–1116, 1996.

[10] A. B. Haimes, R. D. Zimmerman, S. Morgello et al., "MR imaging of brain abscesses," *American Journal of Roentgenology*, vol. 152, no. 5, pp. 1073–1085, 1989.

[11] Y. Bükte, Y. Paksoy, E. Genç, and A. U. Uca, "Role of diffusion-weighted MR in differential diagnosis of intracranial cystic lesions," *Clinical Radiology*, vol. 60, no. 3, pp. 375–383, 2005.

[12] J. D. Eastwood, R. T. Vollmer, and J. M. Provenzale, "Diffusion-weighted imaging in a patient with vertebral and epidural abscesses," *American Journal of Neuroradiology*, vol. 23, no. 3, pp. 496–498, 2002.

[13] S.-C. Chang, P.-H. Lai, W.-L. Chen et al., "Diffusion-weighted MRI features of brain abscess and cystic or necrotic brain tumors. Comparison with conventional MRI," *Clinical Imaging*, vol. 26, no. 4, pp. 227–236, 2002.

[14] P. G. Jorens, P. M. Parizel, H. E. Demey et al., "Meningoencephalitis caused by *Streptococcus pneumoniae*: a diagnostic and therapeutic challenge. Diagnosis with diffusion-weighted MRI leading to treatment with corticosteroids," *Neuroradiology*, vol. 47, no. 10, pp. 758–764, 2005.

[15] K. Weingarten, R. D. Zimmerman, R. D. Becker, L. A. Heier, A. B. Haimes, and M. D. F. Deck, "Subdural and epidural empyemas: MR imaging," *American Journal of Roentgenology*, vol. 152, no. 3, pp. 615–621, 1989.

[16] R. L. Bernays, S. S. Kollias, and Y. Yonekawa, "Dynamic changes during evacuation of a left temporal abscess in open MRI: technical case report," *Neuroradiology*, vol. 44, no. 5, pp. 438–442, 2002.

[17] P.-H. Lai, S.-S. Hsu, S.-W. Ding et al., "Proton magnetic resonance spectroscopy and diffusion-weighted imaging in intracranial cystic mass lesions," *Surgical Neurology*, vol. 68, supplement 1, pp. S25–S36, 2007.

[18] P. H. Lai, J. T. Ho, W. L. Chen et al., "Brain abscess and necrotic brain tumor: discrimination with proton MR spectroscopy and diffusion-weighted imaging," *American Journal of Neuroradiology*, vol. 23, no. 8, pp. 1369–1377, 2002.

[19] G. Luthra, A. Parihar, K. Nath et al., "Comparative evaluation of fungal, tubercular, and pyogenic brain abscesses with conventional and diffusion MR imaging and proton MR spectroscopy," *American Journal of Neuroradiology*, vol. 28, no. 7, pp. 1332–1338, 2007.

[20] M. B. Fukui, R. L. Williams, and S. Mudigonda, "CT and MR imaging features of pyogenic ventriculitis," *American Journal of Neuroradiology*, vol. 22, no. 8, pp. 1510–1516, 2001.

[21] T. Gray, "Streptococcus anginosus group: clinical significance of an important group of pathogens," *Clinical Microbiology Newsletter*, vol. 27, no. 20, pp. 155–159, 2005.

[22] G. Simone, G. Rubini, A. Conti et al., "*Streptococcus anginosus* group disseminated infection: case report and literature review," *Infezioni in Medicina*, vol. 20, no. 3, pp. 145–154, 2012.

[23] B. Sim and D. A. R. Watson, "Pyogenic brain abscess due to *Streptococcus anginosus*," *The Medical Journal of Australia*, vol. 202, no. 5, article 271, 2015.

[24] M. Yassin, G. K. Yadavalli, N. Alvarado, and R. A. Bonomo, "*Streptococcus anginosus* (*Streptococcus milleri* group) Pyomyositis in a 50-year-old man with acquired immunodeficiency syndrome: case report and review of literature," *Infection*, vol. 38, no. 1, pp. 65–68, 2010.

[25] P. C. Y. Woo, H. Tse, K.-M. Chan et al., "'*Streptococcus milleri*' endocarditis caused by *Streptococcus anginosus*," *Diagnostic Microbiology and Infectious Disease*, vol. 48, no. 2, pp. 81–88, 2004.

[26] J. Carpenter, S. Stapleton, and R. Holliman, "Retrospective analysis of 49 cases of brain abscess and review of the literature," *European Journal of Clinical Microbiology and Infectious Diseases*, vol. 26, no. 1, pp. 1–11, 2007.

[27] C. Zhang, B. Xie, F. D. Shi, and J. Hao, "Multiple intracranial arteritis and hypothyroidism secondary to *Streptococcus anginosus* infection," *Journal of Neurology, Neurosurgery & Psychiatry*, vol. 86, no. 9, pp. 1044–1045, 2015.

[28] S. Esposito, S. Bosis, E. Dusi, C. Cinnante, and N. Principi, "Brain abscess due to *Streptococcus intermedius* in a 3-year-old child," *Pediatrics International*, vol. 53, no. 6, pp. 1104–1105, 2011.

[29] M. A. Kirkman, H. Donaldson, and K. O'Neill, "Multiple intracranial abscesses due to Streptococcus anginosus in a previously well individual," *Journal of Neurology, Neurosurgery and Psychiatry*, vol. 83, no. 12, pp. 1231–1232, 2012.

22

Parotid Oncocytoma as a Manifestation of Birt-Hogg-Dubé Syndrome

Kazuki Yoshida [D],[1] **Masao Miyagawa,**[1] **Teruhito Kido,**[1] **Kana Ide,**[1] **Yoshifumi Sano,**[2] **Yoshifumi Sugawara,**[3] **Hiroyuki Takahata,**[4] **Nobuya Monden,**[5] **Mitsuko Furuya,**[6] **and Teruhito Mochizuki**[1]

[1]*Department of Radiology, Ehime University Graduate School of Medicine, Toon, Japan*
[2]*Department of Pulmonary Surgery, Ehime University Graduate School of Medicine, Toon, Japan*
[3]*Department of Diagnostic Radiology, Shikoku Cancer Center, Matsuyama, Japan*
[4]*Department of Pathology, Shikoku Cancer Center, Matsuyama, Japan*
[5]*Department of Head and Neck Surgery, Shikoku Cancer Center, Matsuyama, Japan*
[6]*Department of Molecular Pathology, Yokohama City University Graduate School of Medicine, Yokohama, Japan*

Correspondence should be addressed to Kazuki Yoshida; kn0wn951753@gmail.com

Academic Editor: Bruce J. Barron

Birt-Hogg-Dubé syndrome (BHD) is a rare autosomal dominant disease characterized by skin fibrofolliculomas, pulmonary cysts, spontaneous pneumothoraces, and renal cancers. Oncocytomas are benign epithelial tumors that are also rare. Recently, there have been a few case reports of BHD with a parotid oncocytoma that appears to have a BHD phenotype. Here we document the eighth known case and describe the magnetic resonance imaging features of the parotid oncocytoma, which mimicked Warthin's tumor. Radiologists should be aware of the association between these rare disorders.

1. Introduction

Birt-Hogg-Dubé syndrome (BHD) is a rare autosomal dominant disease characterized by skin fibrofolliculomas, pulmonary cysts, spontaneous pneumothoraces, and renal cancers [1]. In 1977, Birt, Hogg, and Dubé reported on a group of patients from single kindred who had multiple fibrofolliculomas with trichodiscomas and acrochordons [2]. This hereditary condition was later named Birt-Hogg-Dubé syndrome. In 2002, Nickerson et al. identified the BHD gene, which codes a protein called folliculin [3]. The BHD gene is now known as the folliculin gene (*FLCN*).

Patients with BHD often have renal tumors. Pavlovich et al. reported that 34 (27%) of 124 patients with BHD had renal tumors with variable histology, most commonly hybrid oncocytic tumors, and chromophobe renal cell carcinomas [4]. However, there are few reported cases of BHD with parotid gland oncocytoma. Here we present a rare case of a patient with this association.

2. Case History

A 44-year-old woman presented to our hospital complaining of right lower facial swelling and pain around the parotid gland. Her past medical history was unremarkable. However, she had a family history of pneumothoraces in her father and brother. Magnetic resonance (MR) imaging showed a 35 mm diameter mass in the superficial lobule of the right parotid gland. The lesion appeared hypointense on both T1-weighted MR imaging (T1WI) and T2-weighted imaging (T2WI). On fat-saturated T2WI, the masses appeared hyperintense when compared with the native parotid gland tissue but hypointense on contrast-enhanced T1WI with fat saturation. The lesion was hyperintense on axial diffusion-weighted imaging with a b-value of 1000 s/mm^2 and a low apparent diffusion coefficient (Figure 1). The lesion was suspected to be Warthin's tumor, and a right superficial parotidectomy was performed accordingly.

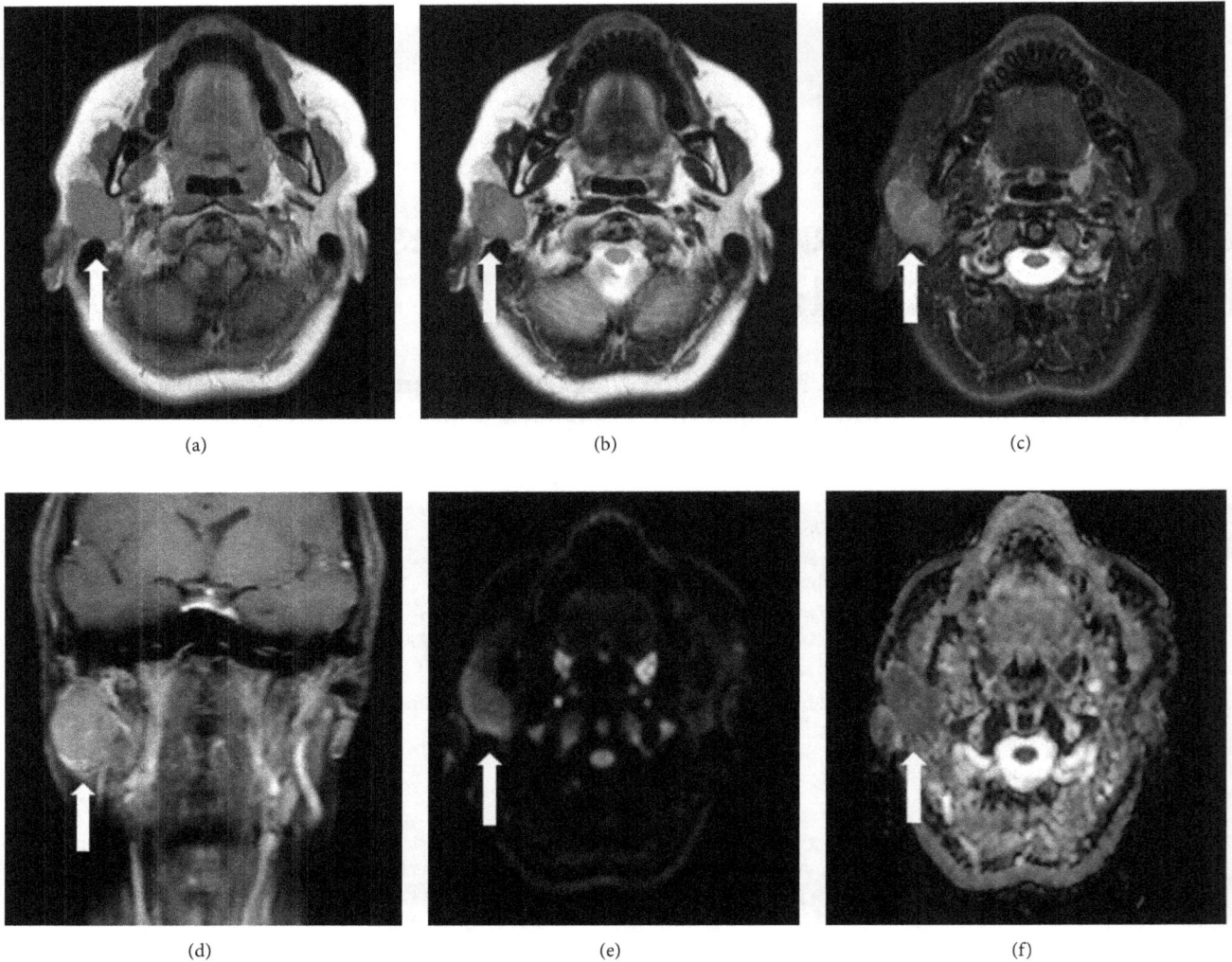

FIGURE 1: Magnetic resonance images of parotid oncocytoma. (a) T1-weighted imaging. (b) T2-weighted imaging. An axial magnetic resonance image shows a well-demarcated mass (white arrow) in the superficial and deep lobe of the right parotid gland. (c, d) The mass appears hyperintense to the native parotid gland tissue on fat-saturated T2-weighted imaging but hypointense on contrast-enhanced, fat-saturated T1-weighted imaging. (e, f) On diffusion-weighted images, the mass appeared hyperintense and hypointense according to the apparent diffusion coefficient.

There was a well-circumscribed, solid, mahogany-colored nodule measuring 3.7 × 2.8 × 2.4 cm in the superficial parotidectomy specimen (Figure 2(a)). Microscopically, the nodule was an encapsulated tumor containing oncocytic cells. These cells formed solid clusters or trabecular patterns, separated by thin strands of fibrovascular stroma (Figure 2(b)) and were round in shape with centrally placed nuclei and clear cytoplasm. Neither necrosis nor capsular invasion was observed. Cytoplasmic granules enriched with glycogen were present but there was no mucin on periodic acid-Schiff staining (data not shown). Although the morphology was comparable to oncocytoma of the kidney, radiologic examination excluded the possibility of a metastatic renal tumor, and immunostaining for PAX-2 and CD10, which

are markers for renal cell carcinoma, was negative (data not shown). The pathologic diagnosis was clear cell oncocytoma.

One and a half years later, the patient presented to hospital again, this time with mild dyspnea. Physical examination revealed decreased breath sounds on the right side. A computed tomography (CT) scan showed a right pneumothorax and multiple cysts. The cysts were located in the medial basilar regions of the lung fields bilaterally and were ellipsoid in shape and variable in size. Some of the cysts abutted the proximal portions of the lower pulmonary arteries (Figure 3). The patient was strongly suspected to have BHD and was subsequently referred for genetic counseling. Informed consent was obtained from the patient for *FLCN* genetic testing, which was performed on genomic DNA extracted from

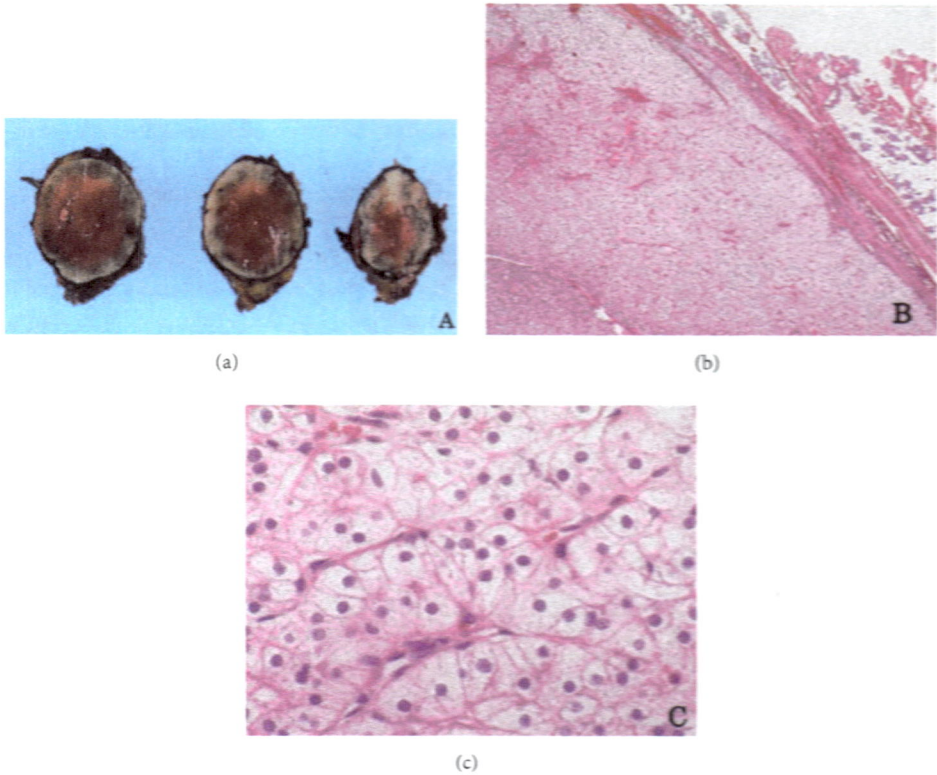

FIGURE 2: Histopathologic findings of parotid clear cell oncocytoma. (a) A well-circumscribed, solid, mahogany-colored nodule. (b, c) Microscopically, fibrous encapsulated nodules containing oncocytic tumor cells with a clear cytoplasm. (hematoxylin-eosin staining; original magnification, 20× in (b) and 400× in (c))

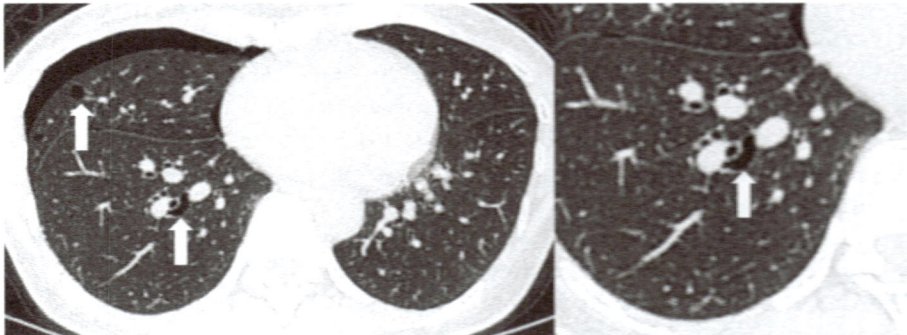

FIGURE 3: Pulmonary computed tomography demonstrating a right pneumothorax and multiple cysts (arrow). An ellipsoidal cyst abuts on the proximal portion of the pulmonary artery.

peripheral leukocytes. Duplication of cytosine was identified in the C8 tract of exon 11 (c.1285dupC), confirming the diagnosis of BHD. The patient's brother, who had an episode of pneumothorax, asked for genetic testing and was found to have the same mutation as the proband (data not shown).

3. Discussion

BHD is a rare disease and there are some reports of its prevalence. In North America, 102 BHD-affected families have been reported by the National Cancer Institute group

[6, 15]. In Asia, 312 affected individuals from 120 Japanese BHD families have been reported [16]. However, information regarding the manifestations of BHD apart from pulmonary cysts, renal tumors, and cutaneous manifestations is limited. Oncocytomas are rare benign epithelial tumors, accounting for only 0.5%–1.5% of all salivary gland tumors. The parotid gland is the site most often affected, accounting for 78%–84% of salivary gland oncocytomas [12]. These tumors are slightly more prevalent in women than in men and usually occur in the seventh to ninth decades of life [14].

Lieu et al. reported the first case of BHD with parotid oncocytoma in a 56-year-old man [5]. Lindor et al. then reported a 45-year-old Caucasian woman with BHD who presented with multiple oncocytic parotid tumors [8]. Pradella et al. also reported a parotid oncocytoma that had arisen in a patient with BHD [9]. In a report by Schmidt et al. in 2005 on 219 BHD-affected individuals, four parotid gland tumors were documented in 2 men and 2 women. Three of those tumors were classified as oncocytoma [6]. In 2011, Maffé et al. reported on 19 BHD-affected individuals, including a 53-year-old man with bilateral parotid oncocytoma. There was a relative reduction of the wild-type signal from the parotid oncocytoma in this patient, who was heterozygous for the *FLCN* mutation. They reported that parotid oncocytoma should be considered as a component manifestation of BHD [7]. To the best of our knowledge, our patient is the eighth reported case of BND with parotid oncocytoma. The previous seven reports and our present case are summarized in Table 1.

FLCN encodes the protein folliculin (FLCN), which acts as a tumor suppressor. FLCN forms a complex with FLCN-interacting protein 1 (FNIP1) and FNIP1 homologue FNIP2, which interacts with 5'-AMP-activated protein kinase and regulates signaling of the mammalian target of rapamycin [17, 18].

The clinical characteristics of BHD are thought to be age-related. Skin lesions usually develop after the age of 20 years. Pulmonary cysts and pneumothoraces are often found in young adult patients aged 20–30 years, and renal cell tumors are more likely to develop after the age of 40 years. Therefore, BHD should be considered in patients who develop repeated spontaneous pneumothoraces and have skin fibrofolliculomas [1, 17]. Abdominal imaging is recommended at least every 3 years in the clinical follow-up of BHD. We prefer MR imaging to CT as it avoids exposure to radiation [18].

The MR imaging reports on parotid oncocytomas suggest that these lesions are hypointense on both T1WI and T2WI [10, 14], are hyperintense on diffusion-weighted imaging, and have low apparent diffusion coefficient values. Dynamic contrast-enhanced MR images show early enhancement with early washout [10]. Parotid oncocytomas also show high uptake of ^{18}F-fluorodeoxyglucose on positron emission tomography (Table 2) [11]. Oncocytomas and Warthin's tumors have very similar features on imaging and therefore are very difficult to differentiate [19]. However, oncocytomas have unique imaging features that have led to them being known as "vanishing tumors" and may help in the diagnosis. Oncocytomas are hard to detect on fat-saturated T2 and T1 postcontrast MR images because of the similarity of intensity of the tumor and the parenchyma, hence the term vanishing tumor. Oncocytes have a low free water content and a lipid-rich membrane. If the amount of lipids in the tumor is approximately equal to that in the normal parotid gland tissue, the vanishing phenomenon occurs [13]. However, in our patient, the lesion appeared slightly hyperintense on fat-saturated T2WI and hypointense on contrast-enhanced and fat-saturated T1WI. The histology was that of a clear cell oncocytoma, which contains large amounts of glycogen and lipid that is different from that of normal oncocytes. A mixed oncocytoma/pleomorphic adenoma is hyperintense on fat-saturated T2WI and contrast-enhanced, fat-saturated T1WI [13]. These different MR imaging features may reflect the histopathologic findings. The previous seven case reports did not describe the imaging findings in detail. It is uncertain if the MR findings of solitary oncocytoma are different from those of oncocytoma associated with BHD; therefore, further investigation is needed.

Tobino et al. reported that multiple, irregular-shaped cysts of various sizes with lower medial lung zone predominance were characteristic features of BHD on CT [20]. Cysts abutting or including the proximal portion of the lower pulmonary arteries or veins are also highly probable in BHD (Figure 3). Our patient had these features, which were very helpful in making a diagnosis of BHD.

In conclusion, we describe the eighth confirmed case in the literature of parotid oncocytoma in BHD, which mimicked Warthin's tumor on MR imaging. Parotid oncocytoma appears to be one of the phenotypes in BHD. It is difficult to distinguish between oncocytomas and Warthin's tumors; however, radiologists should be aware of this association and consider parotid oncocytoma as a differential diagnosis if they detect a parotid mass similar to Warthin's tumor in BHD. If additional imaging is to be recommended, a dedicated renal-protocol MRI would be the choice, alone or in addition to chest CT.

TABLE 1: Reported cases of parotid oncocytoma in Birt-Hogg-Dubé syndrome.

Case	Year	First author and region	Age at diagnosis, years /Sex	Mutation	Pathology of parotid tumor (Age at diagnosis)	Skin	Lung (Age, years)	Kidney (Age, years)	Family history
1	2000	Liu et al. [5] North America	56/M	N/A	Oncocytoma (55)	FF	PTX, LC	-	Sister (PTX, LC)
2-4	2005	Schmidt et al [6] North America	N/A	N/A	Oncocytoma	N/A	N/A	N/A	N/A
5	2011	Maffé et al. [7] Europe	53/M	c.347 dup A (exon 5)	Oncocytoma (32, right; 43, left)	FF or TD	PTX (42)	CCC (37)	Father (hybrid CCC/on with CCC component accounting for 60% of the lesion and sigmoid adenocarcinoma)
6	2012	Lindor et al. [8] North America	45/F	c.779+1 G>T (exon 7i)	Oncocytic neoplasm (45)	FF	-	-	Maternal grandfather (prostate cancer), maternal grandmother (bladder cancer, lung cancer)
7	2013	Pradella et al. [9] Europe	N/A	c.347 dupA (exon 5)	Oncocytoma	FF	PTX LC	N/A	N/A
8	2016	Present report Yoshida et al. Asia	45/F	c.1285dupC (exon 11)	Clear cell oncocytoma (44)	-	PTX (45) LC	-	Father (PTX) Brother (PTX)

M: male; F: female; FF: fibrofolliculoma; TD: trichodiscomas; PTX: pneumothorax; LC: lung cysts; CCC: clear cell carcinoma; N/A: not available.

TABLE 2: Imaging features of parotid oncocytoma.

No.	First author, year	Age/Sex BHD or not	Pathology	Size (mm)	CT enhancement	MRI T1WI	MRI T2WI	MRI DWI/ADC	MRI T2WI FS	MRI T1WI CE FS	FDG-PET
1	Kasai et al. [10] 2007	56/F	Onc	25	N/A	Hypo	Hypo	High /Low	Iso	Iso	N/A
2	Shah et al. [11] 2007	76/M	Onc	20	Homogeneous	N/A	N/A	N/A	N/A	N/A	Intense FDG uptake
3-12	Tan et al. [12] 2010	49–74/ 7 female 3 male	Onc	6–66	6 Heterogeneous 4 Homogeneous	N/A	N/A	N/A	N/A	N/A	N/A
13-21	Patel et al. [13] 2011	N/A 6 female 3 male	8 Onc 1*	13–34	N/A	9 Hypo	N/A	N/A	8 Iso 1 Hyper*	8 Iso 1 Hyper*	N/A
22	Lindor et al. [8] 2012	45/F **BHD**	Onc	9	N/A	Hypo	*Hyper*	N/A	N/A	N/A	N/A
23	Sepúlveda et al. [14] 2014	67/F	Onc	73	Heterogeneous	Iso	N/A	N/A	Hyper	Hyper	N/A
24	This report Yoshida et al.	44/F **BHD**	Clear cell Onc	35	N/A	Hypo	Hypo	High /Low	***Hyper***	***Hypo***	N/A

M: male; F: female; Hypo: hypointense; Iso: isointense; Hyper: hyperintense; N/A: not available; T1WI: T1-weighted imaging; T2WI: T2-weighted imaging; DWI: diffusion-weighted images; ADC: apparent diffusion coefficient; T2WI FS: T2-weighted imaging with fat saturation; T1WI CE FS: postcontrast T1-weighted imaging with fat saturation; FDG-PET: ¹⁸F-fluorodeoxyglucose positron emission tomography; Onc: oncocytes.
*Mixed oncocytoma/pleomorphic adenoma.

Acknowledgments

The authors thank Dr. Shojirou Morinaga, Department of Pathology, Hino Municipal Hospital, Hino, Japan, for making the pathological diagnosis in this patient.

References

[1] F. H. Menko, M. A. van Steensel, S. Giraud et al., "Birt-Hogg-Dubé syndrome: diagnosis and management," *The Lancet Oncology*, vol. 10, no. 12, pp. 1199–1206, 2009.

[2] A. R. Birt, G. R. Hogg, and W. J. Dube, "Hereditary multiple fibrofolliculomas with trichodiscomas and acrochordons," *JAMA Dermatology*, vol. 113, no. 12, pp. 1674–1677, 1977.

[3] M. L. Nickerson, M. B. Warren, J. R. Toro et al., "Mutations in a novel gene lead to kidney tumors, lung wall defects, and benign tumors of the hair follicle in patients with the Birt-Hogg-Dubé syndrome," *Cancer Cell*, vol. 2, no. 2, pp. 157–164, 2002.

[4] C. P. Pavlovich, R. L. Grubb III, K. Hurley et al., "Evaluation and management of renal tumors in the Birt-Hogg-Dubé syndrome," *The Journal of Urology*, vol. 173, no. 5, pp. 1482–1486, 2005.

[5] V. Liu, T. Kwan, and E. H. Page, "Parotid oncocytoma in the Birt-Hogg-Dubé syndrome," *Journal of the American Academy of Dermatology*, vol. 43, no. 6, pp. 1120–1122, 2000.

[6] L. S. Schmidt, M. L. Nickerson, M. B. Warren et al., "Germline BHD-mutation spectrum and phenotype analysis of a large cohort of families with Birt-Hogg-Dubé syndrome," *American Journal of Human Genetics*, vol. 76, no. 6, pp. 1023–1033, 2005.

[7] A. Maffé, B. Toschi, G. Circo et al., "Constitutional FLCN mutations in patients with suspected Birt-Hogg-Dubé syndrome ascertained for non-cutaneous manifestations," *Clinical Genetics*, vol. 79, no. 4, pp. 345–354, 2011.

[8] N. M. Lindor, J. Kasperbauer, J. E. Lewis, and M. Pittelkow, "Birt-Hogg-Dubé syndrome presenting as multiple oncocytic parotid tumors," *Hereditary Cancer in Clinical Practice*, vol. 10, no. 1, article 13, 2012.

[9] L. M. Pradella, M. Lang, I. Kurelac et al., "Where Birt-Hogg-Dubé meets Cowden Syndrome: Mirrored genetic defects in two cases of syndromic oncocytic tumours," *European Journal of Human Genetics*, vol. 21, no. 10, pp. 1169–1172, 2013.

[10] T. Kasai, K. Motoori, T. Hanazawa, Y. Nagai, and H. Ito, "MR imaging of multinodular bilateral oncocytoma of the parotid gland," *European Journal of Radiology Extra*, vol. 63, no. 3, pp. 97–100, 2007.

[11] V. N. Shah and B. F. Branstetter, "Oncocytoma of the parotid gland: a potential false-positive finding on 18F-FDG PET," *American Journal of Roentgenology*, vol. 189, no. 4, pp. 212–214, 2007.

[12] T. J. Tan and T. Y. Tan, "CT features of parotid gland oncocytomas: A study of 10 cases and literature review," *American Journal of Neuroradiology*, vol. 31, no. 8, pp. 1413–1417, 2010.

[13] N. D. Patel, A. Van Zante, D. W. Eisele, H. R. Harnsberger, and C. M. Glastonbury, "Oncocytoma: The vanishing parotid mass," *American Journal of Neuroradiology*, vol. 32, no. 9, pp. 1703–1706, 2011.

[14] I. Sepúlveda, E. Platín, M. L. Spencer et al., "Oncocytoma of the parotid gland: a case report and review of the literature," *Case Reports in Oncology*, vol. 7, no. 1, pp. 109–116, 2014.

[15] J. R. Toro, M.-H. Wei, and G. M. Glenn, "BHD mutations, clinical and molecular genetic investigations of Birt-Hogg-Dubé syndrome: a new series of 50 families and a review of published reports," *Journal of Medical Genetics*, vol. 45, no. 6, pp. 321–331, 2008.

[16] M. Furuya, M. Yao, R. Tanaka et al., "Genetic, epidemiologic and clinicopathologic studies of Japanese Asian patients with Birt–Hogg–Dubé syndrome," *Clinical Genetics*, vol. 90, no. 5, pp. 403–412, 2016.

[17] M. Furuya and Y. Nakatani, "Birt–Hogg–Dubé syndrome: clinicopathological features of the lung," *Journal of Clinical Pathology*, vol. 66, no. 3, pp. 178–186, 2013.

[18] L. Stamatakis, A. R. Metwalli, L. A. Middelton, and W. Marston Linehan, "Diagnosis and management of BHD-associated kidney cancer," *Familial Cancer*, vol. 12, no. 3, pp. 397–402, 2013.

[19] Y. Araki and R. Sakaguchi, "Synchronous oncocytoma and Warthin's tumor in the ipsilateral parotid gland," *Auris Nasus Larynx*, vol. 31, no. 1, pp. 73–78, 2004.

[20] K. Tobino, Y. Gunji, M. Kurihara et al., "Characteristics of pulmonary cysts in Birt-Hogg-Dubé syndrome: Thin-section CT findings of the chest in 12 patients," *European Journal of Radiology*, vol. 77, no. 3, pp. 403–409, 2011.

23

A Case of Ruptured Pulmonary Hydatid Cyst of the Liver and Review of the Literature

Nilufer Bulut[1] and Sevinc Dagıstanlı[2]

[1]Department of Medical Oncology, Kanuni Sultan Suleyman Education and Research Hospital, Istanbul, Turkey
[2]Department of Surgery, Division of General Surgery, Kanuni Sultan Suleyman Education and Research Hospital, Istanbul, Turkey

Correspondence should be addressed to Nilufer Bulut; ferlut@hotmail.com

Academic Editor: Atsushi Komemushi

Background. Hydatid cyst is an endemic disease frequently localized to the liver. It is frequently observed in Southeast Europe, Middle East, and Turkey. Although the cyst rupture can occur spontaneously, it can also occur upon albendazole treatment. Its surgical treatment includes cystotomy, capitonnage, and wedge resection. *Material-Method.* A 56-year-old male immigrant was admitted with fever, pain, and cough. Albendazole treatment was initiated and elective surgery was planned. Upon his admission to emergency service, he was diagnosed with pneumonia, and a spontaneous cyst rupture was detected. *Result.* Thoracotomy and cystotomy were performed. Bile leakage aspiration and lung wedge resection were also performed. *Conclusion.* Different surgical methods are used in the treatment of hydatid cysts depending on the localization and complications. Follow-up with antihelminthic drugs such as albendazole and mebendazole is recommended in medical treatment.

1. Introduction

Hydatid cyst is a parasitic disease caused by *Echinococcus granulosus* which is most frequently seen in the liver, followed by the lungs [1, 2]. It is endemic in Turkey with a prevalence of 50 to 400 in every 100.000 individuals and an incidence of 5.7 in every 100.000 individuals [3, 4].

Cyst hydatid may remain symptomless for years but also may show pressure symptoms due to the growth, rupture to the neighboring structures, or becoming contaminated with infection. In our case, the hepatic cyst hydatid, which did not cause anaphylaxis despite being ruptured, was treated.

2. Case Report

A 56-year-old Syrian male immigrant was admitted with fever, shortness of breath, and abdominal pain in the upper right quadrant for two months. His medical history revealed no known comorbidities. Emergency surgical examination did not reveal any rebound, muscular defence in the abdomen. Laboratory test results were as follows: white blood cell count (WBC) 13.000/mm³ and C-reactive protein

(CRP) 232 mg/dL tomography showed echinococcosis cyst (CE) 102 × 92 mm at the liver segments 6-7. Abdominal ultrasonography revealed a CE-4 type cystic lesion of 106 × 70 mm at the right lobe (Figures 1 and 2). The patient was treated with albendazole (800 mg/day), considering the risk of abscess due to long-term abdominal pain and fever, with metronidazole. However, the patient who admitted to the emergency service the next day with shortness of breath was hospitalized with a diagnosis of pneumonia. In the posteroanterior X-ray examination, a consolidation was observed in the right basal segment of the lung (Figure 3). He was treated with ceftriaxone 2 g/day and clarithromycin 500 mg/day. Oxygen saturation was 97%, arterial blood pressure was 105/65 mmHg, and pulse was 80/min. No rale or rhoncus was determined during the auscultation. We concluded that the rupture of the hydatid cyst was responsible for the clinical condition of the patient and planned drainage of the lung infection with an abdominal incision. As the clinical condition of the patient deteriorated and WBC count gradually increased to 22,700/mm3 and CRP to 314 mg/dl, abdominal cavity was entered with a subcostal incision. A grade 3 hydatid cyst with a solid content and secondary cysts was determined. It was extending to the thoracic cavity through the detachment of the laminar membrane. There was no rupture to the bronchial structures. As the cyst was attached to the diaphragm with a large surface, cystectomy could not be carried out in the liver. Due to the multiple millimetric perforations in the diaphragm, thoracotomy was performed. The cyst was ruptured to the thoracic cavity via diaphragmatic invasion (Figure 4). Cystotomy was also performed to remove the cyst localized in the liver and lung, and millimetric perforations in the diaphragm were primarily repaired. A drainage tube was placed in the abdomen and a thoracic tube was placed in the pleural cavity. The patient was followed in the intensive care unit. During follow-up, his blood chemistry values were as follows: WBC: 28.000/mm^3, CRP: 27 mg/dL, and procalcitonin: 3.38 ng/mL. Twelve days after the operation, biliary content was observed in endotracheal aspiration, and bronchopleural fistula was suspected. Therefore, rethoracotomy and lung wedge resection were performed. One week later, the thoracic tube was removed due to complete expansion of the lung and absence of air leakage (Figure 5). Repeated abdominal ultrasound showed that 120 mL of biliary fluid was aspirated by the injection from the debrided cyst in the liver segment 7. The WBC regressed to 7400/mm^3 and CRP to 15 mg/dL. The patient was discharged on 15th day. In the histopathological examination, necrotic cuticular membrane pieces and bile in the cyst lumen were detected.

3. Discussion

Hydatid cyst rupture can be spontaneous, iatrogenic, or traumatic. Rupture of hydatid cyst in diaphragmatic location is rare. Coughing, dyspnea, fever, hemoptysis, flank pain, chest pain, and secondary pneumonia with pressure to the neighboring bronchi can develop due to the pulmonary rupture of the hydatid cyst on the diaphragmatic surface of the liver [5].

Bronchopneumonia, similar to our case, can develop due to hydatid cyst of the lung, or it can be asymptomatic. While secondary infections develop due to the rupture of the cyst (in 5–8% of the cases), if the cyst fluid enters to the blood circulation of the host, anaphylaxis and allergic reactions can occur [6–8].

Primary lung hydatid cysts can be asymptomatic, as well as spontaneously ruptured or infected. Treatment options include surgical draining of the cyst, peeling of the germinative membrane, and, for inoperable patients, albendazole treatment [1]. Spontaneous or albendazole treatment-related ruptures can be seen in many cases. In some cases, anaphylaxis may not be present due to spontaneous narrowing, although the cyst is ruptured [1, 5]. Moreover, recovery is observed in patients receiving albendazole treatment instead of undergoing surgery, despite the ruptured cyst [1, 4]. For hydatid cyst of the lung, the location of the cyst is critical for the choice of treatment. For centrally localized cysts, as the cyst can lead to bronchopulmonary tree, surgical drainage of the cyst and capitonnage is performed, whereas

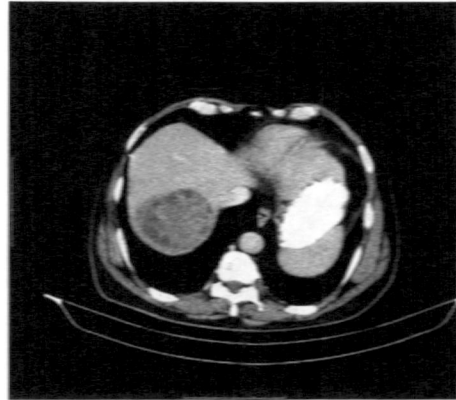

FIGURE 1: Hydatid cyst in right lobe liver on CT scan.

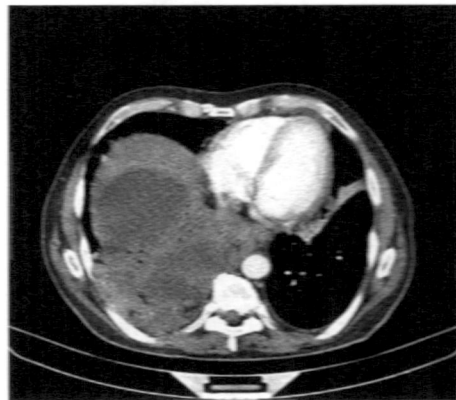

FIGURE 2: Ruptured hydatid cyst in lung via diaphragm on CT scan.

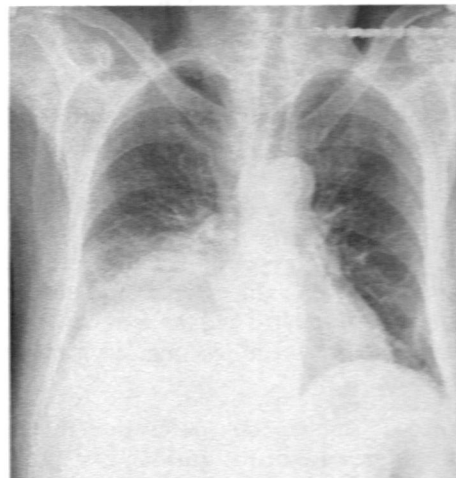

FIGURE 3: Pneumonic infiltration on chest X-ray.

in asymptomatic cysts surrounded by parenchyma, lobectomy, wedge resection, pericystectomy, endocystectomy, or combined albendazole treatment is preferred. Using these for at least 3 months can reduce the number of scolices [1, 5]. However, there is no consensus on the duration of medical treatment. The treatment can last for at least one month to at most one year [7].

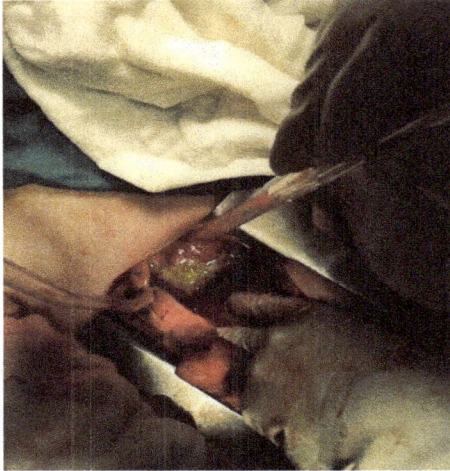

FIGURE 4: Lung cystotomy with subcostal incision.

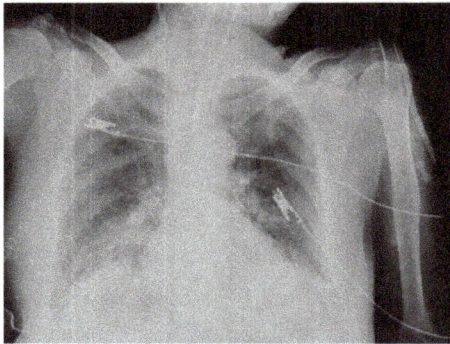

FIGURE 5: Complete expansion of the lung after the thorax tube was removed.

In our case, capitonnage was not performed, since the cyst was close to diaphragmatic surface, and bile leakage was observed during follow-up. Later, wedge resection was performed in the lung, due to the localization of the cyst. Hydatid cyst in the liver was attached to the diaphragmatic surface and cystotomy was performed. Bile accumulation in the liver pouch occurred, as omentopexy was not performed. Bilious content was aspirated and the area was washed with saline. Albendazole treatment (800 mgr/day) was continued for three months.

Spontaneous ruptures can occur due to trauma or increase in the pressure of the growing cyst [2]. Upon its rupture, according to the associated organ, the cyst causes anaphylactic reaction such as pain and fever, hypotension, and fatigue [7, 8]. Although these findings primarily suggest pneumonia, sudden changes in the overall state, persistent cough, and hypotension with the respiratory distress suggest anaphylaxis [3]. In particular, chest X-ray shows cavitary lesions containing air and lotus flower or meniscus appearance. It forms hyperdense areas with an icy glass appearance through direct invasion depending on the defects on the diaphragmatic side of the liver. Diagnosis of the ruptured cyst is made during surgery, via detection of the cavity

and identification of the defect on the degenerate laminar membrane and diaphragmatic surface. In such cases, thoracotomy, laparotomy, or mixed surgical interventions; that is, thoracophrenolaparotomy techniques can be performed. The primary or prosthetic repair of the diaphragm is completed with surgical interventions like cystectomy, cystotomy, and capitonnage or omentoplasty [9].

4. Conclusion

In conclusion, a ruptured hydatid cyst may present asymptomatically or with an anaphylactic reaction. In symptomatic cases, depending on the organ where the cyst is located, surgical excision and capitonnage can be performed. Primary lesion at the dome of the liver can be ruptured into the lung through the diaphragmatic route. In such cases, wedge resection and cystotomy to the primary region are performed, and recurrences are followed with albendazole treatment.

Acknowledgments

The study case was procured from the Department of Medical Oncology and Department of General Surgery of Kanuni Sultan Suleyman Education and Research Hospital.

References

[1] K. Sheikhy, M. Abbassidedezfouli, D. Kakhaki Abolghasem, A. Saghebi, Saghebi Sr., and A. Malekzadegan, "Different and unpredictable clinical outcome of ruptured pulmonary hydatid cysts," Tanaffos, vol. 14, no. 3, pp. 217–221, 2015.

[2] K. Karakaya, "Spontaneous rupture of a hepatic hydatid cyst into the peritoneum causing only mild abdominal pain: a case report," World Journal of Gastroenterology, vol. 13, no. 5, pp. 806–808, 2007.

[3] T. Celik, B. Akçora, M. Tutanç, T. D. Yetim et al., "Ruptured pulmonary hydatid cyst: a case report," Turkey Parasitology Journal, vol. 36, no. 1, pp. 45–47, 2012.

[4] N. Fidan and K. A. Terim Kapakin, "Pathological examinations of lesions seen in the lungs of cows in Ordu and Erzurum provinces," Ataturk Universitesi Veteriner Bilimleri Dergisi, vol. 11, no. 1, pp. 40–46, 2016.

[5] A. Özdemir, Ş. E. Bozdemir, D. Akbiyik et al., "Anaphylaxis due to ruptured pulmonary hydatid cyst in a 13-year-old boy," Asia Pacific Allergy, vol. 5, no. 2, pp. 128–131, 2015.

[6] D. Talwar, A. Dhariwal, B. B. Gupta, and D. Sen Gupta, "Albendazole in pulmonary hydatid disease," The Indian Journal of Chest Diseases & Allied Sciences, vol. 32, no. 4, pp. 237–241, 1990.

[7] B. Victoria and M. B. Rareş, "The management of abdominal hydatidosis after the rupture of a pancreatic hydatid cyst: a case report," Journal of Medical Case Reports, vol. 9, no. 27, pp. 1–5, 2015.

[8] V. Constantin, F. Popa, B. Socea et al., "Spontaneous rupture of a splenic hydatid cyst with anaphylaxis in a patient with multiorgan hydatid disease," Chirurgia (Romania), vol. 109, no. 3, pp. 393–395, 2014.

[9] C. Manterola and T. Otzen, "Hepatic echinococcosis with thoracic involvement. clinical characteristics of a prospective series of cases," Annals of Hepatology, vol. 16, no. 4, pp. 599–606, 2017.

24

Pancreaticoduodenal Artery Aneurysm Associated with Celiac Trunk Stenosis

Jad A. Degheili,[1] **Alissar El Chediak,**[2] **Mohamad Yasser R. Dergham,**[3] **Aghiad Al-Kutoubi,**[3] **and Ali H. Hallal**[1]

[1]*Department of Surgery, Division of General Surgery, American University of Beirut Medical Center, Beirut, Lebanon*
[2]*Department of Internal Medicine, American University of Beirut Medical Center, Beirut, Lebanon*
[3]*Department of Diagnostic Radiology, Division of Interventional Radiology, American University of Beirut Medical Center, Beirut, Lebanon*

Correspondence should be addressed to Ali H. Hallal; ah05@aub.edu.lb

Academic Editor: Daniel P. Link

Pancreaticoduodenal artery aneurysms (PDA) are rare visceral aneurysms. Celiac trunk stenosis represents a common attributable aetiology for those aneurysms. Therefore, an alternative treatment approach, which differs from those isolated aneurysms, is recommended. We hereby present a 77-year-old male patient who was admitted with sudden onset of severe abdominal pain and significant drop in haemoglobin, occurring within a 24-hour interval. Contrast-enhanced computed tomography revealed a ruptured visceral aneurysm arising from the anterior branch of the inferior pancreaticoduodenal artery. A severe stenosis was also noted at the take-off of the celiac trunk. Selective catheterization of the supplying branch of the superior mesenteric artery, followed by coil embolization of the aneurysm, was performed, resulting in cessation of flow within the aneurysm, with preservation of the posterior branch, supplying the celiac territory. PDAs are usually asymptomatic and discovered incidentally at rupture. The risk of rupture is independent of the aneurysmal size and is associated with a 50% mortality rate. The consensus on coping with aneurysms is to treat them whenever they are discovered. Selective angiography followed by coil embolization represents a less invasive, and frequently definitive, approach than surgery. The risk for ischemia mandates that the celiac territory must not be compromised after embolization.

1. Background

Pancreaticoduodenal artery (PDA) aneurysms constitute around 2% of all splanchnic aneurysms [1]. Unlike other aneurysms, the risk of rupture is independent of the aneurysmal diameter. True PDA aneurysms have been linked to hemodynamic alterations in the involved branches caused by altered blood flow due to celiac trunk stenosis [2]. Enlargement of the pancreaticoduodenal arcade is also a contributing factor in the formation of aneurysms secondary to local inflammatory processes such as pancreatitis [3]. Increasing numbers of such cases are being reported due to increased use of cross sectional imaging, and a physiological association between PDA aneurysms and celiac trunk stenosis has been described. However, no unified treatment algorithm has been reported [4]. Transarterial embolization (TAE) is an effective and minimally invasive initial approach [5]. We present a case of a ruptured PDA aneurysm associated with celiac trunk stenosis that was successfully treated with transarterial coil embolization. In this case study, we highlight and discuss the association between celiac trunk stenosis and PDA aneurysms and discuss the various management options.

2. Case Description

A 77-year-old male resident of a nursing home, on antihypertensive and prophylactic anticoagulation medications, sustained a spontaneous postsyncopal head injury that resulted in subarachnoid haemorrhage. He was admitted to our hospital for close observation. One day later, he developed tachycardia with a heart rate reaching 120 bpm,

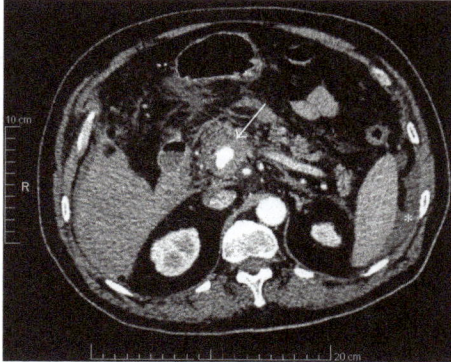

FIGURE 1: *Contrast-enhanced CT scan of the abdomen prior to transarterial embolization*: A 2.0- × 1.4 cm visceral aneurysm *(Arrow)* surrounded by 0.7 cm rim of thrombus arising from the inferior pancreaticoduodenal branch. The hemoperitoneum is also noticed in this figure *(Asterisk)*.

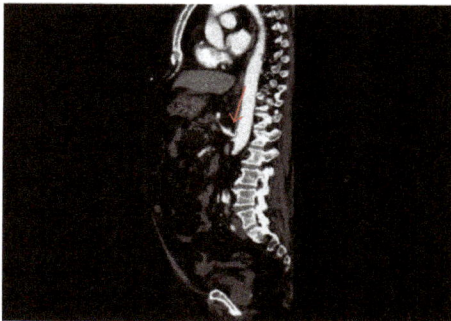

FIGURE 2: *Evidence of celiac trunk stenosis*: Sagittal reconstruction of contrast-enhanced CT scan of abdomen revealing the evidence of stenosis at the take-off of the celiac trunk (Arrow).

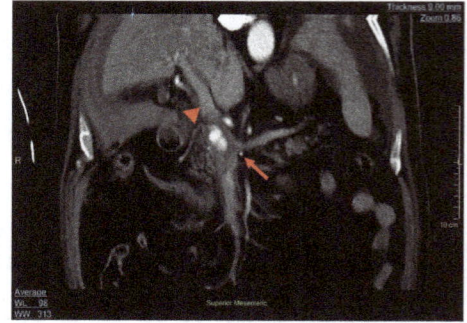

FIGURE 3: *Coronal reconstruction revealing patent portal vein (Arrowhead) with compression of confluence by the aneurysm*: filling of the aneurysm with contrast is shown with its respective compression on the confluence of the superior mesenteric vein and the portal vein *(Arrow)*.

hypotension with a systolic blood pressure of 80 mmHg, and an increase in abdominal girth, distention, and pain. His haemoglobin level decreased from 13.5 to 9.4 g/dL (reference range, 13–18 g/dL). His white blood cell count was 9800/mm^3 (reference range, 4000–11,000/mm^3), and his serum amylase level was 129 IU/L (reference range, 10–120 IU/L). Contrast-enhanced computed tomography (CT) of the abdomen and pelvis showed a 2.0- × 1.4-cm visceral aneurysm, surrounded by a 0.7 cm rim of thrombus, arising from the inferior pancreaticoduodenal branch of the superior mesenteric artery (SMA), in addition to a moderate hemoperitoneum (Figure 1). The aneurysm was associated with severe celiac trunk stenosis (Figure 2). The compression by the aneurysm caused almost complete occlusion of the confluence of the splenic and superior mesenteric veins; however, the portal vein was widely patent (Figure 3). After stabilization with blood and blood product components, selective transarterial embolization was performed.

The right femoral artery was accessed, and a 5-Fr sheath was placed. A 5-Fr SIM2 catheter (Terumo Interventional Systems, Somerset, NJ, USA) was advanced over a 150-cm, 0.035-in Terumo hydrophilic guide wire, and selective catheterization of the inferior pancreaticoduodenal artery from the superior mesenteric artery was performed. Retrograde flow to the common hepatic branch, all the way to the celiac trunk, was demonstrated through the pancreaticoduodenal arcade.

Filling of a single visceral aneurysm originating from an anterior branch of the Inferior PDA (Figures 4(a) and 4(b)) was observed. The aneurysm had an oblong shape arising from the superior aspect of the artery with a very narrow, almost pinpoint neck typical of a false aneurysm. There was also good flow through the posterior branch of the Inferior PDA towards the gastroduodenal arcade and the hepatic artery.

The close proximity of the aneurysm to the pancreas and SMV precluded percutaneous thrombin injection. The parent artery was measured at 4 mm in diameter and was then embolized by advancing a 2.8-F Progate microcatheter (Terumo Interventional Systems) beyond the neck of the aneurysm and the placement of five interlocking 4 mm fibred platinum microcoils (Cook, Bloomingdale, USA), using the "back door-front door" approach, distal, across, and proximal to the origin of the aneurysm (Figure 5(a)). Three millilitres of Gelfoam (Pharmacia and Upjohn Co., Kalamazoo, MI, USA) was introduced into the coiled segment to further promote thrombosis. The posterior branch of the Inferior PDA remained patent and supplying the gastroduodenal arcade and the hepatic artery.

Completion angiography of the SMA revealed cessation of the flow into the aneurysm. The patient's hemodynamic parameters and clinical condition were then stabilized with blood product resuscitation.

3. Outcome and Follow-Up

The patient remained hemodynamically stable, with an almost constant haemoglobin level at 10.4 g/dL and normal liver function test results. His hospital course was protracted with requirement of drainage of ascites and IVC filter insertion. CT angiography performed 5 weeks after embolization for suspicion of bowel ischemia revealed thrombosis of the

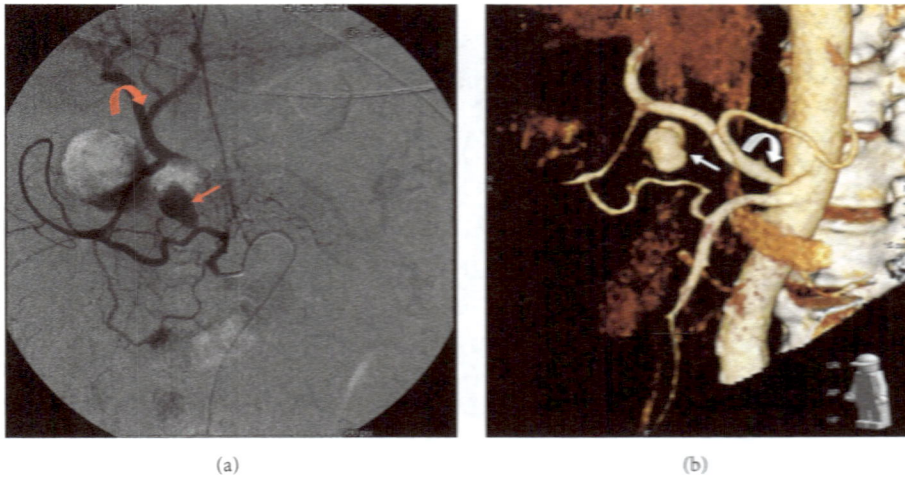

FIGURE 4: *Selective digital subtraction angiography (DSA) of the superior mesenteric artery*: (a) A 2.0- × 1.4-cm narrow neck oblong aneurysm *(Arrow)* originating from the anterior branch of the inferior pancreaticoduodenal artery. Note the several collaterals originating from the superior mesenteric artery and the retrograde filling of the hepatic artery *(Curved arrow)* through the pancreaticoduodenal arcade. (b) Three-dimensional reconstruction of the enhanced CT images revealing the oblong aneurysm *(Arrow)*, originating from a branch of the superior mesenteric artery. Note the stenosis at the origin of the celiac trunk *(Curved arrow)*.

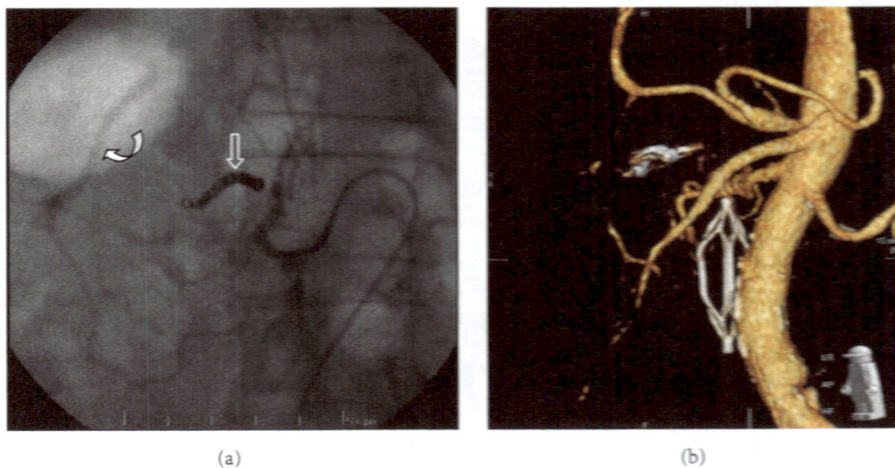

FIGURE 5: *Selective angiography after coil embolization*: (a) Microcoils *(Arrow)* proximal and distal to the origin of the aneurysm. Adequate retrograde collaterals to the liver through the posterior branch are also shown *(Curved arrow)*. (b) Three-dimensional reconstruction on follow-up CT scan, showing persistent occlusion of the aneurysm, with patent hepatic artery.

false aneurysm and a patent hepatic artery and portal vein, with no evidence of liver ischemia (Figure 5(b))

The patient's condition later deteriorated, and he developed severe pneumonia, with acute kidney injury necessitating supportive intubation and initiation of sepsis protocols. He died 40 days after his initial embolization.

4. Discussion and Evaluation

Splanchnic aneurysms can be classified based on their anatomical locations [6]. Visceral aneurysms are the least prevalent among all systematic aneurysms [7]. Several aetiologies have been correlated with their pathogenesis. Infection, trauma from surgeries or endoscopic procedures, collagen diseases, and pancreatitis [8] form the basis for the formation of pseudoaneurysms which becomes more prevalent when the arcade is enlarged secondary to celiac trunk stenosis [9]. True aneurysms have been linked to flow redistribution from celiac trunk stenosis [9]. The high flow rate or kinetics of turbulent blood in the smaller branches of the SMA, as a result of divergent blood from the celiac trunk due to stenosis, increases the shear stress on the single endothelial layer of the intima, resulting in alteration of its biochemical profile, the development of erosion, and increased permeability. These changes reflect deeply into the media layer. The media layer, which maintains the integrity and elasticity of vessels, became itself dysfunctional, resulting in aneurysmal formation [10]. True aneurysms are recognized in 0.09% to 2.00% of the general population [11].

4.1. Celiac Trunk Stenosis and PDA Aneurysms. The pathogenesis behind celiac trunk stenosis may be intrinsic in nature (e.g., caused by atherosclerosis or dysplasia) or extrinsic (e.g., caused by median arcuate ligament compression, which is seen in 10–24% of patients with celiac stenosis) [4]. The first reported case of a PDA aneurysm was published by Ferguson [12]. More than 131 cases of PDA aneurysms have been reported to date; 81 of which were linked to celiac trunk stenosis or occlusion [2]. The initial reported case that correlated PDA aneurysms with celiac trunk occlusion was described by Sutton and Lawton [13]. Since then, the reported incidence has ranged from 45% to 67% [1]. Accumulating evidence reveals that 50% to 80% of PDA aneurysms are associated with celiac artery stenosis [14]. De Perrot et al. [15] reported that 63% of true PDA aneurysms were associated with celiac trunk stenosis. In most reported cases, the cause of this stenosis is ambiguous. Of 12 reported cases with a known aetiology, 9 were attributed to median arcuate ligament compression and 3 were due to atherosclerosis, thrombosis, and agenesis of the celiac trunk, respectively [9].

4.2. Detection of PDA Aneurysms. With the advent and increased utility of imaging techniques, the detection of PDA aneurysms is becoming more frequently reported in the literature. Modalities used for detection include contrast-enhanced multidetector-row CT, three-dimensional contrast-enhanced magnetic resonance angiography [11], and CT angiography with 3D reconstruction from thin section (0.6 mm–3 mm) acquisitions [9, 16]. In addition, CT angiography and magnetic resonance angiography are capable of routinely detecting aneurysms less than 1 cm in diameter [17]. More recently, flow-sensitive four-dimensional magnetic resonance imaging has been implemented in studying chronic hyperkinetic flow in the pancreaticoduodenal arcade secondary to blood shifting from the celiac trunk to the SMA branches in the presence of stenosis. This hyperkinetic flow has been shown to form the basis for the formation of PDA aneurysms [14]. Despite these noninvasive modalities, selective digital subtraction angiography remains the gold standard for the diagnosis of PDA aneurysms because the location of the aneurysm and the supplying artery can be determined, and definitive treatment can be simultaneously performed through TAE.

With the presence of celiac trunk stenosis and the consequent divergent of retrograde blood from the SMA to the celiac territory, the arcade becomes engorged and easily visualized by SMA angiography and MRI kinetic studies [9, 14].

4.3. Management. No treatment guidelines have been established for the management of PDA aneurysms. Consensus states that such aneurysms must be treated, once detected. With a gastrointestinal haemorrhage incidence of 7% to 15% [18], mostly into the retroperitoneal cavity [19], the presence of PDA aneurysms is considered life-threatening. No correlation exists between the size of the PDA aneurysm and the rate of rupture; however, rupture is associated with a significant mortality rate reaching 50% [4, 5, 9] or even higher (up to 75%) [11]. Approximately 17.6% of ruptured aneurysms are ≤10 mm in diameter [1]. Suzuki et al. [20] reported a similar mean diameter (22.2 versus 21.4 mm) among PDA aneurysms that did and did not rupture, respectively. These facts render PDA aneurysms unique with respect to other visceral aneurysms, thus necessitating rigorous planning and implementation of treatment upon recognition.

A major goal in the treatment of aneurysms associated with celiac trunk stenosis revolves around obliteration, resolution of any associated pathologies, and maintenance of adequate blood flow to territories of the celiac trunk. Surgical options oscillate between ligation/resection and aneurysmorrhaphy. These treatments are associated with a high mortality rate and technical difficulties, especially after rupture. Continuous monitoring of the hepatic venous saturation through a right hepatic venous catheter could act as a surrogate approach to ensure an adequate blood supply to the celiac territories after resection or ligation of PDA aneurysms [21]. A saturation level of >60% indicates adequate hepatic perfusion [21].

Less invasive techniques, including TAE with or without recanalization or bypass of celiac stenosis, have recently been predominating. Celiac trunk recanalization promotes stagnation of blood within the aneurysm, resulting in regression in the size of the aneurysm by formation of an intramural thrombus [1]. No cases of PDA aneurysm recurrence after successful endovascular embolization alone have been reported, even without the resumption of adequate celiac flow. Considering this evidence, Suzuki et al. [5] stated that if the ischemic risk to the liver and duodenum is not significant, there is no need to reverse the stenosis. In such an approach, monitoring any ischemic insult to the liver would include serial monitoring of liver function tests, after embolization.

Transcatheter embolization may take the form of total occlusion of the parent artery in cases of fusiform aneurysms, or coil embolization of the aneurysm itself if it is saccular and its neck is accessible. Occlusion of the parent artery beyond and proximal to the neck of the aneurysm (back door/front door technique) is mandatory to prevent retrograde filling. However it is not always successful because vessels can be tortuous and difficult to bypass for deployment of embolic agents. However the development of new trackable microcatheters has improved the ability of the interventional radiologist to reach the target vessel and optimize the embolization procedure. Recurrence of an aneurysm after embolization may occur because collaterals can be excessive and difficult to access and occlude completely [22]. Alternative techniques such as percutaneous thrombin injection (PTI) under CT or ultrasound guidance have been implemented for the treatment of false aneurysms, thus providing patients with more options for minimally invasive procedures before proceeding to surgery [23]. PTI was pioneered by Cope and Zeit in 1986 [24] and since then has shown success in obliterating the aneurysmal sac through thrombin injection and thrombus formation without the need to embolize inflow and outflow vessels. With the use of 21-gauge or smaller calibre needles for percutaneous access, the risk of major organ-specific complications ranges from 0.1% to 2.0% [23]. PTI is characterized by a shorter overall procedure time and lower operational cost than transarterial embolization [22]. We did not attempt this approach because of the patient's moderate hemoperitoneum, fear of bowel perforation, and the proximity of the aneurysmal sac to pivotal organs, as well as the absence of real-time evaluation of the amount of thrombin being administered using CT guidance [23], all of which rendered this approach unfavourable.

The most concerning complication after TAE is reperfusion with subsequent rupture and bleeding into the abdominal cavity. The incidence of this complication depends on the technical success of the procedure and reportedly ranges from 5% to 20% [25], thus necessitating strict radiological follow-up. Some authors suggest imaging initially at 6 months and possibly yearly thereafter [17]. Further complications of embolization include ischemic injury to the liver, pancreas, or duodenum; tissue necrosis with subsequent abscess formation; and possible sepsis [26].

Several reports have described variations in the treatment of PDA aneurysms in conjunction with celiac trunk stenosis. Simultaneous treatment of the stenosis in conjunction with the aneurysm is still a matter of debate. Stambo et al. [27], Lossing et al. [28], and Savastano et al. [29] reported no recurrence of PDA aneurysms after transcatheter embolization and no need for revascularization of the celiac trunk. Others have reported complete obliteration of PDA aneurysms by formation of an intramural thrombus secondary to blood stagnation; only revascularization of the celiac trunk was performed in their situation, without resection or ligation of the aneurysm [1]. A contradictory report published by Suzuki et al. [20] negated the successfulness of embolization alone, without revascularization.

We did not attempt simultaneous celiac trunk recanalization as we considered this additional procedure an extra stress to our already hemodynamically unstable patient. The major concern would be to stop aneurysmal bleeding, at time of presentation. Had it been that our patient was stable enough, celiac trunk recanalization could be considered simultaneously, albeit the different contradictory, yet valid, viewpoints discussed in the pertinent literature.

In addition, Takao et al. [30] had reported the largest series of unruptured true pancreaticoduodenal artery aneurysms, followed without any intervention. Of the five reported patients with a total of eight true PDA aneurysms, four were associated with celiac trunk stenosis. Over a mean follow-up period of 29.4 months, three aneurysms increased in size, but no ruptured aneurysm occurred. Thus, in spite of the small sample size, the authors concluded that the risk of rupture of PDA aneurysms might be lower than predicted from other ruptured aneurysms' series.

Albeit TAE is not a novel approach for the treatment of ruptured PDA aneurysms associated with celiac trunk stenosis, it has been based only on single case series. The present descriptive case, along with the aforementioned literature review, has shed more light on this controversial topic.

5. Conclusion

PDA aneurysms are becoming more frequently reported in association with celiac trunk stenosis. The high mortality rate upon rupture necessitates abrupt treatment. No consensus exists regarding the optimal approach for treatment of these associated conditions. However, TAE is now accepted as an appropriate initial modality. The presence of celiac stenosis complicates the decision-making but is unlikely to require treatment. Whatever approach is used, maintenance of adequate perfusion to the celiac territory is pivotal. The performance of embolization, PTI, or surgery must be decided on a case-by-case basis until a well-established algorithm is devised.

Abbreviations

PDA: Pancreaticoduodenal artery
IPDA: Inferior pancreaticoduodenal artery
TAE: Transarterial embolization
SMA: Superior mesenteric artery
PTI: Percutaneous thrombin injection.

Authors' Contributions

J. A. Degheili carried out the literature review and wrote the initial draft of the manuscript, incorporating all changes and revisions advised, thereafter, by the authors, to achieve the final version of the manuscript. A. El Chediak and M. Y. Dergham assisted with the literature review and initial drafting. A. Al-Kutoubi is the senior interventional radiologist who performed the transarterial embolization and contributed to the detailed description of the procedure, along with major contribution in editing and drafting of the manuscript. A. H. Hallal is the senior surgeon who assisted in drafting and editing the different versions of the manuscript, along with providing his expert opinion.

Acknowledgments

The authors also gratefully acknowledge both Professor Nobuhiro Akuzawa from the Department of Internal Medicine, Gunmachūō General Hospital, Japan, and Professor Katsuyuki Hoshina from the Division of Vascular Surgery,

The University of Tokyo Hospital, Japan, for their expert proofreading and valuable comments.

References

[1] Y.-W. Tien, H.-L. Kao, and H.-P. Wang, "Celiac artery stenting: a new strategy for patients with pancreaticoduodenal artery aneurysm associated with stenosis of the celiac artery," *Journal of Gastroenterology*, vol. 39, no. 1, pp. 81–85, 2004.

[2] M. B. Armstrong, K. S. Stadtlander, and M. K. Grove, "Pancreaticoduodenal artery aneurysm associated with median arcuate ligament syndrome," *Annals of Vascular Surgery*, vol. 28, no. 3, pp. 741-e5, 2014.

[3] T. C. Y. Pang, R. Maher, S. Gananadha, T. J. Hugh, and J. S. Samra, "Peripancreatic pseudoaneurysms: a management-based classification system," *Surgical Endoscopy and Other Interventional Techniques*, vol. 28, no. 7, pp. 2027–2038, 2014.

[4] O. Ikeda, Y. Tamura, Y. Nakasone, K. Kawanaka, and Y. Yamashita, "Coil embolization of pancreaticoduodenal artery aneurysms associated with celiac artery stenosis: report of three cases," *CardioVascular and Interventional Radiology*, vol. 30, no. 3, pp. 504–507, 2007.

[5] K. Suzuki, Y. Tachi, S. Ito et al., "Endovascular management of ruptured pancreaticoduodenal artery aneurysms associated with celiac axis stenosis," *CardioVascular and Interventional Radiology*, vol. 31, no. 6, pp. 1082–1087, 2008.

[6] G. R. Upchurch Jr., G. B. Zelenock, and J. C. Stanley, "Splanchnic artery aneurysms," in *Vascular Surgery*, R. B. Rutherford, Ed., pp. 1565–1581, WB Saunders, Philadelphia, Pa, USA, 6th edition, 2005.

[7] A. Ikoma, "Inferior pancreaticoduodenal artery aneurysm treated with coil packing and stent placement," *World Journal of Radiology*, vol. 4, no. 8, p. 387, 2012.

[8] P. G. Tarazov, A. M. Ignashov, A. V. Pavlovskij, and A. S. Novikova, "Pancreaticoduodenal artery aneurysm associated with celiac axis stenosis: combined angiographic and surgical treatment," *Digestive Diseases and Sciences*, vol. 46, no. 6, pp. 1232–1235, 2001.

[9] M. Katsura, M. Gushimiyagi, H. Takara, and H. Mototake, "True aneurysm of the pancreaticoduodenal arteries: a single institution experience," *Journal of Gastrointestinal Surgery*, vol. 14, no. 9, pp. 1409–1413, 2010.

[10] J. C. Lasheras, "The biomechanics of arterial aneurysms," in *Annual review of fluid mechanics. Vol. 39*, vol. 39 of *Annu. Rev. Fluid Mech.*, pp. 293–319, Annual Reviews, Palo Alto, CA, 2007.

[11] M. Koganemaru, T. Abe, M. Nonoshita et al., "Follow-up of true visceral artery aneurysm after coil embolization by three-dimensional contrast-enhanced MR angiography," *Diagnostic and Interventional Radiology*, vol. 20, no. 2, pp. 129–135, 2014.

[12] F. Ferguson, "Aneurysm of the superior pancreaticoduodenalis, with perforation into the common bile duct," *Proceedings of the New York Pathological Society*, vol. 24, 1895.

[13] D. Sutton and G. Lawton, "Coeliac stenosis or occlusion with aneurysm of the collateral supply," *Clinical Radiology*, vol. 24, no. 1, pp. 49–53, 1973.

[14] Y. Mano, Y. Takehara, T. Sakaguchi et al., "Hemodynamic assessment of celiaco-mesenteric anastomosis in patients with pancreaticoduodenal artery aneurysm concomitant with celiac artery occlusion using flow-sensitive four-dimensional magnetic resonance imaging," *European Journal of Vascular and Endovascular Surgery*, vol. 46, no. 3, pp. 321–328, 2013.

[15] M. De Perrot, T. Berney, J. Deléaval, L. Bühler, G. Mentha, and P. Morel, "Management of true aneurysms of the pancreaticoduodenal arteries," *Annals of Surgery*, vol. 229, no. 3, pp. 416–420, 1999.

[16] A. Horiguchi, S. Ishihara, M. Ito et al., "Multislice CT study of pancreatic head arterial dominance," *Journal of Hepato-Biliary-Pancreatic Surgery*, vol. 15, no. 3, pp. 322–326, 2008.

[17] M. D. Sgroi, N.-K. Kabutey, M. Krishnam, and R. M. Fujitani, "Pancreaticoduodenal artery aneurysms secondary to median arcuate ligament syndrome may not need celiac artery revascularization or ligament release," *Annals of Vascular Surgery*, vol. 29, no. 1, pp. 122–122.e7, 2015.

[18] K. S. Chiang, C. M. Johnson, M. A. McKusick, T. P. Maus, and A. W. Stanson, "Management of inferior pancreaticoduodenal artery aneurysms: a 4-year, single center experience," *Cardio-Vascular and Interventional Radiology*, vol. 17, no. 4, pp. 217–221, 1994.

[19] K. Flood and A. A. Nicholson, "Inferior pancreaticoduodenal artery aneurysms associated with occlusive lesions of the celiac axis: diagnosis, treatment options, outcomes, and review of the literature," *CardioVascular and Interventional Radiology*, vol. 36, no. 3, pp. 578–587, 2013.

[20] K. Suzuki, H. Kashimura, M. Sato et al., "Pancreaticoduodenal artery aneurysms associated with celiac axis stenosis due to compression by median arcuate ligament and celiac plexus," *Journal of Gastroenterology*, vol. 33, no. 3, pp. 434–438, 1998.

[21] M. Tori, M. Nakahara, H. Akamatsu, S. Ueshima, M. Shimizu, and K. Nakao, "Significance of intraoperative monitoring of arterial blood flow velocity and hepatic venous oxygen saturation for performing minimally invasive surgery in a patient with multiple calcified pancreaticoduodenal aneurysms with celiac artery occlusion," *Journal of Hepato-Biliary-Pancreatic Surgery*, vol. 13, no. 5, pp. 472–476, 2006.

[22] R. P. Chan and E. David, "Reperfusion of splanchnic artery aneurysm following transcatheter embolization: treatment with percutaneous thrombin injection," *CardioVascular and Interventional Radiology*, vol. 27, no. 3, pp. 264–267, 2004.

[23] A. Ghassemi, D. Javit, and E. H. Dillon, "Thrombin injection of a pancreaticoduodenal artery pseudoaneurysm after failed attempts at transcatheter embolization," *Journal of Vascular Surgery*, vol. 43, no. 3, pp. 618–622, 2006.

[24] C. Cope and R. Zeit, "Coagulation of aneurysms by direct percutaneous thrombin injection," *American Journal of Roentgenology*, vol. 147, no. 2, pp. 383–387, 1986.

[25] U. Sachdev-Ost, "Visceral artery aneurysms: review of current management options," *Mount Sinai Journal of Medicine*, vol. 77, no. 3, pp. 296–303, 2010.

[26] M. Izumi, M. Ryu, A. Cho et al., "Ruptured pancreaticoduodenal artery aneurysm treated by superselective transcatheter arterial embolization and preserving vascularity of pancreaticoduodenal arcades," *Journal of Hepato-Biliary-Pancreatic Surgery*, vol. 11, no. 2, pp. 145–148, 2004.

[27] G. W. Stambo, M. J. Hallisey, and J. J. Gallagher Jr., "Arteriographic embolization of visceral artery pseudoaneurysms," *Annals of Vascular Surgery*, vol. 10, no. 5, pp. 476–480, 1996.

[28] A. G. Lossing, H. Grosman, R. A. Mustard, and E. M. Hatswell, "Emergency embolization of a ruptured aneurysm of the pancreaticoduodenal arcade," *Canadian Journal of Surgery*, vol. 38, no. 4, pp. 363–365, 1995.

25

Splenorenal Manifestations of *Bartonella henselae* Infection in a Pediatric Patient

Taylor Rising,[1] **Nicholas Fulton,**[2] **and Pauravi Vasavada**[2]

[1]*Northeast Ohio Medical University, 4209 State Route 44, Rootstown, OH 44272, USA*
[2]*University Hospitals Case Medical Center, 11100 Euclid Avenue, Cleveland, OH 44106, USA*

Correspondence should be addressed to Nicholas Fulton; nicholaslfulton@gmail.com

Academic Editor: Yoshito Tsushima

Bartonella henselae is a bacterium which can cause a wide range of clinical manifestations, ranging from fever of unknown origin to a potentially fatal endocarditis. We report a case of *Bartonella henselae* infection in a pediatric-aged patient following a scratch from a kitten. The patient initially presented with a prolonged fever of unknown origin which was unresponsive to antibiotic treatment. The patient was hospitalized with worsening fevers and night sweat. Subsequent ultrasound imaging demonstrated multiple hypoechoic foci within the spleen. A contrast-enhanced CT of the abdomen and pelvis was also obtained which showed hypoattenuating lesions in the spleen and bilateral kidneys. *Bartonella henselae* IgG and IgM titers were positive, consistent with an acute *Bartonella henselae* infection. The patient was discharged with a course of oral rifampin and trimethoprim-sulfamethoxazole, and all symptoms had resolved following two weeks of therapy.

1. Introduction

Bartonella henselae commonly presents as fever and localized lymphadenopathy in children or adolescents with a history of exposure to a scratch from a kitten or cat. The most commonly involved lymph nodes are axillary and epitrochlear, presumably because the majority of contact with cats occurs with the hands [1]. The bacterium is an aerobic Gram-negative bacillus which is thought to disseminate hematogenously, most commonly to the liver and spleen [1, 2]. Hepatosplenic involvement may manifest clinically as periumbilical or upper abdominal pain, weight loss, and hepatosplenomegaly [1]. Typical imaging findings include hypoechoic lesions in the liver and spleen by ultrasound and hypoattenuating lesions on CT [3]. More recently, renal microabscesses as a result of *Bartonella henselae* infection have also been reported [4]. We report a patient with splenorenal involvement of *Bartonella henselae* on ultrasound and CT in a pediatric patient.

2. Case Presentation

A previously healthy six-year-old female presented with complaint of fevers up to 101 degrees Fahrenheit, as well as dry cough, intermittent periumbilical abdominal pain, and night sweat. She was taken to her primary care physician, who diagnosed the patient with a urinary tract infection and she was sent home with a 5-day course of oral amoxicillin. The patient returned to the emergency room 2 days after completion of the antibiotics with persistent fevers and cough. Urinalysis and complete blood count were unremarkable. A chest radiograph showed perihilar peribronchial thickening without focal consolidation. The patient was discharged home with a presumed diagnosis of atypical pneumonia and given a 5-day course of oral azithromycin. Following completion of the azithromycin, the patient had improvement in her cough but still had persistent fevers, prompting another visit to the emergency department the day after completing her antibiotic course. A repeat chest radiograph, complete blood count, urinalysis, and renal function panel were normal. C-reactive protein was elevated at 6.65 mg/dL (normal < 0.80) and erythrocyte sedimentation rate was elevated at 70 mm/h (normal 0–13). The patient was hospitalized for further evaluation of her fever.

Blood and urine cultures on the date of admission did not demonstrate bacterial growth. Further investigation into the patient's history did not demonstrate any sick contacts but did have exposure to a family member who had been recently

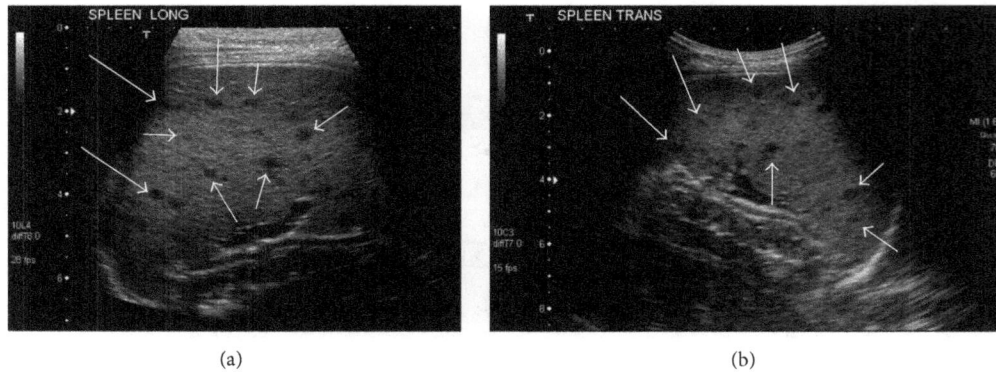

(a)   (b)

FIGURE 1: Longitudinal and transverse ultrasound images of the left upper quadrant demonstrate the spleen to be enlarged. Multiple small hypoechoic lesions (arrows) are seen within the spleen.

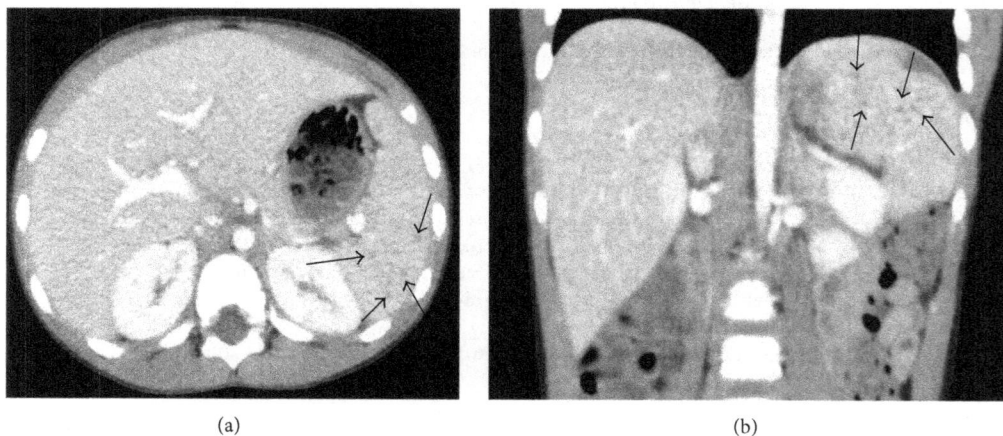

(a)   (b)

FIGURE 2: Axial and coronal contrast-enhanced CT images through the upper abdomen demonstrate an enlarged liver. The spleen demonstrates multiple small hypoattenuating lesions (arrows).

incarcerated. The patient did also have recent contacts with new dogs and kittens prior to the onset of fever, and the patient did suffer a scratch to the chest from one of the kittens which did not require medical care.

On hospital day 2, a complete abdominal ultrasound demonstrated an echogenic liver which was slightly enlarged to 11 cm in the craniocaudal dimension at the right midclavicular line. The spleen was also enlarged to 7.8 cm and contained several small hypoechoic foci (Figures 1(a) and 1(b)). A contrast-enhanced CT was then recommended for further evaluation which showed small, poorly defined hypoattenuating lesions that were seen throughout the spleen suspicious for microabscesses (Figures 2(a) and 2(b)). Additionally, small, poorly defined hypoattenuating lesions were seen within the bilateral renal cortices, findings which were also compatible with microabscesses (Figures 3(a) and 3(b)).

Epstein-Barr virus, cytomegalovirus, toxoplasmosis, HIV, histoplasmosis titers, and purified protein derivative tests were all negative. On hospital day 3, the *Bartonella henselae* IgG was positive at >1 : 1024, consistent with presence of IgG antibody to *Bartonella henselae*, suggestive of current or prior infection. *Bartonella henselae* IgM was also positive at 1 : 128, suggestive of current or recent infection. Given the CT

and ultrasound findings, the constellation of findings was consistent with an active *Bartonella henselae* infection with involvement of the spleen, kidneys, and liver.

Following these results, the patient was treated with oral rifampin 150 mg by mouth twice daily for fourteen days and trimethoprim-sulfamethoxazole 150 mg by mouth twice daily for fourteen days, which was well tolerated. The patient's fever and other symptoms resolved during her hospital stay and she was discharged on continuation of the oral antibiotic regimen. At the time of discharge on hospital day 5, the patient had defervesced and her inflammatory markers had decreased, with a C-reactive protein of 2.61 at the time of discharge. Her subsequent outpatient follow-up in the infectious disease clinic 11 days after discharge demonstrated complete resolution of her symptoms.

3. Discussion

Though well-known as a cause of infection, *Bartonella henselae* is infrequently discussed in the radiology literature. In this report, we present a case of a pediatric patient with *Bartonella henselae* infection resulting in splenomegaly and microabscesses of the spleen and kidneys as noted on CT

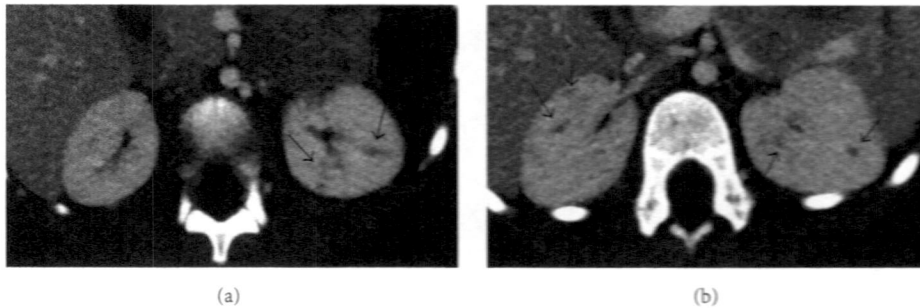

(a)   (b)

FIGURE 3: Two axial contrast-enhanced CT images through the level of the kidneys demonstrate small bilateral hypoattenuating renal lesions.

and ultrasound; to the best of our knowledge, this is the first report of this constellation of findings on radiologic studies in a pediatric patient.

It was only as recently as 1992 that the pathogen *Bartonella henselae* was isolated [5]. *Bartonella henselae* is an intracellular bacillus which can be identified on Warthin-Starry silver stains [1]. The primary reservoir for *Bartonella henselae* is cats, with half of pet cats demonstrating antibodies against the bacteria [1]. Stray cats have an even higher incidence of bacteremia and seropositivity [6]. The infection spreads via an arthropod vector, *Ctenocephalides felis* or the "common" cat flea [1, 5]. The bacterium may be able to survive in the feces of the cat flea for up to a week [5]. Cats associated with cat-scratch disease are highly likely to be bacteremic with *Bartonella henselae*; however, even 28% of cats without known associations to cat-scratch disease or bacillary angiomatosis have *Bartonella henselae* bacteremia [7]. Prevention of infection includes hand-washing after handling pets, with particularly close attention to any bites or scratches one may endure; however, the most effective prevention is the eradication of fleas [8].

Upon infection, the bacteria invade many types of cells; most commonly this involves dendritic cells but also includes CD34+ progenitor cells, erythrocytes, pericytes, and perhaps most importantly endothelial cells [1, 5]. *Bartonella henselae* invasion of endothelial cells in both reservoir and incidental hosts [9] results in vascular proliferation [10]. This is accomplished both by inhibiting endothelial cell apoptosis [10] and by stimulating the host to produce vascular endothelial growth factor (VEGF) which results in endothelial cell stimulation; subsequently, these endothelial cells promote proliferation of *Bartonella henselae* [11].

The clinical presentation of *Bartonella henselae* infection is dependent on the immune status of the affected host. Immunocompetent patients most commonly present with fever which may be protracted, localized lymphadenopathy and hepatosplenic disease. Typical cat-scratch disease includes fever and localized lymphadenopathy usually only affecting a single lymph node [1]. The patient may also note a nontender papule which develops a few days following the inciting scratch [6]. Although it is often thought of as only a self-limiting disease, *Bartonella henselae* can also result in chronic bacteremia, even in immunocompetent patients, who may be entirely asymptomatic or may present with mild or fluctuating symptoms, including fatigue, myalgias,

headaches, generalized pains, and insomnia [5, 12]. *Bartonella henselae* infection is also one of the most common causes of a fever of unknown origin in a pediatric patient, with some studies showing it to be the third most common cause of a prolonged fever in children [13].

Immunodeficient patients may manifest infection as cutaneous bacillary angiomatosis [14], with cutaneous or subcutaneous vascular tumors which may be painful. They may be solitary or few or exist in hundreds [6]. If not adequately treated, these lesions may disseminate throughout the entire body to involve the respiratory tract and gastrointestinal tract, involve the heart to cause endocarditis, and infect the CNS resulting in meningitis and brain abscesses [14]. Bacillary peliosis is another manifestation of *Bartonella henselae* infection in immunocompromised patients, typically those with AIDS [6]. It presents with cystic blood-filled spaces in the spleen and/or liver [6].

Diagnosing *Bartonella henselae* infection can be quite challenging, in part because many of the signs and symptoms are nonspecific and can be seen in other infections including cytomegalovirus infection, HIV infection, toxoplasmosis, and Epstein-Barr virus infection, as well as noninfectious etiologies including malignancies such as lymphoma [15]. The path to diagnosis begins with physical examination and history including presence or absence of lymphadenopathy and any prior exposure to cats [15]. The best test to be performed initially is serology either by indirect fluorescent assay or by enzyme-linked immunosorbent assay [15]. Unfortunately, reports on the efficacy of these are variable, with some showing serology to have a very low sensitivity and others reporting a high sensitivity [15–17]. The sensitivity of serologic tests also varies by which immunoglobulin is evaluated, with the sensitivity of IgM much lower than that of IgG [18]. Conversely, specificity is much higher in IgM tests than IgG [18]. With this in mind, it is important to remember that many patients do not mount a significant antibody response to *Bartonella henselae* infection [16]. An additional possible confounding factor with serology is that results may be positive due to a remote infection. However, immunoglobulin G levels >1 : 256 are strongly suggestive of either an acute or recent *Bartonella henselae* infection. In the setting of equivocal serology but a persistent clinical suspicion for *Bartonella henselae* infection, lymph node biopsy can be performed with polymerase chain reaction to identify DNA fragments [15, 16]. The use of enriched cultures and PCR from a number of sources including blood and CSF can enhance both sensitivity and specificity of diagnosing *Bartonella henselae* infection [5].

On radiologic imaging, hepatosplenic involvement of *Bartonella henselae* is becoming more widely discovered due to increased use and accuracy of radiological studies. These lesions are seen as hypoattenuating on CT and hypoechoic on ultrasound, which correspond to necrotizing granulomata on pathology [1]. These tend to range in size from being as small as 3 mm to being as large as 3 cm [3, 6]. On contrast-enhanced CT, these lesions may remain hypoattenuating or may show mild enhancement [6]. Splenomegaly in the absence of granulomatous involvement may be seen in 12% of patients [3]. In a study of 11 pediatric age patients with hepatosplenic *Bartonella henselae* infection, only two had hepatomegaly and two had splenomegaly; all patients in this series had splenic microabscesses and the majority had hepatic microabscesses [2]. Only 64% of these patients presented with complaints of abdominal pain [2].

The first reported case of bilateral renal microabscesses secondary to *Bartonella henselae* infection was reported in 2010 [4], which presented with a normal urinalysis and renal function panel. Renal microabscesses can be shown by contrast-enhanced CT scan, even in the setting of normal kidney function [4]. However, the most common manifestation of renal involvement of *Bartonella henselae* is glomerulonephritis [1].

In cats, no antibiotic has proven routinely effective for the treatment of *Bartonella henselae*, with incomplete treatment responses to multiple antibiotic agents. Rather, in cats the optimal management is the eradication of fleas [19]. A monthly topical application of 10% imidacloprid-1% moxidectin has been shown to decrease flea transmission [20]. Other drugs including Selamectin have shown to reduce the number of cat fleas [21]. In human infection, the clinical manifestations of *Bartonella henselae* dictate the appropriate therapeutic management; for instance, typical cat-scratch disease is often self-limited and poorly responsive to antimicrobial treatment [1], whereas hepatosplenic disease has been shown to be successfully treated with trimethoprim-sulfamethoxazole, rifampin, ciprofloxacin, and gentamicin. However, it is unknown which antibiotic is optimal for treatment due to a lack of controlled trials [1]. Treatment of *Bartonella henselae* infection is also dependent on the patient's immune status, with immunocompetent patients requiring only 2–4 weeks of therapy. In immunocompromised patients, even prolonged antibiotic courses of 6 weeks may subsequently relapse and demonstrate bacteremia [19]. As a result, these patients may require lifelong antimicrobial therapy [14].

References

[1] T. A. Florin, T. E. Zaoutis, and L. B. Zaoutis, "Beyond cat scratch disease: widening spectrum of *Bartonella henselae* infection," *Pediatrics*, vol. 121, no. 5, pp. e1413–e1425, 2008.

[2] M. W. Dunn, F. E. Berkowitz, J. J. Miller, and J. A. Snitzer, "Hepatosplenic cat-scratch disease and abdominal pain," *The Pediatric Infectious Disease Journal*, vol. 16, no. 3, pp. 269–272, 1997.

[3] D. C. Rappaport, W. A. Cumming, and P. R. Ros, "Disseminated hepatic and splenic lesions in cat-scratch disease: imaging features," *American Journal of Roentgenology*, vol. 156, no. 6, pp. 1227–1228, 1991.

[4] N. Salehi, H. Custodio, and M. H. Rathore, "Renal microabscesses due to *Bartonella* infection," *Pediatric Infectious Disease Journal*, vol. 29, no. 5, pp. 472–473, 2010.

[5] E. B. Breitschwerdt, "Bartonellosis: one health perspectives for an emerging infectious disease," *ILAR Journal*, vol. 55, no. 1, pp. 46–58, 2014.

[6] J. W. Bass, J. M. Vincent, and D. A. Person, "The expanding spectrum of Bartonella infections: II. Cat-scratch disease," *Pediatric Infectious Disease Journal*, vol. 16, no. 2, pp. 163–179, 1997.

[7] D. L. Kordick, K. H. Wilson, D. J. Sexton, T. L. Hadfield, H. A. Berkhoff, and E. B. Breitschwerdt, "Prolonged *Bartonella* bacteremia in cats associated with cat-scratch disease patients," *Journal of Clinical Microbiology*, vol. 33, no. 12, pp. 3245–3251, 1995.

[8] B. B. Chomel, H. J. Boulouis, and E. B. Breitschwerdt, "Cat scratch disease and other zoonotic *Bartonella* infections," *Journal of the American Veterinary Medical Association*, vol. 224, no. 8, pp. 1270–1279, 2004.

[9] C. Dehio, "*Bartonella* interactions with endothelial cells and erythrocytes," *Trends in Microbiology*, vol. 9, no. 6, pp. 279–285, 2001.

[10] C. Dehio, "Recent progress in understanding Bartonella-induced vascular proliferation," *Current Opinion in Microbiology*, vol. 6, no. 1, pp. 61–65, 2003.

[11] V. A. J. Kempf, B. Volkmann, M. Schaller et al., "Evidence of a leading role for VEGF in *Bartonella henselae*-induced endothelial cell proliferations," *Cellular Microbiology*, vol. 3, no. 9, pp. 623–632, 2001.

[12] M. Vayssier-Taussat, S. Moutailler, F. Féménia et al., "Identification of novel zoonotic activity of *Bartonella* spp., France," *Emerging Infectious Diseases*, vol. 22, no. 3, 2016.

[13] R. F. Jacobs and G. E. Schutze, "Bartonella henselae as a cause of prolonged fever and fever of unknown origin in children," *Clinical Infectious Diseases*, vol. 26, no. 1, pp. 80–84, 1998.

[14] K. A. Adal, C. J. Cockerell, and W. A. Petri, "Cat scratch disease, bacillary angiomatosis, and other infections due to rochalimaea," *New England Journal of Medicine*, vol. 330, no. 21, pp. 1509–1515, 1994.

[15] S. A. Klotz, V. Ianas, and S. P. Elliott, "Cat-scratch disease," *The American Family Physician*, vol. 83, no. 2, pp. 152–155, 2011.

[16] K. Chondrogiannis, A. Vezakis, A. Derpapas, A. Melemeni, and G. Fragulidis, "Seronegative cat-scratch disease diagnosed by PCR detection of Bartonella henselae DNA in lymph node samples," *Brazilian Journal of Infectious Diseases*, vol. 16, no. 1, pp. 96–99, 2012.

[17] M. J. Dalton, L. E. Robinson, J. Cooper, R. L. Regnery, J. G. Olson, and J. E. Childs, "Use of Bartonella antigens for serologic diagnosis of cat-scratch disease at a national referral center," *Archives of Internal Medicine*, vol. 155, no. 15, pp. 1670–1676, 1995.

[18] M. J. Vermeulen, H. Verbakel, D. W. Notermans, J. H. J. Reimerink, and M. F. Peeters, "Evaluation of sensitivity, specificity and cross-reactivity in *Bartonella henselae* serology," *Journal of Medical Microbiology*, vol. 59, no. 6, pp. 743–745, 2010.

[19] B. B. Chomel and R. W. Kasten, "Bartonellosis, an increasingly recognized zoonosis," *Journal of Applied Microbiology*, vol. 109, no. 3, pp. 743–750, 2010.

[20] C. A. Bradbury and M. R. Lappin, "Evaluation of topical application of 10% imidacloprid-1% moxidectin to prevent *Bartonella henselae* transmission from cat fleas," *Journal of the American Veterinary Medical Association*, vol. 236, no. 8, pp. 869–873, 2010.

[21] T. L. McTier, R. L. Jones, M. S. Holbert et al., "Efficacy of selamectin against adult flea infestations (Ctenocephalides felis felis and Ctenocephalides canis) on dogs and cats," *Veterinary Parasitology*, vol. 91, no. 3-4, pp. 187–199, 2000.

26

Fractured Ribs and the CT Funky Fat Sign of Diaphragmatic Rupture

Iclal Ocak[1] and Diane C. Strollo[2]

[1]University of Pittsburgh Medical Center, Radiology Suite 200 East Wing, 200 Lothrop Street, Pittsburgh, PA 15213, USA
[2]University of Pittsburgh Medical Center, 200 Lothrop Street, Pittsburgh, PA 15213, USA

Correspondence should be addressed to Iclal Ocak; ocaki@upmc.edu

Academic Editor: Salah D. Qanadli

Traumatic diaphragmatic rupture remains a diagnostic challenge for both radiologists and surgeons. In recent years, multidetector CT has markedly improved the diagnosis of diaphragmatic injury in polytrauma patients. Herein, we describe two cases of subacute presentation of traumatic diaphragmatic rupture from a penetrating rib fracture and subsequent intrathoracic herniation of omental fat, representing the CT "funky fat" sign.

1. Introduction

Traumatic diaphragmatic rupture (TDR) may result from a penetrating injury or blunt thoracoabdominal trauma and results in communication between the pleural and peritoneal cavities [1–3]. A penetrating injury from a displaced rib fracture or a stab or gunshot wound typically causes a small diaphragmatic puncture [1–3]. With blunt trauma, a sudden increase of intra-abdominal pressure may cause a ≥10 cm "blowout" laceration of the diaphragm [4]. In some cases, both mechanisms may occur.

The fractured rib sign was first described by Holland and Quint [5] and later referred to as "presumed laceration of the diaphragm by a fractured rib" in a study by Nchimi et al. [6]. The fractured rib sign is present when a rib fragment points toward and directly penetrates the diaphragm. The reported sensitivity of the fractured rib sign is low, and specificity statistics have not been reported [6]. In patients with severe trauma, this sign should direct attention to the diaphragm and prompt close clinical and imaging follow-up [1].

Numerous signs indicating TDR have been described in the literature [1]. We present two patients with TDR with fractured rib sign on initial trauma multidetector computed tomography (MDCT) and subsequent herniation of omental fat, the funky fat sign, within two weeks of the injury.

2. Case 1

A 59-year-old helmeted man was thrown 20 feet during a motorcycle collision. Because of multiple comminuted left rib fractures with subcutaneous gas on chest radiograph, a left chest tube was placed. Contrast-enhanced trauma MDCT revealed sternal and scapular fractures, left lung contusion, and a small left pneumothorax. In addition, multiple fractures of the 3rd through 10th left ribs were displacement into the chest cavity, with associated chest wall instability. While the 10th-rib fracture abutted the left diaphragm, the diaphragm appeared intact (Figures 1(a) and 1(b)). The patient did not require intubation. Repeat chest MDCT ten days later revealed herniation of omental fat into the left chest, consistent with subacute TDR (Figures 1(c) and 1(d)). However, this finding was missed until a third MDCT was performed three days later. Thoracoscopic exploration confirmed the TDR contained only omentum, which was edematous and viable but could not be completely reduced from the thoracic approach. Therefore, laparoscopic reduction of omental fat and repair of the diaphragm laceration were performed on hospital day 13. Because the left rib fractures had become progressively displaced, side plate fixation of the left 6–8th ribs was performed on hospital day 14. The patient had otherwise uneventful recovery.

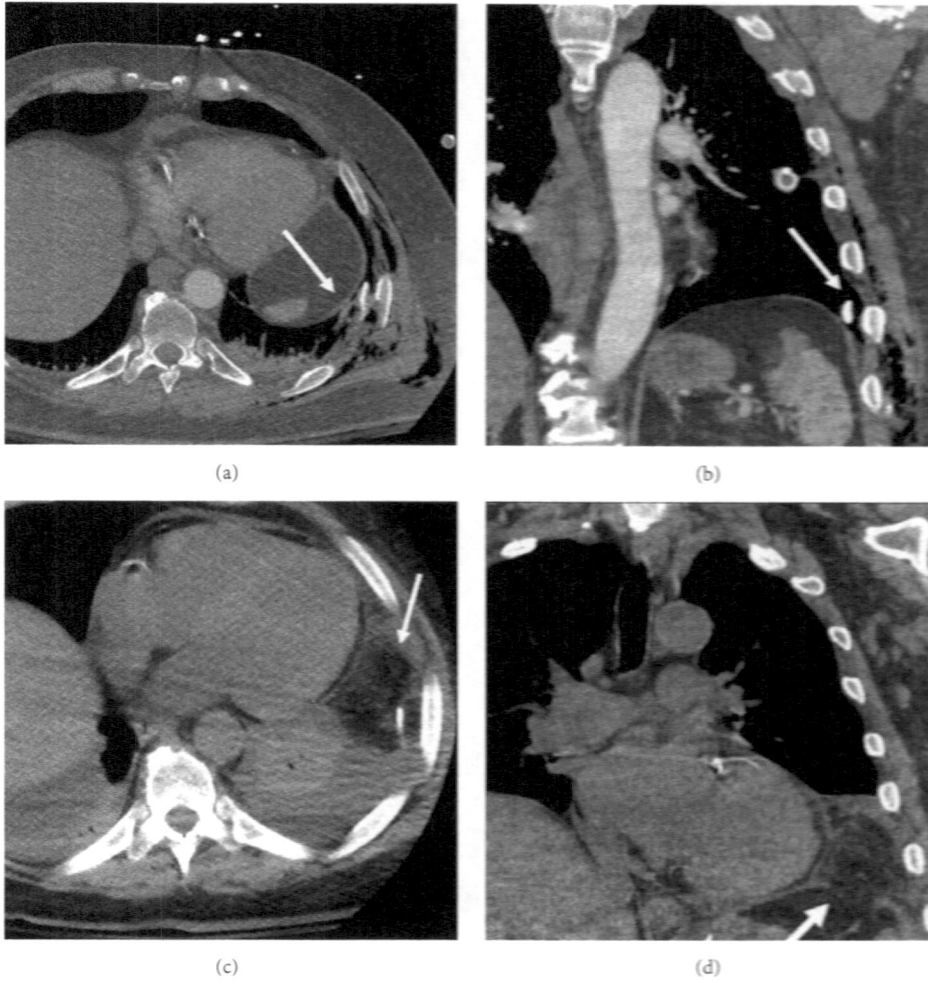

FIGURE 1: Fractured rib sign in a 59-year-old man after blunt trauma. (a, b) Contrast-enhanced chest MDCT (axial and coronal planes) shows left 10th-rib fracture abutting the diaphragm (fractured rib sign). The diaphragm appears intact on initial trauma CT. (c, d) Repeat chest MDCT (axial and coronal planes) 5 days later reveals new herniation of omental fat (arrows) into the left hemithorax due to subacute diaphragm rupture. Note discontinuity of the diaphragm (d).

3. Case 2

A 64-year-old male restrained driver suffered blunt chest and abdominal trauma following a motor vehicle collision. He sustained a large left lung contusion, tiny left pneumothorax, fractures of the pelvis and lumbar transverse processes, and multiple comminuted ribs fractures. He did not require intubation. Contrast-enhanced trauma MDCT better depicted multiple displaced compound fractures of the left 3rd through the 11th ribs in close proximity to the diaphragm, but the diaphragm appeared intact (Figures 2(a) and 2(b)). In addition, left lung contusion and tiny left pneumothorax were present.

A repeat MDCT was performed four days later to evaluate an enlarging left pleural effusion. CT depicted new herniation of omental fat into the left chest, compatible with evolving TDR (Figures 2(c) and 2(d)). Serosanguinous pleura fluid was drained following left chest tube placement. Exploratory

laparotomy on the same day revealed that the left 10th-rib fracture pierced the diaphragm with mildly edematous omental fat herniating through the 5 cm diaphragmatic laceration. The omentum was easily reduced, and the diaphragmatic laceration was surgically closed. His recovery was otherwise uneventful.

4. Discussion

Up to 8% of patients with severe blunt thoracoabdominal trauma develop a traumatic diaphragmatic injury [1]. TDR is rarely an isolated injury, and affected patients typically have a high injury severity score. While TDR represents only 5% of all diaphragmatic hernias, it is responsible for 90% of hernias that eventually become incarcerated, and most manifest within three years of the injury [1]. In a number of cases, TDR can present years later after the trauma and carry a mortality rate of 30–60% [7]. Motor vehicle collisions

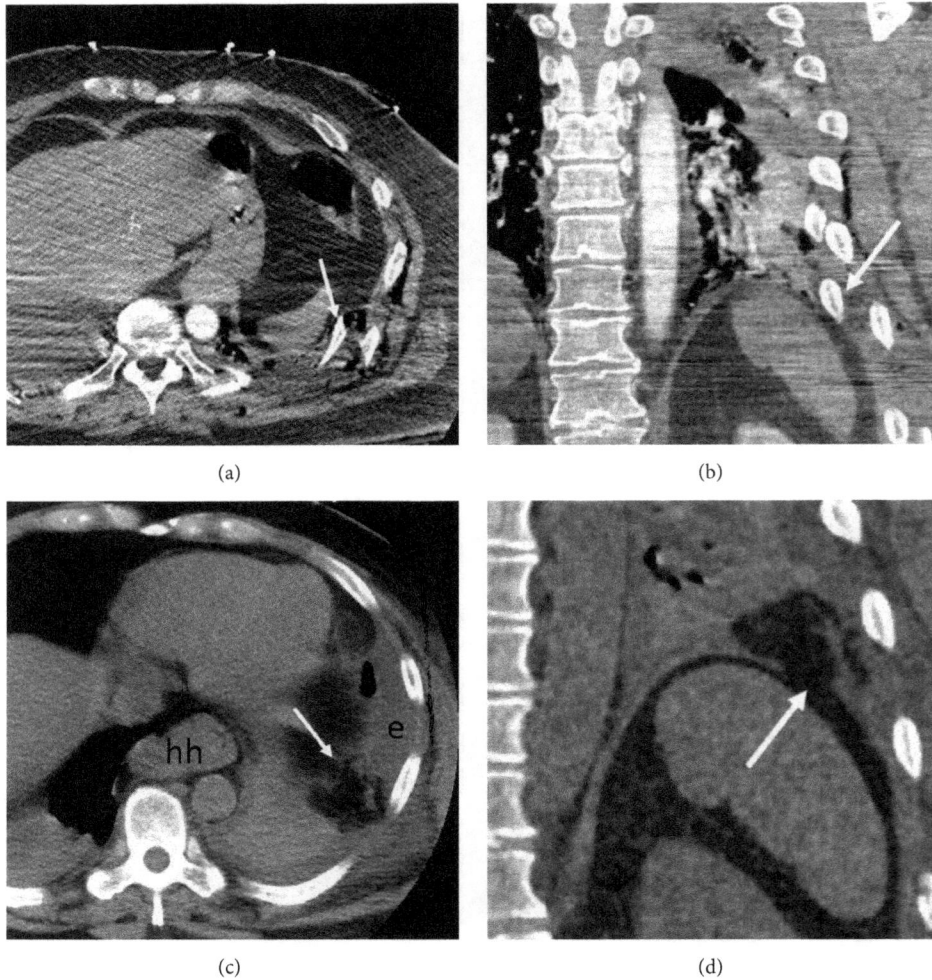

FIGURE 2: Fractured rib sign in a 64-year-old man after blunt trauma. (a, b) Contrast-enhanced trauma chest MDCT (axial and coronal planes) shows displaced left lower rib fractures abutting left diaphragm (arrows). The diaphragm appears intact. A small hiatal hernia is present (hh). (c, d) Repeat MDCT 4 days later (axial and coronal planes) shows new herniated omental fat (arrows) within left hemithorax due to subacute diaphragm rupture. Note discontinuity of the diaphragm (d). The large left effusion (e) is new.

are responsible for up to 90% of TDR, with the remainder due to falls or crush or penetrating injuries [1, 8]. Left-sided diaphragmatic injuries are typically more clinically apparent and symptomatic [8]. The liver likely has a protective effect on the right diaphragm, and right diaphragmatic injuries may be underdiagnosed [1, 8].

It has been reported that the diagnosis of TDR may be missed initially in up to 30% of cases on MDCT [3]. Affected patients typically have severe multisystem injures that may overshadow the diagnosis of TDR, and there may be lack of awareness of the various imaging signs of diaphragmatic injury [1, 8]. In some cases, like ours, herniation of abdominal contents likely develops after the first trauma assessment. While a penetrating diaphragmatic injury from a rib fracture is typically small and may be initially inconspicuous, it will likely enlarge over time as negative intrathoracic pressure with inspiration gradually promotes herniation of abdominal contents into the chest [2, 8]. When a patient with TDR is intubated following blunt trauma, positive pressure ventilation may prevent the herniation of abdominal contents into the chest, and the diagnosis may only become apparent after extubation [1]. Subacute rupture is possible if diaphragmatic tissue is devitalized at the time of injury and subsequently breaks down. Because omentum is pliable and mobile, it may be the first abdominal structure to herniate into the chest [1, 7].

Spontaneous closure of TDR has never been reported, and almost all cases should be surgically repaired [1]. Early diagnosis is important because small laceration is typically easier to repair. A large TDR is inherently more complicated and may be associated with dense thoracic and abdominal adhesions.

In all cases with significant blunt thoracoabdominal trauma, thorough evaluation is required to exclude TDR, and follow-up evaluation should be considered to assess potential delayed development of TDR [9]. In difficult cases,

MR imaging may secure the diagnosis of TDR but may not always be well suited for multitrauma patients [1, 8]. When the index of suspicion is high for diaphragmatic injury but imaging studies are inconclusive, laparoscopic or thoracoscopic exploration may confirm or exclude TDR.

Farboud et al. reported a 77-year-old man who sustained fractures of the left 3rd to 10th ribs and a left diaphragmatic hernia after falling 8 feet from a ladder. Urgent left thoracotomy revealed that the sharp edge of the displaced left 7th-rib fracture had pierced the diaphragm, and omentum had herniated into the chest through the diaphragmatic defect. Following resection of the rib fragment, the omentum was reduced, the diaphragmatic defect was repaired, and the patient recovered uneventfully [10]. Holland and Quint reported a case of left diaphragm laceration adjacent to rib fractures. CT performed 4 hours after the trauma revealed left hemopneumothorax, fractures of several left lower lateral ribs, and herniation of omentum into chest [5].

In our cases, both patients had the subacute diagnosis of a small TDR from a penetrating rib injury, with herniation of omental fat within two weeks following trauma. Both had the fractured rib sign on initial CT, with rib fracture fragments protruding into the chest cavity in close proximity to the diaphragm. In this setting, radiologists should have a high index of suspicion for TDR. Intrathoracic omental fat, the CT "funky fat" sign, should also alert the radiologist to search for TDR and may be a delayed finding. In our cases, early diagnosis of TDR likely prevented strangulation of omentum and subsequent development of a large hernia sac containing abdominal visceral organ.

References

[1] A. Desir and B. Ghaye, "CT of blunt diaphragmatic rupture," *Radiographics*, vol. 32, no. 2, pp. 477–498, 2012.

[2] K. L. Shackleton, E. T. Stewart, and A. J. Taylor, "Traumatic diaphragmatic injuries: spectrum of radiographic findings," *Radiographics*, vol. 18, no. 1, pp. 49–59, 1998.

[3] G. P. Sangster, A. González-Beicos, A. I. Carbo et al., "Blunt traumatic injuries of the lung parenchyma, pleura, thoracic wall, and intrathoracic airways: multidetector computer tomography imaging findings," *Emergency Radiology*, vol. 14, no. 5, pp. 297–310, 2007.

[4] G. Rodriguez-Morales, A. Rodriguez, and C. H. Shatney, "Acute rupture of the diaphragm in blunt trauma: analysis of 60 patients," *Journal of Trauma—Injury, Infection and Critical Care*, vol. 26, no. 5, pp. 438–444, 1986.

[5] D. G. Holland and L. E. Quint, "Traumatic rupture of the diaphragm without visceral herniation: CT diagnosis," *American Journal of Roentgenology*, vol. 157, no. 1, pp. 17–18, 1991.

[6] A. Nchimi, D. Szapiro, B. Ghaye et al., "Helical CT of blunt diaphragmatic rupture," *American Journal of Roentgenology*, vol. 184, no. 1, pp. 24–30, 2005.

[7] R. Kaur, A. Prabhakar, S. Kochhar, and U. Dalal, "Blunt traumatic diaphragmatic hernia: pictorial review of CT signs," *Indian Journal of Radiology and Imaging*, vol. 25, no. 3, pp. 226–232, 2015.

[8] S. Iochum, T. Ludig, F. Walter, H. Sebbag, G. Grosdidier, and A. G. Blum, "Imaging of diaphragmatic injury: a diagnostic challenge?" *Radiographics*, vol. 22, pp. s103–s118, 2002.

[9] T. R. de Nadai, J. C. P. Lopes, C. C. Inaco Cirino, M. Godinho, A. J. Rodrigues, and S. Scarpelini, "Diaphragmatic hernia repair more than four years after severe trauma: four case reports," *International Journal of Surgery Case Reports*, vol. 14, pp. 72–76, 2015.

[10] A. Farboud, H. Luckraz, and E. G. Butchart, "Delayed presentation of diaphragmatic injury secondary to rib fracture," *Respiratory Medicine CME*, vol. 1, no. 2, pp. 158–160, 2008.

27

Contribution of Imaging in Diagnosis of Primitive Cyst Hydatid in Unusual Localization: Pleura—A Report of Two Cases

Fatima Zahra Mrabet ⓘ,[1] **Jihane Achrane,**[1] **Yassir Sabri,**[2] **Fatima Ezzahra El Hassani,**[3,4] **Sanaa Hammi,**[1] **and Jamal Eddine Bourkadi**[1]

[1]*Department of Pneumology, Moulay Youssef University Hospital Center, Rabat, Morocco*
[2]*Department of Parasitology, Ibn Sina University Hospital Center, Rabat, Morocco*
[3]*Department of Cardiology, Military Hospital Mohamed V, Rabat, Morocco*
[4]*Faculty of Medicine and Pharmacy, Mohamed V University, Rabat, Morocco*

Correspondence should be addressed to Fatima Zahra Mrabet; mrabetfatimazahra@gmail.com

Academic Editor: Atsushi Komemushi

Hydatic disease has always been the most common in countries where large amount of sheep and cattle is raised, but increased travel and immigration have made this condition a serious worldwide public problem. Cyst hydatid may affect all parts of the human body like the heart, the bone marrow, the eye, the brain, the kidney, and the spermatic cord. Humans can become infested by accidentally ingesting the eggs that are passed in the feces from definitive hosts (usually a canid, such as a wolf, fox, or dog). Even in endemic countries, the primitive pleural hydatid cyst is exceptional, and it is very difficult to distinguish from other pleural and parietal cystic masses especially that in majority of cases the immunologic tests are negative. We report two cases of pleural hydatid cyst discovered in two young patients, with a nonspecific clinical presentation. The interest of this paper is to raise the primordial role of imaging in the positive diagnosis of primary pleural hydatid cyst.

1. Introduction

Hydatidosis is a parasitic disease caused by the development in humans of the larval form of Echinococcus Granulosus, a small dog tapeworm. Pleural localization is extremely rare even in endemic countries and represents only 1.3% of thoracic locations [1].

We report two observations of primary pleural hydatid cyst by insisting on the fundamental place of different imaging techniques in diagnosis.

2. Case Reports

2.1. Case Report (1). Miss A. L., a 17-year-old girl, with no pathological history and no notion of contact with dogs, reported since 3 months right thoracic pain, stage III of mMRC dyspnea, chest tightness, and some episodes of hemoptysis of low abundance evolving in a context of apyrexia, and conservation of the general state. The clinical examination revealed a right fluid effusion syndrome. The posteroanterior chest roentgenogram showed a homogeneous right basal opacity that effaced the diaphragmatic cupola and merged with mediastinum; its upper limit is convex (Figure 1).

Thoracic ultrasonography revealed an intrapleural cyst with a duplication of its wall suggesting a proliferative membrane without associated pleurisy (Figure 2).

Thoracic CT showed a right basal-thoracic cystic formation, measuring 126 * 93 * 93 mm, with a discreet slope with the adjacent parenchyma; its wall was thickened and enhanced after injection of contrast product. The lung parenchyma was without anomaly with the exception of passive atelectasis adjacent to the cyst, confirming the diagnosis of a right pleural cyst type II of Gharbi classification (Figure 3).

The blood count was normal and the ELISA and Indirect Agglutination serologies were negative. In a second stage, the research for other localizations of the hydatid cyst was negative (abdominal ultrasound, echocardiography,

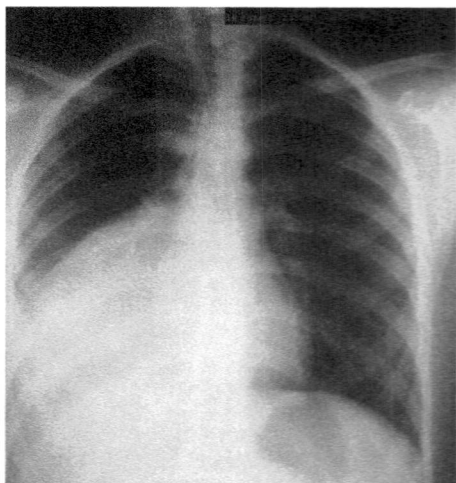

FIGURE 1: Posteroanterior chest roentgenogram showing a right basal opacity of watery tonality with convex upper limit (patient 1).

and cerebral CT), hence the primitive character of pleural hydatidosis in our observation. During surgery, the presence of a cystic formation in the parietal pleural was noted. The delicate dissection had objectified thickened visceral pleura. The cystectomy was successfully performed without rupture and the piece was sent to the parasitology laboratory with evidence of proliferative membrane (macroscopically) and alive scolex in the intracystic fluid (microscopically) (Figure 4).

2.2. Case Report (2). Mr. SF, a 26-year-old man, without any notable pathological history, have a notion of contact with dogs in childhood, asymptomatic on the respiratory plane. The posteroanterior chest roentgenogram was performed for him as a preemployment checkup. It objectified a homogeneous oval opacity, well limited, left hilar, and having the internal edge in intimate contact with the left edge of the heart (Figure 5).

In this context, a chest CT scan revealed a left anterolateral mediastinal mass with a total parietal calcification measuring 70 mm in height and 55 mm in lateral diameter (Figure 6).

Echocardiography confirmed the presence of left-ventricular extracardiac structure without intracavitary lesion or associated pericardial effusion. Likewise, magnetic resonance imaging (MRI) showed a mediastinal cyst next to the anterolateral wall of the left cardiac ventricle, in close contact with the pericardium but with a cleavage plane and no mass effect on the cardiac cavities, measuring 72mm ∗ 53mm. Its tonality was hypointense on T1 and hyperintense on T2 (Figure 7).

The blood count was normal and the ELISA and Indirect Agglutination serologies were negative. In a second stage, the research for other localizations of the hydatid cyst was negative (abdominal ultrasound and cerebral CT), hence the primitive character of pleural hydatidosis in this second observation. In operation, the heart was of normal volume

with no intrapericardial mass. At the opening of the left pleura, the exploration found a solid mass contiguous to the mediastinal pleura and in contact with the left phrenic nerve. The careful dissection and excision of the mass were successfully performed without complications.

3. Discussion

The hydatid cyst is a parasitic disease that is still endemic in several parts of the world, especially around the Mediterranean rim. The lung constitutes the second hydatid localization (20 to 40%) after the liver (75%). The primary pleural localization is exceptional, representing only 1.3% of thoracic locations [2]. It mostly affects the young adult male. We reported two observations of two young patients of different sex.

Once a human has been infested with the taenia eggs, gastric and enteric digestion facilitates the release of larvae, which penetrate the intestinal wall until they reach a small vessel system. Passing through the bloodstream, they arrive at the organ where they can settle and transform into small cysts that increase in size by 2 to 3 cm per year. The usual locations are the liver and lungs; intrathoracic but extrapulmonary locations like the pleura, diaphragm, mediastinum, pericardium, and chest wall are uncommon. Pleural hydatid cysts can develop chiefly as a result of liver or lung cyst rupture into the pleural space with complications of pneumothorax, pleural effusion, or empyema [3]. With taking respiration, eggs settle in the lungs distally. In the humid environment, they become scolexes and pass the alveolocapillary membrane and join systemic circulation by pulmonary veins, form primary isolated cysts in organs such as heart, bone marrow, eye, and brain. But some of scolexes may move into the pleural space with negative pleural pressure and settle there, causing disease. It is probably the case in the patients presented. In the pleura, cysts sit between the parietal pleura and the endothoracic fascia, and the involvement appears to be systemic or lymphatic [4]. The pleural layers are avascular, and a hydatid cyst may form and grow in this region because the structure of the laminated cyst membrane is permeable to calcium, potassium, chloride, water, and urea. Accordingly, these nutritional substances and others that may be useful to the parasite can traverse the membrane via diffusion. Active transport may be involved in this process [5, 6].

The clinical symptomatology is poor and nonspecific; it can simulate any pleural-pulmonary disease (chest pain, dyspnea, and dry cough). The diagnosis can be made in the acute phase in front of a symptomatology of sudden onset of thoracic pain and dyspnea, following the rupture of the cyst. In other cases, this new location may remain asymptomatic for a long time and the diagnosis can be done tardily. Exceptionally, there may exist some signs of mediastinal compression depending on the location. The discovery can also be fortuitous on a chest roentgenogram [1, 7]. As reported in the literature, our first patient had a discreet clinical presentation, while the second patient was outright asymptomatic.

Imaging is a fundamental element for positive diagnosis. The chest roentgenogram shows a homogeneous pleural opacity, well-defined with water tonality. Rarely, it shows

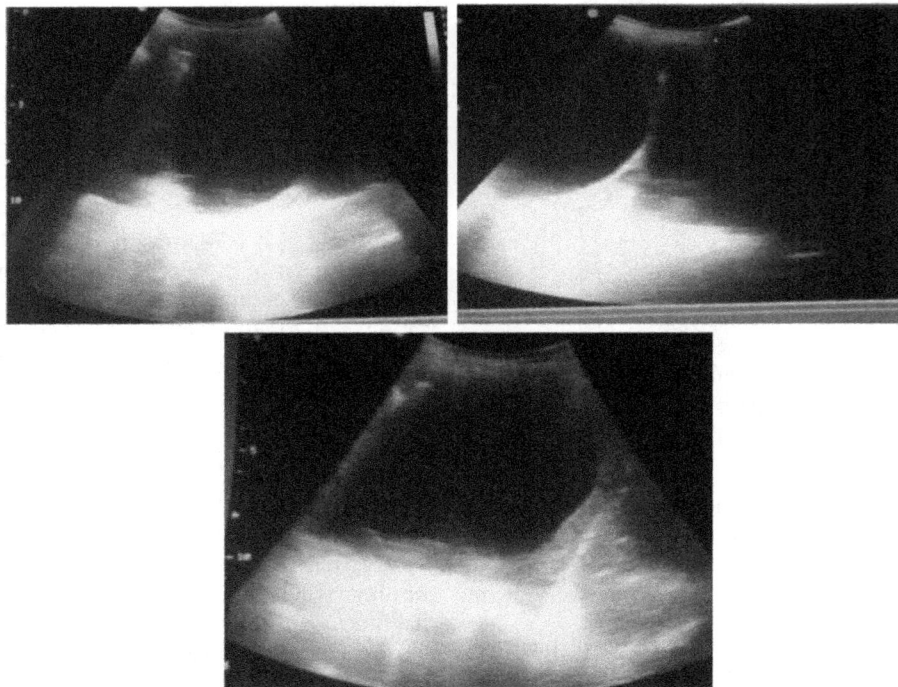

FIGURE 2: Thoracic ultrasound showing an intrapleural cyst with a proliferative membrane (patient 1).

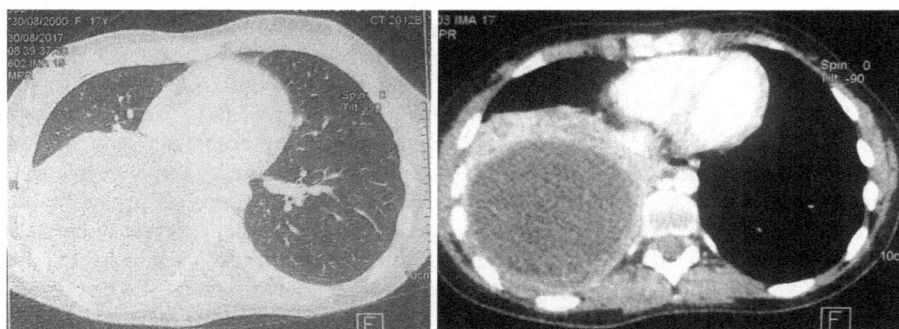

FIGURE 3: Thoracic CT scan showing a right basi-thoracic cystic image and a discreet slope with the pulmonary parenchyma with floating membrane aspect [right: parenchymal window, left: mediastinal window] (Patient 1).

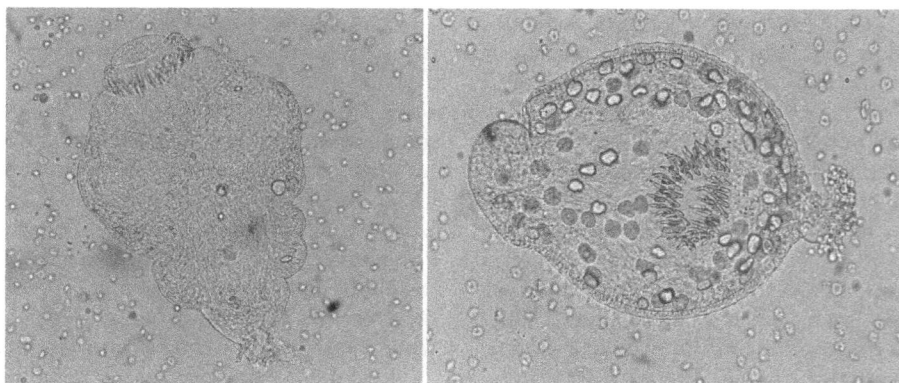

FIGURE 4: Parasitological study of the cyst fluid showing scolex [right: evaginated scolex, left: invaginated scolex] (Patient1).

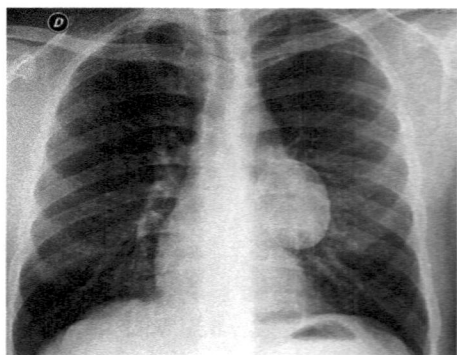

FIGURE 5: Posteroanterior chest roentgenogram elucidating a homogeneous oval opacity of water tonality, well limited, left hilar (Patient 2).

peripheral calcifications that it orientates the diagnosis. In some cases, radiography may guide the diagnosis at an early stage by showing characteristic iconography, such as the presence of cystic formation having a calcified wall, which is rare but represents a strong diagnostic presumption; the chest roentgenogram in our second patient was very characteristic and diagnosis was very probable; the other investigations had an objective to determine the exact localization of the cyst (cardiac or pleural). Ultrasound is a second-line examination; it is very efficient with a diagnostic specificity estimated at 96%. It can make the diagnosis and make the extension assessment and the surveillance, as the case of our first patient [1]; it can show multicystic or hydroaerial images when the hydatid cyst is broken. In many cases, it shows the proliferative membrane doubling inside the pericyst which is pathognomonic of hydatid cyst, confirms the liquid nature of pleural opacity, and evokes positive diagnosis especially in case of multivesicular form; it also makes it possible to detect a possible pleural effusion associated with it and to look for other cystic localizations, particularly abdominal ones. Computed tomography confirms the pleural localization of the well limited fluid mass unmodified by injection of the contrast product; it is more sensitive than the previous modalities and makes it possible to pose the diagnosis with a higher specificity and sensitivity [1]. MRI provides diagnostic support in cases where cysts are not characteristic on ultrasound or CT, especially in pseudo-tumoral forms [1]. MRI makes it possible to better define the topography of the cyst and its relations with the neighboring organs thanks to the multiplanar sections. The T2-weighted sequences visualize small daughter vesicles or a floating membrane in favor of the diagnosis and especially a peripheral T1 and T2 hyposignal in relation to the calcium deposits which is pathognomonic to the diagnosis [2].

Biologically, hypereosinophilia is generally absent in cases of intrathoracic hydatid disease. Immunological tests such as IgG ELISA, indirect hemagglutination, and Western Blot may be helpful, but their sensitivity is only about 60%. A combination of two or more biological tests and radiological imaging should be used for accurate diagnosis [6]. The sensitivity of the immunology increases significantly in case of complication or associated liver cyst. Our two patients had negative immunologic tests similar to that described in the literature.

In the absence of rupture, the puncture of the cyst is formally contraindicated, which explains why cytological or pathological diagnosis is usually performed only after its surgical excision [8].

The positive diagnosis is difficult to establish due to the rarity of the condition and the lack of clinical, radiological, and biological specificity. Indeed, several other causes of cystic lesions can be evoked: bronchogenic cyst, enteric cyst, pleuropericardial cyst, thymic cyst, and lymphangioma. The diagnosis of certitude is almost always operative by visualizing the hydatid membrane and/or daughter's vesicles or after pathological study of the operative piece in case of infected or thickened cyst [3]. The mobilization of the daughter vesicles has been considered as a sign of secondary pleural hydatidosis and, to our knowledge, such mobilization has not been observed in the primitive forms [9]. In our two observations, imaging played a fundamental role and the positive diagnosis was almost certain even preoperatively; otherwise, the surgery and the parasitological study of the operative piece came to set the diagnosis and validate the data already obtained by imaging.

As a general rule, when the presurgical diagnosis of hydatid cyst is suspected, surgeon should take care of four things in order to achieve complete resection and to avoid recurrence of disease from pleural hydatid cysts:

(i) Three days before emergency resection if necessary and 1 week before elective surgery albendazole treatment should start to increase the blood and tissue concentration of the medicine in case of the risk of contamination

(ii) Plan the appropriate surgical approach to prevent cystic rupture or spillage when doing the thoracotomy.

(iii) Inactivate daughter cysts and scolices prior to removal by injecting 20% hypertonic saline solutions into the cyst.

(iv) To give the anticoloidal agent into the cyst, it is necessary to empty some of the cyst content; otherwise, it will leak out of the injection site during surgery. Distention may be reduced by aspiration, and this will ease the manipulation and surgical dissection.

(v) Do not spill cyst contents during surgery to avoid anaphylactoid reaction, recurrence, and multiple hydatidosis.

(vi) Completely remove of the cyst including the innermost germinative layer, which can produce scolices, with en-bloc excision whenever possible; sometimes, to avoid recurrence, it is necessary to resect the affected surrounding tissues completely.

Postoperatively, it may be necessary to place patients on an anthelmintic medical regimen (Albendazol) with appropriate follow-up reevaluations [10].

4. Conclusion

The primary pleural localization of the hydatid cyst is very rare or even exceptional. Its discovery is often fortuitous. More rarely, primary pleural hydatid cyst is symptomatic with a discreet and nonspecific clinical presentation. Imaging

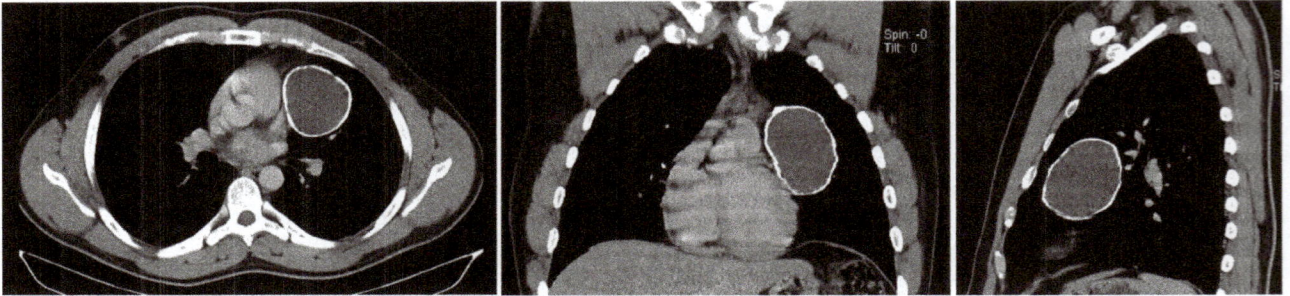

FIGURE 6: Thoracic CT showing left anterolateral mediastinal mass with total parietal calcification (mediastinal window) [right: horizontal section, middle: frontal section, and left: sagittal section] (Patient2).

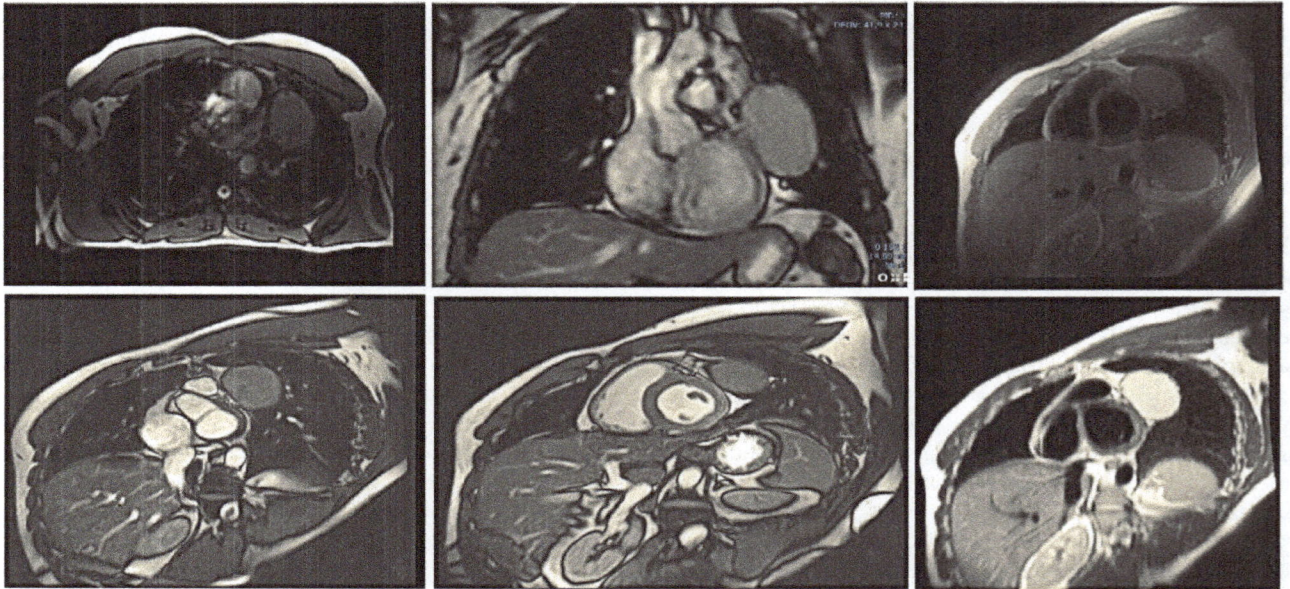

FIGURE 7: Cardiac MRI showing a para cardiac cyst next to the anterolateral wall of the left ventricle in close contact with the pericardium but with a cleavage plane (Patient 2).

especially in section, plays a fundamental role; it makes it possible to guide the diagnosis, to specify the topography and the relations with the neighboring organs, and to look for other localizations.

The feature of our observations, compared to literature, is the similarity to the young age, the discreet clinical presentation, the absence of hypereosinophilia, the negative hydatid serology, and especially the diagnostic orientation thanks to imaging.

Abbreviations

mMRC: Modified Medical Research Council
CT: Computed Tomography
ELISA: Enzyme-Linked Immunosorbent Assay
MRI: Magnetic Resonance Imaging
Ig G: Immunoglobulin G.

Disclosure

This case report was written based on clinical observation without any funding.

Authors' Contributions

Fatima Zahra Mrabet drafted this manuscript under Sanaa Hammi's supervision. Jihane Achrane, Yassir Sabri and Fatima Ezzahra El Hassani have made substantial contributions to acquisition of data. Jamal Eddine Bourkadi had been involved in drafting the manuscript. All authors read and approved the final manuscript.

References

[1] N. F. Badji, B. NDong, G. Akpo, H. Dème, M. H. Touré, and H. Niang El, "Apport de l'imagerie dans le diagnostic de kyste hydatique pleural primitif: à propos de deux cas," *Mali Médical*, vol. 32, no. 4, pp. 33–36, 217.

[2] L. Harzallah, M. Bacha, A. Garrouche et al., "Kyste hydatique pleural primitif : à propos d'une observation," *Revue Medicale De Liege*, vol. 62, no. 7-8, pp. 506–508, 2007.

[3] S. J. Kim, K. H. Jung, W. Jo, Y. S. Kim, C. Shin, and J. H. Kim, "A case of pleural hydatid cyst mimicking malignancy in a non-endemic country," *Tuberculosis and Respiratory Diseases*, vol. 70, no. 4, pp. 338–341, 2011.

[4] A. Zidane, A. Arsalane, F. Atoini, and E. Kabiri, "Les kystes hydatiques thoraciques extra-pulmonaires," *Revue de Pneumologie Clinique*, vol. 62, no. 6, pp. 386–389, 2006.

[5] P. Mardani, M. Y. Karami, K. Jamshidi, N. Zadebagheri, and H. Niakan, "A primary pleural hydatid cyst in an unusual location," *Tanaffos*, vol. 16, no. 2, pp. 166–169, 2017.

[6] S. Gürsoy, A. Ucvet, H. Tozum, A. E. Erbaycu, C. Kul, and O. Basok, "Primary intrathoracic extrapulmonary hydatid cysts," *Texas Heart Institute Journal*, vol. 36, no. 3, pp. 230–233, 2009.

[7] A. Kumar, P. Yadav, S. Pahuja, and S. Chaudhri, "Hydatid disease of pleura: a rare cause of recurrent pneumothorax," *Journal of Case Reports*, pp. 273–276, 2016.

[8] E. Hernandez, P. Eggenspieler, and T. Cruel, "Cas clinique au laboratoire Un épanchement pleural inhabituel," *Revue Française des Laboratoires*, vol. 1998, no. 300, pp. 77-78, 1998.

[9] J. M. Antona Gómez, G. García-Vinuesa, F. Fuentes Otero, J. M. Checa Pinilla, and M. Pérez Miranda, "Giant primitive pleural hydatid cyst," *CHEST*, vol. 76, no. 5, p. 614, 1979.

[10] F. Alakhras Aljanadi, "A rare case of primary hydatidosis; a primary parietal pleural hydatid cyst," *Acta Healthmedica*, vol. 1, no. 4, pp. 93–97, 2016.

28

Cerebral Phosphorus Magnetic Resonance Spectroscopy in a Patient with Giant Cell Arteritis and Endovascular Therapy

Ruth Steiger,[1] Lisa-Maria Walchhofer,[2] Andreas Rietzler(iD),[1] Katherina J. Mair,[3] Michael Knoflach,[3] Bernhard Glodny,[2] Elke R. Gizewski,[1] and Astrid E. Grams[1]

[1]Department of Neuroradiology, Medical University of Innsbruck, Anichstraße 35, 6020 Innsbruck, Austria
[2]Department of Radiology, Medical University of Innsbruck, Innsbruck, Austria
[3]Department of Neurology, Medical University of Innsbruck, Innsbruck, Austria

Correspondence should be addressed to Andreas Rietzler; andreas.rietzler@i-med.ac.at

Academic Editor: Atsushi Komemushi

With phosphorus magnetic resonance spectroscopy (31P MRS) energy metabolites can be visualised. In this case study, we report on a patient with stenosis and wall contrast enhancement in the left internal carotid and the right vertebral artery, due to giant cell arteritis. 31P MRS revealed a decreased inorganic phosphate-to-phosphocreatine ratio (Pi/PCr) in regions with a prolonged mean transit time (MTT). After systemic therapy and angioplasty of the right vertebral artery, the stenosis and the symptoms improved and the area of prolonged MTT became smaller. However, a new decrease in Pi/PCr in areas that developed moderately prolonged MTT was observed.

1. Introduction

Giant cell arteritis (GCA) is a common form of primary arteritis in patients over 50 years, in which supraaortal arteries are predominantly affected [1, 2]. Typical symptoms include headache, jaw claudication, scalp tenderness, pain in the temple, and sudden loss of vision. In rare cases symptoms are night-sweating or weight loss [3]. For the diagnosis of GCA, three out of the five following criteria must be met: age over 50 years, new onset of localised headache, raised erythrocyte sedimentation rate (ESR) of > 50 mm/1st hour, tenderness, nodulesm or reduced pulse of the temporal arterym or abnormal arterial biopsy findings [4]. Brain ischaemia resulting from GCA affecting also intracranial vessels has been reported in 3-4% of the patients. Inflamed vessel segments can be identified by magnetic resonance imaging (MRI) [5], cerebral infarcts can be visualised with diffusion weighted imaging, and brain ischaemia can be estimated by perfusion weighted imaging (PWI), for instance, by measuring a prolongation of the mean transit time (MTT) [6]. First line therapy is drug related to steroids or other immunosuppressant medication. Revascularisation

of the affected vessels can be achieved by angioplasty or stent implantation in selected patients, as the endovascular procedure can be high risk [2, 7].

Phosphorus magnetic resonance spectroscopy (31P MRS) enables the assessment of different energy and membrane metabolites in vivo. The number of metabolites corresponds to the area under the fitted curve of the respective peak (Figure 1). Prior studies reported changes in phosphocreatine (PCr), adenosine triphosphate (ATP), and inorganic phosphate (Pi) in patients with high-grade carotid artery stenosis [8]. However, in order to receive comparable results, it is common to interpret MRS data as ratios between different metabolites [9].

We present the 31P MRS findings in relation to MR perfusion imaging in a patient with severe brain ischaemia and intermittent transient ischaemic attacks due to giant cell arteritis prior to and after systemic and endovascular therapy.

2. Case Report

A 67-year-old Caucasian male presented with recurrent episodes of amaurosis fugax, sharp pain in both temples,

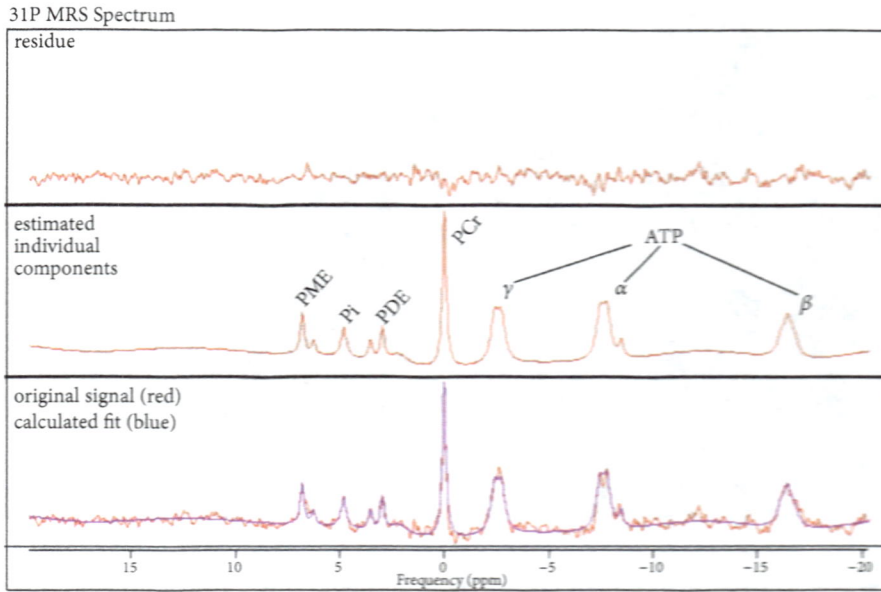

FIGURE 1: Examples for phosphorous magnetic resonance spectrum. The peaks of the different detectable membrane and energy metabolites from left to right: phosphomonoesters (PME), inorganic phosphate (Pi), phosphodiesters (PDE), and phosphocreatine (PCr), as well as three resonances of adenosine triphosphate (gamma-, alpha-, and beta-ATP). The vertical axis displays arbitrary units, while the x-axis shows the chemical shift difference to PCr in ppm. The original signal is depicted as the red curve and the calculated fit can be seen in blue.

masticatory claudication, intermitting paresis of the right arm, and a positive right-sided Babinski sign. ESR was 93 mm/1st hour, C-reactive protein was elevated up to 14.18 mg/dl, and fibrinogen was 1062 mg/dl. The diagnosis of giant cell arteritis was established. Additionally, the patient suffered from arterial hypertension, type 2 diabetes mellitus, hypercholesterinaemia, coronary heart disease, and paroxysmal atrial fibrillation.

The patient received a structural MRI scan with a 3T whole-body system (Verio, Siemens Medical 22 AG, Erlangen, Germany) and a 12-channel reception head coil. MRI angiography revealed short high-grade stenosis of the right vertebral artery (VA) in the V3 segment (Figure 2(a)), a hypoplastic left VA, and a patent posterior communicating artery on the right side. The C6 and C7 segments of the left internal carotid artery (ICA) also showed high-grade stenosis (Figure 2(a)). On a follow-up MRA three weeks later especially the stenosis in the left ICA was longer, but also the stenosis of the right VA (Figure 2(b)). The walls of both ICA (Figure 2(c)), the left temporal artery (TA), and the right VA (Figure 2(d)) were thickened with contrast enhancement, so were the walls of the superficial temporal arteries (Figure 2(e)). In addition, a left-sided pontine infarct was present. Proton emission tomography computed tomography (PET-CT) found no involvement of other noncranial vessels. The diagnosis was based on the 1990 ACR criteria, in which the presence of three out of five points results in a sensitivity of 93.5 % and a specificity of 91.9 % [10]. Even though a halo sign was not seen in color Doppler ultrasound we did not perform a temporal artery biopsy due to the vascular high-risk situation with the need of pronounced antithrombotic therapy. When we retrospectively applied the revised 2016

criteria of the ACR (Sait et al. 2017) for the diagnosis of GCA, we would still confirm the diagnosis with at least four points (three of those in Domain I).

The patient was treated with high-dose corticosteroids as well as acetylsalicylic acid. Within two days, his symptoms had resolved. ESR slowed down to 23 mm/1st hour.

16 days after his first presentation, the patient suffered a new onset of acute aphasia and right-sided facial palsy. In the following days, neurological symptoms fluctuated from mild aphasia to severe aphasia, which could not be stabilised by moderate hypertension, dual antiplatelets, or anticoagulation. A brain MRI revealed a new infarct in the left basal ganglia and the left centrum semiovale. The left ICA stenosis had become more pronounced and extensive compared to the initial MRI, with the C5 segment now involved as well. PWI showed that the MTT, but none of the other perfusion parameters, was inhomogeneously prolonged in the left middle cerebral artery (MCA) territory and in both posterior cerebral artery (PCA) territories (see Figure 3(a)). Additionally to these sequences, a 31P MRS sequence was acquired, with an acquisition time of 10:44, a repetition time of 2000 ms and an echo time of 2.3 ms. The volume of interest was gained with an extrapolated 16 x 16 x 8 matrix and a field of view of 240 x 240 x 200 mm^3, resulting in a voxel size of 15 x 15 x 25 mm^3. For its acquisition the patient had to sit up briefly and the head coil was changed to a double-tuned 1H/31P volume head coil (Rapid 23 Biomedical, Würzburg, Germany). 31P-MRS data was postprocessed offline with the software package jMRUI version 5.0 (current stable version 5.4 available at http://www.jmrui.eu/), utilizing prior knowledge for the nonlinear least square fitting algorithm AMARES [11]. The fitting model was composed of 15 Lorentzian-shaped

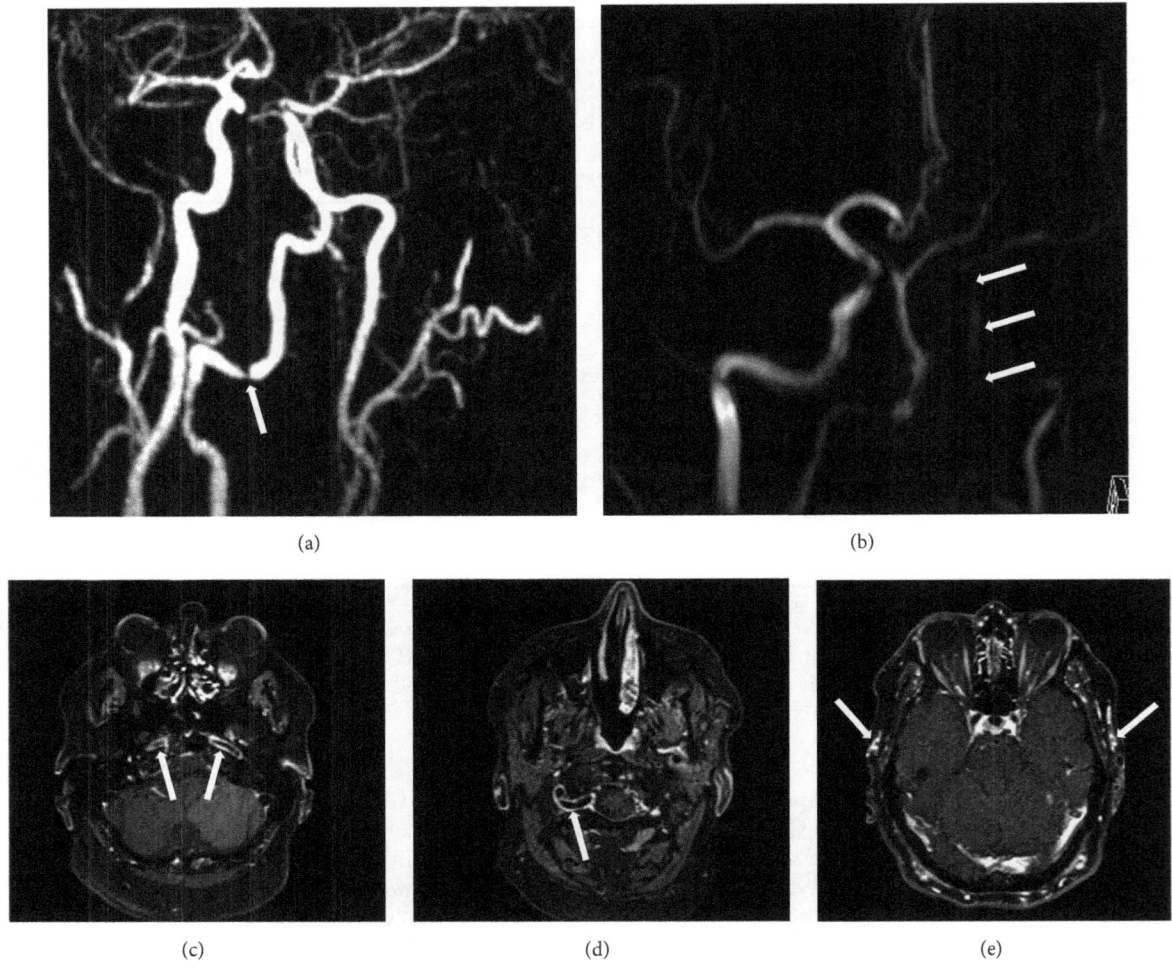

(a)

(b)

(c)   (d)   (e)

FIGURE 2: Magnetic resonance images of the affected arteries. High-grade stenosis of the right vertebral and the left distal internal carotid artery ((a), arrow) on a contrast enhanced magnetic resonance angiography. Long stenosis of the left ICA with tapering aspect ((b), arrows) on a follow-up time of flight MRI. Vessel wall enhancement of both internal carotid arteries ((c), arrows), and the right vertebral artery ((d), arrow), visible on a "dark blood" sequence. Contrast enhancement of both superficial temporal arteries ((e), arrows).

exponentially decaying sinusoids; however, for this patient only the calculation of the metabolite ratio of Pi/PCr was taken into account, as this ratio can be seen as a marker for the energy reserve [12]. 31P MRS revealed a decreased Pi/PCr ratio (Table 1) in both PCA and central left MCA territories in areas which showed a moderately prolonged (3.432 sec) but shorter MTT (Figure 3(a), Table 1) than surrounding areas (3.776 sec, reference value contralateral MCA territory = 3.279 sec). However the adjacent area showed a higher Pi/PCr ratio than the contralateral MCA territory (Table 1).

The therapeutic consequences were an increase of the corticosteroid dosage and initiation of an interleukin-6 receptor blocker therapy. Due to the fluctuating neurological symptoms—with large mismatches between areas with prolonged MTT and the clinical presentation—various potential interventional strategies were discussed. Angioplasty of the left ICA was deemed to be higher risk, because the affected intradural segments were very elongated. With the intention

to improve the perfusion of the ACM territory via the posterior communicating artery, the high-grade stenosis in the V3 segment of the right vertebral artery was corrected via balloon angioplasty and stenting (Figure 4). The intervention was successful, and the aphasia improved rapidly. Follow-up MRI revealed an improved perfusion in parts of the left MCA and both PCA territories. The Pi/PCr ratio in the MCA territory was higher than in the first scan and higher than in the contralateral hemisphere (Table 1). 31P MRS showed a newly decreased Pi/PCr ratio in the border area between MCA and PCA territories and the insular cortex of the left side (Figure 3(b); Table 1), again in an area with a moderately increased MTT (3.462 sec), which was shorter than in the ventral adjacent MCA territory (3.962 sec; Figure 3(b)).

After two months, all mentioned stenoses improved and further clinical improvement was observed. Only a slight aphasia persisted. The corticosteroid dose was able to be reduced to a maintenance level.

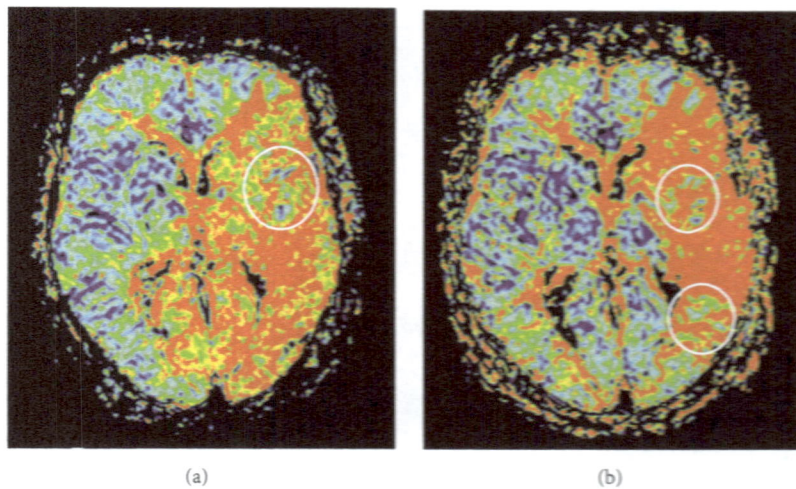

(a)   (b)

FIGURE 3: Perfusion weighted imaging prior to and after therapy. Prior to therapy (a): prolonged mean transit time in large parts of the left middle cerebral artery territory as well as both posterior cerebral artery territories and an area with only moderately elevated transit time was found in the left insular cortex (white circle). After endovascular therapy (b): improvement of cerebral perfusion in both posterior cerebral artery territories and partly the left middle cerebral artery territory, with a now moderately elevated transit time in the border zone between middle and posterior cerebral artery territories, in addition to the area in the insular cortex (circles).

TABLE 1: Results of mean transit time (MTT, given in seconds) and inorganic phosphate to phosphocreatine ratio (Pi/PCr, given in arbitrary units).

Date of exam	Area	MTT	Pi/PCr Mean value
02.07.2015	central MCA	3.43	3.28
02.07.2015	adjacent	3.78	5.49
02.07.2015	contralateral	3.28	4.47
23.07.2015	central MCA	3.96	5.79
23.07.2015	MCA/PCA border	3.46	3.16

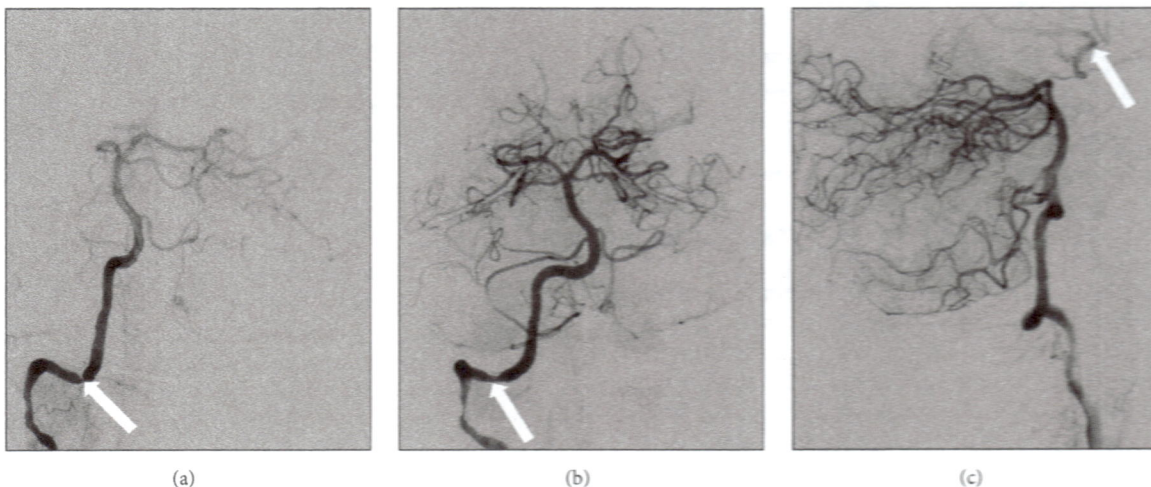

(a)   (b)   (c)

FIGURE 4: Conventional angiography prior to and after angioplasty. Contrast injections in the right vertebral artery, displaying a high-grade stenosis of the right vertebral artery ((a), arrow), which improved after stent angioplasty ((b), arrow). After angioplasty a new cross-flow via the left posterior communicating artery to the left distal internal carotid artery is seen ((c), arrow).

3. Discussion

In this case, energy metabolism changes in a patient suffering from cerebral ischaemia due to giant cell arteritis could be visualised with 31P MRS, both before and after stent angioplasty of the right VA. Prior to the intervention, 31P MRS showed a decreased Pi/PCr ratio in both PCA territories and the central left MCA territory, in areas with a moderately prolonged MTT. In unaffected areas and those areas with a more severely prolonged MTT, no changes of the energy metabolites were found.

Former studies have reported the impact of carotid stenosis or ischaemic stroke on brain energy metabolism [8, 13]. In patients with high-grade carotid artery stenosis, decreased concentrations of ATP and Pi were observed in comparison to healthy controls, and decreased PCr was found in untreated patients [8]. Our findings of a decreased Pi/PCr ratio in less perfused brain tissue are consistent with these results. After VA angioplasty, MTT improved in the left MCA and both PCA territories. However, 31P MRS revealed a newly decreased Pi/PCr ratio in two additional areas with now moderately prolonged MTT. Similar to a prior study [8], we also found a generalised decrease of Pi/PCr in both hemispheres after endovascular therapy.

This is the first time that cerebral 31P MRS has been described to evaluate the therapeutic effects of intracranial angioplasty in giant cell arteritis. Our findings suggest that the fast energy metabolism pathway—the creatine kinase reaction—is a dynamic process, which seems to take place only in areas with moderately but not severely prolonged blood transition. This energy provision seems to be able to adapt according to perfusion circumstances. Our findings imply that the fast energy supply might only take place in less affected areas, which are more likely to survive an ischaemic situation. This in vivo insight into brain energy metabolism might be helpful in selecting patients for interventional treatment of cerebral vessel stenosis in the future. A combination of PWI and 31P MRS could be a novel supplementary method to penumbra imaging, helping to identify tissues at risk of infarction more accurately, for example, in patients with "wake-up" strokes. This could provide more safety for the indication of thrombectomy, intracranial stenting, or bypassing in different situations or display a tool for therapy monitoring. However, the clinical application of this multimodal MRI method has to be evaluated in prospective studies and in a larger sample size of patients with stenosis of the supraaortal or intracranial arteries.

A major limitation of the 31P MRS method is that it is only available in a few centres that the acquisition time is quite long (11 mins) and that the voxel size is larger than most of the regions of interest (15 x 15 x 25 mm). Further modifications with a faster acquisition method and a better spatial resolution could lead to a wider applicability of this method.

4. Conclusion

31P MRS, in combination with the established PWI, is a promising method to evaluate the complex processes that occur during cerebral infarct or ischaemia. Not only does it provide useful information about the perfusion of small vessels, it also gives insight into energy metabolism on a cellular level. Further studies on a larger scale are highly recommended.

References

[1] E. Reinhold-Keller, K. Herlyn, R. Wagner-Bastmeyer, and W. L. Gross, "Stable incidence of primary systemic vasculitides over five years: results from the German vasculitis register," *Arthritis & Rheumatology*, vol. 53, no. 1, pp. 93–99, 2005.

[2] M. Both, T. Jahnke, E. Reinhold-Keller et al., "Percutaneous management of occlusive arterial disease associated with vasculitis: a single center experience," *Cardiovasc Intervent Radiol*, vol. 26, no. 1, pp. 19–26, 2003.

[3] C. Salvarani, F. Cantini, and G. G. Hunder, "Polymyalgia rheumatica and giant-cell arteritis," *The Lancet*, vol. 372, no. 9634, pp. 234–245, 2008.

[4] G. G. Hunder, D. A. Bloch, B. A. Michel et al., "The American College of Rheumatology 1990 criteria for the classification of giant cell arteritis," *Arthritis Rheum*, vol. 33, no. 8, pp. 1122–1128, 1990.

[5] P. Vaith and K. Warnatz, "Clinical and serological findings of giant-cell arteritis," *Z Rheumatol*, vol. 68, no. 2, pp. 124-31, 2009.

[6] M. Reuter, J. Biederer, M. Both et al., "Radiology of the primary systemic vasculitides," *Rofo*, vol. 175, no. 9, pp. 1184-92, 2003.

[7] N. Dementovych, R. Mishra, and Q. A. Shah, "Angioplasty and stent placement for complete occlusion of the vertebral artery secondary to giant cell arteritis," *J Neurointerv Surg*, vol. 4, no. 2, pp. 10–1136, 2012.

[8] E. Hattingen, H. Lanfermann, S. Menon et al., "Combined 1H and 31P MR spectroscopic imaging: impaired energy metabolism in severe carotid stenosis and changes upon treatment," *MAGMA*, vol. 22, no. 1, pp. 43–52, 2009.

[9] X. F. Shi, P. J. Carlson, Y. H. Sung et al., "Decreased brain PME/PDE ratio in bipolar disorder: a preliminary (31) P magnetic resonance spectroscopy study," *Bipolar Disord*, vol. 17, no. 7, pp. 743-52, 2015.

[10] K. Le, L. M. Bools, A. B. Lynn et al., "The effect of temporal artery biopsy on the treatment of temporal arteritis," *Am J Surg*, vol. 209, no. 2, pp. 338–341, 2015.

[11] L. Vanhamme, A. van den Boogaart, and S. van Huffel, "Improved method for accurate and efficient quantification of mrs data with use of prior knowledge," *Journal of Magnetic Resonance*, vol. 129, no. 1, pp. 35–43, 1997.

[12] T. Dinh, J. Doupis, T. E. Lyons et al., "Foot muscle energy reserves in diabetic patients without and with clinical peripheral neuropathy," *Diabetes Care*, vol. 32, no. 8, pp. 1521–1524, 2009.

[13] M. Yamamuro, Y. Katayama, H. Igarashi, and A. Terashi, "1H and 31P-magnetic resonance spectroscopy of cerebral infarction in rats," *Nihon Ika Daigaku Zasshi*, vol. 64, no. 2, pp. 131–138, 1997.

29

A Retroperitoneal Isolated Enteric Duplication Cyst Mimicking a Teratoma

Daichi Momosaka,[1] Yasuhiro Ushijima,[1] Akihiro Nishie,[1]
Yoshiki Asayama,[1] Kousei Ishigami,[1] Yukihisa Takayama,[2]
Daisuke Okamoto,[1] Nobuhiro Fujita,[1] Tetsuo Ikeda,[3] Keiichiro Uchida,[4]
Masaaki Sugimoto,[5] Kenichi Kohashi,[5] and Hiroshi Honda[1]

[1]Department of Clinical Radiology, Graduate School of Medical Sciences, Kyushu University, 3-1-1 Maidashi, Higashi-ku, Fukuoka 812-8582, Japan
[2]Department of Radiology Informatics and Network, Graduate School of Medical Sciences, Kyushu University, 3-1-1 Maidashi, Higashi-ku, Fukuoka 812-8582, Japan
[3]Center for Integration of Advanced Medicine and Innovative Technology, Graduate School of Medical Sciences, Kyushu University, 3-1-1 Maidashi, Higashi-ku, Fukuoka 812-8582, Japan
[4]Department of Medicine and Bioregulatory Science, Graduate School of Medical Sciences, Kyushu University, 3-1-1 Maidashi, Higashi-ku, Fukuoka 812-8582, Japan
[5]Department of Anatomic Pathology, Graduate School of Medical Sciences, Kyushu University, 3-1-1 Maidashi, Higashi-ku, Fukuoka 812-8582, Japan

Correspondence should be addressed to Akihiro Nishie; anishie@radiol.med.kyushu-u.ac.jp

Academic Editor: Roberto Grassi

Enteric duplication cysts lacking anatomic association with the gastrointestinal tract are called isolated enteric duplication cysts (IEDCs). We present an atypical case of a retroperitoneal IEDC with a tortuous tubular complex shape that enfolded the surrounding retroperitoneal fat and mimicked a retroperitoneal teratoma. Multiplanar reconstruction images should be used to evaluate such a lesion correctly. A tortuous tubular complex shape could be a key finding to differentiate from other retroperitoneal cysts.

1. Introduction

Enteric duplication cysts (EDCs) are uncommon congenital anomalies that can be found anywhere along the alimentary tract from the tongue to the anus [1–4]. Essentially they are located in or immediately adjacent to some part of the alimentary tract wall [1]. Histologically, EDCs have a well-developed coat of smooth muscle and an epithelial lining that represents some portion of the intestinal tract mucosa and contain various concentrations of mucus [1]. The incidence rate of EDCs is 1 in every 4000 to 5000 live births [5]. Although the majority of cases are detected in infants, they

can be found in patients of any age [6]. Some cases lack anatomic association with the normal gastrointestinal tract, and they are called isolated enteric duplication cysts (IEDCs) [7]. Prenatal vascular accidents, torsion, and heterotopic tumors may be considered the etiology of IEDCs [8]. This type of tumor has been reported in locations including the tongue [1, 2], pleural space [1], liver [9, 10], pancreas [1, 11], biliary tree [2, 11, 12], and retroperitoneum. Only 17 cases of retroperitoneal IEDCs are found in the literature [8, 13–27]. Most cases have a unilocular or multilocular shape.

Herein, we report a case of a retroperitoneal IEDC that formed a curious shape. This mass was misdiagnosed to

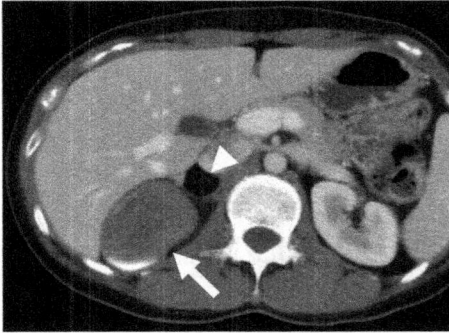

FIGURE 1: Axial contrast-enhanced CT scan. The mass consisted of a cystic component (arrow) and a nodular fatty component (arrowhead) at the right adrenal region. A high-density fluid-fluid level was also seen in the cystic component. No enhancing component was observed.

include a fat component and was difficult to discriminate from a teratoma. We also discuss the radiological findings useful to a correct diagnosis of retroperitoneal IEDC.

2. Case Presentation

A 35-year-old woman visited our institution with an abdominal mass detected on abdominal ultrasound. The patient had no history of parity, drug use, or surgical intervention. On contrast-enhanced computed tomography (CT), a mass with a distorted shape and a diameter of 7.5 cm occupied the region between the right lobe of the liver and the right adrenal gland. The mass, which consisted of nodular fatty components, was well-circumscribed without contrast enhancement. A high-density fluid-fluid level was also seen (Figure 1). On magnetic resonance imaging (MRI), the cystic component showed heterogeneous high intensity on fat-suppressed T2-weighted images (Figure 2(a)) and relatively homogeneous isointensity compared to the muscle on T1-weighted images (Figure 2(b)). On chemical shift images, microscopic fat was not observed in the cystic component. On DWI, the cystic component showed slightly high intensity and its ADC value was 2.0×10^{-3} mm^2/sec, suggesting slightly restricted diffusion compared with the cerebrospinal fluid (Figure 2(c)). A nodular fatty component was again identified on MRI (Figures 2(a) and 2(b)). Based on the presence of a fatty component and possibly calcification or iodine inside the mass, our preoperative diagnosis was retroperitoneal teratoma. The patient underwent laparoscopic surgical intervention. Macroscopically, the mass was a tortuous tubular cyst. The nodular fatty component indicated on preoperative images was not a part of the mass but rather a part of the normal retroperitoneal fat that the complex cyst enfolded. The mass was separated from the colon and the right adrenal gland. Microscopically, the wall of the cyst consisted of well-developed smooth muscle and an epithelial lining representing the large-intestine mucosa. Its content was viscous mucus

(Figure 3). The final diagnosis was retroperitoneal IEDC. Retrospective multiplanar reconstruction (MPR) oblique images revealed the appearance of a tortuous tubular cyst clearly and showed that the nodular fat-density component of the mass was continuous with the normal retroperitoneal fat (Figure 4).

3. Discussion

EDCs are congenital enteric malformations with a cystic appearance, a well-developed coat of smooth muscle, and an epithelial lining representing some portion of the intestinal tract mucosa [28]. On CT and MRI, EDCs are well-circumscribed fluid-filled cysts with a slightly enhanced thin wall, which is located in or adjacent to the normal gastrointestinal wall. The density and intensity of the intracystic fluid can vary depending upon mucous concentration, viscosity, and the existence of intermixed hemorrhage [29, 30]. In our case, the CT and MRI findings on the internal characteristics of the cystic component are consistent with those of the previous reports on EDCs.

The shape of the IEDC in our case is noteworthy. Only 17 cases of retroperitoneal IEDCs have been reported [8, 13–27]. These cases are summarized in Table 1. Retroperitoneal IEDCs demonstrated unilocular (75%) or multilocular to multilobulated (25%) shapes. The present case is the first report of an IEDC with a tortuous tubular complex shape. However, in EDCs that have continuity with the wall of the normal alimentary tract, the shape can be either spherical (80%) or tubular (20%) [31]. That is, 20% of EDCs are tubular. It would be reasonable that IEDCs can also be tubular. To the best of our knowledge, this is the first reported case of a retroperitoneal cyst with such a shape [32].

Another point to discuss is that the mass enfolded surrounding retroperitoneal fat and mimicked a fat-containing tumor. A misdiagnosis resulting from that resemblance may derive from the tortuous tubular complex shape described above. Retrospectively, however, MPR images were useful for differentiating the IEDC from retroperitoneal teratoma because the nodular fat-density component of the mass was continuous with the normal retroperitoneal fat. In addition, MPR images also clearly revealed a tortuous tubular cyst. This image reconstruction technique is of great value for grasping the three-dimensional anatomy of a lesion [33]. For retroperitoneal masses, surgery is basically performed, although ultrasound-guided aspiration and ethanol sclerotherapy can be sometimes performed instead [34]. MPR images can enable surgeons to choose more appropriate operative methods.

In conclusion, a retroperitoneal IEDC can show a tortuous tubular shape and enfold surrounding retroperitoneal fat due to its complex shape. MPR images should be used to evaluate such a lesion correctly. A tortuous tubular complex shape could be a key finding to differentiate from other retroperitoneal cysts.

TABLE 1: Retroperitoneal IEDCs.

Case number	Age/sex	Laterality	Size (cm)	Ectopic gastric mucosa	Ectopic pancreatic mucosa	Shape	Location	Ref.
1	19 y/F	L	11	+	−	Simple	Nearby pancreas	[13]
2	79 y/F	L	5	−	−	Simple	Left adrenal area	[14]
3	34 y/F	R	10	−	−	Simple	Nearby duodenum	[15]
4	19 y/F	L	13	−	−	Multilobular	ND	[16]
5	35 y/F	L	5.5	+	−	Polycystic	Left adrenal area	[17]
6	31 y/F	M	5	−	−	Simple	ND	[18]
7	1 wk/ND	L	3.5	ND	ND	Simple	ND	[19]
8	6 mo/M	R	10	ND	ND	Simple	ND	[19]
9	17 y/M	L	8.6	+	+	Simple	Left adrenal area	[20]
10	27 d/M	L	3	+	−	Simple	Nearby pancreas	[21]
11	28 y/F	L	ND	+	−	Simple	Nearby left kidney	[22]
12	9 d/M	M	5	+	−	Simple	Nearby pancreas	[23]
13	ND/F	Bil	4	−	−	Dumbbell	ND	[24]
14	7 mo/F	R	2	+	−	Polycystic	Right adrenal area	[25]
15	10 mo/F	R	3.8	+	−	Simple	Left adrenal area	[26]
16	9 mo/M	R	ND	+	ND	ND	Nearby right kidney	[27]
17	2 d/M	R	6	−	−	Simple	Nearby extrahepatic bile duct	[8]

IEDCs: isolated enteric duplication cysts, ND: not described, M: male, F: female, Bil: bilateral, R: right, L: left, M: middle, y: years, mo: months, wk: weeks, and d: days.

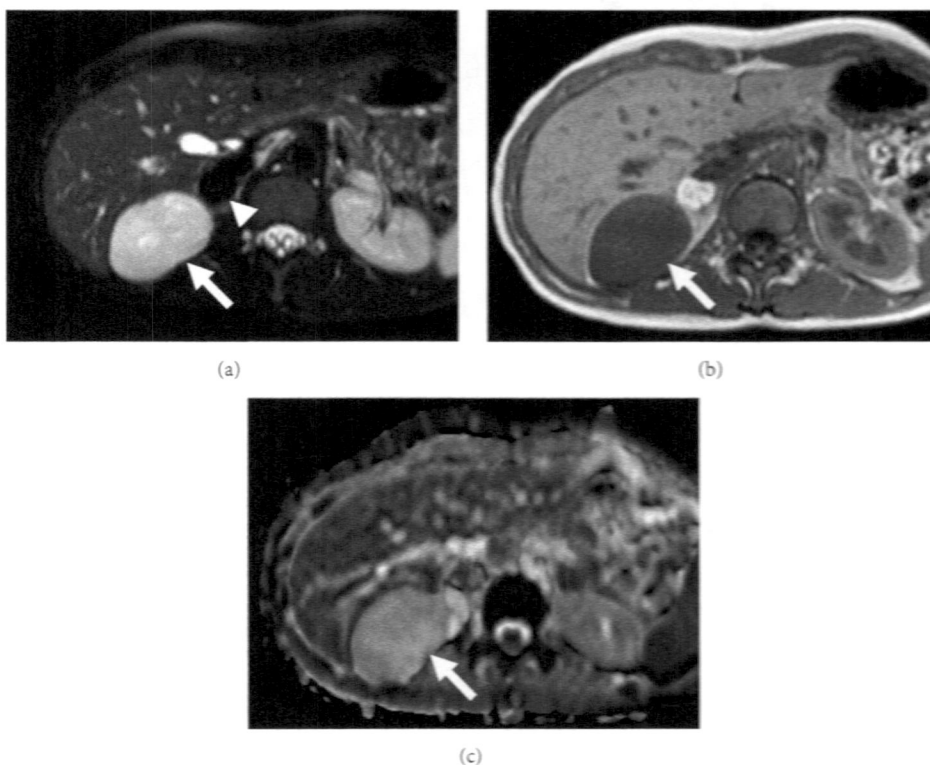

(a)

(b)

(c)

FIGURE 2: Axial MRI. (a) Fat-suppressed T2-weighted image showed heterogeneous high intensity inside the cystic component of the mass (arrow). The fatty component showed low intensity (arrowhead). (b) T1-weighted image showed relative homogeneous isointensity inside the cystic component of the mass (arrow). (c) ADC map showed slightly restricted diffusion compared with cerebrospinal fluid. Its ADC value was 2.0×10^{-3} mm^2/sec.

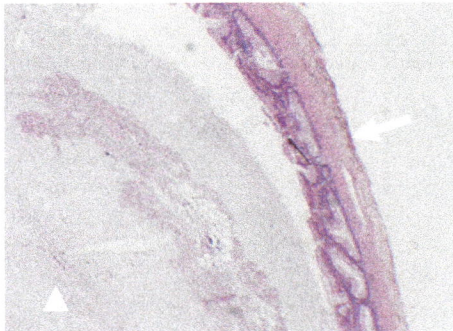

FIGURE 3: Microphotographs of the cystic wall. The wall consisted of large intestinal mucosa, submucosa, muscle layers, and serosa (arrow). The cyst was filled with viscous mucus (arrowhead). A cellular component was not seen (high magnification, hematoxylin and eosin staining).

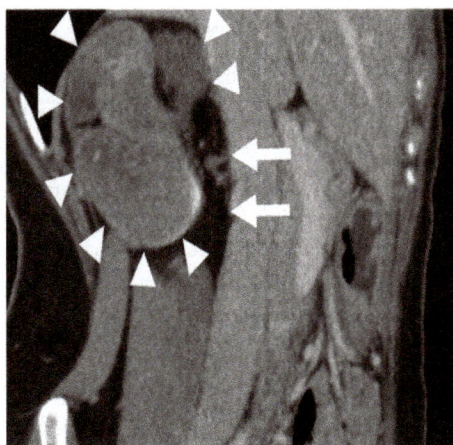

FIGURE 4: Oblique MPR images of contrast-enhanced CT scan. A tortuous tubular complex shape of the mass (arrowheads) was clearly observed. The nodular fatty component of the mass continued to the normal retroperitoneal fat (arrows).

References

[1] R. E. Gross, G. W. Holcomb, and S. Farber, "Duplications of the alimentary tract," *Pediatrics*, vol. 9, no. 4, pp. 448–468, 1952.

[2] B. E. Favara, R. A. Franciosi, and D. R. Akers, "Enteric duplications. Thirty-seven cases: a vascular theory of pathogenesis," *American Journal of Diseases of Children*, vol. 122, no. 6, pp. 501–506, 1971.

[3] R. J. Bower, W. K. Sieber, and W. B. Kiesewetter, "Alimentary tract duplications in children," *Annals of Surgery*, vol. 188, no. 5, pp. 669–674, 1978.

[4] G. W. Holcomb III, A. Gheissari, J. A. O'Neill Jr., N. A. Shorter, and H. C. Bishop, "Surgical management of alimentary tract duplications," *Annals of Surgery*, vol. 209, no. 2, pp. 167–174, 1989.

[5] E. Gilbert Barness, *Potter's Pathology of the Fetus, Infant and Child*, Elsevier, Amsterdam, The Netherlands, 2nd edition, 2007.

[6] S. O. Choi, W. H. Park, and S. P. Kim, "Enteric duplications in children: an analysis of 6 cases," *Journal of Korean Medical Science*, vol. 8, no. 6, pp. 482–487, 1993.

[7] Z. Steiner and J. Mogilner, "A rare case of completely isolated duplication cyst of the alimentary tract," *Journal of Pediatric Surgery*, vol. 34, no. 8, pp. 1284–1286, 1999.

[8] R. Souzaki, S. Ieiri, Y. Kinoshita et al., "Laparoscopic resection of an isolated retroperitoneal enteric duplication in an infant," *Journal of Pediatric Surgery Case Reports*, vol. 1, no. 7, pp. 167–170, 2013.

[9] K. H. Imamoglu and A. J. Walt, "Duplication of the duodenum extending into liver," *The American Journal of Surgery*, vol. 133, no. 5, pp. 628–632, 1977.

[10] J. D. Seidman, A. J. Yale-Loehr, B. Beaver, and C.-C. J. Sun, "Alimentary duplication presenting as an hepatic cyst in a neonate," *American Journal of Surgical Pathology*, vol. 15, no. 7, pp. 695–698, 1991.

[11] D. R. Akers, B. E. Favara, R. A. Franciosi, and J. M. Nelson, "Duplications of the alimentary tract: report of three unusual cases associated with bile and pancreatic ducts," *Surgery*, vol. 71, no. 6, pp. 817–823, 1972.

[12] K. Grumbach, D. H. Baker, J. Weigert, and R. P. Altman, "Biliary tract duplication cyst with gastric heterotopia," *Pediatric Radiology*, vol. 18, no. 4, pp. 357–359, 1988.

[13] N. Upadhyay, D. Gomez, M. F. Button, C. S. Verbeke, and K. V. Menon, "Retroperitoneal enteric duplication cyst presenting as a pancreatic cystic lesion. A case report," *Journal of the Pancreas*, vol. 7, no. 5, pp. 492–495, 2006.

[14] N. E. Terry, C. K. Senkowski, W. Check, and S. T. Brower, "Retroperitoneal foregut duplication cyst presenting as an adrenal mass," *American Surgeon*, vol. 73, no. 1, pp. 89–92, 2007.

[15] H. Hata, N. Hiraoka, H. Ojima, K. Shimada, T. Kosuge, and T. Shimoda, "Carcinoid tumor arising in a duplication cyst of the duodenum," *Pathology International*, vol. 56, no. 5, pp. 272–278, 2006.

[16] Y. S. Lo, J. S. Wang, C. C. Yu et al., "Retroperitoneal enteric duplication cyst," *Journal of the Chinese Medical Association*, vol. 67, pp. 479–482, 2004.

[17] R. D. Laraja, R. E. Rothenberg, J. Chapman et al., "Foregut duplication cyst: a report of a case," *The American Journal of Surgery*, vol. 61, pp. 840–841, 1995.

[18] H. Takiff, J. J. Brems, and M. L. Elliott, "Calcified retroperitoneal enteric duplication cyst," *American Journal of Gastroenterology*, vol. 88, no. 3, pp. 470–471, 1993.

[19] B. W. Duncan, N. Scott Adzick, and A. Eraklis, "Retroperitoneal alimentary tract duplications detected in utero," *Journal of Pediatric Surgery*, vol. 27, no. 9, pp. 1231–1233, 1992.

[20] P.-H. Chen, J.-Y. Lee, S.-F. Yang, J.-Y. Wang, J.-Y. Lin, and Y.-T. Chang, "A retroperitoneal gastric duplication cyst mimicking a simple exophytic renal cyst in an adolescent," *Journal of Pediatric Surgery*, vol. 45, no. 10, pp. e5–e8, 2010.

[21] T. Okamoto, S. Takamizawa, A. Yokoi, S. Satoh, and E. Nishijima, "Completely isolated alimentary tract duplication in a neonate," *Pediatric Surgery International*, vol. 24, no. 10, pp. 1145–1147, 2008.

[22] S. K. Kim, H. K. Lim, S. J. Lee, and C. K. Park, "Completely isolated enteric duplication cyst: case report," *Abdominal Imaging*, vol. 28, no. 1, pp. 12–14, 2003.

[23] N. Nakazawa, T. Okazaki, and T. Miyano, "Prenatal detection of isolated gastric duplication cyst," *Pediatric Surgery International*, vol. 21, no. 10, pp. 831–834, 2005.

[24] D. A. May, S. E. Spottswood, M. Ridick-Young, and B. C. Nwomeh, "Case report: prenatally detected dumbbell-shaped retroperitoneal duplication cyst," *Pediatric Radiology*, vol. 30, no. 10, pp. 671–673, 2000.

[25] H. S. Bal, S. Kisku, S. Sen, and D. Masih, "A retroperitoneal enteric duplication cyst communicating with the right upper ureter in an infant," *BMJ Case Reports*, vol. 2014, 2014.

[26] M. Pachl, K. Patel, C. Bowen, and D. Parikh, "Retroperitoneal gastric duplication cyst: a case report and literature review," *Pediatric Surgery International*, vol. 28, no. 1, pp. 103–105, 2012.

[27] S.-F. Ko, S.-H. Ng, F.-C. Huang, M.-T. Sung, and C.-S. Hsieh, "Postprandial abdominal pain owing to isolated enteric duplication cyst in the superior mesenteric artery root: sonographic and magnetic resonance imaging features," *Journal of Pediatric Surgery*, vol. 46, no. 4, pp. 772–775, 2011.

[28] R. I. Macpherson, "Gastrointestinal tract duplications: clinical, pathologic, etiologic, and radiologic considerations," *Radiographics*, vol. 13, no. 5, pp. 1063–1080, 1993.

[29] T. Berrocal, M. Lamas, J. Gutiérrez, I. Torres, C. Prieto, and M. L. Del Hoyo, "Congenital anomalies of the small intestine, colon, and rectum," *Radiographics*, vol. 19, no. 5, pp. 1219–1236, 1999.

[30] R. L. Teele, C. I. Henschke, and D. Tapper, "The radiographic and ultrasonographic evaluation of enteric duplication cysts," *Pediatric Radiology*, vol. 10, no. 1, pp. 9–14, 1980.

[31] B. Domajnko and R. M. Salloum, "Duplication cyst of the sigmoid colon," *Gastroenterology Research and Practice*, vol. 2009, Article ID 918401, 3 pages, 2009.

[32] D. M. Yang, D. H. Jung, H. Kim et al., "Retroperitoneal cystic masses: CT, clinical, and pathologic findings and literature review," *Radiographics*, vol. 24, no. 5, pp. 1353–1365, 2004.

[33] M. M. Maher, M. K. Kalra, D. V. Sahani et al., "Techniques, clinical applications and limitations of 3D reconstruction in CT of the abdomen," *Korean Journal of Radiology*, vol. 5, no. 1, pp. 55–67, 2004.

[34] G. Gatta, V. Parlato, G. Di Grezia et al., "Ultrasound-guided aspiration and ethanol sclerotherapy for treating endometrial cysts," *Radiologia Medica*, vol. 115, no. 8, pp. 1330–1339, 2010.

30

Dual-Energy CT for Evaluation of Intra- and Extracapsular Silicone Implant Rupture

Katrina N. Glazebrook, Shuai Leng, Steven R. Jacobson, and Cynthia M. McCollough

Mayo Clinic, 200 First Street SW, Rochester, MN 55904, USA

Correspondence should be addressed to Katrina N. Glazebrook; glazebrook.katrina@mayo.edu

Academic Editor: Atsushi Komemushi

Silicone implants are commonly used for both breast augmentation and breast reconstruction. With aging of the implant, the silicone envelope may become weak or may rupture. The technique of choice for evaluation of implant integrity is breast MRI; however this may be contraindicated in some patients or the cost may be prohibitive. Dual-energy CT allows determination of density and atomic number of tissue and can provide material composition information. We present a case of extracapsular implant rupture with MRI and dual-energy CT imaging and surgical correlation.

1. Introduction

Many silicone gel implants placed in the 1990s may be ruptured by now. Robinson et al. performed Kaplan-Meier survival analyses which showed that the proportion of patients with intact implants after 20 years was as low as 5% [1, 2]. We describe the MRI and dual-energy CT (DECT) appearance of intra- and extracapsular rupture with nodal silicone deposits and peri-implant seroma in a patient with long-standing silicone implants following trauma.

2. Case Report

A sixty-two-year-old woman was diagnosed with multicentric ductal carcinoma in situ and invasive ductal carcinoma of the left breast in 1992. She underwent a left modified radical mastectomy and a right simple mastectomy with immediate bilateral reconstruction with subpectoral silicone gel implants. A month before presentation, she fell on the right breast, heard a pop, and had immediate right-sided chest wall pain. She noted progressive enlargement of the right breast with increasing pain. Clinically, the right breast was larger than the left, suspicious for seroma and or hematoma with bilateral animation deformity consistent with subpectoral implants. An MRI exam was performed to evaluate implant rupture using an 8-channel breast coil. The MRI

exam (1.5 T Signa LX Echospeed, General Electric Medical Systems, Milwaukee, WI) consisted of an axial T2 weighted IDEAL sequence, axial and sagittal silicone sensitive series, and pre- and postcontrast Vibrant 3D T1 weighted gradient series. Intracapsular rupture was found on the right with a surrounding seroma, both within the ruptured envelope and mixing with the silicone outside the envelope but within the fibrous capsule (Figures 1(a) and 1(b)). Foci of extracapsular silicone were felt to be present in the right axillary tail. Intracapsular rupture of the left implant was also noted with MRI linguini sign [3] (Figures 1(a) and 1(b)). No axillary adenopathy was identified as the high signal intensity on the silicone sensitive sequence was not appreciated prospectively to correspond to silicone within level I and II axillary nodes (Figure 1(c)). There was no abnormal enhancement to suggest tumor recurrence.

DECT was performed using a dual-source CT scanner (SOMATOM Force, Siemens Healthcare, Forchheim, Germany) using tube potentials of 100 and 150 kV. An additional tin filter was added to the 150 kV beam to increase spectral separation. The patient was scanned prone with a single acquisition using a prototype breast stand modified from a 16-channel breast MRI coil for CT use. No intravenous contrast material was given. Tube current was adjusted to match the radiation dose as that of a routine noncontrast chest CT. The volume CTDI ($CTDI_{vol}$) was 7.19 mGy, and the

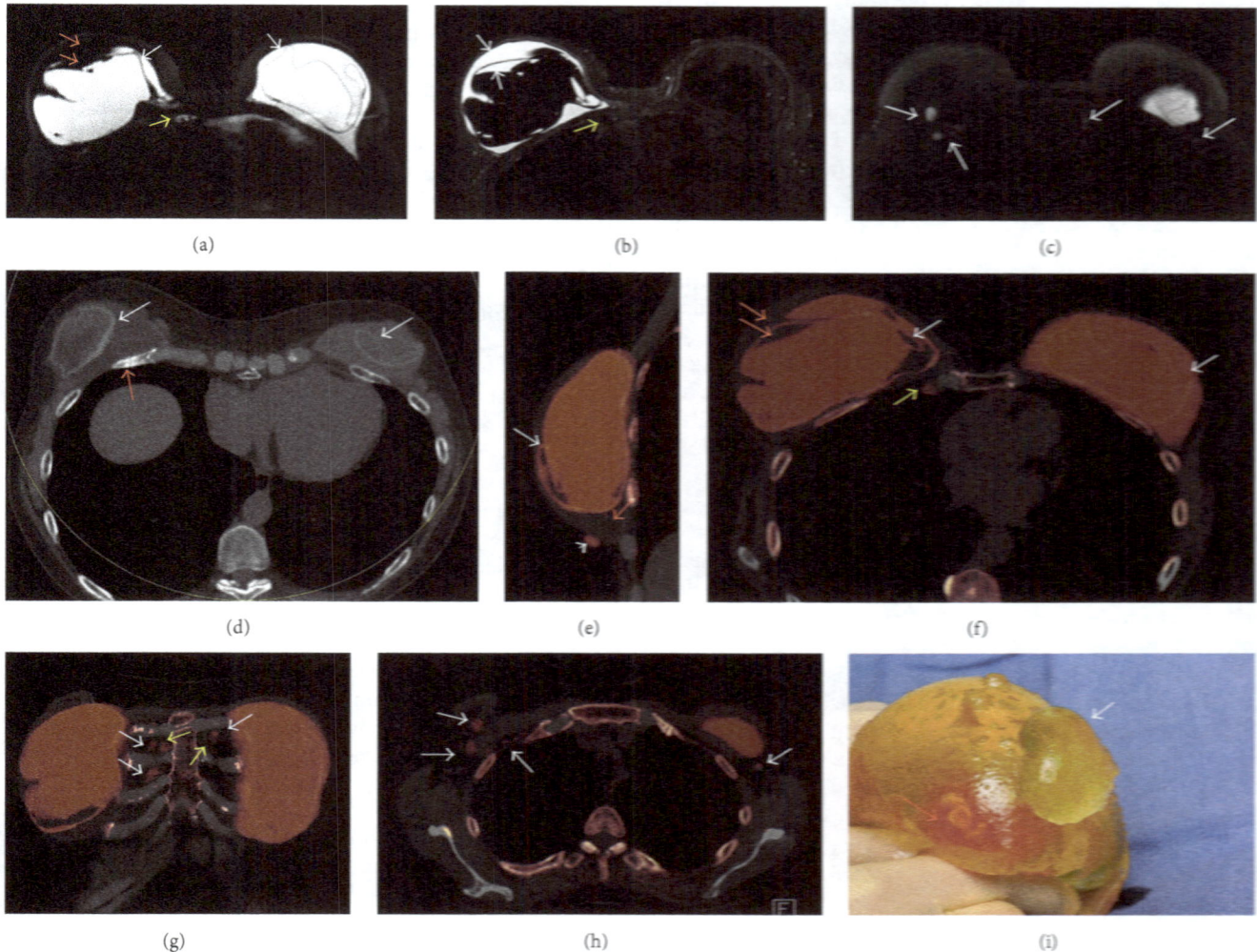

FIGURE 1: 62-year-old woman presents with enlarging right breast following a fall one month earlier and a history of long standing bilateral subpectoral silicone implants following left modified radical mastectomy and simple right mastectomy in 1992. (a) Axial silicone sensitive MR sequence shows bilateral MR linguini sign of intracapsular rupture (white arrows) with low signal intensity material both within the capsule itself, along with high signal intensity silicone, and within the collapsed right envelope (red arrows). The high signal intensity tissue adjacent to the right internal mammary (IM) vessels (yellow arrow) was not appreciated as being likely due to silicone due to the inhomogeneity of the fat and water suppression about heart. (b) T2 weighted axial IDEAL MR sequence shows high T2 signal fluid of intracapsular seroma within the right envelope (white arrows). There is low signal intensity tissue adjacent to the right IM vessels which in retrospect would be in keeping with silicone within an internal mammary node (yellow arrow). (c) Axial silicone sensitive sequence from the high axilla demonstrates high signal intensity material which was not appreciated prospectively as right intranodal silicone in level I and II nodes and left IM nodes (arrows). (d) Axial mixed energy CT shows the CT linguini sign of collapsed implant envelopes bilaterally (white arrows). Healing fractured right anterior rib is clearly seen (red arrow). (e) Sagittal dual-energy noncontrast CT of the right breast with silicone colored as red shows intracapsular rupture with CT equivalent linguini sign of collapsed silicone envelope (white arrow). Water density material (red arrows) is noted surrounding the collapsed right envelope as seen on fluid sensitive MRI sequence (c). Extracapsular silicone is noted inferiorly (arrowhead). (f) Axial dual energy noncontrast CT of the right breast with silicone colored as red shows bilateral intracapsular rupture with CT equivalent linguini sign of collapsed silicone envelopes (white arrows). Water density material (red arrows) is noted surrounding and within the collapsed right envelope as seen on fluid sensitive MRI sequence (c). Silicone is also easily seen within enlarged right IM lymph node (yellow arrows). (g) Coronal DECT with color coding shows silicone within the right and left internal mammary nodes (white arrows). The internal mammary vessels can be seen (yellow arrows). (h) Axial DECT with silicone colored as red show silicone within right level I and II and left level I nodes (arrows). (i) Photograph of the explanted right implant with extruded silicone from the envelope (white arrow). Serosanguineous fluid noted within the envelope as seen on MRI and CT (red arrow).

dose length product (DLP) was 216.8 mGy·cm. Axial images were reconstructed with slice thickness of 1.5 mm. Sagittal and coronal reformats were performed. Images were analyzed using the 3-material decomposition algorithm of the dual-energy CT software (syngo Via Dual Energy, Siemens Healthcare) with "Liver VNC" workflow. The mixed CT images (average if 90 and 150 kV images) demonstrated healing nondisplaced rib fractures of the right 3rd through 6th ribs and left 3rd through 5th ribs anterolaterally (Figure 1(d)). On the silicone color-coded images of the right breast, fluid density material was noted within the right fibrous capsule external to and within the collapsed silicone envelope consistent with seroma and intracapsular rupture (Figure 1(e)). Several foci of extracapsular silicone were noted superficial to the pectoralis muscle in the inferomedial breast that was not appreciated prospectively on the breast MRI exam (Figure 1(e)). Intracapsular rupture was also seen on the left, with fluid noted within the collapsed envelope (Figure 1(f)). No extracapsular silicone was seen in the left breast. There were several enlarged internal mammary nodes bilaterally containing silicone (Figures 1(f) and 1(g)) and level I through III (right) and level I (left) axillary nodes containing silicone (Figure 1(h)). Note was also made of coronary artery calcifications. This led to a cardiac stress test, which the patient failed, and resulted in 3-vessel coronary artery stent placement with good results.

The patient underwent explantation of both implants with bilateral capsulectomies and silicone implant exchange with AlloDerm placement. The explanted right implant showed signs of rupture with seroma actually within the implant itself, as shown on MRI and DECT (Figure 1(i)). Rupture of the left implant was obvious following entry of the capsule.

On follow-up, the patient was very satisfied with her new implants and very grateful for the DECT for identifying the coronary calcifications.

3. Discussion

The prevalence of silicone breast implant rupture in a population-based study has been reported to be as high as 55%, with 22% of ruptured implants showing extracapsular spread of silicone [4, 5]. Local complications and adverse outcomes include capsular contracture, reoperation, and removal. Women may also experience breast pain, wrinkling, asymmetry, scarring, and rarely infection [5]. Breast ultrasound has been used to evaluate implant integrity with sensitivity and specificity ranging from 50 to 77% and from 55 to 84%, respectively [3, 6–9]. The sonographic sign of intracapsular rupture is the "stepladder" sign, with multiple curvilinear reflective lines within the interior of the implant corresponding to the collapsed envelope. Silicone masses within the breast tissue and axillary nodes in cases of extracapsular rupture have a typical "snowstorm" appearance with posterior shadowing [3, 6–9].

Breast MRI is the technique of choice for assessment of implant integrity. The reported sensitivity and specificity for implant rupture range from 74 to 100% and from 55 to 84%, respectively [5, 8–11]. Intracapsular rupture of the implant envelope is seen as hypodense linear lines lying within the hyperintense silicone, the so-called "Linguini" sign [3]. Silicone-specific sequences with 3-point chemical shift techniques show high signal intensity silicone within the breast parenchyma or axillary nodes in extracapsular rupture. The FDA has mandated a surveillance MRI screening examination for silent rupture in patients at 3 years following implantation of silicone breast implants and every 2 years thereafter [4]. MRI may not be possible in patients with contraindications to MRI (e.g., pacemaker, cochlear implant) or claustrophobia. The costs of MRI surveillance are not typically covered by insurance and can be prohibitively expensive for most patients.

Johnson et al. performed DECT for the evaluation of silicone breast implants [12]. Seven silicone implant specimens were evaluated with DECT using 100 and 140 kV tube potentials, with a strong dual-energy signal. Two patients scheduled for implant removal or replacements were examined. In one patient, both implants were intact, while rupture was identified in the other patient. Ultrasound, MRI, surgical findings, and histology confirmed the DECT diagnosis.

CT number (in Hounsfield units) depends on X-ray attenuation, which depends on the physical density (g/cm^3) (electron-density) and atomic number (Z). Different materials may have the same CT number if atomic number differences are offset by density differences [13]. DECT allows determination of density and Z and can provide material composition information. Silicone contains the metalloid element silicon with element number 14. Soft tissue contains lighter elements such as hydrogen and oxygen. Silicon and soft tissue have different slopes in a plot of low-energy CT number versus high-energy CT number and therefore can be differentiated. This technique has been shown to be highly accurate in the characterization of kidney stones and identification of monosodium urate crystals in the extremities in patients with gout [14–17]. In this case, DECT not only defined the intracapsular ruptures and right seroma, but also more clearly identified the extracapsular silicone within chest wall tissue, level I through III axillary nodes, and extra-axillary nodes, compared to the MRI scan. Additional clinically relevant information was also noted on the DECT, including identification of the bilateral rib fractures and coronary artery calcifications.

4. Conclusion

While MRI is the current technique of choice for evaluation of intra- and extracapsular silicone breast implant ruptures, DECT however shows promise in more specifically evaluating the extent of extracapsular rupture and nodal involvement in a single, noncontrast, breathhold scan.

References

[1] O. G. Robinson, E. L. Bradley, and D. S. Wilson, "Analysis of explanted silicone implants: a report of 300 patients," *Annals of Plastic Surgery*, vol. 34, no. 1, pp. 1–6, 1995.

[2] S. L. Brown, B. G. Silverman, and W. A. Berg, "Rupture of silicone-gel breast implants: causes, sequelae and diagnosis," *The Lancet*, vol. 350, no. 9090, pp. 1531–1537, 1997.

[3] N. Yang and D. Muradali, "The augmented breast: a pictorial review of the abnormal and unusual," *American Journal of Roentgenology*, vol. 196, no. 4, pp. W451–W460, 2011.

[4] Center for Devices and Radiological Health, *FDA Update on the Safety of Silicone Gel-Filled Breast Implants*, Center for Devices and Radiological Health, 2011.

[5] C. M. Goodman, V. Cohen, J. Thornby, and D. Netscher, "The life span of silicone gel breast implants and a comparison of mammography, ultrasonography, and magnetic resonance imaging in detecting implant rupture: a meta-analysis," *Annals of Plastic Surgery*, vol. 41, no. 6, pp. 577–586, 1998.

[6] A. Cilotti, C. Marini, C. Iacconi et al., "Ultrasonographic appearance of breast implant complications," *Annals of Plastic Surgery*, vol. 56, no. 3, pp. 243–247, 2006.

[7] C. I. Caskey, W. A. Berg, N. D. Anderson, S. Sheth, B. W. Chang, and U. M. Hamper, "Breast implant rupture: diagnosis with US," *Radiology*, vol. 190, no. 3, pp. 819–823, 1994.

[8] G. Di Benedetto, S. Cecchini, L. Grassetti et al., "Comparative study of Breast implant rupture using mammography, sonography, and magnetic resonance imaging: correlation with surgical findings," *The Breast Journal*, vol. 14, no. 6, pp. 532–537, 2008.

[9] D. M. Ikeda, H. B. Borofsky, R. J. Herfkens, A. M. Sawyer-Glover, R. L. Birdwell, and G. H. Glover, "Silicone breast implant rupture: pitfalls of magnetic resonance imaging and relative efficacies of magnetic resonance, mammography, and ultrasound," *Plastic and Reconstructive Surgery*, vol. 104, no. 7, pp. 2054–2062, 1999.

[10] L. R. Hölmich, I. Vejborg, C. Conrad, S. Sletting, and J. K. McLaughlin, "The diagnosis of breast implant rupture: MRI findings compared with findings at explantation," *European Journal of Radiology*, vol. 53, no. 2, pp. 213–225, 2005.

[11] A. J. Madhuranthakam, M. P. Smith, H. Yu et al., "Water-silicone separated volumetric MR acquisition for rapid assessment of breast implants," *Journal of Magnetic Resonance Imaging*, vol. 35, no. 5, pp. 1216–1221, 2012.

[12] T. R. C. Johnson, I. Himsl, K. Hellerhoff et al., "Dual-energy CT for the evaluation of silicone breast implants," *European Radiology*, vol. 23, no. 4, pp. 991–996, 2013.

[13] T. R. C. Johnson, B. Krauss, M. Sedlmair et al., "Material differentiation by dual energy CT: initial experience," *European Radiology*, vol. 17, no. 6, pp. 1510–1517, 2007.

[14] A. N. Primak, J. G. Fletcher, T. J. Vrtiska et al., "Non-invasive differentiation of uric acid versus non-uric acid kidney stones using dual-energy CT," *Academic Radiology*, vol. 14, no. 12, pp. 1441–1447, 2007.

[15] A. Graser, T. R. C. Johnson, M. Bader et al., "Dual energy CT characterization of urinary calculi: initial in vitro and clinical experience," *Investigative Radiology*, vol. 43, no. 2, pp. 112–119, 2008.

[16] K. N. Glazebrook, L. S. Guimarães, N. S. Murthy et al., "Identification of intraarticular and periarticular uric acid crystals with dual-energy CT: initial evaluation," *Radiology*, vol. 261, no. 2, pp. 516–524, 2011.

[17] T. Bongartz, K. N. Glazebrook, S. Kavrios et al., "Dual-energy CT for the diagnosis of gout: and accuracy and diagnostic yield study," *Annals of the Rheumatic Diseases*, vol. 74, no. 6, pp. 1072–1077, 2015.

31

Osteoblastic Metastases Mimickers on Contrast Enhanced CT

Fahad Al-Lhedan,[1] Sam Samaan,[2] and Wanzhen Zeng[2]

[1]Medical Imaging Department, King Abdullah bin Abdulaziz University Hospital, Riyadh, Saudi Arabia
[2]Department of Nuclear Medicine, The Ottawa Hospital, Ottawa, ON, Canada

Correspondence should be addressed to Fahad Al-Lhedan; fahad_allhedan@hotmail.com

Academic Editor: Samer Ezziddin

Secondary osseous involvement in lymphoma is more common compared to primary bone lymphoma. The finding of osseous lesion can be incidentally discovered during the course of the disease. However, osseous metastases are infrequently silent. Detection of osseous metastases is crucial for accurate staging and optimal treatment planning of lymphoma. The aim of imaging is to identify the presence and extent of osseous disease and to assess for possible complications such as pathological fracture of the load-bearing bones and cord compression if the lesion is spinal. We are presenting two patients with treated lymphoma who were in complete remission. On routine follow-up contrast enhanced CT, there were new osteoblastic lesions in the spine worrisome for metastases. Additional studies were performed for further evaluation of both of them which did not demonstrate any corresponding suspicious osseous lesion. The patients have a prior history of chronic venous occlusive thrombosis that resulted in collaterals formation. Contrast enhancement of the vertebral body marrow secondary to collaterals formation and venous flow through the vertebral venous plexus can mimic the appearance of spinal osteoblastic metastases.

1. Background

Approximately 16% of lymphoma patients will eventually have osseous involvement [1]. Osseous manifestations of lymphoma in the spine commonly occur as a result of direct invasion from adjacent lymph nodes; however, hematogenous osseous metastases are also a possibility [2]. The osseous involvement usually occurs during the course of the disease rather than at the initial presentation. Multiple osseous metastases are more common than a solitary metastasis. Most lymphomatous metastases tend to be osteolytic; however, osteoblastic and mixed metastases may also be encountered [2].

Osseous metastases can cause significant morbidity as a result of pathologic fracture and spinal cord compression [3, 4].

2. Discussion

2.1. Case 1. A 54-year-old male with Hodgkin lymphoma, who was in complete remission, had several new osteoblastic spinal lesions on routine follow-up contrast enhanced CT of the neck (Figure 1) which were worrisome for new

osteoblastic metastases. A whole-body SPECT bone scan was performed to evaluate the extent of osseous disease (Figure 2). In addition, FDG PET/CT was performed to evaluate the disease extent within the body (Figures 3(a) and 3(b)).

The patient has a history of left innominate and left subclavian veins chronic occlusive thrombosis in addition to a partially occlusive thrombosis of the left internal jugular and left axillary veins. The routine follow-up contrast enhanced CT of the neck demonstrated at least partial occlusion of the proximal left subclavian vein with numerous collaterals in the left shoulder and left upper back region.

The lack of uptake on bone scan and FGD PET along with disappearance of lesions on the non-contrast enhanced CT portion of the PET/CT (Figure 3(b)) implies that the apparent osteoblastic lesions were merely vertebral marrow enhancement secondary to collaterals formation and venous flow through the vertebral venous plexus in the cervical and thoracic spine [5].

2.2. Case 2. A 22-year-old male with non-Hodgkin lymphoma, who was in complete remission, had a new osteoblastic lesion at the T5 vertebral body on routine follow-up contrast enhanced CT of the chest (Figure 4) which was

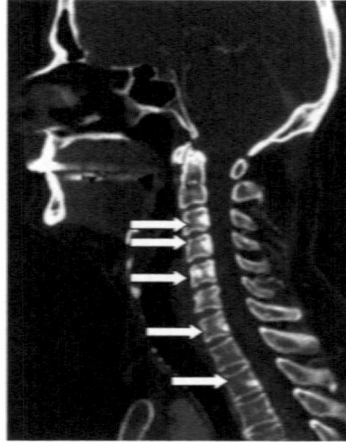

FIGURE 1: *(Neck contrast enhanced CT)*: Sagittal bone window CT demonstrating several osteoblastic lesions at C3, C4, C5, C7, and T2 (white arrows).

FIGURE 2: *(SPECT)*: Sagittal SPECT bone scan does not demonstrate any abnormal focal uptake within the cervical or thoracic spine.

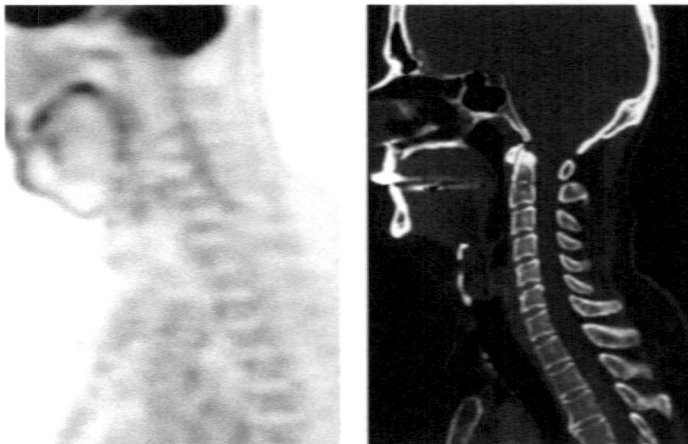

(a) *(PET/CT)*: Sagittal FDG PET does not demonstrate any abnormal focal uptake within the cervical or thoracic spine

(b) *(PET/CT)*: Sagittal non-contrast enhanced CT does not demonstrate any osteoblastic lesions within the cervical or thoracic spine

FIGURE 3

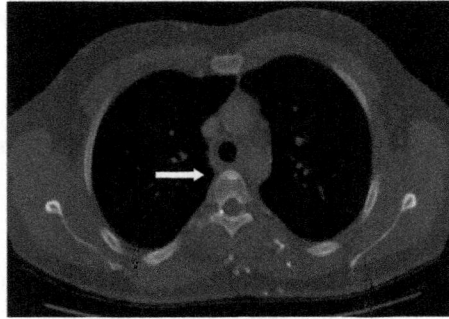

FIGURE 4: *(Chest contrast enhanced CT)*: Axial bone window CT demonstrating a solitary osteoblastic lesion at T5 (white arrow).

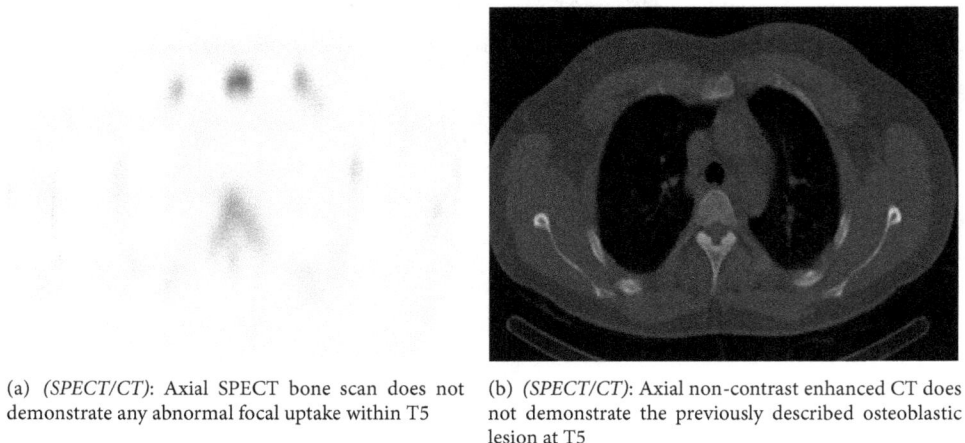

(a) *(SPECT/CT)*: Axial SPECT bone scan does not demonstrate any abnormal focal uptake within T5

(b) *(SPECT/CT)*: Axial non-contrast enhanced CT does not demonstrate the previously described osteoblastic lesion at T5

FIGURE 5

worrisome for a new solitary osteoblastic metastasis. A whole-body SPECT bone scan was performed to evaluate the extent of osseous disease (Figure 5).

The patient has a history of chronic occlusive thrombosis of bilateral brachiocephalic and left subclavian veins which was demonstrated on contrast enhanced CT of the chest.

Again the lack of uptake on bone scan along with disappearance of lesion on the non-contrast enhanced CT portion of the SPECT/CT (Figure 5(b)) implies that the apparent osteoblastic lesion was merely vertebral marrow enhancement secondary to collaterals formation and venous flow through the vertebral venous plexus in the thoracic spine [5].

3. Conclusion

The apparent osteoblastic lesions on both contrast enhanced CTs were merely marrow enhancement of the vertebral bodies. In the presence of chronic subclavian vein occlusion, collaterals often form and may result in venous flow through the vertebral venous plexus after contrast administration.

Contrast enhancement of the vertebral body marrow secondary to venous flow through the vertebral venous plexus can mimic the appearance of spinal osteoblastic metastases. Therefore, focal contrast enhancement of a vertebral body

should be considered as a mimicker of an osteoblastic lesion on contrast enhanced CT in the presence of significant chronic venous occlusion and collaterals formation.

Additional Points

Objective. Identify mimickers of osteoblastic metastases on contrast enhanced CT.

References

[1] P. C. Malloy, E. K. Fishman, and D. Magid, "Lymphoma of bone, muscle, and skin: CT findings," *American Journal of Roentgenology*, vol. 159, no. 4, pp. 805–809, 1992.

[2] J. O'Neill, K. Finlay, E. Jurriaans, and L. Friedman, "Radiological Manifestations of Skeletal Lymphoma," *Current Problems in Diagnostic Radiology*, vol. 38, no. 5, pp. 228–236, 2009.

[3] R. L. Theriault and R. L. Theriault, "Biology of bone metastases," *Cancer Control*, vol. 19, no. 2, pp. 92–101, 2012.

32

Abernethy Malformation Type II and Concurrent Nodular Hyperplasia in a Rare Female Case

Zhen Kang⬤, **Xiangde Min**⬤, **and Liang Wang**⬤

Department of Radiology, Tongji Hospital, Tongji Medical College, Huazhong University of Science and Technology, Wuhan, Hubei 430030, China

Correspondence should be addressed to Liang Wang; wang6@tjh.tjmu.edu.cn

Academic Editor: Atsushi Komemushi

Background. Abernethy malformation is a rare splanchnic vascular abnormality characterizing extrahepatic abnormal shunts that is classified into types I and II. Abernethy malformation type I has a female predilection and is associated with a variety of concurrent hepatic benign or malignant tumours while type II with concurrent tumours is very rare in females. *Case Report.* We report a rare female case of Abernethy malformation type II with concurrent occupying lesion in the right liver, which was successfully transplanted; the occupying lesion was pathologically proven to be nodular hyperplasia. *Conclusion.* This case might provide further knowledge regarding Abernethy malformation. On imaging, the anatomy of portal vein should be carefully investigated to categorize Abernethy malformation, and a wide variety of differential diagnosis of concurrent occupying lesions should be taken into account.

1. Background

Abernethy malformation was first described in 1793, which is characterized by direct drainage of the superior mesenteric vein and splenic vein to the inferior vena cava. The malformation is divided into two types based on the vascular anatomy of the exhepatic shunts. The majority of type I shunts are identified in females, and a variety of benign or malignant tumours may exist simultaneously, whereas type II shunts are almost always recognized in males with a much lower prevalence than type I [1]. Here, we report a rare female case of Abernethy malformation type II with a concurrent mass in the right liver, mimicking a hepatocellular carcinoma or hepatoblastoma on computer tomography (CT), which was ultimately confirmed to be nodular hyperplasia by pathology. This case was successfully transplanted by piggyback method. Through this case report, we might gain further knowledge of Abernethy malformation.

2. Case Report

A 10-year-old female visited the emergency room of our hospital with right upper quadrant abdominal pain for 8 days that was not associated with eating. No fever, jaundice, emesis, diarrhea, or melena was reported. Initial blood tests and liver function test revealed no abnormalities other than mild elevation of cytomegalovirus IgG antibody (23.7 U/ml). A routine urine test was positive for red blood cells (3+, 294.8/μl) and white blood cells (3+, 422.9/μl). Alpha-fetoprotein and cancer embryonic antigen were negative, whereas cancer antigen 19-9 was elevated (247.25 U/ml, normal level 2-37 U/ml).

Ultrasound examination exhibited a mass measuring 78 x 71 mm in the right liver with abundant blood signals. No obvious portal venous trunk and branches were noted in the liver but the superior mesenteric vein and splenic veins drained directly into the inferior vena cava. The left hepatic vein and middle hepatic vein conjoined together and then drained to the inferior vena cava. Abdominal enhanced CT with multiplanar reconstruction revealed that the splenic vein and superior mesenteric vein drained directly into the inferior vena cava after confluence (Figures 1(a) and 1(b)), and only two branches of hepatic veins drained into the inferior vena cava. A barely perceptible small branch of the portal vein measuring 3 mm supplied the left lobe of the liver

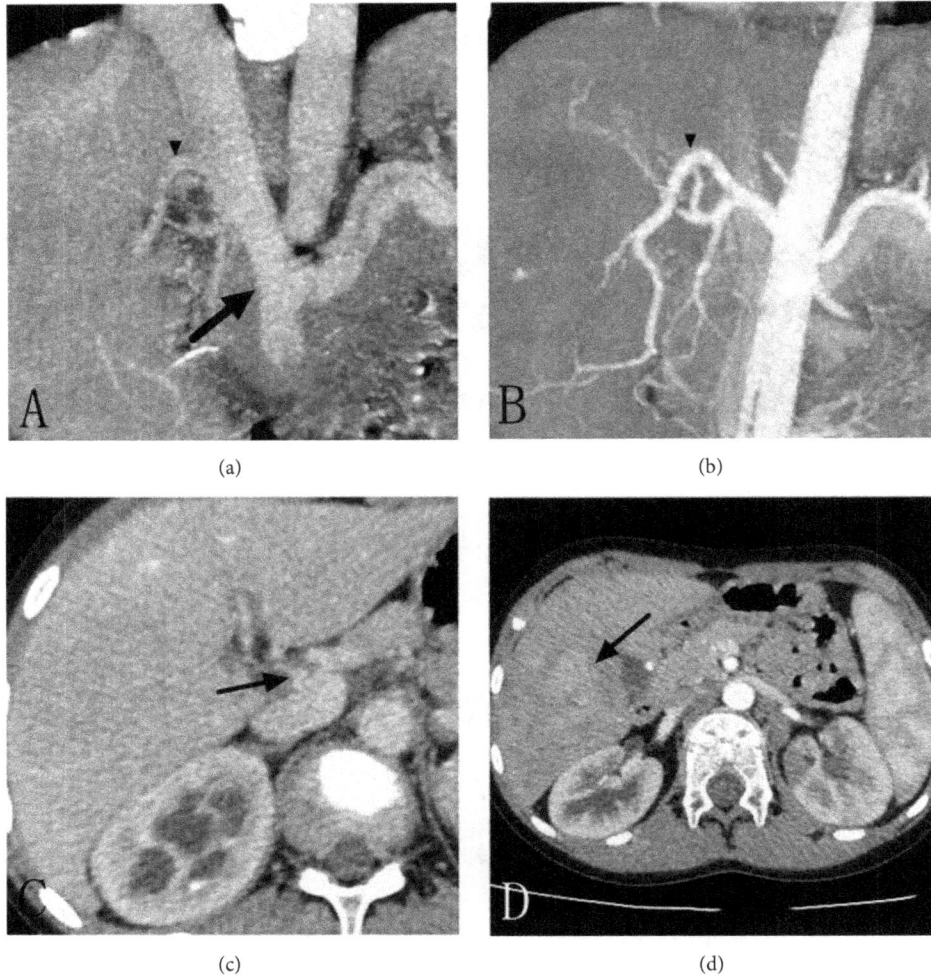

FIGURE 1: (a) The splenic vein and superior mesenteric vein drained directly into the inferior vena cava after confluence (arrow). (b) Note that the branch overlapping with the portal vein in (a) originated from the hepatic artery, not from the portal vein (arrow head). (c) A barely perceptible small branch of the portal vein measuring only 3 mm supplied the left lobe of the liver (arrow). (d) A faintly visible mass in segment 5 approximately 55 mm in size and poorly demarcated with mildly heterogeneous enhancement on arterial phase (arrow), which conferred a diagnosis of tumour initially.

(Figure 1(c)). In addition, a mass was noted in hepatic segment 5, approximately 55 mm in size and poorly demarcated. Arterial phase of enhancement showed mildly heterogeneous enhancement (Figure 1(d)), while portal venous and delayed phases demonstrated iso- to hypoattenuating. Abernethy malformation (type II) and concurrent hepatocellular carcinoma or hepatoblastoma were suspected.

The liver donor from a close relative was examined comprehensively and the transplantation for this suffered child was conducted 1 month after diagnosis. Surgical findings demonstrated a mass occupying the right liver measuring 60 mm in diameter. No remarkable liver cirrhosis was noted. The superior mesenteric vein and splenic vein merged together and drained directly to the inferior vena cava. A faintly visible branch vessel measuring 3 mm dominated the left lobe of the liver; no veins supplied the left medial lobe and the right lobe of the liver. A left lobe of liver was donated from a close

relative. Considering that the donor liver weighed only 250g and the ratio of donor liver to recipient weight was 0.83, an auxiliary piggyback liver transplantation method was conducted. The allogeneic iliac vein was adopted to reconstruct the left branch of the recipient portal vein and the left hepatic vein. Then, the hepatic vein was anastomosed to the recipient vena cava. The right hepatic artery of the recipient and the left lobe of the donor liver were anastomosed with the allogeneic iliac artery. The left hepatic duct, the ductus cysticus, and the biliary tree of the donor liver were also reconstructed by end-to-end anastomosis. Meanwhile, the diseased right lobe of the liver was resected. Two years after the transplantation, physical examination as well as CT and MR imaging showed the patient was in a satisfactory condition (Figures 2(a) and 2(b)); the transplanted liver was practically normal in contour and signal intensity. The child is still alive at present, 4 years after surgery.

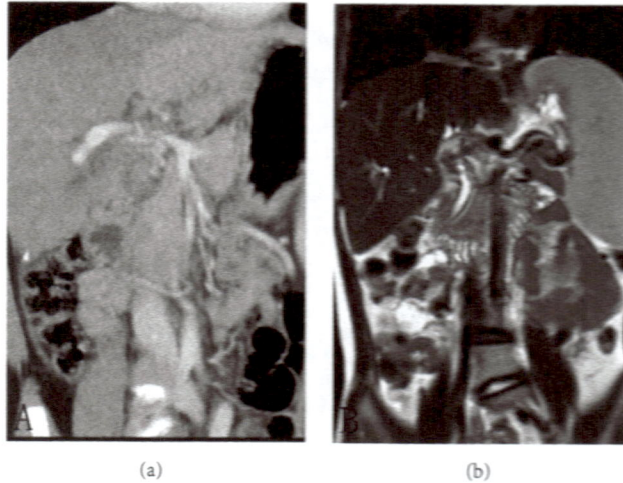

FIGURE 2: Piggyback liver transplantation was conducted. (a) CT imaging illustrated the reconstructed portal vein was in a satisfactory condition. (b) MR showed the transplanted liver was almost normal in signal 2 years following transplantation.

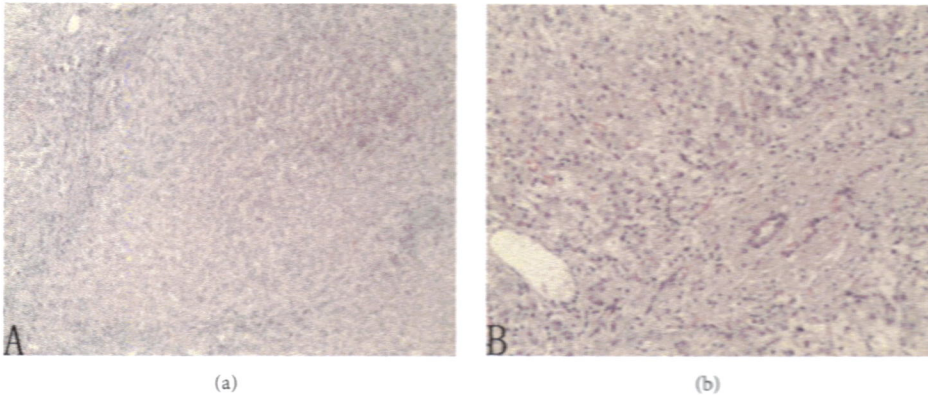

FIGURE 3: (a, b) Pathologically, the lesion was nodular in appearance and lacked a normal hepatic lobule texture, with absence of normal hepatic veins and bile ducts, while multiple fibrous connective tissues and chronic inflammatory cells scattered over.

Microscopically, the lesion was in nodular appearance without feeding hepatic veins. Abnormal hepatic lobules and distorted bile ducts scattered throughout the mass, with increased amount of fibrous connective tissues and chronic inflammatory cells. Some hepatocytes were characterized by hydropic degeneration, cholestasis, and regeneration (Figures 3(a) and 3(b)). Portal spaces exhibited an absence of veins and were replaced by chronic inflammatory cell infiltration, with 3 lymph nodes featuring reactive hyperplasia. The final pathological diagnosis was congenital extrahepatic portal-systemic shunt and coexisting nodular hyperplasia.

3. Discussion

Abernethy malformation was first described in 1793 in an autopsy of a 10-month-old female who died of unknown causes and of whom the portal vein bypassed the liver and directly drained into the inferior vena cava [2]. There have been up to 100 cases currently reported [3]. The abnormalities are believed to result from the persistence of the embryonic

vessels [4]. Type I is characterized by the complete drainage of portal blood into the vena cava congenitally and is further divided into type Ia (drainage of the superior mesenteric and splenic veins into systemic veins separately) and type Ib (drainage of the superior mesenteric and splenic veins into systemic veins after they conjoin to form a short extrahepatic portal vein). Type II is rarer and defined as a side-to-side extrahepatic shunt [5]. Abernethy malformation may be complicated by increased serum ammonia, hepatic encephalopathy, or hepatocellular carcinoma in the absence of cirrhosis [6].

Abernethy malformation type I has a female predilection [1] and predisposes patients to develop hepatic tumours and other abnormalities [7]. Histologically, benign tumours, such as focal nodular hyperplasia, nodular regenerative hyperplasia, and liver adenoma, as well as malignant tumours, such as hepatoblastoma and hepatocellular carcinoma, have all been reported [5]. The majority of reported Abernethy type II in publications were males. Here, we reported a female case of Abernethy malformation type II, which is rare in epidemiology and gender, with coincidental huge nodular hyperplasia confirmed pathologically. To our knowledge, such a case has not been reported to date. On imaging, the mass occupied the right lobe of the liver, showing mildly heterogeneous enhancement on arterial phase on contrast-enhanced CT scan, as well as iso to low density on portal venous phase and delayed phase. Considering the huge size of the mass, the age characteristic, and the enhancement features, hepatocellular carcinoma or hepatoblastoma was suspected. However, the final pathology was confirmed as nodular hyperplasia.

In terms of treatment, the ideal management strategy has not yet been established. There have been reports of other treatments, such as shunt ligation [8] and shunt closure using interventional angiography [9]; however, interventional closure may lead to recurrent hyperammonemia as reported [9]. In this case, taking into consideration that the concurrent mass developed in the right liver, piggyback liver transplantation is the optimal treatment.

4. Conclusion

This case might provide further knowledge regarding Abernethy malformation. On imaging, the anatomy of portal vein should be carefully investigated to categorize Abernethy malformation, and a wide variety of differential diagnosis of concurrent occupying lesions should be taken into account.

References

[1] M. Benedict, M. Rodriguez-Davalos, S. Emre, Z. Walther, and R. Morotti, "Congenital extrahepatic portosystemic shunt (abernethy malformation type Ib) With associated hepatocellular carcinoma: Case report and literature review," *Pediatric and Developmental Pathology*, vol. 20, no. 4, pp. 354–362, 2017.

[2] C. Correa, J. P. Luengas, S. C. Howard, and G. Veintemilla, "Hepatoblastoma and abernethy malformation type I: Case report," *Journal of Pediatric Hematology/Oncology*, vol. 39, no. 2, pp. e79–e81, 2017.

[3] S. Happaerts, A. Foucault, J. S. Billiard, B. Nguyen, and F. Vandenbroucke-Menu, "Combined hepatocellular-cholangio-carcinoma in a patient with Abernethy malformation and tetralogy of Fallot: A case report," *Hepatology*, vol. 64, no. 5, pp. 1800–1802, 2016.

[4] E. R. Howard and M. Davenport, "Congenital extrahepatic portocaval shunts—the Abernethy malformation," *Journal of Pediatric Surgery*, vol. 32, no. 3, pp. 494–497, 1997.

[5] J. Abernethy, "Account of two instances of uncommon formation in the viscera of the human body," *Philosophical Transactions of the Royal Society*, vol. 83, pp. 59-66, 1793.

[6] G. Morgan and R. Superina, "Congenital absence of the portal vein: two cases and a proposed classification system for portasystemic vascular anomalies," *Journal of Pediatric Surgery*, vol. 29, no. 9, pp. 1239–1241, 1994.

[7] R. Sharma, A. Suddle, A. Quaglia et al., "Congenital extrahepatic portosystemic shunt complicated by the development of hepatocellular carcinoma," *Hepatobiliary & Pancreatic Diseases International*, vol. 14, no. 5, pp. 552–557, 2015.

[8] C. Jiang, W. Ye, C. Liu, W. Wu, and Y. Li, "Surgical ligation of portosystemic shunt to resolve severe hematuria and hemafecia caused by type II abernethy malformation," *Annals of Vascular Surgery*, vol. 29, no. 5, pp. 1020–e11, 2015.

[9] H. Li, Z. Ma, Y. Xie, and F. Tian, "Recurrent Hyperammonemia After Abernethy Malformation Type 2 Closure: a Case Report," *Annals of Hepatology*, vol. 16, no. 3, pp. 0–0, 2017.

33

Primary Neuroendocrine Breast Carcinoma in a 13-Year-Old Girl: Ultrasonography and Pathology Findings

**Mazamaesso Tchaou,[1] Tchin Darré,[2] Koué Folligan,[3]
Akomola Sabi,[4] Lantam Sonhaye,[1] Azanledji Boumé,[5] Akila Bassowa,[6]
Solange Adani-Ifé,[7] and Gado Napo-Koura[2]**

[1]Department of Radiology, The University Teaching Hospital of Lomé, Lomé, Togo
[2]Department of Pathology, The University Teaching Hospital of Lomé, Lomé, Togo
[3]Department of Histology-Embryology, The University Teaching Hospital of Lomé, Lomé, Togo
[4]Department of Endocrinology and Nephrology, The University Teaching Hospital of Lomé, Lomé, Togo
[5]Department of Pediatric Surgery, The University Teaching Hospital of Lomé, Lomé, Togo
[6]Department of Obstetrics and Gynecology, The University Teaching Hospital of Lomé, Lomé, Togo
[7]Department of Clinical Oncology, The University Teaching Hospital of Lomé, Lomé, Togo

Correspondence should be addressed to Mazamaesso Tchaou; joseph_tchaou@yahoo.fr

Academic Editor: Samer Ezziddin

Neuroendocrine carcinoma (NEC) of the breast is a rare disease and has been scarcely reported by African authors. The authors report a case of breast NEC in a 13-year-old African girl initially diagnosed as an atypical adenofibroma by ultrasonography. Ultrasound-guided biopsy and conventional histological examination indicated two potential diagnoses: primary malignant non-Hodgkin's lymphoma and undifferentiated carcinoma. According to immunohistochemistry performed on paraffin blocks in France, infiltrating ductal carcinoma with a strong neuroendocrine component was confirmed by CD56, CD57, and chromogranin A markers.

1. Introduction

Neuroendocrine carcinomas (NECs) constitute a very rare entity, affecting mainly the bronchopulmonary system and gastrointestinal tract [1, 2]. Breast localizations are unusual and represent less than 0.1% of mammary cancers and less than 1% of neuroendocrine tumors [1]. Most publications in radiology describe nonspecific suspicious findings and do not indicate consistent imaging characteristics of this particular carcinoma by all the available modalities, including ultrasonography, mammography, and MRI [3]. In pathology, diagnosis is based on morphological criteria and confirmed by the expression of neuroendocrine markers (chromogranin and synaptophysin) in more than 50% of tumor cells [4]. There are two forms of breast NEC: the pure form exclusively composed of neuroendocrine cells and the mixed or composite form that is less well-differentiated [2, 4]. The composite form often poses diagnostic difficulties, especially for laboratories lacking immunohistochemical techniques, which explains the extreme rarity of the cases reported by African authors [5]. We report a primary composite neuroendocrine carcinoma in a 13-year-old Togolese girl confirmed by immunohistochemistry. We detail the epidemiological, morphological, and immunohistochemical aspects of this rare tumor.

2. Case Report

A 13-year-old girl with no remarkable past medical history and no family history of breast cancer presented with a palpable mass in her right breast, which had been evolving

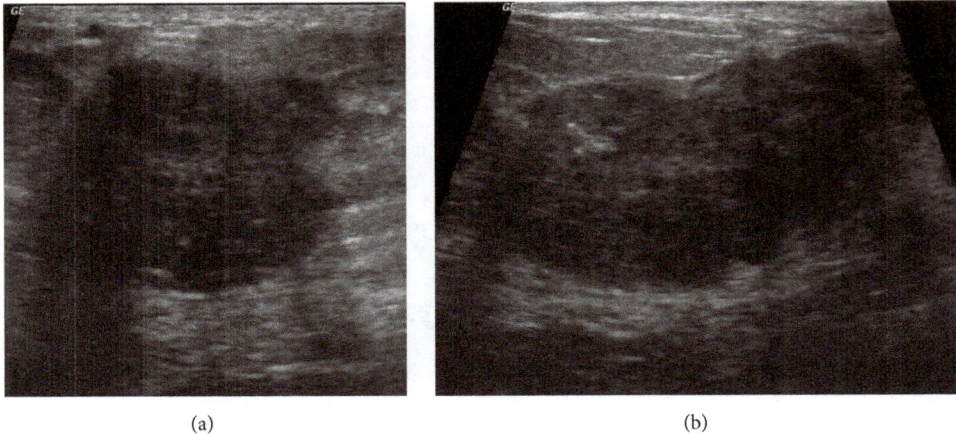

(a)   (b)

FIGURE 1: Transverse (a) and longitudinal (b) ultrasonography images showing a solid hypoechoic and heterogeneous mass with microlobulated contours, measuring 3.6 cm × 2.9 cm × 2.3 cm, with long transversal axis of the external superior quadrant of the right breast (collection of the Department of Imaging, the University Teaching Hospital of Lomé).

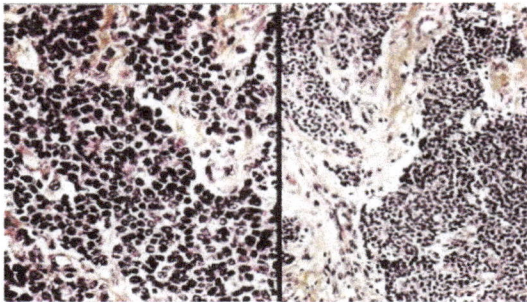

FIGURE 2: Neuroendocrine carcinoma of the breast (HES; ×100): tumoral diffuse proliferation due to medium and large cells invading the gland and oppressing lobules and ducts (collection of the Pathological Anatomy Laboratory, the University Teaching Hospital of Lomé).

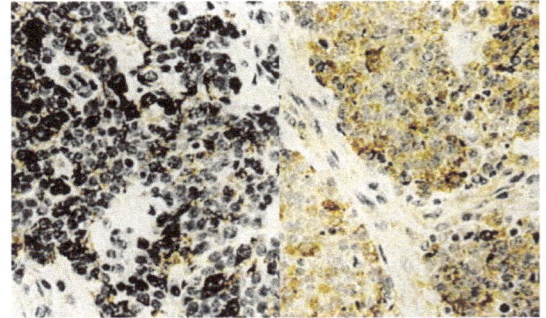

FIGURE 3: Neuroendocrine carcinoma of the breast (IHC; ×100): chromogranin A test positive for neuroendocrine tumor cells and negative for leukocyte markers (collection of the Pathological Anatomy Laboratory, the University Teaching Hospital of Lomé).

for 7 months. On physical examination, an approximately 4 cm firm and mobile nodule was identified. There were no axillary nodes, skin abnormality, nipple retraction, or abnormal nipple discharge. The nodule was located in the superior-external quadrant of the right breast. Due to her young age, the patient underwent only ultrasonography; no mammography was performed. Ultrasonography showed an oval hypoechoic, heterogeneous mass with microlobulated contours, measuring 3.6 cm × 2.9 cm × 2.3 cm, with long transverse axis (Figure 1). The mass was considered as atypical adenofibroma and categorized as ultrasound BI-RADS 3. On the demand of the parents, an ultrasound-guided biopsy with a 16G automatic core needle was performed before surgical ablation. On conventional histological examination, a diffuse tumor proliferation, made up of small round cells with hyperchromatic nucleus and a scant limited amphophilic cytoplasm, massively infiltrating the gland and penetrating the lobules and channels (Figure 2) was observed. This aspect had evoked two diagnoses: primary malignant non-Hodgkin's lymphoma and undifferentiated carcinoma. The

paraffin blocks were sent to France for immunohistochemical analysis. Immunohistochemical studies demonstrated infiltrating ductal adenocarcinoma expressing cytokeratin and membrane epithelial antigen associated with a neuroendocrine population expressing CD56, CD57, and chromogranin A. The leukocyte markers were negative (Figure 3). TTF1 and CDX2 markers were not expressed by the tumor cells. Therefore, the case was a composite form of breast NEC. Abdominal ultrasound and thoracoabdominal CT scan performed excluded any secondary site or any other non-mammary primitive site. The treatment consisted of tumor surgery followed by chemotherapy. The patient died 5 months after diagnosis, owing to local recurrence and metastasis.

3. Discussion

NECs are very rare tumors, with an incidence of approximately 0.7 cases per 100,000 inhabitants, commonly located within the digestive tract. Mammary localization is very rare, accounting for less than 0.1% of all breast cancers and less

Primary Neuroendocrine Breast Carcinoma in a 13-Year-Old Girl: Ultrasonography...

171

than 1% of neuroendocrine tumors [1, 5]. Our observation illustrates the diagnostic difficulties encountered in practice by underequipped pathological laboratories in Black Africa (e.g., absence of immunodetection technology and electron microscopy). Indeed, in our case, if the paraffin blocks were not sent to France, the diagnosis of NEC would be impossible to affirm. The lack of adequate technical equipment and advanced technologies in pathology laboratories noticed in the majority of African countries can explain a large part of the extreme rarity of the cases of NEC of the breast reported by African authors [6]. We reported a case of neuroendocrine carcinoma of the right breast in a young Togolese girl with fatal evolution. NEC occurrence at this age is rarely described in the literature, it is common in elderly patients, specifically between the sixth and seventh decades [2, 4, 6].

In ultrasonography, it is common to misdiagnose breast masses, considering them as benign or probably benign lesions, or adenofibroma [7]. In these situations, fine needle aspiration biopsy or ultrasound-guided core needle biopsy is necessary [3, 7]. In their recent literature review, Collado-Mesa et al. [3] noticed that imaging features of primary neuroendocrine tumor of the breast have been previously described by only a small number of case reports [8–13]. The published cases describe nonspecific suspicious findings and do not indicate consistent imaging characteristics of this particular carcinoma. On ultrasonography, which is the only breast imaging technique performed in this case, lots of aspects have been described. The typical appearance of this cancer has been reported as a hypoechoic or heterogeneous mass, with irregular shape or microlobulated margins and with normal sound transmission [10, 14]. In some situations, especially in older women, mammography can be crucial for the final diagnosis by revealing a distinctive mass with microcalcifications [15]. Imaging such as ultrasound, CT scan, and even PET scan if available can help in excluding another primary site of neuroendocrine tumor [15].

Histologically, the breast NEC is characterized by cell proliferation appearing as pseudorosettes invading the surrounding adipose tissue, with a richly vascularized small or fibrous stroma and small monomorphic cell elements with rather irregular nuclei and weak mitotic activity [1, 16]. Depending on the cell type, grade, degree of differentiation, and presence of mucin production, several subtypes are defined in the WHO classification. Solid NEC, small cell carcinoma, and large cell NEC [6] have been noted. The pure form expresses a high degree of histological differentiation, whereas the composite form, apart from the neuroendocrine expression, presents either a sarcomatous component or an epithelial component [6, 17]. Our case included a composite form of the tumor, and the histological description suggested that an undifferentiated epithelial tumor initially evoked non-Hodgkin's lymphoma. In the absence of immunodetection techniques, it is difficult to confirm the diagnosis of breast NEC, especially in its mixed form, using conventional histology [17]. Even though immunohistochemical techniques have made significant progress in the diagnostic accuracy of the majority of tumors [18, 19], they are still inaccessible for the majority of African countries. The immunohistochemical study in our case presented several diagnostic advantages, including the positivity of the CD56, CD57, and chromogranin markers, for the diagnosis of NEC. Moreover, the immunopositivity of cytokeratin and epithelial membrane antigen indicated the epithelial component of our case, suggesting a composite or mixed form. Lastly, negative result for lymphocytic leukocyte markers made it possible to rule out non-Hodgkin's lymphoma, which was initially evoked. In practice, the distinction between the two forms of breast NEC is necessary because their prognosis is significantly different [20, 21]. Primary neuroendocrine cancer of the breast must be distinguished from a metastatic lesion from other sites. Some markers such as TTF1 even if it is positive in only 55% of primary lung neuroendocrine tumors help to exclude lung primary NEC [15]. Positive nuclear CDX2 expression confirmed intestinal derivation [21]. Thoracic and abdominal imaging screening is helpful.

The treatment of endocrine tumors of the breast mainly comprises surgical tumor removal. The indications of chemotherapy and radiotherapy are the same as for other breast cancers. The indications of hormone therapy and immunotherapy are not coded because their effects remain uncertain [22]. In our case, the patient benefited only from surgery and chemotherapy.

4. Conclusion

Breast NEC is very rare and has poor prognosis. Further, its occurrence in a young patient is unusual. As there are no imaging specific features, the diagnosis of certainty is based on immunohistochemical analysis, which makes it possible to differentiate the pure forms from the composite forms. This case of a composite breast NEC also illustrates the diagnostic difficulties encountered by underequipped pathology laboratories in developing countries, explaining, in part, the extreme rarity of cases reported by African authors.

Additional Points

Availability of Data and Materials. All data generated or analysed during this study are included in this published article.

Authors' Contributions

Mazamaesso Tchaou and Tchin Darré are responsible for the conception of the study, participated in the study design, performed imaging and laboratory exams and interpretation, and wrote the paper. Koué Folligan, Akomola Sabi, Lantam Sonhaye, Azanledji Boumé, Akila Bassowa, and Solange Adani-Ifé were involved in the clinical and therapeutic management of the patient; they have reviewed the paper. Gado Napo-Koura was responsible for the overall scientific management of the study and the preparation of the final paper. All the authors have read and approved the final paper to be submitted for publication.

References

[1] M. Anlauf, M. Neumann, S. Bomberg et al., "Neuroendocrine neoplasms of the breast," *Pathologe*, vol. 36, no. 3, pp. 261–270, 2015.

[2] S. Singh, G. Aggarwal, S. P. Kataria, R. Kalra, A. Duhan, and R. Sen, "Primary neuroendocrine carcinoma of breast," *Journal of Cytology*, vol. 28, no. 2, pp. 91-92, 2011.

[3] F. Collado-Mesa, J. M. Net, G. A. Klevos, and M. M. Yepes, "Primary neuroendocrine carcinoma of the breast: report of 2 cases and literature review," *Radiology Case Reports*, vol. 12, no. 1, pp. 1–12, 2017.

[4] A. Inno, G. Bogina, M. Turazza et al., "Neuroendocrine carcinoma of the breast: Current evidence and future perspectives," *Oncologist*, vol. 21, no. 1, pp. 28–32, 2015.

[5] M. Affane, L. Elmorjani, A. El Omrani, F. Abbadi, H. Rais, and M. Khouchani, "Neuroendocrine carcinoma of the breast: about a case and review of the literature," *Pan African Medical Journal*, vol. 24, article 78, 2016.

[6] W. Bocker and WHO, "classification of breast tumors and tumors of the female genital organs: pathology and genetics," *Verhandlungen Der Deutschen Gesellschaft Fur Pathologie*, vol. 86, pp. 116–119, 2002.

[7] J. I. Malowany, U. Kundu, L. Santiago, and S. Krishnamurthy, "Fine-needle aspiration detects primary neuroendocrine carcinoma of the breast in a patient with breast implants," *CytoJournal*, vol. 12, no. 1, 2015.

[8] H. Ogawa, A. Nishio, H. Satake et al., "Neuroendocrine tumor in the breast," *Radiation Medicine*, vol. 26, no. 1, pp. 28–32, 2008.

[9] H. Ajisaka, K. Maeda, A. Miwa, and K. Yamamoto, "Breast cancer with endocrine differentiation: report of two cases showing different histologic patterns," *Surgery Today*, vol. 33, no. 12, pp. 909–912, 2003.

[10] I. Gunhan-Bilgen, O. Zekioglu, E. Ustun, A. Memis, and Y. Erhan, "Neuroendocrine differentiated breast carcinoma: imaging features correlated with clinical and histopathological findings," *European Radiology*, vol. 13, pp. 788–793, 2003.

[11] A. Irshad, S. J. Ackerman, T. L. Pope, C. K. Moses, T. Rumboldt, and B. Panzegrau, "Rare breast lesions: correlation of imaging and histologic features with WHO classification," *Radiographics*, vol. 28, no. 5, pp. 1399–1414, 2008.

[12] A. Mariscal, E. Balliu, R. Díaz, J. D. Casas, and A. M. Gallart, "Primary oat cell carcinoma of the breast: Imaging features," *American Journal of Roentgenology*, vol. 183, no. 4, pp. 1169–1171, 2004.

[13] J.-Y. Zhang and W.-J. Chen, "Bilateral primary breast neuroendocrine carcinoma in a young woman: report of a case," *Surgery Today*, vol. 41, no. 11, pp. 1575–1578, 2011.

[14] E. D. Chang, M. K. Kim, J. S. Kim, and I. Y. Whang, "Primary neuroendocrine tumor of the breast: Imaging features," *Korean Journal of Radiology*, vol. 14, no. 3, pp. 395–399, 2013.

[15] L. Marinova, D. Malinova, and S. Vicheva, "Primary neuroendocrine carcinoma of the breast: histopathological criteria, prognostic factors, and review of the literature," *Case Reports in Pathology*, vol. 2016, Article ID 6762085, 4 pages, 2016.

[16] Y. Fujimoto, R. Yagyu, K. Murase et al., "A case of solid neuroendocrine carcinoma of the breast in a 40-year-old woman," *Breast Cancer*, vol. 14, no. 2, pp. 250–253, 2007.

[17] O. David and M. Bhattacharjee, "Diffuse neuroendocrine differentiation in a morphologically composite mammary infiltrating ductal carcinoma: a case report and review of the literature," *Archives of Pathology & Laboratory Medicine*, vol. 127, no. 3, pp. 131–134, 2003.

[18] A. Galzerano, N. Rocco, A. Accurso et al., "Medullary breast carcinoma in an 18-year-old female: Report on one case diagnosed on fine-needle cytology sample," *Diagnostic Cytopathology*, vol. 42, no. 5, pp. 445–448, 2014.

[19] S. Frachon, D. Pasquier, I. Treilleux et al., "Breast carcinoma with predominant neuroendocrine differentiation," *Annales de Pathologie*, vol. 24, no. 3, pp. 278–283, 2004.

[20] K. Alva, L. Tauro, P. Shetty, and E. Saldanha, "Primary neuroendocrine carcinoma of the breast: a rare and distinct entity," *Indian Journal of Cancer*, vol. 52, no. 4, pp. 636-637, 2015.

[21] S. C. Wentz, C. Vnencak-Jones, and W. V. Chopp, "Neuroendocrine and squamous colonic composite carcinoma: Case report with molecular analysis," *World Journal of Gastroenterology*, vol. 17, no. 42, pp. 4729–4733, 2011.

[22] S. Kinoshita, A. Hirano, K. Komine et al., "Primary small-cell neuroendocrine carcinoma of the breast: report of a case," *Surgery Today*, vol. 38, no. 8, pp. 734–738, 2008.

34

Testicular Vasculitis: A Sonographic and Pathologic Diagnosis

Anuj Dixit,[1] Cameron Hague,[2] and Simon Bicknell[2]

[1]*Discipline of Radiology, Memorial University of Newfoundland, St. John's, NL, Canada*
[2]*Department of Radiology, University of British Columbia, Vancouver, BC, Canada*

Correspondence should be addressed to Anuj Dixit; anuj.dixit@gmail.com

Academic Editor: Vincent Low

Very little has been published about single-organ vasculitis of the testicle in the radiological literature. Consequently, it is a diagnosis that is unfamiliar to most radiologists. This case report describes the sonographic, pathologic, and laboratory findings of testicular vasculitis and reviews the available literature with regard to this subject.

1. Introduction

Systemic vasculitides can often involve the testes; however, isolated vasculitis of the testes is uncommon [1]. When a clinical history suggesting an underlying vasculitis is not present in the setting of testicular pain, the diagnosis is a difficult one. Although the imaging features are often nonspecific, in the right clinical setting, it is a diagnosis that the radiologist may be able to suggest.

2. Case Presentation

An 84-year-old male presented to the emergency department with testicular pain worsening over a 24-hour period. The patient was otherwise healthy with no significant medical concerns and no other symptoms. A testicular ultrasound was arranged on an urgent basis with the differential diagnosis consisting of epididymoorchitis versus torsion.

Sonographic evaluation revealed a heterogeneous appearance of both testicles with diminished parenchymal Doppler flow (Figure 1). The preliminary diagnosis was testicular infarction secondary to torsion or a neoplastic process such as lymphoma. Surgical excision of the left testicle was arranged.

The specimen was submitted for pathologic evaluation, which revealed a medium vessel vasculitis with associated hemorrhagic infarction of much of the testicular parenchyma (Figures 2 and 3). Vasculitic inflammatory change was also visualized in the regions of the epididymis and spermatic cord. No granulomas were seen and no evidence of lymphomatous or leukemic infiltrates were identified.

Given the pathological findings, additional blood work to assess an underlying vasculitis was obtained. The antinuclear antibody (ANA) screen was negative. The anti-neutrophil cytoplasmic antibody (ANCA) indirect immunofluorescence (IIF) was positive with a perinuclear pattern (p-ANCA). Anti-proteinase 3 (PR3-ANCA) and anti-myeloperoxidase (MPO-ANCA) antibody testing by ELISA (INOVA Diagnostics Inc.) were both negative. Protein electrophoresis revealed decreased albumin and beta 1 (LDL and transferrin) and beta 2 (C3) globulins. Midstream urinalysis was unremarkable. CRP was elevated at 72 mg/L (reference range < 10 mg/L). The patient was hepatitis B and hepatitis C negative. Liver function tests were normal. HIV status was not determined, but the patient had no known risk factors. Given the medium vessel involvement demonstrated on pathological assessment, as well as the blood work, a diagnosis of nongranulomatous testicular vasculitis was made. Clinical workup for the presence of systemic vasculitis was negative and inflammatory markers returned to normal values following orchiectomy and medical management.

Unfortunately, the patient was lost to follow-up following discharge which we acknowledge is a limitation of this case report.

3. Discussion

Various forms of vasculitis can involve the testes. While PAN is the most common form to affect the testicle, granulomatosis with polyangiitis (GPA), Henoch-Schonlein purpura, giant cell arteritis, and vasculitis associated with some

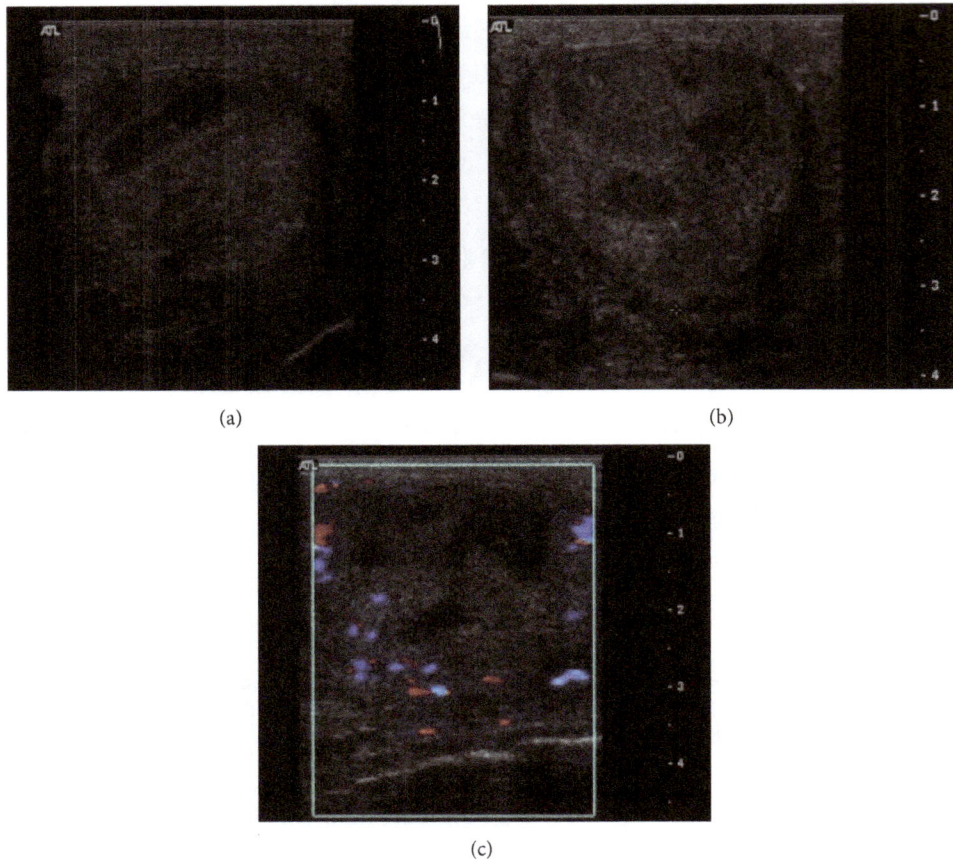

(a)

(b)

(c)

FIGURE 1: Multiple axial and longitudinal sonographic images (a–c) of the left testicle with Doppler color demonstrating a heterogeneous appearance, with multiple hypoechoic mass-like areas, and lack of Doppler flow within the majority of the testis. The right testis (image not provided) had a similar appearance.

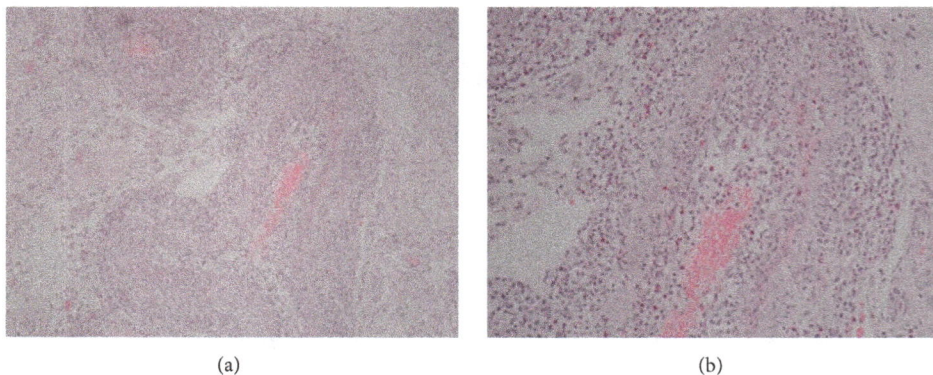

(a)

(b)

FIGURE 2: Histological section (H&E stain) of the left testicle viewed at medium (a) and high (b) power demonstrating a proliferation of neutrophils as well as T and B lymphocytes surrounding one of the intratesticular vessels consistent with vasculitis. No granulomas are present.

autoimmune connective tissue disorders such as Systemic Lupus Erythematosus (SLE) can also involve the testes [1, 2]. A recent study found the majority of cases of testicular vasculitis (TV) involve the testicular parenchyma while a lesser proportion of cases involved the epididymis (44.6%) and spermatic cord (30.6%) [3].

PAN was first described by Kussmaul and Maier in 1866 and commonly affects multiple organs in a patient such as the skin, kidneys, gastrointestinal tract, and peripheral and central nervous systems [4]. PAN is a medium-sized vessel vasculitis predominantly affecting males in their 4th to 6th decade. It is associated with a positive hepatitis B surface

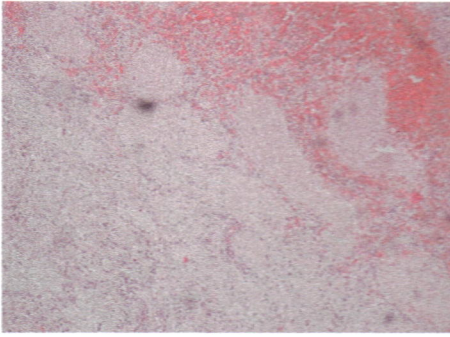

FIGURE 3: Histological section (H&E stain) of the left testicle viewed at medium power demonstrating hemorrhagic infarction of seminiferous tubules secondary to underlying vasculitis.

antigen serology in 10–50% of cases. PAN is also associated with positive HIV serology. Testicular involvement by PAN was first reported in the early 1900s [5]. Since then, isolated PAN has been observed in the gallbladder, uterus, skin, lungs, breast, and kidneys [6].

In 2012, the Chapel Hill consensus for nomenclature of vasculitides added "single-organ vasculitis" as a new category to differentiate PAN which is reserved for the primary systemic form of this medium-sized vessel vasculitis [7]. As such, isolated organ involvement which pathologically shows identical to PAN was to be categorized as single-organ vasculitis (in our case, testicular vasculitis).

Laboratory results, in addition to pathological findings, are crucial in establishing a specific vasculitis as the causative factor for the testicular findings seen on ultrasound. ANA positivity is suggestive of a diagnosis of SLE or other connective tissue diseases. ANCA positivity is helpful in identifying certain small vessel vasculitides. A cytoplasmic pattern (c-ANCA) by IIF and PR-3 positivity by ELISA are suggestive of GPA. A p-ANCA by IIF and MPO positivity by ELISA are suggestive of microscopic polyangiitis (MPA). Our patient had a positive p-ANCA by IIF and a negative PR3 and MPO by ELISA, supporting a diagnosis of PAN-type rather than GPA or MPA. It should be noted that PAN is not classically associated with ANCA [8]. As such, PAN can be diagnosed in patients with c-ANCA, p-ANCA, or nonspecific nuclear ANCA results on immunofluorescence pattern testing. Again, as there was no systemic evidence of vasculitis, our case was classified as testicular vasculitis (TV) or medium-sized vessels vasculitis of the testicle.

There is no consensus regarding the treatment of TV, although it is postulated that the excision of the affected organ is curative [3, 9, 10]. This is important to contrast to a systemic vasculitis with testicular involvement (most commonly PAN) where the treatment involves pharmacologic therapy. The complete absence of systemic symptoms and normal laboratory results suggest no need for further invasive diagnostic procedures, such as renal, skin, or muscle biopsies [4].

As mentioned previously, the sonographic diagnosis of SOV affecting the testicle is a difficult one. Little has been published about the sonographic appearances of testicular

vasculitis in the radiological literature [11], likely contributing to the fact that it is a diagnosis that the radiologist may be unlikely to consider if no history of an underlying vasculitis is provided. Furthermore, the majority of published cases report isolated testicular vasculitis occurring in young patients [4, 12]. Our case is the first describing such findings in a male greater than 80 years of age.

Testicular vasculitis is a great mimic [1, 11, 13]. It can appear sonographically normal or heterogeneous with variable Doppler flow or may present with multiple mass-like intratesticular lesions [1, 13]. Given its variable appearance, it may be prudent to keep vasculitis on the list of differential possibilities when such nonspecific testicular findings are seen. Subacute testicular torsion and neoplastic etiologies including primary testicular malignancy, metastatic disease, and lymphoma should be considered in addition to testicular vasculitis, particularly if sonographic evaluation demonstrates mass-like intratesticular findings and altered Doppler flow.

It should be noted that a limitation of this study was the lack of follow-up of the patient. It is true that, in any medium-sized vessel vasculitis, P-ANCA positivity at IIF (even in the absence of MPO specificity by ELISA) should alert the clinician for the possible systemic extent (even subclinical), which can also evolve to a generalized disease over time.

4. Conclusion

The presented case provides an important teaching point with regard to the differential diagnosis of nonspecific testicular sonographic findings in patients with testicular pain. Due to a relative lack of literature addressing testicular vasculitis, it may be a diagnosis that is overlooked. In the right clinical setting, even given the lack of specific ultrasound findings, testicular vasculitis is an entity that should be considered.

References

[1] K. G. Nigro, F. W. Abdul-Karim, and G. T. MacLennan, "Testicular vasculitis," The Journal of Urology, vol. 176, no. 6, p. 2682, 2006.

[2] T. D. Barber, O. Al-Omar, J. Poulik, and G. A. McLorie, "Testicular infarction in a 12-year-old Boy with Wegener's granulomatosis," Urology, vol. 67, no. 4, pp. 846.e9–846.e10, 2006.

[3] J. Hernández-Rodríguez, C. D. Tan, C. L. Koening, A. Khasnis, E. R. Rodríguez, and G. S. Hoffman, "Testicular vasculitis: findings differentiating isolated disease from systemic disease in 72 patients," Medicine, vol. 91, no. 2, pp. 75–85, 2012.

[4] A. Kussmaul and R. Maier, "Ueber eine bisher nicht beschriebene eigenthümliche Arterienerkrankung (Periarteritis nodosa), die mit Morbus Brightii und rapid fortschreitender allgemeiner Muskellähmung einhergeht," Deutsches Archiv für klinische Medicin, vol. 1, pp. 484–518, 1866.

[5] J. Monckeberg, "Uber periarteritis nodosa," *Beitr Path Anat Allg Pathol*, vol. 380, pp. 101–134, 1905.

[6] H. Pastor-Navarro, L. Broseta-Viana, M. J. Donate-Moreno et al., "Isolated testicular polyarteritis nodosa," *Urology*, vol. 70, no. 1, pp. 178.e7–178.e8, 2007.

[7] J. C. Jennette, R. J. Falk, P. A. Bacon et al., "2012 revised international chapel hill consensus conference nomenclature of vasculitides," *Arthritis and Rheumatism*, vol. 65, no. 1, pp. 1–11, 2013.

[8] C. G. M. Kallenberg, E. Brouwer, J. J. Weening, and J. W. Cohen Tervaert, "Anti-neutrophil cytoplasmic antibodies: current diagnostic and pathophysiological potential," *Kidney International*, vol. 46, no. 1, pp. 1–15, 1994.

[9] J. Hernández-Rodríguez, E. S. Molloy, and G. S. Hoffman, "Single-organ vasculitis," *Current Opinion in Rheumatology*, vol. 20, no. 1, pp. 40–46, 2008.

[10] J. Hernandez-Rodrıguez and G. S. Hoffman, "Updating single-organ vasculitis," *Current Opinion in Rheumatology*, vol. 24, no. 1, pp. 38–45, 2012.

[11] V. S. Dogra, S. Bhatt, and D. J. Rubens, "Sonographic evaluation of testicular torsion," *Ultrasound Clinics*, vol. 1, no. 1, pp. 55–66, 2006.

[12] N. Lintern, N. R. Johnson, I. Mckenzie, and B. Martin, "Testicular vasculitis - literature review and case report in Queensland," *Current Urology*, vol. 7, no. 2, pp. 107–109, 2013.

[13] F. N. Joudi, J. C. Austin, S. A. Vogelgesang, and C. S. Jensen, "Isolated testicular vasculitis presenting as a tumor-like lesion," *Journal of Urology*, vol. 171, no. 2 I, p. 799, 2004.

35

Rhinogenic Optic Neuritis Caused by Sphenoid Mucocele with Sinusitis

Sasitorn Siritho,[1,2] **Weerachai Tantinikorn,**[1]
Paithoon Wichiwaniwate,[1] **and Krit Pongpirul** ⓘ [1,3,4]

[1]*Bumrungrad International Hospital, Bangkok 10110, Thailand*
[2]*Division of Neurology, Department of Medicine, Siriraj Hospital, Mahidol University, Bangkok 10700, Thailand*
[3]*Department of Preventive and Social Medicine, Faculty of Medicine, Chulalongkorn University, Bangkok 10330, Thailand*
[4]*Department of International Health, Johns Hopkins Bloomberg School of Public Health, Baltimore, MD 21205, USA*

Correspondence should be addressed to Krit Pongpirul; doctorkrit@gmail.com

Academic Editor: Atsushi Komemushi

A 59-year-old male who presented with a nonspecific headache at the vertex, resembling retrobulbar optic neuritis, was treated as such but did not show any improvement. Although optic nerve compression from sphenoid mucocele was finally discovered, the delayed diagnosis and improper treatment led to a permanent visual loss. Optic neuritis could be caused by a common problem, "mucocele/sinusitis," but might be easily overlooked in general practice. Rhinogenic optic neuropathy should, therefore, be considered in every case of optic neuritis whenever atypical presentation occurs or is unresponsive to high-dose steroid treatment.

1. Introduction

A 59-year-old healthy male presented with a nonspecific headache at the vertex followed by sudden painless visual loss of the right eye. Ophthalmologic examination showed visual acuity of no light projection with a relative afferent pupillary defect. No diplopia was found. The intraocular pressure, optic disc appearance, and other neurological assessments were normal. Routine blood tests and echocardiogram were unremarkable. Magnetic resonance imaging (MRI) of the brain and orbit performed 2 days after the headache incident revealed minimal fluid surrounding the optic nerve sheath suggestive of mild swelling of the optic nerve sheath papilledema. A follow-up computerized tomography (CT) of the brain 6 days after the onset was unremarkable. With the initial diagnosis of right optic neuritis, he was given intravenous dexamethasone 96 mg/day for 2 days and then switched to oral prednisolone for 3 days.

Because of no improvement within one month, he sought for the second opinion, which revealed persistently no light perception of the right eye with the temporal pallor of the optic disc but normal macula and vascular trunks. Complete blood count, lipid panel, hepatitis profile, anti-HIV, thyroid function, erythrocyte sedimentation rate, C-reactive protein level, autoimmune screening tests, and AQP4-antibody were all unremarkable. Intravenous methylprednisolone 1 g/day for 3 days was given but no improvement was observed.

The previous MRI brain and orbit images were reviewed (Figure 1). Axial STIR T2W showed questionable mucocele of the right sphenoid sinus or sphenoidal sinusitis but the architecture of the optic nerve was not well defined. Coronal FLAIR T2W showed right sphenoidal pathology but the right optic nerve was not well delineated. Thinner slice and contrasted study should have been performed for the definite diagnosis of optic neuritis.

Two weeks later, another MRI of the brain and orbit demonstrated focal hyposignal T1/hypersignal T2 retention cyst, adhering to the right lateral sphenoidal wall, around the right optic nerve canal with mild surrounding enhancement, along the mucosa and around the optic nerve canal, and focal mild swelling with ill-defined hypersignal T2 change and faint enhancement of the corresponding intracanalicular portion of the right optic nerve.

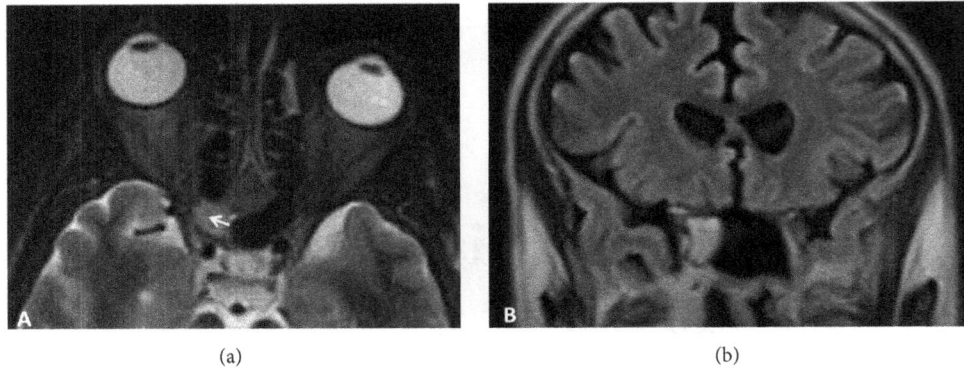

FIGURE 1: Magnetic resonance images of the brain and orbits. (a) Axial STIR T2W showed questionable mucocele of the right sphenoid sinus or sphenoidal sinusitis (white arrow) but the architecture of the optic nerve was not well defined. (b) Coronal FLAIR T2W showed right sphenoidal pathology but the right optic nerve was not well delineated.

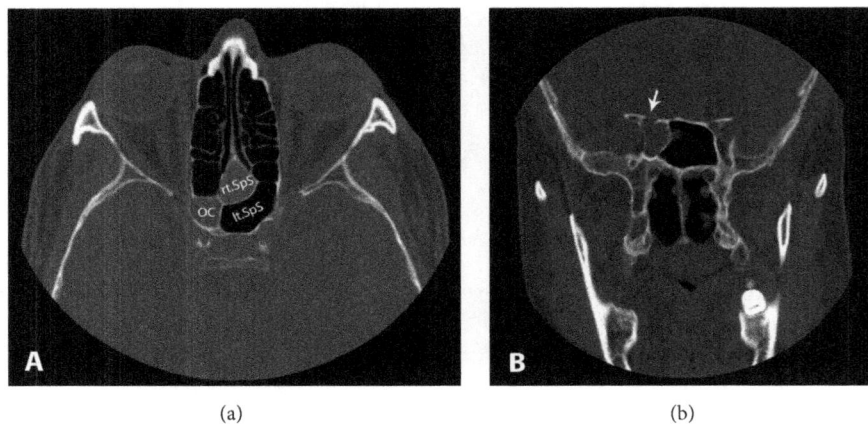

FIGURE 2: Computerized tomography of paranasal sinus. (a) Anatomical location of Onodi cell (OC); right (rt.SpS) and left sphenoid sinus (lt.SpS). (b) Open roof of superior wall of right sphenoid sinus (white arrow).

The multidetector computed tomography (MDCT) of the paranasal sinuses illustrated a nonenhanced soft tissue density lesion in the right sphenoid sinus and no visualization of the superior wall of the right sphenoid sinus, adjacent to the right optic canal, as well as a wall that became thinner by either pressure effect of right sphenoid lesion or bony destruction (Figure 2). There is a loss of perioptic CSF space at the intracanalicular segment of the right optic nerve. Provisional diagnosis with sinusitis-induced optic neuritis was made.

Right posterior ethmoidectomy, sphenoidotomy, and draining the content in the Onodi cell (Figure 3) were performed and found whitish bloody mucoid discharge in the Onodi cell with dehiscence of the superior wall exposing the right optic nerve. The culture of the pus was negative for organisms. Amoxicillin/clavulanic acid was given for 10 days but the visual acuity did not show further improvement. The patient showed no improvement after the surgery, probably because of the long period of compression as evidenced by the presence of optic atrophy.

2. Discussion

The Onodi cell or sphenoethmoidal cell, first described by Adolf Onodi in 1904 [1], is a posterior ethmoid air cell that lies superior to the sphenoid sinus and is in close proximity to the optic nerve or internal carotid artery (ICA) [2–6]. Identification of this structure is useful for minimizing perioperative complication in endoscopic endonasal/transsphenoidal surgery. Because of the anatomical correlation, sinus infection of Onodi cells can lead to inflammation/infection of the optic nerve, so-called "rhinogenic optic neuropathy" [2, 5, 7]. Several mechanisms have been proposed accordingly: (1) direct spreading of the sinus infection through the posterior paranasal sinuses or by osteomyelitis of the sinus wall to the optic nerve [8, 9], (2) compression of the optic nerve caused by ethmoid and/or sphenoid mucoceles [10–14], (3) bacteraemia passing through the mucosa of the sinus [5], (4) vasculitis of the optic nerve [15, 16], and (5) chronic allergic optic neuritis [17].

Our report showed an immunocompetent male patient with atypical unilateral painless retrobulbar neuritis with

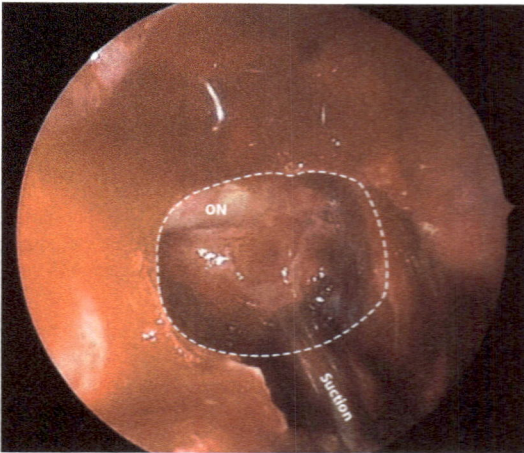

FIGURE 3: Top view of Onodi cell (white dashed line) showed dehiscence of the bone over the naked right optic nerve.

rapid onset but unresponsive to high-dose steroids. MRI of the orbit and CT of paranasal sinus showed soft tissue density lesion with a very thinning wall of the right sphenoid lesion suggestive of bony destruction of the same side of the suspected ON. The surgical drainage of the right Onodi cell revealed noninfectious discharge; the compression optic neuropathy associated with mucoceles with or without direct spreading from suppurative sinusitis should, therefore, be the pathophysiologic causes.

Two cases were reported from Spain [18]: (1) 21-year-old female presenting with typical unilateral ON and central scotoma without eye pain who was unresponsive to steroid treatment and later developed stuffy nose and nasal symptoms and (2) 75-year old male presenting with rapid onset severe unilateral visual loss, severe scotoma, and papillary pallor who underwent surgical opening of suppurative sphenoidal cyst. The former was clinically improved by antibiotic treatment and the latter received successful surgical correction. Another report was a man who presented with right eye pain on the nasal aspect followed by sudden onset of ON secondary to mucocele and had complete recovery after surgical drainage [19].

Even the rare occurrence of rhinogenic optic neuritis, sinusitis of the Onodi cell causing optic neuropathy, should be considered as a possible cause in all cases of atypical optic neuritis. The recommended imaging to analyze the sphenoid sinus and its surrounding structures is the CT paranasal sinuses including axial, coronal, sagittal, and sagittal oblique (parallel to the optic canal) views. Proper treatments with antibiotics and surgical intervention at the right time are crucial in order to prevent the irreversible optic nerve damage.

3. Conclusion

Rhinogenic optic neuritis should be one of the differential diagnoses in patient presenting with atypical neuritis.

4. Learning Points

(i) Optic neuritis could be caused by a common problem "mucocele/sinusitis" but might be easily overlooked in general practice.

(ii) Delayed diagnosis and improper treatment could lead to permanent visual loss.

(iii) The appropriate imaging should be done with fine cut or appropriate sequences if sinusitis of the Onodi cell causing optic neuropathy is suspected.

(iv) Rhinogenic optic neuropathy should be considered in every case of optic neuritis whenever atypical presentation occurs or is unresponsive to high-dose steroid treatment.

Authors' Contributions

Sasitorn Siritho initiated and drafted the paper, obtained inform consent from the patient, and contributed to final approval of the version published and obtaining funding. Weerachai Tantinikorn, Paithoon Wichiwaniwate, and Krit Pongpirul drafted and revised the paper.

References

[1] A. Ónodi, "Die Sehstörungen und Erblindung nasalen Ursprunges, bedingt durch Erkrank-ungen der hinteren Nebenhöhlen," Ophthalmologica, vol. 12, no. 1, pp. 23–46, 1904.

[2] A. Ónodi, "The optic nerve and the accessory sinuses of the nose: a contribution of canalicular neuritis and atrophy of the optic nerve of nasal origin," Annals of Otology, Rhinology & Laryngology, vol. 17, pp. 1–115, 1908.

[3] H. Loeb, "A study of the anatomic relations of the optic nerve to the accessory cavities of the nose," Annals of Otology, Rhinology & Laryngology, vol. 18, p. 243, 1909.

[4] B. Unal, G. Bademci, Y. K. Bilgili, F. Batay, and E. Avci, "Risky anatomic variations of sphenoid sinus for surgery," Surgical and Radiologic Anatomy, vol. 28, no. 2, pp. 195–201, 2006.

[5] R. Jack, R. H. Maisel, N. T. Berlinger, and J. D. Wirtschafter, "Relationship of optic neuritis to disease of the paranasal sinuses," The Laryngoscope, vol. 94, no. 11, pp. 1501–1508, 1984.

[6] J. Lang, Clinical Anatomy of the Nose, Nasal Cavity and Paranasal Sinuses, Thieme Verlag, New York, NY, USA, 1989.

[7] T. Klink, J. Pahnke, F. Hoppe, and W. Lieb, "Acute visual loss by an Onodi cell," British Journal of Ophthalmology, vol. 84, no. 7, pp. 801-802, 2000.

[8] R. W. TEED, "Meningitis from the sphenoid sinus," Archives of Otolaryngology—Head and Neck Surgery, vol. 28, no. 4, pp. 589–619, 1938.

[9] B. A. B. Dale and I. J. MacKenzie, "The complications of sphenoid sinusitis," The Journal of Laryngology & Otology, vol. 97, no. 7, pp. 661–670, 1983.

[10] N. Newton, G. Baratham, R. Sinniah, and A. Lim, "Bilateral compressive optic neuropathy secondary to bilateral sphenoethmoidal mucoceles," Ophthalmologica, vol. 198, no. 1, pp. 13–19, 1989.

[11] M. Ayaki, Y. Oguchi, H. Ohde, and H. Nameki, "Rhinogenous optic neuritis with drastic diurnal variation of visual function," *Ophthalmologica*, vol. 208, no. 2, pp. 110-111, 1994.

[12] K. Yoshida, T. Wataya, and S. Yamagata, "Mucocele in an Onodi cell responsible for acute optic neuropathy," *British Journal of Neurosurgery*, vol. 19, no. 1, pp. 55-56, 2005.

[13] K. Kitagawa, S. Hayasaka, K. Shimizu, and Y. Nagaki, "Optic neuropathy produced by a compressed mucocele in an onodi cell," *American Journal of Ophthalmology*, vol. 135, no. 2, pp. 253-254, 2003.

[14] T. C. C. Lim, W. P. Dillon, and M. W. McDermott, "Mucocele involving the anterior clinoid process: MR and CT findings," *American Journal of Neuroradiology*, vol. 20, pp. 287-290, 1999.

[15] I. Kjœr, "A case of orbital apex syndrome in collateral pansinusitis," *Acta Ophthalmologica*, vol. 23, no. 4, pp. 357-366, 1945.

[16] C. M. Moorman, P. Anslow, and J. S. Elston, "Is sphenoid sinus opacity significant in patients with optic neuritis?" *Eye*, vol. 13, no. 1, pp. 76-82, 1999.

[17] N. A. Rao, J. Guy, and P. S. Sheffield, "Effects of chronic demyelination on axonal transport in experimental allergic optic neuritis," *Investigative Ophthalmology & Visual Science*, vol. 21, pp. 606-611, 1981.

[18] A. C. Pelaz, C. F. Lisa, J. L. L. Pendás, and J. P. R. Tapia, "Reversible retrobulbar optic neuritis due to sphenoidal sinus disorders: two case studies," *Acta Otorrinolaringológica Española*, vol. 59, no. 6, pp. 308-310, 2008.

[19] S. A. Lim, Y. Y. Sitoh, T. C. C. Lim, and J. C. Y. Lee, "Clinics in diagnostic imaging (120)," *Singapore Medical Journal*, vol. 49, no. 1, pp. 84-88, 2008.

36

Brain Herniation into Giant Arachnoid Granulation

Joana Ruivo Rodrigues[1] and Gonçalo Roque Santos[2]

[1]Radiology Unit, Medical Imaging Department, Centro Hospitalar de Tondela-Viseu, EPE, Viseu, Portugal
[2]Neuroradiology Unit, Medical Imaging Department, Centro Hospitalar de Tondela-Viseu, EPE, Viseu, Portugal

Correspondence should be addressed to Joana Ruivo Rodrigues; joana.ruivo130@gmail.com

Academic Editor: Silvio Mazziotti

Arachnoid granulations are structures filled with cerebrospinal fluid (CSF) that extend into the venous sinuses through openings in the dura mater and allow the drainage of CSF from subarachnoid space into venous system. Usually they are asymptomatic but can be symptomatic when large enough to cause sinus occlusion. We report a rare case of a brain herniation into a giant arachnoid granulation in an asymptomatic elderly male patient, which was discovered incidentally.

1. Introduction

Arachnoid granulations represent growths of arachnoid membrane into the dural sinuses through which CSF enters the venous system [1] and are macroscopically visible [2]. They range from a few millimetres to more than 1 cm (giant arachnoid granulations) [3] and consequently may grow to fill and dilate the dural sinuses or expand the inner table of the skull. Rarely, they cause symptoms related with venous hypertension secondary to partial sinus occlusion and usually they are incidental findings [1].

The presence of preexisting arachnoid granulations facilitates the formation of brain herniation into the dural venous sinus (DVS) or adjacent calvarium. The herniations of brain parenchyma into arachnoid granulations are thought to arise spontaneously or as a result of increased intracranial pressure [4].

2. Case Presentation

A 91-year-old man was admitted to our emergency department with an episode of sudden onset of confusion and right arm paraesthesia (without loss of consciousness), lasting less than 3 minutes. Upon admission, neurological exam was normal. Blood pressure was elevated (190/90 mmHg) and

heart rate was stable (90 beats/minute). He had a history of type II Diabetes Mellitus, arterial hypertension, and a left hemicolectomy four years prior, in relation with colon adenocarcinoma (pT2N0M0).

An urgent noncontrast computed tomography (CT) of the brain excluded acute vascular or posttraumatic lesions but showed an area of bone rarefaction localized in the parasagittal left parietal bone, associated with an ipsilateral parietal encephaloclastic cortical and subcortical brain lesion (Figure 1). Magnetic resonance imaging (MRI) of the brain was performed 2 months later (Figures 2, 3, and 4) and demonstrated an unenhancing brain herniation to the incomplete bone defect, with a narrow neck, atrophy, and hyperintensity on T2 FLAIR images. The incomplete bone defect was close to the superior sagittal sinus, posterior and superior to it.

These findings were found to be compatible with a brain herniation into a giant arachnoid granulation, with strangulation and infarction of the herniated brain tissue.

3. Discussion

Arachnoid granulations (AG) were first described by Antonio Pacchioni in 1705 [5]. They represent growths of arachnoid

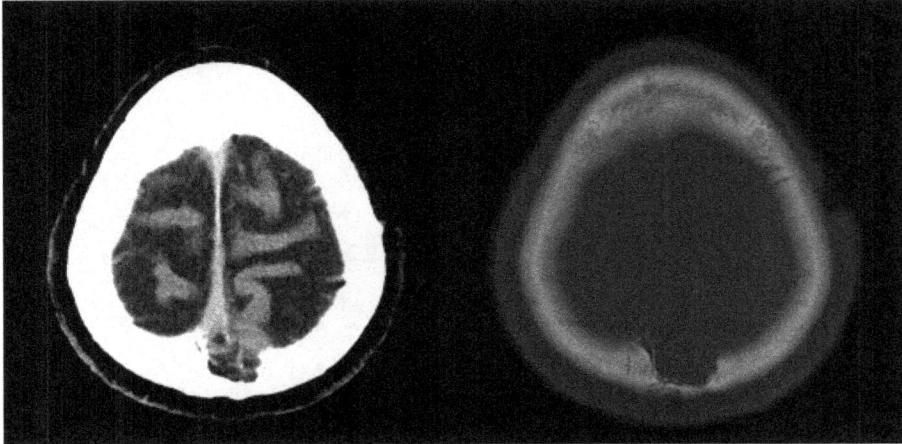

FIGURE 1: Head CT scan images show an incomplete bone defect in the left parietal bone, with heterogeneous density, including calcified areas.

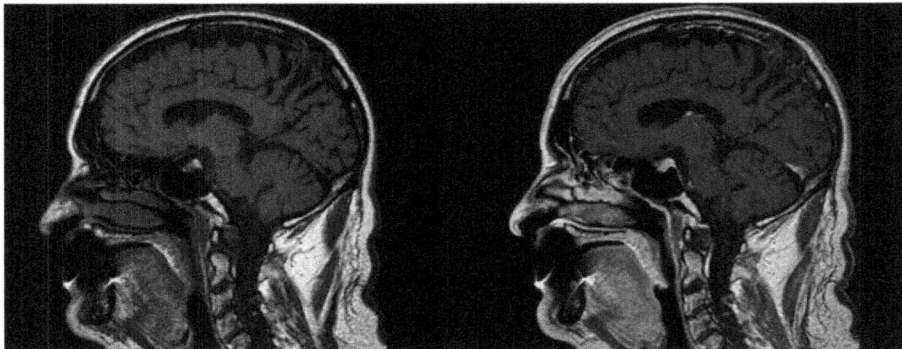

FIGURE 2: Sagittal T1-weighted images without and with contrast show brain herniation to the bone defect, without enhancement.

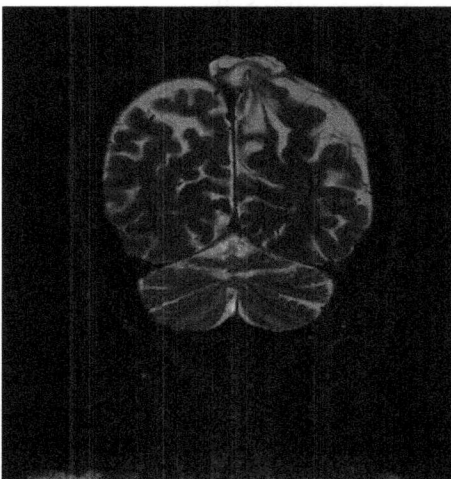

FIGURE 3: Coronal T2-weighted SPAIR image shows narrow neck and atrophy of the brain herniation.

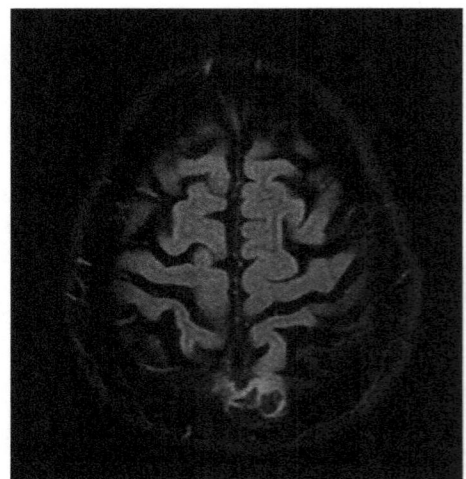

FIGURE 4: Axial T2-weighted FLAIR image shows hyperintensity of the brain herniation, probably due to prior infarction.

membrane into dural sinuses, through which CSF enters the venous system from subarachnoid space [6].

A large AG is called "giant" when larger than 1 cm [3]. However, Kan et al. referred to AG as "giant" when they fill the lumen of a dural sinus, causing local dilatation or filling defects [1].

AG can enlarge with age [7] or in response to an increase in cerebrospinal fluid pressure [8] and they can be found anywhere in the DVS [3]. Mostly giant AG are incidentally discovered on brain studies without any relation to the patients symptoms [1], as occurred in our case.

AG can appear on skull radiography as radiolucent zones or as impressions on the inner table of the calvaria. On CT AG can be hypodense or isodense relative to the brain parenchyma. On MR imaging, they are hyperintense on T2-weighted images and iso- or hypointense relative to brain parenchyma on T1-weighted images [3, 7]. Trimble et al. [3] reported that approximately 80% of giant arachnoid granulations contain CSF-incongruent fluid on at least one MR image and nearly half contain fluid that does not parallel CSF on at least two sequences. FLAIR is the most reliable technique, differing in signal intensity with CSF in 100% of cases [3]. The fluid in the granulations is almost not attenuated on a FLAIR sequence but remains hyperintense, most likely due to pulsation artifacts from the adjacent sinus and differing CSF flow characteristics within the AG [9].

On angiographic studies AG appear as ovoid filling defects in the dural venous sinuses in the venous phase [10]. The internal veins appear as focal linear contrast enhancement on enhanced CT or as linear flow voids on nonenhanced MR [10]. Nonvascular soft tissue can be present in giant AG, and it was interpreted as stromal collagenous tissue, hypertrophic arachnoid mesangial cell proliferation, or invaginating brain tissue [3]. However, it is possible that venous structures and/or connective tissues into the arachnoid granulations may be tamponed by mass effect from the herniated parenchyma, thereby making them invisible [4].

It is important to distinguish giant AG from pathologic processes in the dural venous sinuses, like thrombosis and neoplasia [10]. Thrombosis usually involves an entire segment of sinus or multiple sinuses and can extend into cortical veins, while AG produce focal, well-defined defects [6, 11]. The differential diagnosis with tumour can be made because of the shape, the lack of contrast enhancement, and the lack of diffusion restriction [10]. Brain parenchyma herniations into the calvarium are rare, recently described, and controversial in significance.

Brain parenchyma herniations with surrounding CSF into the DVS and/or calvarium have a prevalence of 0.32% and were encountered more frequently in posterior inferior parts of the intracranial cavity [12].

Malekzadehlashkariani et al. [13] retrospectively analyzed 38 patients with brain herniation into AG and found 68 brain herniations into AG, by order of frequency, in the occipital squama, transverse sinus, lateral lacuna of the superior sagittal sinus, and straight sinus, with cerebellar tissue being the most frequently found in the herniation. Brain parenchyma herniations affect more women than men [13].

Chan et al. [14] described a focal brain herniation into a giant arachnoid granulation located in a DVS which may have developed spontaneously or was induced by elevated intracranial pressure and cerebral edema resulting from prior head trauma.

Çoban et al. [15] reported a symptomatic brain herniation as an occult temporal lobe encephalocele into a transverse sinus. They coined the term "occult encephalocele" due to its distinguishing features from those of more common encephaloceles.

Battal and Castillo [4] observed five patients showing brain herniation into the DVS or calvarium containing various amounts of cerebral or cerebellar parenchyma surrounded by CSF. Battal and Castillo [4] concluded that herniations of brain parenchyma into arachnoid granulations can occur spontaneously or as a result of increased intracranial pressure. Battal et al. [12] in another study referred that the herniations were incidentally detected in all patients and the majority occurred spontaneously, but few were associated with masses that presumably could have increased intracranial pressure. It was suggested that the underlying giant arachnoid granulation might be a predisposing factor.

Brain herniations are best detected with high-resolution T1- and T2-weighted MRI sequences and higher field strengths [12].

Battal et al. [16] believe that brain herniations into the skull have different features from classic encephaloceles, since the brain herniations that were described by the previous referred authors did not occur through a complete calvarial defect but instead occurred through a dural defect into calvarium. Those cases did not show external bony defects that are usually present in encephaloceles.

Brain herniations into the calvarium are very rare, their clinical significance is not well-known, and they should be considered in the differential diagnosis of encephaloceles. The continuity of the external table of the calvarium can be the distinguishing feature of this entity from encephalocele. The presence of preexisting arachnoid granulations facilitates the formation of brain herniation into the DVS or adjacent calvarium [4, 16].

Our case demonstrated such unusual complication of brain herniation into a giant arachnoid granulation, in a rare and unique location: the convexity area (left parasagittal parietal region). Those findings were discovered incidentally without any relation to the patient symptoms. Strangulation (in probable association with a narrow neck) and infarction are present in the herniated parenchyma. Future potential complications, such as brain dysfunction and seizures, should not be ignored. Therefore accurate diagnosis is required and care must be taken not to mistake for a neoplasm. Prompt treatment should be directed at the underlying cause, when it is known.

In conclusion, we have presented a rare and unique case of brain herniation in the convexity area into a giant arachnoid granulation, without any history of trauma or of elevated intracranial pressure.

References

[1] P. Kan, E. A. Stevens, and W. T. Couldwell, "Incidental giant arachnoid granulation," *American Journal of Neuroradiology*, vol. 27, no. 7, pp. 1491–1492, 2006.

[2] D. G. Potts, K. F. Reilly, and V. Deonarine, "Morphology of the arachnoid villi and granulations," *Radiology*, vol. 105, no. 2, pp. 333–341, 1972.

[3] C. R. Trimble, H. R. Harnsberger, M. Castillo, M. Brant-Zawadzki, and A. G. Osborn, ""Giant" arachnoid granulations just like CSF?: NOT!!," *American Journal of Neuroradiology*, vol. 31, no. 9, pp. 1724–1728, 2010.

[4] B. Battal and M. Castillo, "Brain herniations into the dural venous sinuses or calvarium: MRI of a recently recognized entity," *Neuroradiology Journal*, vol. 27, no. 1, pp. 55–62, 2014.

[5] A. Brunori, R. Vagnozzi, and R. Giuffrè, "Antonio Pacchioni (1665–1726): early studies of the dura mater," *Journal of Neurosurgery*, vol. 78, no. 3, pp. 515–518, 1993.

[6] S. C. Chin, C. Y. Chen, C. C. Lee et al., "Giant arachnoid granulation mimicking dural sinus thrombosis in a boy with headache: MRI," *Neuroradiology*, vol. 40, no. 3, pp. 181–183, 1998.

[7] J. L. Leach, B. V. Jones, T. A. Tomsick, C. A. Stewart, and M. G. Balko, "Normal appearance of arachnoid granulations on contrast-enhanced CT and MR of the brain: differentiation from dural sinus disease," *American Journal of Neuroradiology*, vol. 17, no. 8, pp. 1523–1532, 1996.

[8] C. B. Grossman and D. G. Potts, "Arachnoid granulations: radiology and anatomy," *Radiology*, vol. 113, no. 1, pp. 95–100, 1974.

[9] J. L. Leach, K. Meyer, B. V. Jones, and T. A. Tomsick, "Large arachnoid granulations involving the dorsal superior sagittal sinus: findings on MR imaging and MR venography," *American Journal of Neuroradiology*, vol. 29, no. 7, pp. 1335–1339, 2008.

[10] B. De Keyzer, S. Bamps, F. Van Calenbergh, P. Demaerel, and G. Wilms, "Giant arachnoid granulations mimicking pathology: a report of three cases," *Neuroradiology Journal*, vol. 27, no. 3, pp. 316–321, 2014.

[11] H. J. Choi, C. W. Cho, Y. S. Kim, and J. H. Cha, "Giant arachnoid granulation misdiagnosed as transverse sinus thrombosis," *Journal of Korean Neurosurgical Society*, vol. 43, no. 1, pp. 48–50, 2008.

[12] B. Battal, S. Hamcan, V. Akgun et al., "Brain herniations into the dural venous sinus or calvarium: MRI findings, possible causes and clinical significance," *European Radiology*, vol. 26, no. 6, pp. 1723–1731, 2016.

[13] S. Malekzadehlashkariani, I. Wanke, D. A. Rüfenacht, and D. San Millán, "Brain herniations into arachnoid granulations: about 68 cases in 38 patients and review of the literature," *Neuroradiology*, vol. 58, no. 5, pp. 443–457, 2016.

[14] W. C. Chan, V. Lai, Y. C. Wong, and W. L. Poon, "Focal brain herniation into giant arachnoid granulation: a rare occurrence," *European Journal of Radiology Extra*, vol. 78, no. 2, pp. e111–e113, 2011.

[15] G. Çoban, E. Yildirim, B. Horasanli, B. E. Çifçi, and M. Ağıldere, "Unusual cause of dizziness: occult temporal lobe encephalocele into transverse sinus," *Clinical Neurology and Neurosurgery*, vol. 115, no. 9, pp. 1911–1913, 2013.

[16] B. Battal, S. Hamcan, V. Akgun, S. Sari, and B. Karaman, "Brain herniation with surrounding CSF into the skull or encepholecele?" *Journal of Neuroradiology*, vol. 42, no. 3, pp. 187–188, 2015.

37

Unique Presentation of Hematuria in a Patient with Arterioureteral Fistula

Tomas Mujo,[1] **Erin Priddy,**[1] **John J. Harris,**[1] **Eric Poulos,**[2] **and Mahmoud Samman**[1]

[1]*Department of Radiology, University of Louisville Hospital, Louisville, KY 40202, USA*
[2]*University of Louisville School of Medicine, Louisville, KY 40202, USA*

Correspondence should be addressed to Tomas Mujo; t0mujo01@louisville.edu

Academic Editor: Ruben Dammers

Active extravasation via an arterioureteral fistula (AUF) is a rare and life-threatening emergency that requires efficient algorithms to save a patient's life. Unfortunately, physicians may not be aware of its presence until the patient is in extremis. An AUF typically develops in a patient with multiple pelvic and aortoiliac vascular surgeries, prior radiation therapy for pelvic tumors, and chronic indwelling ureteral stents. We present a patient with a left internal iliac arterial-ureteral fistula and describe the evolution of management and treatment algorithms based on review of the literature.

1. Introduction

Arterioureteral fistula is a direct communication between an artery and ureter that presents with hematuria. A fistula may form between the aorta and common iliac, external iliac, and internal iliac arteries. AUF can be seen in the setting of prior oncologic pelvic surgery, pelvic exenteration, vascular surgeries of the pelvis, and pelvic irradiation therapy. Cases can also be seen following urologic diversion procedures and in patients with indwelling ureteral stents requiring frequent stent exchanges.

2. Case Report

We present the case of a 54-year-old woman with past medical history of stage IIB squamous cell carcinoma of the cervix, pelvic radiation therapy 16 years earlier, chronic renal insufficiency, chronic pyelonephritis, solitary left kidney, left ureteral stricture, frequent ureteral stent exchanges, left percutaneous nephrostomy tube, and colovesical and vesicovaginal fistulae. Her past surgical history includes total abdominal hysterectomy and bilateral salpingo-oophorectomy, sigmoid colostomy with Hartmann's pouch, ileostomy, and right nephrectomy.

She developed bright red blood from the left ureteral orifice after removal of a metallic ureteral stent during an exchange procedure at an outside hospital. Subsequently, she had a retrograde pyelogram, which showed no vascular blush or active extravasation of blood. The only positive finding at that time was a large blood clot within the left renal pelvis. Vascular surgery evaluated the patient and was unable to provide the necessary vascular interventions. A technetium pertechnetate tagged red blood cell scan revealed no gastrointestinal source for bleeding. Four days later, the patient began passing bright red blood and clots from her left nephrostomy tube. Her hemoglobin dropped two grams/dL, requiring transfusion of 2 units of packed red blood cells.

Later that day the patient was transferred to our hospital for further management. Active hematuria via the Foley catheter and left nephrostomy tube had resolved at the time of admission to our facility. Her vital signs were within normal limits. Review of the computed tomography (CT) of the abdomen and pelvis without intravenous contrast from the outside institution stated that there was extensive chronic soft tissue thickening at the site of previous surgical operations, likely an area of fistula formation. The patient was placed on continuous bladder irrigation.

Three days later her hemoglobin was stable, and cystoscopy and left retrograde pyelogram were performed. At cystoscopy, a large blood clot was irrigated out and removed. After removal of the clot, a large vesicovaginal fistula was identified in the posterior aspect of the bladder. No other

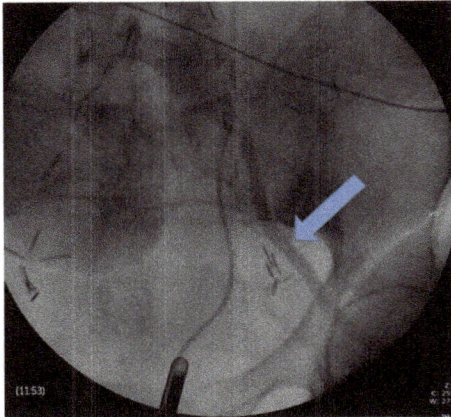

FIGURE 1: Left ureteral retrograde pyelogram demonstrates contrast opacification of the left external iliac artery (solid blue arrow) consistent with left ureteral-arterial fistula. Scattered surgical clips are present throughout the pelvis.

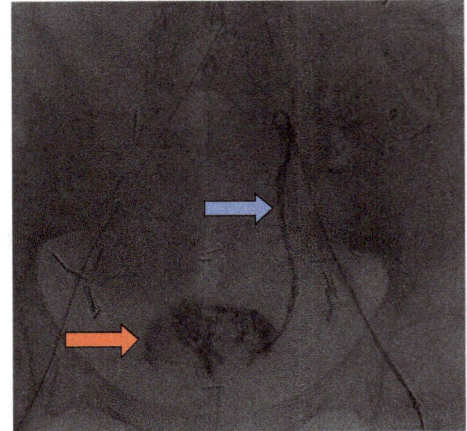

FIGURE 2: Attempted left internal iliac artery selective arteriogram with a 5 F SOS catheter shows the catheter tip in the left ureter and brisk flow of contrast into the left ureter (solid blue arrow) flowing caudad into the urinary bladder (solid red arrow). This confirms the presence of an internal iliac arterioureteral fistula.

lesions were visualized. Subsequently, the left ureteral orifice was intubated with a catheter. Contrast was injected into the ureter and a vascular blush and contrast opacification of the left external iliac artery were identified (Figure 1). At this point, the procedure was aborted and a 3-way Foley catheter was placed for continuous bladder irrigation.

Interventional Radiology was consulted emergently to evaluate and treat the patient. The patient was allergic to contrast media and was premedicated emergently with Benadryl® (West Ward Pharmaceuticals Corp.) and Solu-Cortef® (Pharmacia and Upjohn Company). Visipaque® (GE Healthcare Inc.) contrast was used for the procedure. Initially the left groin was accessed and a hand injection performed via a five-French measuring pigtail catheter in the left proximal common iliac artery. The left proximal external iliac artery demonstrated severe stenosis. At this point the right groin was accessed and a six-French vascular sheath was placed over a 0.035-inch wire, and an arteriogram of the pelvic arteries from the right common femoral artery approach was performed. This arteriogram was helpful to estimate the length and the severity of the stenosis of the left iliac arteries. It also aided to minimize traumatic injury to the left iliac arteries during their recanalization.

Then, after successful recanalization of the left iliac arteries, a pigtail catheter was inserted over the wire into the distal abdominal aorta and a pelvic arteriogram was performed. Subsequently, the left pigtail catheter was exchanged for a 5 Fr SOS selective catheter and attention was directed to the left internal iliac artery. An attempt was made to select the left internal iliac artery, and then contrast was injected which demonstrated the catheter to be in the left ureter instead of the left internal iliac and the arteriogram demonstrated brisk flow of contrast into the left ureter, confirming the presence of an internal iliac arterioureteral fistula (Figure 2). The arterial catheter ending up into the ureter is presumably dangerous and can cause a sudden enlargement of the fistula.

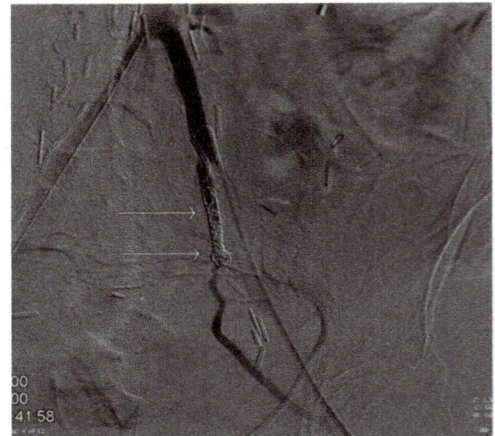

FIGURE 3: Four platinum coils were used to embolize the left proximal internal iliac artery (white arrows).

The SOS catheter was advanced to the common iliac artery and another attempt was made utilizing the road map technique to select the left internal iliac artery for the purpose of embolization planning. Essentially, embolization of the left internal iliac artery was planned to prevent formation of a type II endoleak after placement of a covered stent from the common to external iliac arteries. The left internal iliac artery was successfully selected. Subsequently, four platinum coils were used to embolize the left internal iliac artery (Figure 3) and the postembolization arteriogram demonstrated no flow into the left internal iliac artery or left ureter (Figure 4).

Attention was then directed to the severe stenosis of the left proximal external iliac artery. A 7 mm × 4 cm Dorado® (BARD Peripheral Vascular) angioplasty balloon was utilized to dilate the stenosis followed by placement of an 8 mm × 5 cm Gore Viabahn Endoprosthesis which extended from the mid left common iliac artery to the proximal external

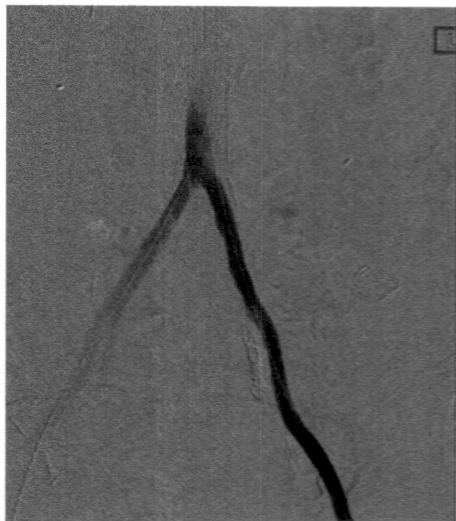

FIGURE 4: Left iliac arteriogram after left internal iliac embolization and placement of an 8 mm × 5 cm Gore® Viabahn® stent from the mid left common iliac artery to the proximal external iliac artery shows no opacification of the left internal iliac artery and elimination of the left internal iliac arterioureteral fistula.

iliac artery (Figure 4). The patient was discharged home in stable condition three days after the procedure. No follow-up documentation or imaging is available in our electronic medical system.

3. Discussion

The arterioureteral fistula is a rare and often life-threatening complication that requires a high index of suspicion for early detection and to discern it from other causes of hematuria. Virtually all patients present with gross hematuria, the timing of which is variable: some develop episodic hematuria requiring intermittent transfusion, others demonstrate no clinical signs prior to massive hematuria with hemodynamic instability, and others only present with hematuria at the time of stent exchange [1]. Increasing numbers of case reports have emerged over the recent years describing the expansion of endovascular interventions and the rare endoureteral intervention as treatment algorithms have evolved [2, 3].

There is a prominent identifiable subset of patients most commonly affected. Of the 118 patients studied in multi-institutional analysis, 73.7% had a history of chronic indwelling ureteral stents, 70.3% had a history of malignancy, and 69.5% had undergone prior pelvic surgery [4]. Women were more commonly affected than males (approximately 3 : 2 ratio), which is logical considering many of these patients have had prior pelvic surgery or radiation [4]. A smaller single-institution study [2] added prior radiation therapy to the predisposing factors. Of the patients identified with AUF, 75% had undergone prior radiation treatment.

The increase in incidence of AUF is almost exclusively related to the increased number of secondary fistulae. Primary fistulae are overwhelmingly related to inflammation from an adjacent aneurysm or pseudoaneurysm. In one review series, of the ten primary fistulae identified, nine were related to aortoiliac aneurysm, while the additional patient had an arteriovenous malformation [5]. Rare instances of primary fistulae during pregnancy have been identified postmortem secondary to ureteropelvic obstruction from advanced pyelonephritis. This etiology has disappeared, likely because of aggressive antibiotic use in the setting of urinary tract infection during pregnancy [5].

While the pathophysiology of fistula formation is still not fully understood, various contributing factors are thought to play a role. External beam radiation and pelvic surgery may disrupt the vasa vasorum, making larger vessels more prone to necrosis and subsequent fistula formation [1, 6]. A similar phenomenon is believed to occur in ureters, causing ureteral devascularization in patients subjected to the same interventions [7, 8]. A mechanical component also is thought to contribute: arterial pulsations cause chronic microtrauma to the overlying ureter, particularly those having been chronically stented. This mechanical microtrauma may then progress to pressure necrosis and arterial-ureteral communication [7]. Krambeck et al. state that ureteral stent size plays a role in fistula formation. For instance, the smaller diameter 7 F ureteral stent provides the same flow rate as a 12 F stent, with potential reduced risk of ureteral dilatation, wall compression, and ischemic changes.

Even with high clinical suspicion, diagnostic confirmation is challenging for a variety of reasons, including the same basis of episodic symptoms: intermittent thrombus occlusion of the fistulous tract [9]. According to the most recent review of the literature, selective arteriography is the most effective modality for diagnosis [4, 10]. An older study cites arteriography sensitivity of 50% [11]. Interestingly, in our case, we did not see the arterial-ureteral communication during our initial pelvic arteriogram. Rather, we saw the arterial-ureteral communication only when we tried to select the left internal iliac artery and unintentionally selected the left ureter. Thus, even when not visualized with contrast, an arterioureteral fistula may be present. The friable, fragile nature of the tissues separating the left ureter from the left iliac arteries likely explains the inconsistent visualization of the AUF and the intermittent nature of the patient's hematuria.

Several studies [2, 4, 5, 12–14] suggest that provocative angiography, by way of stent removal or manipulation of urinary catheter, improves sensitivity, although with obvious needed preparation for potential massive hemorrhage. Computed tomography is not typically revealing, although Krambeck et al. stated it was diagnostic in half of the patients who underwent the aforementioned imaging modality in their study cohort [1]. Bleeding from a ureteral orifice during cystoscopy is a nonspecific finding, but should heighten suspicion in the appropriate clinical scenario. Suggestion has been made for anterograde or retrograde ureterogram to demonstrate contrast extravasation from the ureter into the arterial circulation [4]; however, the pressure gradient makes this visualization less likely.

Open surgical repair, via local reconstruction, ligation with or without extra-anatomic reconstruction, or ligation of the internal iliac artery was commonly used prior to the advent of endovascular techniques [5]. Most often, femoral crossover bypass was utilized given the potential infection risk [5]. Common complications from open surgical management include enterocutaneous fistulae or wound infection [1]. In addition to the morbidity, which can often accompany open surgical repair, these complex patients are often not ideal operative candidates, due to a mummified pelvis from prior surgeries and radiation, thus allowing for advantageous utilization of endovascular techniques.

With the first use of iliac artery stent grafts in 2004, many patients have been treated definitively with endovascular stenting. Despite this utilization of iliac artery stents for over 10 years, standardized antibiotic prophylaxis has yet to be established. Other endovascular treatment options include isolated coil embolization of the iliac artery or most commonly exclusion of the fistula with a covered stent graft [12]. A few studies describe successful treatment with isolated coil embolization and self-expanding polytetrafluoroethylene covered stent graft via the ureter by Inoue et al. and Horikawa et al., respectively [3, 15]; however, the efficacy of this method has not been demonstrated in other instances.

A single center retrospective review of arterioureteral fistulas by Fox et al. did not demonstrate a clear morbidity or mortality advantage for open or endovascular management [6]. In the majority of patients, endovascular intervention is preferred given patient comorbidities, hostile surgical abdomen given multiple prior surgeries and radiation therapies, and complication rate. Prior to 1980 the mortality rate associated with AUF was estimated at 69% [1], and a mortality rate of 100% without treatment. The most recent review of the literature gives an acute mortality rate of 10–38% [4]. True long-term follow-up is still yet to be determined as the longest duration of median follow-up in studies to date is 26 months [2].

Hematuria is a nonspecific finding which should raise suspicion for arterioureteral fistula in patients with history of prior pelvic vascular or oncologic surgery, radiation therapy, and multiple ureteral stents. Diagnostic confirmation can be elusive as seen in our case, but lack of imaging confirmation should not deter diagnosis and treatment of AUF in cases of high clinical suspicion. Increased utilization of endovascular treatment in the past decade has proven useful given the complexity of the majority of patients who present with AUF.

In conclusion, continued attention to long-term morbidity and mortality is needed, particularly in reference to hematuria or fistula recurrence. Mortality due to AUF has markedly decreased over the past 30 years and may continue to decrease with increased clinical awareness and minimally invasive treatment options, which will be seen with ongoing long-term follow-up.

Abbreviations

AUF: Arterioureteral fistula
CT: Computed tomography
IRB: Institutional review board.

Disclosure

This case report does not meet the common rule definition of human subjects' research. Therefore, it did not require institutional review board (IRB) review prior to conducting the research.

References

[1] A. E. Krambeck, D. S. DiMarco, M. T. Gettman, and J. W. Segura, "Ureteroiliac artery fistula: diagnosis and treatment algorithm," *Urology*, vol. 66, no. 5, pp. 990–994, 2005.

[2] R. D. Malgor, G. S. Oderich, J. C. Andrews et al., "Evolution from open surgical to endovascular treatment of ureteral-iliac artery fistula," *Journal of Vascular Surgery*, vol. 55, no. 4, pp. 1072–1080, 2012.

[3] M. Horikawa, H. Saito, H. Hokotate, and T. Mori, "Treatment of ureteroarterial fistula with an endoureteral stent graft," *Journal of Vascular and Interventional Radiology*, vol. 23, no. 9, pp. 1241–1243, 2012.

[4] A. Das, P. Lewandoski, D. Laganosky, J. Walton, and P. Shenot, "Ureteroarterial fistula: a review of the literature," *Vascular*, vol. 24, no. 2, pp. 203–207, 2016.

[5] D. Bergqvist, H. Pärsson, and A. Sherif, "Arterio-ureteral fistula—a systematic review," *European Journal of Vascular and Endovascular Surgery*, vol. 22, no. 3, pp. 191–196, 2001.

[6] J. A. Fox, A. Krambeck, E. F. McPhail, and D. Lightner, "Ureteroarterial fistula treatment with open surgery versus endovascular management: long-term outcomes," *Journal of Urology*, vol. 185, no. 3, pp. 945–950, 2011.

[7] P. F. Escobar, J. L. Howard, J. Kelly et al., "Ureteroarterial fistulas after radical pelvic surgery: pathogenesis, diagnosis, and therapeutic modalities," *International Journal of Gynecological Cancer*, vol. 18, no. 4, pp. 862–867, 2008.

[8] R. C. van den Bergh, F. L. Moll, J. P. de Vries, K. K. Yeung, and T. M. Lock, "Arterio-ureteral fistula: 11 new cases of a wolf in Sheep's clothing," *The Journal of Urology*, vol. 179, no. 2, pp. 578–581, 2008.

[9] S. P. Quillin, M. D. Darcy, and D. Picus, "Angiographic evaluation and therapy of ureteroarterial fistulas," *American Journal of Roentgenology*, vol. 162, no. 4, pp. 873–878, 1994.

[10] R. C. van den Bergh, F. L. Moll, J. P. de Vries, and T. M. Lock, "Arterioureteral fistulas: unusual suspects-systematic review of 139 cases," *Urology*, vol. 74, no. 2, pp. 251–255, 2009.

[11] D. R. Vandersteen, R. R. Saxon, E. Fuchs, F. S. Keller, L. M. Taylor Jr., and J. M. Barry, "Diagnosis and management of ureteroiliac artery fistula: value of provocative arteriography followed by common iliac artery embolization and extraanatomic arterial bypass grafting," *Journal of Urology*, vol. 158, no. 3, part 1, pp. 754–758, 1997.

[12] A. Copelan, M. Chehab, C. Cash, H. Korman, and P. Dixit, "Endovascular management of ureteroarterial fistula: a rare potentially life threatening cause of hematuria," *Journal of Radiology Case Reports*, vol. 8, no. 7, pp. 37–45, 2014.

[13] D. C. Madoff, B. D. Toombs, M. D. Skolkin et al., "Endovascular management of ureteral-iliac artery fistulae with Wallgraft endoprostheses," *Gynecologic Oncology*, vol. 85, no. 1, pp. 212–217, 2002.

38

Magnetic Resonance Enterography Findings of Intestinal Behçet Disease in a Child

Tommaso D'Angelo,[1] **Romina Gallizzi,**[2] **Claudio Romano,**[2] **Giuseppe Cicero,**[1] **and Silvio Mazziotti**[1]

[1]Section of Radiological Sciences, Department of Biomedical Sciences and Morphological and Functional Imaging, University of Messina, Policlinico "G. Martino", Via Consolare Valeria 1, 98100 Messina, Italy
[2]Department of Pediatrics, University of Messina, Policlinico "G. Martino", Via Consolare Valeria 1, 98100 Messina, Italy

Correspondence should be addressed to Silvio Mazziotti; smazziotti@unime.it

Academic Editor: Stefania Rizzo

Behçet's disease (BD) is a multisystem disorder of unknown aetiology, characterized by recurrent oral ulcers, genital ulcers, uveitis, skin lesions, and pathergy. Gastrointestinal disease outside the oral cavity is well recognized and usually takes the form of small intestinal ulcers, with the most significant lesions frequently occurring in the ileocaecal region. Symptoms usually include nausea, vomiting, colicky abdominal pain, and change in bowel habit and it is not unusual that patients may present late, with life-threatening complications requiring surgery. Diagnosis has been hindered for many years by limitations in imaging the small bowel and it is usually achieved by means of endoscopy and CT of the abdomen. Magnetic resonance enterography (MRE) is a relatively new technique, which has a high diagnostic rate in patients with Crohn's disease (CD). Although many similarities between CD and intestinal BD have already been described in literature, the role of MRE in the evaluation of intestinal BD has never been defined up to now. We report a case of a 12-year-old female patient with diagnosis of BD who presented at our institution for recurrent colicky abdominal pain and diarrhoea. The patient underwent MRE that demonstrated the gastrointestinal involvement.

1. Introduction

Gastrointestinal Behçet's disease (BD) outside the oral cavity is well recognized and may lead to life-threatening complications, such as bowel perforation or fistulization, which may require bowel resection surgery. It usually takes the form of deep, circular ulcers that can potentially occur in all the gastrointestinal tract, with the most common involvement of the ileocaecal region.

Patients usually complain of colicky abdominal pain, nausea, and change in bowel habit. Differential diagnosis with Crohn's disease (CD) may be difficult, with several extraintestinal manifestations, such as oral ulcers and arthralgia, which can be recognized in both diseases.

Imaging of intestinal BD has been hindered for many years by the inherent limitations of the radiological techniques used to study the small bowel. In fact, only wireless capsule endoscopy and CT have been able to adequately image the gastrointestinal involvement of BD up to now [1, 2]. MR-Enterography (MRE) is a relatively new imaging technique, which has already demonstrated a good diagnostic value for small bowel pathology, particularly in CD. However, no data has ever been reported regarding the usefulness of MRE in patients with intestinal BD [1].

We describe a case in which a 12-year-old female patient with intestinal BD underwent MR-Enterography that allowed a better evaluation of the small bowel involvement.

2. Case Presentation

A 12-year-old female patient, with history of recurrent oral aphthosis, was originally admitted at our university hospital in September 2014 for progressively increasing pancranial headache, stiff neck, left oculomotor paresis, left facial nerve paresis, unstable gait, and left hemiparesis. The laboratory findings showed neutrophilic leukocytosis, increased C-reactive protein (CRP), and raised erythrocyte sedimentation rate (ESR). She presented papilledema and retinal vasculitis, which led to loss of visual acuity in both eyes. The lumbar puncture revealed high pressure of the cerebrospinal fluid (CSF) and oligoclonal bands of IgG, which were found also in the blood serum. MRI of the brain confirmed the papilledema, showing a bilateral budging of the optic disc into the vitreous chamber. It showed prominent subarachnoid space around the optic nerves and a partial empty sella turcica, compatible with diagnosis of pseudotumour cerebri.

The patient also presented gastrointestinal symptoms such as colicky abdominal pain, nausea, and diarrhoea and she reported an episode of gastrointestinal bleeding. A colonoscopy was performed, revealing round-shaped ulcers along the caecal region. In addition, esophagogastroduodenoscopy showed several aphthous ulcerations along the gastric body.

Diagnosis of BD with neurological and gastrointestinal manifestations was made and the patient was treated with intravenous bolus injection of methylprednisolone for the first three days, followed by oral administration of prednisone and cyclophosphamide. Three months later, the latter was replaced with mycophenolate mofetil.

After relief of neurological and gastrointestinal symptoms, steroids dose was progressively reduced with a concomitant relapsing of gastrointestinal symptoms, such as abdominal pain and fever.

In December 2015, the same patient presented to our institution to undergo MRE, which was suggested to evaluate the small bowel.

To perform the magnetic resonance of the small bowel, the patient was asked to fast from solids and liquids for 4–6 hours prior to the study and to assume a 1200 mL of polyethylene glycol (PEG) solution within 45 minutes before the beginning of the scan. MRE was performed using a 1.5 T MR imaging system (Gyroscan Achieva, Philips, Best, Netherlands) and phased-array abdominal coils. The standard protocol that we applied included various T2-weighted pulse sequences along axial and coronal planes. True fast imaging with steady-state (True-Fisp; TR/TE: 4.20/2.10 ms, FA: 60°) and half-Fourier acquisition single-shot turbo spin echo (HASTE; TR/TE: ∞/80 ms) with and without fat-suppression were also performed as well as diffusion-weighted sequences, using a diffusion factor b fixed at 0, 400, and 800 s/mm^2.

MRE showed marked inflammatory changes of the terminal ileum and caecum, characterized by diffuse bowel wall thickening with polypoid appearance. Diffusion-weighted images demonstrated restricted diffusion of water within the affected bowel wall. Mesenteric vascular engorgement was also seen (Figure 1).

These findings were also confirmed by the endoscopic evaluation of the ileocaecal region that showed oedema, erythema, and round-shaped ulcers (Figure 2). The patient started therapy with adalimumab, with rapid relief of symptoms.

In March 2016, gastrointestinal symptoms relapsed again and the patient was planned for elective terminal ileum

TABLE 1: International criteria for Behçet's disease, point-score system: scoring $\geq$ 4 indicates Behçet's diagnosis [4].

Sign/symptom	Points
Ocular lesions	2
Genital aphthosis	2
Oral aphthosis	2
Skin lesions	1
Neurological manifestations	1
Vascular manifestations	1
Positive pathergy test*	1*

*Pathergy test is optional and the primary scoring system does not include pathergy testing. However, where pathergy testing is conducted one extra point may be assigned for a positive result.

resection surgery. However, a few days prior to the expected day of surgery, the same patient came to emergency room with intense and acute abdominal pain. Emergency CT scan showed bowel perforation in adjacencies of the diseased terminal ileum and the girl rapidly underwent bowel resection surgery.

The surgical specimen revealed multiple punched out ulcerations along a diffusely thickened terminal ileum (Figure 3). Histological examination of the surgically resected intestinal specimens showed thickening of the blood vessel wall and infiltration of inflammatory cells in the vascular wall and perivascular area (i.e., neutrophils and mononuclear cells), findings consistent with submucosal phlebitis.

3. Discussion

BD is classified among inflammatory vascular diseases, affecting arterial and venous vessels of all kinds and sizes. The etiopathogenesis of the disease remains still unknown, but the combination of genetic factors affecting the immune regulation and undetermined environmental triggers may explain the variability of disease expression. The disease distribution is worldwide, even if there is a higher prevalence in populations living along the ancient "Silk-Route," which extends from eastern Asia to the Mediterranean basin [3]. Although the disease rates and the clinical expression vary to some extent by ethnic origin, recurrent mucocutaneous lesions, skin lesions, ocular findings, and reactivity of the skin to needle prick or injection (pathergy test) constitute the most common clinical hallmarks of BD. In addition, a wide spectrum of findings can be observed, with frequent involvement of neurologic, genitourinary, mucocutaneous, and gastrointestinal systems. Diagnosis is made when typical symptoms occur, and it is primarily based on sets of clinical criteria. Recently, the International Team for the Revision of the International Criteria for Behçet's disease (ICBD) has suggested a new criteria/point score, which markedly improved sensitivity and diagnostic accuracy over the older International Study Group (ISG) criteria still maintaining a comparable specificity [3, 4]. Our clinical case encompassed one major and one minor ISG criteria for diagnosis of BD and had a score of five according to ICBD point-score system, allowing diagnosis of BD (Table 1).

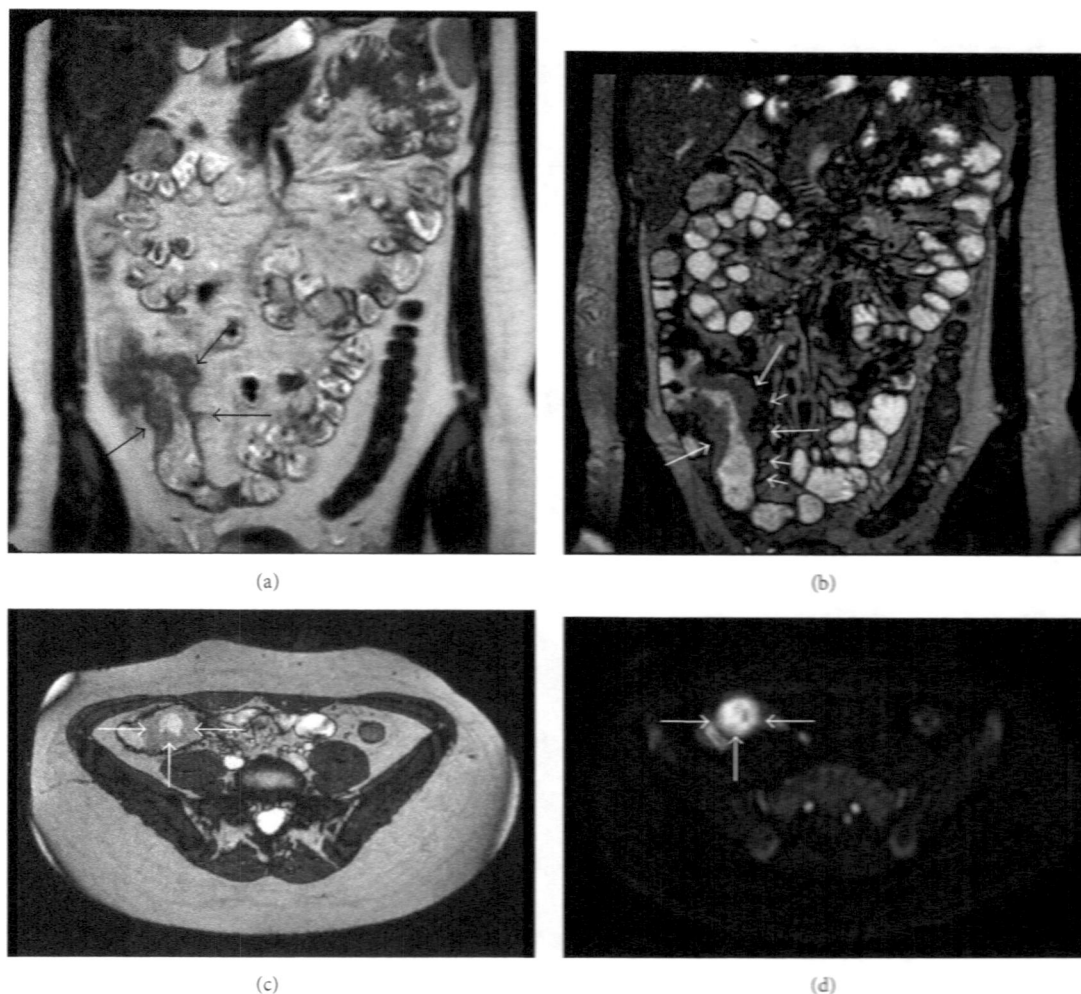

FIGURE 1: Diffuse bowel wall thickening with polypoid appearance *(long arrows)* and ectatic vasa recta *(short arrows)* are well demonstrable on coronal HASTE (a) and True-FISP (b) MR images as well as on axial True-FISP MR image (c). Axial diffusion-weighted MR image, performed at *b*-value of 800 s/mm^2, shows restricted diffusion of water within the affected bowel wall (d).

In addition, several studies have demonstrated that gastrointestinal symptoms (i.e., chronic diarrhoea and proctorrhagia) are significantly correlated with BD and when associated with typical intestinal lesions (e.g., round-shaped ulcers, deep ulcers with discrete margins, and inflammatory pseudopolyps) they configure a clinical condition also known as intestinal BD. Intestinal BD is more frequent in East Asia (i.e., Japan, Taiwan, and Korea) than in other areas of the world [3]. Factors associated with poor patient's prognosis are younger age, high ESR and CRP, and low albumin concentration at diagnosis [5].

Clinical course of intestinal BD is usually mild and patients uncommonly experience a disease flare-up. However, some patients may have a severe clinical course, undergoing frequent disease flare-ups and requiring corticosteroid therapy, immunosuppressant therapy, or surgical treatment [5].

The most typical form of gastrointestinal manifestation of BD is the ulceration of the ileocecal region, which may result in a high frequency of complications, such as fistulization, perforation, bleeding, and peritonitis. Up to now, diagnosis and monitoring of patients with intestinal BD have been done with endoscopy, while radiological examinations have had a limited role [2].

In fact, although conventional enteroclysis is able to detect mucosal abnormalities of the small bowel, this fluoroscopic technique is not widely used since it is relatively invasive and often unpleasant for the patient. Since the last decade, Computed Tomography Enterography (CTE) has been widely used in patients with gastrointestinal manifestations of BD, showing a significant diagnostic value and depicting most of intra- and extraluminal abnormalities [6].

At CTE, the typical findings of gastrointestinal BD are a thickening of bowel wall, consisting in mural edema with ulcer penetration, or a "tumor-like" polypoid mass, both usually showing a marked mural hyperenhancement due to the vascular pathologic substrate [7].

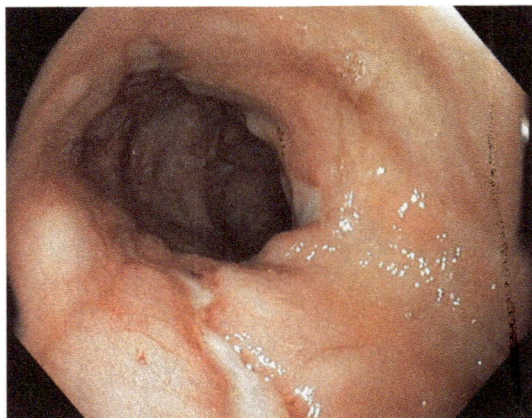

FIGURE 2: Endoscopic image of the caecal region shows deep, large, and oval ulcerations.

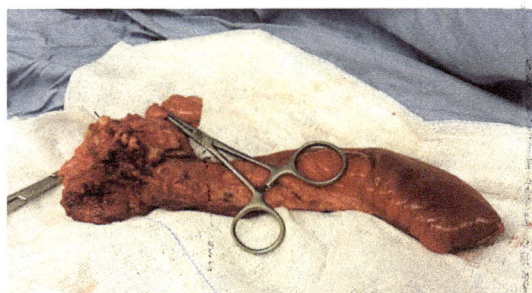

FIGURE 3: Surgical specimen reveals multiple punched out ulcerations along a diffusely thickened terminal ileum.

The chronic relapsing nature of intestinal BD requires patients to undergo repeated diagnostic imaging sessions due to high frequency of complications. Several studies have recently demonstrated that patients with intestinal BD are exposed to high levels of radiation owing to an excessive use of abdominal CT scans [1].

MRE is a relatively new imaging technique, which has demonstrated a good diagnostic value for small bowel pathology [8–10]. In particular, it is currently considered part of the standard evaluation algorithm in the follow-up of CD patients, thanks to its capacity to finely depict pathological alterations of the small bowel and to detect structured segments. In addition, MRE is a radiation-free, noninvasive technique, well-tolerated also by younger patients.

In our case, MRE showed a diffuse bowel wall thickening of the terminal ileum and caecum with polypoid appearance and restricted water diffusivity on DWI, a characteristic feature of active inflammation.

Due to similar CTE and MRE findings, the main differential diagnosis includes CD.

Although a multitude of authors have already assumed that intestinal BD may mimic CD clinically, pathologically, endoscopically, and even radiologically, the role of MRE in detecting intestinal BD has never been reported up to now [1, 3, 11].

MRE may have a primary role in patients with history of BD for staging, follow-up, and also depicting extraluminal complications.

In conclusion, gastroenterologists and radiologists should be aware of the similar CTE and MRE findings of CD and intestinal BD. Although CTE has some advantages compared to MRE, such as a better spatial resolution and a lower acquisition time, if validated by larger studies MRE may represent a useful diagnostic tool to evaluate patients with gastrointestinal manifestations of BD, sparing them an excessive and inappropriate radiation exposure.

References

[1] Y. S. Jung, D. I. Park, C. M. Moon et al., "Radiation exposure from abdominal imaging studies in patients with intestinal Behçet disease," *Gut and Liver*, vol. 8, no. 4, pp. 380–387, 2014.

[2] T. Hisamatsu, M. Naganuma, K. Matsuoka, and T. Kanai, "Diagnosis and management of intestinal Behçet's diseas," *Clinical Journal of Gastroenterology*, vol. 7, no. 3, pp. 205–212, 2014.

[3] F. Davatchi, F. Shahram, C. Chams-Davatchi et al., "Behcet's disease: from east to west," *Clinical Rheumatology*, vol. 29, no. 8, pp. 823–833, 2010.

[4] International Team for the Revision of the International Criteria for Behçet's Disease, "The International Criteria for Behçet's Disease (ICBD): a collaborative study of 27 countries on the sensitivity and specificity of the new criteria," *Journal of the European Academy of Dermatology and Venereology*, vol. 28, pp. 338–347, 2014.

[5] Y. S. Jung, J. H. Cheon, S. J. Park, S. P. Hong, T. I. Kim, and W. H. Kim, "Clinical course of intestinal Behçet's disease during the first five years," *Digestive Diseases and Sciences*, vol. 58, no. 2, pp. 496–503, 2013.

[6] H. K. Ha, H. J. Lee, S.-K. Yang et al., "Intestinal Behcet syndrome: CT features of patients with and patients without complications," *Radiology*, vol. 209, no. 2, pp. 449–454, 1998.

[7] M. J. Park and J. S. Lim, "Computed tomography enterography for evaluation of inflammatory bowel disease," *Clinical Endoscopy*, vol. 46, no. 4, pp. 327–336, 2013.

[8] S. Mazziotti, T. D'Angelo, S. Racchiusa, I. Salamone, A. Blandino, and G. Ascenti, "Peritoneal inclusion cysts in patients affected by Crohn's disease: Magnetic resonance enterography findings in a case series," *Clinical Imaging*, vol. 40, no. 1, pp. 152–155, 2015.

[9] S. Mazziotti, A. Blandino, E. Scribano et al., "MR enterography findings in abdominopelvic extraintestinal complications of Crohn's disease," *Journal of Magnetic Resonance Imaging*, vol. 37, no. 5, pp. 1055–1063, 2013.

[10] S. Mazziotti, G. Ascenti, E. Scribano et al., "Guide to magnetic resonance in Crohn's disease: from common findings to the more rare complicances," *Inflammatory Bowel Diseases*, vol. 17, no. 5, pp. 1209–1222, 2011.

[11] E. Rodrigues-Pinto, F. Magro, S. Pimenta, J. Guimarães, and G. Macedo, "Mimicry between intestinal Behçet's disease and inflammatory bowel disease," *Journal of Crohn's and Colitis*, vol. 8, no. 7, pp. 714-715, 2014.

39

CT Findings of Axillary Tuberculosis Lymphadenitis: A Case Detected by Breast Cancer Screening Examination

Hiroko Shojaku,[1,2] **Kyo Noguchi,**[1] **Tetsuya Kamei,**[2] **Yasuko Tanada,**[3] **Kouichi Yoshida,**[3] **Yasuko Adachi,**[4] **and Kazuhiro Matsui**[5]

[1]*Department of Radiology, University of Toyama, 2630 Suginani, Toyama, Toyama 930-0194, Japan*
[2]*Department of Radiology, Saiseikai Takaoka Hospital, 387-1 Futazuka, Takaoka, Toyama 933-8525, Japan*
[3]*Department of Surgery, Saiseikai Takaoka Hospital, 387-1 Futazuka, Takaoka, Toyama 933-8525, Japan*
[4]*Department of Respiratory Internal Medicine, Saiseikai Takaoka Hospital, 387-1 Futazuka, Takaoka, Toyama 933-8525, Japan*
[5]*Department of Pathology, Saiseikai Takaoka Hospital, 387-1 Futazuka, Takaoka, Toyama 933-8525, Japan*

Correspondence should be addressed to Hiroko Shojaku; hiroshojaku@yahoo.co.jp

Academic Editor: Amit Agrawal

We report the first description of CT findings of axillary tuberculous lymphadenitis confirmed by the pathological specimen. The breast cancer screening examination is one of the prime methods of detection of axillary tuberculous lymphadenitis. The most common site of axillary tuberculous lymphadenitis is the deep axilla. Screening mammography often fails to cover the whole axilla. The presence on the contrast-enhanced CT of unilateral multiple circumscribed dense nodes, some of which have large and dotted calcifications, might suggest tuberculous lymphadenitis in axillary region.

1. Introduction

Peripheral tuberculous lymphadenitis occurs predominantly in females and develops in people at ages younger than those of other tuberculous conditions [1, 2]. The clinical features and the age distribution between peripheral tuberculous lymphadenitis and breast cancer with axillary metastasis overlap and may initially lead to misdiagnosis [2]. Due to the high prevalence of breast carcinoma and tuberculosis especially in Asia, the coexistence of breast carcinoma and axillary tuberculous lymphadenitis and concomitant tuberculosis and metastasis in axillary lymph nodes has been reported without any radiological examination being performed, so histology and microbiology are essential [3, 4]. The treatment of both diseases is quite different, so it is imperative to perform necessary radiological examinations to expose the possibilities for differential diagnosis.

Although the discovery of axillary tuberculous lymphadenitis has been reported chiefly via mammography [1, 2, 5], there are no reports of the condition being identified by a contrast-enhanced CT [6]. We herein report a case of axillary tuberculous lymphadenitis mainly from the viewpoint of the contrast-enhanced CT findings and their relation to the other radiological findings and pathologic correlations.

2. Case Report

A 67-year-old woman presented with a left palpable, fingertip-sized axillary mass that was found by palpitation during breast cancer screening. No breast mass or cervical lymph adenopathy was palpable. The patient was 147 cm in height and weighed 45 kg. Her white blood cell count and C-reactive protein levels were within normal limits.

Screening mammography revealed no abnormalities in the breast or axilla. Ultrasound (US) showed unilateral multiple, markedly hypoechoic, ovoid lymph nodes, some of which had multiple coarse calcifications (Figure 1(a)). We then tried left axillary tail mammography, which revealed irregularly shaped macrocalcifications and dense tissue nodes partially at the corner of the films (Figure 1(b)).

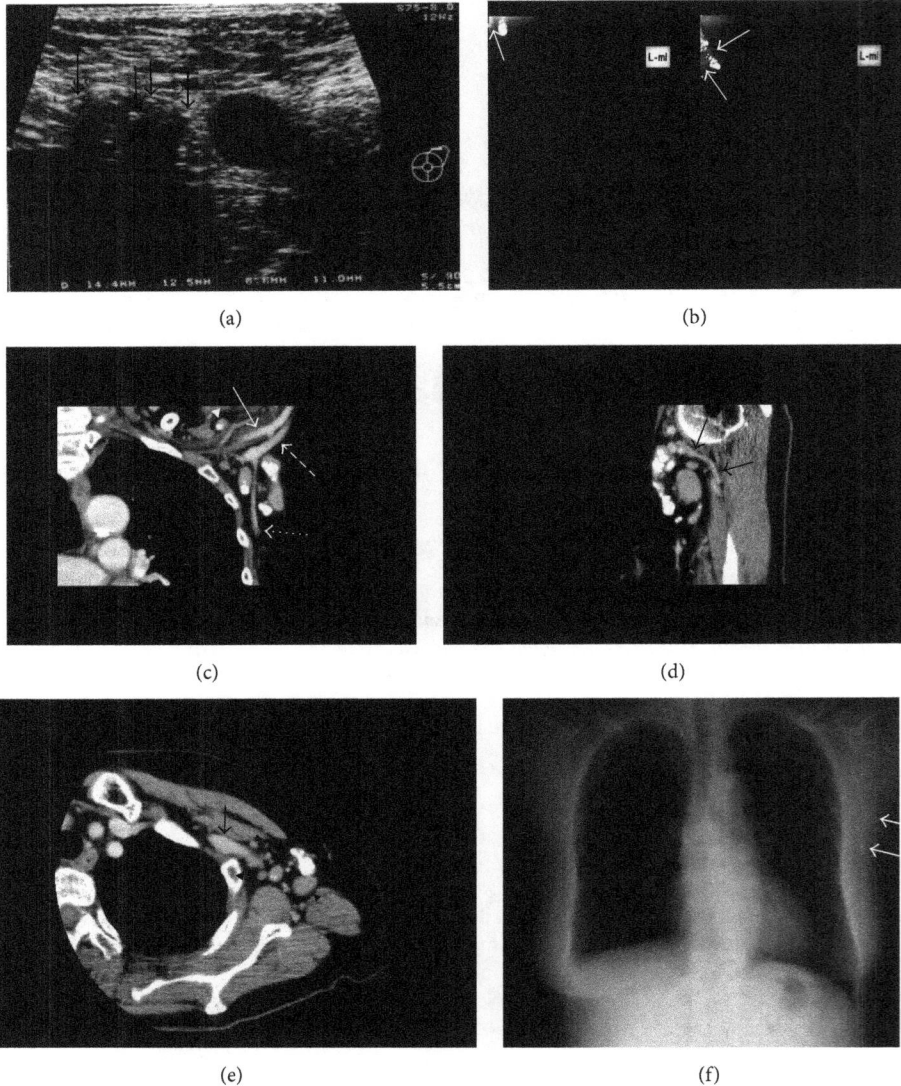

FIGURE 1: Images from a 67-year-old woman with a left axillary mass. (a) Axillary US shows well-circumscribed markedly hypoechoic ovoid lymph nodes. Every node lacks a hilum. Note the multiple rough calcifications (arrows) within the lymph node. (b) Axillary tail mammographic films show irregularly shaped macrocalcifications and matted, slightly dense homogeneous lymph nodes (arrows). Screening mammographic films show normal findings (not shown). (c) Coronal CT image after administration of intravenous contrast shows the presence of multiple well-circumscribed lymph nodes spread throughout the axillary vein (long dotted line) and dorsal thoracic vein (dotted line) in the left axilla and supraclavicular (arrowhead) and infraclavicular regions. Most nodes have large and dotted calcifications. Arrow indicates axillary artery. (d) Sagittal CT image shows the largest node, which has no calcification itself but is surrounded by nodules rich in calcification. Arrows indicate dorsal thoracic vein. (e) Axial CT image shows the lymph nodes are distributed extending from the axilla to the infraclavicular region around the axillary vein (arrow). (f) Chest radiograph shows clustered calcifications in the left deep lower axilla (arrows).

Non-contrast-enhanced CT showed a cluster of 12 well-circumscribed dense nodes spread around the vessels throughout the left deep axilla and supraclavicular and infraclavicular regions. Some of the nodes had large and dotted calcifications. The noncalcified sites of the lymph nodes showed 54–62 Hounsfield units (HU) on the non-contrast-enhanced CT. On contrast-enhanced CT, they appeared to be highly homogenous and showed 98–102 HU. The lymph nodes were partially to completely calcified, and both the calcified and noncalcified nodes were randomly distributed. The largest node measured 3.0 × 2.8 × 1.7 cm (Figures 1(c)–1(e)). The CT images also showed right parabronchial lymph node calcification and small pulmonary nodes in the apex of the right lung, that is, the primary tuberculous complex. A chest radiograph before biopsy showed clustered calcifications in the left deep lower axilla (Figure 1(f)).

Because we could not rule out malignancy as the cause of the pathological nodes, an excisional biopsy was performed.

(a)

(b)

(c)

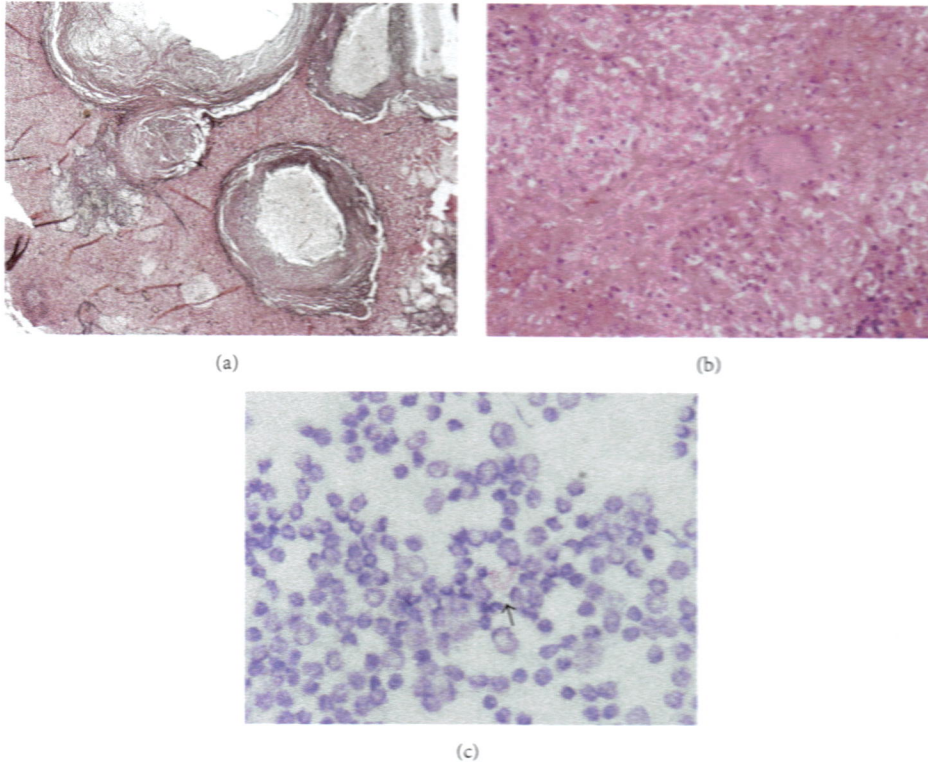

FIGURE 2: Photomicrography findings. (a) Photomicrograph in low-power view shows caseating necrotizing granulomas with calcification in the central areas (Ag stain, ×40). (b) Photomicrograph shows many necrotizing granulomas that occupy the whole area of the lymph node. Note the Langhans-type multinucleated giant cell (HE stain, ×400). (c) A stamp cytological section stained for acid-fast bacilli reveals Gram-positive rod-shaped bacilli (arrow) (×1000).

The excised fibroadipose tissue specimen measured 8 × 8 × 1.5 cm and contained several lymph nodes of various fingertip sizes. The histological sections revealed necrotizing granulomas within the architecture of the lymph nodes resulting in the fact that the disease was diagnosed to be active. Calcification was observed in areas of caseating central necrosis (Figures 2(a) and 2(b)). Stamp cytology was performed according to the Ziehl-Neelsen method, which revealed clusters of positive rod-shaped bacilli on the Ziehl-Neelsen-stained slide (Figure 2(c)).

A tuberculin skin test performed after surgery was positive (0 × 0/30 × 28). Smear staining from the bronchial aspiration fluid resulted in a Gaffky scale rating of 0. Culture and PCR of the left axillary lymph nodes and gastric juice were performed, and both were negative. The patient underwent antituberculosis treatment for 9 months. She has remained symptom- and disease-free for over 6 years.

3. Discussion

We present the contrast-enhanced CT findings of axillary tuberculous lymphadenitis characterized by unilateral multiple circumscribed well enhanced dense nodes around the vessels in the deep axilla, some of which had large and dotted calcifications. The pathological section showed necrotizing granulomas, which could result from tuberculous

lymphadenitis found beyond the proliferative lesions [7]. The contrast-enhanced CT diagnosis of axillary tuberculous lymphadenitis and its differential diagnosis from other disease have not been reported.

To our knowledge, there is only one CT report of the axilla that mentions calcifications, which was of pilomatrix carcinoma. This rare malignant soft-tissue tumor had a well-circumscribed mass with multiple small calcifications and cystic components [8]. The CT diagnosis of axillary tuberculous lymphadenitis and its differential diagnosis from other diseases may be difficult. However, the presence on CT of unilateral multiple circumscribed dense nodes around vessels, some of which have large and dotted calcifications, might suggest axillary tuberculous lymphadenitis. We propose this description as a new CT finding suggestive of axillary tuberculous lymphadenitis because it corresponds to the findings from other modalities.

Calcification in the axillary lymph nodes is uncommon on mammography [9–11] although approximately 5–20% of primary breast carcinomas have mammographically detectable calcifications. In our case, axillary tail mammography showed unilateral irregularly shaped macrocalcifications and dense tissue nodes. Similar findings were reported in 3 of 10 cases in which lymph nodes pathologically confirmed as axillary tuberculous lymphadenopathy showed macrocalcification [2]. Axillary lymphadenitis caused by a prior tuberculous infection may include large, coarse calcifications [5]. Contrastingly, microcalcifications detected on mammography were described in cases of metastasis from breast carcinoma [11], ovarian carcinoma [12], metastatic papillary carcinoma [10], and fat necrosis [9]. Axillary macrocalcification might be helpful in suggesting a diagnosis of tuberculous lymphadenitis.

The US findings in our case showed multiple markedly hypoechoic axillary lymph nodes with multiple coarse calcifications without visualization of a normal central fatty hilum. This US finding was described previously in only one patient who had three nodes of relatively hyperechoic and hypoechoic conglomerate nodes that lacked normal fatty hilum, and no calcification was detected, even though dense calcification was detected on the mammogram [1].

Tuberculosis lesions are histologically characterized as proliferative or exudative. Proliferative lesions are granulomas composed of compact aggregates of epithelioid cells, lymphocytes, and Langhans-type giant cells with variable degrees of central necrosis and relatively few acid-fast bacilli. Exudative lesions consist of an amorphous exudate of mononuclear cells, neutrophils, fibrin, and usually extensive necrotic debris [7]. We showed in the histological section that calcifications were observed in areas of caseating central necrosis, and these calcifications were revealed in every radiological examination on our patient that contained a chest radiograph. The extreme enhancement of the nodes seen on CT in our case reflects the granulomas and aggregate inflammatory cells. From the histopathological viewpoint, our case and the other reported cases of axillary tuberculous lymphadenitis were thought to be of proliferative lesions [1, 2, 4].

Screening mammography often fails to cover the deep axilla, and US for breast cancer screening does not survey the infraclavicular and supraclavicular regions. Associated supraclavicular lesions were reported in 7 of 10 cases of axillary tuberculous lymphadenitis, and all were unilateral [2]. None of the reports on tuberculous lymphadenitis mentioned the infraclavicular region. The axilla and nearby lymph nodes regions can be easily surveyed and objectively viewed with CT, so we can observe the distribution of the lesions and the unique calcification within the nodes, which may help in reaching a diagnosis of axillary tuberculous lymphadenitis.

We described the first enhanced CT findings, to our knowledge, of axillary tuberculous lymphadenitis correlated by other radiologic modalities and pathologic studies. The breast cancer screening examination can be a useful tool for discovering findings of axillary tuberculous lymphadenitis and mammary tuberculosis. We found multiple unilateral circumscribed and well-enhanced dense nodes around the vessels in the deep axilla, some of which had large and dotted calcifications indicative of tuberculous lymphadenitis.

In this case, the axillary lymphadenitis was diagnosed as the active disease because the pathological specimen by the biopsy of the node shows the necrotizing granulomas. Her axillary lymph nodes were calcified and homogenously enhanced in the contrast CT. So, we think that the calcification of the axillary nodes might be one of characteristic findings on contrast CT in tuberculous lymphadenitis, but it might not be helpful for diagnosing the activity of the disease. Further studies are necessary to evaluate this possibility.

Early diagnosis and treatment of active tuberculosis decrease the opportunities for the spread of the disease in the community. Thus, physicians should familiarize themselves with the radiological findings associated with axillary tuberculous lymphadenitis.

4. Conclusion

The breast cancer screening examination is one of the prime methods of detection of axillary tuberculous lymphadenitis. On US the notice of coarse calcifation within the nodes helps in the suspicion of the disease. Further examination on the contrast-enhanced CT and the presence of unilateral multiple circumscribed dense nodes, some of which have large and dotted calcifications, might suggest tuberculous lymphadenitis in axillary region. Early diagnosis and treatment of tuberculosis decrease the opportunities for the spread of the disease in the community.

References

[1] W. T. Yang, M. Suen, and C. Metreweli, "Mammographic, sonographic and histopathological correlation of benign axillary masses," *Clinical Radiology*, vol. 52, no. 2, pp. 130–135, 1997.

[2] M. Muttarak, S. Pojchamarnwiputh, and B. Chaiwun, "Mammographic features of tuberculous axillary lymphadenitis," *Australasian Radiology*, vol. 46, no. 3, pp. 260–263, 2002.

[3] M. Khurram, M. Tariq, and P. Shahid, "Breast cancer with associated granulomatous axillary lymphadenitis: a diagnostic and clinical dilemma in regions with high prevalence of tuberculosis," *Pathology Research and Practice*, vol. 203, no. 10, pp. 699–704, 2007.

[4] E. D. Babu, N. Tariq, F. A. Aref, and R. Vashisht, "Axillary gland involvement in breast carcinomas is not always metastatic: a case report," *International Surgery*, vol. 89, no. 3, pp. 150–151, 2004.

[5] A. J. Leibman and R. Wong, "Findings on mammography in the axilla," *American Journal of Roentgenology*, vol. 169, no. 5, pp. 1385–1390, 1997.

[6] J. W. Lee, S. M. Lee, and J. H. Choi, "18F-FDG PET/CT findings in a breast cancer patient with concomitant tuberculous axillary lymphadenitis," *Nuclear Medicine and Molecular Imaging*, vol. 45, no. 2, pp. 152–155, 2011.

[7] J. Tomashefski and C. Farver, "Tuberculosis and nontuberculous mycobacterial infections," in *Dail and Hammar's Pulmonary Pathology*, J. Tomashefski, C. Farver, P. Cagle, and A. Fraire, Eds., pp. 316–348, Springer, New York, NY, USA, 3rd edition, 2008.

[8] T. Niwa, T. Yoshida, T. Doiuchi et al., "Pilomatrix carcinoma of the axilla: CT and MRI features," *British Journal of Radiology*, vol. 78, no. 927, pp. 257–260, 2005.

[9] R. Hooley, C. Lee, I. Tocino, N. Horowitz, and D. Carter, "Calcifications in axillary lymph nodes caused by fat necrosis," *American Journal of Roentgenology*, vol. 167, no. 3, pp. 627–628, 1996.

[10] S. W. Chen, G. Bennett, and J. Price, "Axillary lymph node calcification due to metastatic papillary carcinoma," *Australasian Radiology*, vol. 42, no. 3, pp. 241–243, 1998.

40

Primary Pulmonary Anaplastic Large Cell Lymphoma: A Rare Malignancy and Rare Cause of the Luftsichel Sign

Elizabeth Von Ende,[1] **Travis Kauffman** ⓓ,[2]
Philip A. Munoz,[3] **and Santiago Martinez-Jiménez** ⓓ[3]

[1]Kansas City University, 1750 Independence Ave, Kansas City, MO 64106, USA
[2]University of Missouri-Kansas City, 4401 Wornall Rd, Kansas City, MO 64111, USA
[3]Saint Luke's Hospital of Kansas City, University of Missouri-Kansas City, 4401 Wornall Rd, Kansas City, MO 64111, USA

Correspondence should be addressed to Travis Kauffman; kauffmant@umkc.edu

Academic Editor: Roberto Iezzi

Primary pulmonary lymphomas are rare with primary pulmonary non-Hodgkin lymphoma accounting for only 0.3% of primary lung neoplasms. Of these, the large majority are made up of marginal zone B-cell lymphoma and diffuse large B-cell lymphoma. We present a case of a very rare primary pulmonary anaplastic large cell lymphoma presenting as the luftsichel sign on chest radiograph. Pertinent imaging and pathology findings are discussed.

1. Introduction

Primary pulmonary lymphomas (PPL) are overall rare neoplastic malignancies. Notably primary non-Hodgkin lymphoma (P-NHL) accounts for only 0.3% of primary lung neoplasms and primary pulmonary Hodgkin lymphoma is even rarer. Among P-NHL, marginal zone B-cell lymphoma of mucosa-associated lymphoid tissue and diffuse large B-cell lymphoma are responsible for over 95% of all PPL [1, 2]. Other types of lymphoproliferative processes such as primary lung plasmacytoma and lymphomatoid granulomatosis account for a good proportion of the remaining PPL [3]. Anaplastic large cell lymphoma (ALCL), which most often occurs in lymph nodes and skin, is an exceedingly rare type of PPL [1, 2, 4]. Here we describe a case of primary ALCL of the lung in a patient who made a complete recovery after presenting with complete left upper lobe atelectasis.

2. Case Presentation

A 42-year-old male presented with nonproductive cough, shortness of breath, 15-pound weight loss, and night sweats for one month in duration. There was no history of smoking, upper respiratory symptoms, or chest pain. Physical exam showed shortness of breath and mildly decreased breath sounds in the left upper lung zone. Initial chest radiograph showed the luftsichel sign (i.e., complete atelectasis of the left upper lobe) and trace left pleural effusion (Figure 1). Subsequent CT scan showed complete left upper lobe atelectasis with a distinct central left upper lobe mass measuring 4.5 × 3.5 cm obstructing the left upper lobe bronchus (Figure 2). The patient eventually underwent further lab work, bronchoscopy, and PET-CT for further testing.

The PET-CT demonstrated focally increased metabolic activity within the left upper lobe which was favored to represent lung cancer or, less likely, metastatic disease (Figure 3). Bronchoscopy revealed a large tumor obstructing the left upper lobe segmental bronchus. Bronchoscopic biopsies of the mass showed neoplastic cells with large nuclei, scant cytoplasm, and vesicular nuclear chromatin, suggestive of a poorly differentiated malignant process. Due to lack of definitive immunohistochemical staining characteristics, additional percutaneous biopsy was performed which was indeterminate for malignancy. Eventually the patient underwent left pneumonectomy with final histology including immunohistochemistry demonstrating anaplastic large cell lymphoma positive for CD30, Ki-67, CD45, and ALK-1

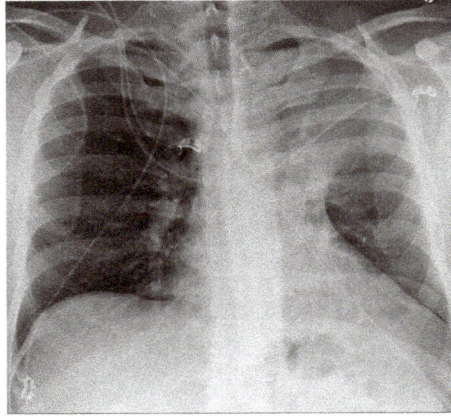

FIGURE 1: PA chest radiograph demonstrates complete left upper lobe atelectasis and the luftsichel sign.

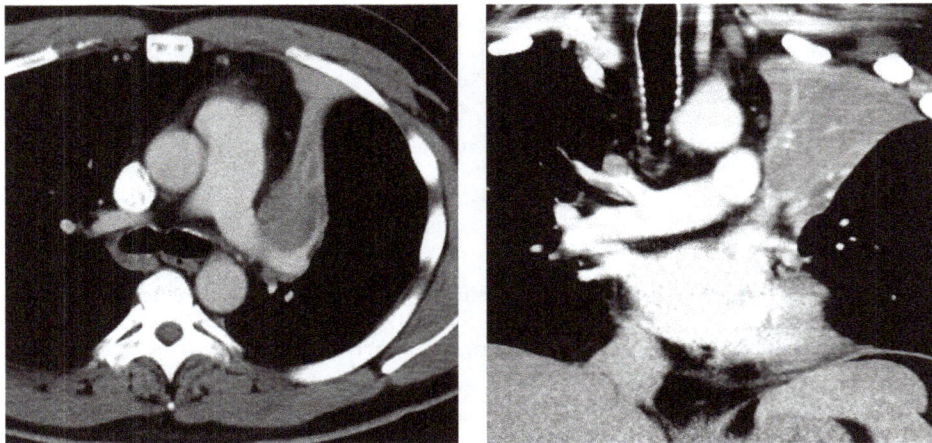

(a) Axial CT image of the same patient at the level of the pulmonary trunk demonstrates a central mass obstructing the left upper lobe bronchus with resultant collapse of the left upper lobe

(b) Corresponding coronal CT image better demonstrates complete left upper lobe atelectasis. The obstructing central mass is again seen

FIGURE 2

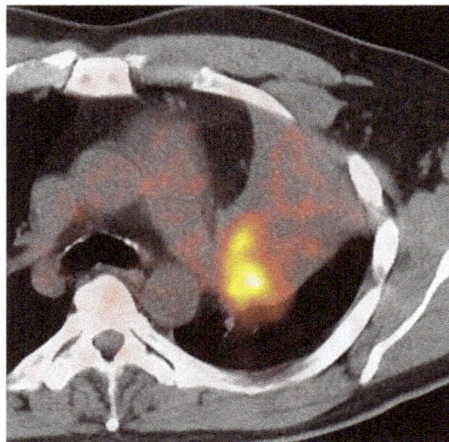

FIGURE 3: Axial PET/CT image demonstrates avid FDG uptake by the left upper lobe mass.

TABLE 1: Results of immunohistochemistry.

Antibody	Results
Pan cytokeratin	Negative in neoplastic cells
CAM 5.2	Negative in neoplastic cells
S-100	Negative in neoplastic cells
Vimentin	Positive, strong diffuse cytoplasmic staining
CD45	Negative to weakly positive in scattered neoplastic cells
CD20	Negative in neoplastic cells
PAX5	Negative in neoplastic cells
CD3	Negative in neoplastic cells
CD30	Positive, strong membrane and heterogeneous cytoplasmic staining
Alk-1	Positive, strong membrane and cytoplasmic staining
CD7	Negative in neoplastic cells
CD4	Negative for neoplastic cells
CD8	Weakly positive in neoplastic cells
CD10	Negative in neoplastic cells
CD15	Negative in neoplastic cells
CD56	Negative in neoplastic cells
CD68	Negative in neoplastic cells
MUM-1	Negative in neoplastic cells
Ki-67	Positive in greater than 90% of neoplastic cells

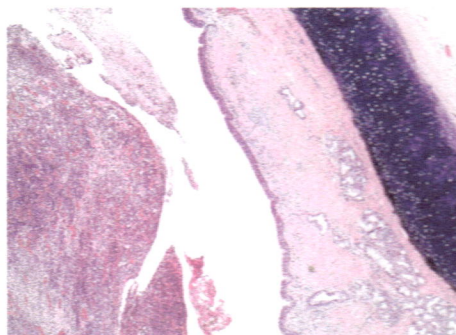

(a) Endobronchial polypoid mass

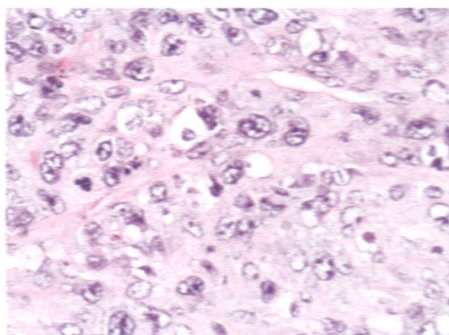

(b) Large cells with pleomorphic nuclei, some with folded nuclear contours. Arrows depict so-called "clue cells"

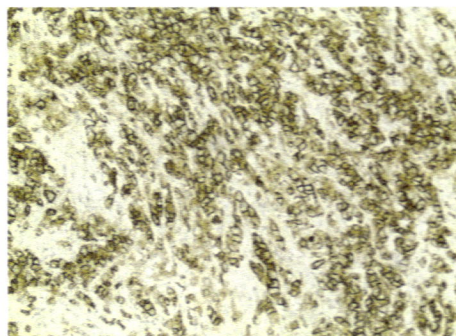

(c) Neoplastic cells show strong, diffuse reactivity to CD30

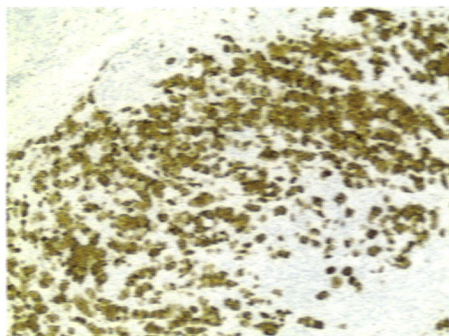

(d) Neoplastic cells showing strong reactivity to ALK-1

FIGURE 4

(Figure 4). A full list of antibodies tested and results are listed in Table 1. An excised left hilar lymph node was free of tumor.

The patient recovered satisfactorily. A follow-up CT performed after 6 months showed no signs of recurrent disease.

3. Discussion

Imaging appearance of primary pulmonary non-Hodgkin lymphoma is varied. One retrospective study described the computed tomography findings in multiple cases of primary and secondary pulmonary lymphoma which included consolidation, ground-glass opacification, air-bronchograms, lymphadenopathy, CT-halo sign, lung nodules, reticular opacities, and pleural effusions [3]. In the majority of pulmonary non-Hodgkin lymphoma cases, patients presented with a combination of multiple CT findings. The most common combination of findings in primary and secondary non-Hodgkin's lymphoma included consolidation with air bronchogram, ground-glass opacities, and lymphadenopathy. As described, the CT findings are often nonspecific, and therefore could resemble various pathological processes. The differential diagnosis may include infectious processes, other neoplasms, inflammatory processes, or autoimmune processes. Interestingly, left upper lobe atelectasis, which is present in this case, seen on chest radiographs in the form of the luftsichel sign, is not a common imaging description in PPL. On the contrary, the luftsichel sign is almost always indicative of central primary lung cancer. Due to the variety and lack of specificity of the presenting imaging findings, definitive diagnosis requires tissue for histopathological examination.

In this case, the initial diagnosis of anaplastic large cell lymphoma was challenging. Ultimately, the resected specimens proved diagnostic. Morphologically, this neoplasm lacked typical glandular, squamous, or neuroendocrine differentiation characteristic of most primary lung cancers. Immunohistochemical studies further substantiated this morphologic impression by negative reactivity for pancytokeratin and CAM 5.2. The possibility of metastatic melanoma was effectively ruled out by negative reactivity for S-100 protein. The morphologic consideration of lymphoma was substantiated by patchy reactivity for CD45. The possibility of B-cell lymphoma was effectively ruled out by negative reactivity for CD20 and PAX5. Strong immunoreactivity for CD30 and ALK-1 are considered diagnostic in this morphologic and immunophenotypic context. Expressions of other T-cell markers are variable, as demonstrated in this case.

Anaplastic large cell lymphoma (ALCL) is a rare non-Hodgkin T-cell lymphoma characterized by large lymphoid cells with abundant cytoplasm, pleomorphic, horseshoe-shaped nuclei, and uniform CD30/Ki-1 expression [1, 4, 5]. ALCL was first described and reported by Stein et al. in 1985, as a neoplastic proliferation of lymphoid cells that are anaplastic in appearance, grow cohesively, invade lymph node sinuses, and consistently express CD30 [6, 7]. The morphology and patterns of tissue invasion commonly mimic those seen with non-small cell carcinoma or melanoma. Two types of ALCL with lung involvement have been described: primary pulmonary ALCL and secondary pulmonary ALCL. ALCL most commonly presents as a primary cutaneous lymphoma and may secondarily involve the lungs. Primary pulmonary ALCL is exceedingly rare and most often presents with mediastinal lymphadenopathy. Zhao et al. conducted a search of Medline and PubMed databases, in an attempt to identify all reported cases of primary ALCL of the lung, and found only 10 published cases between 1990 and 2015 [1].

ALCL is further subclassified based on positive or negative ALK protein expression. This protein is a tyrosine kinase encoded by a unique gene rearrangement. The most common gene rearrangement is t(2;5) which approximates the ALK gene located on chromosome 2 with the nucleophosmin NPM1 gene located on chromosome 5. Other ALK translocation partners occur with lesser frequency but most are involved with ALK protein upregulation. ALK is an important prognostic indicator for ALCL, as patients with ALK(+) staining tend to have a more favorable prognosis with a 5-year survival rate of 70–90%, compared to a 5-year survival rate of 40–60% in ALK(−) staining patients [8, 9]. Diagnosis of ALCL is based on the World Health Organization classification and includes typical histopathology and immunohistochemistry staining, with strong immunoreactivity for CD30.

There is no standardized treatment for ALCL of the lung; however, current first-line therapy includes anthracycline-based regimen, such as CHOP [10, 11].

References

[1] Q. Zhao, Y. Liu, H. Chen et al., "Successful chemo-radiotherapy for primary anaplastic large cell lymphoma of the lung: A case report and literature review," *American Journal of Case Reports*, vol. 17, pp. 70–75, 2016.

[2] S. H. Han, Y. H. Maeng, Y. S. Kim et al., "Primary anaplastic large cell lymphoma of the lung presenting with acute atelectasis," *Thoracic Cancer*, vol. 5, no. 1, pp. 78–81, 2014.

[3] S. Piña-Oviedo, A. Weissferdt, N. Kalhor, and C. A. Moran, "Primary pulmonary lymphomas," *Advances in Anatomic Pathology*, vol. 22, no. 6, pp. 355–375, 2015.

[4] W. L. Rush, J. A. W. Andriko, J. K. Taubenberger et al., "Primary anaplastic large cell lymphoma of the lung: A clinicopathologic study of five patients," *Modern Pathology*, vol. 13, no. 12, pp. 1285–1292, 2000.

[5] J.-F. Cordier, E. Chailleux, D. Lauque et al., "Primary pulmonary lymphomas; a clinical study of 70 cases in nonimmunocompromised patients," *CHEST*, vol. 103, no. 1, pp. 201–208, 1993.

[6] H. Stein, D. Y. Mason, J. Gerdes et al., "The expression of the Hodgkin's disease associated antigen Ki-1 in reactive and neoplastic lymphoid tissue: evidence that Reed-Sternberg cells and histiocytic malignancies are derived from activated lymphoid cells," *Blood*, vol. 66, no. 4, pp. 848–858, 1985.

[7] M. A. Pletneva and L. B. Smith, "Anaplastic large cell lymphoma: Features presenting diagnostic challenges," *Archives of Pathology & Laboratory Medicine*, vol. 138, no. 10, pp. 1290–1294, 2014.

[8] H.-B. Yang, J. Li, and T. Shen, "Primary anaplastic large cell lymphoma of the lung: Report of two cases and literature review," *Acta Haematologica*, vol. 118, no. 3, pp. 188–191, 2007.

[9] D. Sibon, M. Fournier, J. Brière et al., "Long-term outcome of adults with systemic anaplastic large-cell lymphoma treated within the Groupe d'Étude des Lymphomes de l'Adulte Trials," *Journal of Clinical Oncology*, vol. 30, no. 32, pp. 3939–3946, 2012.

[10] K. J. Savage, N. L. Harris, J. M. Vose et al., "ALK—anaplastic large-cell lymphoma is clinically and immunophenotypically different from both ALK + ALCL and peripheral T-cell lymphoma, not otherwise specified: report from the International Peripheral T-Cell Lymphoma Project," *Blood*, vol. 111, no. 12, pp. 5496–5504, 2008.

41

Ectopic Thyroid Tissue in the Mediastinum Characterized by Histology and Functional Imaging with I-123 SPECT/CT

Jed Hummel,[1] Jason Wachsmann,[1] Kelley Carrick,[2] Orhan K. Oz,[1] Dana Mathews,[1] and Fangyu Peng[1,3]

[1]Department of Radiology, UT Southwestern Medical Center, Dallas, TX, USA
[2]Department of Pathology, UT Southwestern Medical Center, Dallas, TX, USA
[3]Advanced Imaging Research Center, UT Southwestern Medical Center, Dallas, TX, USA

Correspondence should be addressed to Fangyu Peng; fangyu.peng@utsouthwestern.edu

Academic Editor: Atsushi Komemushi

Ectopic thyroid tissue is a rare entity and when discovered it is typically along the pathway of embryologic migration of the thyroid. We present a case of incidental finding of ectopic thyroid tissue within mediastinum in a 61-year-old female patient with a history of total thyroidectomy for thyroiditis and nodules. The patient presented to emergency room with cough and right chest pain and underwent a chest computed tomographic angiogram (CTA) to exclude pulmonary embolism as part of chest pain workup. One right paratracheal mediastinal soft tissue nodule was visualized on the images of CTA. This right paratracheal soft tissue mass was found to be ectopic benign thyroid tissue by histological analysis of the biopsied tissue samples. The function of this ectopic thyroid tissue was characterized by I-123 radioiodine uptake and single photon emission computed tomography/computed tomography (SPECT/CT) imaging. This case illustrates that ectopic thyroid tissue should be included for differential diagnosis of a hyperdense soft tissue mass located within mediastinum. I-123 SPECT/CT is useful for guiding tissue biopsy of ectopic thyroid tissue distant from orthotopic thyroid gland and functional and anatomic characterization of mediastinal ectopic thyroid tissue for surgical resection when it is medically necessary.

1. Introduction

Ectopic thyroid tissue is a rare entity and when discovered it is typically along the pathway of embryologic migration of the thyroid [1]. The embryologic development of the thyroid provides an anatomic roadmap for the typical locations of ectopic thyroid tissue. The thyroid gland originates from an endodermal thickening between the first and second pharyngeal arches. There is a caudal migration of the thyroid primordium from the foramen cecum of the tongue to the thyroid bed at the pretracheal neck base, typically positioned anterolateral from the second to the fourth tracheal cartilage. This pathway of descent is marked by the thyroglossal duct. The thyroid primordium passes anterior to the hyoid bone and then loops inferiorly and posteriorly to the hyoid bone. There are circumstances in which thyroid tissue may be found outside the gland due to faulty embryogenesis related

to genetic factors, mechanical implantation secondary to surgical intervention or trauma, a sequestered thyroid nodule adjacent to the gland but without anatomic connection, or thyroid tissue as a component of a teratoma [1–4]. The lingual area is the most common location of ectopic thyroid tissue [3, 5, 6], and ectopic thyroid tissue is occasionally localized within mediastinum [7–13]. Most of patients with ectopic thyroid presented with symptoms of hypothyroidism, and rare cases of hyperthyroidism with histological features similar to Graves' disease have been reported [14–18]. Some patients may present with symptoms such as cough, dysphagia, dyspnea, stridor, and dysphonia related to mass effect on regional structures [16]. Ultrasound of neck was often used for detection of ectopic thyroid and radioiodine uptake and radioiodine scan could be used for functional characterization of suspected ectopic thyroid tissues [19]. Hybrid SPECT-CT imaging is a useful imaging modality

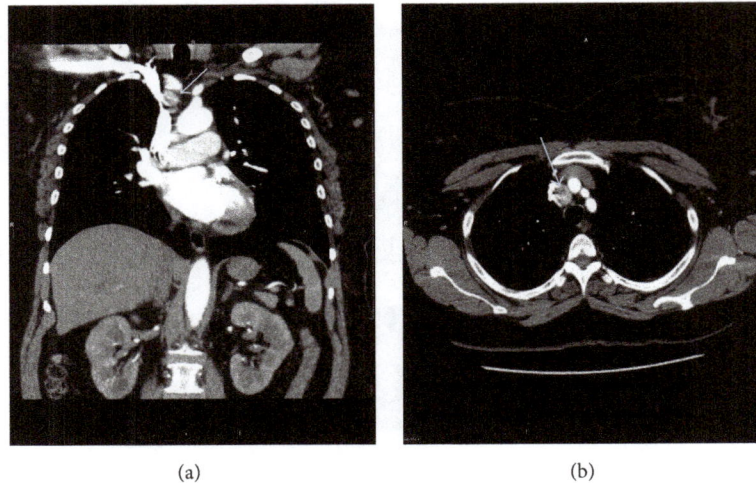

(a)   (b)

FIGURE 1: A right paratracheal mass in the upper mediastinum on the images of CTA from a patient presented with cough and right chest pain. A 61-year-old female presented to emergency room with cough and right chest pain. Chest CTA was performed to exclude pulmonary embolism as part of chest pain workup. One 2.0×1.7 cm right paratracheal mediastinal mass was visualized on the images of CTA, as indicated by a blue arrow on coronal (a) and axial (b) view images of CTA. This right paratracheal mass appeared hyperdense or showed mild contrast enhancement relative to other small mediastinal lymph nodes.

for both functional and anatomic evaluation of suspected ectopic thyroid tissue, particularly those located at an unusual location distant from expected location of ectopic thyroid tissues along the thyroglossal duct [6, 20, 21]. Herein, we report a case of incident finding of an ectopic thyroid tissue mass within the mediastinum in a patient with remote history of total thyroidectomy for thyroiditis and nodules. The patient presented to emergency room with cough and chest pain and underwent CTA to exclude pulmonary embolism as part of workup of right chest pain. On CTA images, one right paratracheal nodular soft tissue mass was visualized, which was found to be ectopic thyroid tissue by histological analysis of tissue sample from endoscopic bronchial ultrasound-guided biopsy and functional imaging with I-123 SPECT/CT.

2. Case Presentation

A 61-year-old female presented to the emergency room with cough and chest pain. A CTA of the chest was performed to exclude pulmonary embolism as part of chest pain workup. The result of CTA was negative for pulmonary embolism. The patient was diagnosed with acute bronchitis and the patient's symptoms of cough and chest pain were resolved after antibiotic treatment of acute bronchitis. On the images of CTA, a 2.0×1.7 cm right paratracheal mediastinal mass was noted which appeared slightly hyperdense or showed mild contrast enhancement (Figure 1).

Differential diagnosis for this upper mediastinal mass included an enlarged lymph node reactive to infection or a chronic inflammatory process, sarcoidosis, or nodal metastasis from occult malignancy. The patient underwent an endobronchial ultrasound-guided biopsy of the right paratracheal mass, which was found to be benign ectopic thyroid tissues by histological analysis of the biopsied tissue samples (Figure 2).

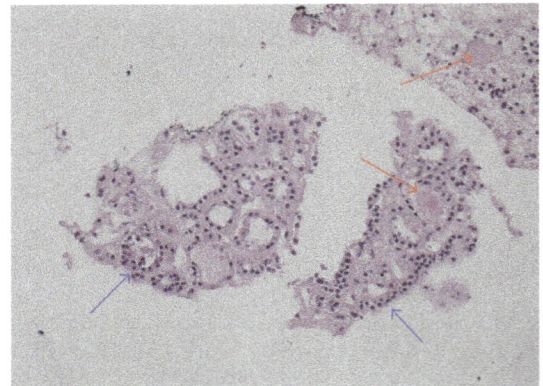

FIGURE 2: Histological image of ectopic thyroid tissue in mediastinum. Fragments of benign thyroid tissue identified in the cell block prepared from the transbronchial needle aspiration. Bland, uniform thyroid epithelial cells are disposed in the follicular units characteristic of benign thyroid tissue (blue arrow), with focal presence of intrafollicular colloid (red arrow). There is mild perifollicular hyalinization (Hematoxylin and Eosin, 20x).

A 24 hours I-123 uptake and scintigraphic scan were performed for further functional characterization of this ectopic thyroid tissue mass within the mediastinum. The 24-hour radioiodine uptake by residual thyroid tissue in the thyroidectomy bed was measured at 1.5% and no thyroid tissue with I-123 uptake was visualized on the surgical bed, compatible with the patient's history of prior total thyroidectomy 10 years ago. One focus of increased I-123 radioiodine accumulation was identified in the region of upper mediastinum on the planar images of I-123 scan. For further anatomic localization of the focal uptake in the upper mediastinum visualized on planar scintigraphic images, a SPECT/CT was performed

Ectopic Thyroid Tissue in the Mediastinum Characterized by Histology and Functional...

203

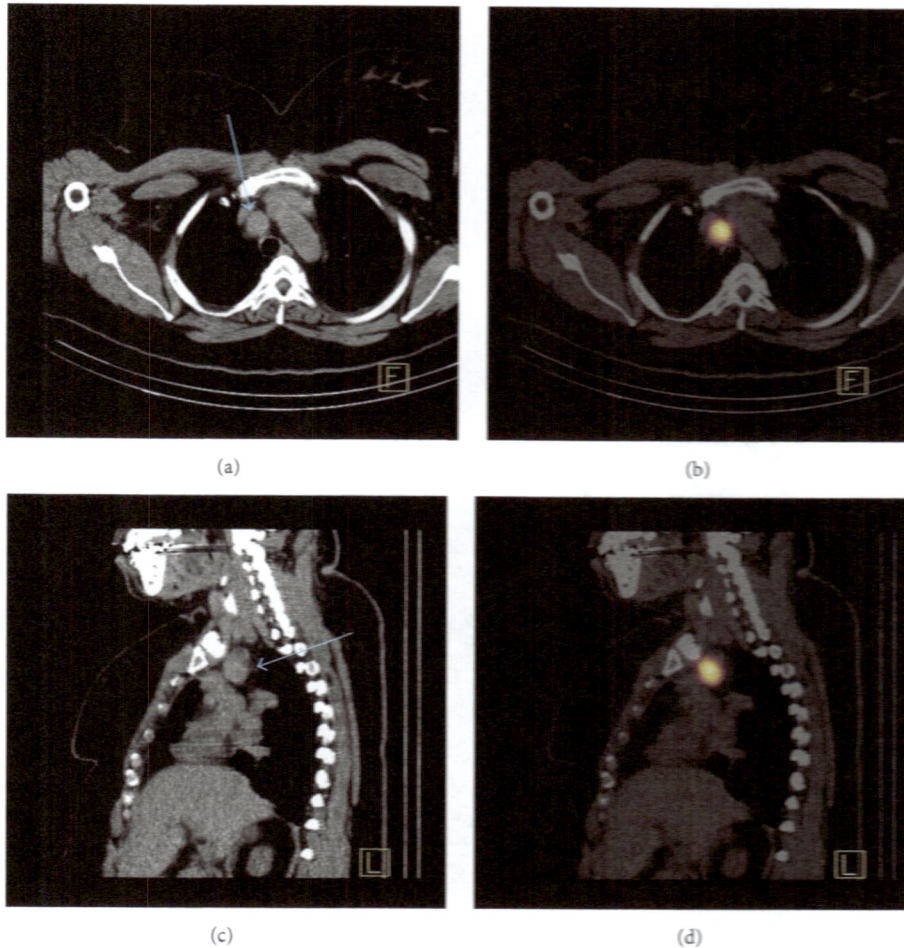

FIGURE 3: Functional and anatomic imaging of ectopic thyroid tissue in the upper mediastinum by I-123 SPECT/CT. A 2.0 × 1.7 cm right paratracheal mediastinal mass was visualized on the axial (a) and sagittal (c) view images of noncontrast enhanced low dose CT component of I-123 SPECT/CT as indicated by a blue arrow. On the axial (b) and sagittal (d) view images of coregistered I-123 SPECT/CT images, I-123 radioiodine uptake by the right paratracheal mass was visualized as indicated by orange color presentation of I-123 radioiodine activity. The findings of I-123 SPECT/CT further confirmed that the upper mediastinal mass seen on the CTA images represented functional ectopic thyroid tissue, consistent with the results of histological analysis of the tissue samples obtained by endobronchial ultrasound-guided biopsy.

using a dual headed Siemens Symbia T2 SPECT/CT camera in a method as previously described [22]. On SPECT/CT images, the focus of increased radioiodine uptake in the mediastinum seen on planar imaging was localized to the 2.0 × 1.7 cm right paratracheal mediastinal mass visualized on CTA (Figure 3).

The patient had a history of total thyroidectomy for thyroiditis and nodules 10 years ago. Post-total thyroidectomy hypothyroidism was treated with oral administration of 125 to 137 mg Levoxyl daily for one year. Subsequently, the dose of Levoxyl was reduced to 100 mg/daily and the results of thyroid functional tests were normal with a TSH level of 2.63 mIU/L (normal reference range 0.40–4.50 mIU/L) and a free T4 of 1.6 ng/dL (normal reference range of 0.8–1.8 ng/dL) at 4 years after the patient was maintained on 100 mg of Levoxyl daily for treatment of post-total thyroidectomy

hypothyroidism. However, TSH level was low at 0.09 mIU/L and a free T4 level was high at 2.58 ng/dL when a thyroid functional test was performed at 6 days after mediastinal ectopic thyroid tissue was diagnosed with histological analysis of the biopsied tissue samples. In preparation for I-123 SPECT/CT, Levoxyl was stopped for 4 weeks and the patient developed symptoms of hypothyroidism (fatigue, constipation, and hair loss). In view of possible functional activity of benign mediastinal ectopic thyroid tissue confirmed by I-123 SPECT/CT, a reduced dose of 75 mg of Levoxyl daily was prescribed for this patient upon completion of I-123 SPECT/CT. One year later, the results of thyroid functional test were normal with a free T4 level 1.4 ng/dL and a TSH level 2.11 mIU/L. Follow-up CT of chest one year after I-123 SPECT/CT revealed no significant interval changes of the size and morphology of the ectopic thyroid tissue, supporting a conservative management without rebiopsy or surgical resection of the ectopic thyroid tissue within the upper mediastinum.

3. Discussion

Ectopic thyroid tissue that coexists with a normally located orthotopic thyroid gland has been reported at equal incidences with ectopic thyroid occurring without a normally located gland. There are rare reports of dual ectopia or two separate foci of ectopic tissue in different locations [23]. Ectopic thyroid tissue should be suspected when a hyperdense mass within mediastinum is detected by CT chest. On CT ectopic thyroid tissue is typically identical in appearance to orthotopic thyroid tissue, a well-circumscribed homogeneous mass with increased attenuation (70 HU ± 10) relative to adjacent skeletal muscle due to iodine content and avidly enhancing on postcontrast images [23, 24]. In the clinical cases of suspected ectopic thyroid tissue, radionuclide imaging with technetium-99m pertechnetate, iodine-123, or iodine-131 is useful for functional assessment of radioiodine uptake by the suspected ectopic thyroid tissues. Retrospectively, it might be desirable to determine whether the hyperdense right paratracheal mass seen on CTA represented an ectopic thyroid tissue mass with I-123 SPECT/CT prior to biopsy. Prebiopsy I-123 SPECT/CT could be used for both functional and anatomic characterization of the right paratracheal mass and guiding tissue biopsy of the mass. Malignancy may occur within ectopic thyroid tissue with a variety of cell types reported. In contrast to ectopic thyroid gland neoplasms, the majority of tumors reported in lingual thyroid tissue are follicular while papillary forms are reported to comprise 23% [2]. Histological analysis of the biopsied tissue samples is desirable to determine benign versus malignancy of ectopic thyroid tissue. In the absence of significant symptoms of hyperthyroidism or mass effect from large ectopic thyroid tissue, management is typically conservative. Ectopic thyroid may be excised when mass effect becomes symptomatic or clinically significant or there is suspicion of malignancy [5, 15].

Majority of patients with ectopic thyroid presented with hypothyroidism which can be medically managed with thyroid hormone supplement. Rare cases of hyperthyroidism with histological features similar to Graves' disease have been reported [17, 18]. The patient of this case report was initially treated with 125 to 137 mg Levoxyl daily after total thyroidectomy. Because the patient's TSH and free T4 levels were not normalized or unstable, the dose of Levoxyl was reduced to 100 mg daily. However, the patient's TSH (0.09 mIU/L) and free T4 levels (2.58 ng/dL) were still abnormal prior to stopping oral administration of Levoxyl (100 mg, daily) in preparation for I-123 SPECT/CT. The results of thyroid functional tests were normal after the dose of Levoxyl was reduced to 75 mg daily after I-123 uptake by the benign mediastinal ectopic thyroid tissue was demonstrated by I-123 SPECT/CT. The results of abnormal thyroid functional tests (low TSH and high free T4) when the patient received 100 to 135 mg of Levoxyl might be related to combined effects of Levoxyl and thyroid hormone produced by the benign ectopic mediastinal thyroid tissues.

In summary, the findings from this case suggested that ectopic thyroid tissue should be suspected for differential diagnosis of incidental finding of hyperdense mediastinal mass on CT. I-123 SPECT/CT can be useful for both functional and anatomic characterization of suspected ectopic thyroid tissue to guide medical management of hypothyroidism, rarely hyperthyroidism, in the patients with ectopic thyroid tissues.

References

[1] J. Rosai, L. V. Ackerman, and J. Rosai, *Rosai and Ackerman's Surgical Pathology*, Mosby, Edinburgh, Scotland, 2011.

[2] N. A. Ibrahim and I. O. Fadeyibi, "Ectopic thyroid: etiology, pathology and management," *Hormones*, vol. 10, no. 4, pp. 261–269, 2011.

[3] G. Noussios, P. Anagnostis, D. G. Goulis, D. Lappas, and K. Natsis, "Ectopic thyroid tissue: anatomical, clinical, and surgical implications of a rare entity," *European Journal of Endocrinology*, vol. 165, no. 3, pp. 375–382, 2011.

[4] V. Triggiani, V. A. Giagulli, B. Licchelli et al., "Ectopic thyroid gland: description of a case and review of the literature," *Endocrine, Metabolic & Immune Disorders Drug Targets*, vol. 13, no. 3, pp. 275–281, 2013.

[5] M. P. Abdallah-Matta, P. H. Dubarry, J. J. Pessey, and P. Caron, "Lingual thyroid and hyperthyroidism: a new case and review of the literature," *Journal of Endocrinological Investigation*, vol. 25, no. 3, pp. 264–267, 2002.

[6] A. Gandhi, K. K. Wong, M. D. Gross, and A. M. Avram, "Lingual thyroid ectopia: diagnostic SPECT/CT imaging and radioactive iodine treatment," *Thyroid*, vol. 26, no. 4, pp. 573–579, 2016.

[7] H. Dominguez-Malagon, J. Guerrero-Medrano, and S. Suster, "Ectopic poorly differentiated (insular) carcinoma of the thyroid. Report of a case presenting as an anterior mediastinal mass," *American Journal of Clinical Pathology*, vol. 104, no. 4, pp. 408–412, 1995.

[8] S. Karapolat and I. Bulut, "Ectopic posterior mediastinal thyroid: a case report," *Cases Journal*, vol. 21, no. 1, p. 53, 2008.

[9] G. D. Gorur, S. Isgoren, Y. Z. Tan, Z. Utkan, H. Demir, and F. Berk, "Graves' disease in a patient with ectopic mediastinal thyroid," *Clinical nuclear medicine*, vol. 36, no. 11, pp. 1039–1040, 2011.

[10] A. D. Mace, A. Taghi, S. Khalil, and A. Sandison, "Ectopic sequestered thyroid tissue: an unusual cause of a mediastinal mass," *ISRN Surgery*, vol. 2011, Article ID 313626, 5 pages, 2011.

[11] F. Thuillier and J. Venot, "Ectopic thyroid tissue in the anterior mediastinum with a normally located gland: a case report," *Annales d'Endocrinologie*, vol. 73, no. 1, pp. 34–36, 2012.

[12] J. Wang and J. Fang, "Ectopic thyroid mass in the left lateral neck and anterior mediastinum: a case report," *Journal of Medical Case Reports*, vol. 8, article 351, 2014.

[13] K. K. Kamaleshwaran, F. Rajan, P. Asokumar, V. Mohanan, and A. S. Shinto, "Mediastinal ectopic benign colloid goitre detected using iodine-131 whole body scintigraphy and single-photon emission computed tomography-computed tomography," *Indian Journal of Nuclear Medicine*, vol. 30, no. 2, pp. 180–182, 2015.

Ectopic Thyroid Tissue in the Mediastinum Characterized by Histology and Functional...

205

[14] F. M. Cunha, E. Rodrigues, J. Oliveira, A. Saavedra, L. S. Vinhas, and D. Carvalho, "Graves' disease in a mediastinal mass presenting after total thyroidectomy for nontoxic multinodular goiter: a case report," *Journal of Medical Case Reports*, vol. 10, no. 1, article 878, 2016.

[15] C. Adelchi, P. Mara, L. Melissa, A. De Stefano, and M. Cesare, "Ectopic thyroid tissue in the head and neck: a case series," *BMC Research Notes*, vol. 7, article no. 790, 2014.

[16] M. Salvatori, V. Rufini, S. M. Corsello et al., "Thyrotoxicosis due to ectopic retrotracheal adenoma treated with radioiodine," *The Journal of Nuclear Biology and Medicine*, vol. 37, no. 2, pp. 69–72, 1993.

[17] S. Basaria and D. S. Cooper, "Graves' disease and recurrent ectopic thyroid tissue," *Thyroid*, vol. 9, no. 12, pp. 1261–1264, 1999.

[18] K. Kamijo, "Lingual thyroid associated with Graves' disease and Graves' ophthalmopathy," *Thyroid*, vol. 15, no. 12, pp. 1407–1408, 2005.

[19] D. A. Zander and W. R. K. Smoker, "Imaging of ectopic thyroid tissue and thyroglossal duct cysts," *Radiographics*, vol. 34, no. 1, pp. 37–50, 2014.

[20] C. N. B. Harisankar, G. R. Preethi, and M. George, "Hybrid SPECT/CT evaluation of dual ectopia of thyroid in the absence of orthotopic thyroid gland," *Clinical Nuclear Medicine*, vol. 37, no. 6, pp. 602–603, 2012.

[21] P. Joshi, V. Lele, and J. Kapoor, "Dual ectopic thyroid—non-invasive diagnosis on radionuclide thyroid scan with SPECT/CT correlation: a case report and brief review of literature," *Indian Journal of Endocrinology and Metabolism*, vol. 17, no. 2, pp. 359–361, 2013.

[22] J. C. Bulman, J. Wachsmann, and F. Peng, "Asymmetric radiotracer activity of enlarged cerebral spinal fluid space on radionuclide cisternography with SPET/CT," *Hellenic Journal of Nuclear Medicine*, vol. 19, no. 3, pp. 269–271, 2016.

[23] R. J. Hammond, K. Meakin, and J. E. Davies, "Case report: lateral thyroid ectopia—CT and MRI findings," *The British Journal of Radiology*, vol. 69, no. 828, pp. 1178–1180, 1996.

[24] D. Radkowski, J. Arnold, G. B. Healy et al., "Thyroglossal duct remnants: preoperative evaluation and management," *Archives of Otolaryngology—Head and Neck Surgery*, vol. 117, no. 12, pp. 1378–1381, 1991.

42

A Rare Case of Penile Metastases as a Harbinger of Primary Pulmonary Adenosquamous Carcinoma

Partha Hota ⓘ,[1] **Tejas N. Patel,**[2] **Xiaofeng Zhao,**[3] **Carrie Schneider,**[3] **and Omar Agosto**[1]

[1]*Division of Abdominal Imaging, Department of Radiology, Temple University Hospital, Philadelphia, PA, USA*
[2]*Division of Abdominal Imaging, Atlantic Medical Imaging, Galloway, NJ, USA*
[3]*Department of Pathology, Temple University Hospital, Philadelphia, PA, USA*

Correspondence should be addressed to Partha Hota; partha.hota@tuhs.temple.edu

Academic Editor: Suayib Yalcin

Although lung cancer has a high propensity for distant metastatic disease, penile metastases from primary lung neoplasms are considered particularly rare. A 71-year-old male presented to our hospital with a rapidly enlarging hard palpable penile mass. MR imaging demonstrated two penile masses centered in the left and right corpus cavernosa. Subsequent CT imaging revealed a spiculated pulmonary mass in the right upper lobe with PET/CT, MRI, and surgical exploration, demonstrating evidence of metastases to the left adrenal gland, right subscapularis muscle, brain, and small bowel. Tissue sampling of lesions in the small bowel, right subscapularis muscle, and penis demonstrated histopathology consistent with an adenosquamous carcinoma which in combination with the appearance of the right upper lobe mass on PET/CT imaging suggested the patient's lung cancer as the primary lesion. Prior to our case, pulmonary adenosquamous carcinoma metastasizing to the penis has only been reported once in the literature. Herein, we report a rare case of penile metastases as the presenting sign of metastatic pulmonary adenosquamous carcinoma characterized with PET/CT and MR imaging.

1. Introduction

Lung cancer is currently the leading cause of cancer mortality in the United States, contributing to approximately 25% of all cancer-related deaths [1]. Early diagnosis of metastatic disease is crucial for staging and treatment planning with common sites of extrathoracic metastases including the adrenal glands, liver, brain, and bone via hematogenous dissemination [2]. Although the penis has an abundant and highly complex vascular and lymphatic supply, metastatic disease to the penis is quite rare with just over 500 reported cases of penile metastases reported to date [3–5]. Of these patients, there have been only a total of 40 cases of penile metastases from primary lung cancer with the following histopathologic incidences: squamous cell carcinoma (63%), adenocarcinoma (18%), and a single reported case of adenosquamous carcinoma (ASC) [3]. Approximately one-third of penile metastases from lung cancer are detected at the time of primary tumor detection with the vast remainder detected several months later, often in end-stage disease [3].

Penile metastases are rarely detected prior to the diagnosis of a primary lung cancer with only four cases reported in the current literature [2, 3, 6, 7]. We report a rare case of penile metastases as the presenting sign of a metastatic ASC of the lung.

2. Case Report

A 71-year-old male with a history of hypertension and hyperlipidemia and a 70 pack-year smoking history presented to our hospital with a two-month history of an enlarging penile mass at the base of the shaft as well as a 10-pound weight loss. Physical examination demonstrated a hard, smooth, approximately 2 cm mass surrounding the base of the penile shaft. In addition, both testes were distended. No palpable pelvic lymphadenopathy was found and vital signs and laboratory data were within normal limits.

Penile magnetic resonance (MR) imaging demonstrated a 6 × 2.5 × 6 cm (AP × TV × CC) irregularly shaped mass centered in the left corpus cavernosum involving the

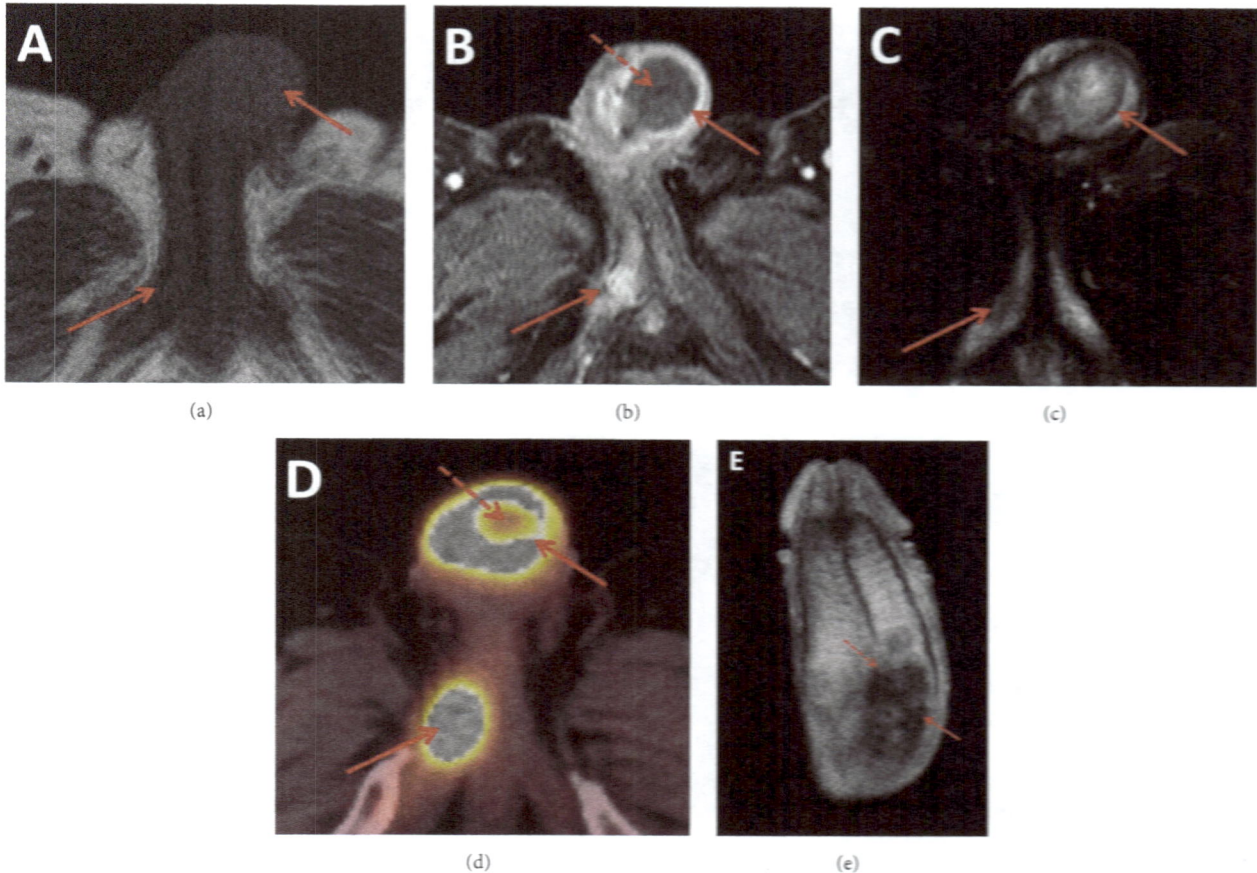

FIGURE 1: Axial T1-weighted MR images (a) demonstrate two isointense penile masses involving the left paracentral shaft base and proximal right corpus cavernosum (solid arrows) demonstrated with enhancement (solid arrow) on axial contrast enhanced T1-weighted MR images (b). The larger lesion at the base of the shaft demonstrates peripheral enhancement and central nonenhancement (dashed arrow), while the smaller posterior lesion homogeneously enhances. Both lesions demonstrate intrinsic heterogeneous hyperintensity (solid arrow) on axial T2-weighted MR imaging (c). Corresponding to the areas of enhancement on postcontrast MR images, PET/CT imaging (d) demonstrates hypermetabolic activity (solid arrow). A central area of decreased metabolic activity within the larger lesion (dashed arrow) corresponds to the area of central nonenhancement on MRI in keeping with central necrosis. Coronal contrast-enhanced T1-weighted MR image (e) demonstrates the larger lesion (solid arrow) extending across the intercorporal septum (dashed arrow).

proximal and mid aspects of the penile shaft extending across the intercorporal septum (Figure 1). This lesion demonstrated intrinsic isointensity on T1-weighted images and hyperintensity on T2-weighted images with peripheral enhancement and central hypovascularity. No extension into the corpus spongiosum or penile urethra was identified. A second lesion measuring 3×1 cm (AP $\times$ TV) demonstrating similar T1 and T2 signal characteristics but with more homogeneous enhancement was present in the more proximal right corpus cavernosum. Transcoporeal biopsy of the larger penile lesion was performed and subsequent immunostaining was positive for squamous cell markers (p63 and CK5/6) as well as an adenocarcinoma marker (mucicarmine) in keeping with an ASC with a predominantly squamous cell pattern (Figures 2(a) and 2(b)) [8].

Subsequent whole-body positron emission tomography/computed tomography (PET/CT) imaging demonstrated hypermetabolism of both penile lesions with a maximum standard uptake value (SUV$_{max}$) of 19.9 (Figures 1(d) and 3). In addition, a 4.3×2.1 cm (AP $\times$ TV) hypermetabolic spiculated pulmonary mass (SUV$_{max}$ = 10.7) was identified in the right upper lobe abutting the mediastinum with areas of central cavitation. The overall imaging features of this lung lesion were suggestive of a primary lung squamous cell carcinoma. Additional mass lesions with similar hypermetabolic activity were identified in the right subscapularis muscle (SUV$_{max}$ = 15) and left adrenal gland (SUV$_{max}$ = 15.3). No hypermetabolic lymphadenopathy was identified. Subsequent contrast-enhanced MR imaging of the brain demonstrated multiple supratentorial and infratentorial ring enhancing lesions in keeping with metastases (Figure 4). Percutaneous biopsy of the right subscapularis mass was performed demonstrating positive immunostaining with p63 and mucicarmine in a pattern similar to the penile masses. Given the similar immunostaining profile of the penile and right subscapularis lesions, the overall distribution of

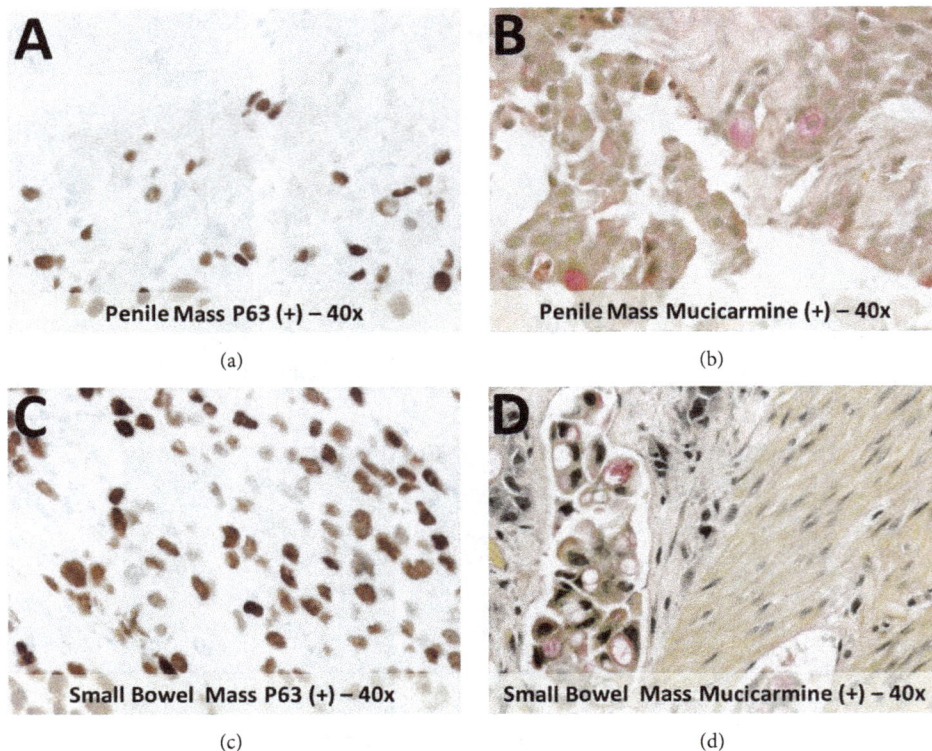

FIGURE 2: Immunohistochemical analysis of the penile mass demonstrates positive staining with an indicator for squamous cell carcinoma, p63 (a), and positive staining with an indicator for adenocarcinoma, mucicarmine (b). The small bowel mass demonstrates a similar immuno-staining profile and is positive with both p63 (c) and mucicarmine (d) immunostains.

metastases, and appearance of the right upper lobe mass, a unifying diagnosis of stage IV primary ASC of the lung with distant metastases was made.

Six months later, following chemotherapy, the patient returned to our hospital with severe abdominal pain with CT imaging demonstrating a perforated small bowel obstruction with a transition point in the mid-small bowel. The patient underwent explorative laparoscopy and partial small bowel resection. Immunostaining of the resected segment of small bowel demonstrated positive staining with p63 and muci-carmine strongly resembling the staining profile of the penile and subscapularis masses in keeping with metastatic spread to the small bowel (Figures 2(c) and 2(d)). The patient was subsequently discharged from the hospital and died 1 month later from cardiac arrest.

3. Discussion

Since secondary penile metastases were first reported by Eberth in 1870, a total of just over 500 cases of this rare entity have been reported in the literature [3, 9]. Of the primary malignancies attributed to penile metastases, approximately 70–75% originate from regional genitourinary or lower gastrointestinal organs of the pelvis [7, 10]. The remainder arise from extrapelvic organs and include tumors of the lungs (4–6.2%), upper gastrointestinal tract, kidneys, hematological system, and osseous structures [3, 4, 7]. Although lung

cancer is the second most common malignancy with the estimated 222,500 new cases diagnosed in 2017, secondary penile metastases from lung cancer are particularly rare with only 40 reported cases to date, with our case being the 41st [1, 3, 4, 7].

The mean age of secondary penile metastases is dependent on age of incidence of the primary malignancy and in the setting of the lung cancer is approximately 61 years of age [3, 10]. Approximately one-third of penile metastases from lung cancer are detected at the same time as the primary tumor with the remainder detected in advanced disease at a mean interval of 18 months following primary tumor detection [3, 11]. Penile metastases are rarely detected prior to the diagnosis of the primary lung cancer with only four cases with this presentation reported in the current literature, with our case being the fifth [2, 3, 6, 7].

Clinical manifestations vary widely with the most common symptom being the presence of a palpable penile mass which has been reported in 45% to 80% of patients with an average size of 3.5 cm [3, 5, 6, 10]. Low flow priapism secondary to occlusion of the venous plexus by tumor cells, the so-called "malignant priapism," has been reported in 20% to 53% of patients with secondary penile metastases and may be a particularly useful clinical tool in differentiating secondary penile metastases from primary penile malignancies as it is almost never observed in the latter [6, 10, 12]. Although penile pain and obstructive uropathy

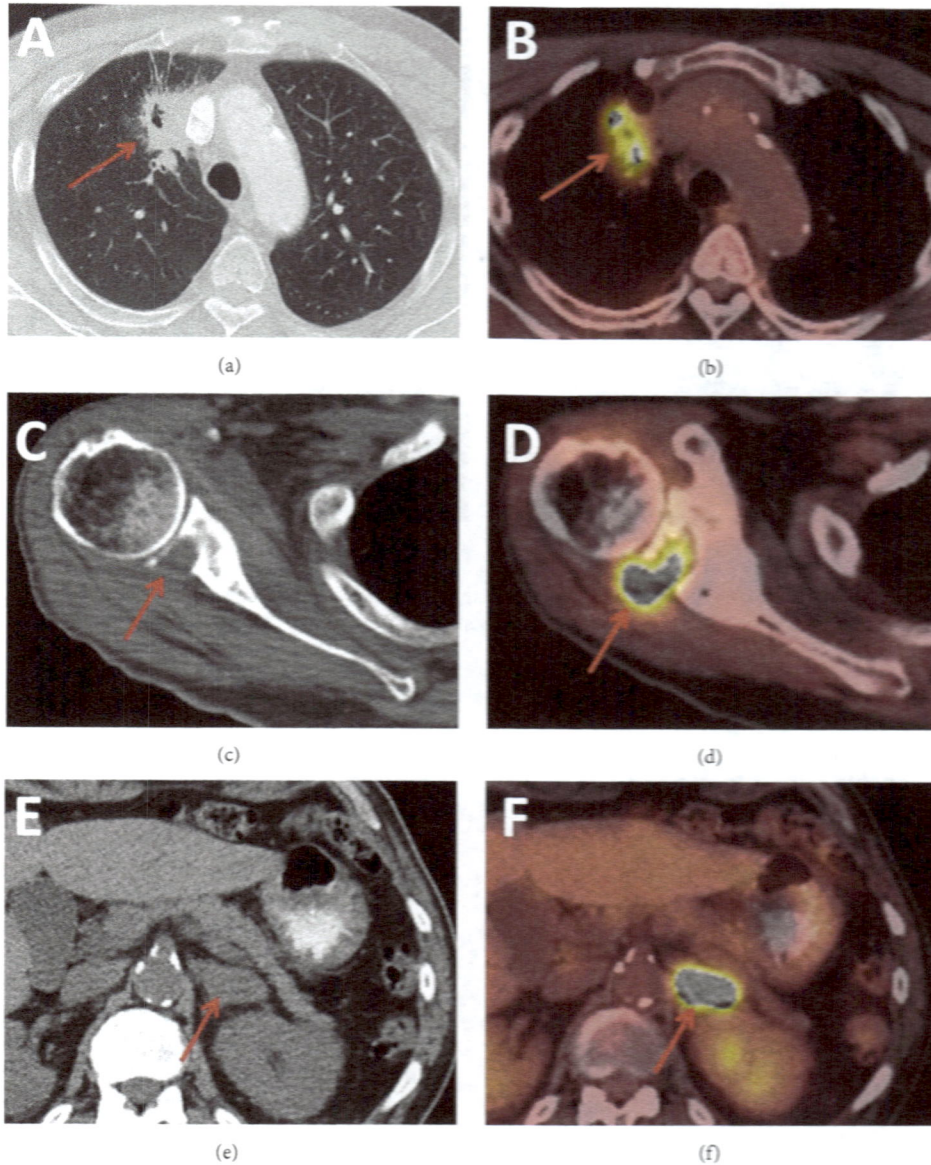

FIGURE 3: Axial noncontrast CT image in the lung window (a) demonstrates a spiculated mass with an area of central cavitation in the right upper lobe abutting the mediastinum (solid arrow) with imaging features suggesting a primary pulmonary squamous cell carcinoma. The corresponding PET/CT image (b) demonstrates hypermetabolic activity (solid arrow) within this lesion. Axial noncontrast CT image in the bone window (c) demonstrates a soft tissue lesion centered in the right subscapularis muscle with destruction of the adjacent bony glenoid (arrow). Corresponding PET/CT imaging (d) demonstrates hypermetabolic activity (solid arrow) within this lesion. Axial noncontrast CT image in the soft tissue window (e) demonstrates a solid lesion in the left adrenal gland (solid arrow) with the corresponding PET/CT image (f) demonstrating hypermetabolic activity (solid arrow) within this lesion.

are infrequently reported as initial symptoms, they may be present in advanced disease secondary to increased mass effect and infiltration, notably into the corpus spongiosum [10]. As patients presenting with penile metastases often have widespread metastatic disease, symptoms specific to the primary tumor are common, including dyspnea, cough, and weight loss in the setting of a primary lung malignancy [2, 3, 6].

Distant metastatic disease from primary lung cancer predominantly spreads secondary to hematogenous dissemination of malignant cells via the arterial route [10]. Counterintuitively, despite a robust vascular supply even during flaccidity, extensive vascular communication between the penis and the adjacent pelvis organs, and its location as an end organ of arterial, venous, and lymphatic systems, the overall incidence of secondary penile metastases remains quite low

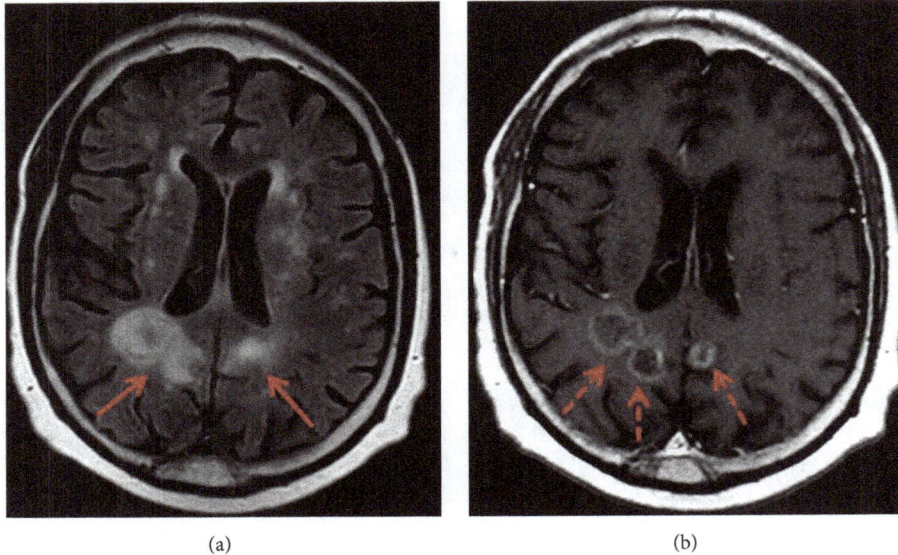

(a)   (b)

FIGURE 4: Axial T2-weighted FLAIR (a) and T1-weighted postcontrast MR images (b) demonstrate ring enhancing lesions in the bilateral parietal lobes (dashed arrows) at the level of the basal ganglia with surrounding vasogenic edema (solid arrows) in keeping with brain metastases.

[3, 4]. Several hypotheses have been proposed to explain the pathogenesis behind this enigma including an imperfect penile microenvironment as well as high flow through arterial and venous communications which may lead to difficulty in neoplastic seeding via hematogenous spread [10]. Decreased or retrograde venous flow may facilitate tumor seeding and may explain the relatively high propensity for penile metastases in the setting of neoplasm in neighboring pelvic organs which may cause obstruction of the pudendal plexuses and sluggish or reversed flow [10, 13].

The majority of the reported secondary penile metastases from lung cancer are of squamous cell carcinoma origin (23 of 40 cases) followed by adenocarcinoma origin (7 of 40 cases) [3]. In contrast, the incidence of pulmonary adenocarcinoma is greater than pulmonary squamous cell carcinoma with an incidence of 40% and 25–30% in the lung cancer population, and this discrepancy may be the result of a female-predominant incidence of pulmonary adenocarcinoma [6, 14]. Penile metastasis from pulmonary adenosquamous carcinoma is even rarer, which up until now has only been reported once in the literature, with our case being the second [3].

MR imaging is a particularly advantageous imaging modality in evaluating secondary penile metastases due to the superior soft tissue resolution, multiplanar functionality, and ability to accurately characterize disease extension [15]. In the setting of lung cancer, penile metastases most commonly involve the corpora cavernosa in the shaft with one meta-analysis reporting 85% of lesions in this location [3, 6, 10]. As was demonstrated in our case, both cavernosa are involved in the majority of cases which may be attributed to free communication through an incomplete intercorporal septum [3]. Penile metastases generally show hypointensity on T1-weighted imaging with a variable appearance on T2-weighted

imaging ranging from hypointense to slightly hyperintense compared to the cavernosa [10, 15]. Following administration of intravenous contrast, metastatic lesions typically avidly enhance [10, 15]. In lesions with increased metabolic activity, central areas of hypoenhancement may be present, reflecting areas of central necrosis, as was demonstrated in our case. In addition to a discrete mass, an infiltrative appearance has also been described [15].

PET/CT has been described as a valuable imaging tool in the setting of secondary penile metastases for detection of the primary malignancy, identification of additional sites of metastases, and the facilitation of staging, which may all be provided in a single examination [3, 16, 17]. Moreover, in asymptomatic patients, PET/CT may incidentally detect small penile metastases that would otherwise be difficult to assess on conventional CT imaging or physical examination [17]. Secondary penile metastases demonstrate similar metabolic activity to the primary neoplasm, as was seen in our case, where the penile metastases, primary lung lesion, adrenal lesion, and right subscapularis lesion all demonstrated similar hypermetabolic activity. To the best of our knowledge, there has been no study investigating SUV_{max} as a measure of distinguishing a primary penile malignancy versus metastases from a synchronous neoplasm; however, a meta-analysis may be difficult given the rarity of cases. In the posttreatment setting, PET/CT is particularly advantageous for disease restaging when assessing for treatment modification [16, 17].

Choice of treatment in the setting of metastatic penile cancer is multifactorial and is based on the type of cancer, size and number of metastatic lesions, age, and patient constitution [3, 6]. With the vast majority of patients with penile metastases presenting with widespread metastatic disease and a mean survival time of 4.5 to 5.5 months, treatment is usually palliative, involving local resection or radiation therapy to improve the overall quality of life [3, 6, 18]. Additional treatment strategies include chemotherapy as was chosen by the patient in our case [3, 6]. In the setting of urinary obstruction, suprapubic catheterization may be performed and in cases of a large mass producing intractable pain, penectomy has been reported [6].

In conclusion, this is a case of secondary penile metastases as the initial presentation of metastatic pulmonary ASC and is, to the best of our knowledge, only the second described case of this entity. Penile metastases are rarely identified prior to the diagnosis of the primary neoplasm. Given the rarity of secondary penile metastases from primary lung cancer and the poor prognosis following identification, knowledge of this atypical heralding lesion to prompt further assessment for end-stage extrapelvic metastatic malignancy is crucial when evaluating these patients.

References

[1] American Cancer Society, *Cancer Facts & Figures 2017*, American Cancer Society, Atlanta, GA, USA, 2017.

[2] S. Ozkaya, S. Findik, and A. G. Atıcı, "Penile metastasis as a first sign of lung cancer," *International Medical Case Reports Journal*, vol. 2, pp. 19–21, 2009.

[3] L.-C. Guo, G. Li, X.-M. Wang, M. Zhang, J.-A. Huang, and Y.-B. Chen, "Penile metastases from primary lung cancer: case report and literature review," *Medicine*, vol. 96, no. 26, Article ID e7307, 2017.

[4] F.-F. Zheng, Z.-Y. Zhang, Y.-P. Dai, Y.-Y. Liang, C.-H. Deng, and Y. Tao, "Metastasis to the penis in a patient with adenocarcinoma of lung, case report and literature review," *Medical Oncology*, vol. 26, no. 2, pp. 228–232, 2009.

[5] K. Zhang, J. Da, H.-J. Yao et al., "Metastatic tumors of the penis: a report of 8 cases and review of the literature," *Medicine (United States)*, vol. 94, no. 1, p. e132, 2015.

[6] D. E. Du Plessis, A. Van Der Merwe, and C. F. Heyns, "Penile metastases from primary bronchus carcinoma—a case report and literature review," *African Journal of Urology*, vol. 21, no. 1, pp. 57–61, 2015.

[7] F. Hizli and F. Berkmen, "Penile metastasis from other malignancies: a study of ten cases and review of the literature," *Urologia Internationalis*, vol. 76, no. 2, pp. 118–121, 2006.

[8] N. Rekhtman, D. C. Ang, C. S. Sima, W. D. Travis, and A. L. Moreira, "Immunohistochemical algorithm for differentiation of lung adenocarcinoma and squamous cell carcinoma based on large series of whole-tissue sections with validation in small specimens," *Modern Pathology*, vol. 24, no. 10, pp. 1348–1359, 2011.

[9] C. J. Eberth, "Krebsmetastasen des Corpus cavernosum penis," *Archiv für Pathologische Anatomie und Physiologie und für Klinische Medicin*, vol. 51, no. 1, pp. 145-146, 1870.

[10] L. Mearini, R. Colella, A. Zucchi, E. Nunzi, C. Porrozzi, and M. Porena, "A review of penile metastasis," *Oncology Reviews*, vol. 6, no. 1, article e10, pp. 80–87, 2012.

[11] U. Maier and M. Grimm, "Transitional cell carcinoma of the bladder with solitary metastasis to the penis 4 years after successful heart transplantation: a case report and review of the literature," *Transplantation*, vol. 58, no. 7, pp. 861–863, 1994.

[12] B. S. Abeshouse and G. A. Abeshouse, "Metastatic tumors of the penis: a review of the literature and a report," *The Journal of Urology*, vol. 86, pp. 99–112, 1961.

[13] W. T. Hayes and J. M. Young, "Metastatic carcinoma of the penis," *Journal of Chronic Diseases*, vol. 20, no. 11-12, pp. 891–895, 1967.

[14] The American Cancer Society, "What is non-small cell lung cancer?" 2017, https://www.cancer.org/cancer/non-small-cell-lung-cancer/about/what-is-non-small-cell-lung-cancer.html.

[15] T. Kendi, E. Batislam, M. M. Basar, E. Yilmaz, D. Altinok, and H. Basar, "Magnetic resonance imaging (MRI) in penile metastases of extragenitourinary cancers," *International Urology and Nephrology*, vol. 38, no. 1, pp. 105–109, 2006.

[16] H. Öztürk, "Evaluation of the response chemotherapy for penile metastasis of bladder cancer using 18F-fluorodeoxyglucose-PET/CT," *International Journal of Surgery Case Reports*, vol. 11, pp. 33–36, 2015.

[17] E. G. Spinapolice, C. Fuccio, B. Rubino et al., "Penile metastases from bladder and prostate cancer detected by pet/ct: a report of 3 cases and a review of literature," *Clinical Genitourinary Cancer*, vol. 12, no. 4, pp. e155–e159, 2014.

[18] N. Fujimoto, A. Hiraki, H. Ueoka, and M. Harada, "Metastasis to the penis in a patient with squamous cell carcinoma of the lung with a review of reported cases," *Lung Cancer*, vol. 34, no. 1, pp. 149–152, 2001.

43

Atypical Growth Pattern of an Intraparenchymal Meningioma

Zhen Zeng,[1] **Tijiang Zhang,**[1] **Yihua Zhou,**[2] **and Xiaoxi Chen**[1]

[1]Department of Radiology, Affiliated Hospital of Zunyi Medical College, Zunyi, China
[2]Department of Radiology, Saint Louis University School of Medicine, St. Louis, MO, USA

Correspondence should be addressed to Tijiang Zhang; tijzhang@163.com

Academic Editor: Salah D. Qanadli

Meningiomas are the most common primary nonneuroglial extra-axial neoplasms, which commonly present as spherical or oval masses with a dural attachment. Meningiomas without dural attachment are rare and, according to their locations, are classified into 5 varieties, including intraventricular, deep Sylvain fissure, pineal region, intraparenchymal, or subcortical meningiomas. To the best of our knowledge, intraparenchymal meningioma with cerebriform pattern has never been reported. In this paper, we report a 34-year-old Chinese male patient who presented with paroxysmal headaches and progressive loss of vision for 10 months and blindness for 2 weeks. A thorough physical examination revealed loss of bilateral direct and indirect light reflex. No other relevant medical history and neurologic deficits were noted. Computed tomography and magnetic resonance imaging scans showed an irregular mass with a unique cerebriform pattern and extensive peritumoral edema in the parietal-occipital-temporal region of the right cerebral hemisphere. The initial diagnosis was lymphoma. Intraoperatively, the tumor was completely buried in a sulcus in the parietal-occipital-temporal region without connecting to the dura. The histological diagnosis was intracranial meningioma based on pathological examination. Therefore, when an unusual cerebriform growth pattern of a tumor is encountered, an intraparenchymal meningioma should be considered as a differential diagnosis.

1. Introduction

Meningiomas are the most common primary extra-axial neoplasms in adults, accounting for up to 30% [1] of all primary intracranial tumors and as high as 35.5% in Asian and African people [2]. Generally speaking, meningiomas have a dural attachment. Meningiomas without dural attachment are rare and are more likely to occur in young males and are most commonly seen in the intraventricular and pineal regions and the Sylvain fissure [3–5]. Intraparenchymal meningioma is extremely rare with only few reported cases [6–8]. To the best of our knowledge, an intraparenchymal meningioma with a cerebriform pattern has never been reported. In this paper, we present a rare case of primary intraparenchymal meningioma with a cerebriform pattern, which was preoperatively misdiagnosed as lymphoma based on conventional CT and MR imaging features.

2. Case Presentation

A 34-year-old male patient presented with paroxysmal headaches and progressive vision loss for 10 months and blindness for 2 weeks. Physical examination revealed loss of bilateral direct and indirect light reflexes. No other neurologic deficits and laboratory abnormalities were noted. CT imaging showed an irregular isodense mass with a cerebriform appearance and extensive peritumoral edema in the parietal-occipital-temporal region of the right hemisphere (Figure 1). MR imaging demonstrated a cerebriform, irregular but well-defined mass in the parietal-occipital-temporal region of the right hemisphere. It was isointense on T1-weighted images and slightly hyperintense on T2-weighted images. There was intense homogeneous enhancement after contrast administration. The mass measured 46 cm × 80 cm × 51 cm and demonstrated significant mass effect, resulting in a 13 mm midline shift, compression of brainstem, and effacement of the occipital horn of the right lateral ventricle and ambient cistern (Figure 2). Unfortunately, advanced MRI imaging such as perfusion weighted imaging (PWI) and MR spectroscopy (MRS) was not performed on the initial MRI examination and the patient denied the suggestion of further imaging. Intraoperatively, the tumor was completely buried in a sulcus in the parietal-occipital-temporal region.

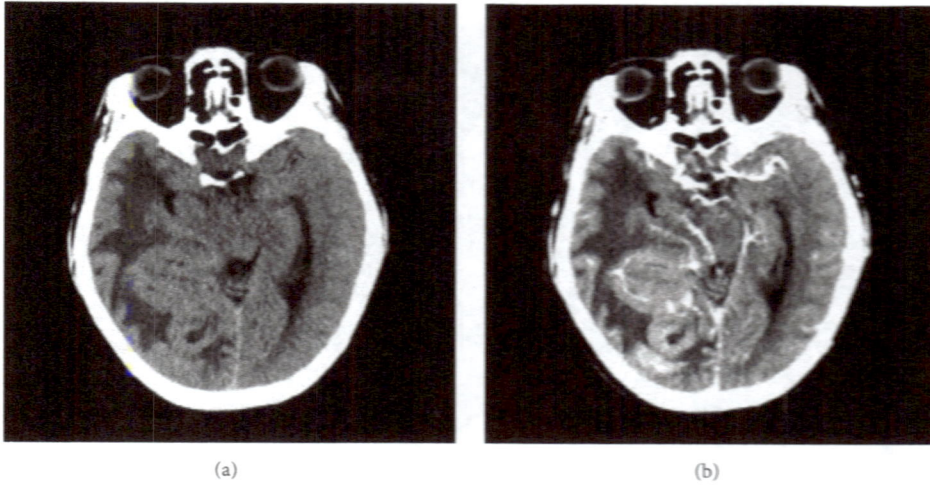

FIGURE 1: Axial CT images showed a cerebriform and irregular isodense mass with prominent perilesional vasogenic edema (a-b) in the right parietal-occipital-temporal region.

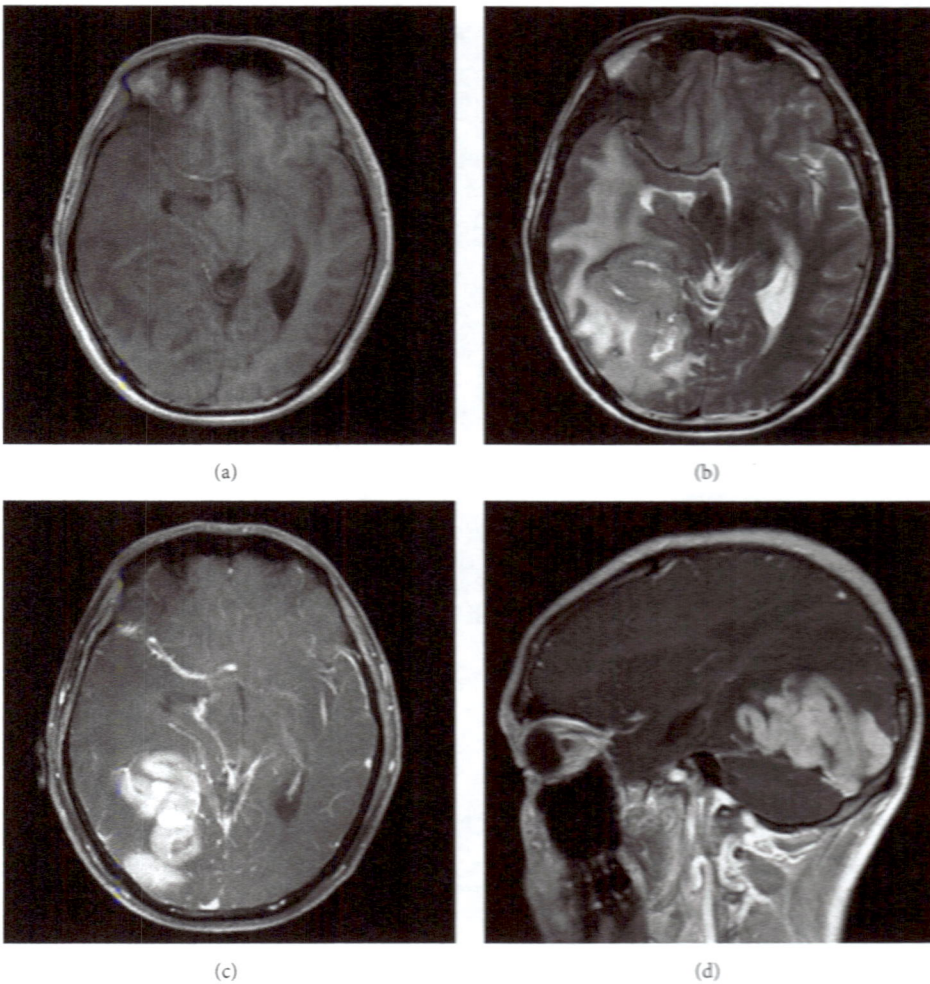

FIGURE 2: MR images from the same patient as shown in Figure 1. The mass was isointense on T1WI and slightly hyperintense on T2WI. Contrast-enhancement T1WI images showed prominent homogeneous enhancement. The tumor was buried in the cortical sulci without any dural attachment.

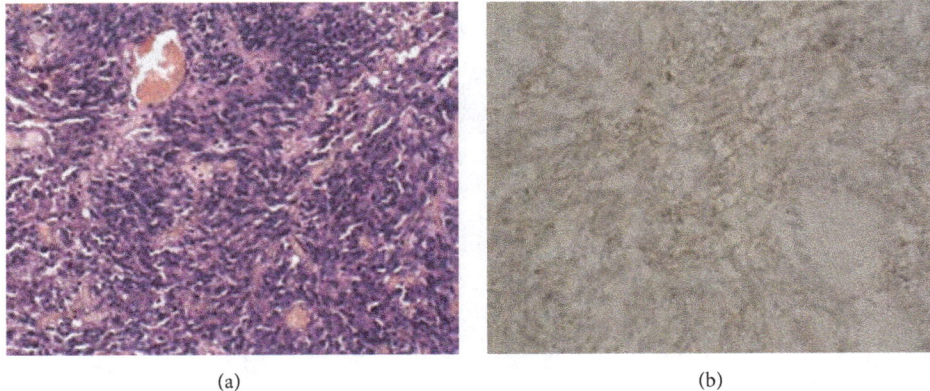

(a)   (b)

Figure 3: Histopathological manifestations of meningioma with cerebriform pattern. Hematoxylin-eosin staining (×100) revealed a large number of spindle cells rich in *Vascellum*, without distinct karyokinesis (a). Immunohistochemical staining (diaminobenzidine ×20) showed cells positive for epithelial membrane antigen (b).

The tumor was completely resected with careful dissection from the surrounding brain tissues. Macroscopically, the tumor showed rich blood supplies but without a dominant feeding artery. Histopathologically, the tumor showed a large number of spindle cells. There was no distinct karyokinesis. On immunohistochemical staining, the pathology specimen of this tumor was positive for epithelial membrane antigen (Figure 3) but negative for glial fibrillary acidic protein. Ki-67 was 1-2%. Based on the pathological findings, the tumor was diagnosed as a grade II intracranial meningioma according to classification criteria of World Health Organization (WHO). Postoperative radiotherapy was offered to the patient. However, the patient and his family denied the treatment. Twelve months after surgery, the patient returned to the hospital with progressive headaches and a CT examination (Figure 4) revealed recurrence. After surgical resection, the recurrent tumor was determined to be meningioma (WHO grade III).

3. Discussion

In this report, we presented a case of a large intraparenchymal meningioma that likely developed in a long period of time. Despite the significant mass effect caused by its large size and the extensive surrounding edema, no cognitive deficit was detected at presentation, which was likely due to compensation/accommodation to the slow growth of the tumor. However, the patient had lost both direct and indirect pupil light reflexes which may be secondary to optic atrophy resulting in slowly but progressively elevated intracranial pressure.

This case showed an unusual growth pattern of an intraparenchymal meningioma which was characterized by its cerebriform appearance and the involvement of cortical sulci without dural attachment. The etiology of meningiomas without dural attachment is unclear. Some authors presume that such meningiomas arise from arachnoid cells of the pia mater that stretches into the surface of brain or sulcus with perforating blood vessels, while others suggest that the arachnoid cells rest in the brain during the migration progress [3].

The meninges comprise three membranous layers: pia dura, arachnoid membrane, and dura mater. The arachnoidal villi, which are capped by the arachnoidal cells, may stretch into the sulci and the brain parenchyma along with the pia mater. We believe this may explain the cerebriform appearance of the meningioma in our case.

To the best of our knowledge, only 31 cases of intraparenchymal meningiomas have been reported [3–8]. This is the first case of intraparenchymal meningioma with a cerebriform pattern. It has been known that intraparenchymal meningiomas are more likely to occur in young males and can be located in all regions of the supratentorial brain. The most common presenting symptom is seizure followed by headache. Most intraparenchymal meningiomas are of fibrous type, whereas the extra-axial meningiomas are of the meningothelial type. The most common location is the frontal lobe followed by the parietal lobe. The imaging findings of intraparenchymal meningioma are nonspecific and most cases are misdiagnosed preoperatively.

Due to the absence of typical features on CT and MRI scans, the preoperative diagnosis of this disease is extremely difficult. Our case showed several findings leading to the diagnosis of lymphoma: the periventricular location, an isodense or slightly hyperdense mass on CT scans, and isointensity with intense enhancement on T1WI and T2WI. It has been reported that intraparenchymal meningioma may be misdiagnosed as glioma [3], metastatic tumor [6], and cavernoma [4]. Our case and others reported in the literature indicate that conventional imaging technology may not always be reliable in distinguishing intraparenchymal meningiomas from other intra-axial tumors. Advanced MRI techniques are important supplemental means in improving the reliability of preoperative diagnosis and differential diagnosis.

Given their vastly different treatments and prognosis, intraparenchymal cerebriform meningiomas should be differentiated from lymphoma and glioblastoma multiforme (GBM). Brain lymphoma accounts for 1–5% of all brain tumors [9]. It occurs in the brain parenchyma nearly

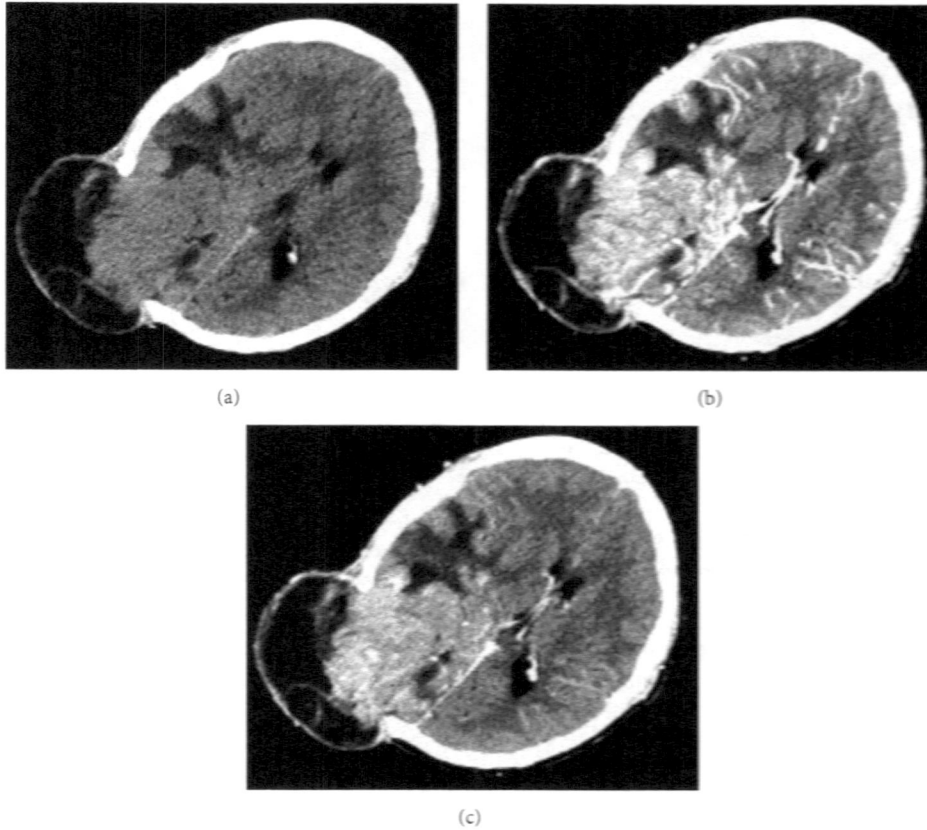

(a)

(b)

(c)

Figure 4: Follow-up CT images showed recurrent tumor herniating through the craniectomy defect. An axial CT image showed an irregular isodense mass with prominent perilesional vasogenic edema located in the right parietal-occipital-temporal lobe (a). Contrast-enhancement CT images showed intense homogeneous enhancement (b, c).

exclusively. Lymphomas have a predilection for the basal ganglia and periventricular and superficial regions. They appear to be isodense or hyperdense on CT and isointense on T1WI and T2WI images with intense enhancement but relatively mild surrounding edema. In immune compromised patients, CNS lymphomas usually have cystic changes and central necrosis. They may show the "clench fist sign" and "cusp angle sign" on contrast-enhancement images. CNS lymphomas typically demonstrate diffusion restriction on diffusion weighted imaging and hypoperfusion on MR PWI. An abnormal lipid peak and increased choline/creatine ratios can be seen on MRS [10]. As CNS lymphoma may be cured by chemotherapy and/or radiotherapy for its sensitive response to chemotherapeutic agent or radiation, when it is difficult to differentiate an intraparenchymal meningioma from a CNS lymphoma, stereotactic biopsy of lesion should be considered to obtain tissue diagnosis before proceeding to gross total resection of the tumor. GBM is the most common primary malignant brain tumor in adults. It usually occurs in the supratentorial white mater, particularly in the frontal and temporal lobes. Imaging characteristics of GBM include invasion of the corpus callosum and involvement of the contralateral hemisphere, producing the so-called "butterfly sign." Necrosis and cystic changes are common in GBM. Therefore, contrast-enhanced images often demonstrate

prominent irregular enhancement in the periphery of the tumor. The aggressive growth pattern of GBM is commonly accompanied by marginal tumor infiltration and extensive surrounding edema. The relative cerebral blood volume in meningioma is remarkably higher than that in GBM on PWI. In addition, meningiomas typically display the type III curve (postenhancement part is below the baseline) on signal intensity-time curve analysis, whereas GBM shows the type II curve (postenhancement part is above the baseline). On MRS, Alanine is commonly observed in meningioma. Based on these imaging features, the cerebriform meningioma may be differentiated from GBM and lymphoma.

In summary, in this paper, we present the cerebriform imaging characteristics of a rare intraparenchymal menin- gioma and review the imaging findings that could help us differentiate atypical meningiomas from other types of brain tumors such as CNS lymphoma and GBM. Advanced MRI techniques such as MRS and PWI can effectively improve the preoperative diagnosis of this tumor.

Ethical Approval

All procedures performed in studies involving human participants were in accordance with the ethical standards of the institutional research committee and with the 1964

Helsinki Declaration and its later amendments or comparable ethical standards.

Acknowledgments

This study was supported by National Natural Science Foundation of China (Grant no. 81360218) and Programs from Science and Technology Department of Guizhou Province (Grant no. LKZ(2011)38). The authors thank Heng Liu, M.M., for his help with revising the manuscript.

References

[1] M. J. Riemenschneider, A. Perry, and G. Reifenberger, "Histological classification and molecular genetics of meningiomas," *The Lancet Neurology*, vol. 5, no. 12, pp. 1045–1054, 2006.

[2] A. Das, W.-Y. Tang, and D. R. Smith, "Meningiomas in Singapore: demographic and biological characteristics," *Journal of Neuro-Oncology*, vol. 47, no. 2, pp. 153–160, 2000.

[3] T. Wada, M. Suzuki, T. Beppu et al., "A case of subcortical meningioma," *Acta Neurochirurgica*, vol. 142, no. 2, pp. 209–213, 2000.

[4] S. Jadik, A. C. Stan, U. Dietrich, T. A. Pietilä, and A. E. Elsharkawy, "Intraparenchymal meningioma mimicking cavernous malformation: a case report and review of the literature," *Journal of Medical Case Reports*, vol. 8, article 467, 2014.

[5] D. Bansal, P. Diwaker, P. Gogoi, W. Nazir, and A. Tandon, "Intraparenchymal angiomatous meningioma: a diagnostic dilemma," *Journal of Clinical and Diagnostic Research*, vol. 9, no. 10, pp. ED07–ED08, 2015.

[6] I. H. Tekkök, L. Cinel, and S. Zorludemir, "Intraparenchymal meningioma," *Journal of Clinical Neuroscience*, vol. 12, no. 5, pp. 605–608, 2005.

[7] X.-B. Jiang, C. Ke, Z.-A. Han et al., "Intraparenchymal papillary meningioma of brainstem: case report and literature review," *World Journal of Surgical Oncology*, vol. 10, article 10, 2012.

[8] K. Nayil, R. Makhdoomi, R. Malik, and A. Ramzan, "Intraparenchymal anaplastic meningioma in a child: a rare entity," *Asian Journal of Neurosurgery*, vol. 10, no. 2, pp. 111–113, 2015.

[9] I. S. Haldorsen, B. K. Krossnes, J. H. Aarseth et al., "Increasing incidence and continued dismal outcome of primary central nervous system lymphoma in Norway 1989–2003: time trends in a 15-year national survey," *Cancer*, vol. 110, no. 8, pp. 1803–1814, 2007.

[10] I. S. Haldorsen, A. Espeland, and E.-M. Larsson, "Central nervous system lymphoma: characteristic findings on traditional and advanced imaging," *American Journal of Neuroradiology*, vol. 32, no. 6, pp. 984–992, 2011.

44

Bilateral Simultaneous Pseudoangiomatous Stromal Hyperplasia of the Breasts and Axillae: Imaging Findings with Pathological and Clinical Correlation

Afsaneh Alikhassi,[1] Fereshteh Ensani,[2] Ramesh Omranipour,[3] and Alireza Abdollahi[2]

[1]Department of Radiology, Cancer Institute, Tehran University of Medical Sciences, Tehran, Iran
[2]Department of Pathology, Cancer Institute, Tehran University of Medical Sciences, Tehran, Iran
[3]Division of Surgical Oncology, Cancer Institute, Tehran University of Medical Sciences, Tehran, Iran

Correspondence should be addressed to Afsaneh Alikhassi; afsanehalikhassi@yahoo.co.uk

Academic Editor: Stefania Rizzo

Pseudoangiomatous stromal hyperplasia (PASH) of the breast is a pathology that is usually diagnosed by accident during pathological examination of other breast lesions. PASH is an uncommon and benign tumoral lesion of the mammary stroma that can be pathologically mistaken for other tumours, such as phyllodes, fibroadenoma, and sometimes even angiosarcoma. We report the case of a 45-year-old woman with complaints of huge bilateral breast enlargement. This is a rare case of PASH presenting with gigantomastia and involving bilateral breasts and axillae simultaneously. Mammography, ultrasonography, and MRI features are illustrated with histopathological correlation.

1. Introduction

First described in 1986 by Vuitch et al. [1], pseudoangiomatous stromal hyperplasia (PASH) of the breast is an uncommon pathology that is usually diagnosed incidentally upon pathological examination of other benign lesions, and it rarely occurs as a growing lump [2, 3]. Huge nodular PASH in the breast and simultaneous axillary tumoral PASH are extremely rare [4, 5]. Only rarely, like in this case, PASH may present with a simultaneous bilateral enlargement of the breast [6, 7]. It is usually found in women, but male cases have been reported [8, 9]. Although its etiology and pathogenesis are still unclear, it is generally thought that PASH represents a neoplastic process of myofibroblasts and the correlations between PASH and hormonal stimuli have been widely discussed in the literature [10].

Upon gross examination, PASH is usually well encapsulated, sometimes lobulated, and usually oval. Histologically, PASH can pose differential diagnostic problems, especially with benign and malignant vascular lesions. As PASH may occasionally adopt a solid fascicular growth pattern, it can be confused with pure mesenchymal spindle cell lesions (myofibroblastoma; leiomyoma; and fibromatosis) or fibroepithelial lesions (fibroadenoma; hamartoma; and phyllodes tumors) containing a spindle cell component [11, 12]. Most patients are premenopausal women, but there are also reports of PASH cases in adolescent girls [13]. PASH in males has also been reported with incidence in as many as 47% of gynecomasty cases [8, 9]. Final diagnosis requires biopsy, and the treatment of choice is excision with wide margins because the rate of local recurrence is quite high [14]. However, it seems to have no association with malignancy, and it does not seem to be a premalignant lesion [15].

2. Case Presentation

A 45-year-old woman was referred to our breast clinic due to slow growing bilateral breast enlargement since 2 years priorly, with more growth on the left, causing significant asymmetry (Figure 1). In clinical examination, both breasts were enlarged with multiple lumps which were soft and mobile upon palpation. The overlying skin was thickened and

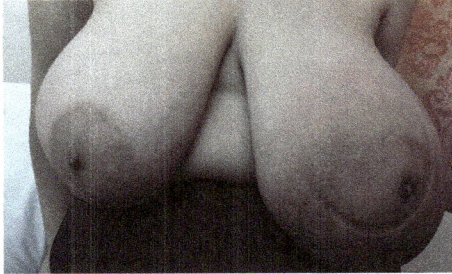

FIGURE 1: Breasts asymmetric enlargement in a patient with bilateral pseudoangiomatous stromal hyperplasia of the breasts.

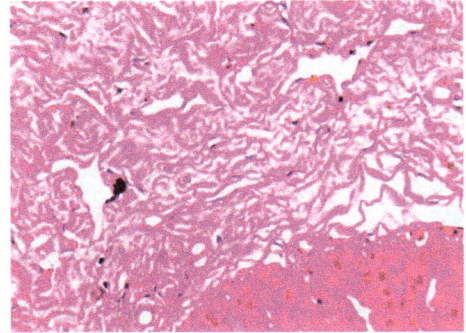

FIGURE 2: Breast tissue with expanded stroma and scattered ducts in our patient with PASH.

erythematous on the left side. Soft lumps were also palpable in both axillae. The patient did not have any systemic disease, was premenopausal, had not received any hormonal therapy, and had no family history of breast cancer.

The patient underwent mammography, ultrasound examination, and MRI of the breasts. To obtain samples for pathology, we performed a 14-gauge core needle biopsy of the right breast under ultrasound guidance and excisional biopsy of the left breast. PASH was diagnosed in both methods. Breast tissue with expanded stroma and scattered ducts were seen and immunohistochemical analyses were performed using antibodies against vimentin, CD34, CD31, and alpha-smooth muscle actin. The cells lining the pseudovascular spaces were stained with vimentin and alpha-smooth muscle actin, while they were negative to CD31. These immunohistochemical findings were consistent with the myofibroblastic nature of the cells, supporting the diagnosis of PASH. The patient was informed about the standard treatment of local excision with free margin. The conditions warranted left breast skin-sparing mastectomy, so she refused to undergo surgery, but she wanted to have more time to think about it and is now under observation. Informed consents and permissions were obtained from the patient.

3. Discussion

3.1. Clinical Presentation. PASH patients can present with mass palpation, abnormality in screen mammography, or incidental pathology finding [16, 17].

3.2. Etiology and Risk Factors. The etiological factors of PASH are unknown. A hormone-dependent etiology of PASH has been strongly suggested [11], mostly because of its occurrence in premenopausal women and postmenopausal females receiving hormone replacement therapy, as well as the presence of estrogen and progesterone receptors in most cases of PASH. A well-accepted hypothesis is that the stromal hyperplasia in PASH results from an exaggerated responsiveness of myofibroblasts to hormonal stimuli [1, 10, 12].

3.3. Pathology. PASH is not a rare pathological diagnosis [15] and can accompany other breast pathologies. However, the tumoral form of PASH is rare. The microscopic appearance of PASH is similar to endothelial spindle cells forming vessel-like slits within the stroma, which are not true vessels covered

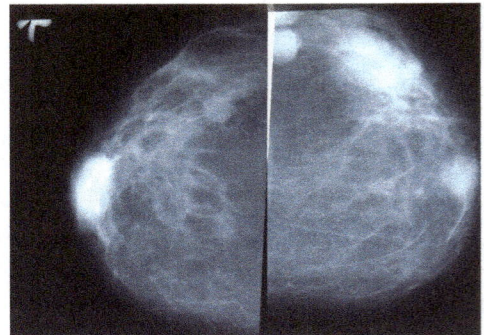

FIGURE 3: Analogue mammography of the patient two years priorly showing multiple well-defined oval masses or partially defined masses in both breasts.

with endothelia but are vacant spaces bordered by myofibroblasts (Figure 2). The slit-like channels may be mistaken for a low-grade angiosarcoma. However, angiosarcoma can be differentiated based on malignant cytologic features and positive immunohistochemical staining of endothelial markers [14]. Coexisting or subsequent development of carcinoma at the site of PASH and subsequent diagnosis of malignancy in the opposite breast are all possible [1].

3.4. Imaging. Jones et al. reported mammographically detected abnormality in 78% of cases [17], but another study [18] reported that 69% of patients with PASH presented no mammographic abnormality. However, that series included cases in which PASH was incidentally noted histologically. Mammographic features of PASH can be quite nonspecific and can show a wide range of variation. In mammography, PASH can appear as a partially or well-circumscribed mass or as an asymmetric density. Mass is the most prevalent sign in mammography [13, 16, 19]. In the present case, multiple well-defined and partially ill-defined masses had been seen in both breasts two years priorly, which increased significantly in size in subsequent mammography (Figures 3 and 4).

3.5. Sonography. Most patients with dominant PASH pathology have detectable lesions in sonography [17]. Jones et al. reported that the most common sonographic sign is

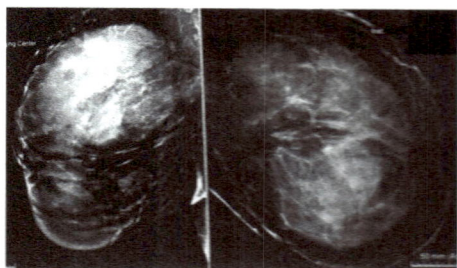

FIGURE 4: Recent digital mammography of the same patient. Bilateral masses were significantly enlarged compared with previous mammography.

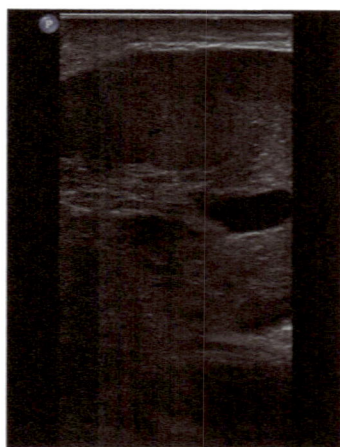

FIGURE 5: Sonography of the patient's left breast showing hypoechoic mass with solid appearance and multiple cystic changes.

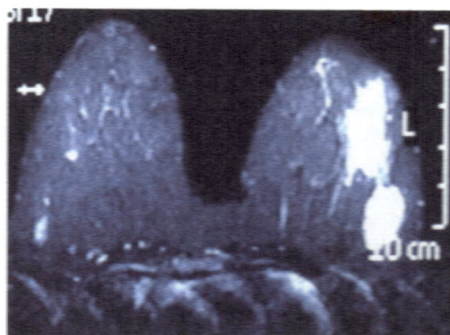

FIGURE 6: MRI with contrast of the patient shows two intensely enhanced masses with irregular borders in left breast lateral part.

a circumscribed oval hypoechoic mass [17]. PASH findings in sonography are usually nonspecific and are more suggestive of a benign lesion [20–22]. Coexisting fibrocystic changes can result in more heterogeneous appearance [5]. A less commonly reported morphology is a heterogeneous or echogenic area with hypoechoic central areas [21]. Piccoli et al. described this finding in association with a developing focal asymmetry in mammography [21]. Another reported PASH morphology is a mass with irregular or poorly defined borders, which mandates biopsy to rule out malignancy [17].

In our patient, multiple similar hypoechoic masses with solid appearance and prominent internal cystic changes were seen bilaterally and in both axillae (Figure 5). The simultaneous bilateral and biaxillary involvement makes this case very notable and rare. It seems that cystic changes inside these solid masses could be an important sonographic sign for nodular PASH.

3.6. MRI. The appearance of PASH has been described in very few cases of MRI [23]. Our patient shows multiple masses with low signal in T1 sequences and high signal in T2 sequences with irregular border or microlobulated border throughout both breasts. This was observed with early homogenous and intense enhancement (Figure 6) with all three types of enhancing curves that are more common for

types I (persistent) and II (plateau). Malignant disease could not be excluded in this patient with MRI, which prompted biopsy.

3.7. Treatment. Most often, PASH is stable over time [24], but it may increase rapidly in size or recur. High PASH recurrence without complete surgery has been reported [12, 25]. Surgery may warrant a lumpectomy or even a skin-sparing mastectomy. Tumor histology, cosmetic results, and patient preferences should be considered before performing the operation or referring the patient for follow-up at a reference center for breast cancers.

4. Conclusion

PASH is a rare benign lesion that can have variable imaging features and may mimic malignant or other benign conditions. Diagnosis requires biopsy. Follow-up or surgery may be warranted, depending on tumoral histology, cosmetic results, risk factors, and patient preference.

References

[1] M. F. Vuitch, P. P. Rosen, and R. A. Erlandson, "Pseudoangiomatous hyperplasia of mammary stroma," *Human Pathology*, vol. 17, no. 2, pp. 185–191, 1986.

[2] N. Taira, S. Ohsumi, K. Aogi et al., "Nodular pseudoangiomatous stromal hyperplasia of mammary stroma in a case showing rapid tumor growth," *Breast Cancer*, vol. 12, no. 4, pp. 331–336, 2005.

[3] E. Bowman, G. Oprea, J. Okoli et al., "Pseudoangiomatous stromal hyperplasia (PASH) of the breast: a series of 24 patients," *Breast Journal*, vol. 18, no. 3, pp. 242–247, 2012.

[4] A. C. Jordan, S. Jaffer, and S. E. Mercer, "Massive nodular pseudoangiomatous stromal hyperplasia (PASH) of the breast arising simultaneously in the axilla and vulva," *International Journal of Surgical Pathology*, vol. 19, no. 1, pp. 113–116, 2011.

[5] T. R. Shimpi, V. Baksa Reynolds, S. Shikhare, S. Srinivasan, M. J. Clarke, and W. C. Peh, "Synchronous large tumoral pseudoangiomatous stromal hyperplasia (PASH) in the breast and axilla with subsequent carcinoma in the contralateral breast: routine and strain imaging with histopathological correlation," *BJR Case Reports*, vol. 1, no. 3, Article ID 20150017, 2015.

[6] E. M. Ryu, I. Y. Whang, and E. D. Chang, "Rapidly growing bilateral pseudoangiomatous stromal hyperplasia of the breast," *Korean Journal of Radiology*, vol. 11, no. 3, pp. 355–358, 2010.

[7] H. S. Go and S. K. Jeh, "Radiologic imaging findings of bilateral infiltrating pseudoangiomatous stromal hyperplasia of the breasts: a case report," *Journal of the Korean Society of Radiology*, vol. 68, no. 4, pp. 329–332, 2013.

[8] S. Badve and J. P. Sloane, "Pseudoangiomatous hyperplasia of male breast," *Histopathology*, vol. 26, no. 5, pp. 463–466, 1995.

[9] M. F. G. Milanezi, F. P. Saggioro, S. G. Zanati, R. Bazan, and F. C. Schmitt, "Pseudoangiomatous hyperplasia of mammary stroma associated with gynaecomastia," *Journal of Clinical Pathology*, vol. 51, no. 3, pp. 204–206, 1998.

[10] C. Anderson, A. Ricci Jr., C. A. Pedersen, and R. W. Cartun, "Immunocytochemical analysis of estrogen and progesterone receptors in benign stromal lesions of the breast: evidence for hormonal etiology in pseudoangiomatous hyperplasia of mammary stroma," *The American Journal of Surgical Pathology*, vol. 15, no. 2, pp. 145–149, 1991.

[11] Y. Huang and F. Chen, "Review of myofibroblastoma of breast and its most common mimickers," *American Chinese Journal of Medicine and Science*, vol. 5, no. 1, pp. 38–42, 2012.

[12] C. M. Powell, M. L. Cranor, and P. P. Rosen, "Pseudoangiomatous stromal hyperplasia (PASH): a mammary stromal tumor with myofibroblastic differentiation," *The American Journal of Surgical Pathology*, vol. 19, no. 3, pp. 270–277, 1995.

[13] K. W. Gow, J. K. Mayfield, D. Lloyd et al., "Rapidly growing nodular pseudoangiomatous stromal hyperplasia of the breast in an 18-year-old girl. Case report," *The American Surgeon*, vol. 70, pp. 605–609, 2004.

[14] D. J. Spitz, V. B. Reedy, and P. Gattuso, "Fine-needle aspiration of pseudoangiomatous stromal hyperplasia of the breast," *Diagnostic Cytopathology*, vol. 20, no. 5, pp. 323–380, 1999.

[15] A. C. Degnim, M. H. Frost, D. C. Radisky et al., "Pseudoangiomatous stromal hyperplasia and breast cancer risk," *Annals of Surgical Oncology*, vol. 17, no. 12, pp. 3269–3277, 2010.

[16] M. R. Polger, C. M. Denison, S. Lester, and J. E. Meyer, "Pseudoangiomatous stromal hyperplasia: mammographic and sonographic appearances," *American Journal of Roentgenology*, vol. 166, no. 2, pp. 349–352, 1996.

[17] K. N. Jones, K. N. Glazebrook, and C. Reynolds, "Pseudoangiomatous stromal hyperplasia: imaging findings with pathologic and clinical correlation," *American Journal of Roentgenology*, vol. 195, no. 4, pp. 1036–1042, 2010.

[18] G. C. Hargaden, E. D. Yeh, D. Georgian-Smith et al., "Analysis of the mammographic and sonographic features of pseudoangiomatous stromal hyperplasia," *American Journal of Roentgenology*, vol. 191, no. 2, pp. 359–363, 2008.

[19] M. A. Cohen, E. A. Morris, P. P. Rosen, D. D. Dershaw, L. Liberman, and A. F. Abramson, "Pseudoangiomatous stromal hyperplasia: mammographic, sonographic, and clinical patterns," *Radiology*, vol. 198, no. 1, pp. 117–120, 1996.

[20] R. Salvador, J. L. Lirola, R. Dominguez, M. Lopez, and N. Risueno, "Pseudo-angiomatous stromal hyperplasia presenting as a breast mass: imaging findings in three patients," *Breast*, vol. 13, no. 5, pp. 431–435, 2004.

[21] C. W. Piccoli, S. A. Feig, and J. P. Palazzo, "Developing asymmetric breast tissue," *Radiology*, vol. 211, no. 1, pp. 111–117, 1999.

[22] A. T. Stavros, "Ultrasound of solid breast nodules: distinguishing benign from malignant," in *Breast Ultrasound*, A. T. Stavros, C. L. Rapp, and S. H. Parker, Eds., pp. 445–527, Lippincott Williams & Wilkins, Philadelphia, Pa, USA, 2004.

[23] R. K. Virk and A. Khan, "Pseudoangiomatous stromal hyperplasia: an overview," *Archives of Pathology and Laboratory Medicine*, vol. 134, no. 7, pp. 1070–1074, 2010.

[24] J. Jędrys, L. Rudnicka-Sosin, and W. Nowak, "Pseudoangiomatous stromal hyperplasia of the breast—case reports of two patients treated conservatively with long follow up and a literature review," *Nowotwory Journal of Oncology*, vol. 58, no. 2, pp. 100–102, 2008.

[25] C. L. Mercado, S. A. Naidrich, D. Hamele-Bena, S. A. Fineberg, and S. S. Buchbinder, "Pseudoangiomatous stromal hyperplasia of the breast: sonographic features with histopathologic correlation," *Breast Journal*, vol. 10, no. 5, pp. 427–432, 2004.

45

Acute Abdominal Pain Caused by an Infected Mesenteric Cyst in a 24-Year-Old Female

Davy R. Sudiono,[1] **Joep B. Ponten,**[2] **and Frank M. Zijta**[1]

[1]*Department of Radiology, Medical Center Haaglanden, Lijnbaan 32, 2512 VA Den Haag, Netherlands*
[2]*Department of Surgery, Medical Center Haaglanden, Lijnbaan 32, 2512 VA Den Haag, Netherlands*

Correspondence should be addressed to Davy R. Sudiono; d.sudiono@mchaaglanden.nl

Academic Editor: Yoshito Tsushima

A mesenteric cyst is a rare cause for abdominal pain. This umbrella term includes cystic entities which reside in the mesentery. We present a case of an infected false mesenteric cyst in a 24-year-old female patient without prior surgery or known trauma. Mainstay of treatment involves surgical resection, although less invasive treatments have been described. Prognosis depends on the origin of the cyst.

1. Introduction

The mesenteric cyst is a rare entity. It has been reported as an umbrella term that comprises any cystic mass which manifests within the mesentery, including lymphangioma, benign and malignant cystic lymphangioma, enteric duplication cyst, dermoid cyst, and pseudocyst [1]. Mesenteric cysts are often asymptomatic; however, if infected or ruptured patients may present with acute abdominal pain. This lack of specific symptoms and moreover its rare occurrence make diagnosis and consequently clinical decision challenging. Ultrasound (US) and contrast-enhanced computed tomography (CT) are decisive in distinguishing a wide variety of potential disorders that may result in acute abdominal pain. In case of a symptomatic mesenteric cyst a complete enucleation or resection either with laparoscopy or with laparotomy surgery is considered treatment of choice [2, 3]. We report a case of an infected mesenteric cyst in a 24-year-old woman presenting with severe abdominal pain and signs of infection, without previous abdominal surgery. Additionally we discuss relevant literature regarding histopathological differentiation and categorization as well as potential treatment options.

2. Case Presentation

A 24-year-old female without previous abdominal surgery or relevant medical history presented at her general practitioner because of severe abdominal pain, principally in the right lower quadrant, gradually increasing over the last 4 days. In addition she complained about lower back pain and feeling feverish. Consequently, she was admitted to the emergency department for advanced assessment. On physical examination at the time of admission, the patient experienced severe pain if palpitated in the right lower abdomen. Her vital signs were stable and cardiac and pulmonary examination were normal. She was afebrile; however, this could also be attributed to the use of anti-inflammatory drugs (NSAIDs). Laboratory results confirmed inflammation with an elevated C-reactive Protein (CRP) of 162 mg/L and a normal leucocytes amount of 9.8×10^9/L. Urinalysis was normal. As a result acute appendicitis was suspected and imaging evaluation was performed.

Ultrasound revealed a large, moderately anechoic, multilocular mass in the right hemiabdomen with echogenic components extending from the caudal border of the liver into the

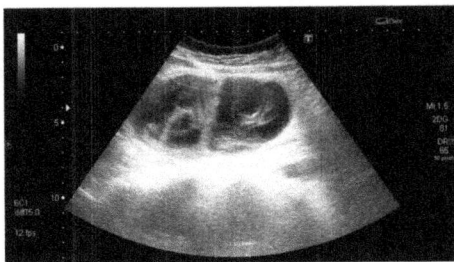

FIGURE 1: Ultrasound demonstrated a large abdominal, multilocular cystic mass. Notice the thick septations and echogenic content suspicious of a complicated cyst.

right lower quadrant (Figure 1). Due to difficulties in defining the origin of the cystic mass, a complementary contrast-enhanced CT of the complete abdomen was performed. CT confirmed a well-encapsulated, intramesenteric, multilocular cystic mass in close proximity to the ascending colon, inferior part of the duodenum, and segmental mesenteric vessels (Figure 2(a)). The cystic mass with a mean CT-attenuation value of 10 Hounsfield Units (HU) showed no enhancing, solid components. The superior border of the mass showed distinctive, perifocal fat stranding. In accordance with ultrasound findings, a normal appearing right ovary was observed in the right lower abdomen, as well as a normal appearing appendix. Consequently, based on the radiologic findings an infected mesenteric cyst was suggested.

The patient retained severe, unceasing abdominal pain; consequently laparotomy was proposed. At surgery, enucleation of the cystic mass from the mesentery was attempted; however, due to the intraoperative findings of enlarged lymph nodes and the diffuse mesenteric infiltration in close relationship with adjacent ascending colon, an ileocecal resection was performed. Intraoperatively the cyst was partially opened and did not contain pus but rather greasy, brown, hemorrhagic-like material (Figure 2(b)). Macroscopically there seemed to be no relation with any muscular layer of the bowel wall. Fluid material was not sent for examination; therefore, bacterial growth was not determined.

Microscopic examination of the specimen showed lining of the cystic wall with a flatted cell layer. Additional staining with mesothelial markers and endothelial markers was negative (Figure 2(c)). The peripheral border of the mass contained a fibrous moderately cell-rich wall with inflammatory, infiltrating changes. The cystic fluid consisted of protein rich fluids, combined with fibrin, inflammatory cells, and blood components. Furthermore multiple smaller cysts were noted in the mesenteric fat (Figure 3). Based on these findings a false infected mesenteric cyst was histopathologically confirmed. The patient recovered uneventfully and was discharged in good health three days after surgery.

3. Discussion

Mesenteric cysts are one of the rarest abdominal tumors with less than thousand cases reported in literature, with a reported incidence rate of approximately 1/100.000 admissions in adults [2, 4]. They seem to occur slightly more frequently in females and are found in almost all age groups with the highest incidence in the fourth decade of life [5]. Mesenteric cysts are often asymptomatic and are frequently incidental findings in routine abdominal imaging studies. However, if acute symptoms occur, that is, in case of complications, surgical resection is often required [5].

Symptoms are variable and rather nonspecific in case of a complicated cyst. Complications include infection, rupture, hemorrhage, and torsion. If large, the cyst can cause intestinal obstruction or bowel ischaemia on account of bowel compression or arterial compression, respectively [6, 7]. Although very rare, malignant degeneration has been described [7].

Ultrasound and contrast-enhanced abdominal CT are crucial in the diagnostic workup. The US appearance is reported to be diverse but should be considered if an avascular oval mesenteric mass is visualized [8]. CT is helpful for detecting signs of infection, rupture, or internal bleeding: a thickened enhancing cyst wall with perifocal fatty stranding is suggestive of such complications. On the other hand, US is superior to CT in demonstrating the internal nature of the cyst. Echogenic content, a thickened capsule, and septations indicate hemorrhage or infection [9].

Mesenteric cysts can occur anywhere in the mesentery along the gastrointestinal tract but most frequently appear in the small bowel mesentery, particularly the ileum [10]. Nonetheless determining the origin of an abdominal cystic structure remains a diagnostic challenge, whereas parapancreatic cysts and ovarian related pathology (e.g., ectopic endometriosis) should be ruled out. In addition, a fine needle aspiration has been reported to be helpful in distinguishing cysts of different origin [11]. MRI might be complementary in the characterization of the cysts content and its origin. As an alternative, preoperative laparoscopy could also be helpful in localizing and characterizing.

Numerous classifications have been reported to categorize mesenteric cysts based on pathology and etiologic origin [12, 13]. In 2000 de Perrot et al. proposed a classification based on histopathology [1], where distinction has been made between true and false cysts or pseudocysts. True cyst possesses an endothelial or mesothelial lining located on the inner wall, whereas a false cyst does not. In case of an endothelial inner lining it is classified as a cyst of lymphatic origin. On the other hand, a mesothelial cell lining defines a cyst of mesothelial origin. Hence, immunohistological analysis helps differentiating between these endothelial and mesothelial cells [14]. If the cyst arises from adjacent bowel wall, it is consequently classified as an enteric cyst or enteric duplication cyst. Lastly, cysts of urogenital origin and mature, cystic teratomas are the residual classes in this classification system [1].

Although the exact etiology is uncertain, an accepted theory for cysts of lymphatic origin entails a congenital benign proliferation of ectopic lymphatic tissue in the mesentery, which does not communicate accurately with the residual lymphatic system. By contrast, mesothelial cysts are believed to be a result of congenital failure of the mesenteric leaves to fuse [14]. False cysts are caused by trauma or other injuries

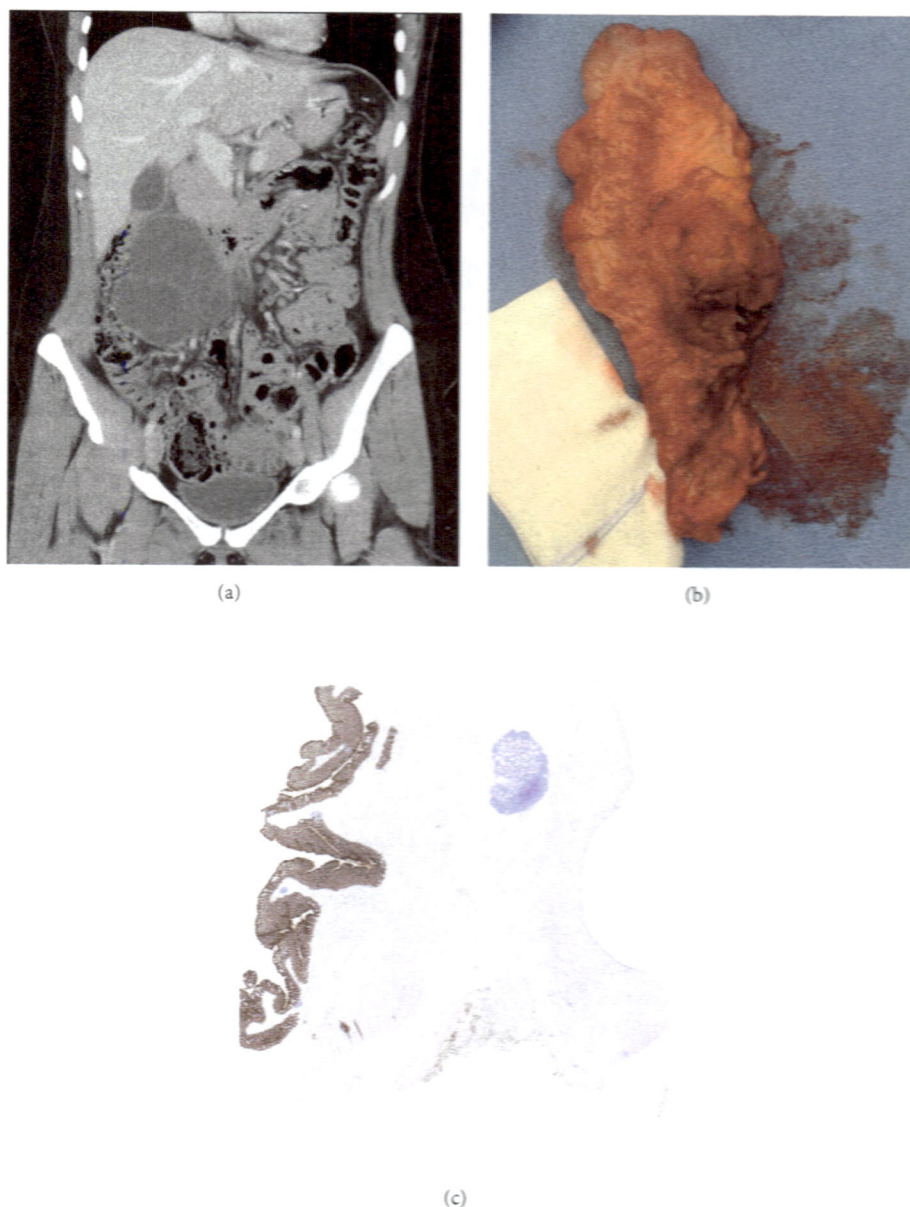

(a)

(b)

(c)

FIGURE 2: Mesenteric cyst in coronal plane in different viewings. (a) Contrast-enhanced CT. Cystic mass in close relation with surrounding bowel structures. (b) Macroscopic appearance of the resected ascending colon and adjacent mesenteric cyst in the mesenteric fat. (c) Histological appearance of the resected colon (left) and cyst (right). Notice the mesothelial cell specific AE1/AE3 positive (brown colour) cells lining the colon wall, which lacks at the cystic site. This confirms the lack of mesothelial cells in the cyst wall. AE1/AE3 stain, ×10.

like previous abdominal surgery or infection. In our case patients' medical history was unremarkable, in particular no history of previous abdominal surgery or trauma. Still the cyst was classified as a false cyst based on the lack of an endothelial or mesothelial cell lining. Furthermore, the presence of multiple smaller cysts in the mesentery indicated areas of fat necrosis supporting the classification of a false cyst. A report by Kim et al. [15] presented a similar case of a young female with an infected false mesenteric cyst without a history of predisposing factors.

Treatment of choice for complicated mesenteric cysts is surgery with complete resection of the cystic mass. This can be accomplished by either laparotomy or laparoscopy. An advantage of laparoscopy includes the shorter hospital stay but comes with longer procedure time and technical difficulty when the cyst is large or if infected [3]. In such cases laparotomy is advised. Depending on the nature of the cyst, anatomic relationships, and intraoperative findings, partial bowel resection might be necessary. This is mainly required when severance of important arteries in the mesentery cannot

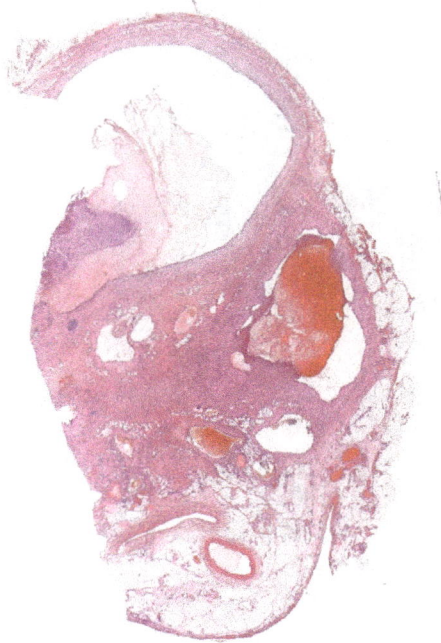

FIGURE 3: Histological image shows multiple smaller cysts spread out through the mesenteric fat. These represent areas of fat necrosis, which favor the diagnosis of a false cyst. Hematoxylin and eosin stain, ×10.

be avoided [16]. Simple aspiration or drainage is not advised due to a high recurrence rate and the chance of infection [2]. Three case reports have described a successful attempt to sclerose the cyst with ethanol after percutaneous drainage by the interventional radiologist [17, 18]. Complete regression of the cyst maintained after follow-up of up to 16 months. Although promising, this technique should be reserved for simple unilocular cysts, with no signs of malignant transformation.

Prognosis depends largely on the nature of the mesenteric cyst. Most are benign and therefore have a general good prognosis. Even so, mesotheliomas and lymphangiomas especially have a tendency to recur if resected incompletely [1].

4. Conclusion

Mesenteric cysts are rare abdominal tumors and encompass cysts from different origins with different etiologies. Consequently they can occur at diverse sites in the abdominal cavity. Symptomatic mesenteric cysts are preferably treated by surgical resection, through either laparotomy or laparoscopy. Also sclerosis of the cyst might be a feasible alternative, though with limitations. Diagnosis might be challenging and is mainly obtained with CT and US; therefore, radiologists should be aware of the existence of the (complicated) mesenteric cyst.

References

[1] M. de Perrot, M.-A. Bründler, M. Tötsch, G. Mentha, and P. Morel, "Mesenteric cysts. Toward less confusion?" *Digestive Surgery*, vol. 17, no. 4, pp. 323–328, 2000.

[2] A. Sardi, K. J. Parikh, J. A. Singer, and S. L. Minken, "Mesenteric cysts," *American Surgeon*, vol. 53, no. 1, pp. 58–60, 1987.

[3] J.-H. Vu, E. L. Thomas, and D. D. Spencer, "Laparoscopic management of mesenteric cyst," *American Surgeon*, vol. 65, no. 3, pp. 264–265, 1999.

[4] V. W. Vanek and A. K. Phillips, "Retroperitoneal, mesenteric, and omental cysts," *Archives of Surgery*, vol. 119, no. 7, pp. 838–842, 1984.

[5] J. S. Burkett and J. Pickleman, "The rationale for surgical treatment of mesenteric and retroperitoneal cysts," *American Surgeon*, vol. 60, no. 6, pp. 432–435, 1994.

[6] T. S. Chang, R. Ricketts, C. R. Abramowksy et al., "Mesenteric cystic masses: a series of 21 pediatric cases and review of the literature," *Fetal and Pediatric Pathology*, vol. 30, no. 1, pp. 40–44, 2011.

[7] J. J.-Y. Tan, K.-K. Tan, and S.-P. Chew, "Mesenteric cysts: an institution experience over 14 years and review of literature," *World Journal of Surgery*, vol. 33, no. 9, pp. 1961–1965, 2009.

[8] M. Sato, H. Ishida, K. Konno et al., "Mesenteric cyst: sonographic findings," *Abdominal Imaging*, vol. 25, no. 3, pp. 306–310, 2000.

[9] B. Vargas-Serrano, N. Alegre-Bernal, B. Cortina-Moreno, R. Rodriguez-Romero, and F. Sanchez-Ortega, "Abdominal cystic lymphangiomas: US and CT findings," *European Journal of Radiology*, vol. 19, no. 3, pp. 183–187, 1995.

[10] A. Hebra, M. F. Brown, K. M. McGeehin, and A. J. Ross, "Mesenteric, omental, and retroperitoneal cysts in children: a clinical study of 22 cases," *Southern Medical Journal*, vol. 86, no. 2, pp. 173–176, 1993.

[11] F. K. Baddoura and V. A. Varma, "Cytologic findings in multicystic peritoneal mesothelioma," *Acta Cytologica*, vol. 34, no. 4, pp. 524–528, 1990.

[12] P. R. Caropreso, "Mesenteric cysts: a review," *Archives of Surgery*, vol. 108, no. 2, pp. 242–246, 1974.

[13] B. G. Moynihan, "I. Mesenteric cysts," *Annals of Surgery*, vol. 26, no. 1, pp. 1–30, 1897.

[14] P. R. Ros, W. W. Olmsted, R. P. Moser Jr., A. H. Dachman, B. H. Hjermstad, and L. H. Sobin, "Mesenteric and omental cysts: histologic classification with imaging correlation," *Radiology*, vol. 164, no. 2, pp. 327–332, 1987.

[15] E.-J. Kim, S.-H. Lee, B.-K. Ahn, and S.-U. Baek, "Acute abdomen caused by an infected mesenteric cyst in the ascending colon: a case report," *Journal of the Korean Society of Coloproctology*, vol. 27, no. 3, pp. 153–156, 2011.

[16] M. F. O'Brien, D. C. Winter, G. Lee, E. J. Fitzgerald, and G. C. O'Sullivan, "Mesenteric cysts—a series of six cases with a review of the literature," *Irish Journal of Medical Science*, vol. 168, no. 4, pp. 233–236, 1999.

[17] G. Pozzi, A. Ferrarese, A. Borello et al., "Percutaneous drainage and sclerosis of mesenteric cysts: literature overview and report of an innovative approach," *International Journal of Surgery*, vol. 12, supplement 2, pp. S90–S93, 2014.

[18] T. Irie, M. Kuramochi, N. Takahashi, and T. Kamoshida, "Percutaneous ablation of a mesenteric cyst using ethanol: is it feasible?" *CardioVascular and Interventional Radiology*, vol. 33, no. 3, pp. 654–656, 2010.

46

Unusual Malposition of a Chest Tube, Intrathoracic but Extrapleural

Alqasem Fuad H. Al Mosa ⓘ,[1] **Mohammed Ishaq,**[2]
and Mohamed Hussein Mohamed Ahmed[3]

[1]*ATA, King Faisal Specialist Hospital & Research Centre (KFSH&RC), Riyadh, Saudi Arabia*
[2]*Surgery, King Faisal Specialist Hospital & Research Centre (KFSH&RC), Riyadh, Saudi Arabia*
[3]*Organ Transplant Centre, King Faisal Specialist Hospital & Research Centre (KFSH&RC), Riyadh, Saudi Arabia*

Correspondence should be addressed to Alqasem Fuad H. Al Mosa; alqasem.almosa@gmail.com

Academic Editor: Atsushi Komemushi

Chest tube malpositioning is reported to be the most common complication associated with tube thoracostomy. Intraparenchymal and intrafissural malpositions are the most commonly reported tube sites. We present a case about a 21-year-old patient with cystic fibrosis who was admitted due to bronchiectasis exacerbation and developed a right-sided pneumothorax for which a chest tube was inserted. Partial initial improvement in the pneumothorax was noted on the chest radiograph, after which the chest tube stopped functioning and the pneumothorax remained for 19 days. Chest computed tomography was done and revealed a malpositioned chest tube in the right side located inside the thoracic cavity but outside the pleural cavity (intrathoracic, extrapleural). The removed chest tube was patent with no obstructing materials in its lumen. A new thoracostomy tube was inserted and complete resolution of the pneumothorax followed.

1. Introduction

Tube thoracostomy has become a common surgical procedure with relatively safe outcomes [1]. The complications include malposition, infection, and organ injury with chest tube malposition being the most common complication [2]. Not all malpositioned chest tubes are clinically significant. Different studies described the locations of malpositioned chest tubes by means of CT imaging. In one study looking at complications after emergency tube thoracostomy utilizing CT scan, chest tube malposition was the most common complication, which was observed in 26% of the patients (20 out of 77 tubes). Eighteen intrathoracic (5 intraparenchymal, 9 intrafissural), and two extrathoracic malpositioned tubes were seen by CT imaging [3].

2. Case Presentation

The patient is a young male with cystic fibrosis, combined obstructive and restrictive lung disease with bilateral bronchiectasis with poor compliance to medications and chest physiotherapy.

The patient has a chronic respiratory illness with declining lung function. He is on CF treatment and was following in the pulmonology and lung transplant clinic for his disease.

On admission the patient had a right-sided pneumothorax on chest X-ray (Figure 1). He had chronic respiratory symptoms with cough, sputum, and shortness of breath that was associated with pleuritic chest pain. On examination, the patient was stable with oxygen saturation of 91% on room air. He had absent breath sounds on right side with normal cardiovascular and abdominal examination.

His laboratory work was unremarkable. His chest X-ray (Figure 1) showed significant right-sided pneumothorax with partial collapse of the right lung. Severe destruction of the lungs with bronchiectasis and atelectasis was noted. The mediastinum and heart were shifted to the left side.

The patient was admitted and was given broad-spectrum antibiotics based on previous culture results. A right-sided

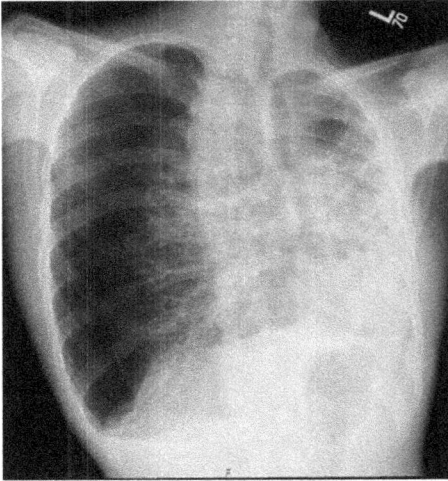

FIGURE 1: Chest X-ray on admission showing a right-sided pneumothorax prior to insertion of the chest tube.

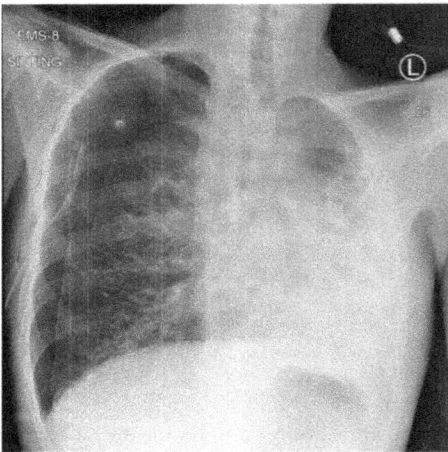

FIGURE 2: Chest X-ray after insertion of the chest tube showing improvement in the right-sided pneumothorax with subcutaneous emphysema.

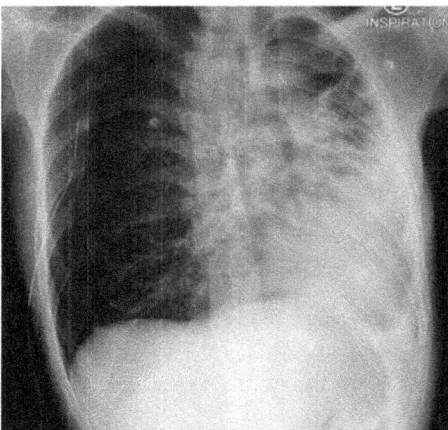

FIGURE 3: Chest X-ray showing worsening of the pneumothorax despite presence of a chest tube.

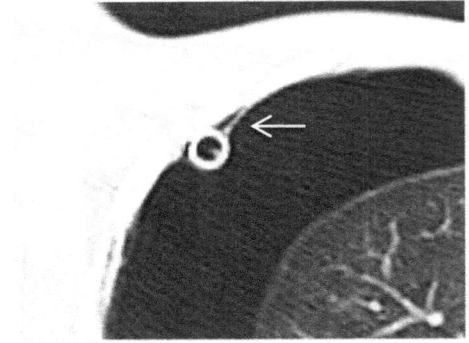

FIGURE 4: Chest CT axial cut magnified showing the location of the chest tube in relation to the chest wall and parietal pleura. White arrow: parietal pleural.

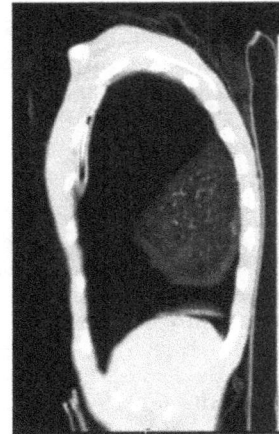

FIGURE 5: Chest CT scan sagittal cuts showing the chest tube location.

chest tube of size 24 Fr was inserted in the emergency room under aseptic measures.

The chest X-ray (Figure 2) after the chest tube insertion showed interval improvement of the pneumothorax with subcutaneous emphysema near the chest tube insertion site.

Repeated chest X-rays showed interval improvement of the pneumothorax, and the pneumothorax remained relatively unchanged for about a week, but then it worsened (Figure 3).

After worsening, the pneumothorax remained stable for 10 days. During that time, a chest CT was done which showed diffuse bronchiectasis in the right lung and bronchial wall thickening with partial atelectasis of the right lower lobe. The left lung was collapsed and showed diffuse bronchiectasis with cystic changes. The large right-sided pneumothorax was seen. The chest tube was noted to be intrathoracic but extrapleural with the parietal pleura shown as thin layer in some of the transverse cuts (Figure 4). Sagittal and coronal planes show the chest tube inside the chest cavity but in the extrapleural space adherent to the inner side of the right chest wall (Figure 5).

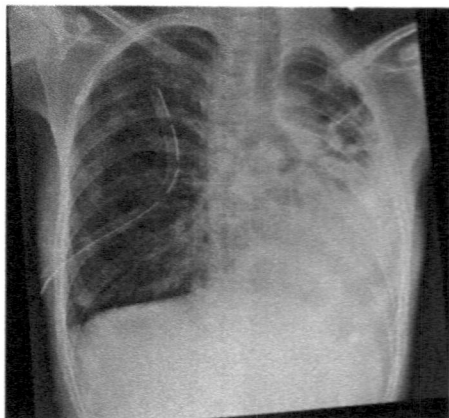

FIGURE 6: Chest X-ray after insertion of a new chest tube and resolution of the right-sided pneumothorax.

Manipulation of the chest tube (stripping, flushing, and aspirating) resulted in partial regain of tube function for 24 to 48 hours and observation of oscillation with breathing. The tube stopped functioning afterwards.

The chest tube remained for nineteen days before it was changed. The old tube was removed, and it was noted that it does not have any obstructing materials in its lumen. A new 28 Fr rigid chest tube was inserted under aseptic technique and connected to underwater seal with wall suction. A chest X-ray obtained after insertion revealed total resolution of the pneumothorax (Figure 6).

Four days later, the patient was discharged home after removal of the chest tube in a stable condition with follow-up in the clinic.

3. Discussion

Chest tube malposition is the most common complication related to chest tube insertion [2], with intraparenchymal and intrafissural locations being the most common. Different studies have demonstrated the ability of computed tomography imaging to describe the location of a malpositioned chest tube. We report on a rare location of a malpositioned chest tube that is intrathoracic but extrapleural.

Different studies described the locations of malpositioned chest tubes by means of CT imaging. In one study, CT imaging of emergency tube thoracostomy was positive for malpositioned chest tubes in 28 out of 76 chest tubes inserted. The locations described were as follows: 23 tubes in the intrathoracic location (20 intraparenchymal, 3 intrafissural) and 5 tubes in the extrathoracic location (4 in mediastinum, 1 in chest wall) [4].

In a CT pictorial review of tube thoracostomy, malpositioned chest tubes were found to be placed in intraparenchymal, intrafissural, mediastinal, and abdominal locations [5].

In a prospective study, chest tube malposition in critically ill patients was identified in 35 patients (30% of 106 chest tubes in total). Twenty-two chest tubes (21%) were diagnosed as being intrafissural, and ten (9%) were diagnosed as being intraparenchymal (9%). One tube was described as being intrathoracic but extrapleural which was inserted by a trocar and was replaced by another chest tube because it was not draining a pneumothorax [6].

In a 2013 paper enrolling 42 patients studying the results of emergency placement of chest tubes, malpositioning of chest tubes was detected by CT scan in 9 cases [7].

Multiple case reports have described rare malpositioning and complications of chest tube placement. An example of such complications is heart puncture [8, 9].

An example of a life-threatening intra-abdominal chest tube related malpositioning and complication is presented in a case report published in 2015 in which a life-threatening hemoperitoneum and liver injury occurred after chest tube insertion [10].

The authors of the paper were able to find one study describing a similar location [6] with brief description of the clinical presentation of the patient.

We believe that the initial improvement in the pneumothorax resulted from opening the pleural space during the insertion of the chest tube causing the release of an initial gush of air and hence initial temporary improvement of the pneumothorax; however, the chest tube was inserted subpleurally and not through the pleural opening. There could also have been some temporary communication between the drain and the pleural space which gradually closed off.

A couple of thoracic surgeons in our unit have noted possible similar previous observations of this malpositioning in a patient with thickened pleura. It is known that there are varying degrees of pleural involvement in cystic fibrosis patients due to chronic pleural inflammation secondary to chronic pulmonary infections. Pulmonary involvement in this patient with cystic fibrosis could explain the unusual malpositioning of the chest tube despite proper insertion technique.

4. Conclusion

We present a case about a malpositioned chest tube in a young male patient suffering from cystic fibrosis. The tube was intrathoracic but extrapleural in location. Patients with pleural pathology, like cystic fibrosis patients, could be prone to unusual malpositioning of a chest tube secondary to pleural inflammatory involvement in chronic pulmonary infections. Care and proper techniques are paramount to preventing such complications. It is important to determine why a chest tube is not functioning and to insert a new chest tube as opposed to keeping a nonfunctioning one in situ for prolonged periods. Keen and prompt review of CT chest scans can provide great insight on whether or not a chest tube is malpositioned.

References

[1] L. Chan, K. M. Reilly, C. Henderson, F. Kahn, and R. F. Salluzzo, "Complication rates of tube thoracostomy," *The American Journal of Emergency Medicine*, vol. 15, no. 4, pp. 368–370, 1997.

[2] C. G. Ball, J. Lord, K. B. Laupland et al., "Chest tube complications: how well are we training our residents?" *Canadian Journal of Surgery*, vol. 50, no. 6, pp. 450–458, 2007.

[3] J. J. Curtin, L. R. Goodman, E. J. Quebbeman et al., "Complications after emergency tube thoracostomy: Assessment with CT," *Radiology*, vol. 198, no. 1, pp. 19-20, 1996.

[4] K.-E. Lim, S.-C. Tai, C.-Y. Chan et al., "Diagnosis of malpositioned chest tubes after emergency tube thoracostomy: Is computed tomography more accurate than chest radiograph?" *Clinical Imaging*, vol. 29, no. 6, pp. 401–405, 2005.

[5] G. Gayer, J. Rozenman, C. Hoffmann et al., "CT diagnosis of malpositioned chest tubes," *British Journal of Radiology*, vol. 73, no. 871, pp. 786–790, 2000.

[6] F. Remérand, V. Luce, Y. Badachi, Q. Lu, B. Bouhemad, and J.-J. Rouby, "Incidence of chest tube malposition in the critically ill: A prospective computed tomography study," *Anesthesiology*, vol. 106, no. 6, pp. 1112–1119, 2007.

[7] R. Schupfner, W. Wagner, and A. Schneller, "Results of thoracic drainages placed in air rescue," *Interventional Medicine and Applied Science*, vol. 5, no. 4, pp. 168–174, 2013.

[8] D. Kim, S.-H. Lim, and P. W. Seo, "Iatrogenic perforation of the left ventricle during insertion of a chest drain," *The Korean Journal of Thoracic and Cardiovascular Surgery*, vol. 46, no. 3, pp. 223–225, 2013.

[9] N. Anitha, S. Kamath, E. Khymdeit, and M. Prabhu, "Intercostal drainage tube or intracardiac drainage tube?" *Annals of Cardiac Anaesthesia*, vol. 19, no. 3, pp. 545–548, 2016.

[10] J. M. Bae, "Life threatening hemoperitoneum and liver injury as a result of chest tube thoracostomy," *Clinical Medicine Insights: Case Reports*, vol. 26, no. 8, pp. 15–17, 2015.

47

Painful Os Peroneum Syndrome: Underdiagnosed Condition in the Lateral Midfoot Pain

Francisco Abaete Chagas-Neto,[1,2,3] **Barbara Nogueira Caracas de Souza,**[1] **and Marcello Henrique Nogueira-Barbosa**[4]

[1]*Division of Radiology, Antonio Prudente Hospital, Fortaleza, CE, Brazil*
[2]*School of Medicine, Division of Radiology, University of Fortaleza, Fortaleza, CE, Brazil*
[3]*School of Medicine, Division of Radiology, Christus University Center, Fortaleza, CE, Brazil*
[4]*Division of Radiology, Internal Medicine Department, Ribeirao Preto Medical School, Sao Paulo University, Ribeirão Preto, SP, Brazil*

Correspondence should be addressed to Francisco Abaete Chagas-Neto; abaeteneto@yahoo.com.br

Academic Editor: Atsushi Komemushi

Os peroneum is an accessory ossicle located within the peroneus longus tendon. The painful os peroneum syndrome (POPS) results from a wide spectrum of conditions, including fractures, diastases, and other causes. POPS can result in tenosynovitis or discontinuity of the peroneus longus tendon with a clinical presentation of pain in the lateral aspect of the midfoot. Authors report a typical case of POPS, illustrating this entity through different imaging methods (radiographs, ultrasound, and magnetic resonance imaging). We emphasize the prevalence of this ossicle and discuss painful complications.

1. Introduction

Os peroneum is an accessory ossicle located within the substance of the peroneus longus tendon. Os peroneum is identified in 4.7–30% of normal feet [1] and is bipartite in approximately 30% of cases and unilateral in 40%. Its fully ossified form is found in about 26% of population [2].

Painful os peroneum syndrome (POPS) results from a wide spectrum of conditions, including fractures or diastases, and may result in tenosynovitis or even rupture of the peroneal tendon [1].

This syndrome should be considered in patients with pain in the lateral aspect of the midfoot. A positive physical examination reveals pain during palpation of the ossicle; however, it is easily overlooked.

Imaging, such as radiographs, ultrasonography, and magnetic resonance imaging (MRI), plays an important role in the diagnosis and in the other associated conditions.

This report aims to illustrate, using different imaging methods, a typical case of POPS, to raise the degree of suspicion of this entity and highlight possible related complications.

2. Case Report

A 60-year-old female patient presented with progressive pain in the lateral aspect of the right midfoot. She denied any history of recent trauma, sprain, or high-impact sport activity. She was evaluated by an orthopedist who requested plain films of the right foot (Figure 1(a)). The plain film showed the presence of an accessory ossicle in the lateral aspect of the midfoot, located in the path of the peroneal tendons with cortical discontinuity, fragmentation, irregular margins, and heterogeneous density. Simple contralateral comparative radiograph of the left foot also showed the same accessory bone; however, there was intact margins and homogeneous density (Figure 1(b)).

Following the plain film, an MRI was performed for soft tissue evaluation. The accessory ossicle was identified within the peroneus longus tendon in the lateral aspect of the midfoot. It showed diffuse marrow edema, irregular margins, and cortical discontinuity. Also, there was edema and intense enhancement in the adjacent soft tissues (Figure 2). The peroneus longus tendon was thickened and heterogeneous, consistent with tendinopathy.

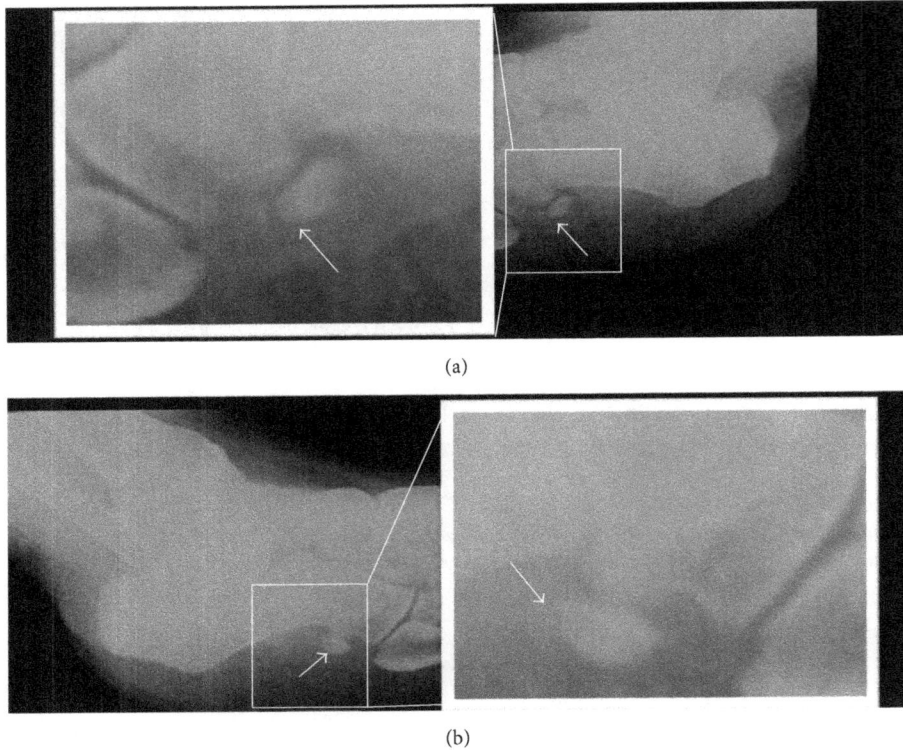

(a)

(b)

FIGURE 1: 60-year-old female plain film of the feet in an oblique view. (a) Right foot: complaint side, showing an irregular and fragmented os peroneum with heterogeneous density (arrow). (b) Left foot: comparative contralateral side, showing a regular and complete os peroneum with regular contours and homogeneous density (arrow).

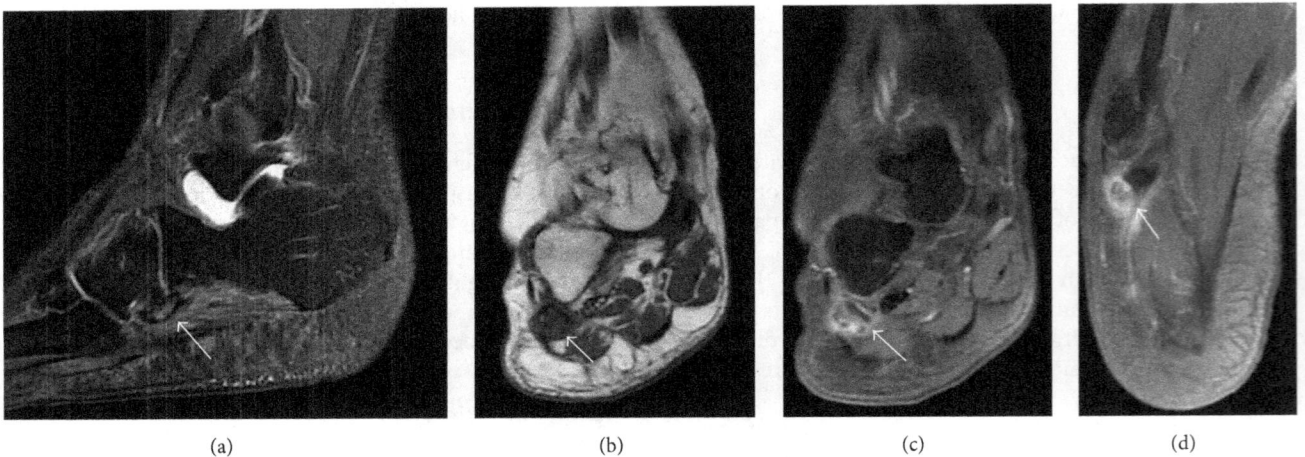

(a)   (b)   (c)   (d)

FIGURE 2: 60-year-old female right midfoot MRI. (a) Sagittal T2; (b) coronal T1; (c) fat suppressed T1 postgadolinium; and (d) axial fat suppressed T1 postgadolinium. Arrows show the os peroneum within the peroneal tendon, with irregular contours, bone marrow edema, and intense enhancement of the adjacent soft tissues, characterizing inflammatory changes. The peroneus longus tendon is thickened and heterogeneous, consistent with associated tendinopathy.

We also performed comparative ultrasonography of the feet (Figure 3). It also identified irregularity and discontinuity of the right os peroneum (Figure 3(a)) and regular shape of this ossicle in the left (Figure 3(b)). We were also able to identify edema of the adjacent soft tissues and tendinopathy of the peroneus longus only on the right side (Figure 3(a)).

We emphasize that the right os peroneum region coincided with the exact painful area.

The patient follows conservative treatment with medication and physical therapy. However, she reports only partial improvement of symptoms and most recently began a slightly painful condition on the left foot. It was not symptomatic

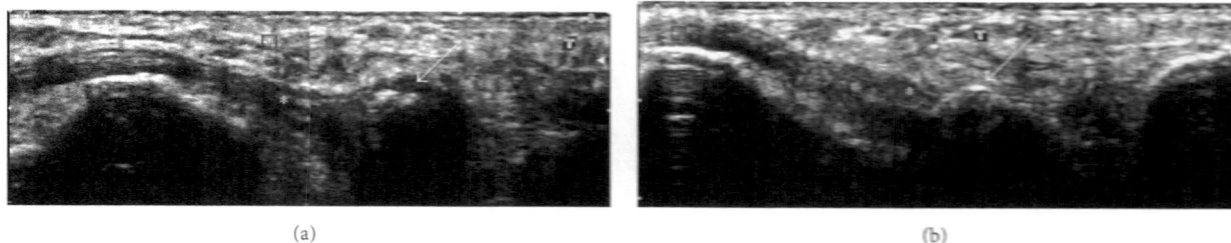

(a)   (b)

FIGURE 3: 60-year-old female ultrasonography of the long axis of the peroneus longus tendon. (a) Right foot: complaint side demonstrating a thickened and heterogeneous peroneus longus tendon (asterisk) and irregular and fragmented os peroneum, associated with swelling of the surrounding soft tissues. (b) Left foot: contralateral side for comparison, demonstrating a preserved echotexture of the peroneus longus tendon (asterisk) and regular contours of the os peroneum without changes in the surrounding soft tissues.

when the imaging studies presented in this report were performed.

3. Discussion

There are different sesamoids and accessory ossicles in the skeleton. Some of them are known to be associated with painful syndromes, such as os trigonum, os navicular, and fabela. These syndromes may be caused by different etiologies such as trauma, infection, impact, and degenerative changes [3].

The os peroneum is an accessory ossicle, round or oval, within the substance of the peroneal tendon [1], and can be classified accordingly to Nwawka et al. and Blitz and Nemes as a sesamoid [2, 4]. Its histological structure is composed of different degrees of ossification and fibrous tissue [5].

The peroneus longus tendon is located proximal and posteriorly to the lateral malleolus on the lateral surface of the calcaneus, cuboid (along the midfoot), and distally inserting at the base of the first metatarsal and medial cuneiform [1, 6].

There are several causes for pain in the lateral aspect of the foot, including dislocation or subluxation of the peroneal tendon, injury, to the talofibular ligament or calcaneofibular ligament, or fractures in the fifth metatarsal, anterior process of the calcaneus, or cuboid [1].

The os peroneum fracture may be complicated by rupture of the peroneus longus tendon. The most common mechanism occurs with a strong contraction of the peroneus longus muscle in response to a sudden inversion or supination. Such contraction can compress the os peroneum against the cuboid, resulting in fracture and rupture of the peroneus longus tendon. It has been suggested that the presence of this ossicle can predispose to its distal rupture due to potential increased friction with adjacent structures [7]. Physical examination can reveal swelling over the cuboid, with pain in this area during palpation. The pain is usually exacerbated by plantar flexion and heel elevation stage during gait [7].

POPS has two main forms: acute and chronic. The acute form occurs as a result of trauma, commonly with ankle sprain or supination movement, resulting in fracture or diastasis of the os peroneum, which may or may not be associated with peroneus longus tendon rupture. Chronic presentation

is closely linked to a healing process of a fracture with subsequent calcification, remodeling, or chronic diastasis of the os peroneum with a variable frequency of tenosynovitis of the peroneus longus tendon.

With MRI, the ossicle is usually isointense to bone marrow and presents with increased intrasubstantial signal within the peroneus longus tendon, typically close to the cuboid. Under ultrasonography, its identification is easily appreciated because of its typical bone appearance, as a curved echogenic focus with posterior acoustic shadow [8].

On radiographs, it is better identified in an oblique view of the foot. Both radiography and computed tomography may demonstrate displacement of the os peroneum from its usual position, fracture, or diastasis of a bipartite sesamoid. The displacement of the os peroneum is an indirect sign of a peroneal tendon rupture [2].

The radiographic differentiation between a fractured or split os peroneum may be difficult. In an acute event, fracture margins seem relatively nonsclerotic and bone fragments generally fit together, as "pieces of a puzzle." In the bipartite sesamoid, margins become rounded and sclerotic. It is possible that over time due to remodeling, the edges of the fracture resemble the appearance of a split os peroneum. Brigido et al. suggested that a diastasis between fragments of os peroneum, greater than five millimeters, must indicate the diagnosis of fracture. US and MRI can also be used, especially to evaluate other possible associated abnormalities.

In the same study by Brigido et al., all bone fragments identified with US were hyperechogenic [7]. The evaluation of sesamoid fractures with MRI is difficult because of their small size and low signal. Bone marrow swelling can also complicate evaluation of fractures due to abnormal marrow signal intensity.

Therefore, early diagnosis and correct characterization of POPS are essential for an adequate management of these patients. Knowledge of its presentation through different imaging methods is very important during training of specialists in Radiology and Diagnostic Imaging.

References

[1] S. J. Oh, Y. H. Kim, S. K. Kim, and M.-W. Kim, "Painful os peroneum syndrome presenting as lateral plantar foot pain," *Annals of Rehabilitation Medicine*, vol. 36, no. 1, pp. 163–166, 2012.

[2] O. K. Nwawka, D. Hayashi, L. E. Diaz et al., "Sesamoids and accessory ossicles of the foot: anatomical variability and related pathology," *Insights into Imaging*, vol. 4, no. 5, pp. 581–593, 2013.

[3] A. R. Barreto, F. A. Chagas-Neto, M. D. Crema et al., "Fracture of the fabella: a rare injury in knee trauma," *Case Reports in Radiology*, vol. 2012, Article ID 390150, 3 pages, 2012.

[4] N. M. Blitz and K. K. Nemes, "Bilateral peroneus longus tendon rupture through a bipartite os peroneum," *The Journal of Foot and Ankle Surgery*, vol. 46, no. 4, pp. 270–277, 2007.

[5] C. M. Sofka, R. S. Adler, G. R. Saboeiro, and H. Pavlov, "Sonographic evaluation and sonographic-guided therapeutic options of lateral ankle pain: peroneal tendon pathology associated with the presence of an os peroneum," *HSS Journal*, vol. 6, no. 2, pp. 177–181, 2010.

[6] K. L. Moore, A. F. Dalley, and C. L. C. de Araújo, *Anatomia Orientada Para a Clínica*, Guanabara Koogan, 2006.

[7] M. K. Brigido, D. P. Fessell, J. A. Jacobson et al., "Radiography and US of os peroneum fractures and associated peroneal tendon injuries: initial experience," *Radiology*, vol. 237, no. 1, pp. 235–241, 2005.

[8] A. Donovan, Z. S. Rosenberg, J. T. Bencardino et al., "Plantar tendons of the foot: MR imaging and US," *Radiographics*, vol. 33, no. 7, pp. 2065–2085, 2013.

Permissions

List of Contributors

Savitha Srirama Jayamma, Seema Sud and TBS Buxi
Department of CT and MRI, Sir Ganga Ram Hospital, New Delhi 110060, India

VS Madan and Ashish Goyal
Department of Neurosurgery, Sir Ganga Ram Hospital, New Delhi 110060, India

Shashi Dhawan
Department of Pathology, Sir Ganga Ram Hospital, New Delhi 110060, India

Gernot Rott and Frieder Boecker
Radiological Department, Bethesda-Hospital Duisburg, Heerstr. 219, 47053 Duisburg, Germany

Sayyad Yaseen Zia and Richard L. Bakst
Department of Radiation Oncology, Icahn School of Medicine at Mount Sinai Hospital, New York, NY, USA

Qiusheng Si
Department of Pathology, Icahn School of Medicine at Mount Sinai Hospital, New York, NY, USA

Mike Yao
Department of Otolaryngology, Icahn School of Medicine at Mount Sinai Hospital, New York, NY, USA

Peter M. Som
Department of Radiology, Icahn School of Medicine at Mount Sinai Hospital, New York, NY, USA

Umairullah Lodhi, Uzair Sarmast, Saadullah Khan and Kavitha Yaddanapudi
Department of Radiology, Stony Brook University Hospital, 101 Nicolls Road, Stony Brook, NY 11794, USA

Takashi Takahashi, Jutta Ellermann and Shelly Marette
Department of Radiology, University of Minnesota Medical Center, 420 Delaware Street SE, MMC 292, Minneapolis, MN 55455, USA

Lauren MacCormick and Denis Clohisy
Department of Orthopaedic Surgery, University of Minnesota Medical Center, 2450 Riverside Avenue, Suite R200, Minneapolis, MN 55454, USA

Puneeth Kumar, Amit Kumar Dey, Kartik Mittal, Rajaram Sharma and Priya Hira
Department of Radiology, Seth G. S. Medical College and KEM Hospital, Mumbai 400012, India

Willian Schmitt, Marta Baptista and Ana Germano
Department of Radiology, Hospital Prof. Doutor Fernando Fonseca, Amadora, Portugal

Marco Ferreira
Department of Pathology, Hospital Prof. Doutor Fernando Fonseca, Amadora, Portugal

António Gomes
Department of Surgery, Hospital Prof. Doutor Fernando Fonseca, Amadora, Portugal

Gernot Rott and Frieder Boecker
Radiological Department, Bethesda-Hospital Duisburg, Heerstr. 219, 47053 Duisburg, Germany

Marie Tominna and Sayf Al-Katib
Beaumont Hospital, Oakland UniversityWilliam Beaumont School of Medicine, Department of Diagnostic Radiology and Molecular Imaging, 3601W13 Mile Rd, Royal Oak, MI 48073, USA

M. Hamard, S. P. Martin and S. Boudabbous
Department of Imaging and Medical Information Sciences, Division of Radiology, Geneva University Hospitals, Geneva, Switzerland

Takashi Takahashi and Jeffrey Rykken
Department of Radiology, University of Minnesota Medical Center, Minneapolis, MN, USA

Jawad M. Qureshi, Jeffrey Mueller and Matthew S. Hartman
Department of Radiology, Allegheny Health Network, Pittsburgh, PA, USA

Brian Pagano
Temple University School of Medicine, Philadelphia, PA, USA

Lana Schumacher
Department of Surgery, Allegheny Health Network, Pittsburgh, PA, USA

Claudia Velosa
Department of Pathology, Allegheny Health Network, Pittsburgh, PA, USA

Laura Geraci, Massimo Midiri, and Cesare Gagliardo
Section of Radiological Sciences, Department of Biopathology and Medical Biotechnologies, University of Palermo, Palermo, Italy

Alessandro Napoli and Carlo Catalano
Radiology Section, Department of Radiological, Oncological and Anatomopathological Sciences, "Sapienza" University of Rome, Rome, Italy

Mehmet Barburoglu, Bulent Acunas, Yilmaz Onal
Istanbul Medical Faculty, Istanbul University, Department of Radiology, Millet Caddesi, Capa, Fatih, 34390 Istanbul, Turkey

Murat Ugurlucan, Omer Ali Sayin, and Ufuk Alpagut
Istanbul Medical Faculty, Istanbul University, Department of Cardiovascular Surgery, Istanbul, Turkey

Kyung Ryeol Lee
Department of Radiology, Jeju National University Hospital, Aran 13gil 15 (Ara-1-Dong), Jeju-si, Jeju Special Self-Governing Province 63241, Republic of Korea

Young Hee Maeng
Department of Pathology, Jeju National University Hospital, Aran 13gil 15 (Ara-1-Dong), Jeju-si, Jeju Special Self-Governing Province 63241, Republic of Korea

William E. Haley and Joseph G. Cernigliaro
Mayo Clinic, Jacksonville, FL 32224, USA

El-Sayed H. Ibrahim
 Mayo Clinic, Jacksonville, FL 32224, USA
University of Michigan, Ann Arbor, MI 48109, USA

Mingliang Qu and Cynthia H. McCollough
Mayo Clinic, Rochester, MN 55905, USA

David S. Goldfarb
New York Harbor VA Healthcare System, Brooklyn, NY 11209, USA

William Ryan
Lewis Katz School of Medicine at Temple University, USA

Farouk Dako, Gary Cohen, David Pryluck, Joseph Panaro, Emily Cuthbertson, and Dmitry Niman
Temple University Hospital, USA

Tuan Phung
Radiology Department, Military Hospital 103, Vietnam Military Medical University, Hanoi, Vietnam

Thach Nguyen
Anesthesiology Department, National Institute of Burns, Vietnam Military Medical University, Hanoi, Vietnam

Dung Tran and Hung Nguyen
Histopathology Department, Military Hospital 103, Vietnam Military Medical University, Hanoi, Vietnam

Nga Phan
Neurology Department, Military Hospital 103, Vietnam Military Medical University, Hanoi, Vietnam

Mujtaba Mohammed and Andrij Wojtowycz
Department of Radiology, SUNYUpstate University Hospital, Syracuse, NY 13210, USA

Mary Allen-Proctor
Department of Pathology, SUNY Upstate University Hospital, Syracuse, NY 13210, USA

Farzaneh Mosavat, Roxana Rashtchian, Negar Zeini, Daryoush Goodarzi Pour, and Shabnam Mohammed Charlie
Oral and Maxillofacial Radiology Department, School of Dentistry, Tehran University of Medical Sciences, Tehran, Iran

Nazanin Mahdavi
Oral and Maxillofacial Pathology Department, School of Dentistry, Tehran University of Medical Sciences, Tehran, Iran

K. O. Kragha
Department of Radiology, University of Louisville, 530 S. S. Jackson Street, CCB C07, Louisville, KY 40202, USA

Kazuki Yoshida, Masao Miyagawa, Teruhito Kido, Kana Ide and Teruhito Mochizuki1
Department of Radiology, Ehime University Graduate School of Medicine, Toon, Japan

Yoshifumi Sano
Department of Pulmonary Surgery, Ehime University Graduate School of Medicine, Toon, Japan

Yoshifumi Sugawara
Department of Diagnostic Radiology, Shikoku Cancer Center, Matsuyama, Japan

Hiroyuki Takahata
Department of Pathology, Shikoku Cancer Center, Matsuyama, Japan

Nobuya Monden
Department of Head and Neck Surgery, Shikoku Cancer Center, Matsuyama, Japan

Mitsuko Furuya
Department of Molecular Pathology, Yokohama City University Graduate School of Medicine, Yokohama, Japan

Nilufer Bulut
Department of Medical Oncology, Kanuni Sultan Suleyman Education and Research Hospital, Istanbul, Turkey

Sevinc Dagistanli
Department of Surgery, Division of General Surgery, Kanuni Sultan Suleyman Education and Research Hospital, Istanbul, Turkey

Jad A. Degheili and Ali H. Hallal
Department of Surgery, Division of General Surgery, American University of Beirut Medical Center, Beirut, Lebanon

Alissar El Chediak
Department of Internal Medicine, American University of Beirut Medical Center, Beirut, Lebanon

Mohamad Yasser R. Dergham and Aghiad Al-Kutoubi
Department of Diagnostic Radiology, Division of Interventional Radiology, American University of Beirut Medical Center, Beirut, Lebanon

Taylor Rising
Northeast Ohio MedicalUniversity, 4209 State Route 44, Rootstown, OH44272, USA

Nicholas Fulton and Pauravi Vasavada
University Hospitals Case Medical Center, 11100 Euclid Avenue, Cleveland, OH 44106, USA

Iclal Ocak
University of Pittsburgh Medical Center, Radiology Suite 200 East Wing, 200 Lothrop Street, Pittsburgh, PA 15213, USA

Diane C. Strollo
University of Pittsburgh Medical Center, 200 Lothrop Street, Pittsburgh, PA 15213, USA

Fatima Zahra Mrabet, Jihane Achrane, Sanaa Hammi, and Jamal Eddine Bourkadi
Department of Pneumology, Moulay Youssef University Hospital Center, Rabat, Morocco

Yassir Sabri
Department of Parasitology, Ibn Sina University Hospital Center, Rabat, Morocco

Fatima Ezzahra El Hassani
Department of Cardiology, Military Hospital Mohamed V, Rabat, Morocco
Faculty of Medicine and Pharmacy, Mohamed V University, Rabat, Morocco

Ruth Steiger, Andreas Rietzler, Elke R. Gizewski, and Astrid E. Grams
Department of Neuroradiology, Medical University of Innsbruck, Anichstraße 35, 6020 Innsbruck, Austria

Lisa-Maria Walchhofer and Bernhard Glodny
Department of Radiology, Medical University of Innsbruck, Innsbruck, Austria

Katherina J. Mair and Michael Knoflach
Department of Neurology, Medical University of Innsbruck, Innsbruck, Austria

Daichi Momosaka, Yasuhiro Ushijima, Akihiro Nishie, Yoshiki Asayama, Kousei Ishigami, Daisuke Okamoto, Nobuhiro Fujita and Hiroshi Honda
Department of Clinical Radiology, Graduate School of Medical Sciences, Kyushu University, 3-1-1 Maidashi, Higashi-ku, Fukuoka 812-8582, Japan

Yukihisa Takayama
Department of Radiology Informatics and Network, Graduate School of Medical Sciences, Kyushu University, 3-1-1 Maidashi, Higashi-ku, Fukuoka 812-8582, Japan

Tetsuo Ikeda
Center for Integration of Advanced Medicine and Innovative Technology, Graduate School of Medical Sciences, Kyushu University, 3-1-1 Maidashi, Higashi-ku, Fukuoka 812-8582, Japan

Keiichiro Uchida
Department of Medicine and Bioregulatory Science, Graduate School of Medical Sciences, Kyushu University, 3-1-1 Maidashi, Higashi-ku, Fukuoka 812-8582, Japan

Masaaki Sugimoto and Kenichi Kohashi
Department of Anatomic Pathology, Graduate School of Medical Sciences, Kyushu University, 3-1-1 Maidashi, Higashi-ku, Fukuoka 812-8582, Japan

Katrina N. Glazebrook, Shuai Leng, Steven R. Jacobson, and Cynthia M. McCollough
Mayo Clinic, 200 First Street SW, Rochester, MN55904, USA

Fahad Al-Lhedan
Medical Imaging Department, King Abdullah bin Abdulaziz University Hospital, Riyadh, Saudi Arabia

Sam Samaan and Wanzhen Zeng
Department of Nuclear Medicine, The Ottawa Hospital, Ottawa, ON, Canada

Zhen Kang, Xiangde Min, and Liang Wag
Department of Radiology, Tongji Hospital, Tongji Medical College, Huazhong University of Science and Technology, Wuhan, Hubei 430030, China

Mazamaesso Tchaou and Lantam Sonhaye
Department of Radiology, The University Teaching Hospital of Lom´e, Lom´e, Togo

Tchin Darré and Gado Napo-Koura
Department of Pathology, The University Teaching Hospital of Lom´e, Lom´e, Togo

Koué Folligan
Department of Histology-Embryology, The University Teaching Hospital of Lom´e, Lom´e, Togo

Akomola Sabi
Department of Endocrinology and Nephrology, The University Teaching Hospital of Lom´e, Lom´e, Togo

Azanledji Boumé
Department of Pediatric Surgery, The University Teaching Hospital of Lom´e, Lom´e, Togo

Akila Bassowa
Department of Obstetrics and Gynecology, The University Teaching Hospital of Lom´e, Lom´e, Togo

Solange Adani-Ifé
Department of Clinical Oncology, The University Teaching Hospital of Lom´e, Lom´e, Togo

Anuj Dixit
Discipline of Radiology, Memorial University of Newfoundland, St. John's, NL, Canada

Cameron Hague and Simon Bicknell
Department of Radiology, University of British Columbia, Vancouver, BC, Canada

Weerachai Tantinikorn and Paithoon Wichiwaniwate
Bumrungrad International Hospital, Bangkok 10110, Thailand

Sasitorn Siritho
Bumrungrad International Hospital, Bangkok 10110, Thailand
Division of Neurology, Department of Medicine, Siriraj Hospital, Mahidol University, Bangkok 10700, Thailand

Krit Pongpirul
Bumrungrad International Hospital, Bangkok 10110, Thailand
Department of Preventive and Social Medicine, Faculty of Medicine, Chulalongkorn University, Bangkok 10330, Thailand
Department of International Health, Johns Hopkins Bloomberg School of Public Health, Baltimore, MD 21205, USA

Joana Ruivo Rodrigues
Radiology Unit, Medical Imaging Department, Centro Hospitalar de Tondela-Viseu, EPE, Viseu, Portugal

Gonçalo Roque Santos
Neuroradiology Unit, Medical Imaging Department, Centro Hospitalar de Tondela-Viseu, EPE, Viseu, Portugal

Tomas Mujo, Erin Priddy, John J. Harris, and Mahmoud Samman
Department of Radiology, University of Louisville Hospital, Louisville, KY 40202, USA
Eric Poulos
University of Louisville School of Medicine, Louisville, KY 40202, USA

Tommaso D'Angelo, Giuseppe Cicero, and Silvio Mazziotti
Section of Radiological Sciences, Department of Biomedical Sciences and Morphological and Functional Imaging, University of Messina, Policlinico "G. Martino", Via Consolare Valeria 1, 98100 Messina, Italy

Romina Gallizzi and Claudio Romano
Department of Pediatrics, University of Messina, Policlinico "G. Martino", Via Consolare Valeria 1, 98100 Messina, Italy

Kyo Noguchi
Department of Radiology, University of Toyama, 2630 Suginani, Toyama, Toyama 930-0194, Japan

Hiroko Shojaku
Department of Radiology, University of Toyama, 2630 Suginani, Toyama, Toyama 930-0194, Japan
Department of Radiology, Saiseikai Takaoka Hospital, 387-1 Futazuka, Takaoka, Toyama 933-8525, Japan

Tetsuya Kamei
Department of Radiology, Saiseikai Takaoka Hospital, 387-1 Futazuka, Takaoka, Toyama 933-8525, Japan

Yasuko Tanada and Kouichi Yoshida
Department of Surgery, Saiseikai Takaoka Hospital, 387-1 Futazuka, Takaoka, Toyama 933-8525, Japan

Yasuko Adachi
Department of Respiratory Internal Medicine, Saiseikai Takaoka Hospital, 387-1 Futazuka, Takaoka, Toyama 933-8525, Japan

Kazuhiro Matsui
Department of Pathology, Saiseikai Takaoka Hospital, 387-1 Futazuka, Takaoka, Toyama 933-8525, Japan

Elizabeth Von Ende
Kansas City University, 1750 Independence Ave, Kansas City, MO 64106, USA

Travis Kauffman
University of Missouri-Kansas City, 4401Wornall Rd, Kansas City, MO 64111, USA

Philip A. Munoz and Santiago Martinez-Jiménez
Saint Luke's Hospital of Kansas City, University of Missouri-Kansas City, 4401Wornall Rd, Kansas City, MO 64111, USA

Jed Hummel, Jason Wachsmann, Orhan K. Oz and Dana Mathews
Department of Radiology, UT Southwestern Medical Center, Dallas, TX, USA

Fangyu Peng
Department of Radiology, UT Southwestern Medical Center, Dallas, TX, USA
Advanced Imaging Research Center, UT Southwestern Medical Center, Dallas, TX, USA

Kelley Carrick
Department of Pathology, UT Southwestern Medical Center, Dallas, TX, USA

Partha Hota, and Omar Agosto
Division of Abdominal Imaging, Department of Radiology, Temple University Hospital, Philadelphia, PA, USA

Tejas N. Patel
Division of Abdominal Imaging, Atlantic Medical Imaging, Galloway, NJ, USA

Xiaofeng Zhao and Carrie Schneider
Department of Pathology, Temple University Hospital, Philadelphia, PA, USA

Zhen Zeng, Tijiang Zhang, and Xiaoxi Chen
Department of Radiology, Affiliated Hospital of Zunyi Medical College, Zunyi, China

Yihua Zhou
Department of Radiology, Saint Louis University School of Medicine, St. Louis, MO, USA

Afsaneh Alikhassi
Department of Radiology, Cancer Institute, Tehran University of Medical Sciences, Tehran, Iran

Fereshteh Ensani and Alireza Abdollahi
Department of Pathology, Cancer Institute, Tehran University of Medical Sciences, Tehran, Iran

Ramesh Omranipour
Division of Surgical Oncology, Cancer Institute, Tehran University of Medical Sciences, Tehran, Iran

Davy R. Sudiono and Frank M. Zijta
Department of Radiology, Medical Center Haaglanden, Lijnbaan 32, 2512 VA Den Haag, Netherlands

Joep B. Ponten
Department of Surgery, Medical Center Haaglanden, Lijnbaan 32, 2512 VA Den Haag, Netherlands

Alqasem Fuad H. Al Mosa
ATA, King Faisal Specialist Hospital and Research Centre (KFSH&RC), Riyadh, Saudi Arabia

Mohammed Ishaq
Surgery, King Faisal Specialist Hospital and Research Centre (KFSH&RC), Riyadh, Saudi Arabia

Mohamed Hussein Mohamed Ahmed
Organ Transplant Centre, King Faisal Specialist Hospital and Research Centre (KFSH&RC), Riyadh, Saudi Arabia

Barbara Nogueira Caracas de Souza
Division of Radiology, Antonio Prudente Hospital, Fortaleza, CE, Brazil

Francisco Abaete Chagas-Neto
Division of Radiology, Antonio Prudente Hospital, Fortaleza, CE, Brazil
School of Medicine, Division of Radiology, University of Fortaleza, Fortaleza, CE, Brazil
School of Medicine, Division of Radiology, Christus University Center, Fortaleza, CE, Brazil

Marcello Henrique Nogueira-Barbosa
Division of Radiology, Internal Medicine Department, Ribeirao Preto Medical School, Sao Paulo University, Ribeir͠ao Preto, SP, Brazil

Index